Schuster

# Reinforcement of Elastomers by Nanoscaled Fillers

Robert H. Schuster

# Reinforcement of Elastomers by Nanoscaled Fillers

## Carbon Black, Silica/Silane, Carbon Nanotubes, and Layered Silicates

HANSER

Print-ISBN: 978-1-56990-913-3
E-Book-ISBN: 978-1-56990-930-0



Bibliographic information of the German National Library:
The German National Library lists this publication in the German National Bibliography; detailed bibliographic data are available on the Internet at http://dnb.d-nb.de.


Kolbergerstraße 22 | 81679 Munich | info@hanser.de
*www.hanserpublications.com*
*www.hanser-fachbuch.de*
Editor: Dr. Mark Smith
Production Management: Cornelia Speckmaier
Cover concept: Marc Müller-Bremer, www.rebranding.de, Munich
Cover design: Max Kostopoulos
Cover picture: © Max Kostopoulos
Typesetting: Eberl & Koesel Studio, Kempten

# The Author

Robert H. Schuster studied Organic Chemistry at the "A. I. Cuza" University in Jassy, Romania, and obtained his doctorate degree at the Institute of Macromolecular Chemistry in Freiburg, Germany. From early 1984, he worked at the German Institute for Rubber Technology in Hanover, which he headed from 1992. During this time, he enjoyed good cooperation with Prof. Jacobi from the State University Porto Alegre (Brazil). The author has been a professor at the University of Hanover since 1995. The institute is involved in national and international projects with raw material manufacturers, the rubber industry and customers. The institute has six departments through which more than 70 doctoral students find their way into the rubber industry. Topics covered range from filler reinforcement, rubber blending, vulcanization, physical properties, magnetic elastomers, rubber-filler composites, electrospinning fibers, magnetic elastomers, investigation of physical properties and fatigue. An essential part of the work is the study of carbon black, silica, silanization of silica, wet-mixing of layered silicates and carbon nanotubes. After leaving the DIK, the author collaborated with Lanxess AG and with the EVE Institute of Dr. M.-J. Wang in Qingdao. Over the years, the author has participated in DKG Meetings and ACS Meetings and Conferences around the world. Over his career in rubber research, Robert H. Schuster has published more than 300 scientific papers, two books and five book chapters and is coauthor in several German Patents. In 1995 he received from the DKG the Carl Dietrich Harries medal and in 2011 from the Rubber Division the Melvin Mooney Award.

# Preface

Elastomers are fascinating and indispensable materials for society. They make a huge contribution to our safety and comfort in the world of mobility. Based on the unique principle of entropy elasticity, elastomers cannot be replaced by any other class of material. Their performance, functionality and quality depend on the selected raw materials, especially rubber and filler, on mechanical mixing to form homogeneous compounds, and on efficient crosslinking. Some topics have been studied intensely and have led to technological progress. There is no doubt that the reinforcement of elastomers with nanoscaled fillers is one such topic. Therefore, the nanoscaled fillers are important due to their availability to improve the physical properties of elastomers. The use of nanoscaled fillers in rubber compounds is a multidisciplinary task that requires a knowledge of how to produce the filler particles, of how to mix them with the rubber, of how to disperse them and of how they enhance physical and dynamic-mechanical properties while reducing elastomer rupture.

Reinforcing fillers increase abrasion resistance, ensure low wear and a longer service life, as well as confer useful dynamic-mechanical properties on elastomers. In order to be able to describe and predict these properties, it is necessary to understand that the degree to which a filler is dispersed during mixing is a quality parameter. With the dispersion of each filler, rubber-filler interaction is improved, an effective phase connection between filler and rubber is achieved and thus the mechanical properties and the service life of elastomers are increased. The importance of filler dispersion as a material-determining parameter of elastomers increasingly became the focus of technological and scientific interest as high-performance elastomers were developed: The properties of rubber and fillers were improved and harmonized, more-efficient mixing technologies were developed and the rubber-filler interaction was studied extensively and systematically to deepen our understanding of it.

The book compares how different fillers are used to provide reinforcement for rubber: carbon black, silanized silica, layered silicates, and multi-walled carbon nanotubes. The fillers offer a variety of reinforcement options that can be used and expand the applications of the elastomers. Regarding the impact of the different fillers on the aspect of rubber reinforcement, filler properties, such as aggregates morphology and the aspect ratio play an outstanding role in filled elastomers. The book explores how combining conventional fillers such as carbon black or silanized silica with small amounts of high-aspect-ratio fillers creates hybrid fillers that offer new properties which cannot be achieved with conventional fillers. It is explained in great detail how the rubber-filler interaction affects the physical, dynamic and fatigue properties of rubber. The presented chapters emphasize a consistency, providing an overall picture of rubber reinforcement.

This book draws on the author's longstanding experience with the global rubber community and is aimed at rubber technology experts, developers, compounders, process engineers, and physicists. It offers a wealth of information for anyone interested in pursuing a career in this fascinating industry. Students of polymer chemistry and materials science looking for an elastic focus outside of thermoplastics will find this to be a complex and interesting subject. Finally, the book will provide inspiration and be a reliable resource for managers who care about improved product quality and reduced costs. The author would like to express his thanks to the German Rubber Institute that provide interesting contributions to the presented book. Without their firm backing and understanding, this effort could not have been accomplished.

Hanover, Germany *Robert H. Schuster*

November 2024

# Contents

# 1 Introduction to Filled Elastomers

There is probably no industry more adept at using and varying raw materials than the rubber industry. In addition to the main components of each mixture – rubber, filler and plasticizer – the usual formulations contain a number of ingredients that have a certain functionality and interact with the main components. These include the ingredients that enable crosslinking to form a three-dimensional network or the anti-aging agents that are intended to reduce the rubber chains' susceptibility to oxygen. Assuming materials of satisfactory quality have been selected for the formulation, producing a high-quality mix represents the heart of the rubber processing. Subsequently the downstream processing and crosslinking to the end product are prerequisites for obtaining a high performance vulcanizate with the desired physical properties. With a few exceptions, the added fillers lead to a significant improvement in the rheological and, above all, the vulcanizate properties. In particular, the stress-strain behavior, the dynamic-mechanical properties and, above all, the fracture behavior, are significantly improved. The morphological features of the filler particles and the physical interaction with the rubber chain play a decisive role, and can effect significant improvements in the properties of the elastomers.

An elastomer is characterized by the crosslinking density of the rubber and the physical properties that result from the presence of a particulate filler and the oil added. Few elastomer applications can be envisaged without some level of reinforcing fillers. The effect of fillers on rubber reinforcement depends on the improvement of the properties, which depend on the primary particles, the shape of the filler aggregates, the loading, the aspect ratio of the aggregate, the interaction with the rubber chains and the chemical nature of the rubber. Four different types of filler have been developed for rubber technology that improve the physical properties of the elastomer due to their reinforcing properties:

- carbon black
- amorphous silica
- layered silicates
- carbon nanotubes

A characteristic of these types of fillers is that the particles are usually in the nanoscale when they are mixed into the rubber matrix. Nanocomposites are defined as composite materials characterized by the presence of dispersed filler particles whose size is in the nanoscale region, defined as having one or more dimensions in the order of 100 nm or less. Among these fillers, there are two groups which can be identified by the size of the particles and their shape. Carbon black and amorphous silica have a small aspect ratio, while layered silicates and especially multi-walled carbon nanotubes have a high specific surface area and a high aspect ratio. The first two fillers are incorporated into a rubber matrix in much higher concentrations, while the last two are used at low concentrations on account of their high aspect ratio.

Carbon black and precipitated silica are iso-dimensional fillers used in the production of filled elastomers. When properly dispersed in the rubber matrix, their particle size, aggregate structures and special surface properties make them unique among filled elastomers. The two-dimensional layered silicates and uni-dimensional multi-walled carbon nanotubes in particular are used, because their high aspect ratio and high specific surface area deliver physical improvements at small volume fractions.

Carbon black is the most important and the most used filler in the rubber industry. The generic term "carbon black" refers to a group of industrial products composed essentially of elemental carbon that is coalesced mainly into aggregates and agglomerates in an industrial manufacturing process. Their physical effects in natural rubber (NR) were published over 100 years ago. The first study on fillers in rubber was produced in 1891 [1] and true reinforcement was first reported in 1906 [2]. The discovery of the reinforcing effect of carbon black by S. Ch. Mote in 1904, who realized that the tensile strength and abrasion resistance of natural rubber are significantly improved by the addition of carbon black, opened the door for modern rubber compounds [3]. From that time on, carbon black been produced to improve the performance of rubber products manufactured from natural rubber (NR) initially and later from synthetic rubber (SR).

Carbon black is manufactured in different ways. The most common is thermo-oxidative decomposition of aromatic mineral oil. The second is the thermal decompositions of low hydrocarbons. The phenomenon of reinforcement remained defined and was only partially understood for the first few decades, due to the chemical and physical complexity of commercial important carbon blacks and the difficulty of measuring the complex mechanical properties. Reinforcement was defined as an increase in stiffness, modulus of elasticity, rupture energy, tensile strength, and resistance to cracking, fatigue and abrasion. Improvements were made to the specific surface area and the geometry of the aggregates, as ascertained initially by determining the iodine adsorption and the tinting strength. Surface activity can promote better compatibility between the filler surface and the rubber chains, which improves the reinforcement of the rubber. The production technology was improved and the yield of the processes significantly increased. Contrary to pessimistic assumptions, the carbon black industry survived the oil crisis without major losses and continued to develop.

Modified carbon black types were created that exhibit much-improved morphology in the rubber compounds, and leads to a significant improvement in reinforcement. The so-called "magic triangle" which reduces rolling resistance, friction and grip, and abrasion resistance was improved by certain types of rubber. Although the same grade may be prepared by different manufacturers, it may not be identical in terms of processing or physical properties after the final curing step. Global production was 5 million metric tons per year but, due to its polluting nature, carbon black depends on petroleum feedstock for manufacturing. Nowadays, compounders can choose the right carbon back for a specific application from a wide range of grades available on the market. Carbon black is the most important rubber filler for tires, as it can increase the speed of cars and extend the tires' life time. More than 80 percent is used in the automotive market alone, with 20 percent finding use in non-automotive applications, such as printing inks and plastics. By 2022, the volume of carbon black was estimated to have risen to more than 15 million tons. In rubber compounds, they are used in a loading of 10–60 wt.% and offer an advantage in terms of physical properties. Carbon black manufacturers strive to produce products that meet the analytical and performance specifications for each grade, but owing the difference in the furnace geometry and the origin of the feedstock. In the second half of the 20th century, the reinforcing properties that carbon black imparts to rubber were the subject of a large number of books, of which a small selection can be found in [4–9].

Non-black fillers have always been an important part of the rubber industry. They represent the principal fillers originally used in rubber technology, which almost exclusively uses natural rubber (NR). The early non-black fillers were mainly naturally occurring minerals, such as clay, whiting, zinc oxide, calcium carbonate and others. But these non-black fillers merely diluted the amount of the rubber in the formulation and failed to produce any significant improvement in the performance of the vulcanized NR.

With the Second World War came the industrial production of amorphous silica. Fumed silica, which is obtained from $SiCl_4$ in a gas-phase reaction is less polar, but has silanol groups on its surface. Today, this material is mainly used to boost the physical properties of silicone rubber. Some years later, an aqueous solution of $Na_2SiO_3$ and $H_2SO_4$ was used to synthesize amorphous precipitated silica, which has a better quality than fumed silica and is particularly useful in polar rubber, such as acrylonitrile-butadiene rubber (NBR). In addition to carbon black, mention should be made of precipitated silicas, which is the second largest class for rubber applications todays.

These fillers contain a high proportion of silanol groups on their surface and exhibit strong filler-filler interactions. They are differentiated according to their activity, although the transitions between the different activity levels are fluid. Precipitated silica emerged in 1950, when amorphous silica demonstrated better reinforcing effects than fumed silica, but at a lower price [10–13]. Initially, precipitated silica was used for shoe soles and for off-road tire-tread compounds to improve chipping and chunking resistance. The reason for this discrepancy is the high polarity of precipitated silica which makes it less compatible with non-polar rubbers, such as NR and styrene-butadiene rubber (SBR). Up to 1970, precipitated silica was used for shoe soles and truck tires, where abrasion resistance was low. However, the great advance made in precipitated silica was the use of organo-silanes, which significantly reduced the polarity of the silica surface, enabling its use in unsaturated non-polar rubber. With the discovery of organo-silanes [10, 14], which can transform a polar surface into a non-polar surface, precipitated silica became a competitive material that provides non-polar rubber with a similar level of reinforcement as carbon black. This treatment renders the product less hydrophilic and produces a typical hydrophobic effect, where filler-filler interaction is at a low level. The initial concept was developed at Bridgestone [15] and commercially introduced by Michelin [16] with its "green tire" technology with a tire tread based on silanized silica in S-SBR/cis-BR blends that improves wet grip and rolling resistance. The high concentration of silanized precipitated silica poses special technological problems in achieving an excellent silica dispersion. Some silanized silica rubber treads must be mixed several times in an internal mixer to deliver a high tire quality. Global consumption is running at more than 1 million metric tons and growing significantly.

Layered silicates are hydrophilic fillers that cannot be dispersed in rubber compounds by a mechanical process, as the tactoids cannot be destroyed in an internal mixer and remain behind as hard agglomerates. Interest in layered silicates increased after 1990 when Toyota researchers used quaternary ammonium salts to produce organo-clays that widened the diameter of the layers and could be mixed with polar polymers [17]. It was found that tactoids can be intercalated and exfoliated by treating the suspension with quaternary ammonium salts and blending it with a polar polymer. Consequently, the high specific surface area and the high aspect ratio can be used to improve the physical properties of rubber [18, 19]. After transformation, the fillers were used in different types of polar rubber. Wet compounding proved to be the best method for using layered silicates without any transformation to improve the barrier properties of different polar polymers [20–22]. Layered silicates are environmentally-friendly fillers and pose no a risk to humans, as they occur naturally in pristine form in large quantities.

At the end of the last century, allotropic carbon forms were synthesized and have had an enormous impact on various applications. Carbon nanotubes have attracted the interest of academia and industry [23, 24]. A large number of reports have heightened interest in understanding the structure-property relationship and finding useful applications. Multi-walled carbon nanotubes are characterized by a high specific surface area and a high aspect ratio and exhibit remarkable mechanical and electrical properties that cannot be achieved with other fillers. The mechanical properties of carbon nanotubes in a rubber matrix makes them candidates for high reinforcement, due to their low density, high aspect ratio and high specific surface area. Rubber nanocomposites reinforced with carbon nanotubes can be considered a kind of particulate composite that varies extensively with the state of the dispersion and the alignment. Carbon nanotubes can be used in very low concentrations (less than 5 wt.%) on account of their high specific surface area and high aspect ratio. The effect of the fillers described is enhanced by the reinforcement they exert in the soft rubber and by the improvement in activity arising from the interaction with the rubber chains.

## 1.1 Reinforcement of Elastomers

An important property is the dispersion of the filler in the rubber matrix and the uniform distribution within it. This ensures good interaction between the filler and the rubber. The dispersion of the filler has always been an important factor for the quality of the rubber compound. Therefore, the mixing of the filler and the choice of mixing tool are predicated on achieving good dispersion at short mixing times. To an extent depending on the degree of dispersion achieved in the rubber compound, the physical properties of the elastomer are significantly improved. This reinforcing effect is described by the attachment of the rubber chains to the filler surface, which leads to reduced chain mobility and is characterized overall by a high level of reinforcement.

The term "rubber reinforcement" refers to the sum of the rubber-filler interactions in the interface between the filler surface and the rubber chains, especially in the crosslinked state of the rubber. The term "reinforcement" was introduced in 1920 by Wiegand, a pioneer in rubber science, who saw the effect of the filler as a particular contribution to significant increase in elastomer properties [25]. The importance of adequate surface bonding in filled rubber compounds was addressed by Wiegand, when he wrote in an article published in 1925: "... with a reinforcing pigment the surface energy is sufficient to prevent the tensional stress in the rubber phase from separating the rubber-pigment surface...". This statement contains the essence of rubber-filler interaction and filler dispersion. It is on one hand the degree of filler dispersion, which indicates the reduced particle size, and on the other hand the bonding of the rubber chains to the filler surface, which converts this geometric requirement into an energetic factor, that enables the material to withstand large dynamic loads. The increase in fracture energy (integral of the stress-strain curve) was introduced as a criterion for the reinforcing effect of fillers. Comparison of the fracture energy of different fillers reveals a steady decrease with increase in concentration of inactive fillers and an increase in the case of reinforcing fillers until the property reaches a maximum [26].

It is worth noting that the term “reinforcement” does not mean the enhancement of a specific physical property, but rather comprises the beneficial changes in elastomer properties generated by the use of active fillers. Thus, “reinforcement” is a technical-commercial value and not a defined property. Reinforcement is to be understood as the sum of the beneficial improvements conferred on the product by the presence of the filler, which should be well dispersed and homogenized in the elastomer [27]. It is ultimately the physical properties demonstrated to use the filler reinforcement. That will determine the composite utility, while its proliferation will be determined by the cost of the product. One can refer to the filled rubber materials as nanocomposites, even though this term was only introduced three decades ago, but the reinforcement of elastomers has been studied for many decades before that. Before the beneficial effects of the reinforcing fillers were known, naturally occurring minerals, such as clay, zinc oxide, mica and others that are above the nano size limit, were used to occupy space in the rubber compound and dilute the expensive natural rubber imported from Brazil and Southeast Asia. The fillers lowered the price of the natural rubber vulcanizate, but did not reinforce the filled rubbery material. If the development of reinforcements in rubber composites is considered in terms of the use of carbon black, parallels with car development can be seen, both in tire technology, where abrasion resistance and rolling resistance play a role, but also under the hood, where heat resistance, aging resistance and fuel transport play a role. In all other areas, filler-reinforced elastomers play an important role in our society, be it in bridge construction, earthquake protection or home comfort.

It was found that the physical properties of such nanocomposites exceed those of unfilled rubber vulcanizates many times over. The beneficial improvement in the use of carbon black as a nanoscale filler was recognized for synthetic rubbers, which show a significant tremendous increase in performance. Over the last century, it has become clear that carbon black as a filler can significantly enhance the property of rubber while lowering costs [5, 6]. Besides its environmental benefits, the manufacturing process allows for greater flexibility, as it is able to manufacture more different grades of carbon blacks in terms of quality, type and quantity.

In addition to the class of carbon blacks, significant improvements have resulted from the use of silanized silica in non-polar rubber, which in S-SBR/cis-BR effects a substantial improvement in wet grip and rolling resistance in automotive treads. More knowledge is also being gained in the use of fillers of high aspect ratio and high specific surface to improve reinforcement. The physical basics can be seen in the use of exfoliated layered silicates or isolated carbon nanotubes, which depend on the high aspect ratio and the increased specific surface area as well as the dispersion of the filler, and manage with a low concentration in the vulcanizate. Although the use of carbon nanotubes leads to a high degree of reinforcement in rubber compounds, it also greatly affects the electrical conductivity needed in the rubber matrix.

One option that is being used more and more frequently is the use of hybrid fillers, i. e. the use of a large quantity of conventional fillers alongside a small quantity of carbon nanotubes or layered silicates.

## 1.2 Rubber Melts and Elastomers

To keep the terminology consistent, the term "rubber" will be reserved for the raw rubber obtained by polymerization reactions or by tapping the *Hevea brasiliensis* tree, while "elastomer" will be used for the Brasiliense's vulcanized rubber or elastic engineering material. The unfilled rubber melt is an assembly of polymer chains of small cross-sectional diameter that possesses a glass transition temperature below 0 °C, where the melt takes on a glassy consistency and can easily break. Above the glass transition temperature, the rubber melt can be deformed by an external force, but it will not return to its original dimensions when the force is removed, on account of the entanglement of the rubber chains and residual elongation in the rubber melt. The rubber melt is a quasi-elastic assembly arising from the entangled and elastic rubber chains. Due to the entanglements in the raw rubber melt, the material exhibits viscoelastic behavior; in other words, it behaves in between the deformation of an ideal solid and an ideal viscous material. The degree of viscoelasticity depends on the length of the rubber chain, the number of entanglements and finally the crosslinking density.

Chemical crosslinking causes the rubber melt to enter a vulcanized state that is characterized by a wide-meshed network, such that the large distance between the network nodes makes the crosslinked chains elastic. Such a vulcanizate is also known as an elastomer. When a force is applied to the elastomer, rapid deformation takes place and when the force is removed, the elastomer quickly returns to its original size. The difference between the behavior of the elastomer and that of the non-crosslinked rubber compound is shown in Figure 1.1.

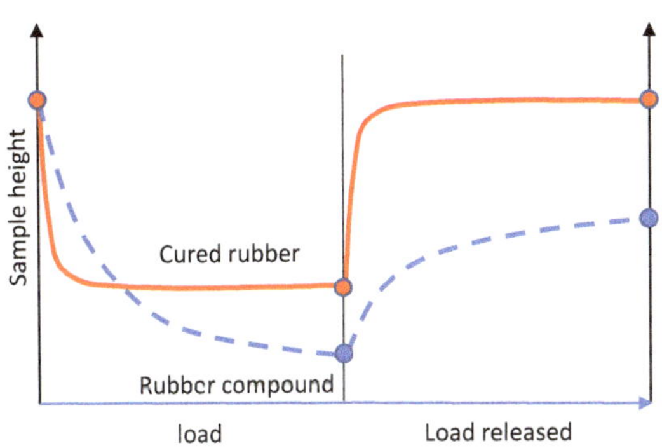


**Figure 1.1**
Response to a filler force acting on the mixed compound and the vulcanizate

Crosslinking of the rubber chains transforms the system from a free-flowing rubber melt into an elastic material, the elastomer. The chain sections between adjacent crosslinks are long enough to take on a coil-like shape and make the elastomer highly elastic.

The manufacture of elastic rubber chains is characterized by the polymerization reaction of monomers, i. e. olefins and dienes. During polymerization, the monomers are linked together to form long-chain macromolecules and, due to the low glass transition temperature, highly elastic rubber chains. The synthesis of rubbers uses different monomers to generate the polymerization mechanism [28–30]:

- radical mechanism
- anionic mechanism
- cationic mechanism
- coordination catalysis

The resulting rubber is divided into homo- and copolymers in accordance with the composition of the monomer. Copolymers use the properties of both monomers and therefore change the physical behavior. They can form linear or branched polymers, depending on the mechanism of polymerization. The macrostructure of the chain molecules has a significant influence on the processing behavior and crosslinking as well as the physical properties of the rubber. One exception is anionic polymerization, which yields linear polymers of different chain lengths and a narrow molecular weight distribution. The other strategies for polymerizing the monomers, lead to long-chain, branched polymers with a broad molar mass distribution. According to ISO 1629, the classification of rubber is based on the chemical structure shown in Table 1.1.

**Table 1.1** Classification of Rubber

| Chemical Composition | Rubber Grade | Type |
|---|---|---|
| Unsaturated chains (diene rubber) | NR, SBR, CR, NBR, IR und IIR | R |
| Saturated chains (polymethylene chain) | EPDM, EPM, CM, ACM | M |
| C-O bonds in the main chain | ECA, GPO | O |
| Silicone rubber | MQ, MVQ, MPVQ, MFQ | Q |
| Polyurethane rubber | AV, EV | U |
| Sulfur in the main chain | Polysulfide rubber | T |

Finding the right choice of rubber, which is essential for reinforcement, can be an important task for a compounder, especially when mixing with appropriate fillers of a certain activity. The next task is chemical crosslinking, for which reliable chemicals are sought and in turn is another task for the compounder. Crosslinking of the unsaturated rubber is carried out with sulfur and accelerators or with peroxides at elevated temperatures, above 120 °C, in the case of saturated chains which cannot be vulcanized with sulfur. For saturated rubbers the peroxides or the resins are used. To an extent depending on the type of crosslinking, the elastomer acquires defined properties that determine its physical behavior.

## 1.3 Entropy Elasticity

When an ideal rubber network is extended by an external force, the network chains are elongated in the direction of the force vector and the entropy of the entire network decreases. When the force is released, the network quickly returns to its initial state of higher system disorder. The elasticity of a rubber network is thermodynamically based on the change in conformational entropy and is referred to as entropy elasticity or rubber elasticity [31]. This reversible process corresponds to a quantitative transformation of mechanical energy into a lower state of entropy. The principle of entropy elasticity is shown in Figure 1.2.

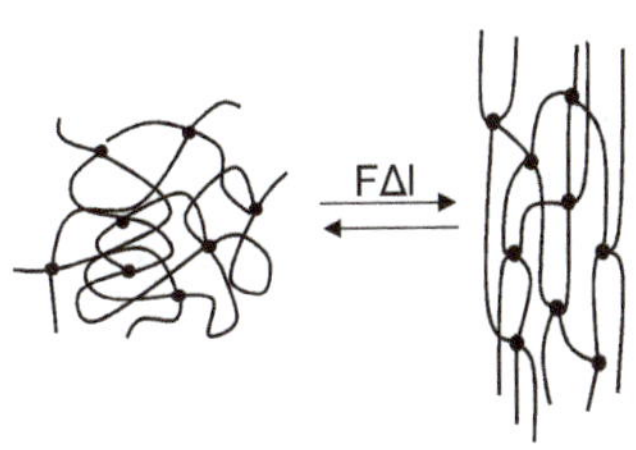


**Figure 1.2**
Concept behind entropy elasticity

The concept of entropy elasticity applies to ideally elastic bodies for which the theory does not provide for intra- and intermolecular interactions between the chains and internal friction processes of adjacent chains during deformation. The increased mechanical work is compensated for by a lower entropy of the polymer chains. For a system in thermodynamic equilibrium under isothermal conditions, the work done, $W$, is equal to the change in internal energy, $U$, and entropy, $S$:

$$dW = dU - TdS \tag{1.1}$$

where $W$ is the work done, $U$ is the internal energy, $S$ is the entropy of the system and $T$ is the absolute temperature

Assuming that the enthalpy does not change during deformation, the restoring force under strain increases with increase in temperature. This is known as the Joule effect:

$$f = -T\,(dS/dl) \tag{1.2}$$

where $f$ is the restoring force, $l$ is the length of deformation, $S$ is the entropy, $T$ is the absolute temperature

The calculation leads to an equation for an ideal unfilled elastomer that represents the applied mechanical energy as a reduction in the entropy of the elastic system:

$$F\Delta l = -T\Delta S \tag{1.3}$$

This relationship is describing the behavior of an ideal rubber network. For real networks, some corrections are needed that are related to the structure of the unfilled elastomer:

- presence of chain entanglement
- free chain ends
- inhomogeneities of the crosslinks
- different lengths of rubber chains

Generally, the equation describes the mechanical energy that is transformed into a higher-order network structure. Rapid deformation and recovery are important components of the elasticity of a rubber network.

Elastomers are soft and compressible materials with moduli of elasticity between $10^{-1}$ and $10^{2}$ MPa. They are characterized by high elongation of up to 1000%, behave hyper-elastically when deformed and have a positive temperature coefficient [32]. However, elastomers are limited in their application due to mechanical fracture failure. Fillers can significantly improve the physical properties and, especially, significantly reduce failure under extreme deformation [33]. Although real rubber networks deviate from this ideal state, this unique material concept is guides the behavior of both unfilled and filled elastomers. From a purely phenomenological point of view, elastomers differ from the other polymeric materials in four properties:

1. soft materials
2. highly elastic
3. dimensionally stable, and
4. small residual elongation after extensive elongation

In contrast, thermoplastic elastomers, which are crosslinked by physical interactions, are not crosslinked, but contain a hard phase and exist in a glassy or a semi-crystalline state. As opposed to this, elastomers exhibit highly elastic behavior from a temperature above of the glass transition temperature up to an elevated temperature at which thermal decomposition of the rubber sets in. The residual elongation of elastomers is used to distinguish between elastomers, thermoplastic elastomers and thermoplastics. Elastomers have very low residual elongation (1–2%), while thermoplastic elastomers have much higher residual elongation (> 30%), with thermoplastic materials having the highest value, as shown in Figure 1.3.

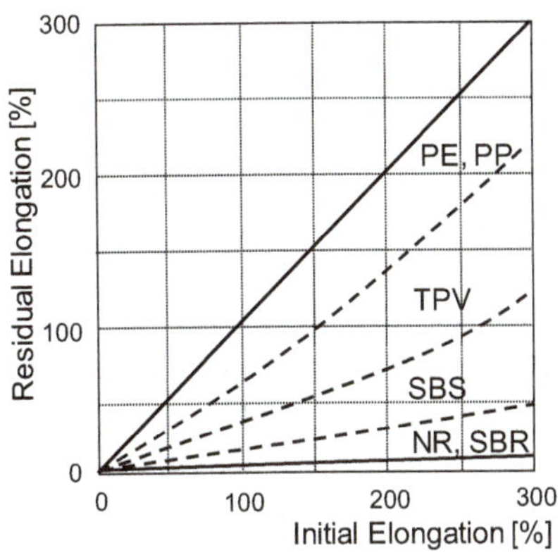


**Figure 1.3**
Residual elongations after stretching different classes of polymers

For any rubber network, the most important molecular characteristic is the crosslink density, which can be expressed either as the number of crosslinks or network chains per unit volume. The higher the crosslink density, the lower is the extensibility of the network chains and consequently the higher the force required to extend the network. Above a certain crosslink density, the entanglements are permanently locked or trapped between the network chains and contribute to the elastomer's resistance to uniaxial deformation. While the stiffness increases proportionally to the crosslink density and the hysteresis decreases inversely proportional to it, the ultimate properties, such as tensile strength and fatigue resistance show a characteristic maximum at a certain crosslink density that depends on network formation during vulcanization.

## 1.4 Advantages of Fillers

After the invention of vulcanization by Charles Goodyear (1839), the incorporation of a solid filler to dilute the expensive natural rubber became standard practice in rubber technology. The filler increases the viscosity of the filled rubber melt and the hardness is increased, the stress-strain behavior, the dynamic – mechanical moduli and the crack propagation is strongly reduced. The role of the filler in the interaction with the rubber chains is shown in Figure 1.4.

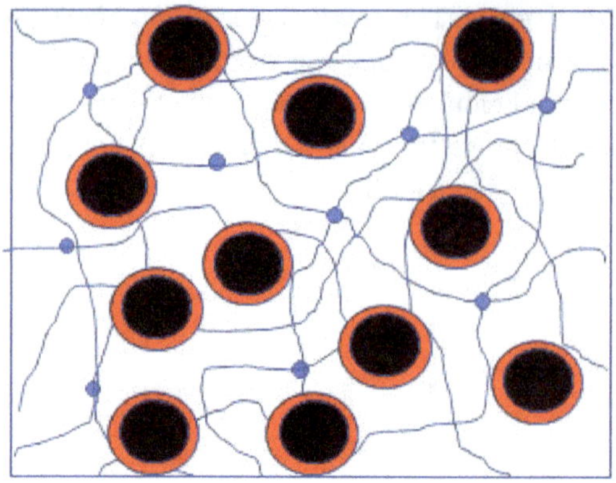

**Figure 1.4**
Filler in a rubber vulcanizate

Attempts were made to reduce the size of the particles to the filler aggregates, which are important for the physical properties of the elastomers. The incorporation of carbon black into the rubber compound and the subsequent dispersion of the aggregates reduces the fracture mechanical properties, such as tensile strength, tear energy and cut growth resistance and leads to a continuous improvement in the life time of the elastomer. The high-activity carbon blacks were used for the development of high-performance tire treads, while the semi-active carbon blacks were used for carcasses. In addition, a number of semi-active to active carbon blacks were used for technical rubber goods. Soon it was recognized that the dispersion of carbon black could be improved by providing better mixing elements for carbon black in the synthetic rubbers. Dispersion was an important challenge for the rubber industry if the performance of the rubber was to be improved. It can be assumed that dispersion is related to the interaction between the filler surface and the rubber material as the filler volume fraction increases (see Figure 1.4). The specific properties of the filled compounds change almost proportionally, such as the increase in tensile strength, the reduction in crack growth and the increase in tear energy [5–7]. The carbon blacks of highest surface activity are used comparatively little for tread compounds, because dispersion during mixing was not as good as for the grades of lower specific surface area.

## 1.5 The Role of Filler Activity

The degree to which the physical properties of an elastomer are increased by particulate fillers is qualified by the specific surface area, the "structure" of the filler aggregates and the filler activity. The term "activity" refers to the degree of interaction between rubber and filler during processing and in the vulcanized final state. Fillers of higher activity significantly increase the reinforcement of the elastomer. The activity range of the main groups of particulate fillers used in the rubber industry is shown in Figure 1.5.

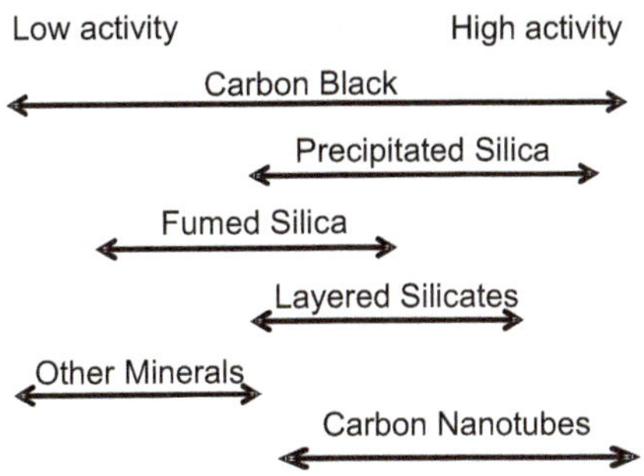


**Figure 1.5**
Activity of fillers used in rubber technology

The carbon blacks are produced under low reaction times to afford small primary particles that lower the price of the carbon black on account of the short residence time and low oil consumption. The carbon blacks are used in tread compounds and to enhance the vulcanizate properties and to increase abrasion resistance. The active fillers of high specific surface area yield reinforced elastomers that possess excellent physical properties, but are difficult to efficiently disperse. The activity of the fillers is associated with significant changes in rheological properties, such as an increase in the compound's viscosity and a reduction in extruder die swell. They provide major improvements in the physical properties of the rubber vulcanizate, especially enhancement of the ultimate properties, such as tensile strength, and a reduction in dynamic crack growth. The semi-active carbon blacks of larger primary particle size are produced at longer reaction times and are used at high loading capacity and are easy to disperse and process. Low-activity fillers result in less interaction between rubber and filler, allowing for better dispersion and downstream processing, which is reflected in extrusion or flexural strength, lower heat buildup, but their effect in terms of cut growth and abrasion resistance is rather limited. Carbon black imparts excellent flexural strength and good resilience to rubber and is preferred for mechanical goods, belts, hoses and cable jackets. As can be seen, carbon blacks cover the entire range from low to high activity and underline the importance in rubber technology and the wide range of technical applications in filled elastomers.

After suitable silanization, precipitated silicas exhibit medium to high activity, which improves dispersibility in non-polar rubber. Silanized silicas are now used in an increasing number of products, due to their good dispersibility, and ensure low rolling resistance and good wet grip of car tires, while interacting very well with the rubber chains. Fumed silicas are an active filler used for silicone rubber in medical applications.

Layered silicates can serve as pristine clays or can be transformed to organo-clay and are mainly used in polar rubbers that interact well with the two-dimensional platelets. They are important fillers that are aligned with the direction of flow and exhibit enhanced reinforcement in rubber components. Their main application in barrier properties demonstrates effective influence on the transport properties of gases and liquids. They are important fillers for reducing the explosive decomposition of polar rubber compounds used in oil production.

Multi-walled carbon nanotubes exhibit a high activity in all rubbers at small volume fractions, mainly due to their exceptionally high specific surface area and high aspect ratio. Certainly, combination with iso-dimensional fillers lead to practical applications. Due to the high aspect ratio, multi-walled carbon nanotubes lead in combination with carbon black or precipitated silica to an improvement in important physical properties that cannot be achieved with a conventional filler. However, the price of carbon nanotubes is still prohibitive and their availability for broad industrial applications is far from satisfactory.

There is practically no physical property of elastomers that is not influenced to some extent by the interaction between rubber and filler and the affinity of the filler for rubber. Active fillers are able to build up strong physical interactions with polymer chains, which are adsorbed on the filler surface and form a dense polymer interface that causes physical phase adhesion. In contrast to this class of fillers, inactive fillers can only build up weak interactions with chain segments of the rubber and so rubber-filler phase adhesion is poor. The difference between an active and a non-active filler can be illustrated by the loading shown in Figure 1.6. In this illustration, it is assumed that the filler is actually completely dispersed in the rubber matrix. Active fillers increase the tensile strength and the abrasion resistance and pass through a maximum value when the filler loading is increased. This maximum value is characteristic of each type of carbon black in a rubber. After reaching the maximum, the ultimate properties decrease continuously to lower values.

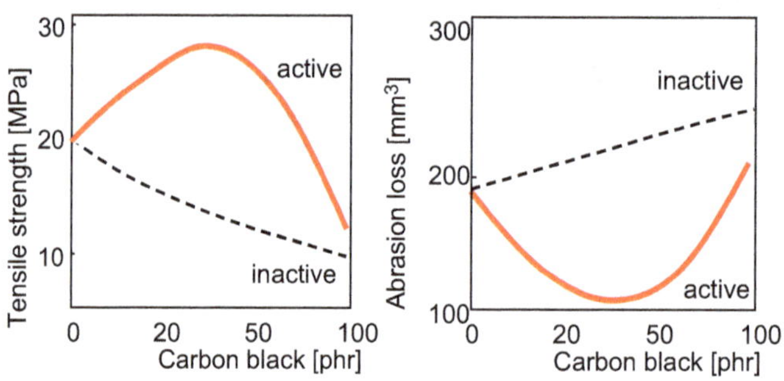


**Figure 1.6** Effect of particle size on the enhancement of the mechanical properties

In the case of active carbon black, the property under consideration is improved up to loadings corresponding to a volume fraction of 0.17–0.19. However, the maximum values depend on the type of rubber and the activity of the filler as well as on the mixing conditions and the carbon black dispersion. However, the maximum values depend on the type of rubber and the activity of the filler as well as on the mixing conditions and the dispersion of the filler. Important vulcanizate properties, such as stress-strain behavior, tensile strength, dynamic-mechanical properties, cut growth, abrasion resistance and tear resistance, are improved by active fillers, due to the presence of a rubbery interface layer that is physically bonded to the filler surface. In the case of semi-active carbon black, the tensile strength and abrasion resistance values decrease steadily as filler loading increases. It has been found that decreasing the particle size down to the nanoscale range boosts reinforcement of the mechanical properties according to an exponential function, as shown in Figure 1.7.

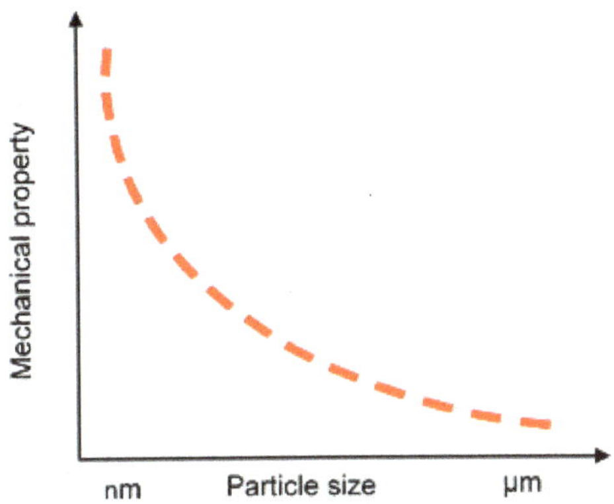


**Figure 1.7** The properties of an elastomer increase with decrease in particle size

Since the fillers used in rubber technology are supplied by the suppliers in pelletized form, the filler particles have to be broken down during mixing in order to achieve reinforcement effects in elastomers. For this purpose, various internal mixer concepts with specific mixing rotors and cooling systems have been developed to improve dispersive and distributive mixing in the rubber compound at a certain temperature. While large particles disperse relatively quickly, fine particles can be difficult to disperse and may require more than one pass through the mixer. When the rubber compound is extruded through an extrusion die, it exhibits "elastic memory". The fillers improve the extrusion properties by suppressing the elastic memory, and this ability to reduce swelling of the extrudate in the die is a direct function of the increased specific surface area and structure.

When using precipitated silica, hydrophobization of the polar silica surface with organo-silanes is required in order to achieve good dispersion in non-polar rubbers. Such treatment of the silica leads to low rolling resistance and excellent wet grip on tire treads. For layered silicates, it is important to use intercalated or, preferably, exfoliated layered silicates produced by wet compounding or organic clays to obtain good barrier properties. The use of carbon nanotubes is made possible by commercially oriented mixing processes, in which small volumes are incorporated into a rubber, mostly in combination with carbon black or silanized silica. In this way, the special effect of the high aspect ratio can be used to produce elastomers with very good strength, reduced dynamic crack growth and good abrasion resistance.

## 1.6 The Role of Filler Dispersion

The quality of elastomers containing reinforcing fillers depends not only on the physico-chemical properties of the rubber matrix, but also on the type and activity of the filler and dispersion of the filler during the mixing process. The dispersion of the filler is increased by the high elongational and shearing forces and the low mixing temperature. The greater the polymer-filler contact area and the interfacial interaction at the boundary of the rubber phases, the more bonding between the filler and the rubber leads to an improvement in dispersion. The key improvement is the progressive dispersion of the filler and the increase in beneficial contacts between the filler surface and the chemical chains of the rubber. Overall, the volume fraction of the rubber-filler interface providing the phase adhesion increases during dispersion and the proportion of non-dispersed filler agglomerates decrease. The fraction of non-dispersed filler leads to defects in the compound and premature failure of the vulcanizate, especially under continuous dynamic load. The effects of filler dispersion are shown in Figure 1.8.

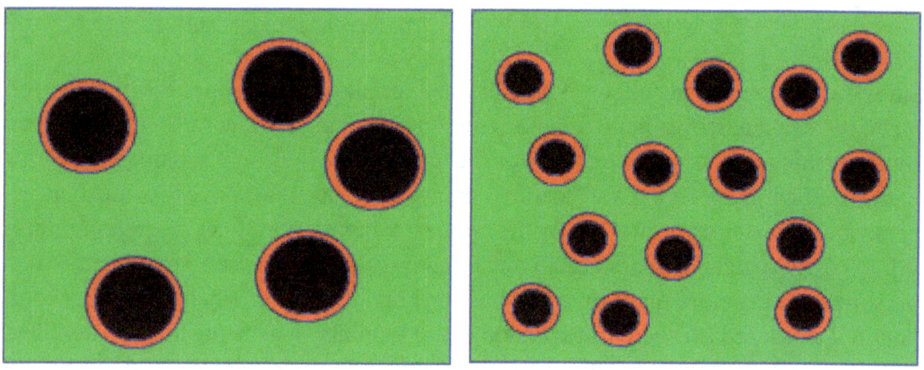

**Figure 1.8** Dispersion increases the interaction between the filler surface and the rubber matrix

The bonding forces on the filler surface and the chemical nature of the rubber as well as the efficiency of the mixing tools contribute to better dispersion. The parameters of the mixing tool are largely responsible for good dispersion of the filler particles in the nanocomposite. It is important to know the degree of dispersion of the filler and the distribution of the filler during the mixing and downstream processes, as well as the influence of high temperatures during the induction time of the crosslinking process. It is advisable to routinely check the filler dispersion in the mixing chamber by means of user-friendly test methods and thus to derive the quality of the mixed rubber compound.

The effects arising from the dispersion of fillers must be understood as an interplay of material parameters that support the interaction between rubber and filler and the size reduction of particulate fillers. Thus, compounders should have a detailed knowledge of filler morphology and its role in the process of filler dispersion. This knowledge should exceed the level of commercial product specifications. In the author's opinion, it is important that compounders know in detail and are able to justify the choice of a particular type of filler and the loading used for the product.

The problem of selecting the right raw material for manufacturing a commercial elastomer falls to the compounder, who should be well versed not only in the performance of the raw rubbers, the type of fillers and the other ingredients of the formulation, but also in the multidisciplinary material science so as to be able to assess the multiple interactions and sometimes the conflicting interferences between the compound ingredients during mixing and downstream processing and final curing process curing to obtain the desired and specified properties of a competitive product.

The information about the particular functionalities of compound ingredients, their interactions in the compounds and in the vulcanizates or the interferences and interdependencies during mixing and further processing cannot be conveyed in sufficient detail to compounders, process engineers and experts active in research and development in companies that do not have access to specialist libraries.

That reinforcement is still at the center of research and that innovations are still being made in fillers and polymers are an indicator of its importance. In recent decades, the topic has gained much interest in connection with the development of high-performance elastomers and the more efficient use of raw materials, including the reduction of waste. The concepts developed for understanding the reinforcement of elastomers by particulate fillers consider the colloidal properties of the fillers, especially the specific surface area and the shape (or "structure") of the particles (i. e. aggregates) and assume an "ideal" dispersion, i. e. the size of the filler particles is reduced to the smallest level possible. However, an "ideal" dispersion is not achieved in real compounds manufactured on two-roll mills or internal mixers. A real, filled compound, even one used for a scientific study, contains particles whose size distribution is broad or narrow to an extent depending on the mixing technique employed. Thus, the particle size distribution in real compounds ranges from nanoparticles to pellet fragments.

As rubber manufacturers today are increasingly faced with economic pressures and rising demands on product quality and durability, it seems necessary to identify the potential of compound ingredients for transforming them into quality elastomers that meet the stringent requirements of the market. A look at the main components of rubber compounds – rubber, filler and plasticizer – reveals that they account for 90% of the total weight, but only 80% of product performance and 80% of product cost. The remaining 20% is covered by the vulcanization system and the antioxidants. Ownership of the rubber formulation makes it possible to control material cost at the expense of running a materials development department and a mixing chamber [35]. Raw materials must therefore be used consciously in order that production efficiency may be maximized and waste minimized, as shown in Figure 1.9.

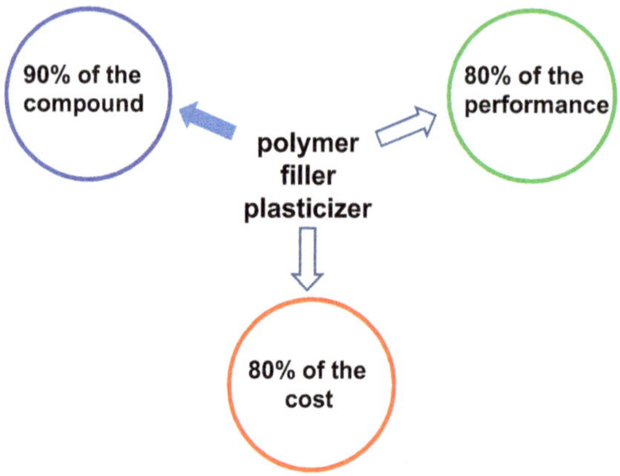


**Figure 1.9**
Importance of the main compounding ingredients

This requires the potential of all ingredients to be identified and fully exploited at a competitive cost. This task calls for complex knowledge of the functionality of each ingredient, the order in which they are added to the mixing tool, the mutual interactions of the ingredients during mixing, the degree of filler dispersion achieved during mixing, the changes that occur during extrusion, calendaring, shaping and molding and the curing state achieved during vulcanization.

As a side-effect, the better the knowledge of the material functionalities during processing and in the product, the less effort is needed to identify superfluous compound ingredients. In recipes still used in the rubber industry today, there are a number of ingredients that are intended to serve the same purpose or those that have no purpose. The elimination of such useless ingredients contributes substantially to the reduction of manufacturing costs.

In order to meet the requirements for an end product with specific properties, the compounder must select the rubber with the best performance. This requires a consistent knowledge of the property set of the rubber, its interaction with reinforcing fillers and the impact of the impact of the interaction on flow behavior of the compound, and finally the curing behavior. Any problem that arises because of an unsatisfactory property of the product or unsatisfactory processability requires a definitive answer on how to change the manufacturing process or to find a better formulation. The relationship between the formulation ingredients and the manufacturing process must be background knowledge for daily operations and problem solving.

The nature of the intimate mechanism is still not completely elucidated. More than 40 years ago Gerard Kraus wrote: “One might think that in such a mature field, most of the important questions would have been settled long ago. This is not so; few topics in rubber science still continue to be the subject of more controversy. To understand why, one only needs to keep in mind the great complexity of the system" [36]. The complex behavior of a filled elastomer results from the different types of filler and rubber chains of specific elasticity. To an extent depending on the type of mixing, the internal mixer, the order of addition of the ingredients, the mixing temperatures, the mixing time and the geometry and speed of the rotors, the loading factor, the dispersion and distribution of the filler in the rubber can be increased. The morphology of the rubber compound is determined solely by the interfacial tension between the rubber and the filler during the mixing process. In addition, the plasticizers, the antioxidants and, above all, the crosslinking system contribute to the formation of the elastomer.

The rubber-filler influences determine the range of properties of the filled elastomers in extreme temperatures and make it possible to further improve their performance limit. Without filled elastomers, both comfort and strategic importance in the mobile world would be inconceivable. An understanding of filled elastomers is an important prerequisite for further improving their properties, especially through the use of conventional fillers and new nanofillers.

## References

[1] Heinzerling, C., Pahl, W., Verk. Ver. Behördl. Gewerbefl., 1981, 415

[2] Ditmar, H., Gummi Ztg., (1906), 20: 394

[3] "Vignettes from the Science Hall of Fame", ACS Rubber Division (1989), p. 193

[4] Kraus, G., in "Reinforcement of Elastomers", Kraus, G., Ed., Interscience Publishers, New York (1965)

[5] Donnet, J.-B., Bansal, R. C., Wang, M.-J., (Eds.) in "Carbon Black Science and Technology", Marcel Dekker Inc. New York (1993)

[6] Gerspacher, M., Wampler, W. A., in "Basic Elastomer Technology", Baranwal, K. C., Stephens, H. L., Eds., Rubber Division (2001), Ch. 4a

[7] Röthemeier, F., Sommer, F., Ed., in "Kautschuktechnologie", Hanser Verlag, Munich (2001)

[8] Mark, J. E., Erman, B., Roland, M. C., Eds., in "The Science and Technology of Rubber", Academic Press, Waltham, (2013)

[9] Wang, M.-J., Morris, M., in "Rubber Reinforcement with Particulate Fillers", Hanser Verlag, Munich (2022)

[10] Wagner, M. P., Rubber Chem. Technol., 1978, 49: 703

[11] Görl, U., in "Handbuch der Kautschuktechnologie", Hofmann, W., Gupta, H., Eds., Dr Gupta Verlag, Ratingen, (2001), Ch. 8b

[12] Meon, W., Blume, A., Luginsland H.-D., Uhrland, S., in "Rubber Compounding, Chemistry and Applications", B. Rodgers, Ed., Marcel Dekker, Inc. New York, Basel (2004)

[13] Byers, J. T., in "Basic Elastomer Technology", Baranwal, K. C., Stephens, H. L., Eds., Rubber Division (2001), Ch. 4b

[14] Thurn, F., Wolff, S., Kautsch. Gummi Kunstst., 1971, 28: 733

[15] Takashito, M., Makai, U., Sugawara, S., US Patent 4820751A (1988)

[16] Rauline, P., US Patent 5227425A (1992)

[17] Fujiwara, S., Sakamato, T., Japan. Kokai Pat. Appl. 109998 (1976)

[18] Giannelis, E. P., Krishnamoorti, R., Manias, E., Adv. Polym. Sci., 1999, 138: 108

[19] LeBaron, P. C., Wang, Zh., Pinnavia, T. J., Applied Clay Sci., 1999, 15: 11

[20] Wu, Y.-P., Zhang, L-Q. Comp. Sci. Technol., 2005, 65: 1195

[21] Klafke de Azeredo, L., Schuster, R. H., Jacobi, M. M., Kautsch. Gummi Kunstst., 2010, 62: 130

[22] Schellenberg, M., Klafke de Azeredo, L. Schuster, R. H., 184th ACS Rubber Division Cleveland 2013

[23] Ma, P-Ch., Kim, J.-K., Eds., in "Carbon Nanotubes for Polymer Reinforcement", CRC Press, Taylor & Francis Group, Boca Raton, USA (2011)

[24] Grady, B. P., in "Carbon Nanotube-Polymer Composites, Manufacture, Properties and Applications", J. Whiley & Sons, Hoboken, USA (2011)

[25] Wiegand, W. B., Ind. Rubber J. (1920) 60

[26] Wiegand, W. B. Trans. Inst. Rubber Ind. (1925) 1

[27] Studebaker, L., in "Reinforcement of Elastomers", G. Kraus, Ed., Interscience Publishers, New York (1965)

[28] Quirk, R. P., Gomochak-Pickel, D. L., in "The Science and Technology of Rubber", (3rd ed., Mark, J., Erman, B., Eirich, F. R., Eds.) Elsevier Acad. Press, (2005) Ch. 2

[29] Odian, G., "Principles of Polymerization", Wiley-Interscience, 4th ed., New York, (2004) Ch. 4

[30] Carraher, Jr., Ch. E., in "Polymer Chemistry", CRC Press, (2011), Ch. 7

[31] Kuhn, W., Kolloidzeitschrift, 1934, 34

[32] Treloar, L. R. G., in "The Physics of Rubberlike Elasticity", 3rd ed., Clarendon Press, Oxford (1975)

[33] Gent, A., in "Science and Technology of Rubber", (Mark, J. E., Erman, B., Eirich, F. R., Eds.) Elsevier, 3rd ed., (2005), Ch. 1

[34] Coran, A. Y., in "The Science and Technology of Rubber", Mark, J., Erman, B, Eirich, F. R., Eds., 3rd ed., Elsevier Academic Press (2005)

[35] Graf, H.-J., in "Cost Reduction in Rubber Processing" (Graf, H.-J., Ed.) TechnoBiz Communications Co., Ltd. (2014), Ch. 7

[36] Kraus, G., Angew. Makrom. Chem., 60/61 (1977) 215

# 2 Active Fillers

Developing an efficient formulation and finding the best conditions for flawless processing on the equipment available in the factory requires knowledge of raw materials and their interactions during processing. Compounding requires a thorough understanding of all raw materials and a deep knowledge of the functionality of the compound ingredients during processing. Considering that the reinforcement of elastomers provided by nanoscale fillers is based on the size and morphology of the fillers interacting with the rubber, it is appropriate to first introduce the manufacturing processes for the filler classes used in the rubber industry. As the origin of all reinforcement effects lies in the interaction of nanoscale fillers and the rubber chains, the formation of the filler particles is presented in this chapter. The technology for producing the filler particles is described in the following technical processes.

## 2.1 Carbon Black

Carbon black is the generally accepted term for the most important class of fillers for the rubber industry. Carbon black consists of elemental carbon aggregates, which are discussed in textbooks and relevant reports [1–4]. There are different types of carbon black covering a wide range of specific surface area and "structure" and exhibiting different activity in their interaction with rubber, ranging from low-activity to semi-reinforcing carbon blacks used in tire carcasses and rubber goods and high-activity carbon blacks mainly used in tire tread compounds. The manufacturing process is divided into two groups of technologies. The main group utilizes the incomplete or partial combustion of aromatic hydrocarbons. The chemical reaction for carbon black production by thermal-oxidative decomposition of hydrocarbons can be formulated as follows:

$$C_xH_y + O_2 \longrightarrow \text{Carbon black} + CO + CO_2 + H_2 + H_2O$$

This process can be ensured by various methods and process parameters and leads to carbon black grades of different particle sizes, structures and surface activities. This group includes the furnace blacks, the gas blacks, the channel blacks and the lamp blacks, as mentioned in Table 2.1, which lists the different chemical processes, the manufacturing process and the feedstock that lead to carbon black. The most commonly used technology is the furnace black process, which is based on incomplete combustion of crude aromatic oil and tar oil products. Furnace production has a yield of 70% and covers the production of 95% the world's carbon black [2, 3].

The second technological process is the thermal decomposition of low molecular weight hydrocarbons in the absence of oxygen. This is called thermal black technology and it yields larger carbon black particles. The thermal decomposition process yields carbon black and hydrogen:

$$C_xH_y \longrightarrow \text{Carbon black} + H_2$$

There are two processes for the thermal decomposition of hydrocarbons, which differ significantly both in terms of the feedstock and the properties of the resulting carbon black. Thermal blacks are obtained from natural gas and low-boiling oil hydrocarbons, while acetylene blacks are obtained from acetylene; both have a special morphology and in the case of acetylene black a high electrical conductivity. These thermal blacks account for up to 5% of global production [2]. A summary of the different methods used to produce carbon black along with the feedstock and global production is given in Table 2.1.

**Table 2.1** Chemical Process (Types of Carbon Black and Feedstock Used)

| Chemical Process | Manufacturing Process | Percentage of Production | Feedstock |
|---|---|---|---|
| Partial combustion | Furnace black | 95% | Aromatic oils, coal tar oils<br>Natural gas |
| | Gas black | | Coal tar oils |
| | Channel black | | Natural gas |
| | Lamp black | | Petrochemical oils, coal tar oils |
| Thermal decomposition | Thermal black | 5% | Natural gas, oil |
| | Acetylene black | | Acetylene |

The various carbon black processes were named for their industrial use. The first industrial carbon black produced was channel black, but after 1889 the furnace process was introduced, which guaranteed a higher yield and wide variety of specific surface areas [5].

## 2.1.1 Thermo-Oxidative Process

This process is based on the transformation of the aromatic feedstock, which decomposes at high temperatures to produce a carbon black that interacts well with the rubber. The main argument for using the thermo-oxidative process is to ensure economical production and to meet customer expectations regarding this wide variety of carbon black. However, the manufacturing companies need space for:

- storage of the oil feedstock
- the reaction for forming carbon black
- filtration/separation from the tail gas
- pelletizing, and
- drying and shipping

Today the technology has been refined to produce different carbon blacks with different specific surface areas and structures. However, carbon black has fundamental properties that cannot be obtained with other groups of fillers. In addition to being inexpensive and relatively easy to produce, they also offer a wide range of possible applications in rubber compounds.

### 2.1.1.1 Channel Black Process

The first carbon black process was the channel black, which was introduced in 1872 in Texas using natural gas from the oil industry [6]. The carbon source in the channel process is the natural gas from which the lower-boiling gasoline was used. The name channel black stems from the iron channels used to collect the carbon black deposited by small natural gas flames striking the surface of the channels. The channels are moved in a slow, reciprocating motion over stationary scrapers. The carbon black is collected in large containers and conveyed to packaging by a longer shift. The channel black process is shown in Figure 2.1.

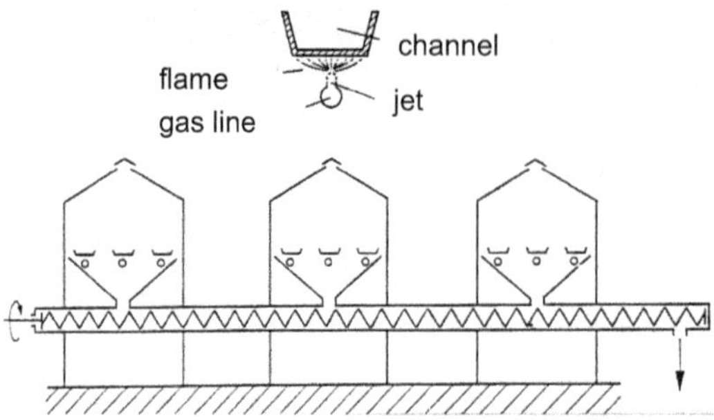


Figure 2.1
Channel black process

For most non-rubber applications, the carbon black is then packed into paper bags. For rubber applications, it is occasionally dry-pelletized in large horizontally rotating pelletizing drums. The process of channel black has improved over the decades, but it remains a very-low-yield process and has not been able to compete with the increase in market size in the tire industry. The oxidation causes the oxygen content of the carbon black to increase from 3% to as much as 10%, reducing the particle diameter from 30 nm to much lower values. The particle size of channel carbon black was rather small and the surface of the particles was oxidized, a fact affected the reinforcement of rubber compounds and especially slowed down vulcanization [7]. The acidic nature of channel blacks also contributes to delaying scorch time and the cross-linking process in the rapid vulcanization of NR. Up until the 1940s, channel black played a role as a filler in the rapidly developing tire industry, but, due to oxygen functionalization, the extent of vulcanization was very low and so it was replaced by the more efficient furnace black, which does not affect vulcanization isotherms. The limited yield of channel black (3–6%) and the environmental hazard posed by the emission of very fine carbon black particles into the environment were the reasons for terminating the process.

### 2.1.1.2 Furnace Black Process

The gas furnace process was introduced industrially on a large scale in 1920, when the yield of the channel process was too low and the environmental conditions were no longer sufficient to produce large quantities of reinforcing carbon black for the tire industry [8]. Production in a gas furnace yields large particle sizes, with diameters over 50 nm. Manufacturing of small carbon black particles is carried out by the oil-furnace process, which was developed in 1942 [9]. During production in gas furnaces, a diffusion flame is generated by burning part of the natural gas with added air.

The furnace reactors are designed to operate as trouble-free as possible over a long period of time and to support aggressive conditions with short reaction times. The wide variety of furnace blacks requires different reactor designed and sizes to cover the full range of carbon blacks. Reactors for high surface area carbon black used for tire treads operate with small reaction times and large turbulence to ensure rapid mixing of the feedstock with the hot environment. For low surface area carbon black, the reactors are larger and operate at lower temperatures and longer reaction times.

The furnace process offers a high yield of carbon black (approx. 70%) and a wider production of different types of carbon blacks. Furnace blacks have almost forced the others types of carbon black processes out of the market. The most common type of carbon black is furnace black (> 95%), which is produced in a continuous furnace process. Chemically speaking, highly aromatic hydrocarbons of molecular weight of up to 500 g/mol fulfill the conditions for use as feedstock. In order that oils with a high aromatic content may be used, recourse is made petrochemical mineral oils produced during cracking or those resulting from the distillation of coal tar. Suitable oils are selected on the basis of their density, boiling behavior, viscosity and asphaltene content. In particular, polynuclear aromatics are very useful, while short side chains with many double bonds enhance the quality of the feedstock. Long paraffinic side chains are not useful in this process. In contrast to the channel process, the aromatic oil feedstock, which should contain as few sulfur compounds as possible, is introduced into a reactor heated by a flame to a high temperature and is burned incompletely. Ash and sulfur accumulate in the carbon black and must be avoided. In Europe, naphthalene and anthracene oils are economically viable and frequently serve as feedstocks; coal tar is also used. Reports on feedstocks for carbon black production have been published [10, 11].

A furnace reactor consists of a refractory lined tube arranged either in horizontally or vertically. A linear diameter is used for preparing the semi-reinforcing carbon blacks of lower specific surface area or larger primary particle size that are used for carcass blacks in tire manufacturing and for technical rubber goods. For the reinforcing carbon blacks of high specific surface, the reactors are used in either horizontal or vertical forms of different length, and may be rectangular or cylindrical. Many producers operate with smaller reactors in parallel to increase the production of carbon black. Air and natural gas are introduced to increase the temperature in the combustion zone either axially, radially or tangentially. The feedstock can be introduced either axially or radially into the high-velocity section of the furnace.

A special model is the biconcentric reactor, where, in the first half, the diameter decreases linearly and the residence time decreases with the square of the distance, while thereafter the diameter increases and the residence time increases. Inside the reactor, at high velocities of the feedstock and the carbon black that forms, erosion takes place at the edges and can contribute to contamination of the carbon black with some impurities. A reactor geometry consisting of two conical halves is schematically shown in Figure 2.2 [3].

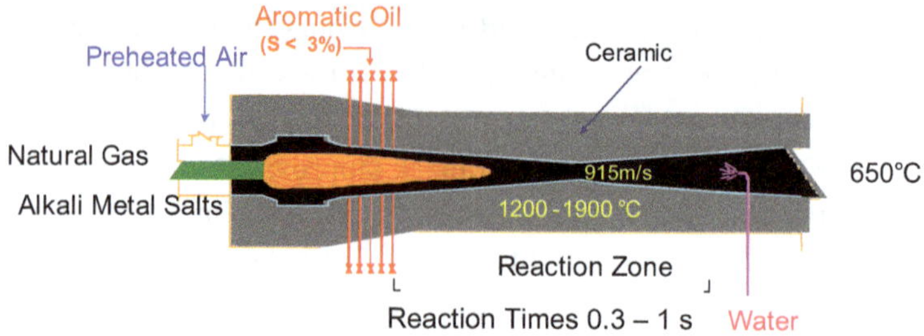


**Figure 2.2** Biconcentric furnace for producing furnace carbon black. The alkali metal salts are introduced to obtain carbon black with reduced aggregate structure

The first zone of the reactor is the combustion zone in which the natural gas and air are completely combusted to generate the high temperature in the reactor. In the reaction zone, the gas flow is accelerated to higher velocities that allow the formation of irregular structured carbon black aggregates, resulting in intense to high velocities and turbulent mixing. Refinery heavy oil with a high fraction of aromatics, but with fewer sulfur compounds in the feedstock, ethylene cracker tar and coal tar distillates, which are produced during coke production, serve as the feedstock for the production of carbon blacks of smaller size. The hydrocarbon feedstock, generally preheated to about 250 °C, is injected into the end of the flame zone. The feedstock is atomized by introduction under pressure through specially designed nozzles. The turbulence required for the process is achieved via the chosen furnace design. The carbon black formed in the flame zone is fed through the furnace section at a high, pre-calculated speed. The heavy aromatic oils and tars give better yields than lighter oils with a higher fraction of hydrogen. At the high temperatures, the remaining oxygen is consumed by burning a small portion of the feedstock (less than 25–30%). Therefore, a portion of the feedstock is consumed in the reaction zone. If the temperature in the furnace is increased, a larger portion of the feedstock is burned and does not participate in the carbon black formation, a fact which increases the price of the fine furnace black.

The entropic nature of the furnace process determines the physical properties of the carbon black aggregates. The added feedstock is quickly decomposed and begins to dehydrogenate and form the carbon black in a turbulent process. In the reaction zone, the decisive portion of the feedstock, 70–75%, is very quickly converted into form carbon black aggregates in a reaction time of 0.3 to 1 s. The shorter the reaction time, the smaller is the primary particle size of the carbon black and the specific surface area and the higher the activity formed in the furnace process.

The hydrocarbon molecules are broken down into smaller units, which now form highly viscous aggregates in the turbulent atmosphere and simultaneously undergo dehydrogenation and carbonization, while the diameter of the primary particle increases during the reaction time. In a few fractions of a second, highly viscous droplet aggregates are produced which carbonize in a very short time. The first section of the reactor comprises the combustion zone, where the fuel consisting of natural gas and an excess of process air is ignited to ensure a high temperature in the reactor. The combustion of natural gas generates the high temperatures in the furnace reactor that are required for thermally oxidative cracking of the oil feedstock. The temperatures reached are crucial for the type of carbon black to be produced. Depending on the primary particle size, temperatures of 1400 to 1900 °C are used.

For active carbon blacks that have a small primary particle size, a very short reaction time is chosen, while, for larger primary particles, these times are much higher. Consequently, a higher temperature of 1600 to 1800 °C is achieved by using more feedstock, but it leads to a smaller amount of carbon black and higher production costs. If less active carbon black is to be produced, for instance carcass carbon black, the temperature in the furnace reactor is kept lower, at around 1300–1400 °C, and that reduces the price of carbon black.

Reaction times vary with the choice of natural gas and air for the flame and the quantity of oil feedstock. The reaction times range from 0.3 s to 1 s before dehydrogenation is complete at the end of the reactor. With a suitable reactor design and a good oil/air ratio and a good residence time in the reactor zone, the particle size is distributed over a fairly wide range, as shown in Table 2.2.

**Table 2.2** Particle Size and Reaction Time in the Reactor [12]

| Carbon Black | Particle Size (nm) | Spec. Surface ($m^2/g$) | Speed (m/s) | Residence Time (s) |
|---|---|---|---|---|
| N 220 | 20 | 125 | 200–400 | 0.01 |
| N 550 | 35 | 80 | 30–40 | 1 |
| N 762 | 65 | 40 | 0.5–1.5 | 2 |

With increase in the residence time or reduction in the speed of the decomposing hydrocarbon, there is an increase in the particle diameter of the primary particles. Another representation combines the oil flow rate at constant fuel and air flow rates with the reaction temperature and the particle size of the final carbon black and is given in Table 2.3 [13]. This shows the relationship between the oil rate (kg/h) and the primary particle size at different set temperatures. If the oil rate decreases by half and the reaction temperature increases at the same time, the radius of the primary particles decreases by half. This is certainly associated with higher oil consumption. The particle size is primarily controlled by the feedstock rates and the temperature in the reactor. A higher combustion ratio increases the temperature in the furnace and reduce the particle size as well as the yield.

**Table 2.3** Influence of Oil Rate at Constant Fuel and Air Rates on the Primary Particle Size [13]

| Oil Rate (kg/h) | Reaction Temperature (°C) | Primary Particle Size (nm) |
|---|---|---|
| 4060 | 1450 | 44 |
| 3660 | 1500 | 35 |
| 3170 | 1570 | 26 |
| 2740 | 1630 | 21 |
| 2370 | 1680 | 19 |

In order to reduce the "structure" of the carbon black aggregates, it was discovered that alkali metal salts additives can reduce the "structure" of the aggregates of carbon black. The positive charge of the potassium ions prevents the attachment of other charged aggregates and thus the number of primary particles in an aggregate can be increased. The alkali cations reduce the degree of aggregation in the reaction zone and produce carbon black with a smaller "structure" [14]. This may be done in the form of aqueous salt solutions injected directly into the furnace or through the use of an oil-soluble alkali metal compound [15]. Quantities of 5–25 ppm potassium chloride are used to significantly reduce the particle "structure". However, the process is also characterized by the geometry of the furnace and other inexactly given factors, which tend to make quality control difficult. High "structure" blacks, characterized by a high proportion of particles that clump together to form persistent aggregates are advantageously produced by controlling the turbulence of the decomposing feedstock in a desired pattern [16]. The particle size of the produced carbon black depends largely on the combustion ratio, i. e. the ratio of the total air present to air needed for the complete combustion of all hydrocarbons present. A high combustion ratio increases the flame temperature and reduces the particle size as well as the yield of the carbon black formed. The average diameter of the carbon black made by the furnace process varies from 85 nm for low specific area to 15–18 nm for high-activity carbon black. After the carbon black has formed in the reaction zone, it must be quenched by injecting water at a distance dependent on the type of carbon black to be produced. If the reaction time is shortened, a large quantity of water is required. The temperature of the product stream of the carbon black process is lowered to 650 °C, which also prevents any undesirable secondary reactions which reduce the yield of carbon black production.

The wet slurry is transported for filtration and separation of the carbon black and tail gas (mainly $H_2$ and small amounts of CO, $CO_2$, and $N_2$). Separation is done using commercial filter bags. The collected carbon black is removed from the filter bag, usually into a pneumatic conveying system that transports the carbon black until a secondary filter separates the gas. The gases produced during the reaction are combustible and, in most cases, are fed into an afterburning stage where the heat is used to dry the carbon black, or are burnt in a boiler to generate steam. Since the filtered gas contains 30–50% water vapor, most filters are operated at elevated temperatures to avoid condensation. Two different types of backwash filter are used, namely repress back filters and pulse-jet filters. The filter bags are usually made of special surface-treated glass fibers and function for up to two years; afterwards, the filter has to be changed. The heat from the vapor stream is used to preheat the feedstock and air and to generate steam that can be used within the company or is discharged externally.

The separated carbon black consists of a mixture of loose aggregates and agglomerates with a very low apparent density of 20–60 g/l. This mixture is called "fluffy black" and must be compacted for handling. The fluffy black is densified by a pelletizing process or densified to facilitate onward handling. For the pelletization, two different technologies have been developed. The most commonly used is wet-pelletization, which has been employed industrially since 1922. It consists in mixing equal weight ratios of water and fluffy black in pin mixers. To promote the formation of easily dispersible pellets and to minimize the presence of undesirable fines, molasses from sugar refining or lignosulfonic salts, byproducts of paper manufacturing, are used. The liquid content of the charge has a profound effect on the rate of pellet growth [17]. The wet-pelletization process yields a pellet size of 1–2 mm that contains 50% of water and must then be dried. The wet-pelletizing process is commonly employed with all types of furnace carbon to form a user-friendly product [7, 18, 19]. The bulk density of the pelletized carbon black is 350 g/l. The wet-pelletizing process affords a technically and economically favorable process for the production of carbon black modifications. The size distribution of the pelletized carbon black is important for its behavior in automated transportation. The formation and quality of the pellets are dependent on the pelletizing speeds and the feed rates of carbon black and water. The pelletizing increases the bulk density for easy handling.

The second technology is dry-pelletization, in which the fluffy black is pelletized in inclined drums in the presence of an already pelletized material (seed) [19, 21]. Dry-pelletization is a simple, energy-saving process that produces softer pellets of smaller diameter. Special carbon blacks produced by the furnace black process are either loosely compacted and packaged as powder carbon blacks or converted into easily dispersible pellets by the dry-pelletizing process. Such carbon blacks are used to reinforce polyolefins, such as PE and PP.

The pelletized carbon black is dried in rotating gas-fired dryers. The water content must be reduced from the initial 50% to less than 1% before the product is delivered to the customer. Finally, the furnace black is passed over magnetic separators to remove metallic impurities. The carbon black compounds contain small amounts of solid impurities, such as coke particles, extraneous particles, and mostly grit, which is discharged via the grit separators.

The energy consumption in the production of one kilogram of furnace black is in the range of 9–16 × $10^7$ J and the yields are 300–600 kg/$m^3$, the exact figure depending on the type of carbon black. Furnace black has a variable surface area of 25 to 145 $m^2$/g and a wide variety of structures. The pellets are fed into rotating gas-fired dryer and dried in an air stream at a temperature of 200 °C. The material is screened to remove material of an undesirable size and is transported to the shipping container.

The carbon black pellets must be able to withstand transportation, storage in silos and pneumatic conveying into the internal mixers. On the other hand, they must be soft enough to be easily crushed and dispersed during mixing. The complex manufacturing technology is shown schematically in Figure 2.3.

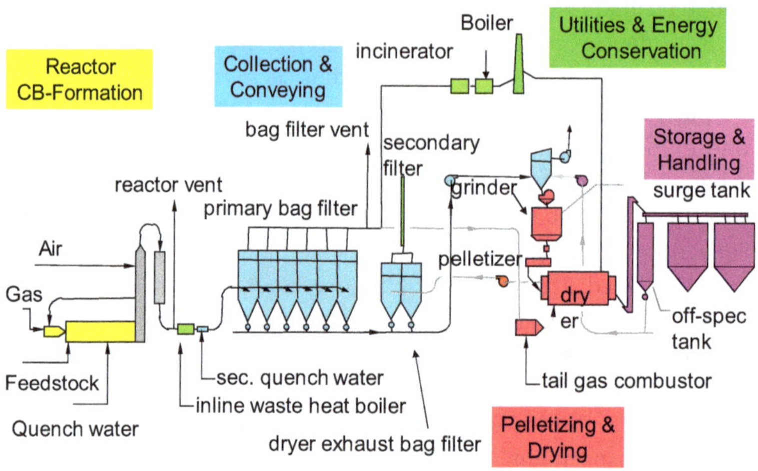


**Figure 2.3** View of a furnace carbon black plant

After the production of carbon black, there are several necessary installations that prepare it for shipment to the customer. Carbon black can vary enormously in specific surface area and structure, and its surface activity, as a result of a functionalization process, ranges from low to very high values. It is produced in a very capital intensive and energy intensive process, making it inherently imperative to reduce energy costs throughout the production process. Therefore, in the furnace process the furnace blacks of varying particle size are shown in Figure 2.4 [27].

In addition, many different specific surfaces and structures of the filler aggregates can be achieved with furnace carbon black process. This method of production is expressly endorsed. The type of carbon black chosen for a given compound can greatly influence the properties of the compound, and also a major impact on the methods used to process the compound. With the development of emulsion SBR, the furnace carbon black demonstrates a similarly high reinforcement as the NR/carbon black vulcanizates.

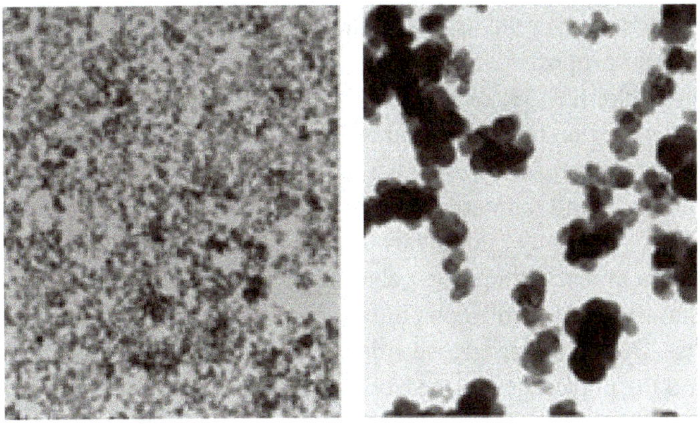

Figure 2.4 Aspects of furnace carbon blacks

#### 2.1.1.3 Formation of Furnace Carbon Black

One of the most interesting aspects of the manufacture of carbon black is the mechanism by which it forms in the furnace reactor. Due to the high rate of carbon black formation, it is difficult to establish a simple correlation with what is happening in the reactor. However, many studies have unraveled the mechanism to some extent. It is beyond the scope of this treatise to review all reports, but the authors conclude that there are two distinct processes:

1. nucleation of the carbon particle, and
2. subsequent growth that leads to the aggregates

During the oxidative decomposition of hydrocarbons, it can be assumed that oxygen diffuses in the gaseous hydrocarbon stream and leads to formation of carbon black, namely the aggregates containing primary particles. This process was studied during the last century, but is not completely understood. No single mechanism can explain the formation from different feedstocks and different processes or how some is also produced on the walls of the reactor [23]. Some of the studies of the mechanism of nucleation and particle growth show that there are two phases in the formation of carbon black [24–26]. It is assumed that carbon black occurs in several stages in its formation [27]. It is also assumed that the condensation of degraded feedstock leads to small $C_2$ radicals that can form polyacetylene, which is not the only precursor for the forming carbon black [28]. The intermediate products are unsaturated and it appears that the formation of carbon black entails a polymerization process. Initially, liquid, highly viscous droplets are formed by polymerization of the unsaturated molecules. This hypothesis is consistent with the results on nucleation and growth of viscous particles [29–32].

At the beginning of the process, the feedstock is introduced as a gas or in finely dispersed form into the reaction zone at high temperature, initially forming $C_2$ and

$C_2H_2$ radicals, which then form polycyclic aromatics having a molar mass of 150 to 600 g/mol. Theories have been proposed which, on one hand, assume growth of free radicals on condensation nuclei [29, 33] or condensation of gaseous hydrocarbons and the continuing dehydrogenation leads to liquid droplets with high viscosity from which the solid particles are finally formed [34–38]. If the saturation limit is exceeded, they condense into droplets of small diameter (2–5 nm). As a result, further radicals and polycyclic hydrocarbons accumulate in the turbulent system of the reactor zone. Progressive dehydrogenation and growth of the polycyclic layers lead to an increase in droplet viscosity, so that on further collisions the droplets adhere to one another and form fused aggregates. Due to the progressive dehydrogenation of the particles, the viscosity increases and the polycyclic hydrocarbons are carbonized and form imperfect graphitic layers at the surface of the particles. Aggregates are formed which take on a solid form after a reaction time in which the dehydrogenation process occurs. The final solids are carbon black aggregates that flocculate to build larger agglomerates, due to van der Waals forces. As dehydrogenation of the particles and carbonization progress, the aggregates are formed through ballistic collision of the particles. Besides aggregate growth, resulting from the collision of neighboring aggregates, surface growth occurs due to the deposition of carbon atoms on the aggregate during the formation of primary carbon black aggregates according to an ordering principle that allow the superior graphitic layers to be more organized than those from the middle of the primary particles. [39–43].

On the basis of transmission electron micrographs, the concept of concentration seems to prevail when the final shape of the particles is considered. All carbon blacks have some formation features in common [44]:

- The particles generated at high temperature are generated in a supersaturated vapor phase and have a related micro-structure the micro-structure does not depend on the aggregate state of the feedstock, and
- the carbon black particles are formed from graphitic layers that are built into the particle according to an ordering principle.

The resonance-stabilized aromatic compounds arrange themselves, following the lowest energy in the $sp^2$ hybridization, in layers that are preferably oriented parallel with the surfaces of the droplets. Impurities, ions, and hydrocarbon molecules of high molar mass serve as nuclei. As aromatization progresses by dehydrogenation and carbonization, the agglomerated droplets become less and less likely to break apart, and a transition to a solid carbonized phase takes place. This means that the growth in size of the carbon aggregates is as good as complete [45]. The nucleation and growth mechanism demonstrate that the growth of the particle increases in very short reaction times. The formation of the droplets leads to a rapid decrease in vapor pressure, such that no formation of additional nuclei takes place. The growth of the particles continues on the existing nuclei. The continuity of the planar carbon microstructure in fused carbon black aggregates also indicates coalescence of the liquid droplets prior to carbonization into solid carbon black aggregates. It has been shown that the temperatures required for growth of the aggregates are lower than those needed for nucleation of the carbon black particles. The various stages of formation of the carbon black aggregate are summarized in Table 2.4.

**Table 2.4** Formation Components of the Carbon Black Aggregate [46]

| Structural Unit | Time (ms) | Particle (nm) | Molar Mass (g/mol) |
|---|---|---|---|
| Organic radicals | < 0.5 | - | 25–50 |
| Polycyclic aromatics | 0.5–3.0 | 1–2 | 150–600 |
| Cond. of aromatics | 3–5 | 2–5 | 2400–40,000 |
| Growth of particles | 5–10 | 10–20 | > $10^5$ |
| Aggregates | > 10 | > 20 | > $10^6$ |

The aggregates formed in this way accumulate under van der Waals forces to form large agglomerates, which form what is referred to as "fluffy" carbon black. The fast and complicated process can be represented schematically by a simple model that shows disintegration of the feedstock into molecular fragments and the formation of viscous droplets, which subsequently split off hydrogen and undergo a process of carbonization to form the furnace carbon black aggregates. This is shown in Figure 2.5.

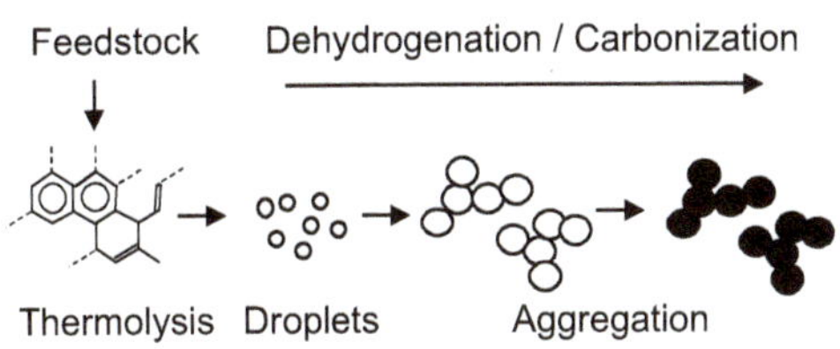


**Figure 2.5** Mechanism behind furnace carbon black formation by dehydrogenation and carbonization of the mineral oil

Such aggregated particles subsequently end up agglomerating, producing what is known as fluffy carbon black. The fluffy black is free-flowing pellets of 0.3–1 mm in diameter that are difficult to handle in industry. The particle sizes and the shape of aggregates can be controlled by changing the processing parameters, such as feedstock flow rate, gas rate, air rate, air-to-gas ratio, quenching position and furnace geometry. The biconical reactor geometry allows a quadratic increase in the flow rate in the first section of the reactor and a decrease after the smallest part of the reactor. In this way, the residence time for the formation of the carbon black particles is shortened, while the time available for cooling the hot material is extended.

In addition to the influence of geometric parameters, the usual technique for adjusting the primary particle size of carbon blacks is to change the ratio of air to natural gas. An increase in the air-to-gas ratio leads to an increase in both the temperature and the overall flow rate, which has a strong influence on the size of the carbon black aggregates. If this time is shortened, particle growth is reduced and the size of the primary particles is small (15–20 nm). If growth of the particle has more time to increase the size of the primary particle, the aggregate size also increases. According to some publications, the increase in the particle diameter is proportional to the residence time in the reactor [47].

$$\Delta d = c\, d \Delta t \tag{2.1}$$

where $c$ is a constant, $d$ is the size of the primary particle and $t$ is the residence time in the reactor.

Integration of this equation leads to an exponential increase in the size of the primary particle with residence time in the reactor. The nuclei generally have a normal size distribution; particles formed at lower feedstock concentrations without significant growth should have a normal size distribution. It is not possible for carbon black aggregates consisting of coalesced nuclei to change to the normal particle size distribution. The higher the reactor temperature, the smaller the primary particle size becomes. As shown in Figure 2.6, increasing the furnace temperature leads to a substantial reduction in primary particle and aggregate size [1].

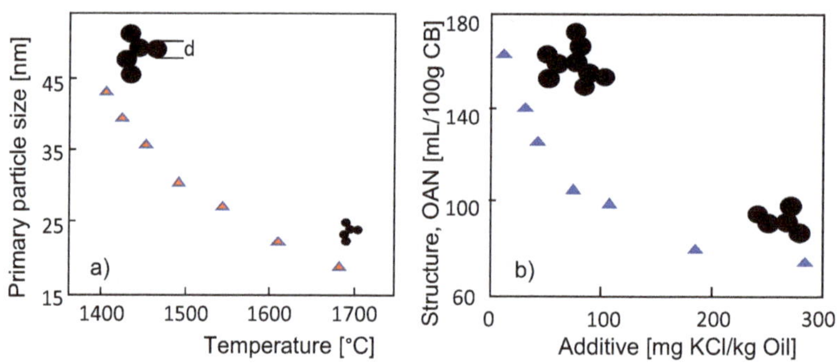


**Figure 2.6** Morphology of primary particle formation: a) change in aggregate size with temperature and b) reduction in aggregate structure through incorporating potassium salt

Increasing the temperature in the reaction zone leads to a reduction in the diameter of the primary particles. Therefore, the average size of the primary carbon black particles can be adjusted via the process parameters. Ballistic collisions of carbonizing droplets in the turbulent flow prevailing in the reaction zone lead to the formation of branched aggregates. It is assumed that the irregular shape of the aggregates, the so-called "structure," is more dependent on the turbulence at the feedstock injection point. In order to reduce the degree of branching during the formation of the aggregates, a solution of potassium chloride is injected into the furnace. The positive ions act as nucleation centers for the nanodroplets formed in the first stages of the process. Due to electrostatic repulsion between the similarly charged droplets, aggregation is largely prevented [48]. The reduction in the degree of branching triggered by the potassium salt is shown in Figure 2.6b. The presence of potassium cations reduced the predilection to form larger aggregates and the increase in structure is reduced to a lower level that can be measured experimentally.

The yield of furnace black depends on the feedstock used and the reactor temperature. Aromatic oils, which consist of polynuclear, unsubstituted aromatics, give rise to a higher yield. The primary particle sizes and the structure of carbon black can be controlled independently of each other, such that there are carbon blacks with increasing particle size, but the same aggregate branching and the case where the particle size is constant, but the aggregates have a different shape. Product quality is therefore determined by the process control, in which a large number of parameters must be taken into account, but are not changed frequently.

#### 2.1.1.4 Gas Black Process

A further development of the channel process was operated by Degussa in Germany in 1930s, whereby the feedstock is obtained from tar distillation and hydrogen or air serves as a carrier gas on small flames under rotating drums that are cooled [2]. Gas black production is an open process with a constant air flow. The feedstock ranges widely from gases to renewable sources to highly aromatic oils, such as coal tar distillate or bottoms from the cracking process of naphthene oils for constant high quality products. The gas black apparatus consists of a burner pipe approximately 5 m long, which is fitted with 30–50 diffusion burners. The flames impinge on the surface of a water-cooled drum, where about half of the carbon black is deposited. The carrier gas is used to generate the energy required for the thermal decomposition of the hydrocarbon. The carbon black is continuously scraped off and transported to a filter system and deposited to be shifted to the rubber industry. Onward processing is then similar to the furnace black process. The gas black is collected in a sealed filter system that exceeds official emission standards by a significant margin.

Due to the continuous contact with oxygen in the flame and the decreasing temperature, the gas black process has special properties on the size of the primary particles and the size of the aggregates. The gas black is characterized by a loose structure and very good dispersibility in rubber compounds and plastics. The production rate is 7–9 kg/h and the observed yield is approx. 60%. The gas-phase process allows the production of a small primary particle size of 10–30 nm. A schematic representation of this technology is shown in Figure 2.7.

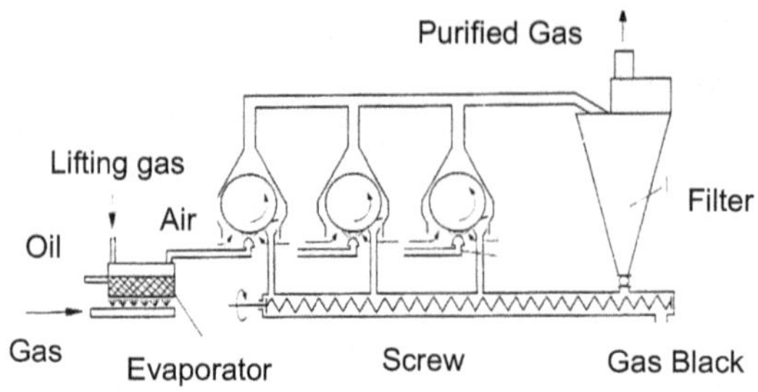


**Figure 2.7**
Scheme of the gas black process

The average primary particle diameter is 10–30 nm and the specific surface area are approx. 100 $m^2/g$. Gas blacks are characterized by their loose structure and good dispersibility. As a result of contact with oxygen at high temperatures, acidic functionalities are formed that can react with polar rubbers (i. e. NBR). As a result of contact with oxygen at high temperatures, acidic oxides are formed on the surface. Carbon black produced by this technology has small primary particles in the lower to medium particle size range and is used in rubber compounding, but also in in the paint industry. Oxidative post-treatment using nitrogen dioxide, ozone or other oxidants makes it possible to NBC further increase the acidic function for use in inks and coatings production.

#### 2.1.1.5 Lamp Black Process

The lamp black process is the oldest and most primitive method of producing carbon black, but it is still used today. This type of carbon black is made by burning aromatic oils in shallow, open pans in a restricted area with a limited air supply. The process falls into the category of thermal-oxidative decomposition of the oil feedstock. The difference is that the temperature is relatively low and the temperature distribution is relatively large. The feedstock, oil with a high aromatic hydrocarbon content, is burned in shallow iron pans (diameter 1.5 m), which is surmounted by a fire-proofed flue hood that is lined with refractory bricks. The air gap between the pan and the hood, as well as the vacuum present in the system, help to regulate the air supply and thus to tune the carbon black's properties. The smoke from the burning pans passes though the ceramic chamber. The off-gas containing carbon black is sucked into an exhaust pipe, which is coated with a ceramic inner liner and leads to the cooling and collecting system. The smoke from the burning pans passes through low velocity settling chambers at the end of the simple reactor, from which the carbon black is cleaned by motor-driven plows. By varying the size of the burner pans and the amount of combustion air, the particle size can be controlled within relatively narrow limits, but the particle size is always significantly higher than in the furnace process. The carbon black resulting from the incomplete and slow combustion is then separated in chambers or cyclones [7]. The lamp black process is shown in Figure 2.8.

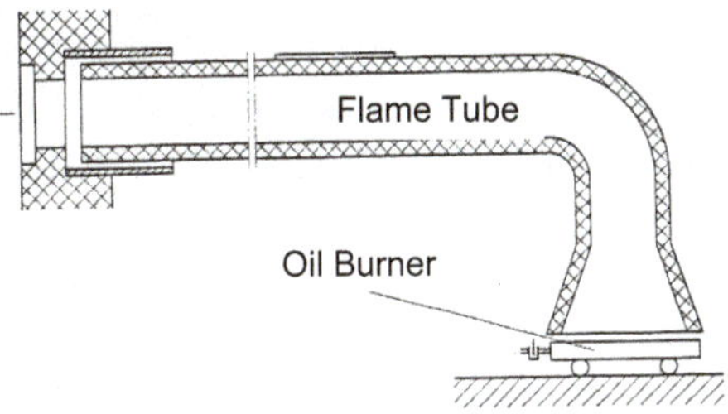


**Figure 2.8** Lamp black process

Onward processing of lamp black is similar to that of the furnace blacks. A lamp black process can produce 100 kg/h of carbon black, because the oil burns slowly. Lamp carbon black has an average primary particle diameter greater than 65 nm, a specific surface area of 22 $m^2$/g and an aggregate structure of 130 mL/100 g.

Unlike the furnace carbon black process, the lamp black process does not allow any major changes in primary particle size, surface area and structure. In the past, different types of lamp black were produced. Typically lamp carbon black shows a very broad particle size distribution ranging from approx. 60 to 200 nm.

A more extensive description of the process and the properties of lamp carbon black can be found in the literature [1, 2]. Nevertheless, the typical lamp black process leaves a unique fingerprint of carbon black properties. It yields large quasi-spherical carbon black particles that have large particle size and small specific surface area and wide particle size distribution. This carbon black has a very low activity and is used to increase the hardness in EPDM or FKM and for special applications, where increasing the rubber properties is not as important as for active carbon blacks. Thermal blacks have good extrusion behavior and ensure a smooth extrudate surface.

## 2.1.2 Thermal Decomposition Process

Special carbon blacks are made by thermal decomposition, without using any oxygen in the process. Depending on the hydrocarbon used and the manufacturing technology one can use two types of carbon black manufacturing:

- thermal black process
- acetylene black process

In both cases, high temperatures are set in the process of forming carbon black at which the feedstock thermally decomposes into carbon black and hydrogen at 1000–1300 °C. Thermal black processes produce the carbon black from natural gas and the acetylene process use the exothermic decomposition of acetylene.

#### 2.1.2.1 Thermal Black Process

At the beginning of the 20th century this process was used to produce hydrogen, but later the emphasis was placed on the production of a semi-active carbon black [1, 2]. This thermal carbon black is made by thermal decomposition of natural gas or low boiling liquid hydrocarbons in the absence of oxygen. The thermal black process is carried out at relatively low temperatures (about 1300 °C) and requires longer residence times than does the furnace process. The process is carried out in two cylindrical furnaces lined with an open checker-work of silica brick. While the first generator is heated with a stoichiometric mixture of the gaseous fuel and air, the second hot generator is charged with natural gas. Thermal decomposition takes place and about half of the carbon black is formed is swept out with the gases, while the rest remains on the bricks. After 4 to 5 minutes, natural gas is turned off and air and fuel are introduced for reheating the generator, first burning off the carbon remaining from the previous cycle. Since the reaction is endothermic, the temperature decreases and at 900 °C a new heating process starts. The material stream is finally quenched with water and filtered in a bag house. The distinction from other processes is that, in the thermal black process, energy generation and decomposition of the feedstock are not simultaneous. The fact that the carbon black produced occurs in the absence of oxygen and at decreasing temperatures, results in large carbon black particles with a wide agglomerate distribution. The existing carbon black may be further processed and pelletized, screened and packaged for shipment to customers. The process for producing thermal black is shown in Figure 2.9.

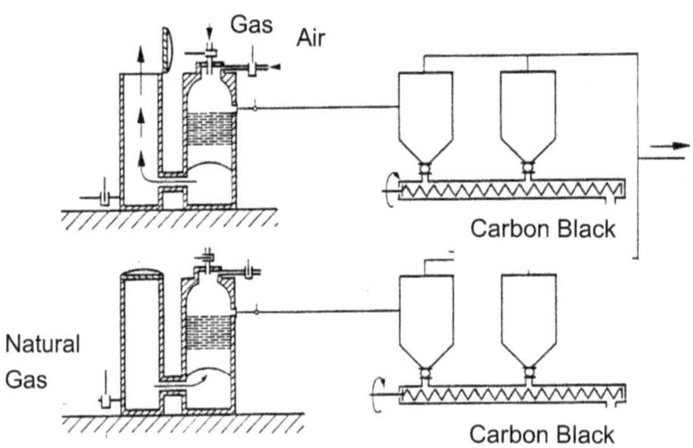


Figure 2.9 Thermal black process

Due to the long residence time at elevated temperatures, thermal blacks consist of nearly spherical particles that demonstrate a broad particle size distribution, the largest form of industrial carbon blacks. Thermal blacks form relatively slowly and yield coarse primary particles ranging from 300 to 500 nm. These carbon blacks have the lowest reinforcement of all carbon blacks.

The fraction of the smaller particles (approx. 0.5 µm average diameter) can reinforce moderately amorphous elastomers. However, since thermal blacks have a wide particle size distribution, their largest particles approach the limit of non-reinforcement. If the feedstock is only natural gas, it is possible to reduce the particle size to 120–200 nm, in which case it is called fine thermal black.

Medium thermal blacks (MT blacks) with a larger particle size of 300–500 nm are still produced by using an undiluted feedstock and the yield of MT blacks is 40%. Fine thermal blacks have diameter of 120–150 nm. Diluting the natural gas with hydrogen leads to a decrease in particle size. These carbon blacks are characterized by a high tar content (> 1%). The production and the properties of thermal black are described in several publications [7, 49]. Thermal carbon blacks are much more expensive than furnace carbon blacks due to their high energy requirements. Thermal carbon blacks have very good processing properties, but a low reinforcing effect and produce vulcanizates with low hardness, low compression set and low hysteresis. They can be used at high loadings to give benefits in some compounding applications. They have a high loading capacity and are used in mechanical goods, particularly in insulation jackets, as the electrical conductivity is relatively low. This type of carbon black is used in O-rings and seals due to its low hardness, high extensibility and low compression set, and low dynamic hysteresis. Thermal carbon blacks are employed as fillers for special rubbers such as FKM, CO, ECO, CSM.

### 2.1.2.2 Acetylen Black Process

Acetylene blacks are a special group produced by thermal decomposition of acetylene at high temperatures in a continuous process. In principle, there are two methods for producing acetylene black. The older method is based on the decomposition of compressed acetylene in steel pressure vessels, initiated by an electrical discharge [50]. The process used today works at normal pressure and is based on the thermal decomposition of acetylene in a preheated reactor (Shawinigan technology), which leads to yield in water-cooled jacketed cylindrical retorts [51].

The apparatus consists of a "burner" which serves to introduce the gases into the decomposition chamber, a refractory to form a retaining wall for the reacting gases and auxiliary equipment to remove the acetylene black from the hydrogen. The decomposition reaction starts by burning a mixture of acetylene and air. Thereafter, the air is closed off and decomposition continues, as acetylene decomposition is an exothermic reaction ($\Delta H_{293}$ = – 230 kJ/mol). The formation of aggregates requires more primary particles to join the formation of the aggregate and that leads to a much higher aggregate structure. Other hydrocarbons are usually added to acetylene to prevent the reactor temperature from rising due to the exothermic reaction. The temperature is maintained at the desired level by water cooling. The production of acetylene black is shown in Figure 2.10.

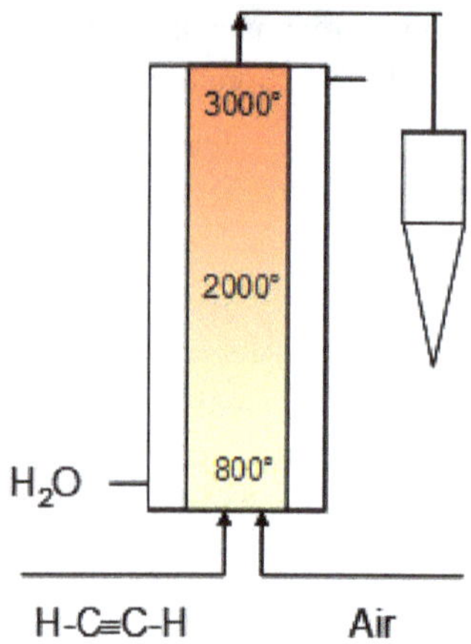


**Figure 2.10**
Acetylene black process

The carbon black is separated from the hydrogen, degritted and packed in paper bags, uncompressed or compressed. The tail gas, hydrogen, is flared or used as a fuel. Continuous processes were developed with a production rate of 500 kg/h. Although the medium primary particle size of acetylene black is in the same range as that of the furnace blacks, the structure diverges from the spherical form. Acetylene black consists of carbon particles with a high content of $sp^2$ hybridization (graphitic layers) that provide quite high electrical conductivity. To achieve comparable electrical conductivity in a compound, only small levels of these types are needed, compared with standard furnace blacks. Acetylene carbon blacks are normally fluffy and have a low bulk density of 19 kg/m$^3$. They are difficult to compress and resist the process of pelletization. They can be compressed to a bulk density of up to 200 kg/m$^3$. Acetylene carbon black is very pure and has a carbon content of 99.7%. It has an average particle diameter of 40 nm and a surface area of 65 m$^2$/g and a high structure of 250 mL/100 g. It has unconventional laminar particles that are larger than those of furnace black N 330. Acetylene carbon blacks are much more expensive than furnace carbon blacks. The high proportion of $sp^2$ hybridization makes acetylene carbon black suitable for highly electrically conducting applications and also thermal applications.

### 2.1.3 Pellet Size and Hardness

The hardness of individual pellets is related to the size of the pellets and the average pellet hardness. The pellet hardness stated in product certificates only provides information about the average value. In most applications, the carbon black cannot be used in its original fluffy form as produced in the furnace process. The filler particles need to be compacted in the form of pellets to survive transportation, storage in silos and handling in the mixing chamber, without generating dust. Special tests have been developed for measuring the ability of the pellets to disintegrate and pack during bulk handling and for measuring the degree of dispersion in a polymer [52, 53] The ASTM method involves crushing a total of 20 pellets from which the results obtained at high and the low force are eliminated, because they are not as relevant. For pellets of the same matrix, there is a direct relationship between size and hardness. Therefore, pellet size must be specified when describing pellet hardness. Individual pellet hardness is measured according to a standard procedure [54]. Wet-pelletized and dry-pelletized carbon black are shown in Figure 2.11.

**Figure 2.11** Pelletized carbon black: a) wet-pelletized and b) dry-pelletized

The pellet hardness test [55] involves applying an increasing force to a single pellet unit until it breaks. The breaking force varies with pellet size, so the carbon black industry uses the –12/+14 mesh fraction. Compounds that use low-viscosity polymers usually require soft carbon black pellets. The pellet hardness is related to pellet fragility and therefore abrasion resistance. Pellets that are too hard can cause problems with carbon black dispersion, especially in soft compounds. Softer pellets make for better dispersion, but their inherent propensity towards fineness may create handling issues. Wet-pelletized carbon blacks are usually characterized by greater pellet hardness than dry-pelletized carbon blacks. Pellet strength and hardness should not exceed a certain level, which is determined by the ease of fragmentation and dispersion during mechanical mixing. The recommended hardness for carbon blacks is in the range of 5–20 g for pellets of 1 mm or less than 15–30 g for larger pellet sizes. Pellet hardness becomes critical if the viscosity of the compound is low or if the plasticizer oil content is high. Such compounds may require a pellet hardness of less than 30 g. Pelletized carbon blacks offer advantages over powder carbon black from ecological and application-related aspects, such as dust loading, good flow behavior, ease of filling and meterability.

The pellet size distribution is a parameter that affects the flow characteristics of pelletized carbon black. A uniform pellet size means optimum flow behavior, which is recommended for automatic feeding of carbon black into an internal mixer. The fines content of carbon black pellets is also determined. Fines content in carbon black is undesirable for several reasons, not least because of problems associated with dusty material. Excessive fines can lead to problems with dustiness and flowability.

The production process, especially in the vertical furnace, also produces carbonaceous impurities, known as grit, which are larger than the carbon black aggregates and present certain challenges when carbon black is used as a filler in rubber compounds. The presence of grit adversely affects the polymer matrix into which the carbon black is dispersed and significantly lowers the tensile strength and other mechanical properties. The grit content must be removed or significantly reduced to become an integral part of the matrix. The amount of grit present in a sample of carbon black is measured by washing a given quantity of the carbon black on a specified screen size [56]. The insoluble residue, expressed in mg/kg (or ppm) of the original sample, provides indication of the quality of the carbon black. The smaller the grit, the greater is its ability to improve the properties of the rubber compounds. However, the presence of grit can dramatically shorten the life of the screens and the extrusion dies used in tire manufacturing. Grit size reduction can be achieved by grinding in long-gap mills or air-classifying mills.

### 2.1.4 Purity of Carbon Black

The purity of carbon blacks is generally assessed on the basis of toluene discoloration [57], ash content [58], grit residue [59] and sulfur content [60]. Toluene discoloration gives a rough indication of extractables present on the surface of carbon blacks. In carbon black with a lot of tar, the toluene extract is dark yellow in color. The main sources of grit are coke that forms in the reactor, pieces of the refractory that have eroded from the reactor and metals from the process equipment. The aim is to produce carbon black without the various impurities.

### 2.1.5 Classification of Carbon Black

As a large number of carbon blacks are used, the development of a more rational nomenclature for new types of carbon black was inevitable. The classification system introduced for carbon black is based on a nomenclature consisting of four characters. However, the classification is far from perfect, particularly because the last two digits are arbitrarily assigned. The ASTM classification and the type code used in the past are shown in Table 2.4. The classification is used to distinguish between the:

- influence on vulcanization speed
- primary particle size, and
- structure of the carbon blacks

The first character in the classification system is a letter that indicates the effect of carbon black on the curing rate of a compound containing the carbon black. Thus, N stands for carbon blacks that do not change the curing rate and S is used for blacks that slow down the curing rate (e. g. channel black). The second character refers to the nitrogen specific surface area, which is arbitrarily divided into nine classes, each of which is assigned a digit that describes the range of the specific surface area. The third and fourth characters are digits that describe the "structure" of the filler aggregate. A normal structure corresponds to 100 mL oil/100 g carbon black, as is the case for N330. A carbon black with a higher structure is indicated by a higher number (e. g. N339) and a carbon black with a lower structure by a smaller number (e. g. N326).

**Table 2.5** Classification of Carbon Blacks

| ASTM Classification | Generic Name | Type | Nitrogen Spec. Surface Area ($m^2/g$) | Average Particle Size (nm) |
|---|---|---|---|---|
| N 110 | SAF | Super-abrasion furnace | 130 | 11–19 |
| N 220 | ISAF | Intermediate-abrasion furnace | 115 | 20–25 |
| N 330 | HAF | High-abrasion furnace | 79 | 26–30 |
| N 550 | FEF | Fast-extrusion furnace | 41 | 40–48 |
| N 660 | GPF | General-purpose furnace | 35 | 49–60 |
| N 762 | SRF | Semi-reinforcing furnace | 28 | 61–100 |
| N 990 | MT | Medium thermal | 9 | 200–500 |

The table also shows a use of carbon black, taken from an older classification, which is still used today in many companies, like high abrasion, fast extrusion or general purpose furnace black. Although the classification could be improved to include the processability of carbon blacks, it has been used in rubber technology for half a century and there is sign yet of any improvements being made.

### 2.1.6 Global Production of Carbon Black

Carbon black is the most commonly employed reinforcing filler in rubber technology. Most than 92% of global production of carbon black is used in the rubber industry, especially in tire production. The remaining 8% is used for non-rubber applications, for pigments, plastics, and coatings. About 13.9 million metric tons of carbon black were produced worldwide in 2016, placing it among the top 50 industries. A typical tire contains 30–35% carbon black, and there are several different grades in the tire, depending on the reinforcement requirements. An almost equally large amount, 2–3 million metric tons, is used for the manufacture technical products, owing to its ability to reinforce other rubber components and reduce production costs. The leading producers of carbon black in the world are:

- Cabot Corporation 43.5%
- Birla Carbon 15.7%
- Orion Engineering 15%
- Black Cat 8.6%
- Tokai Carbon 6.8%
- Others 6.4%

The use of carbon black in various end-use rubber industries is important for the progress of production in tires and technical rubber goods. The global market has been expanded through the use of carbon black for tire construction. Despite the climate-mitigation measures being discussed today, demand for carbon black remains strong, as an outstanding filler that keeps the properties of tires and technical rubber goods at a high level. The growth in demand for carbon black in the automotive industry is driving the overall carbon black market. It will be interesting to see how things develop over the next 20–30 years, once the discussions on climate protection have taken hold, especially when electric cars are introduced, which can reach high speeds more quickly.

## 2.2 Amorphous Silica

Precipitated silica is a polar filler that is synthesized in an aqueous slurry from water glass and a strong mineral acid. About 70% of synthetic precipitated silica is processed in the rubber industry. It is an important nanofiller that is becoming increasingly important in passenger tire production and technical rubber goods. It has evolved from a simple filler used for colored rubber products to a functionalized filler with organosilane that is able to improve the reinforcement of unsaturated rubbers, especially to reduce rolling resistance and increase the wet grip of car tire treads. The use of amorphous silica has increased exponentially over the last three decades since it was first used in tire-tread compounding [61–66]. By manufacturing it in a slurry solution, it is possible to obtain the desired result with a defined morphology and different particle size and high structure in a rapid process. As the process is fast enough to prevent the formation of silica crystallites, the silica produced has an amorphous structure. There are two methods for the rapid production of amorphous silica:

- precipitated silica made in a solution reaction
- fumed silica produced by pyrohydrogenation

Both yield amorphous silica, but the essential difference is the amount of polar silanol groups that are present on the silica surface. The first method is a solution process while the second is much faster and takes place in the gas phase. As the starting chemicals are different for precipitated silica and fumed silica, therefore the manufacturing processes are also different. The difference between the two methods is the size of the primary particles and the number of silanol groups on the aggregates.

The use of organo-silanes converts the "water friendly" or hydrophilic precipitated silica into "water-repelling" or hydrophobic silica that can be used for much better interaction with the rubber chains of unsaturated non-polar rubber [61, 67]. A variant is the *in situ* manufacturing of precipitated silica in the rubber network from a soluble form of silica in an organic solvent [68, 69]. However, this type of well dispersed silica nanoparticle is only of academic interest. Of course, the use of the same particle size description for precipitated silicas and carbon blacks does not imply the same level of reinforcement. It is known that silicas and carbon black of the same particle size normally do not provide the same level of reinforcement, because the interaction with the rubber is based on different principles. The difference disappears when coupling agents, such as organo-silanes, are used with silica. Organo-silanes improve rubber-filler interaction and decrease polar filler-filler interaction. The silica and the carbon black of similar size then provide equivalent reinforcement, as measured by modulus, tensile strength and tread wear.

### 2.2.1 Manufacture of Precipitated Silica

Precipitated silica is produced by precipitating an aqueous solution of sodium silicate (water glass) with a mineral acid, usually sulfuric acid. The sodium silicate is produced by alkaline fusion of high-purity quartz sand and sodium carbonate at high temperatures, a process which is energy intensive and expensive. As a salt, water glass is soluble in water [61, 62, 65, 66]. The chemical reaction for the synthesis of amorphous silica is:

$$SiO_2 + Na_2CO_3 = Na_2SiO_3 + CO_2$$
$$Na_2SiO_3 + H_2SO_4 = SiO_2 + Na_2SO_4$$

Once a defined pH value has been set, the components are fed continuously into the reactor. This process takes place simultaneously over a certain period of time in order to ensure consistency of product morphology. Another option is to first feed in a certain quantity of water glass and an initial dose of sulfuric acid and then to dose both simultaneously under defined reaction conditions. Precipitation, which occurs on overly rapid addition of concentrated acids, produces a small amount of silica gel. This gel content has no adverse effect on reinforcement, but can be a significant source of non-dispersed silica for the customer. The resulting filler has a higher proportion of silanol groups on the filler surface which make the surface of the aggregates acidic.

The primary particles consist of amorphous $SiO_2$ and a small number of acidic silanol groups (Si-OH) situated on the surface of the particles. Subsequently, the primary particles react with each other by eliminating water, forming siloxane bonds (Si-O-Si) on the surface of the aggregate. Formation of the agglomerates depends on the reaction time, the pH and the electrolyte. A state of equilibrium is reached between the aggregates and the agglomerates which depends on the process conditions and can be influenced by the process operator. Depending on the intended application of the precipitated silica, the process conditions such as feed rates, stirring, precipitation parameters, temperature, pH, alkali content etc. can be adjusted to provide a wide range of products of different properties. Between pH 3 and 6–7, a silica gel is formed, but it is not easily dispersed in elastomers. The usual pH range for commercial precipitated silica is 8.0–8.5, so that precipitated silica can be dispersed in rubber compounds [65, 67].

The resulting suspension is transferred to a homogenizer to reach the end point of the chemical reaction to produce amorphous silica. Precipitation parameters involved in setting the particle size include temperature, reaction rate, reactant concentrations, and the presence of additives in the silica. The precipitation temperature correlates with particle size; low temperatures produce small silica particles. Slow rates of neutralization reduce gel formation [71]. The aqueous suspension of silica particles is filtered and washed to remove traces of the mineral acid and sulfates. Since the solids content of the filter cake is 15–25% water, this must be removed from the filter cake.

Several types of dryers are utilized in the production of silica: rotary drum dryers, spray dryers and spin flash dryers. For the production of 100 kg of precipitated silica, 400 kg of water must be evaporated. The dried product still contains 10–14% water and needs to be finely milled. In line with the drying step employed, the precipitated silica may be milled and then granulated or granulated directly into a low-dust product [72, 73]. Special product properties can be achieved by spray-drying the filter cakes after redispersion in water or acid [74]. Both the size of the particles and the surface activity (see below) can be controlled within relatively narrow limits via the choice of processing parameters [75].

After the homogenizer, the product is sent for washing and filtering. Acid precipitation produces a slurry of silica and residual salts (sodium sulfate). The salts are removed by washing in a counter-current decanter or by means of a filter press. Washing lowers the salt content to 1 or 2%. Further concentration in rotary or plate and frame filters yields a solid wet cake which contains only 15–25% silica. Due to the high water content, the final drying step, whether performed with rotary, tray, belt or spray-dryers, is a major energy consumer. Owing to lower investment and operating costs, spray-drying has become the dominant drying process [66]. The formation of precipitated silica is shown in Figure 2.12.

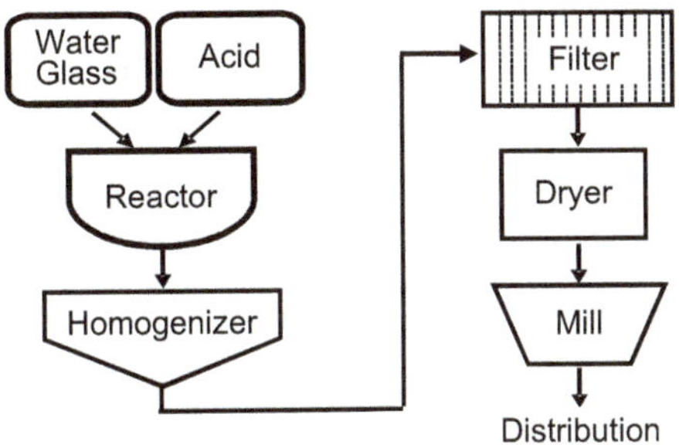


**Figure 2.12**
Process of precipitated silica production

In all cases, the end product still contains about 6% free water, because the silanol groups on the surface of the silica attract water molecules. The free water content reflects the hydrophilicity nature of silica and its tendency is to adsorb moisture to attain equilibrium with the relative humidity of the environment. At temperatures above 105 °C, the content of free water decreases. In contrast to the ephemeral nature of free water, bound water is held firmly in place in the form of silanol groups until the temperature rises above 250 °C. Silanols are responsible for the hydrophilic nature of precipitated silica and its unique reactivity with water. The end product is often milled and compacted to achieve an optimum balance between the absence of visible particles and minimal dustiness during rubber mixing.

A grade with a specific surface area of 50 to 230 $m^2/g$ is required for the production of precipitated silica in elastomers. The smaller the aggregate size, the greater is the change in the physical properties of the rubber compound. The reinforcing properties therefore depend entirely on the primary particle size or the specific surface area and the content of the silanol groups.

Silica has stronger interparticle interactions, due to the hydrogen bonds between the silanol groups, and preferentially interacts with polar rubber. Compounding leads to low interaction with non-polar rubber, bestowing poor abrasion resistance on non-polar rubber, compared to carbon black reinforced compounds. The use of silanized precipitated silica increased the use of amorphous synthetic silica and turned silica into a reinforcing filler in passenger tires that is growing in importance today. Precipitated silica initially offers more options in polar rubber compounds, such as NBR or HNBR, because of a strong interactions with the polar groups of the rubber. Precipitated silica is used also in non-tire products, especially in EPDM products for domestic appliances, such as washing-machine seals, sports shoes and others.

Precipitated silica was only able to gain a foothold in tire manufacturing when it became possible to use a second component, namely organo-silane. This is how silica/organo-silane became a high-performance filler for tire treads, which is excellently dispersed in the rubber matrix and significantly improves the physical properties. Silanization with a bifunctional organic silane led to a decrease in the number of silanol groups on the aggregate and to bonding of the silica aggregate to the non-polar rubber chains which influences the dynamic-mechanical and the abrasion resistance. As precipitated silica became an active, high-performance filler for tire treads, all producers of precipitated silica developed so-called highly dispersible silica grades that can be dispersed efficiently in S-SBR/cis-BR blends and provide a reduction in rolling resistance and an increase in the wet grip of passenger tires.

## 2.2.2 Formation of Precipitated Silica

The formation of precipitated silica in a reaction between $Na_2SiO_3$ and $H_2SO_4$ leads first to ortho-silicic acid, which can dissociate and form longer chains in the subsequent steps which, depending on the temperature and the concentration, lead to silica aggregates [61, 65]. The process of the polymerization of ortho-silicic acid is shown in Figure 2.13.

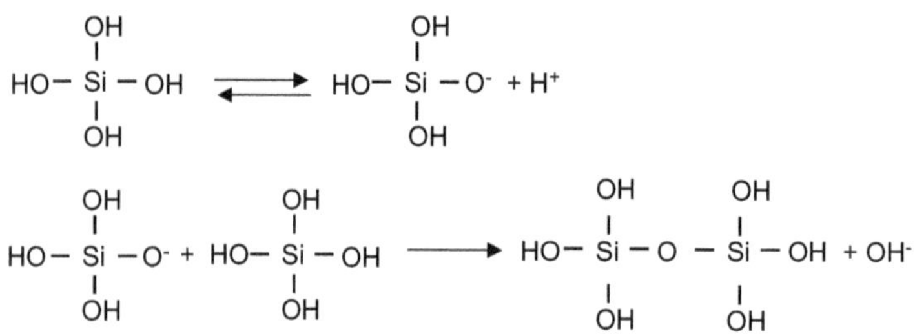

**Figure 2.13** Polymerization of ortho-silicic acid

In the first phase of the reaction, small silica particles are formed from ortho-silicic acid at nucleation centers. The duration of particle growth can be controlled by adding sodium sulfate and correlates with the temperature in the reaction chamber. The size of the primary particles at low temperatures leads to smaller silica particles. The parameters that govern the growth of the concentration of the sodium silicate, the concentration of the acid, the reactor temperature, the addition of the acid and other compounds are specified in patents and reports from the producers.

However, aggregation of precipitated silica can occur both during the production of the silica and during compounding when the filler is distributed throughout the polymer matrix and increases the viscosity. The high specific surface area of the fillers makes surface silanol groups available for interaction with those others silica particles, leading to larger aggregates [76, 77]. Large agglomerates can diminish the properties of the reinforced materials by disrupting the activity of the silanol groups when interacting with the polymer; they are considered to be mechanical defects. Under high pH conditions, the colloidal particles are negatively charged and repel each other. The electrostatic repulsion is reduced by the presence of sodium cations, resulting in growing colloidal silica aggregates. The particles collide with each other and form aggregates, which are consolidated by further deposition of $SiO_2$ on the surface of the aggregates, as shown in Figure 2.14.

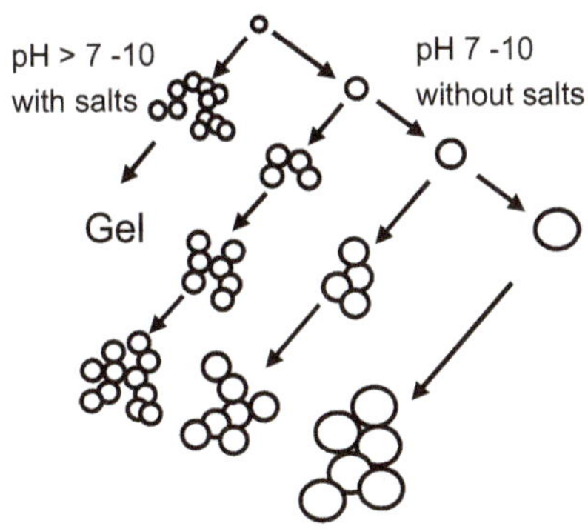


**Figure 2.14**
Silica agglomeration during precipitation

Silanol groups are present both on the silica surface and inside the silica skeleton. For the production of precipitated silica, the pH value varies from 5–9 in accordance with the conditions of silica production. Compared with carbon black, precipitated silica is less compatible with general-purpose rubbers (NR, SBR and BR). They provide lower cohesive bonding forces with non-polar rubbers. Precipitated silicas tend to link with each other to form a strong filler-filler network that is active in uncured rubbers and has a high viscosity. The silica-filled compound is difficult to be extrude, because of the higher viscosity. For this reason, organo-silanes are used to reduce the concentration of silanol groups and to bring the precipitated silica into a state that reduces the formation of filler-filler networks. The use of organo-silanes leads to a decreased intermolecular interaction of the silanol groups that provides lower viscosity in the rubber compound and excellent dynamic properties in the crosslinked non-polar rubber.

## 2.2.3 Surface-Treated Silicas

For use in a tire-tread compound, the precipitated silica is combined with an organo-silane in the mixing process, which leads to the emission of ethanol. This reaction will be discussed in the next chapter. The manufacturers of precipitated silica have identified the problem with the reaction in the internal mixer and ethanol emission. This has been corrected and customers are offered a pre-silanized silica that does not release ethanol in the mixers of the tire industry.

The surface-treated silicas contain the result of the reaction with organo-silanes in the molecule and can be introduced directly into the mixing process. The coupling with organo-silanes is performed in the silica producer's facilities, where the ethanol emission takes place in special reactors. The customer can use the silanized silica directly in the mixing process and reduce the emission of volatile ethanol to zero and shorten the silica mixing process [78]. The use of surface-silanized silica has so far been limited, due to the higher price of the product. Pre-silanized silica is commercially available as the products:

- Coupsil® (Evonik)
- Ciptane® (PPG Ind.)

They reduce mixing problems in a tire company, which is not forced to deal with the emitted ethanol, and considerably improve environmental conditions by avoiding the multiple silanization steps during mixing. Of course, this improvement results in an increase in the price of silanized silica.

### 2.2.4 Classification of Precipitated Silica

Since the 1960s, there has been a tendency to classify silica in the same ways as carbon black. It was proposed to classify silica by manufacturing method, particle size and vulcanizate properties. The ASTM designation consists of two letters and a three-digit number code [79]. The first letter describes the manufacturing process (A = fume process and H = precipitation process), the second letter stands for the curing behavior (S = slow curing, M = moderate curing), and the numerical code indicates the particle size, which is given in Table 2.6.

**Table 2.6** Classification of Precipitated Silicas

| Precipitated Silica | Specific Surface Area ($m^2/g$) |
|---|---|
| A | > 191 |
| B | 161–190 |
| C | 136–160 |
| D | 106–135 |
| E | 71–105 |
| F | < 70 |

A proposal was made for the determined fraction of the nitrogen surface area, which is an important property of the rubber-filler interaction. The higher the specific surface area, the greater is the interaction with the rubber.

Easy incorporation of the silica into the rubber mixture and its dispersion are critical parameters as they have a considerable effect on processing costs and rubber performance. The silanol groups and the possibility of quantitative coupling are crucial for good processing and enhancement of physical properties. Historically, three major groups of precipitated silica have been developed to improve the dispersion in unsaturated rubbers:

- first generation of precipitated silica (e. g. VN2 and VN3)
- second generation of easily dispersible silica (semi-HD silica)
- third generation of highly dispersible precipitated silica (HD silica)

The type produced depends on the process parameters. For types 1 and 2 precipitated silica, long-retention dryers (rotary dryers) are used. For the more advanced HD silica, short-retention dryers (flash or spray-dryers) are employed. This classification has gained acceptance in the literature and is observed in practice. The trade names of HD silicas are Ultrasil® for silica from Evonik, Zeozil® for silica from Rhodia SA, Hubersil® for silica from Huber Inc. and Hi-Sil® for silica from PPG Inc.

The choice of process parameters can influence the surface properties and therefore improve the reactivity of the precipitated silicas towards the organo-silanes. The silanol groups on the silica surface should be present in sufficient concentration. The highly dispersible silica makes for a faster and more efficient reaction with the organo-silanes, owing to the greater reactivity of the silanol groups on the silica surface. As the coupling efficiency with these silicas is higher, it is possible to use a smaller quantity of silane without having to compromise on technical properties [80]. One way to produce silica with new technical properties is to modify the manufacturing process. Reports have been published on highly dispersible reactive silicas (HDRS™) which enable a faster reaction with organo-silanes and a higher yield. As coupling with the silanes is more extensive, the lower is the quantity of organo-silanes required. Evonik has developed a new class of silicas of high specific surface area using a modified process. These afford a new type of structure that enhances the reinforcing properties, without raising the viscosity of the rubber compound. Micro-dispersion of the precipitated silica in rubber is under development by the silica producer.

## 2.2.5 Global Production

Precipitated silica for the production of tires and technical rubber is offered by several companies. Due to the growth in the passenger car industry, market demand for precipitated silica has risen sharply. The chief producers of silica are:

- Evonik Industries (Germany)
- Gujaret Multi Gas Base Chemicals (China)
- Industrial Quimicas del Ebro, S. A. (Spain)
- PPG Industries (USA)
- Mahdu Silica Ltd. (India)
- Pa Corporation (USA)
- Quechen Silicon Chem. Co. (China)
- Tosoh Silica Corporation (Japan)

The demand for precipitated silica surged, due to the growth of the automobile tire industry across the globe. The increase in high-quality precipitated silica is expected to drive the global market, especially in the passenger tires application where it provides better wet grip and the lower rolling resistance.

## 2.3 Fumed Silica

The smallest particle sizes of silica are produced in a high-temperature reaction between $SiCl_4$ and water vapor. The technology for producing "white carbon black" was developed by Degussa in 1942 [81]. The product was considered a curiosity, due to its multiple properties. However, fumed silica was not able to establish itself as a general filler for rubber compounds, if only for reasons of price. Fumed silicas, also called "pyrogenic", "colloidal" or "anhydrous" silica, are produced by high-temperature flame hydrolysis of $SiCl_4$ in a mixture of hydrogen and oxygen around 1800 °C, according to the global reaction:

$$SiCl_4 + 2H_2 + O_2 \longrightarrow SiO_2 + 4HCl$$

Fumed silicas, produced by flame hydrolysis of silicon tetrachloride, are relatively "anhydrous" fillers, on account of the less than 1.5% bound water on the surface [82]. Several other processes have been developed for the production of fumed silica. The "electrothermal" and the "arc" processes produce low-moisture forms of anhydrous silica [83, 84].

The oxygen in the fuel is a hyper-stoichiometric component. Alkylchlorosilanes, which are a by-product of silicone production, are often added to the feedstock. The fumed silica, which is finely distributed in the HCl-containing gas stream, is separated with cyclones and filter chambers. Most of the HCl is then removed in eddy current reactors using steam and hot air. The primary particles of 7–50 nm formed in the burner immediately melt at the high temperature to form large aggregates 100–500 nm in size. During storage, van der Waals forces and hydrogen bonds form agglomerates with a size of 100 µm [84]. The production of fumed silica is shown in Figure 2.15.

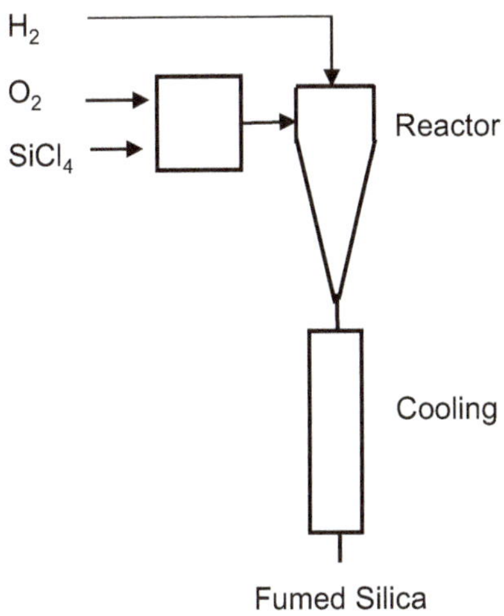


**Figure 2.15**
Manufacture of fumed silica

Both the normal moisture level and the affinity for moisture are reduced. The resulting density of fumed silica is similar to that of precipitated silica. The bulk density is extremely low, however, at about 40 g/l, although it can be increased through compression to reach 100 g/l. Fumed silica with a low specific surface area and low "structure" is produced in an electric arc. The silicon dioxide is reduced to silicon and reacts with oxygen to form fumed silica with a high surface area, but it does not play a role in rubber technology. Fumed silica is characterized by small particles 7–15 nm in size, high purity (99%) and a reduced number of silanol groups on the aggregate surface. In rubber processing, fumed silicas with a specific surface area between 130 and 400 $m^2/g$ are used. Due to their low bulk density and high fluidity, they are difficult to mix into solid rubbers. This type of silica is used in silicone rubber, where it offers a combination of heat stability, low electrical conductivity and transparency. In view of the manufacturing process, its high price is acceptable. Fumed silica with a low specific surface and low structure is produced using the arc process. These products have not been used in rubber technology.

## 2.3.1 Formation of Fumed Silica

It is assumed that small nanoscale particles of amorphous silica form in the initial phase, the growth of which depends on the processing conditions. At high temperatures, collisions and coalescence of proto-particles lead to the formation of primary particles. The rate of coalescence depends on the viscosity of the molten oxide, which collides and form the three-dimensional, branched chair-like aggregates [85]. The silica aggregates are growing at high temperatures and at the end of the reactor could down at the end of the process. The fumed silica now becomes solid and interact with each other through surface interactions of the few silanol groups.

The aggregate of a fumed silica particle can form large clusters, whose size is generally inversely proportional to the primary particle size or proportional to the specific surface area. The agglomerate size also increases in line with the extent to which the fumed silica is compacted. The binding forces holding these agglomerates together are relatively weak, with some silanol groups (3–4.5 silanol groups per $nm^2$) providing hydrophilic interaction of with a weakly acidic character that creates hydrogen bonds. Agglomerates are less tightly clustered than those of carbon black or precipitated silica. This is due to the smaller surface concentration of silanol groups on the particle surface.

Normal moisture levels and affinity for moisture are reduced in fumed silica. Commercial fumed silica is available in a small particle size and with a relatively high specific surface area. Compared with precipitated silica, fumed silica is hard to disperse in rubber. Fumed silica is most widely used in silicone rubber, where incorporation is relatively tricky and takes a long time.

## 2.3.2 Global Production

The main use of fumed silica is the reinforcement of silicone rubber, as its cost precludes general application in rubber. Commercial fumed silicas are available in average particle sizes of 7–50 nm and with specific surface areas ranging from 50 to 400 $m^2/g$. Some fumed silica production processes include compacting, which can increase the bulk density by a factor of 2–2.5. The main producers of fumed silica are:

- Evonik Industries (Germany)
- Cabot Industries (USA)
- Wacker Chemie (Germany)
- Tokyama Coorporation (Japan)
- Application Material Solutions (USA)

Demand for fumed silica has risen exponentially to over 250,000 metric tons and it is expected to grow further.

## 2.4 Layered Silicates

For several decades, attempts have been made to successfully incorporate the layered silicates into a rubber matrix as an active filler. However, the success was only achieved when a Japanese research group found a way to separate the platelets and incorporate them into polar polymers, especially polyamides. It was found that the strong interparticle attractions can be reduced if the layered silicate is treated with a quaternary ammonium salt that makes it possible to increase the distance between the particles and obtain an intercalated or exfoliated layered silicate.

The silicates are the largest and one of the most complex class of minerals in the world. Approximately 30% of all minerals are silicates and geologists estimate that 90% of the earth's crust is made up of silicates. Clay is a naturally occurring material composed primarily of fine minerals that can become plastic with the right amount of water and that harden when dried, according to the Clay Mineral Society. Clays can be divided into the following groups on the basis of their electrical charge layer:

- neutral layers
- negatively charged layers
- positively charged layers

In the last two cases, the layer charge is compensated for by an equal amount of opposite charges located in the interlayer space. The subclass contains four subgroups, among them a subgroup known as 2:1 phyllosilicates or 2:1 smectites. The 2:1 phyllosilicates are a subclass of silicates, with a structure of silicate tetrahedrons arranged in two sheets and linked to aluminum oxide or magnesium oxide, generating a sheet-like structure. The excess negative charge within the layers is saturated by the positive charges of the cations between the layers. This subclass consists of materials made up planar agglomerates which can further break into small platelets of layered silicates that are 1 nm thick and form horizontal sheets of several hundred nm thickness. Dispersion of the layered silicates into the simple two-dimensional platelets requires electrostatic forces that cannot be generated by mechanical mixing. However, the large platelets offer a high aspect ratio and even a high specific surface area when they are well dispersed in the exfoliated state in a rubber matrix. Since the specific area and the aspect ratio are present, the advantage of layered silicate over conventional iso-dimensional fillers becomes apparent:

- lighter weight due to low loading
- low cost due to a lesser amount of filler
- improved properties at low loadings
- combination of specific properties

The color of 2:1 smectites is white, gray or yellow. The material increases its original volume when added to water; layered silicates can swell in water and exfoliate in water. Their real chemical composition depends on the region in which they were formed.

A characteristic is the presence of aggregates, consisting of 20–30 platelets adhering perpendicularly to the orientation of the platelets. The interaction between the individual platelets is very strong and cannot be destroyed by mechanical mixing.

Layered silicates were discovered in Vienne, the commune of Montmorillon, in the year 1849 and were thus called montmorillonite [86]. Some years later, in 1897, they were discovered in Fort Benton in Montana, where they are called bentonite. More were discovered in numerous other locations. They are a cheap material, yet provide a substantial reinforcing effect in polymers. In the following, the term layered silicate or montmorillonite (MMT) is used instead of 2:1 smectites or 2:1 phyllosilicates.

Montmorillonite is the most commonly used clay for the production of polymer nanocomposites, due to its widespread availability, low cost, high aspect ratio and high specific surface area. However, this clay form leads to multiscale organization on the microscale because many of the sheets are forming aggregates or tactoids that are not easily dispersed. The dispersion that is important in rubber requires the opening of the tactoids and allow the single platelets to be dispersed in polymers. The material is considered safe and is handled like a standard powder. It can yield two-dimensional platelets when it is properly dispersed in rubber.

## 2.4.1 Chemical Composition of Montmorillonite

Montmorillonite (MMT) is a sodium aluminum silicate with a variable proportion of water of crystallization. The chemical composition of this layered silicate is a hydrated sodium, calcium aluminum magnesium silicate hydroxide of the approximate formula:

$$(Na, Ca)_{0.33}\,(Al, Mg)_2\,(Si_4O_{10})\,(OH)_2\,.nH_2O$$

The simple-looking formula does not reflect the complex nature of this material. The structure of MMT is derived from the original pyrophyllite structure in that the $Al^{3+}$ cations in the octahedral layer have been partially replaced by the bivalent $Mg^{2+}$ cation. Oxygen atoms from each aluminum oxide from the octahedral layer also belongs to the tetrahedral layer of the silica. Due to the difference in charge between $Al^{3+}$ and $Mg^{2+}$, the center layer is negatively charged and this is balanced by alkali ($Na^+$ or $K^+$) or alkaline earth ions ($Ca^{2+}$ and $Mg^{2+}$) between the layers [87]. These ions do not fit into the tetrahedral layer and are located between the layers, with the result that 2:1 phyllosilicate can adsorb water between the charged layers, due to this electrostatic bonding and the space between the layers increases [88]. Figure 2.21 shows the model crystal structure of a montmorillonite layer.

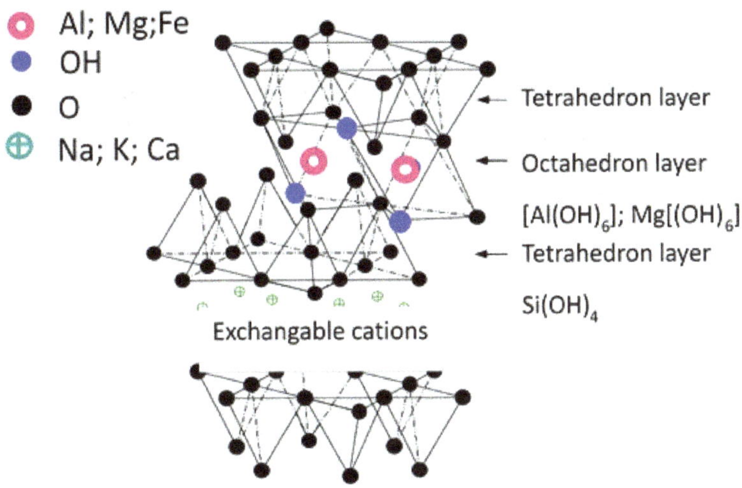


**Figure 2.16** Structure of a 2:1 phyllosilicate

The peculiar form of the layers shows a significant difference in length relative to thickness. They are two-dimensional platelets with a thickness of 1 nm and a lateral dimension with a diameter from 300 to 1000 nm, depending where they come from and how they are produced. Therefore, the aspect ratio of montmorillonite is very important when it comes to its use for reinforcement in rubber matrixes and thermoplastic polymers. Particles with a planar shape have more surface area available for contact with the rubber chains than spherical, iso-dimensional particles. The higher the aspect ratio, the higher are the values obtained for dynamical mechanical moduli and reduction in gas and fuel permeation. The total amount of cations is expressed in milliequivalents per 100 g of dry material and is called the "cation exchange capacity" (CEC) of the clay. The CEC is generally determined by saturating the clay with $NH_4^+$ and determining the amount held at pH 7 by conductometric titration [90]. Another method consists in saturating the clay with alkali-ammonium ions and evaluating the quantity of ions intercalated by ignition of the sample [91]. Comparison of sodium montmorillonites obtained from different mining areas reveals significant differences in mineralogical compositions. Due to the high content of alkaline earth ions in some layered silicates, these demonstrate a significantly lower ability to exfoliate in water, compared with $Na^+$cloisite.

A high aspect ratio enables layered silicates to be dispersed and oriented in the processing direction. In this direction, the stress-strain properties have the highest values. Clays are planar particles that align with the rubber chains during mixing and processing and enable orientation in the processing direction. Perpendicular to the processing direction, the orientation of the platelets represents a hindrance for any gas or fluid that has to surround the solid platelets. Therefore, there is a barrier to the transport of gases or liquids in the direction perpendicular to the surface of the platelets and a longer flow distance is apparent.

The layered silicates are not easily dispersed in most polymers, as they prefer to be in the tactoids (agglomerates). Dispersion of the tactoids into discrete monolayers is further hindered by the intrinsic incompatibility of the hydrophilic layered silicates and the hydrophobic non-polar polymer. For this reason, a purely mechanical attempt to disperse the montmorillonite with available means of dispersion in an internal mixer proves to be unsuccessful. The idealized structures of the 2:1 layered silicates in rubber are shown in Figure 2.17.

**Figure 2.17** Possible structures of layered silicates in a rubber matrix: non-dispersed, intercalated and exfoliated platelets

In conventional composites, the clay layer remains stocked in the form of the diminished tactoids, that cannot dispersed further. The polymer chains cannot intercalate into the galleries and remain attached to the large tactoids. The second system corresponds to the intercalated layered silicates, in which a single, extended polymer chains penetrate between the clay layers to produce a very ordered multilateral layer with alternating polymer and layered clay. When the polymer chains are intercalated into the gaps in the clay galleries, they are unable to break the layered structure. The mixing entropy prevents larger intercalation of the clay. The third system is that of exfoliated layered silicates, in which the clay is dispersed as monolayers in the polymer matrix [91]. The intercalation is so extensive that the distance between the layers increases to a large extent and they are no longer in the tactoid structure. Through modifying the clay surface, they can be rendered organophilic to make them compatible with hydrophobic organic polymers [92].

In the past, the pristine layered silicates were used to reduce the cost of the elastomer, but no attempt was made to achieve an interesting dispersion of the filler by using two-roll mills or an internal mixer. Pristine montmorillonite can be compounded by wet-mixing [93]. A process for exfoliating the platelets by more than 80% was developed in the German Institute of Rubber Technology and called continuous dynamic latex compounding (CDLC) [94, 95]. These compounds have very high dispersion in rubber compounds and provide excellent results in terms of dynamic properties and a significant reduction in the explosive decomposition of polar NBR and HNBR products used in oil exploration. The advantage of using exfoliated platelets of layered silicates is the possibility of producing rubber compounds offering improvements in physical performance:

- mechanical properties
- low permeability of gases and fuels
- thermal stability
- flame retarders

These dispersed layered silicates show high dispersion in rubber compounds and provide excellent results in dynamic properties of the filled rubber vulcanizate. The explosive decomposition of NBR and HNBR used in oil exploration is strongly reduced.

The use of pristine montmorillonite can lead to significant reinforcement in rubber compounds, especially of the barrier properties, which are increased significantly by the use of well-exfoliated layered silicates at a very low concentration in rubber compounds.

### 2.4.2 Organo-Clays

In practice, it has been found that the functionalization of layered silicates to organo-clays with several organic onium bases leads to the reinforcement of latex-based elastomers [96, 97]. It was Toyota researchers who began a detailed study of the interaction of layered organo-silicates and nylon-6. At only 4.2 wt.% organo-clay, the modulus doubles, the tensile strength increases by more than 50%, the heat deflection temperature increases by 100% and the rate of combustion heat release decreases by 63% [98–100]. Organo-clays have been widely used for dispersion in polar polymers to take advantage of the specific surface area and the high aspect ratio.

The pristine layered silicates usually contain alkali cations which can be replaced by other cationic surfactants, especially by quaternary ammonia ions that turn the normally hydrophilic silicates into organophilic silicates, a fact which facilitates interaction with many polymers. Organo-clay can be produced from normally hydrophilic clay by exchange with an organic ammino cation. The way in which this is done has a major influence on the formation of organo-clay. The most widely used alkylammonium cations are based on primary amines introduced into an acidic medium to protonate the amine function. It is important to note that the length of the ammonium ions can have a strong impact on the resulting structure of the nanocomposite. Alkyl ammonium ions with a chain length larger than eight carbon atoms favors the synthesis of efficient organo-clays [101]. The intercalation of ammonium cations leads to an increase in interlayer spacing in the direction perpendicular to the structural layers, which can be measured by X-ray diffraction. Compared with the interlayer distance of sodium montmorillonite (d = 122 nm), the dimethyl-talloyl-dihydroxyethyl group and the dimethyl-diatalloyl group lead to an increase in the interlayer distance to 1.85 nm and 3.5 nm. It should be noted that one and the same ammonium cation can give rise to different interlayer distances, as a function of its weight percent and of its bond with the inorganic layer. Table 2.7 shows the interlayer spacing values determined for commercially available organoclays with dimethyl-talloyl-dihydroxy-ethyl ammonium cations.

**Table 2.7** Organoclays with Dimethyl-talloyl-dihydroxyethyl Ammonium Cation [101]

| Organoclay | Organic (wt.%) | d (nm) |
|---|---|---|
| Dellite®72T | 36–38 | ~ 2.5 |
| Cloisite®20A | 38 | ~ 2.5 |
| Cloisite®15A | 43 | ~ 3.0 |
| Dellite®67G | 45 | ~ 3.6 |

Ion-exchange reactions with cationic surfactants, including primary, tertiary and quaternary ammonium or phosphonium ions, transform the hydrophilic layered silicate into an organophilic layer that enables intercalation and exfoliation in engineering polymers [102, 103]. The role of alkylammonium cations in organo-silicates is to lower the surface energy of the inorganic platelets and improve the wetting characteristics with the polymer. This exchange reaction increases the interlayer spacing to higher values. The organo-silicates are much more compatible with the majority of engineering plastics. In addition, organic cations can contain different functional groups that react even more strongly with the polymer and improve the adhesion between the inorganic phase and matrix. Polymeric nanocomposites using layered silicates as nanofiller have gained particularly in importance over the last three decades, once dispersion of the layered silicates was solved by making various changes to the nature of the filler. Dispersing of the layered silicates into exfoliated platelets requires special wet-mixing techniques [104]. Nanocomposites with exfoliated layered silicates have several advantages, including:

- They are lighter than conventional filled polymers, because of the low concentrations.
- The mechanical properties are potentially better than those of the conventional filler because of the high aspect ratios and the high specific surface, and
- they exhibit outstanding diffusion barrier properties, without the need for a special design.

Methods that use pristine layered silicates and organo-clays are suitable for industrial development, whereas solution blending can be employed in laboratories. One can consider the advantageous role of layered silicates as transport barriers in elastomers and their reduction of explosive decompression of polar rubbers. This filler materials offer scope for greatly enhancing the elastomer properties. Layered silicates have an advantage in the rubber industry and especially in the plastics industry which is characterized by their reinforcement of the mechanical properties and especially by their barrier properties, which greatly reduce the permeation of gases and liquids through the filled material. It is of some advantage if the composite is lighter than materials reinforced with carbon black and precipitated silica. In addition, the layered silica is readily available but it needs to be selected carefully and the price is relatively low.

### 2.4.3 Global Production of Layered Silicates

In 2018, China was the top producer of layered silicates, accounting for almost a quarter of global production, followed by the United States and India. Total global was 400,000 metric tons, but only a small proportion finds application in the rubber field as a reinforced filler. Although organic pretreatment drives up the cost of organo-clay, the overall cost is still low because of the cheapness of layered silicates.

## 2.5 Carbon Nanotubes

Chemists and physicist may prefer to think on carbon nanotubes as linearly extended fullerenes. But carbon nanotubes are more complicated and has a huge application in science and industrial applications. To those interested in nanocomposites carbon nanotubes are ultrafine carbon fibers capable of a high elasticity to withstand stress and strain in polymer compounds. Carbon nanotubes are one-dimensional fillers with the diameter of several nm and length of microns. They consist of one or more concentric graphitic layers and are smaller carbon fibers and have diameters similar to the fullerenes. The variation in the diameter of graphitic materials is shown in Figure 2.18 [105].

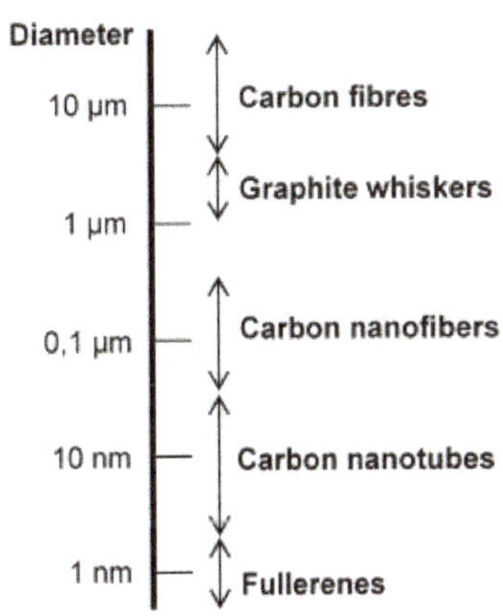


**Figure 2.18**
Comparison of diameters of various fibrous carbon-based materials

Between the allotropic forms of carbon atoms, the carbon nanotubes have received during the last two decades an important interest in improving the reinforcement of elastomers. Carbon nanotubes have very high mechanical, electrical and thermal properties. In contrast to commercial carbon fibers with a much larger diameter of 7 to 20 µm and the vapor grown fibers with a diameter of a few hundred nanometers to approx. 1 mm carbon nanotubes have diameters which is from 1 to 50 nm. They are considered as very small fibers or heavy molecules, since the diameter of the smallest molecules are in similar to those of common polymer molecules. One distinguishes between *single-walled carbon nanotubes* (SWCNTs) and *multi-walled carbon nanotubes* (MWCNTs), which have gained importance as fillers in polymers.

There are several ways to roll the graphene plane depending on the angle and therefore there are different types of carbon nanotubes that can be formed. The orientation of the graphite lattice relative to the axis defines the chirality of the carbon nanotubes [106] A single-walled carbon nanotube is formed by rolling the sheet of graphene into a cylinder along the m, n lattice vectors in the graphene plane. The m, n indices determine the diameter and the chirality, which are key parameters for carbon nanotubes. The molecular models for SWCNTs show different geometric arrangements depending on how they are rotated in the graphene plane, which is shown in Figure 2.19.

**Figure 2.19** Molecular models of SWCNTs with different chiralities:
a) armchair arrangement, b) zig-zag arrangement and c) chiral conformation

Depending on how the tubes are curled up, one can obtain metallic or semiconducting properties of SWCNTs [107]. Dependent on the manufacturing procedure, carbon nanotubes can have different structures, lengths and thicknesses. SWCNTs appear as bundles, containing, because of the van der Waals attractions, many different tubes that cannot be dispersed easily in a polymer [108, 109]. MWCNTs are formed by a concentrically arranged number of tubes separated by an interlayer spacing of 0.134 nm which is the same as in graphite. This concentric structure of the MWCNTs enables them to be used to significantly increase the properties of an elastomer and to significantly improve its fracture properties. They may have a more complex array, because each concentric single-walled nanotube can have different structures, and hence there are a variety of sequential arrangements, are either regular layering or random layering.

However, the first CNTs were produced in the western world in 1976 by Endo as a result of the synthesis of tubular nanotubes using hydrocarbon chemical vapor decomposition [110]. The first recognition of single walled CNTs was the observation of graphitic tubular structure in an arc discharge apparatus in a report published by Ijima in the year 1991 [111, 112] This beginning has significantly accelerated scientific interest in carbon nanotubes, both from a scientific point of view and in terms of industrial application. But the first publication of carbon nanotubes was published by Raduchevich in 1952 in a Russian publication, which has not been read in the western world [113]. The wider the diameter of the multi-walled carbon nanotubes, the stiffer is the carbon nanotube and the lower the interaction with rubbers. However, the type of the rubber plays an important role. If the rubber is unsaturated and polar it shows a better interaction with the carbon nanotubes than those who are saturated. The narrower the diameter of the carbon nanotube, the more its intrinsic properties depend upon its specific type. On a molecular level, SWCNTs are 100 times stronger than steel at one-sixth the weight and have a very large aspect ratio making them very useful as a mechanical property enhancing filler reinforcing properties in rubber compounds.

Carbon nanotubes have attracted particular interest in scientific and technology because they show remarkable physical properties at low concentrations in polymer matrices. The combination of these properties obtained at low concentrations suggest that carbon nanotubes are ideal candidates for high performance polymer composites.

Although significant advances have been made in recent years to overcome difficulties in the manufacturing of the polymer nanocomposites, processing of compounds that contain carbon nanotubes remains a challenging problem.

Carbon nanotubes have already found commercial applications in the fields of engineering plastics, polymers, displays, anticorrosion paints, thin films and coatings, transparent and nontransparent conductive electrodes, super hydrophobic coatings and antistatic packaging while active research is ongoing in fields such as batteries, fuel cells, solar cells, advanced devices, optics, water desalination and many others. They are attracting the particular interest in rubber technology. A large aspect ratio of the carbon nanotubes and the high specific surface area fillers are requirement to allow these fillers, to orient during processing in the rubber matrix and produce an anisotropic elastomer that has a much higher physical properties in the processing direction.

Multi-walled carbon nanotubes were shown to be attractive nanofillers for rubber compounds in order to reach high electrical conductivity and excellent mechanical properties at low filler loadings. Although hundreds kilogram of multi-walled carbon nanotubes are currently produced per day, the development of high stiffness elastomers based on this filler has been hampered so far by the lack of quality assurance.

The quality and yield of carbon nanotubes strongly depend on the synthesis and the purification techniques and the specific growth conditions used.

### 2.5.1 Manufacturing of Carbon Nanotubes

The production techniques lead to carbon nanotubes with various characteristics such as diameter, length and different entanglements. Both SWCNTs and MWCNTs can be produced by different processes, which can be divided in two categories. The high temperature evaporation processes of the arc discharge and the laser ablation of graphite is in one group. The next group occurs at much lower temperature and consists of the chemical vapor deposition or the catalytic growth process. In the high temperature methods, the SWCNTs requires the presence of metallic catalysts, while the MWCNTs are produced by the evaporation of pure carbon. The chemical vapor deposition method requires for both of carbon nanotubes a metallic catalyst.

#### 2.5.1.1 Arc Discharge Process

The arc discharge technology is based on the generation of a DC arc plasma process between two carbon electrodes in a helium atmosphere [111, 114–116]. By separating the contact of the graphite electrodes to 1–2 mm and applying arc discharge between the pure graphite electrodes a high temperature (3000–4000 °C) is created in a pure hydrogen or helium stream and that causes sublimation of carbon and the deposition of CNTs on the anode. The SWCNTs form bundles of varying lengths that cannot be dispersed in a rubber matrix. However, the carbon nanotubes are of high quality with low amounts of defects [117, 118]. While the anode is consumed, a soft fibrous deposit from the cathode.

It was shown that with this technology it is able to produce MWCNTs if both electrodes contain only graphite. If the graphite anode is doped with metal catalyst, i. e. Fe, CO, Ni, Ni-Co, Co-Y, Ni-Y in a concentration of 3% or greater, on the anode and on the walls are deposited SWCNTs [119]. The MWCNTs produced by this method are generally entangled by the van der Waals interaction. A schematic diagram of the arc discharge process is shown in Figure 2.20.

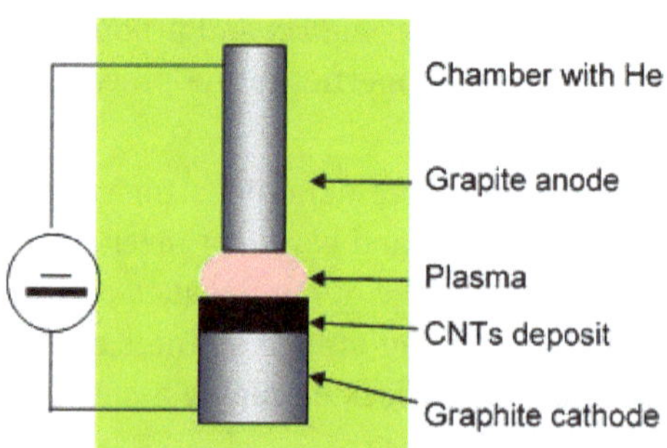


**Figure 2.20**
Preparation of CNTs by arc discharge

However, when cobalt is added as a catalyst, the nature of the product changes to form a web, with large strands. The mechanism by which cobalt changes this process is unclear, however one possibility is that such metals affect the local electric fields and hence the formation of the five-membered rings. The mass production of MWCNTs by using the arc discharge method is realized soon after the first publication [120].

Usually, carbon nanotubes synthesized by arc discharge show a high degree of structural perfection; however, a number of variables (temperature of the chamber, the composition and concentration of the catalyst, the presence of hydrogen, etc.) influence the size and structure of the carbon nanotubes. The carbon arc discharge method is common and perhaps easiest way to produce carbon nanotubes, as it is from the technology rather simple. The reaction times resulting from this technology are very short and the error in the formation of the carbon nanotubes is not very high. The deposit contains nanotubes, fullerenes, amorphous carbon and catalyst particles. The technique requires purification to separate the carbon nanotubes from the by products in the crude material. However, it is a technique that produces a complex mixture of components, and therefore requires further purification to separate the carbon nanotubes from the other carbonic materials and the residual catalytic metals present in the crude product will complicate the result of the purification process.

#### 2.5.1.2 Laser Ablation Process

Using a different technique to generate the carbon vapor the laser ablation is used to produce both types of carbon nanotubes. A graphite target, containing 1–2% metal catalyst, is held in a furnace at 1200 °C in an inert atmosphere and is evaporated using a high-power laser Nd-YAG [121]. At the cooled collector the carbon nanotubes were deposited. In order to obtain SWCNTs the transition metal particle (Co and Ni) must be added to the graphite target. The result was annealed at 1000 °C to remove the fullerenes, which are also formed in this reaction. The material produced by this method appears as a mat of “ropes” of 10–20 nm in diameter and up to 100 µm or more in length. The SWCNTs will self-assemble and condense on the walls of the flow tube. The products are highly crystalline, with low defect concentrations, but are not pure and contain carbonaceous impurities. The diameter of the SWCNTs is around 1.0–1.6 nm. The carbon nanotubes produced by laser ablation has a purity of 90% [122]. Each rope is found to consist primarily of a bundle of single walled nanotubes, aligned along a common axis but this SWCNTs cannot be dispersed in a polymer matrix. By varying the growth temperature, the catalyst composition, and other process parameters, the average nanotube diameter and size distribution can be varied. The quantity and the quality of the carbon nanotubes depends on several factors such as the amount of catalyst, laser power and wavelength, pressure, type of inert gas. With this type of production, too, the carbon nanotubes have to be freed from the other carbon-containing materials in order to obtain a carbon nanotube that is as pure as possible. The arc discharge and the laser ablation are usually work on the gram scale and are relative expensive.

### 2.5.1.3 Chemical Vapor Deposition Process

The carbon nanotubes that are used in polymer nanocomposites require large amounts and a relative low cost and are handled without a complicated purification. Undoubtedly the most common method to produce of multi-walled carbon nanotubes is catalyzed chemical vapor deposition (CVD) of low molecular weight carbon molecules over a metal catalyst [123, 124]. Only this method satisfy the requirements to provide the quantity of carbon nanotubes in the scientific research and industrial application This technology holds the greatest promise, since it allows the production of larger quantities of carbon nanotubes under more easily controllable conditions and at lower cost, a fact which is important to use the MWCNTs to enhance the performance of elastomers.

In the CVD process, the metal catalyst, such as iron, cobalt or nickel, is combined with carbon-containing reaction gases, such is acetylene and argon or carbon monoxide and hydrogen to form carbon nanotubes at lower temperature. However, the quality of carbon nanotubes shows structurally some defects. a high-temperature furnace. The vapor chemical deposition is shown in Figure 2.21.

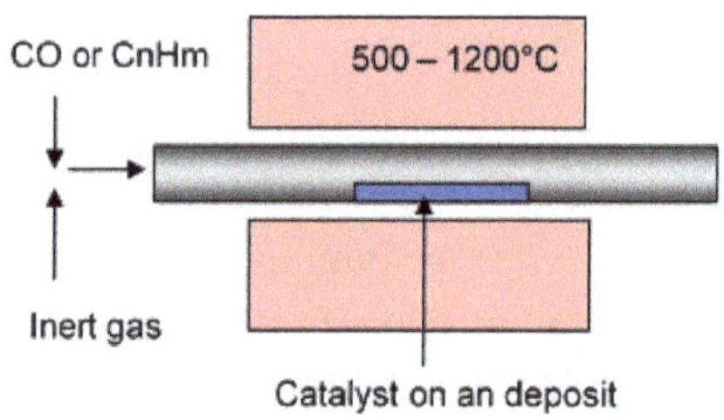


**Figure 2.21**
Schematic view on the manufacturing to carbon nanotubes by chemical vapor deposition

The CVD process can be purely catalytic or plasma-supported. The latter process requires slightly lower temperatures (200–500 °C) than the catalytic process (up to 750 °C) and aims at producing 'lawn-like' carbon nanotubes growth. There are several options for introducing the catalyst in the furnace, ranging injection of the organometallic vapors and metallic colloids, to pre-depositions of metal films on ceramic supports. Injecting the catalyst into the gas stream allows for the continuous production of carbon nanotubes. In addition, CVD process has the capability to control the size, shape and alignment of the multi-walled carbon nanotubes through a carefully designed patterning of the catalysts on the substrate surface [125]. Frequently used substrates for bulk synthesis are ceramic particles such as magnesium oxide, aluminum oxide [126]. The synthesis of well-aligned and straight MWCNTs can be produced during a plasma growth process at low temperatures of 120 °C [127, 129].

The size of the catalyst determines the thickness of the MWCNT and the alignment. Vertically aligned carbon nanotube arrays also grown by thermal chemical vapor deposition and can easily remove from the substrate and constitute a carbon nanotube for nanocomposite applications. Typically, that layer is iron and is deposited via sputtering to a thickness of 1–5 nm. A 10–50 nm underlayer of alumina is often also put down on the substrate first. The thickness of the MWCNT obtained is crucial for the interaction forces between the polymeric material and the multi-walled carbon tube.

The smaller the diameter, the higher the interactions between the π-orbitals of the $sp^2$-hybridized carbon and the polymer. This imparts controllable wetting and good interfacial properties. When the substrate is heated to the growth temperature (~700 °C), the continuous iron film breaks up into small islands with each island then nucleating a carbon nanotube. The sputtered thickness controls the island size and this in turn determines the nanotube diameter. Thinner iron layers drive down the diameter of the nanotubes grown. The amount of time the metal island can sit at the growth temperature is limited as they are mobile and can merge into larger (but fewer) islands. Annealing at the growth temperature reduces the site density while increasing the catalyst diameter. At present catalytic MWCNTs is appear an optimum choice, because the materials are of the large quantities and a high purity.

The unannealed carbon nanotubes manifest dual weight losses in thermoanalytical gravimetry. The first thermogravimetric peak is located at 442 °C corresponds to a loss of 22.3% may be attributed to the degradation of less stable carbon. The second weight loss at 554 °C can be attributed to the degradation of the carbon nanotubes [129]. A newer variant is the Hipco technology carried out at 900–1000 °C under high pressures (30–40 bars). It was developed for creating SWCNTs from the gas-phase reaction of iron carbonyl with high-pressure carbon monoxide gas [130]. Iron pentacarbonyl is used to produce carbon nanotubes that provide a nucleation surface for the transformation of carbon monoxide into carbon nanotubes [131, 132]. It is commercially exploited by Carbon Nanotechnologies Inc., USA.

MWCNTs can also be produced by diffusion flame synthesis, electrolysis, use of solar energy, heat treatment of a polymer, and low-temperature solid pyrolysis. In flame synthesis, combustion of a portion of the hydrocarbon gas provides the elevated temperature required, with the remaining fuel conveniently serving as the required hydrocarbon reagent. The technology of catalytic vapor deposition can be driven at lower temperatures, which means an advantage in the production of larger quantities of multi-walled carbon nanotubes and a lower energy consumption. However, this technology can lead to both variations in the orientation of graphite planes with respect to the axis of the tubes and an increased amount of structure defects. Even in the simplest case, it is necessary to purify the mixture in order to separate the nontarry materials and the metal.

Different methods are used to determine the number of walls in a MWCNTs. Usually, it is estimated by transmission electron microscopy (TEM), which reveal the length of the tubes. The X-ray diffraction analysis can be used to determine the number of the tubes. At present, only CVD-grown nanotubes satisfy the use to produce elastomer nano composites. The main contaminant in CVD materials are the residual catalysts which are incorporated into the multi-walled carbon nanotubes. The multi-walled carbon nanotubes produced in CVD process show higher quality in terms of defects than those prepared by shorter reaction time.

The quality and yield of multi-walled carbon nanotubes depend on the synthesis technique and grow conditions used. Catalytic process involves growth temperatures which lead to both variations in the orientation of the graphitic planes with the tube axis and to an increased concentration of defects [133]. Multi-walled carbon nanotube may not produce replacements of existing fillers like carbon black or precipitated silica, but they are able to increase the mechanical and electrical properties together with the traditional fillers.

## 2.5.2 Purification of Carbon Nanotubes

As deposited multi-walled carbon nanotubes are frequently contaminated with metal catalysts, ceramic particles, fullerenes and amorphous carbon. The first step of purification includes the removal of the impurities by wet chemical oxidation ($HNO_3$). Additional steps include decanting, centrifugation or filtration followed by rinsing in de-ionized water [133, 134]. The advantage of this method is that unwanted nanoparticles and amorphous carbon are removed simultaneously and the carbon nanotubes are not chemically modified. A common method is treatment with a strong acid to remove the material, which is amorphous carbon. The addition of 2–3 mol nitric acid can be helpful to remove the carbonaceous impurities. Oxidative purification and filtration techniques a shown to purify carbon nanotubes in Figure 2.22.

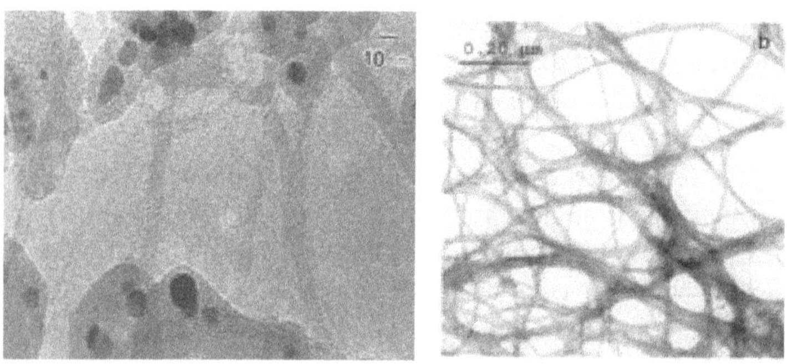


Figure 2.22 Oxidative agents ($HNO_3$, $H_2O_2$, $KMnO_4$) remove impurities form the carbon nanotubes: a) original sample and b) purified sample

Even though synthetic techniques have been improved to obtain high-purity multi-walled carbon nanotubes, the formation of byproducts containing impurities such as metal encapsulated nanoparticles, metal particles in the tip of a carbon nanotube, and amorphous carbon has been an unavoidable phenomenon, because the metal nanoparticles are essential for the nanotube growth.

These foreign nanoparticles, as well as structural defects that occurred during synthesis, have the unfortunate implication that they modify the physico-chemical properties of the produced carbon nanotubes. That is why carbon nanotubes need to be purified with the help of various methods such as acid treatment or ultrasound at the end of the production process.

### 2.5.3 Use of Carbon Nanotubes

The large aspect ratio and the excellent electric conductivity recommend the single walled carbon nanotubes especially for high electrical and electronic applications. They are separated according to the angle of rotation of the graphite planes. The high thermal properties lead them to increase thermal conductivity when properly aligned.

For use in the rubber sector and in the thermoplastic sector, the high specific surface paired with the high aspect ratio are of great importance in order to provide the polymer materials with better physical properties. The lead to exceptional mechanical properties, high tensile strength, high cut growth resistance and tear energy or elastomers. Their use to substantially increase the reinforcement of elastomers, a fact which is experienced in the German Institute of Rubber Technology (DIK) and Universities. Through cooperation with the industry, some products have come onto the market that use this advantage revealed by the multi-walled carbon nanotubes that give much better performance in the area of abrasion resistance, tear resistance and fracture mechanics with only small fractions of multi-walled carbon nanotubes in combination with a conventional filler, such as an active carbon black or a silanized silica. The improved properties are attained at lower carbon nanotubes content in comparison with the conventionally filled system and will find an advantage in the market this way.

### 2.5.4 Producer of Carbon Nanotubes

Among the many new companies, eight have been selected that produce multi-walled carbon nanotubes:

- Nanocyl (Belgium)
- CHASM (USA)
- Arkema (France)
- Cabot Corporation (USA)

- Nano Shell (GB)
- Showa Denko (Japan)
- Carbon Solutions (USA)

It is found that the production is not accompanied by uniform rules relating to the purity and impurities of the carbon nanotubes.

## 2.6 Human Risk of the Filler

Increased environmental awareness today accompanies the growing production of rubber compounds and thus also the use of fillers in their various forms. In recent years, the topic of the environment has evolved from being a question that people interested in ecology have to become a cross-cutting topic in society. The discussion about the toxicity of carbon black due to the polycyclic hydrocarbons adsorbed on the surface has largely receded since no demonstrable consequences were shown in animal tests. The hazard from used black and white fillers has been regulated by the containment of air pollution laws and other legislations. Today the present carbon black production and processing using a sure technology do not significant influence the risk effects for humans. Detailed investigations have revealed that carbon black does not cause structural damage on lungs or irritations of eyes and mucous membranes. However, during the production and processing the general dust limits must be adhered to.

The oral toxicity $LD_{50}$ of carbon black is of more than 8000 mg/kg, which is really high. Based on the findings or the National Toxicologic Program (NTP) and the Legislation regarding chemicals (OSHA) of the United States as well as the European Legislation carbon black is not exhibiting mutagenic, teratogenic or carcinogenic risk.

Carbon black is classified by the International Agency for Research on Cancer as having the potential to cause cancer in humans and is classified as Category 2B. It is pointed out that this filler incorporated in rubber does not pose any risk to humans.

The studies on precipitated silica did not show any health risk. Epidemiological studies do not support the hypothesis that amorphous silica have any relevant potential to induce fibrosis in workers with high occupational exposure to these substances. Because the state of precipitated silica and the fumed silica have no crystalline structure it is classified as nontoxic material and can be handled in rubber compounds without safety precautions. However, the use of organo-silanes to reduce the concentration of silanol groups on the surface of precipitated silica is questionable due to the ethanol emissions. In the rubber industry, is eliminated by an incoming ventilation in the internal mixer.

Care must be taken with handling and processing multi-walled carbon nanotubes. Animal studies have shown that if inhaled, multi-walled carbon nanotubes cannot only enter the lungs, but can circulate on vital organs. Early scientific studies have indicated that nanoscale particles may pose a greater health risk than bulk materials. For carbon nanotubes that have a surface area and are very thin objects the studies investigated are not giving a clear answer. According to EU definition of 2011, carbon nanotubes are classified as a group of materials with high aspect ratios that should be used as "preparations" produced in special departments, so they cannot be inhaled when they are mixed in the polymer. As of October 2016, SWCNTs have been registered through the European Units Registration, Evaluation and Restrictions of Chemicals (REACH) regulation on the potential of dangerous materials. They can damage the body's natural defense or cause increased response o common allergens. The biological interaction of carbon nanotubes is not well understood and the field is open to continued toxicological studies.

## References

[1] Kraus, G., in "Reinforcement of Elastomers", (Kraus, G., Ed.) Interscience Publ. (1965)

[2] Donnet, J.-B., Bansal, R. C., Wang, M.-J., Eds., in "Carbon Black Science and Technology", Marcel Dekker Inc., New York, (1993)

[3] Gerspacher, M., Wampler, W., in "Basic Elastomer Technology" (Baranwal, K. C., Stephens, H. L., Eds.) Rubber Division (2001), Ch. 4

[4] Wang, M.-J., Maurice, M., Eds. in "Rubber Reinforcement with Particular Fillers", Carl Hanser Verlag, Munich (2021)

[5] Thalwitzer, O., German Pat. 50605 (1889)

[6] McNutt, L. J., US Pat. 481240 (1982)

[7] Studebaker, M., L. Rubber Chem. Technol., 1957, 30: 1400

[8] General Atlas, US Pat. 1,902,753 (1927)

[9] Krejci, J. C., US Pat. 2,375,759 (1945)

[10] Powell, R., in "Carbon Black Technology" Noyes Development Corp., Park Ridge, New Jersey (1968)

[11] Beede, A. F., Stokes, C. A., Petrol Processing, 1954, 9: 1415

[12] Kühn, K., Voll, M. in "Carbon Black Science and Technology", Donnet, J.-B., Bansal, R. C., Wang, M.-J., Eds., Marcel Dekker Inc., New York, (1993), Ch. 1

[13] Vohwinkel, K., Kautsch. Gummi Kunstst., 1986, 39: 810

[14] Krejci, J. C., Johnson, P. H., US Pat. 3,250,794 (1966)

[15] Voet, A., Iannicelly, J., Teter, A. C., US Pat. 3,240,200 (1965)

[16] De Land, C. L., De Cuir, G. L., Wiggins, L. E., US Pat. 3,253,890 (1966)

[17] Voet, A., US Pat. 2,960,413 (1960)

[18] Venuto, L. J., US Pat. 2,500,968 (1950)

[19] Heller, G. L., Snow, C. W., US Pat. 2131686, 27. Sept. 1938

[20] Price, H., US Pat. 2146164, 27. June 1939

[21] Studebaker, M. L., US Pat. 2503361, 11. April 1950

[22] Orion booklet "What is carbon black"

[23] Gordon, M. D., Lavroskii, K. P., Doc. Akad. Nauk USSR, 1969, 19: 1289

[24] Aten, C. F., Greene, E. F., Disc. Faraday Soc., 1956, 22: 126
[25] Bradley, J. N., Kistiakowski, G. B. J., Compt. Rend. Acad. Sci., Paris, 1972, 247: 569
[26] Prado, G., Lahae, J. J., Chem. Phys., 1973, 70: 1078
[27] Palmer, H. P. J., Chem. Phys. Spec. Issue 18th Annual Meeting (1969) p. 87
[28] Cole, D. J., Minkoff, G. F., Fuel, 1956, 35: 135
[29] Porter, G., in "Combustion Researches and Reviews", Butterworths London (1955)
[30] Schweitzer, C. W., Heller, H. H., Rubber World, 1956, 134: 855
[31] Prado, G., Lahaye, J., C. R. Acad. Sci., Paris, 1972, 274: 1880
[32] Cullis, C. F., Minkoff, G. J., Nettleton, M. A., Trans. Faraday Soc., 1962, 58: 1117
[33] Smith, S. R. Proc. Roy. Soc., London, A, 1940, 174: 122
[34] Parker, W. G., Wolfhard, H. G., J. Chem. Soc., 1950, 2038
[35] Prado, G., Lahae, J., J. Chem. Phys., 1973, 70: 1078
[36] Arthur, J. R., Napier, D. N., 5th Symp. Combustion, Pittsburgh (1955) p. 303
[37] Thorp, N., Long, R., Garner, F. H., Fuel, 1955, 34: 51
[38] Gaydon, A. G., Whittingham, G., Proc. Roy. Soc., London, A, 1947, 139: 313
[39] Hess, W. M., et al., Kautsch. Gummi Kunstst., 1968, 21: 689
[40] Medalia, A. I., Heckman, F. A., J. Colloid Interface Sci., 1971, 36: 173
[41] Meakin, P., Donn, B., Mulholland, G. W., Langmuir, 1989, 5: 510
[42] Negaridis, M., Dobbins, R. A., Combust. Sci. Technol., 1990, 71: 95
[43] Thomas, A., Combustion and Flames, 1962, 6: 46
[44] Lahae, J., Prado, G., Donnet, J.-B., Carbon, 1974, 12: 77
[44] Wolff, S., Rubber Chem. Technol., 1993, 66: 163
[45] Donnet, J.-B., Voet, A., in "Carbon Black, Physics chemistry and Elastomer Reinforcement", Dekker, M., New York and Basel (1977), p. 76
[46] Friauf, G. F., Thorley, B., US Pat. 3010794 (1961)
[47] Grisdale, R. O., J. Appl. Phys. 1953, 24: 1082
[48] Bosch, K., German Pat. 270,199 (1913)
[49] Kaufmann, C., Hull, R. H., US Pat. 2,453,440 (1944)
[50] Holmes, J. M., Beebe, R. A., Can. J. Chem., 1957, 35: 1542
[51] Buckley, B. P., in Kirk Othmer "the Encyclopedia of Chemical Technology", 3 (1949) 66
[52] ASTM D 1937
[53] ASTM D 2663
[54] ASTM D 1508
[55] ASTM D 3313
[56] Patnick, T., Brown, B., Rubber Plastik News, 2010, 13: 16
[57] ASTM 1618
[58] ASTM 1506
[59] ASTM 1514
[60] ASTM 1619
[61] Wagner, M. P., Rubber Chem. Technol., 1973, 49: 703
[62] Iler, R. K., in "The Chemistry of Silica", J. Wiley & Sons, New York (1979)
[63] Wolff, S., Kautsch. Gummi Kunstst., 1988, 41: 674
[64] Ferch, H., Toussant, H. E., Kautsch. Gummi Kunstst., 1996, 49: 589

[65] Görl, U., in "Handbuch der Kautschuk-Technologie", Hofmann, W., Gupta, H., Eds., Gupta Verlag, Ratingen (2001) Ch. 8.3

[66] Meon, W., Blume, A., Luginsland, H.-D., Uhrland, S., in "Rubber Compounding, Chemistry and Applications", Rodgers, B., Ed., Marcel Dekker, Inc., New York, Basel (2004) Ch. 7

[67] Thurn, F., Wolff, S., Kautsch. Gummi Kunstst., 1971, 28: 733

[68] Ikeda, Y., Tanaka, A., Kohjiya, Sh., J. Mater. Chem., 1997, 7: 1497

[69] Grady, B. P., Start, R. P., Mauritz, K. A., J. Polym. Sci., B, Polym. Phys., 2001, 39: 197

[70] Coudurier, R., Rhone-Poulenc, EP 0 018 866 (1980)

[71] Brandt, B., Degussa, DE 2 505 191 (1975)

[72] Dorigito, A. Dzenis, Y. Pegeotti, A., Mechanics of Materials, 2013, 12: 79

[73] Byers, J. T., in "Basic Elastomer Technology", Baranwal, K. C., Stephens, H. L., Eds., Rubber Division, American Chemical Society, (2001), Ch. 4

[74] Cochet, P., Int. Conf. "Functional Fillers", Fort Lauderdale, (2001)

[75] Cocket, P., Intertech. Conf. "Functional Tire Fillers", Fort Lauderdale, (2001)

[76] Görl, U., et al., Kautsch. Gummi Kunstst., 1995, 48: 407

[77] Chevalier, Y., Moravski, J. C., US Patent 4590052 (1960)

[78] Kudo, M., et al., US Patent 6521290 B1 (1999)

[79] ISO 5794/1E

[80] Uhrland, S., Blume, A., ACS Meeting, Savannah, GA (2002) p 15

[81] Klöpfer, H., Deutsches Pat. 1,208,741 (1942)

[82] Bachmann, J. H., Sellers, J. W., Wagner, M. P., Wolf, R. F., Rubber Chem., 1959, 32: 1286

[83] Barthel, H., Rösch, L., Weiß, J., J. Organo-silicon Chemistry II, From Molecules to Materials. Weinheim: VCH 1996

[84] Pratsinis, S. E., Progr. Energy Combi. Sci. 1998, 47: 197

[85] Ulrich, G. D. J. Colloid Interf. Sci., 1982, 87: 257

[86] Anthony, J. W., Bideaux, R. A., Bladh, K., Nichols, M. C., eds. "Montmorillonite" Handbook of Mineralogy, Vol. II (1995)

[87] Giannelis, E. P., et al., Adv. Polym. Sci., 138 (1999) 108

[88] Hoffmann, U., Endell, K., Wilm, D. Z., Kristallogr. 1933, 86: 340

[89] Mortland, M. M., Mellos, J. L., Soil Sci. Soc. Proc., 1954, 18: 363

[90] McAtee, J. L. Am. Miner., 1958, 44: 1230

[91] Krishnamoorti, R. K., Vaia, R. A., Giannelis, E. P., Chem. Mater. 1996, 8: 29

[92] Johamardi, C., et al., J. Nanostr. Chem., 2013, 3: 82

[93] Wu, Y.- P., Zhang, L-Q., Comp. Sci. Technol., 2005, 65: 1195

[94] Klafke de Azeredo, L., Schuster, R. H., Jacobi, M. M., Kautsch. Gummi Kunstst., 2010, 62: 130

[95] Schellenberg, M., Klafke de Azeredo, L., Schuster, R. H., 184$^{th}$ ACS Rubber Division Cleveland (2013)

[96] Carter, L. W., US Pat. 2.531.396 (1950)

[97] Hauser, E. A., Kallman, R. C., US Pat. 2.951.087 (1960)

[98] Fujiwara, S., Sakamato, T., Jap. Kokai Pat. Appl. 109.998 (1976)

[99] Usuki, A., et al., J. Mater. Res. 8 (1993) 119

[100] Usuki, A., et al., J. Appl. Polym. Sci., 55 (1995) 119

[101] Lau, T., Kavinata, P. D., Pinnavia, T. J., Chem. Mater., 1995, 7: 2144

[102] Messersmith, P. B., Giannelis, E. P., Chem. Mater. 1994, 6: 1719

[103] Lagaly, M., Angew. Chem., 1976, 15: 575

[104] Schuster, R. H., Klafke de Azeredo, L., DE Patent 048995A1 (2007)

[105] Shaeffer, M. S. P., Sandler, J., in "Processing and Properties of Nanocomposites", Advani, S., Ed., World Science Ltd. Ch.1

[106] Harris, P. J. F., in "Carbon nanotubes and related structures", Cambridge University Press (2001)

[107] Charlier, J. C., Acc. Chem. Res., 2002, 35: 1063

[108] Jourmet, C., et al., Nature, 1997, 388: 756

[109] Tess, A., Lee, R., Nikolaev, P., Doi, H., Petit, P., Robert, C., Xu, C., Lee, Y. H., Kim, S. G., Rinzler, A. G., Colbert, D. T., Scusenia, G. E., Tomaned, D., Fischer, J., Smally, R. E. Science, 1996, 273: 438

[110] Oberlin, A., Endo, M., Koyama, T. J., Crystal Growth. 1976, 32: 335

[111] Ijima, S., Nature, 1991, 354: 56

[112] Ijima, S., Ichihouchi, T., Nature, 1998, 365: 603

[113] Raduchevich, L. W., J. Physical Chemistry (Russian), 1952, 26: 88

[114] Ebbesen, T. W., Ajayan, P. M., Nature, 1992, 358: 320

[115] Guo, T., Nikolaev, A. N., Tess, T. D., Smalley, R. E., Chem. Phys. Lett., 1995, 243: 49

[116] Tess, A., et al., Science, 1996, 273: 483

[117] Bethune, D. S., Kiang, C. H., de Vries, M. S. Gorman, G., Savoy, R., Velasques, R., Beyers, R., Nature, 1993, 364: 605

[118] Seito, Y., Nishibuko, K., Kawabata, K., Matsumoto, T., J. Appl. Phys., 1996, 80: 3062

[119] Demency, N., et al., Europ. Phys. J. B Condens. Mater. Compl. Systems, 1998, 4: 147

[120] Guo, T., Nikolaev, P., Tess, A., Colbert, D. T., Smalley, R. E., Chem. Phys. Lett. 1995, 243: 49

[121] Smith, D., in "Thin Film Deposition: Principles and Practice", 1st ed., McGraw-Hill Incorp., New York (1995)

[122] Tibbets, G. G., et al., Carbon, 1993, 31: 809

[123] Hitchman, M., Jensen, K., in "Chemical Vapour Deposition: Principles and Applications", Academic Press (1997)

[124] Liu, J., Fan, S., Dai, H., MRS Bulletin, 2004, Vol. 29, 4: 224

[125] Li, W., Wan, J., Renz, Z., J. Appl. Phys., A, 2002, 74: 397

[126] Huan, Z. N., et al., Appl. Phys. Lett., 1998, 73: 3945

[127] Brown, C., et al., Appl. Phys. Lett., 2000, 77: 830

[128] Perez, L. D., et al., Polym. Eng. Sci., 2009, 49: 866

[129] Bornikowski, M., et al., J. Vacuum Sci. 2001, 19: 1800

[130] Takeuchi, T., et al., Nanojournal, ifmo. 2015, 5: 15

[131] Nikolaev, P., J. Nanosci. Nanotechnol., 2004, 4: 304

[132] Itkis, M. E., et al., Nano Letters, 2003, 3: 309

[133] Parro, S., et al., Physica E., 2007, 37: 58

[134] Meyer-Plath, A., et al., Carbon, 2006, 50: 3943

[135] Park, T.-J., et al., J. Mater. Chem., 2006, 16: 14

# 3 Characterization of Fillers

Physical and chemical methods are available for detecting and examining the properties of reinforcing fillers in rubber compounds and elastomers. They are differentiated according to their microscopic properties, their interaction with the rubber, their aspect ratio, but also detect the chemical composition of the elastomer. The size of the primary particles, the shape of the aggregates and the surface activity have a major influence on the dispersion and the intrinsic interactions with the rubber chains and the degree of dispersion obtained during the mixing process. Industry standards exist for characterizing the filler properties, defined in ASTM, ISO or DIN standards, which can be used not only as a test benchmark but also as quality assurance for the daily production process.

In the case of iso-dimensional fillers, a distinction is made between the geometric parameters, which differ in terms of the shape of the structure and the degree of aggregate branching, and the surface activity. These properties can influence the strength of the filled compound and the service life of the product under extreme conditions. Not only the composition of the surface of the filler but also the type of rubber chains and the process conditions during mixing and downstream processing play a major role in the dispersion and the final properties of the vulcanizate. Due to the wide range of carbon blacks and precipitated silica, the compounder has the choice of using a filler that provides the necessary reinforcement with a specific rubber or of using a rubber blend, where the distribution in the blend can determine the properties. In the case of precipitated silica, the size of the primary particles is even smaller, but the effect on non-polar rubber is diminished, due to the interaction of the silanol groups, which increase filler agglomeration and reduce dispersibility. Much better performance is obtained in non-polar rubbers (NR, SBR, BR), as precipitated silica can react with an organo-silane and is coupled to the rubber network. Layered silicates are suitable fillers for boosting the barrier properties of any rubber. Multi-walled carbon nanotubes and layered silicates can lead to special physical effects in terms of electrical properties and anisotropic effects, due to the high specific surface area and the high aspect ratio. Multi-walled carbon nanotubes are often used in combination with carbon black or silanized silica to achieve a reduction in crack growth and an increase in tearing energy.

Surface activity describes the physical adsorption that occurs to an extent depending on the particle surface and the structure of the molecular chains of the rubber. A key point for the quality of the interaction of the filler with the rubber is the degree of filler dispersion during mixing, and is a clear measure of the quality of the interaction between the filler and the rubber. From the wide range of fillers that can be used, the compounder selects the filler with the optimal effect on down-loading and final elastomeric properties. This chapter discusses the chemical composition of fillers first and then the physical methods used to characterize them.

## 3.1 Chemical Composition of Fillers

Chemical characterization of the filler is important, because it allows conclusions to be drawn about the degree of interaction with the respective rubber, yet it is not of great importance from a commercial point of view, because the changes in properties of the filled elastomers can rarely be attributed solely to the composition of the filler.

### 3.1.1 Carbon Black

The chemical nature of carbon black is quite complex and differs in its adsorptive capacity with the rubber chains and influence the distribution in rubber blends. Carbon black consists of 97–99.5% elemental carbon in the form of imperfect graphitic layers and amorphous carbon, depending on the feedstock used and the manufacturing process. The elemental carbon content increases from the furnace grades (97.3%) to the medium thermal grades (99.4%) and reaches a maximum value with acetylene black (99.8%).

The content of hydrogen, oxygen, sulfur and ash is of some importance and has been studied in detail [1–3]. Hydrogen atoms are distributed at the periphery of the basal layers and their number depends on the feedstock used and on the reaction times during production and the duration to quenching with cold water. It has been shown that the hydrogen content depends on the reaction conditions and the quenching behavior and is located to the margins of the graphitic layer of carbon black. In the case of carbon black with a low surface area and low dispersive efficiency that is observed in fine thermal black (FT) and medium thermal black (MT), the concentration of hydrogen is high, which means that these particles still have this element inside the primary particles and not only on the margins of the graphitic layers at the surface of the primary particles. The amount of hydrogen directly bound to the carbon skeleton is determined by the proportion of micro-crystallites and can be considered as an indication of the extent of the mechanical reinforcement of the vulcanizate [3]. The hydrogen content decreases from furnace black (up to 0.80%) to medium thermal black (up to 0.50%) and acetylene black (up to 0.10%). Carbon black from tire treads contains a lower proportion of hydrogen atoms than carbon blacks from carcasses.

The oxygen that comes into contact with the carbon black at elevated temperatures leads to a variable content of oxygen functional groups, which are located at the margin of the graphitic micro-crystalline structure of channel black. Today, furnace blacks contain 0.2 to 1.5 wt.% that does not influence the dispersion and the reinforcement. Gas black, which is formed in gas burners, contain up to 4 wt.% of oxygen on the primary particle surface, and can change the reinforcement of the filler in the rubber compound. The oxygen-containing functional groups are roughly divided into four groups [4–6]:

- carboxylic acids
- quinones
- lactones
- phenols

The oxygen-containing groups can be detected by pyrolysis reactions, redox reactions, ion exchangers, pH measurement, and neutralization with various alkaline solutions. To provide some examples, carboxyl groups are detected with $NaHCO_3$, lactone groups are titrated against $Na_2CO_3$, phenol groups are determined with NaOH and quinones are determined with sodium ethoxide. The oxygen-containing functional groups influence the chemical reactivity of the carbon black. However, the increase in oxidation lowers the pH and the vulcanization speed, which is a great obstacle for channel black [7–9]. Rubbers of a polar nature, such as nitrile rubber or polychloroprene, will interact more strongly with carbon blacks containing -OH and -COOH groups on the surfaces of their particles. In contrast to the previously published amount of oxygen that is made for channel black, there has been found for furnace blacks a significant reduction in the number of oxygen functionalities on the primary particle size [7].

Sulfur atoms are not so important, as this element is not located on the filler surface and cannot interfere with the rate of vulcanization. The content of sulfur is inside the filler aggregate. However, the sulfur is present as a sulfonyl group, as sulfonic acid and mercaptan or thiophene and cannot participate on the reinforcement of carbon black. [10, 11]. It has been shown that sulfur [12] is unable to participate in sulfur crosslinking processes [7]. A reduction in vulcanization speed was previously observed with channel carbon black. The furnace carbon black used today does not exhibit this reduction in vulcanization speed.

Although the carbon black production process is controlled in such a way that the dehydrogenation of the hydrocarbons occurs at almost 100%, residues of hydrocarbons in carbon black can be detected as tar. The amount of tar on the surface of carbon black depends on the formation temperature, the efficiency of the quenching process, and the extent of oxidation or thermal treatment. However, even if the amount of tar is low in carbon black, there are indications that it may be responsible for its carcinogenic activity. The extractable tar is determined by means of toluene extraction [13]. The tar contains polycyclic aromatic hydrocarbons, which are carcinogenic.

Inorganic impurities are summarized in the ash content, which can analyzed in detail by chemical methods [14]. Inorganic components are added to the carbon black during quenching. The chemical analysis shows the ash content decreases with specific surface area and structure of furnace carbon blacks. In the ash, they make up less than 0.5%. The ash content indicates the level of inorganic impurities stemming primarily from the feedstock. Gas blacks and acetylene blacks are characterized by a low ash content, due to their production process. Carbon black contains trace metals originating from the feedstock [15]. Typical concentrations of trace metals are: Ba (< 10 ppm). Cd (< 1 ppm), Cr (< 5 ppm), Co (< 5 ppm), Cu (< 5 ppm), Pb (< 10 ppm), Ni (< 10 ppm), Hg (< 10 pm), Se (< 10 ppm) and Zn (10 ppm). The concentrations of these metals are chemically analyzed by the carbon black manufacturer and should remain below the indicated limit.

### 3.1.2 Precipitated and Fumed Silica

The purity of precipitated silica varies with the silica content over the range 88 to 92%. Fumed silica has a purity of 98% or more. Amorphous silicas consist of silicon and oxygen tetrahedrally bound into the imperfect three-dimensional structure. The degree of lattice imperfection, which is dictated by the synthesis conditions, has several consequences. There is no long-range crystal order, only short-range, with a random arrangement of more-distant neighbors, and the imperfections within the lattice structure leave free silanol groups. The primary particles are firmly bound to one another and form aggregates that do not break up when mixed into the rubber. On the surface of the aggregates, there are silanol groups between the siloxane groups (Si-O-Si). The silanol groups on the silica surface are [16]:

- isolated silanol groups
- geminal silanol groups
- vicinal silanol groups

The primary particles are formed by the condensation of silicic acid. Subsequent condensation of silanol groups (Si-OH) from the aggregates leads to siloxane groups

(Si-O-Si) that links the aggregates together. The presence of the silanol groups and the siloxane dipoles is shown in Figure 3.1.

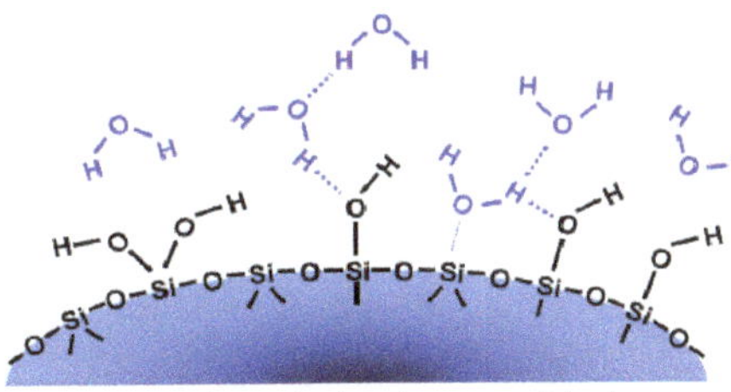


**Figure 3.1**
Surface structure of a precipitated silica

The main impurities on the surface of precipitated silica are the silanol groups that can exist in different forms. Because their strong interaction with water molecules they can bound water on the surface of the silica [17–19]. Depending on the amount of silanol groups and their structure the lattice imperfections on the surface permit more than 4.6 -OH groups per $nm^2$. Precipitated silicas have 15% of silanol groups on the surface. The fumed silicas possess fewer silanol groups on the filler surface, which is due to the different manufacturing process.

The number of silanol groups, their distribution, and even the conformation of the surface siloxanes also depends on the synthesis method and the thermal treatment of the precipitated silica. Therefore, it is important to quantify the number of silanol groups on the surface of precipitated and fumed silica to establish the degree of filler-filler interaction, which is important for the dynamic-mechanical properties. The silanol groups are analyzed using the following methods:

- hydrogenation of silanol groups
- substitution with alkyl lithium
- thermogravimetric analysis (TGA)
- IR spectroscopy
- NMR studies

One chemical method is hydrogenation of the silanol groups with $LiAlH_4$ [20] and substitution with alkyl lithium (LiR), which leads to the determination of the volatile alkane (R-H) [21]. Quantitative analysis by means of IR spectroscopy can discriminate between the three types of silanol groups, as well as trapped water adsorbed on the silica surface. The amount of silanol groups and the accessibility to water molecules have been quantified in the IR-spectroscopic reports. The strong absorption band at 3750 $cm^{-1}$ are found to occur mainly on extensively dehydrated silica. The unequivocal assignment of this band to isolated vibrating silanol is generally accepted. The hydrogen-bonded infrared band is at 3740 $cm^{-1}$. The absorption bands for vicinal (adjacent) silanol groups are located at 3660 $cm^{-1}$ and the corresponding water band at

3400 $cm^{-1}$. Hydrogen bonding of vicinal silanols is assigned to the bands at 3540 $cm^{-1}$ and 3607 $cm^{-1}$. Geminal silanol have been identified on freshly prepared precipitated silica. The broad band at 3500 $cm^{-1}$ was assigned to geminal hydroxyls, as their thermal behavior differs significantly from that of the other types of silanol. They can be reversibly eliminated by heating *in vacuo* at 200 °C. The types of silanol groups identified on silica surfaces and their infrared adsorption assignments are shown in Table 3.1.

**Table 3.1** Characteristic Adsorption Bands of Silanols

| Type | Structure | Infrared band, $cm^{-1}$ | References |
|---|---|---|---|
| Isolated silanol | 1a | 3750 | 20, 21 |
| | 1b | 3740 | 20 |
| Vicinal silanol | 2a | 3600 | 23, 24 |
| Water H-bonded | 2b | 3607 | 20 |
| Geminal silanol | 3 | 3500 | 25 |

Solid-state NMR spectroscopy is a method for measuring the total number of silanol groups as well as the isolated and geminal silanol groups.[29] Si NMR has been used to find out how the silanol groups react on the surface of precipitated silica [26, 27]. The isolated and the geminal silanol groups and the siloxane bonds were identified as active silanol groups. The total number of silanol groups and the amount of surface silanol have been calculated for meso-porous silicates. However, the method is experimentally difficult and has additionally high experimental costs [28, 29].

A simple method is thermogravimetric analysis, which can distinguish between dehydrogenation and dehydroxylation. A method has been developed to determine the silanol concentration from precipitated silica using a simple equation which employs TGA data [30]:

$$OH = 20\ [WL\,(T_0) - WL\,(T_{final})]\ /MH_2O \tag{3.1}$$

where $WL(T_0) - WL(T_{final})$ is the weight-percent loss in the temperature range where the silanol group decomposes

This equation can be used to readily determine the concentration of silanol groups. When the thermal stability of the silanol groups was determined, it was found that the concentration of silanol groups per square nanometer decreases with increase in temperature under similar heating conditions. The value decreased rapidly from 190 to 400 °C and then more slowly from 400 to 780 °C. For precipitated silica, it was found that there are 5 to 7 silanol groups per $nm^2$ silica surface. The silanol content was determined to be 4.6 OH groups per $nm^2$ surface area. This corresponds to the available silicon atoms permitted on the surface by the lattice configuration. The silanol concentration on the surface depends on the number of silicon atoms per unit area on the surface and the number of hydroxyl groups present on each silicon atom.

Commercial silicas have varying degrees of imperfections in their lattice structure and some, notably the precipitated types, are not ideally pure. Impurities, even at low concentrations, can have a marked effect on the surface properties of silicas. Reports indicate a difference between a virgin surface of silicas after outgassing at room temperature and the non-treated precipitated silica. Re-adsorption of water can vary with the type of silica [31]. The silanol groups in fumed silicas are sufficiently isolated to permit equivalent site adsorption by water. There are rules that must be followed regarding the water content of precipitated silica. The humidity, measured at 105 °C, should not be more than 4–8% [32] and the dry-ignition loss at 1000 °C should not exceed 7% [33]. The disappearance of silanol groups with increase in temperature demonstrates a significant change in surface properties of precipitated silica. The sieve residue in a 45 µm sieve should not contain more than 0.5% [34]. The pH of silica suspensions is variable and varies with manufacturing method from 5 to 8 [35]. The content of metals such as Cu, Mn and Fe is determined by wet-chemical methods, but should not exceed 30, 50 and 500 ppm respectively [36].

### 3.1.3 Layered Silicates

The 2:1 smectites are isomorphic tetrahedral and octahedral substitutions that induce a permanent negative layer which is compensated by hydrated alkali cations or in some cases alkaline earth cations in the interlayer. The presence of alkali cations in the interlayer is important for the use of layered silicates in rubber technology. The ability of 2:1 smectites to incorporate water and the subsequent change in basal spacing has been studied in several reports [37]. Layered silicate is a product which depends on factors related to the evolution of the mineral. A chemical analysis of the composition of the layered silicates is required to check the water swelling. If the content of $Ca^{2+}$ and $Mg^{2+}$ ions is low, good swelling in water can be achieved, because the interlayer distance increases significantly and the two-dimensional layer can be combined with a rubber latex to form a nanocomposite. If the $Ca^{2+}$ and $Mg^{2+}$ content is high, swelling in water is significantly reduced because the alkaline earth ion provides a link between the layered silicate and can form bridges, and so the layered silicate cannot serve as an dispersed filler. Chemical analysis of the composition of several layered silicates was used to separate out the unfavorable layered silicates. The layered silicates offered by suppliers have different chemical compositions, as indicated in Table 3.2 [38].

**Table 3.2** Chemical Composition in wt.% of Layered Silicates [38]

| Element | Nanofil 116 | Na⁺Cloisite | PGN |
|---|---|---|---|
| Silicon | 27.49 ± 0.70 | 28.14 ± 0.70 | 27.86 ± 0.56 |
| Aluminum | 8.44 ± 0.14 | 10.46 ± 0.13 | 5.69 ± 0.15 |
| Calcium | 1.16 ± 0.12 | 0.41 ± 0.13 | 0.64 ± 0.10 |
| Iron | 2.19 ± 0.09 | 2.85 ± 0.09 | 1.99 ± 0.01 |
| Magnesium | 2.99 ± 0.08 | 1.37 ± 0.09 | 0.61 ± 0.02 |

The analysis showed that the sample with a high content of alkaline earth ions cannot be used for wet compounding, as the swellability in water is impaired. The products show different interlayer distances, which are expressed by the CEC value. This value indicates how many milliequivalents of layered silicates are exchanged. This provides a quality feature of this filler to absorb water molecules and is important for the dispersion of the layered silicates in the rubber. [38] For $Na^+$Cloistite (Süd Chemie), the CEC is 92 mequ./100 g and the interlayer distance is 1.17 nm. For Nanocor, produced by PGN, the CEC is 120 mequ/100 g and the interlayer distance is 1.20 nm. Therefore, layered silicates with a high CEC value are used for achieving good filler exfoliation.

## 3.1.4 Carbon Nanotubes

Multi-walled carbon nanotubes contain $sp^2$-hybridized carbon atoms, but also the catalyst metals from production and, to an extent depending on the purification of the product, amorphous carbon. For this group of fillers, there are no regulations or quality requirements that are taken into account by the different manufacturers. The compaction technology affects the density of the material, but has no influence on the chemical analysis. One type of less highly compacted MWCNT, Nanocyl 7000, has a powder density of 70 kg/$m^3$ and a carbon purity of 90%, while Baytubes C150P, which is more compacted, has twice the powder density, but a higher carbon purity of 99%. The difference in catalyst metal concentration is greater in the multi-walled carbon nanotubes from Nanocyl than in those from Bayer Material Science [39, 40]. For this reason, despite all their advantages as a filler, multi-walled carbon nanotubes are not at the same quality level as carbon black or precipitated silica.

## 3.2 Primary Particle Size

The primary particle size is the relevant value that describes the activity of a filler in a rubber compound. For iso-dimensional fillers, such as carbon black, precipitated silica and fumed silica, literature reports discuss the structure of the aggregates, which consist of primary particles sintered together during manufacture [7]. The particles of a given filler do not all have the same diameter, but have a size distribution curve that is characteristic of that filler. The relevant data are obtained by direct measurements of the particles' dimensions by transmission electron microscopy. The primary particle size is the fundamental property of iso-dimensional fillers to determine the specific surface of the particle and, therefore, the reinforcement of elastomers through the rubber-filler interaction. However, as the determination of primary particles is relatively complex, the specific surface area is determined in practice by means of easier gas adsorption measurements.

### 3.2.1 Hierarchical Organization of Aggregates

This section describes the organization of the iso-dimensional fillers that form aggregates during manufacture. As a result of the reaction zone, the primary particles formed at the beginning continue to form aggregates, which increase in size with the duration of manufacturing time and a function of the process conditions. Finally, the aggregates are the solid units of the iso-dimensional fillers and cannot be broken up during the mixing process. However, with advances in analytical capabilities, our understanding of the morphological characteristics of particulate fillers has expanded, especially as a result of the development of transmission electron microscopy (TEM). Unfortunately, these measurements are relatively complicated and initially were not very accurate:

- Particles are not perfectly spherical.
- Particles are connected to one another by chemical bonds.
- Some particles cannot be seen in TEM because they are hidden by other particles of similar aggregate size.

The use of high-resolution TEM served as the starting point for gaining an understanding of the morphology of carbon black aggregates.

#### 3.2.1.1 Carbon Black

The hierarchical organization of carbon black was described by visualizing the size of the primary particles inside the aggregates and the shape of the aggregates. The carbon black aggregates are formed by sintering a number of primary particles, which is dependent on the irregular flow in the reaction zone. Especially the short residence time in the reaction zone is relevant for the size of the primary particles. A short residence time at high temperatures is necessary for obtaining small primary particles that deliver small aggregates which interact well with the rubber chains and lead to good dispersion in the rubber compound. Conversely, longer reaction times at low temperatures allow growth in the reaction zone. The surface of the particles becomes more homogeneous and they can easily be dispersed in rubber compounds [41, 42]. An exception is the thermal blacks, which consist mostly of single spherical particles that do not form aggregates and are used to improve the hardness.

For a representative particle size distribution, many particles must be counted. Large particles are counted with more precision than small ones. The simplest method is manual measurement, where each diameter is measured on the TEM image at a known magnification. The use of semi-automatic and automatic particle counters reduces the time required and increases the accuracy of the determination. It is important to realize that the counting of objects, whether manually or automatically, is subject to significant operator bias. A schematic representation of the organization of the filler aggregates is shown in Figure 3.2.

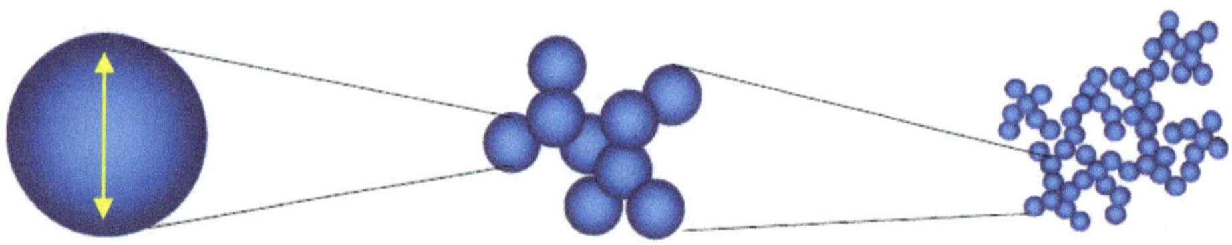

**Figure 3.2** Organization of a primary particle, an aggregate and an agglomerate

The aggregate size can be changed by the number of primary particles and the diameter of these particles. Like the aggregates, the primary particles have a size distribution. The average diameter of primary particles determines the specific surface area per gram of material. The major limitations in determining the size of the primary particles are related to the operational speed of the measurement and the complexity of the aggregates. The aggregates attract each other via van der Waals forces and form large agglomerates that are present in the pelletized filler. How strongly the aggregates interact is important for the degree of dispersion during the mixing process.

The surface area of a carbon black is crucial, because it determines how much surface area is available for interaction with the rubber compound. Generally, carbon black particles of small particle size have a higher surface area for interaction with the rubber phase. The potential of higher-quality TEM, which allows the determination of morphological parameters, are enhanced nowadays by the use of automated image analysis that permits the capture of structural details and information about specific differences in filler types. The technology makes it possible to conduct scientific analysis based on particle diameter, particle microstructure and aggregate branching, which is related to the structure of the filler. Such an analysis is used to evaluate the reinforcement potential of the filler in primary particles and the shape of the aggregates. The diameter of the quasi-spherical primary particles is a fundamental property of carbon black fillers that affects the specific surface area, dispersibility when carbon black is being mixed with rubber, and the performance of the filler in rubber compounds. The characteristic shape of carbon black aggregates is illustrated in Figure 3.3, which shows TEM micrographs of different furnace blacks (N220, N330 and N550). The variation in size of the primary particles, which increases from left to right, is apparent [43].

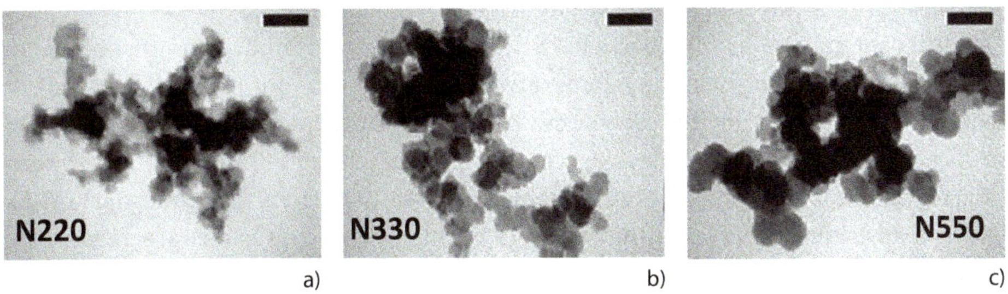


**Figure 3.3** Aggregates increasing in primary particle size:
a) CB N220, b) CB N330 and c) CB N550

The size of the primary particles increases from N200 (21 nm) to N330 (30 nm) and to N550 (56 nm). The residence time of carbon black formation increases in the direction of the growing primary particles. The turbulence in the reactor during the formation of carbon black gives rise to a size distribution of the primary particles and the aggregate size, which are important for elastomer properties. The primary particle size varies with the type of carbon black from N110 (18 nm) to N660 (67 nm) and leads to an increase in the average aggregate diameter of 135 nm for N110 and 535 nm for N660 [44].

However, a closer look at carbon black aggregates reveals a planarity; in other words, their width in a particular plane is greater than their thickness. Due to the different residence times in the furnace reactor, the primary particles in the aggregates have different primary particle diameters. The different thickness and size of the primary particles is shown in Figure 3.4.

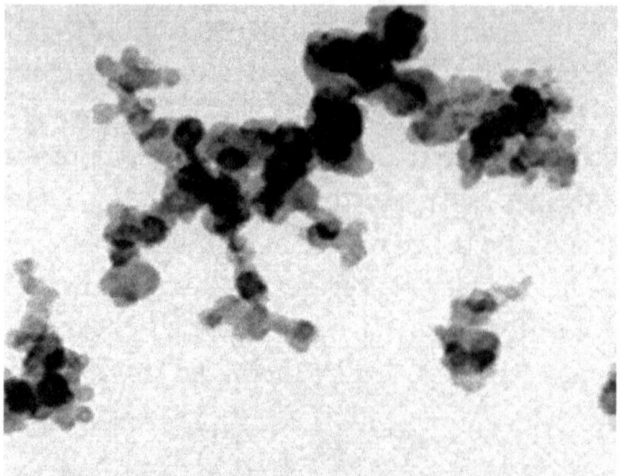

Figure 3.4
Influence of the thickness of the primary particle

The different grayscale of the particles within the aggregates indicates a three-dimensionality of the aggregates or agglomerates that is due to more primary particles in the center of the aggregates. The semi-quantitative information delivered by TEM studies indicate that carbon black aggregates contain a large number of quasi-spherical particles. This contradicts results obtained by small angle neutron scattering (SANS), which lead to the conclusion that average carbon black aggregates are formed by fewer than 10 primary particles and the inner density is smaller than the outside shell density. These aggregates have been likened to a "peanut shell" because the interior of the agglomerate has a lower density than the material as a whole [45]. These changes in the density of an aggregate may result from the rapid quenching of the carbon black. The size of the primary particles can be determined by means of TEM and AFM. Although previous reports on carbon blacks have questioned the fragility of the aggregates during mixing, it is now evident that the aggregates are the smallest solid units of carbon black and cannot break during mechanical mixing. The fraction of the agglomerates that is not dispersed in rubber compounds can be visualized by microscopic methods.

### 3.2.1.2 Precipitated Silica and Fumed Silica

The precipitated silica fillers consist of clusters with a porous structure, which creates an inner surface that generates higher aggregate branching [16]. The properties of the precipitated silica depend on the manufacturing process. The level of purity of the silica and the impurities are determined by the conditions of the manufacturing process. X-ray diffraction studies have demonstrated the absence of crystalline structures [48]. The formation of precipitated silica by the intermolecular effect of silanol groups is more rapid than the initiation of a long-lasting crystallization process. Electron microscopic studies show an approximately spherical primary particle size with a diameter of 5 to 50 nm, which depends on the conditions in which the precipitated silica was produced. The differences in aggregate formation, size of the primary particles and tendency to form agglomerates have their origins in the inter-aggregate interaction of silanol groups of the precipitated silica formed in an aqueous reaction. The primary particles become aggregates through the formation of strong siloxane bonds between the primary particles. TEM shows that the primary particles in precipitated silica are smaller than those of carbon black. They consist of clusters whose porous structure creates an inner surface which generates a larger number of primary particles in an aggregate and make the aggregate geometry more complex. Precipitated silica is more highly structures than carbon black. The agglomerates are formed by hydrogen bonds between the silanol groups of the aggregates and the Si-O-Si bonds have a high bonding energy.

The size of the precipitated silica aggregates depends on the production conditions (concentration of the reactants, pH, temperature and reaction time) and thus has a different primary particle size of 5–100 nm. The aggregate size of 1–40 µm is also not very large, resulting in an agglomerate size of 3–100 µm. In the case of precipitated silica, in addition to the average primary particle size and the aggregate size, the content of silanol groups on the surface of the filler plays a decisive role in the interaction with a rubber and the degree of dispersion achieved during mixing [49, 50].

In addition, changes in the specific surface area alter the total available area and so the concentration of silanol groups is a determining factor. An increase in silanol groups leads to a decrease in the Young's modulus and abrasion index of the rubber vulcanizate. Since the adsorption energies of the non-polar rubber are expected to be low, the lower modulus and abrasion index can be attributed to less rubber-filler interaction.

Silanol groups can interact with the backbone or other silanol groups through hydrogen bonding and form larger, elliptical aggregates that can reinforce the rubber. The silanol groups on the precipitated silica show a group density of 5–6 per $nm^2$. The concentration is less than previously reported for precipitated silica. The average number of silanol groups on the aggregate surface varies with the manufacturing process from 13 to 16% [51]. The same mechanism applies to the aggregates of precipitated silica, which are held together by strong chemical siloxane bonds that cannot be broken during mechanical mixing. The aggregates formed are more or less elliptical in shape, indicating a second property provided by the fillers.

Branching of the filler occurs where both carbon black and silica have a solids content that is spatially distributed, thus creating a voids volume between the solid particles. When the filler is incorporated into a rubber matrix, the voids of the aggregates and agglomerates are filled with rubber, a fact which leads to a change in the rubber/filler composition. The experimental methods for determining the concentration of silanol groups were discussed in the previous section.

Due to the interactions of the aggregates, large agglomerates of precipitated silica are formed that are more stable than those of carbon black. Fumed silica, which is formed in a gas reactor, has a smaller primary particle size of 7–50 nm. The high temperature permits the formation of silica aggregates with fewer silanol groups than in precipitated silica. These silanol groups provide weaker interaction in the agglomerates that form at the end of the manufacturing process. The acidic reaction of fumed silica is less strong, due to the lower number of isolated silanol groups on the aggregate surface.

#### 3.2.1.3 Layered Silicates

Layered silicates have two-dimensional, large primary particles and to separate them into exfoliated particles requires a wet-mixing method. The aggregates of layered silicates are stacked vertically on top of each other and form tight agglomerates. It is not possible to separate the individual layers from the tactoids by simply physically mixing them in an internal mixer. Since the layered silicates are often used as pure layered silicates, latex compounding is the best way to incorporate them into rubber [52–55]. The organo-clays which require a separate chemical conversion that involves the incorporation of quaternary ammonium cations must be added to the rubber by wet compounding. In both cases, the layered silicates are intercalated or exfoliated in the rubber vulcanizate. This is described in the section on mechanical mixing and dispersion in rubber compounds [56, 57].

#### 3.2.1.4 Carbon Nanotubes

The one-dimensional carbon nanotubes can be understood as single carbon molecules extended in one length. Therefore, carbon nanotubes have a high aspect ratio and a large specific surface area [58–60]. SWCNTs form large aggregates or undergo bundling by van der Waals forces and cannot easily be dispersed in polymers by mechanical mixing. Chemical conversion can transform the filler-filler interaction and the bundled system of functionalized carbon nanotubes can be dispersed in a polymer. After purification, the MWCNTs are present in a type of special aggregate that is related to the compaction conditions [61]. If they are not additionally compacted, the bulk density is low and mixing easily leads to good dispersion in the rubber matrix [62]. The decisive factor is the length of the MWCNTs after the incorporation in the rubber compound, which is determined by the effective mechanical mixing.

## 3.3 Morphology of Primary Particles

The primary particles of the iso-dimensional fillers determine the direct interaction with the rubber chains and ultimately the reinforcing properties in rubber compounds. Therefore, the morphological structure of the primary particles is important for a better understanding of the interaction between the surface of the filler and the rubber. Thanks to modern methods, such as X-ray diffraction and transmission electron microscopy (TEM), these interactions have been better understood in recent decades. In the case of precipitated silicas, besides the size of the primary particle, the number of silanol groups has a decisive impact on the behavior of the particles. For this reason, the abundance of the silanol groups and their type can have a decisive influence on dissociation of the agglomerates. The more silanol groups there are on the silica surface, the more difficult it becomes to disperse the precipitated silica in a non-polar rubber compound.

### 3.3.1 Microstructure of Primary Particles of Carbon Black

X-ray diffraction studies of carbon black have demonstrated very clearly that it is a semi-crystalline material that consists of graphite-like crystals and amorphous carbon [63–65]. The X-ray diffractograms show a major structure of $sp^2$-hybridized carbon atoms arranged in condensed aromatic rings. The interplanar distances are a bit higher than in the graphitic layer; instead of 0.334 nm, the distance is 0.35–0.365 nm, as a consequence of the spherical orientation of the particles. It is clear that a longer residence time and later quenching leads to larger graphitic micro-crystallites. For carbon black, the average crystallite diameter, La, is about 1.7 nm and the stacking height, Lc, is 1.5 nm, which corresponds to the formation of planar layers per microcrystallite [66]. It was found that the primary particles of furnace blacks consist of graphitic crystallites with Lc values ranging from 12 to 18 Å, depending on the type of carbon black. The stacking height of the crystalline layers La is roughly the same, but is slightly higher than the layers of pure graphite. This has led to a para-crystalline model of the primary particles which cannot explain the existence of the quasi-spherical shape of the primary particles.

TEM images show that the primary particle is quasi-spherical [67–69]. It can be understood that the growth of the primary particles takes place in the reaction zone where the carbon layers are deposited on top of each other such that the diameter of the primary particles increases. The ballistic approach causes the primary particles to come together and form aggregates. Therefore, besides the $sp^2$ hybridization of the carbon atoms, which yields planar graphite crystals, the formation of $sp^3$-hybridized carbon atoms is necessary to describe the curved shape of the primary particles. The para-crystalline model was therefore replaced by a turbostratic model [70]. Between the graphitic micro-crystallites are introduced a number of $sp^3$ carbon atoms to represent the spherical shape of the primary particles. This representation is far removed from the actual image of the primary particles. This is shown with the aid of a model of a carbon black aggregate, in which the dark areas represent the micro-crystallites and the gray areas correspond to the domains covered by amorphous carbon [42].

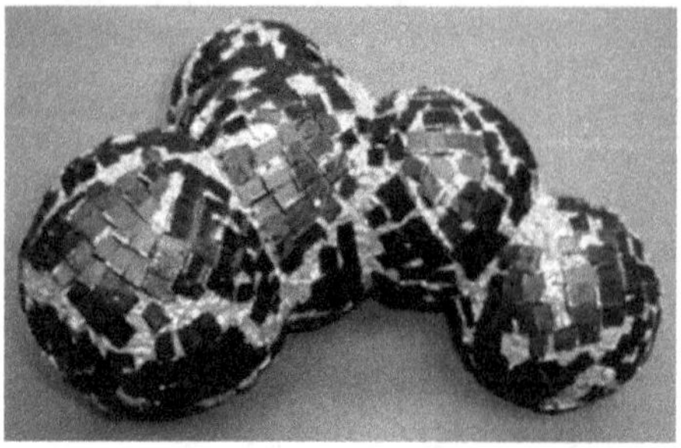

**Figure 3.5**
Model of a carbon black aggregate

The quasi-spherical primary particles observed in all types of carbon black is underscored by scanning tunneling microscope (STEM) images, which also show the distribution of surface roughness, or electrical energy differences [71]. The disordered arrangement of the micro-crystallites on the surface of the primary particles decreases with the increase in particle diameter that results from a longer residence time in the reaction zone. The difference in surface properties can be seen by comparing carbon black N115 and N234. Carbon black N115 with the smaller surface area exhibits a significant increase in surface roughness. Figure 3.6 shows a clear transition to a uniform distribution of the micro-crystallites on the surface of the carbon black as the diameter of the primary particles increases.

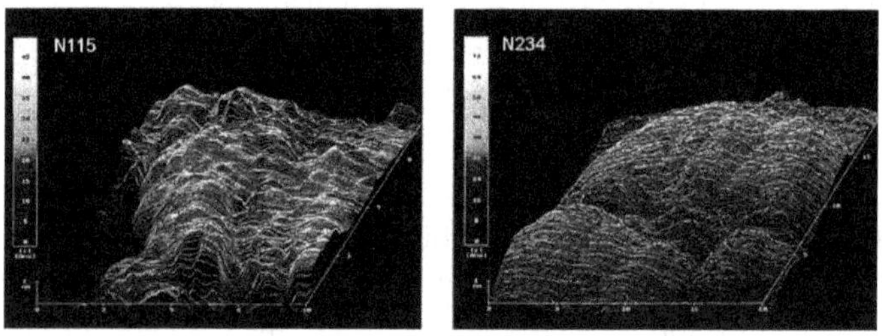


**Figure 3.6** STEM images of primary particles of carbon black N115 and N234 [71]

The surface of N115 show clear differences from the surface of N234. High-resolution AFM and STEM methods were used to profile the topology of carbon black at resolutions down to the nanometer range. The surface of the primary particles of carbon black reveals an irregular arrangement, which could correspond to a certain surface roughness. Accordingly, the surface defects on carbon blacks which occur at short residence times and high manufacturing temperatures can be seen. There are deviations that are reflected in the heterogeneity of the surface of the primary particles, which are much more pronounced than on coarse primary particles. This method captures an electrically conductive carbon black aggregate in that the topology is not directly correlated to the surface roughness, but rather is a function of the electric field. In STEM the electron beams are focused to a fine spot which is scanned over the sample in a raster illumination. The rastering of the beam across the sample makes STEM suitable for electron loss spectroscopy. It images the surface energy of the carbon black surface. Technical changes during carbon black production lead to surface energy changes and show strong changes in the distribution of the micro-crystals of the primary particles, especially in process-functionalized carbon black production, which will be presented at the end of this chapter.

Through the use of Raman spectroscopy, the existence of the $sp^3$-hybridized carbon was experimentally proven by a special band in the spectrum that provides the "amorphous" carbons [72]. The existence of "amorphous" carbon has been demonstrated by static gas adsorption measurements of propene and propane on carbon black samples [73]. It was shown that the "amorphous" carbon decreases exponentially with increase in size of the primary particles. The appearance of the "amorphous" carbon between the graphitic carbon crystallites is associated with the short reaction time in the furnace reaction zone. With short residence times in the reaction zone, only a portion of the carbon atoms are able to undergoes $sp^2$ hybridization. As the reaction time increases, more carbon atoms are able to undergo the lower-energy $sp^2$ hybridization. As a result, the adsorption behavior of the surface of the primary particles is greatly reduced and the interaction with the rubber decreases, as will be described later. This primarily affects the dispersion in the rubber compound and the reinforcement behavior of the carbon black. This aspect is dealt with in the section on surface activity. Our understanding of the architecture of the primary particles is based on instrumental effort and expertise in interpreting the micro-structure of the primary particles (see Figure 3.7) [74].

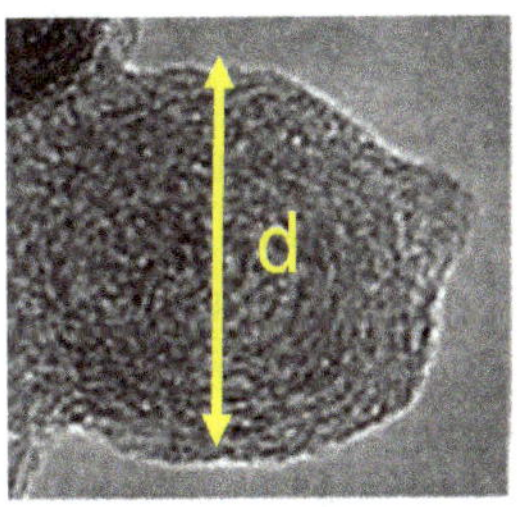


**Figure 3.7**
High-resolution TEM image of a primary particle N220 [74]

The outer crystalline layers are deposited in a more ordered manner, because the reaction time leads to a reduction in internal stresses, and a more uniform orientation corresponding to an onion-like arrangement of graphitic micro-crystallites is organized by the $sp^2$ hybridization, which has a lower energy than the existing $sp^3$-hybridized carbon, to effect the rounding of the primary particle. In the case of carbon blacks with increasing primary particle diameter, the graphitic layers show more order relative to the smaller primary particles that have a shorter reaction time and less time to establish an order in the graphitic layer, resulting in the energetical defects. In all primary particles, however, there is a central nucleation center in the particle, from which the structure of the particle results [75]. More quantitative details on the morphology of carbon blacks can be obtained from the review given in [76].

### 3.3.2 Microstructure of Primary Silica Particles

The size of the primary particles of precipitated silicas depends on the reaction conditions. The primary particle size depends on the concentration of the reactants, the pH of the reaction medium, the reaction time and important the temperature at which the reaction proceeds. The smallest precipitated silica primary particles is obtained when the concentration is optimal and the reactor temperature is low. TEM images deliver exact information about the particle size distribution. The primary particle size of the usual precipitated silica used in rubber technology is 5–30 nm, depending on the manufacturer's conditions. The resulting silica aggregates can reach up to 40 µm, depending on the number and distribution of the silanol groups [16, 47].

For precipitated silica, not only the size of the primary particles is important, but also the type and concentration of the silanol groups, which are crucial for the interaction with the rubber chains. The silanol groups can be determined by thermogravimetric analysis, IR spectroscopy and solid-state NMR spectroscopy, as described in the previous section. The microstructure of the precipitated silica is thus determined by the chemical nature of the acidic silanol groups, which make the silica hydrophilic or "water friendly". Consequently, water molecules adhere physically to the silanol groups through hydrogen bonds. Heating to 105 °C releases the free water, whereas the hydrogen-bonded water is only released by higher temperatures up to 250 °C. Above 200 °C the silanol groups condense to form dipolar siloxane bridges (Si-O-Si). One method for determining the silanol groups is to titrate the acidity of the Si-OH groups against sodium hydroxide [77]. The measured Sears number is equivalent to the quantity of 0.1 N NaOH required to titrate a suspension of 5 g silica and indicates the number of silanol groups present on the silica surface; it is not related to the primary particle size, but depends on the amount of the three silanol groups and which group is the most reactive. The Sears number cannot answer the question; it delivers only the answer as to the total acidity of the primary particles. The concentration of the silanol groups determines the interparticle interaction and correlates with the dispersion of untreated silica in non-polar compounds.

Another test examines the thermal stability of silanol groups. The weight loss of precipitated silica heated to 900 °C corresponds to the amount of water released during the condensation of silanol groups to siloxane groups [78]. Several authors have found that the number of silanol groups per unit surface area decreases to a minimum when heated to 900 °C. Such a thermally treated silica has a minimum Sears number and behaves similarly to fumed silica. The effect of thermal heating is shown in Figure 3.8.

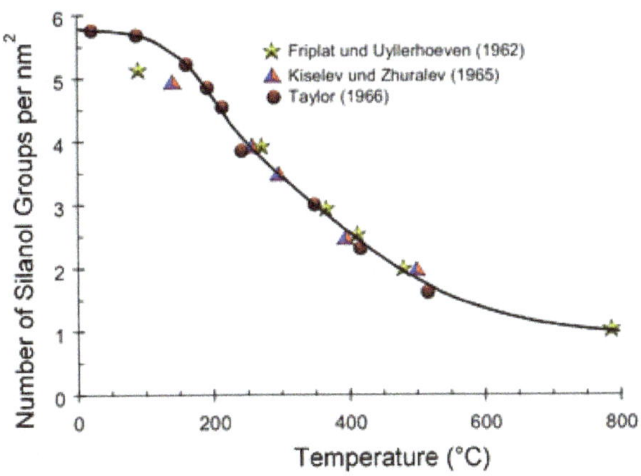


**Figure 3.8** Influence of temperature on the disappearance of silanol groups

The concentration of silanol groups was found to drop during a calcination weight-loss experiment from initially 6 silanol groups per $nm^2$ to 1 silanol group per $nm^2$. This change underlines the loss in surface activity through heat treatment of the precipitated silica.

Quantification of the type of silanol groups (individual, geminal and vicinal) and reactivity remains a difficult analytical task. Such methods led to the concept that there are only 5 to 10 silanol groups per square nanometer of precipitated silica, and fewer than 2.5 g of silanol groups in fumed silica. Silanol groups can undergo various chemical reactions that change the surface. The higher the concentration of silanol groups on the surface unit is, the more difficult it is to realize a good dispersion of precipitated silica in non-polar rubber. The greatest opportunity for precipitated silica is reaction with organo-silanes that reduce the strong filler network and reduce the Payne effect of the rubber compound. The reaction with the organo-silanes will be treated in a separate section of this chapter.

However, the primary particle diameters of silica also depend on the conditions of the TEM measurement. The different energy input that leads to the breakdown of bond states can have a decisive effect on the particle size and shape [79, 80]. Bonds are found to be cleaved, with the weak bonds being most affected [81, 82]. This reversibility of the hydrogen bonds is, however, responsible for the fact that the silica particles can react again after mixing and lead to a strong silica network. However, one should be aware that this formation only results in a morphological structure of the precipitated silica. Other preparative conditions, such as changing the energy input or application from a suspension, can lead to a significantly different morphology. This is because the different energy input or other conditions that lead to the breakdown of the bond states can have a decisive effect on the particle size and shape [83].

## 3.4 Size of Primary Particles

The dependence of the filler reinforcement on particle size is explained by two contributions: the primary particle size and the particle mass. The iso-dimensional fillers are characterized by the size of the primary particles, which determines the end-use application in rubber compounds. Particle size is an extensive parameter in rubber compounds and determines the interfacial interaction between the filler and the rubber chains. However, the different types of fillers can differ in terms of the uniformity of the primary particles in different aggregates.

Considering the manufacturing conditions, it is obvious that primary particles of a given carbon black, precipitated silica and fumed silica have an average diameter and a diameter size distribution that depends on the manufacturing technology. From 2D projections of aggregates the average diameter d of the primary particles can be determined by image analysis techniques. However, the method requires not only expensive instrumentation, but also personnel and a great deal of time.

Particle size is defined as the arithmetic mean (average) diameter, d, of the primary particles, that make up the aggregates. Any sample of a filler has a distribution of particle sizes that varies with the manufacturing conditions. [84]:

$$d = \Sigma\, d_i^{\,3} N_i \,/ \Sigma d_i^{\,2} N_i \tag{3.2}$$

where $d_i$ is the diameter of a primary particle and $N_i$ is the number of the primary particle with the diameter $d_i$

In the case of carbon nanotubes, this method leads to the determination of the tube diameter and to the length of the tubes. In the case of layered silicates, the size of the two-dimensional platelets can be measured, but the results are a function of the location where the mineral was found. The important property for filler grade classification, the production control and the rubber reinforcement are included in the specific surface area.

The relationship between the particle size and the aggregate volume is discussed in scientific reports [85–88]. The number of particles in an aggregate results from the assumption that the particles are of the same size according to the theory of flocculation. If this assumption is taken into account, an average particle size, d, determines each individual aggregate using the mean chord length, π A/P [89]:

$$d = \alpha\, \pi\, A/P \tag{3.3}$$

where α is an aggregate factor to correct for the fact that the mean chord is usually a measure of the average distance across the particle, A is the projected area of the aggregate and P is the perimeter of the aggregate

For non-porous carbon blacks of well-established size and surface area, the factor α decreases hyperbolically with the ratio $P^2/A$, so that the equation becomes:

$$\alpha = 13.092\ \left(P^2/A\right)^{-0.92} \tag{3.4}$$

For carbon blacks where the surface area is determined by TEM the results are in reasonable agreement with the surface areas determined by nitrogen adsorption measurements.

The primary particle concept is useful because it relates the size and surface properties of the fillers to the conditions of the production process. The primary particles of highly reinforcing carbon blacks range consistently from 15 to 30 nm, depending on the residence time of the material in the furnace. The semi-reinforcing carbon blacks have an average particle size of 30 to 200 nm. The thermal process with its long residence time at high temperatures and without rapid cooling stage yields large spherical particles with diameters of 100 to 500 nm and a very broad particle size distribution.

However, the size and the size distribution of primary particles is related to the residence time in the furnace during which particle growth occurs. The following parameters have to be considered:

- The particles are not perfectly spherical.
- The particles are fused together.
- Many particles cannot be observed well.

When the size of the primary particles and their size distribution are taken into account, channel black has the smallest primary particle distribution, followed by furnace black and acetylene black, while the flame black and thermal black have the broadest particle size distribution. The particle size distribution for several carbon blacks is shown in Figure 3.9 [90].

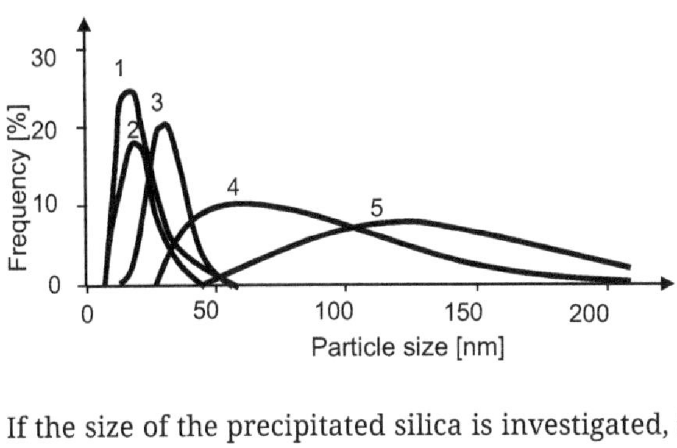


Figure 3.9
Variation in primary particle size distribution: 1) channel black, 2) furnace black, 3) acetylene black, 4) flame black and 5) thermal black [90]

If the size of the precipitated silica is investigated, it is found that the primary particles and their size distribution depend on the process parameters. The concentration of the reactants, agitation, the value of pH, salt concentration and dwell time determine the size of the primary particles in precipitated silicas. These values can specifically change the size of the primary particles and therefore also the size of the aggregates. Compared with furnace black, the size of primary particles is significantly decreased.

Electron micrographs show that the precipitated silica is made up of almost spherical primary particles with a diameter of 5 to 50 nm. The lower the reactor temperature and the higher the concentration of the reactants in the reactor, the smaller is the primary particle size of the precipitated silica. The primary particles cluster together to form aggregates and have a variable diameter of 1 to 40 µm. Both the size of the primary particles and the size of the aggregates depend on the selected production conditions [91]. For fillers of high aspect ratio, carbon nanotubes and layered silicates, TEM studies reveal the size of the large particle diameter, the average length of the carbon nanotubes and the area of the layered silicates if they are located perpendicular to the electron beam.

## 3.5 Specific Surface Area

Specific surface area is a geometric characteristic that can be measured by molecular adsorption of a gas or liquid. This phenomenon is influenced by the surface energy and activity of the filler, which is inhomogeneous across the surface, due to the surface structure. To perform gas adsorption measurements, it is necessary to establish the average number of layers of gaseous molecules adsorbed on the surface of the filler under the conditions of the test method as well as the area of one molecule of the adsorbate as adsorbed on the filler. Generally, a partial or complete adsorption isotherm is determined, i. e. the quantity of gas adsorbed at a selected temperature in a desired pressure range.

The specific surface area of the fillers is one of the most frequently used parameters for characterizing the interaction with the rubber and the reinforcement of the elastomer. It almost goes without saying that the specific surface area directly impacts the level of interfacial contact area with the rubber and the magnitude of the reinforcement of the elastomer. Therefore, the specific surface area is important for determining how fillers affect the properties of the rubber compound.

### 3.5.1 TEM Determination of Specific Surface Area

Of the various approaches developed for determining the average primary particle size, measuring the diameter of the primary particles of the filler and generating an average value for the primary particles leads to the specific surface area per gram of filler. If the primary particles are considered as spheres at the same size, the specific surface area per gram is expressed by a simple equation:

$$\text{Specific surface area (SSA)} = \pi d^2/(1/6)\,\pi d^3\rho = 6/d\rho \qquad (3.5)$$

where d is the diameter of the primary particle and ρ is the density of the particle

The specific surface area is inversely proportional to the average diameter of the primary particles and the density of the material. This equation shows how important the size of the primary particles is. However, the deviations from the specific surface area yielded by TEM may be due to fusion of the primary particles, which does not reveal the full size of the primary particles. The different viscosity of the carbonizing particles allows a different overlap of the primary particles in the furnace process. This is directly related to temperature and the reaction time during manufacture of the filler and offers a way to determine the size distribution of the filler. A good TEM instrument and a specialist are required to perform these complex measurements. There is always error, which is characterized by a lack of clarity regarding some of the primary particles. This method was introduced quite late, when the recording technique with this method was already well established. There is a whole series of specific surface areas of carbon black depending on the mean size of the primary particles that agree very well with the gas adsorption results.

### 3.5.2 Brunauer–Emett–Teller Method

Since it is not possible to measure the primary particles without using an expensive TEM or AFM method to determine the diameter d, it was necessary to resort to simpler methods for determining the specific surface area in square meters per unit mass of filler ($m^2/g$). This method, which appears simple at first glance, is based on the adsorption of a quasi-ideal gas molecules on the cleaned surface of the filler. The specific surface area is determined from the number of adsorbed molecules and the molecular cross-section of the gas as the product of the two values, which are determined experimentally. On the one hand, it is necessary to determine the number of adsorbed molecules on the filler surface and, on the other, the cross-section of the adsorbed molecule should be known. The surface covered by adsorbed molecules corresponds to the product of the number of molecules in the monolayer ($N_m$) that cover a unit mass and the cross-section of the probe molecule ($\sigma_m$):

$$\text{Specific Surface Area (SSA)} = N_m \cdot \sigma_m \left(m^2/g\right) \qquad (3.6)$$

where $N_m$ is the number of molecules in the monolayer and $\sigma_m$ is the cross-section of the adsorbed gas

In this way, the specific surface area of the different filler types can be determined. The method is much easier to carry out than the time-consuming microscopic examinations. Using gas adsorption on a gram of filler to form an adsorbed monolayer at the surface of the filler is quite a simple technique to perform the specific surface area measurements. However, the technique requires a physical phenomenon that depends on the surface structure of the filler and the gas to be adsorbed. When the specific surface area is measured, the result of the specific surface areas contains the contribution of the energetic interaction between the filler surface and the adsorbed molecules.

The method for determining the specific surface area is based on the adsorption of an ideal gas (e. g. nitrogen) on a filler surface, which has previously been cleaned to minimize the influence of foreign matter and impurities on the result. The adsorption isotherms of gases show two regions. At low pressure, the isotherm is concave; at higher pressures it is convex, typical of multi-modality adsorption [92, 93]. According to the theory, the conditions of gas adsorption depend on certain properties of adsorbate and adsorbent, particularly with respect to the filling of the monolayer and the beginning of subsequent layer.

Adsorption takes place on fillers on which two monolayers are adsorbed in order to compensate for the non-uniformity of the adsorbent assumed by the inventors of the method. The test begins with very small portions of the surface of the filler covered by the adsorbed gas. The number of adsorbed gas molecules is counted using the ideal gas law. The degree of adsorption is increased at intervals until the pressure of the gas at that temperature is reached. With nitrogen, which is assumed to behave like an ideal gas and has a cross-section of 1.62 $nm^2$, the micropores of the filler surface can also be detected. The adsorption is carried out until two monolayers cover the filler. For the second adsorbed layer, it is assumed that all adsorption molecules are in contact with the same type of molecules and are not influenced by the solid surface of the filler. On this basis, the specific surface layer can be determined. The derived adsorption isotherm is used to determine the specific surface area of the filler:

$$p/V\,(P_0 - P) = 1/(CV_m) + (C-1)\,/(CV_m) \cdot (Pd/P_0) \tag{3.7}$$

where P is the pressure of the gas at temperature T,
$P_0$ is the vapor pressure of temperature T,
V is the volume of the adsorbate,
$V_m$ is the volume of the monolayer,
and C is an experimental constant

In this way, the specific surface area of the fillers can be determined by gas adsorption. The experiment is carried out in a special apparatus. The procedure is described in industrial standards [94–96]. For evaluation of the Brunauer–Emmet–Taylor (BET) method, the $P/V(P_0 - P)$ is plotted against $P/P_0$ and a linear intermediate range with the slope $(C-1)/V_m$ and the intercept is $1/V_mC$ is obtained. The constant C depends on the adsorbate-adsorbent system and on the temperature and on the difference between the heats of liquefaction and adsorption of the adsorbate. The value $V_m$ and C can be calculated from the second monolayer. If the cross-section of a nitrogen molecule is known, the specific surface area of a filler can be calculated from $V_{m.}$

For routine measurements on fillers, a single-point measurement is carried out for the case that the material constant C is known. It was recognized that the multi-point nitrogen surface area was the most accurate way of measuring the total surface of the primary particles. A representation of the results of the BET method for a filler having an inner and an outer surface is shown in Figure 3.10.

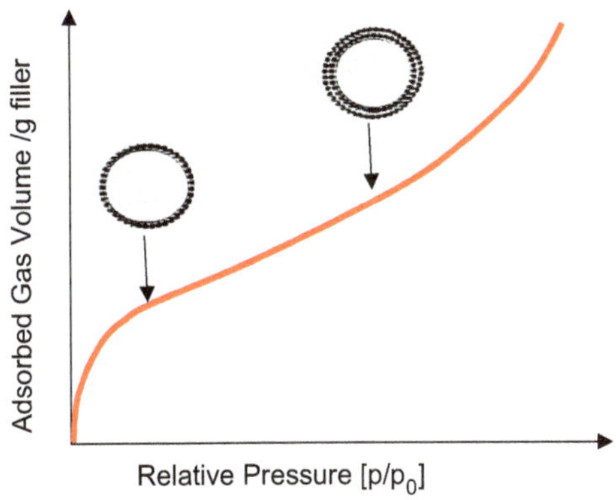


**Figure 3.10**
Specific surface area determined by the BET method

For determining the specific surface area of carbon black, the nitrogen adsorption isotherms are usually determined at −196 °C, the temperature of liquid nitrogen at atmospheric pressure. The measurements are performed at relative pressures of 0.05 to 0.35. Generally, six points are required to establish the linear relationship of Equation 3.10.

Occasionally a single-point determination is used to quickly check the specific surface area of a carbon black. In that event, the slopes of the straight line connecting one pressure point to the origin of the straight-line plots can be used to determine $V_m$. Standard measurements of the surface area use a pressure of 0.30 kPa and the time to run a single-point test is less than for the multi-point test, associated with some inaccuracies. However, since the equipment is simpler, it may be adequate for the filler manufacturer's quality control. Another method for determining the specific surface area is continuous gas chromatography with a mixture of nitrogen/helium gas flowing through the filler in the sample cell immersed in liquid nitrogen [97]. Of the many proposed modifications to the BET theory, none is more satisfactory than the classical theory from 1938. The BET theory strikes the best compromise between precision and simplicity.

### 3.5.2.1 Carbon Black

In absence of significant micro-porosity, which includes almost all rubber carbon black compounds, the specific surface area exhibits an inverse correlation with the size of primary particles. This means that the smaller the primary particles of carbon black, the larger is the specific surface area. The BET-nitrogen specific area ($N_2SA$) determined for carbon black, which does not possess an "inner" surface, ranges from a few $m^2/g$ for thermal blacks to 140 $m^2/g$ or more for special tire tread grades. However, it is questionable whether the adsorption process delivers a true measure of the specific surface area or is significantly affected by the nature of the surface and its energetic heterogeneity. There is a good agreement between the electron microscopic data derived from the diameter of the primary particles [98] and the BET specific surface area determined with the aid of Equation 3.8. A comparison of the two methods of determining the specific surface area is shown in Figure 3.11.

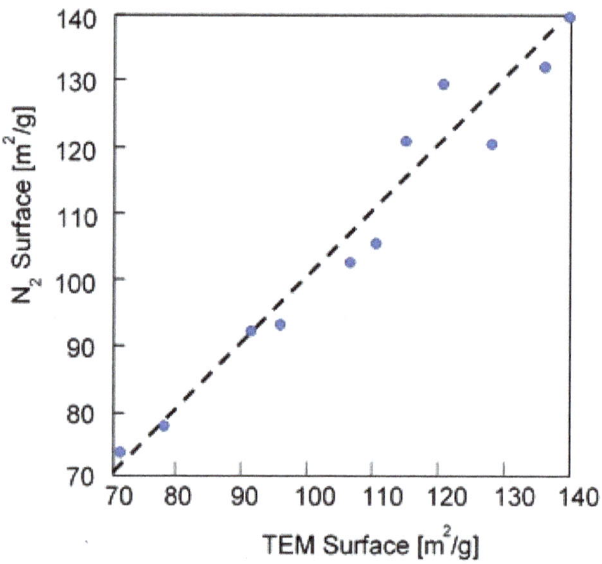


**Figure 3.11** Comparison of the TEM and the BET surfaces for carbon black [98]

The specific surface area determined from the average size, d, of the primary particles is somewhat larger than the corresponding nitrogen specific surface area. The difference can be attributed to the proportion of surface area that is lost when the primary particles coalesce to form the aggregates. For reasons of time, the entire adsorption isotherm is not recorded, but only measured at a degree of occupancy, for example, of 0.15, at which experience has shown that the second monolayer forms. The specific surface area of carbon blacks is shown in Table 3.3.

**Table 3.3** Specific Surface Area of Carbon Blacks

| CB Type | Specific Surface Area ($m^2/g$) |
|---|---|
| N110 | 143 |
| N220 | 119 |
| N330 | 83 |
| N550 | 42 |
| N660 | 35 |
| N770 | 29 |
| N990 | 9 |

Another method for creating isotherms is based on selective adsorption of other gases, such as helium, which does not require vacuum systems and is time saving. Due to the small volume of the nitrogen molecules, they are able to penetrate the pores of the surface. This is usually important for electrically conductive carbon blacks that have macro-pores, and in this case the specific surface area is larger. In the case of precipitated silica, which often forms porous aggregates, this aspect is particularly evident from the size of the pores that can be experimentally measured.

Besides nitrogen gas, other number of gases are required to measure the specific surface area of carbon blacks. The BET tests have been conducted with different gases whose molecular size is known: argon [99, 100], krypton, which is used for materials with low specific surface area [101], xenon [102] and propane [103]. On Graphon Black, the surface area was measured with $N_2$, Ar, Ne and $H_2$. The results show a specific difference compared with nitrogen adsorption. Nitrogen adsorption gives the smallest value of 85.9 $m^2/g$, followed by Ar at 86.1 $m^2/g$, Ne at 93.4 $m^2/g$ and $H_2$ at 169.3 $m^2/g$. Due to the small volume of the hydrogen molecule, it can be adsorbed inside the pores. Therefore, the question of the correct value of the area of adsorbate molecule is important, particularly for the nitrogen specific area, which serves as the reference for all surface area measurements [104].

The deviations from the standard isotherms were interpreted as being the result of micropores in the fillers. The BET standard tests for determining the specific surface area of particulate fillers provide the total of the "outer" and "inner" surface, which led to some ambiguity in the early literature on carbon black. The adsorption and desorption of nitrogen isotherms are used to estimate the microporosity in the filler aggregates [105, 106]. BET adsorption has been used to calculate pore size and size distribution in carbon blacks [107]. The analysis compared adsorption isotherms from fillers with a porosity with those from nonporous fillers.

It has been proven that the surfaces of most particulate fillers, including carbon blacks, are energetically heterogeneous and that the surface energy site distribution depends on the manufacturing process. It can be stated that the nitrogen specific surface area for carbon blacks is not a measure of a real surface area in square meters per unit mass, but rather a reliable indication of about the density of active sites on the filler surface and the filler activity [109]. Consequently, it has been suggested that a simple reporting of the amount of nitrogen adsorbed per gram of filler should be used as a useful indicator of filler activity, even if it does not measure the "true" specific surface area in $m^2/g$. Carbon blacks with high specific surface area tend to form defects by aggregation of aggregates into larger clusters. Such behavior must be taken into account when mixing the carbon black.

### 3.5.2.2 Precipitated Silica and Fumed Silica

The small silica aggregates yield a high specific surface area. By varying the precipitation conditions (concentration, pH, temperature and time) silicas with a variable specific surface area of between 25 and 250 $m^2/g$ are obtained. The specific surface area of the precipitated silica is measured by the BET method, but removal of the adsorbed water is somewhat more complicated. The $N_2SA$ values for conventional precipitated silica from Evonik are listed in Table 3.4 [46].

**Table 3.4** Specific Surface Area of Silicas Produced by Evonik

| Precipitated Silica | Specific Surface Area ($m^2/g$) |
|---|---|
| VN3 | 175 |
| VN2 | 125 |
| Durosil | 50 |

Silicas with a large surface area can affect the cure rate and crosslinking density of rubber compounds. The higher the specific surface area per gram of precipitated silica, the greater the filler-filler interaction and the greater the density of the resulting silica filler network. However, the addition of the correct amount of organo-silanes can reduce the intensity of the filler interaction from hydrogen bonding. There is a difference in particle size distribution between a conventionally produced precipitated silica and a HD silica [46]. The conventionally precipitated silica has a high fraction of large particle sizes while HD silicas show a group of smaller fractions of diameter 70 to 100 nm. This difference may explain some of the different properties of both type of silicas. The HD silica can much easy dispersed in unsaturated and saturated rubber during the mixing process. The semi-HD silicas have a larger primary particle size (40 nm) and a smaller specific surface area (35–100 $m^2/g$). The HD silica 7000G has a narrower aggregate size distribution than VN3. The aggregates of precipitated silica are larger than those of fumed silica and consist of clusters with a porous surface structure, a fact which indicates an inner surface. For precipitated silica, such methods return higher specific surface areas of 120 to 200 $m^2$/g. If the value obtained for the surface using the BET method is compared with that of TEM images, the ratio of "outer" and "inner" surface areas can be estimated. HD silicas, which are of interest in tire technology, show a specific surface area of 175 $m^2$/g and are used for high-performance tire treads, the lower specific area of 165 $m^2$/g is used for standard passenger treads, while the 125 $m^2$/g is used for winter tires [108].

### 3.5.2.3 Layered Silicates

The natural size of the specific surface area of layered silicates is very large, due to the width of the platelets and their small wall thickness of 1 nm. Using a latex compounding process, the large specific surface area of the 2:1 phyllosilicate can be dispersed into a rubber matrix of high specific surface area of 700 $m^2$/g, whether in the pristine MMT or in the organo-clay, which has been made to react with an additional quaternary ammonium salt [55, 56, 110–112]. Chemical composition plays a role in the specific surface area of layered silicates. If the exchange ions consist of $Na^+$, the isolated two-dimensional platelets have a high specific surface area of more than 500 $m^2$/g. If $Ca^{2+}$ and $Mg^{2+}$ are present in the exchange layer, the natural product cannot disperse well and the different bridges of the platelets connect with one another, yielding a reduced specific surface area in practice.

### 3.5.2.4 Carbon Nanotubes

For the cleaned multi-walled carbon nanotubes, the BET determination yields a high specific surface area of 300–450 $m^2$/g [113, 114]. In the case of one-dimensional multi-walled carbon nanotubes, the specific surface is determined by the diameter of the tube. The smaller the diameter, the larger is the specific surface area of the multi-walled carbon nanotube. The diameter of the multi-walled carbon nanotube depends on the way it is manufactured. In CVD processes, the diameter of the multi-walled carbon nanotube is determined by the diameter of the metal catalyst, with a certain distribution of the catalyst playing a role [115, 116]. TEM analysis is very helpful in this assessment, because it determines the average thickness of the carbon nanotubes and the size distribution. The specific surface area of the multi-walled carbon nanotubes measured in practice is 300–450 $m^2$/g, i. e. twice as high as that of carbon black with the largest specific surface area. The tubes of the smallest diameter have a large specific surface area and at the same time the strongest interaction with the rubber. A look at the one-dimensional multi-walled carbon nanotubes and the two-dimensional layered silicates reveals an increase in specific surface area for both fillers, which is twice or three-times the value measured for carbon black and precipitated silica.

### 3.5.3 External Surface Area (STSA)

A particularly fruitful approach to the measurements of the specific surface area is the statistical thickness surface area (STSA). This method measures just the external surface area via the film thickness of the adsorbed film. Known as the t-method, it is based on an adsorption isotherm developed in a published report [117, 118]. Valuable information about the specific surface area, the size and shape of pores, the onset of reversible capillary condensation, and complete filling of pores may be obtained by plotting the experimental volumes of adsorbed nitrogen, $V_a$, as a function of statistical thickness, $t$, of the adsorbed layer. This method is based on the finding that gas molecules do not arrange themselves in monolayers during adsorption, but accumulate in stacks of different heights around the active centers of the surface. The statistical thickness of the stack "t" is derived from comparison with non-porous adsorbents [119]. A plot of the volume of adsorbed nitrogen as a function of the average layer thickness is assumed to provide a more accurate assessment of the "external" surface area (free of micropores). The density of the adsorbed layer of nitrogen is equal to that of the bulk liquid. The statistical thickness, t, in A can be calculated as follows [120]:

$$t = 15.47\ V_a/N_2SA \tag{3.8}$$

where $V_a$ is the adsorbed volume of nitrogen and $N_2SA$ is the surface area of the adsorbent measured by nitrogen in $m^2/g$

Equation 3.8 suggests that for nonporous adsorbents, a line runs through the origin. With pores that do not show capillary condensation, the surface area available for adsorption decreases abruptly at the point where pores of a given size are filled. The STSA test indicates that micro-porosity is present in tread blacks with relatively low specific surface area that could not be detected by other methods; this difference has not yet been resolved. The method is cited as a better alternative for evaluating the external surface area than CTAB measurements, which is discussed in the section 3.5.5. Today, the STSA method has basically replaced CTAB surface area and is recommended for determining the specific surface area of carbon blacks and precipitated silicas [121, 122] and is standardized in the automated ASTM method [123]. With the introduction of the new automated test method, it has become possible to obtain the results of BET multi-point and STSA results with minimum effort in a much shorter test time.

There are some uncertainties that require a non-porous adsorbent to serve as a standard for the purpose of comparison with other porous filler samples. In Figure 3.12, the STSA values obtained from the regression of Equation 3.8 over the range of relative pressure up to 0.5 correspond to the statistical thickness of N660, N326 and N121 at 0.431 nm, 0.643 nm and 0.122 nm [120].

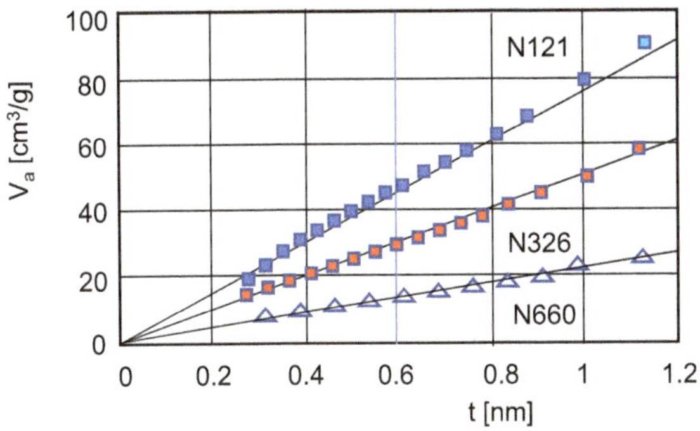


**Figure 3.12**
t-plots for furnace carbon blacks [120]

The STSA values obtained from the regression of $V_a$ versus t-plots over the relative pressure range of 0.2 to 0.5 corresponds to a statistical thickness of 0.43 nm to 0.64 nm or 1.22 and 1.82 times the monolayer adsorption, respectively. Adding potassium chloride to increase the structure during manufacture has an effect on the number of contact points during nitrogen condensation at relatively high pressure. With increase in the amount of the potassium salt, structure is reduced, which should lead to greater condensation of the nitrogen gas. In the rubber industry, the t-method is preferred to the CTAB method for measuring the external surface area of carbon black, because of the following advantages:

- precise results that are not affected by surface oxidation
- measurement of specific surface area
- shorter times

Several methods have been proposed for calculating the pore size distribution from adsorption data on porous solids [124]. Usually, the desorption loop of an adsorption isotherm relates the amount of adsorbate lost in the desorption step to the size of the existing pores [125].

### 3.5.4 Iodine Adsorption

One method that was developed in the early stages of carbon black characterization is the iodine adsorption test, which provides undefined information about the specific surface area. Correlations on iodine absorption have been published which indicate a linear relationship with the surface area of carbon black [126, 127]. Elemental iodine is released from an aqueous solution of KI and absorbed on the carbon black surface. The excess of the non-adsorbed iodine is determined by titration. The method is included in industrial standards and is often used in practice [128–130]. Carbon blacks show an increase in iodine adsorption that correlates well with the $N_2SA$ for carbon blacks [131]. The iodine adsorption is shown in Table 3.5.

**Table 3.5** Iodine Adsorption of Carbon Blacks [131]

| Carbon Black | Iodine Adsorption (g/kg) |
|---|---|
| N110 | 145 |
| N220 | 121 |
| N330 | 82 |
| N550 | 42 |
| N660 | 36 |
| N774 | 29 |

Although the units of measurement are completely different, the numerical values agree well with those returned by the $N_2SA$. However, the tar content in carbon blacks has the effect of decreasing the measured iodine number [132].

Although the method is very simple, its accuracy is poor, as the results are affected by the surface chemistry of the carbon black and these properties depend on the type of carbon black and how it has been processed. The result does not indicate the surface area per unit mass, but rather grams of iodine adsorbed per kg of carbon black, which can vary with the manufacturing conditions. The iodine number determined is somewhat sensitive to the surface chemistry of the carbon black. Like other tests proposed in the past, iodine test fails at predicting in-rubber performance, although it may be adequate for quality control purposes. Several reports show that the iodine number is sensitive to the porosity and surface chemistry of carbon black.

It is still a very common test for process control in carbon black production and rubber factories. However, as iodine has an oxidizing effect, it can react with active groups on the surface of carbon black, reacting with the polycyclic aromatics from the tar that has accumulated. Thus, although iodine adsorption is a standard test, it is only recommended as a comparative measurement of the different types of carbon black.

### 3.5.5 CTAB Method

One method of eliminating the influence of the porosity of the filler is to use surfactant compounds whose molecules are larger than the average pore size and are therefore not accessible to them [133]. With a view to avoiding influences exerted by pores on the filler surface, a method has been developed that uses a cationic surfactant hexadecyl-trimethylammonium bromide (CTAB) adsorbed on the filler from an aqueous suspension. For rapid adjustment of the adsorption equilibrium, the carbon black is stirred in a solution of CTAB and the carbon black sample is then separated by ultrafiltration. The amount of CTAB remaining in the filtrate is determined by titration

[134]. Calibration with a reference carbon black is required in the CTAB method. Carbon black N330 with a surface area of 83 $m^2/g$ is generally selected as a reference. CTAB adsorption by carbon black is independent of oxygen-containing groups and residual tars. Chemical inertness toward the filler surface and the adsorption must be sufficiently high on the filler, according to reports. Due to the large size of the CTAB, which has a cross-section of 6.16 $nm^2$, it is assumed that it does not enter into the micropores and thus represents a measure of the external specific surface area. The adsorption of CTAB on carbon black is characterized by a long horizontal plateau of the adsorption isotherm, which has a "Langmuir" shape. CTAB adsorption is not influenced from functional groups on the filler surface. Rapid equilibrium is achieved through the use of ultrasonic vibrations.

Comparison of STSA and CTAB results for 13 carbon blacks used in tire production revealed good correlation between the two groups of data, especially for the semi-active carbon blacks [135]. The data show that CTAB returns a larger surface area than STSA, which has to be understood as more evidence of low surface area furnace carbon blacks and, concomitantly, an overestimation on the part determined by the CTAB method.

The use of CTAB to determine the specific surface area of precipitated silica reveals monolayer adsorption that is independent of the surface chemistry of the precipitated silica [136]. However, the pH conditions have to be adjusted with buffer solutions [137]. If the particle size of the precipitated silica is decreased, adsorption of liquid CTAB increases. The specific surface area is calculated by multiplying the amount of CTAB adsorbed by the area occupied by the CTAB molecule. The relative method is often used in practice in many reports on carbon black and precipitated silica, because the method is easy to apply.

It has been reported that CTAB leads to misinterpretations of specific surface areas [138] because the surface activity per unit mass of carbon black, which is the basis for the adsorption of low molecular weight molecules, is inversely proportional to the size of the primary particles. CTAB values deviate from the linear correlation, especially for the fillers with high specific surface area of 20 $m^2/g$. It can be seen that, while the BET surface area increases, the CTAB surface area does not increase in proportion. Due to the simpler feasibility of the method, the CTAB method was used in many reports, even if errors occurred in the upper range. A comparison of the CTAB method with the BET method is provided in Figure 3.13.

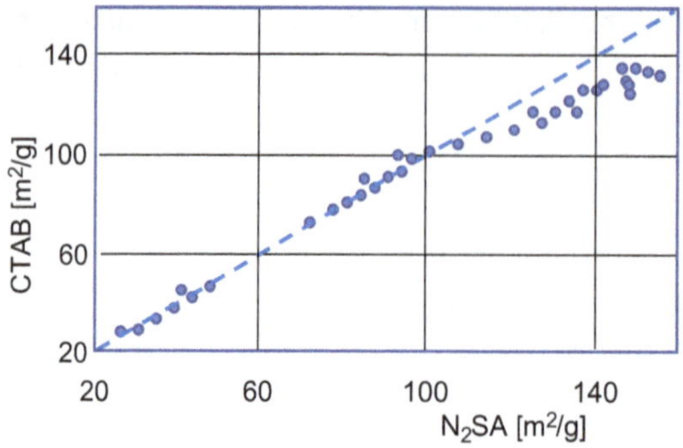


**Figure 3.13**
Comparison of the CTAB and BET methods for a group of carbon blacks [138]

However, the deviation of the CTAB tests becomes evident for carbon blacks with a nitrogen surface area greater than 100 $m^2/g$. This is due to the hyperbolic relationship between the filler activity and the radius of the particles. In particular, this deviation invalidates the hypothesis that larger molecules (CTAB) cannot enter micropores on the carbon black surface like the smaller nitrogen molecules do. In addition, this method has problems with test reproducibility between laboratories. The measured values often coincide with those determined from TEM images which capture the outer surface.

Specific surface area greatly affects the reinforcing potential of the filler in the filled elastomer. Suffice it to say that the surface activity is related to the specific surface area and that affects the ability of the filler surface to immobilize the polymer chains and establish elastic interactions that leads to higher tensile strength, Young's modulus and higher tear resistance. It should be noted that many reports cite the specific surface area measured using CTAB instead of BET.

### 3.5.6 Disc Centrifuge Photosedimentometry

The aggregate size distribution can be determined using specialized techniques, such as sedimentation, ultra-centrifugation and light refraction. One of the most convenient methods for determining the aggregate size distribution is based on centrifugal sedimentation of the filler in a liquid, i. e. a disc centrifuge equipped with an optical detector that detects the size of the filler. Disc centrifuge photosedimentometry (DCP) is probably the most reliable method and is used to determine the aggregate size distribution [139–141]. The method is based on Stokes' law that is modified for centrifugal sedimentation. Disc centrifuge photosedimentometry is used for isodimensional fillers, like precipitated silica. At the start, a small volume is injected onto the meniscus of a rotating hollow disc filled with liquid, called spin fluid. After injection, the particles of various sizes are forced outward by centrifugal forces and settle at rates determined by their size and densities. The diameter of the particles can be derived from the sedimentation rates in the gravity field. A DCP analysis for a precipitated HD silica is shown in Figure 3.14 [108].

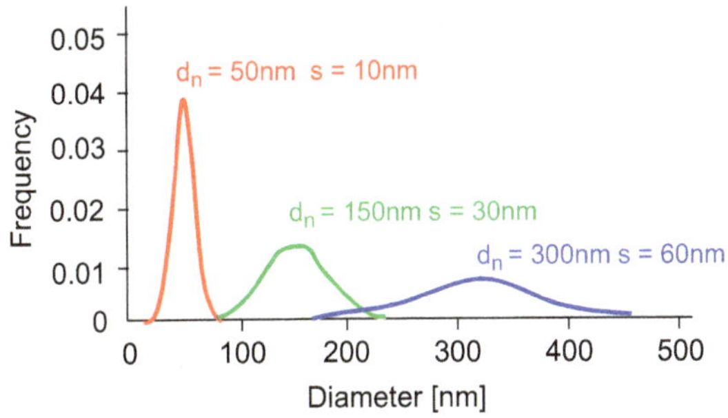


**Figure 3.14**
Aggregate distribution of HD silica by the DCP method [108]

It should be noted that, although large aggregates sediments at a faster rate than smaller ones, the sedimentation rate is also influenced by the void volume of the aggregates. The concentration of a certain size range of aggregates is measured using the absorbance of a light beam during centrifugal sedimentation.

## 3.5.7 Tinting Strength of Carbon Black

Another method for obtaining information about the tinting strength of carbon black consists in measuring it against a white pigment (zinc oxide). The tinting strength of carbon black has long been used for quality control of grades of carbon black. It is assumed that tinting strength is influenced by the particle size of the carbon black [142, 143]. However, it is also accepted that carbon black structure measured by oil-absorption tests has a significant influence on tinting strength as well. For a given size of primary particles, the structure of the aggregates provides an explanation of this property. The finer the carbon black, the greater is the tinting strength. For the tinting strength test, carbon black is mixed with white zinc oxide. The gray paste is applied to a thin film whose reflectance is measured. Tinting strength is measured according to standard methods. The reflectance depends not only on the optical properties pf the carbon black and the white pigments but also on the test conditions such as the concentration and the degree of dispersion of the pigments, the refractive index of the vehicle, the wavelength of the light and the optics of the reflectance measurement [144, 145]. It is assumed that the test mixture of carbon black and white fillers is influenced by the adsorption of the carbon black [146]. The effect of the morphological properties of the carbon black is described by the tinting strength:

- Tinting strength increases with decrease in particle size.
- For small particles, the tinting strength is less sensitive to the primary particle size.

The tinting strength can be calculated from the particle diameter and the refractive index using Mie theory. It was found that the Mie theory gives a good estimate of the carbon black aggregate size below which the tint strength is independent of the aggregate size. The tinting strength decreases with the diameter of carbon black particles, which have different structures. It has been shown that high-structure blacks lead to higher tinting strength. For furnace blacks, the tinting strength increases with the specific surface area. Tinting strength is one of the commonly used quality control tests and is expressed as a function of the DBP number [147]:

$$Da = [2540 + 71\,(DBP)]\,/S \tag{3.9}$$

where Da is given in nanometers, DBP is the DBP number and S is the specific surface area

The theoretical curve describes the tinting strength for all grades of carbon black supplied to the rubber industry. The high values of the tinting strength found with carbon blacks depends on the narrow aggregate size distribution [148]. The tinting strengths plotted against the solid-sphere diameter calculated from Equation 3.9 shows a decreasing curve with the specific area of the carbon black, which represents all the grades supplied in the rubber industry.

## 3.6 Aggregate Structure

One way to characterize the iso-dimensional fillers is to determine the void volume in the branched aggregates, which is related to the free voids of the aggregate. The properties of the branched aggregates depend on the number of primary particles in the aggregate. Therefore, the free voids in the aggregate, which are larger than when spheres lie on top of each other, have a considerable size compared to the volume of the solid aggregate. The presence of these more complex shape creates internal voids within the aggregates, which are greater than those that occur in simple packing spheres. It is not surprising then that the most used techniques for measuring the filler structure have been based on void volume using absorption measurements. The aggregates thus show a structure in carbon black, precipitated silica and fumed silica. Due to their morphological structure, layered silicates and carbon nanotubes do not belong to the type of fillers and cannot form branched aggregates.

The fact that carbon black forms a void volume has been observed by rubber scientists since the 1940s when it was found that the processing properties of carbon black can change. The incorporation time into a rubber compound, the viscosity of the filled rubber compound and the mechanical properties can vary as a result of carbon black production [142]. The not precisely defined term "structure" was initially chosen to describe this property. Later, the property of "structure" was changed to mean the void volume of the aggregates, which describes the property much better. Many of the mechanical properties of the rubber composite are influenced by the rubber content in the void volume of the aggregates.

The aggregate structure is a geometric parameter that is related to the bulkiness of the individual aggregates. It affects the aggregate size, the number of primary particles sintered together in one aggregate and the void volume of the aggregate [148]. As can be seen from TEM images, the aggregates are not Euclidean particles describing a simple geometric aggregate, but a fractal object in which the perimeter is larger than the square root of the area. The aggregates are branched objects that are disordered in their shape and size. The criteria for classifying the degree of branching of the aggregates are related to the determination of the void volume or the "structure" of the filler aggregate that can be shown in more branched aggregates. TEM images of aggregates of carbon black N300 are shown in Figure 3.15 [148].

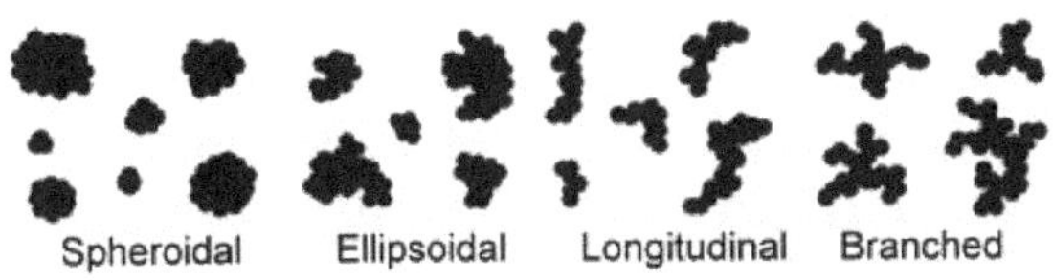


**Figure 3.15** Aggregate shapes found in carbon black N300

The existence of these criteria was known, as it was found that a filler of similar primary particle size is incorporated differently in rubber compounds. The viscosity of the rubber melts and the dispersibility of the rubber are different and the Young's modulus depends on the fact that the processing of carbon black, especially the reaction time, differs [149–151]. In the mid-1960s, it was discovered that the addition of alkaline metal salts could be used to reduce the structure of carbon blacks in an industrial process. This gave the carbon black industry an opportunity to select the desired structure of the filler in the furnace process and to influence the mixing process and dispersion. There is a correlation between the number of primary particles in an aggregate and the structure of the filler. A carbon black aggregate with a high "structure" always requires a higher number of primary particles in the aggregate, while a carbon black with a lower "structure" requires a smaller number of primary particles in the aggregate. A filler aggregate can have different types of branching, depending primarily on the concentration of alkali metal salts which is added in the furnace process and the processing conditions in the reactor zone.

There are different methods for determining the void volume in reinforcing fillers. The first determines the space required to fill the empty space between the aggregates when the aggregate is drawn together by the surface tension of an organic liquid. The second is used to measure the void volume in the filler that supported a high pressure previously. This method measures the "in-rubber morphology" of the fillers. The third is based on exploring the size of the pore using a liquid that can be introduced into the pores. The last approach uses the projected area of TEM images to calculate the amount of void volume from the projected area of the filler aggregates. The "structure" or the void volume of the aggregates can be determined under the following conditions:

- the absorption of a low molecular weight compound [151–154]
- the change in absorption upon prior compression of the filler, 24M4 DBP [155, 156]
- depression of a liquid introduced into the pores of the filler
- interpretation of the projected TEM images

These four methods deliver the size of the aggregates of the void volume as measured under different conditions. The results can explain the increase in the compound viscosity and the increase in Young's modulus in filled elastomers [157]. The last method is related to an interpretation of the void volume observed in TEM micrographs to understand the importance of the presence of branching the aggregates from the two-dimensional projections. The role of the void volume in understanding the importance of the fractal dimensions of the filler is elusive. A summary of the methods to determine the structure of the filler is given in Table 3.6.

**Table 3.6** Methods for Determining the "Structure" of Iso-dimensional Fillers

| Method | Specific Determination |
|---|---|
| Oil and DBP absorption | Void volume |
| Compression measurement | Decrease in void volume |
| Thermoporosimetry | Distribution of void volume |
| TEM evaluation | Ratio of void volume to solid volume |

The information yielded by these methods is discussed below and the advantages of their application are shown.

## 3.6.1 Determination of Void Volume

The oldest method for determining the void volume of iso-dimensional fillers is based on filling the cavities of the aggregates with a liquid and determining the volume of liquid per 100 g of filler. In the past, dibutyl phthalate was used, but was replaced by paraffin oil for toxicological reasons. Measurement of the oil absorption test is an indirect method, as the degree of wetting of the voids of the aggregates is not precisely clarified. Nevertheless, the test still remains the most meaningful and reliable test and it is performed according to the specifications described in the standard methods [158–160]. The void volume of the filler is determined using volumetric apparatus at specific pressures that can be used for quality control tests by absorption. In the standard procedure, the paraffin oil is added from a constant-rate burette to a sample of carbon black in the mixer chamber of an absorptometer. As the sample absorbs the oil, the mixture changes from suddenly from a free-flowing powder to semi-plastic mass of increased viscosity, which is transmitted to the torque system of the absorptometer. The end point of the test is given by a pre-defined torque level. The more liquid is added to the filler, the lower the displayed torque becomes. The result is expressed as the oil adsorption number, measured in mL/100 g filler. As a measure of "structure", the oil absorption test still remains the most meaningful and reliable test and is performed to the specifications described in the standard methods. The oil absorption test is illustrated in Figure 3.16.

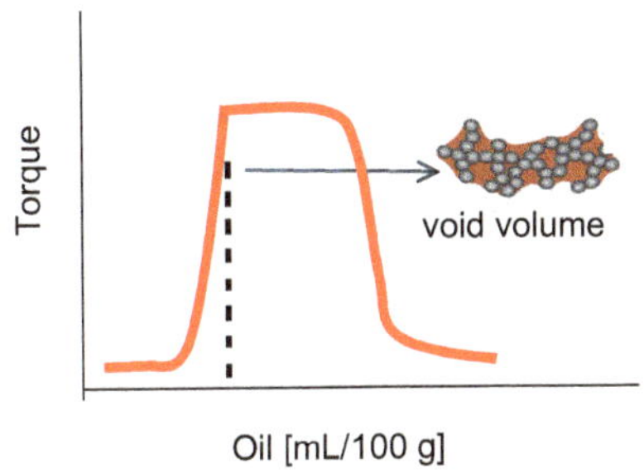


**Figure 3.16**
Determination of filler structure

A distinction is made between three filler structures. A normal structure is determined when the absorption is 100 mL per 100 g filler. A low structure absorbs less liquid and a high structure can absorb more than the value stated above. An amount of 100 g of a normal "structured" carbon black, with a solid volume of 54 mL, absorbs roughly twice the volume of dibutyl phthalate (DBP) or, nowadays, a paraffin oil. The experimental results show some different absorption numbers for the two types of liquids. A high "structured" carbon black is able to absorb more DBP or paraffin oil, namely 120–130 mL/100 g, while a low "structured" carbon black can absorb a smaller amount, around 72–73 mL/100 g. However, the result of the oil absorption measurement is influenced by the degree of pelletization and the rate at which the liquid is absorbed by the filler.

#### 3.6.1.1 Carbon Black

Measurement of the void volume does not provide unequivocal information about the voids between the aggregates of pelletized filler. The results of the test can be understood if the aggregates of a similar primary particle size are considered in order of increasing number of primary particles, resulting in an increased hydrodynamic volume of the aggregate. This can be exemplified with carbon black N326 to N347. Starting from a carbon black with a small number of particles, N326 is more condensed and has a small void volume. The micrographs show a complex agglomeration of more structures that overlap each other. By adding more primary particles to the aggregate the degree of branching increases, the void volume increases and this is observed in the higher void volume of N330 and even more in N339. The opening of the aggregates with more void volume is shown in Figure 3.17.

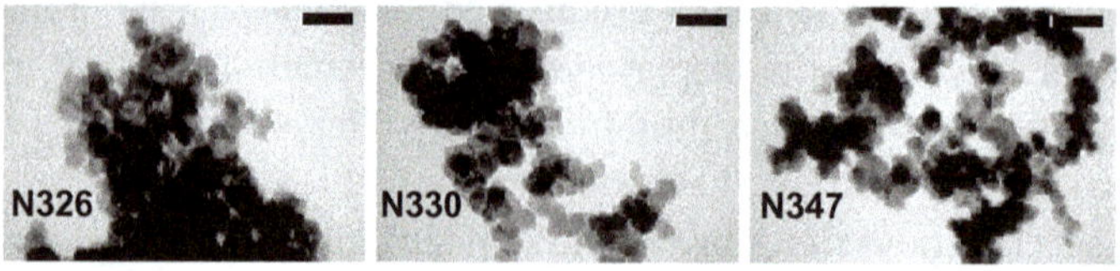


**Figure 3.17** Structure of carbon black aggregates: N326, N330 and N339

For those carbon blacks which exhibit increased viscosity in rubber compounds and Young's modulus, the DBP number changes in a manner similar to that shown in Table 3.7 [161].

**Table 3.7** Void Volume of Active Carbon Blacks [161]

| ASTM Carbon Black | DBP Number (mL/100 g) |
|---|---|
| N110 | 112 |
| N121 | 122 |
| N220 | 114 |
| N234 | 125 |
| N326 | 75 |
| N330 | 102 |
| N339 | 120 |
| N539 | 111 |
| N550 | 121 |

It can be seen that carbon blacks having the same degree of structure have a similar DBP number. This is important for compounding, which also requires an increase in the viscosity of the rubber compound and an increased strength of the vulcanized compound.

Due to the special conditions in the production of furnace blacks, not only the primary particle size, but also the number of primary particles that combine to form aggregates is subject to a distribution that increases significantly with the structure. TEM investigations performed on different carbon blacks reveal that the branching of the aggregates of a carbon black type is not uniform, but corresponds to a distribution ranging from less branched to highly branched aggregates [162]. It has been shown that the degree of branching increases with the number of primary particles of that aggregate. According to this scheme, low-structure carbon blacks contain more spheroidal and ellipsoidal aggregates, while high-structure carbon blacks are characterized by a larger proportion of linear and branched aggregates.

#### 3.6.1.2 Precipitated Silica

The precipitated silica agglomerates have a strong tendency to agglomerate because of the silanol groups on the silica surface. Therefore, the void volume is more pronounced than with carbon black. Triethanolamine is used for precipitated silica that has an outer and inner surface because the test results are more reproducible. The void volume determined is higher than for carbon black [46]. With conventional precipitated silica, VN2 has a void volume of 192 mL/100 g silica and VN3 has a void volume of 209 mL/100 g silica. In precipitated silica, the aggregates are cluster-shaped and the pores within the clusters have a diameter of 10–60 nm. With precipitated silica, measurement of the structure is less satisfactory. Precipitated silicas and fumed silicas give very high structure, apparently because of the high interaction with the polar DBP during the manufacturing process. The chemical reaction conditions determine the number of primary particles that aggregate in precipitated silica [47].

### 3.6.2 Influence of Pressure on Void Volume

After mixing of the filler into the rubber compound, the void volume in the filler is occupied with the polymer. Therefore, the aggregate has another hydrodynamic volume that includes a part of the rubber that is no more elastically active. But when the pressure is increased the agglomerates can accept only a part of the rubber in the branched aggregates. The mechanical stress acting on the agglomerates when they are incorporated into the rubber matrix was simulated in a preliminary condensation test in order to obtain information about the state of oil adsorption. A promising test procedure consists in compressing the filler multiple times to bring the morphology of the aggregates into a state of densification prior to measurement of the oil adsorption. The equipment and the methodology are not complicated. The main component of the compression system is a cylinder with two plugs that fit loosely and so permit air to escape and create relatively little friction. The filler is compressed with a mechanical press that crushes the filler sample four times in the cylinder at 165 MPa. After this, the oil adsorption test is performed to yield what is called the crushed oil absorption number (COAN). The test results refer to the decrease in the void volume after the crushing process. The test conditions are standardized [163]. The test results vary from one producer to the other and are dependent on the pelletizing process. That is why the difference between the normal determination of the void volume and the compressed oil adsorption test is described [164]. The difference indicates the stability of the void volume test. The compressed volume is calculated from the following equation:

$$V_A = \pi D^2 h/4000 \tag{3.10}$$

where $V_A$ is the actual compressed volume of the weighted carbon black, h is the height of the compressed carbon black and D is the diameter of the cylinder

The calculated void volume of carbon black is a value that is not affected by particle size and is expressed by the equation:

$$e = (V_A - V_T)/V_T \tag{3.11}$$

where $V_A$ is the compressed volume of carbon black and $V_T$ is the theoretical volume of the aggregate

This equation formed the basis of several studies on the influence of pressure on filler. A larger study was performed on the test conditions and an extension of the pressure to higher values [165].

#### 3.6.2.1 Carbon Black

The measured compressed oil adsorption numbers are significantly lower than the oil adsorption numbers measured for carbon blacks at atmospheric pressure. Therefore, a process of multiple compression is recommended to bring carbon black into a morphology that can be more easily incorporated into the compound [166]. A direct comparison between the DBP numbers and the CDBP numbers of different carbon blacks shows a sharp drop in CDBP values for all carbon blacks used in tires, as shown in Table 3.8 [164].

**Table 3.8** DBP Numbers and 24M4 DBP Numbers for Different Active Carbon Blacks Used in the Tire Industry [164]

| Carbon Black | DBP (mL/100 g) | 24M4 DBP (mL/100 g) |
|---|---|---|
| N110 | 112 | 98 |
| N115 | 113 | 97 |
| N121 | 132 | 120 |
| N326 | 72 | 69 |
| N330 | 102 | 88 |
| N339 | 120 | 101 |

The data can be plotted semi-logarithmically and for straight lines for the logarithm of OAN versus the applied pressure. The decrease in the volume as a function of the actual pressure on the sample can be easily determined from such a plot. The values show a strong decrease in volume for acetylene blacks, which have a high oil adsorption number of 337 mL/100 g carbon black. It is assumed that compression leads to a breakdown of filler agglomerates. The specific volume of carbon black as fixed pressure of 5.06 MPa show good correlation with the mineral oil adsorption [166]. The influence of pressure was extended to other types of carbon blacks and measured between 0.07 MPa and 69 MPa [167]. The higher-structure carbon blacks ranging from N200 to N700 clearly show the sharpest drop-off in volume with increase in pressure.

#### 3.6.2.2 Precipitated Silica

The structure of precipitated silica is more strongly affected by the silanol groups on the surface of the aggregates. Precipitated silica delivers higher values in the DBP test, apparently because of inter-aggregate agglomeration as a result of silanol groups forcing hydrogen bonding between the aggregates. The agglomeration is apparently not broken down in the test procedure at 1 atmosphere. But increasing the test pressure the COAN values for precipitated silica decrease after the high pressure is applied, as shown in Table 3.9 [46].

**Table 3.9** Comparison of the 24M4-DBP Values of Precipitated Silica [46]

| Precipitated Silica | DBP ($cm^3$/100 g) | 24M4-DBP ($cm^3$/100 g) |
|---|---|---|
| Prec. silica VN2 | 225 | 96 |
| Prec. silica VN3 | 220 | 99 |
| HD silica 7000G | 265 | 103 |

The DBP number increases in the order VN2 > VN3 > HD silica at normal pressure. But at high pressures the value strongly decreases at the values of between 96 and 103 $cm^3$/100 g. HD silica shows a significant increase in DBP numbers at low pressure, exceeding the DBP numbers of the conventional silicas VN2 and VN3, and showing greater aggregate stability. The void volume of conventional silicas is decreased, but the value is twice that obtained with an active carbon black. If the test pressure is increased the void volume decreases in the same way for all the different types of precipitated silicas. Comparison shows the 24M4-DBP values are significantly reduced and comparable to those of active carbon blacks. Therefore, under pressure the occluded rubber is reduced in silica and the large structure is comparable to carbon black grades.

During mixing with rubber, the initial stress caused by the compression corresponds to the shear stress that occurs, which also causes the agglomerates to break down. In contrast to carbon black, where the void volume and the specific surface area are independent of each other, there is a connection between the two properties in the case of precipitated silica. This is illustrated for the COAN of precipitated silica which decreases with the applied pressure. Figure 3.18 compares new silica (Zeosil, 1165 MP, Ultrasil 7000GR) and the conventional silica VN3 of same specific surface area [47].

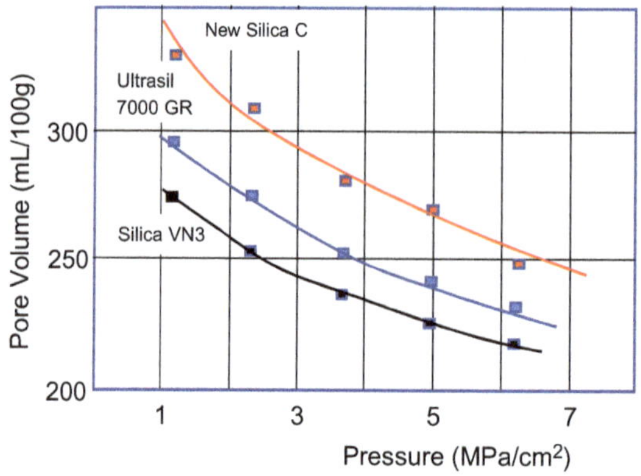


**Figure 3.18**
Decrease in the structure of silica as pressure increases [47]

Comparing the HD silica with the conventional products, the starting position at normal pressures demonstrates high stability of the agglomerates, due to the hydrogen bonding in the aggregates. On repeated compression of the sample, the COAN value decreases sharply to lower values, but they remain above the respective references provided for the conventional precipitated silica, which are slightly higher than for active carbon blacks. In the case of precipitated silica, the number of pores and the pore distribution are criteria for assessing the quality of the filler grade. It shows a decrease in the oil adsorption value determined at normal pressure. In terms of application, the differences can be interpreted as follows. In the case of silica with a high structure, penetration of the polymer is facilitated and this is reflected in the dispersion.

### 3.6.3 Porosimetry

Mercury porosimetry is used to describe the pore size distribution in iso-dimensional fillers. The principle behind porosimetry is based on the pressure required to force mercury into small pores in the filler [168, 169]. The basic equation for conducting the porosimetry is related to the surface tension of the mercury, the contact angle between the liquid and the capillary diameter. The pore size is determined from the equation:

$$d = 1.5/p \qquad (3.12)$$

where d is the pore size in μm and p is the pressure in 0.1 MPa

The procedure measures the pore size distribution by pressing the mercury into the pores of a filler. The test uses an outgassed filler placed in an evacuated steel cylinder. Pure mercury is admitted to the filler in a series of increasing pressures. The measurements distinguish between intra-aggregate, inter-aggregate and inter-agglomerate void volumes. In this way mercury porosimetry fulfills an important task in evaluating the type of the filler. The porosity of the nano-fillers describes the quantity of pores introduced into a solid filler during the manufacturing process.

In Figure 3.19, the steep part of the porosity curve is attributed to the pores in the aggregates and is considered to be the real pore volume [170].

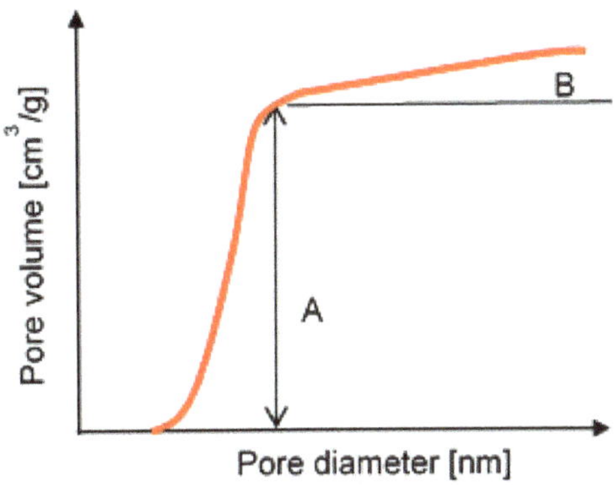


**Figure 3.19** Porosity curve from mercury porosimetry [170]

A small part is attributed to the pores between the aggregates. A mixture of two carbon blacks N234 and N550 shows a pore distribution between the two fillers [171]. Mercury porosimetry has been described as the most useful tool for characterizing precipitated silica. Measuring the pore distribution of precipitated silica is another way of determining the void volume in the filler. For a conventional silica and HD silica of similar specific surface area (165 $m^2/g$), it was found that the pore size and volume in terms of voids within the aggregates are similar for both silicas, while the void volume between the aggregates and between the agglomerates is much higher for the HD silica [47]. This is certainly the root cause of the better dispersion of silica in compounds. The method offers the possibility of drawing conclusions about the size of the voids from the consumption of mercury. It is noticeable that the pores are distributed differently. If mercury porosimetry is compared for conventional silica and HD silica with similar surface area, the pore size and volume are related to the intra-aggregated voids, the real structure of the filler, is similar for the silicas, the inter-aggregate and intra-agglomerate void volumes are much higher in the HD silicas.

### 3.6.4 Calculated Void Volume

Attempts have been made to classify the shape of the iso-dimensional aggregates by studying the void volume obtained with TEM images. In other words, the DPB number should have some physical meanings related to the size and shape of the filler aggregates. One finding was the asymmetric structure of the aggregates that exist between the solid primary particles and the voids which are empty when the filler is incorporated into the rubber and which are filled by the rubber chains in the incorporation phase, resulting in a shift in the apparent ratio of filler and the rubber. In the case of carbon black containing more highly branched aggregates, the region of the void volume is much higher than for carbon black with fewer primary particles and small void volume.

Such an approach has been carried out using the theory of flocculation. The total volume is calculated from the two-dimensional projection of the filler aggregate. The volume of the projected area corresponds to the number of primary particles, $N_p$, and the volume, $V_p$, when the particles are the same size. The total volume of an aggregate corresponds to an equivalent spherical volume, $V_g$, which is determined by the volume of the primary particles and the number of primary particles, $N_p$ [172]:

$$V_g = V_p N_p^{1.305} \tag{3.13}$$

The void volume occupied by the aggregate results from the difference between the total volume $V_g$ and the solid volume $V_a$ of the filler:

$$\left(V_g - V_a\right) / V_a = N_p^{0.305} - 1 = \left(A/A_p\right)^{0.35} - 1 \tag{3.14}$$

Thus, the void volume increases with increase in number of primary particles per aggregate and occluded rubber. The larger void volume is related to the number of primary particles in one aggregate [173]. Therefore, the hydrodynamic volume of the aggregate includes the solids fraction of the primary particles and the occluded rubber, as shown in Figure 3.20.

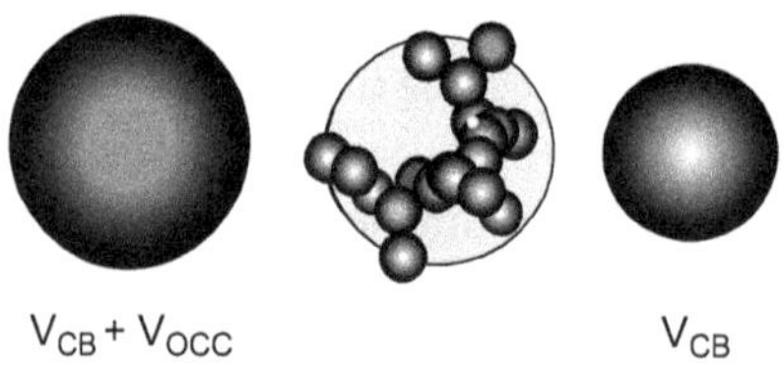


**Figure 3.20**
Hydrodynamic volume includes the solids fraction and the occluded volume of the aggregate

For a filler that has no void volume, $V_g = V_a$, there is a void of 31.5%, which corresponds to the space between a collection of spheres of the same size. The validity of this equation can be confirmed by comparing the experimentally determined DBP numbers with the calculated ones. For this the structure of the filler has to be introduced in the estimation. Such a comparison is shown in Table 3.10 [173].

**Table 3.10** Comparison of Calculated and Experimentally Determined DBP Numbers [173]

| Carbon Black | $V_g/V_a$ | DBP exp. | DBP calc. | Ratio |
|---|---|---|---|---|
| N219 | 3.233 | 72.4 | 66.6 | 0.92 |
| N220 | 4.959 | 117.4 | 127.1 | 1.08 |
| N242 | 5.602 | 139.5 | 149.6 | 1.07 |
| N326 | 3.390 | 73.1 | 90.6 | 1.09 |
| N330 | 4.385 | 97.9 | 106.9 | 1.09 |
| N347 | 5.039 | 128.9 | 129.9 | 1.01 |

Within a group of similar specific surface area, TEM imaging clearly shows an increase in void volume distribution with increase in structure, which is expressed by the ratio $V_g/V_a$. The ratio is quite similar for carbon blacks that have different surface areas. This leads to the calculated structure of the aggregates, which agrees with the experimental values. As a result of this direct effect of the void volume, the following equation provides a basis for calculating the hydrodynamic volume of an aggregate from the determined DBP number and the filler volume of the aggregate [173]:

$$V_g = (1 + 0.02139\,DBP)\;0.685\,V_a \tag{3.15}$$

where DBP is the experimentally determined DBP absorption, $V_g$ is the hydrodynamic volume of the aggregate, and $V_a$ is the solid filler volume

Image analysis is based on a large number of measured particles and is therefore unsuitable for routine investigations of the carbon black aggregates

There are some reasons for considering the void volume of the aggregate as a generic term for the imprecisely defined "structure". If a filler is used in the rubber, the polymer penetrates into the void volume and is excluded from the elastic process of the elastomer. The reduction in the elastic properties of the occluded rubber has a fundamental influence on the elasticity of a filled vulcanizate. It can be seen that if the oil absorption is high and more rubber is occluded, the elasticity of the rubber vulcanizate corresponds to a higher amount of filler. In order to keep the elasticity constant, a smaller amount of filler can be used. The proportion of rubber occluded in agglomerates was represented by adjusting stress values as a function of DBP number [174]. Of particular importance is the principle of equivalence of structure and carbon black concentration. It states that the stress value achieved with a lower concentration of a carbon black of higher structure is the same as that achieved with a higher concentration of carbon black of lower structure [175]. This allows the compounder to select the desired filler concentration. In many cases, the higher structure and lower filler concentration will be chosen.

### 3.6.5 Fractal Approach to Aggregates

It has been recognized that many structural units of many objects that appear in nature can be described using fractal sizes. This is true also for filler aggregates that are manufactured. The aggregates of the fillers that have been combined with fractal geometry. They are particularly suitable for the self-similarity of the carbon black and silica aggregates. The fractal approach was used to replace the structure of fillers with the corresponding fractal dimensions through the self-similarity of the filler aggregates. There were two approaches used to classify the geometry of the filler aggregates by using fractal dimensions of the perimeter or of the mass fractals of the aggregates. Taking into consideration that aggregates and agglomerates of fillers have a self-similarity, the use of perimeter fractal dimensions $\alpha_p$ and the mass fractal dimensions $\alpha_m$ was attempted to bring more consistency than the measured structure from the oil absorptiometer [176]. Carbon black and precipitated silica are non-Euclidean objects, in that the perimeter increases more than the corresponding area. The expressed condition for the perimeter fractal dimension is:

$$P \sim A^{\alpha p/2} \text{ or } \log P \sim \alpha_{p/2} \log A \tag{3.16}$$

where P is the perimeter of the aggregate, A is projected area of the aggregate and $\alpha_p$ is perimeter fractal dimension

For a sphere, the circumferential dimension is, by definition, 1. For an iso-dimensional filler, it is between 1 and 2. Starting from the approximately spherical particles that are typical of thermal carbon blacks, the aggregate shape of furnace black changes independently of the mean primary particle size. These aggregates are given a perimeter fractal dimension greater than 1. For furnace blacks, the perimeter fractal dimension is in the range of 1.30–1.40 [177]. The self-similarity results in a structural principle, which states that in an agglomerate of branched aggregates with a higher fractal perimeter dimension, a void volume is formed that is significantly higher than that which occurs in the usually dense packing of the primary particles. To determine the perimeter fractal dimensions $\alpha_p$, a large number of aggregates are measured with the aid of TEM micrographs, by plotting the perimeter and evaluating the area. The perimeter fractal dimension is determined from the slope of log P vs. log A. The perimeter fractal dimension $\alpha_p$ increases with increase in structure of the carbon black. The measured perimeter fractal dimension for carbon black N330 of 1.32 indicates the normal branched nature of the carbon black aggregates.

The other approach leads to the mass fractal dimensions, which are also related to the structure of the aggregates. The mass fractal dimension uses the mass of the aggregates as a function of radius, as shown in the following equation [178]:

$$M \sim R^{\alpha m/2} \text{ or } \log M \sim \alpha_{m/2} \log R \tag{3.17}$$

where M is the mass of the aggregate, R is the mean radius of the aggregate and $\alpha_m$ is the mass fractal dimension

In contrast to a sphere, where the mass increases with the third power of the radius, iso-dimensional fillers are objects whose mass increases with a smaller power, and for branched filler aggregates it decreases to lower values. In order to determine the mass fractal dimension, the aggregates are divided into circle fractions and the proportion of solids in these fractions is measured. It turns out that there is a good correlation between the structure measured by absorption and the mass fractal dimensions measured with the aid of TEM micrographs. With decrease in structure of the filler, the mass fractal dimension increases, reaching a value of 3 for spheres. For branched aggregates, the mass fractal dimension of the filler is therefore between 2 and 3. This shows that branched aggregates have a smaller mass fractal dimension than a continuous solid, because of the cavities. However, since the aggregates have a certain degree of branching, the mass fractal dimension does not reach the upper limit corresponding to a spherical object, but remains at a lower level. As can be seen from experimental results on perimeter fractal dimensions and mass fractal dimensions, the structure of the aggregates can be followed by the fractal approach, as shown in Figure 3.21 [179].

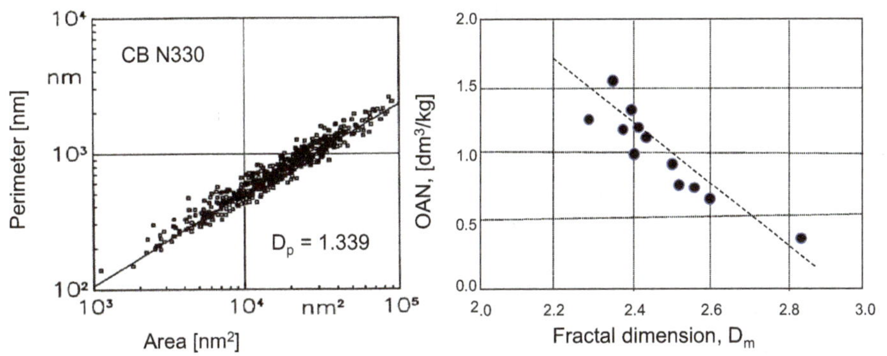


**Figure 3.21** Determining the void volume from fractal dimensions of the filler [179]

For mass fractal dimensions, the OAN decreases with decrease in branching of the aggregates.

In the case of precipitated silica, the structure can be described precisely by the mass fractal dimension. However, these scientifically consistent approaches, which are intended to determine the structure of aggregates as a fractal geometrical effect, have generated only limited interest in rubber science. Modern knowledge of fractal geometry offers consistent criteria for unequivocally assigning the shape of aggregates. However, this requires considerable effort, which is usually met with resistance in practice. Despite all the advantages of fractal dimensions, void volume continues to be determined using standardized methods.

### 3.6.6 Independent Parameters: Specific Surface Area and Void Volume

The specific surface area and the structure of the aggregates are two independent variables that are not related to each other. Different parameters govern the primary particle size and the aggregation tendency when the filler is formed. The manufactured products form a large group of different products from which the compounders can select the size of the fillers and the tendency to form the aggregate structure. The carbon black manufacturers can decrease the specific surface area by increasing the turbulence and the temperature in the furnace, and can reduce the structure by adding potassium chloride into the reactor. The two parameters are therefore relatively independent of each other. An overview of the carbon black types supplied by one manufacturer is shown in Figure 3.22.

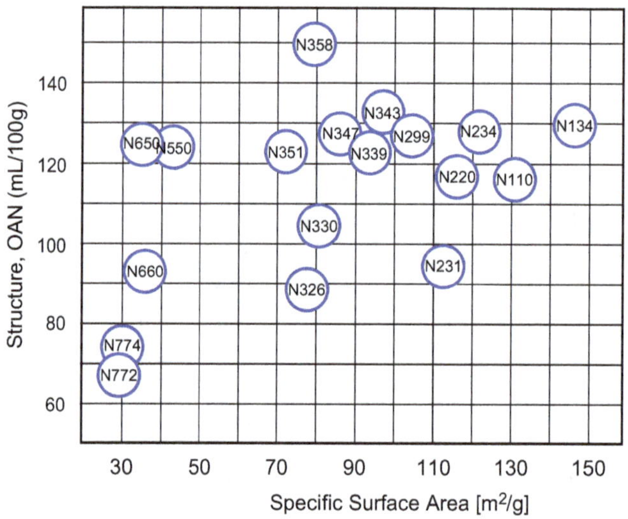


**Figure 3.22** Representation of the $N_2SA$ determined by the BET method and the structure of carbon blacks determined by oil adsorption

The choice of carbon black depends on the application in tire compounds and in technical rubber goods. For tire treads, the products used have a large surface area and a high structure to disperse them easily, while the side-wall requires carbon black that increases the dynamic properties at low frequencies. The semi-reinforcing carbon blacks show a large variety to change the aggregate structure. This helps to achieve good dispersion and high reinforcement in saturated rubbers, such as EPDM. However, there are differences in the mixing processes and the flow properties of the rubber compounds and finally, of course, in the increase of the reinforcement of the rubber vulcanizates.

## 3.7 Aspect Ratio

In the discussion of the reinforcing effect of fillers, the ratio of length to width of the filler particles plays a special role in determining filler loading and filler dispersion as well as the effects on the elastomer properties. In order to be able to make predictions about the mechanical properties of filled vulcanizates, the aspect ratio or the anisometry of the filler needs to be evaluated to ensure the reinforcement is sufficient by having a good dispersion in the rubber compound. The aspect ratio is particularly important when considering the properties of layered silicates and multi-walled carbon nanotubes that naturally possess a high aspect ratio and are present in rubber compounds at very small loadings.

### 3.7.1 Fillers with Low Aspect Ratios

The aspect ratio in iso-dimensional fillers has small values of 5–10. It has been observed in TEM images that carbon black aggregates tend to have an elongated shape which influences the viscosity of the rubber compound and the Young's modulus or dynamic storage modulus in the elastomer. Realistically, the aspect ratio is evaluated by approximating the silhouette of an aggregate by a geometric shape that approximates the fractal object.

Usually, the TEM image of an active carbon black can be used to evaluate the average length and the diameter from which the aspect ratio can be estimated [180–182]. In order to obtain reliable results that serve to explain the increase in the rubber reinforcement, it is necessary to compare the geometrically determined aspect ratio with the results of other methods, e. g. light scattering. It has been found that the average diameter of the aggregate is half of the average length of the carbon black aggregates, resulting in an aspect ratio of 1.8 to 3.5, with the exact value depending on the type of filler. However, a certain amount of agglomeration, due to sample preparation, cannot be ruled out in TEM images.

Further possibilities for estimating the irregular structure of the aggregates are obtained by expressing the third power of the aggregate circumference as a ratio of the microscopically determined total volume. With this imprecise method it is still possible to determine the difference in the aspect ratio. With increase in aggregate structure, the aspect ratio increases from 3.3 for N326 to 6.4 for N330 and 9.3 for N347 [183, 184]. The increase in the aspect ratio is related to the number of primary particles that are growing the aggregate. A similar discussion for precipitated silica relates the aspect ratio to the improved properties of the elastomers. Although there is still no recognized method for determining the aspect ratio of iso-dimensional fillers, the question arises as to how to explain the physical properties related to the aspect ratio, such as the dynamic moduli in elastomers Therefore, in addition to the specific surface area and the structure of aggregates, the surface activity is used, which clearly shows the difference in the rubber-filler interaction.

### 3.7.2 Fillers with High Aspect Ratios

The excellent mechanical properties of layered silicates and multi-walled carbon nanotubes, in particular, such as the high modulus, tensile strength and elongation at break of the composite, are explained by the high aspect ratio of these fillers. In order to obtain optimal property enhancement, it is essential to resolve the key issues of improved dispersion and alignment of the high aspect ratio fillers [185]. Unlike conventional fillers, which form a filler network at high concentrations, layered silicates and multi-walled carbon nanotubes can have a much higher aspect ratio of 100 and more. Layered silicates in the exfoliated state have a platelet length of 300 to 800 nm and, given a thickness of 1 nm, the aspect ratio is in the region of 300 to 800. Therefore, the layered silicate can occupy a small volume, but realize a high aspect ratio that provides a good interaction with the rubber chains [186–190].

Due to the size of the aspect ratios, a significantly larger area of the rubber chains is brought into interaction with the large surface area of the layered silicate. The substitution of inorganic ammonia cations for organo-clays not only serves to make the rubber matrix and the organo-clay more compatible, but also causes higher interlayer spacing of layered silicate platelets. In an intercalated structure, the aspect ratio is lower, because several layers overlap and reduce the value of the aspect ratio. The maximum aspect ratio is found in exfoliated layered silicates, which affects the level of physical properties and barrier function of these fillers. In the organo-clay form as intercalated platelets with a lower aspect ratio or in the exfoliated form having maximum interaction with the polymer, the mechanical properties of a rubber vulcanizate are fundamentally changed [191, 192]. For this type of filler, X-ray diffraction is largely used to assess the distance between the layers. Constructive interference occurs in X-rays scattered from the planes of the layered silicates in accordance with the Bragg equation:

$$d = n\lambda/2\sin\theta \qquad (3.18)$$

where d is the distance between the layers, λ is the wavelength of the X-ray and θ is the measured scattering angle

The structure of the interfaces in montmorillonite can be determined from the values of the θ angles determined during X-ray scattering. The distance between the layers is measured for the pristine montmorillonite, and is about 1.2 to 1.4 nm for different samples of montmorillonite. If wet-mixed layered silicates are exfoliated, the individual platelets are present without neighbors, so that the size of the aggregates can be determined. Of course, the polarity of the rubber also plays a crucial role: the greater the polarity and viscosity of the rubber, the further apart the exfoliated platelets are.

The equilibrium layer spacing for an organically modified layered silicate depends both on the cation exchange capacity of the layered silicates and on the chain length of the organic cation. Traditional structural characterization to determine the orientation and arrangement of the alkyl chain involves X-ray diffraction [193, 194]. An advantage in improving the physical properties of the nanocomposite is obtained when the polymer chains are intercalated into the galleries of the montmorillonite. In the exfoliated structure, the layers are dispersed in the rubber matrix and the d-spacing is larger. The organo-clay structure has been determined in NR [195], SBR and XSBR [196–198].

Multi-walled carbon nanotubes are a second example of a filler with extremely high aspect ratio of 100 to 1000. Single-walled carbon nanotubes form bundled aggregates and have less of a tendency to disperse in rubber compounds [199]. Very strong van der Waals forces hold the tubes together and dispersion will not occur in such a filled system. The length of multi-walled carbon nanotubes formed during CVD manufacturing is reduced during the mechanical processing and dispersion in the rubber matrix. The cleavage of multi-walled carbon nanotubes reduces the aspect ratio to lower values. A comparison of the diameter and the mean length of the chemical vapor deposition multi-walled carbon nanotubes has been carried out on Nanocyl 7000 and Baytubes C150 HP [200, 201]. The former has a diameter of 9.5 nm and a length of 1.5 μm while the latter has a larger diameter of 18 nm and a length of 1 μm. If the diameter of the carbon nanotube increases, the aspect ratio is decreasing accordingly. The former has an aspect ratio of 158 and the latter of 55. The length of the multi-walled carbon nanotubes in a rubber depends on precautions taken during mechanical incorporation and the breaking strength of the filler, as shown in Figure 3.23 [202].

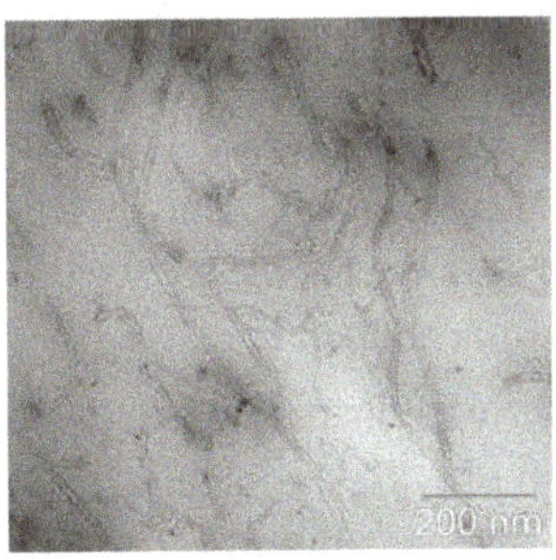


**Figure 3.23**
Aspect ratio for 2.5 vol.% multi-walled carbon nanotubes in HNBR (34% ACN)

The length of 400 to 800 nm achieved leads to a high aspect ratio, as defined by the length of the multi-walled carbon nanotubes divided by the diameter of the tubes. Multi-walled carbon nanotubes exhibit a variety of high aspect ratios that vary with the diameter of the external tube and the length of the nanotube. This is also reflected in the reinforcing properties on the rubber matrix that contains the multi-walled carbon nanotubes. The degree of multi-walled carbon nanotube alignment in the composite is governed by two factors:

- the diameter of the carbon nanotubes
- the content of the carbon nanotubes

A smaller diameter can enhance the degree of multi-walled carbon nanotube alignment, due to greater extensional flow, while a higher multi-walled carbon nanotube content reduces alignment, due to agglomeration and restriction of movement of neighboring carbon nanotubes [203–205]. Multi-walled carbon nanotubes tend to aggregate during the manufacturing process. Homogenous dispersion of the nanofillers can be crucial for obtaining high reinforcement, because of the orientation in the elastomer. The mechanical properties may be enhanced along the aligned direction, but sacrificed in the perpendicular direction. Nanocomposites that contain multi-walled carbon nanotubes aligned with the direction of loading will display the best physical properties in these directions. It has been observed that, after being subjected to acid and ball-milling, carbon nanotubes disperse well in the rubber matrix.

## 3.8 Electrical Properties of Carbon Blacks and Carbon Nanotubes

The chemical structure of carbon blacks and multi-walled carbon nanotubes allows the transport of electrical charge carriers through the filler network that forms at a higher filler concentration in a rubber compound. The formation of $sp^2$-hybridized carbon atoms allows high electrical conductivity. Carbon black and multi-walled carbon nanotubes are generally electrically conductive materials because of the highly conjugated bonding scheme present in the crystalline region of carbon atoms. In the case of carbon black, increasing the surface area and the structure both contribute to increasing the electrical conductivity. If the filler is formed with a longer reaction time, more $sp^2$-hybridizations can arise and higher electrical conductivity is achieved. In general, the electrical conductivity of these types of fillers depends on the graphitization of the carbon atoms, which is a property of the filler manufacturing.

The higher this proportion of hybridization is, the higher is the electrical conductivity of the material. Single-walled carbon nanotubes are able to provide a high electrical conductivity on a par with metallic copper. Multi-walled carbon nanotubes exhibit slightly lower electrical conductivity, but it is significantly higher than that of active carbon black or acetylene carbon black.

Many applications require electrical conductivity of the rubber vulcanizate in order to prevent electrostatic charging or to ensure higher electrical conductivity in an elastic system. Electrical conductivity, while commonly treated as an end-use property of carbon-black-filled compounds, has been proven as valuable for many applications. Electrical resistivity measurements have tended to be a subject of scientific interest rather than a routine method for determining the degree of dispersion of rubber compounds [206–209].

In general, the electrical conductivity (or, conversely, the electrical volume resistivity) exhibits a non-linear increase (or a decrease) with change in filler concentration. At low filler concentrations, the conductive filler particles are separated from each other and the electrical conductivity is dominated by the insulating properties of the rubbers. With increase in carbon black or multi-walled carbon nanotube concentration, local clusters form in the rubber matrix. These clusters form a three-dimensional filler network that shows an exponential increase in electrical conductivity known as the electrical percolation threshold. The stepwise changes in electrical resistivity (or conductivity) are commonly described by the percolation theory [210]. It is assumed that a carbon black network forms at a certain concentration of the filler (percolation threshold). At this critical point, the first conductive path through the compound forms, making the compound significantly less resistive. The theory predicts a correlation between the conductivity of compounds, σ, and the volume fraction of the conductive filler:

$$\sigma = \sigma_0(\Phi - \Phi_c)^{\alpha} \tag{3.19}$$

where $\sigma_0$ is the conductivity of the filler particles, Φ the volume fraction of the filler, $\Phi_c$ is the volume fraction of the filler at the electrical percolation threshold and α is a quantity determining the power of the conductivity increase above $\Phi_c$

With this power-law equation, the electrical conductivity can be described as a function of filler concentration. The conductivity of a filled polymer increases with the type of conductive filler. In order to ensure the transport of charge carriers, it is not necessary for the carbon black aggregates to touch. At the percolation threshold, the charge carriers responsible for conducting the current through the rubber sample can "tunnel" through isolating polymer layers that separate the individual aggregates. It has been theoretically deduced that the current can be transported through the tunnel effect at an inter-aggregate distance of around 2.5 nm [211]. This distance corresponds quite well to the value deduced from 3D-TEM studies of filled compounds. The tunneling current, measured in DC or low-frequency AC, is exponentially proportional to the inverse gap width between neighboring aggregates [212]. The electrical threshold of a conductive filler is shown in Figure 3.24.

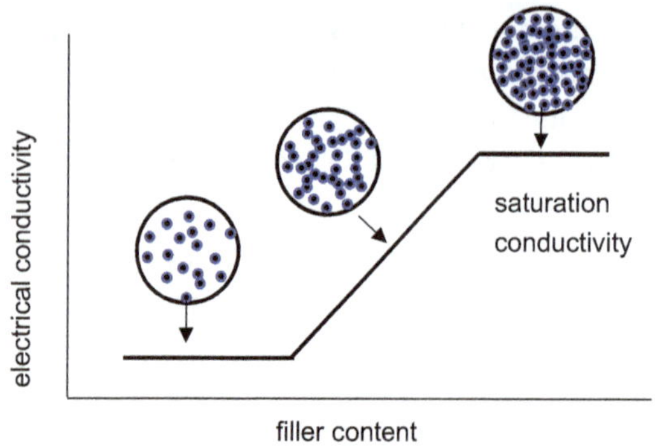


**Figure 3.24**
Electrical percolation threshold versus filler loading

Theoretical concepts state that, at constant filler loading, greater mixing efficiency leads to a lower conductivity range, provided that the concentrations are below the percolation threshold. Efficient mixing leads to a more uniform distribution of the dispersed particles. This is the result of fewer filler-filler interactions and larger spacing between the particles, which directly translates into lower electrical conductivity. The formation of filler networks, as evidenced by the sharp stepwise increase in electrical conductivity, can be considered a true indicator of the micro-dispersion of conductive fillers. As a rule of thumb, the greater the volume fraction at which percolation takes place, the smaller the agglomerates and the better the micro-dispersion of carbon black. It is worth noting that, at filler loadings greater than the percolation threshold almost all physical properties of the filled compound change significantly (i. e. the storage modulus increases exponentially.)

The experiments can be performed either using a DC current or an AC current source. For DC experiments, usually only the current source and a voltmeter are needed. The electrical conductivity can be measured through the vertical thickness of thin specimens. For good contact between the sample and the copper electrodes, silver paste should be used. AC methods require more sophisticated instrumentation. However, the DC method is considered difficult and unreliable, due to the large variation in measurements and their sensitivity to contact conditions. Fixed-frequency AC methods have been used to avoid difficulties encountered in measuring DC resistance. In order to avoid the effects of parasitic currents and to minimize the errors associated with the measurements, it is advisable to use the "alternate polarity" method [213].

Impedance spectroscopy with a frequency sweep appears to be a more interesting technique. This perturbation-response technique wherein the test system (rubber sample) is perturbed (from an initial steady-state condition) by a periodic electrical stimulus and the electrical response is monitored provides reliable values on conductivity. In general, the relationship between input and response is described by the system transfer function, which contains information about the system under study. The analysis is usually performed in the frequency domain, and the transfer function in this case becomes the electrical impedance Z* [214]:

$$Z^*(\omega) = Z'(\omega) + iZ''(\omega) = V(t)/I(t) \tag{3.20}$$

where ω is the angular frequency (in rad $s^{-1}$), V(t) is the periodic voltage perturbation, and I(t) is the corresponding periodic current response

The instrumentation used for this type of measurements consists of a broadband frequency generator of variable signal amplitude. Cylindrical test specimens with carefully cleaned surface are used. For perfect electrical contact, gold contacts are sputtered on the previously cleaned surface. The resonance frequency can be determined as a frequency at the maximum of Z″(ω). The resonance frequency correlates with the degree of carbon black dispersion.

### 3.8.1 Electrical Conductivity of Carbon Black

In contrast to TEM and AFM, which provide information about the size and shape of the nanoparticles (aggregates and agglomerates), the measurement of electrical conductivity (or the electrical resistivity) provides information about the separation of individual aggregates. The better the separation of the conductive particles by isolating rubber layers, the greater the electrical resistivity of the compound and vice versa. The following parameters of carbon black contribute to a high electrical conductivity of filled compounds [215]:

- high structure (OAN)
- low primary particle size
- high specific surface area
- high porosity (difference between $N_2SA$ and CTAB-SA)

Electrical resistivity measurements have been the subject of scientific interest rather than a routine method for determining the degree of dispersion of rubber compounds [216, 217]. The experiments can be performed either using a DC or an AC current source [218, 219]. For the DC experiments usually the current source and a voltmeter is all that one needs. The electrical conductivity can be measured through the vertical thickness of thin specimens. In order to ensure good contacts of the sample and the copper electrodes, silver paste should be used. AC methods require more sophisticated instrumentation. Essentially, three types of measurements can be performed by the DC technique, namely volume resistivity, surface resistivity, and current vs. voltage measurements. Simple volume resistivity measurements can provide information on dispersion and can classify the polymers in order of efficiency of carbon black dispersion. Measurements over a wide range of temperatures and strains have been reported [220]. For carbon black, three distinct zones are observed in the evolution of resistivity as a function of carbon black loading:

- the insulating zone with very high resistivity (approx. $10^{18}$ Ω cm)
- the percolation zone with the resistivity sharply decreasing over orders of magnitude
- the conductive zone with a resistivity lower than $10^{6}$ Ω cm

It has been reported that, irrespective of the chemical nature of the polymer, the electrical percolation threshold for carbon blacks increases with the size of primary particles and aggregates and decreases with the surface activity, as shown in Figure 3.25 [221].

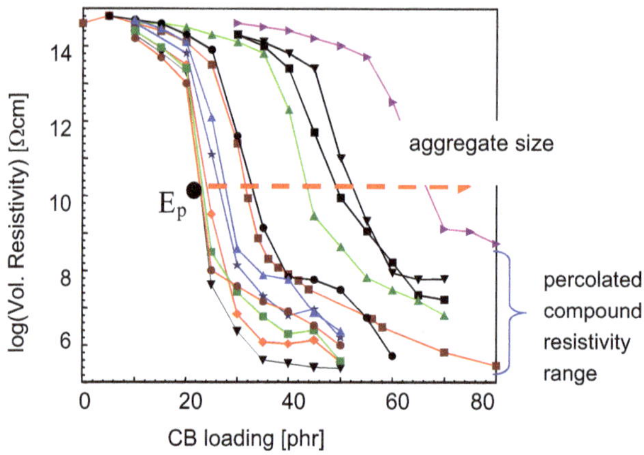


**Figure 3.25** Electrical percolation for carbon blacks [221]

The lowest percolation threshold for the active carbon blacks with the highest specific surface area is found at 20 phr, increasing to 60 phr for high concentrations of semi-reinforcing carbon blacks with low specific surface area. The exponential curve changes with the type of carbon black. Thus, coarse blacks (N550–N770) are easier to disperse than fine blacks (N330–N110). It is noteworthy that the electrical conductivity is some orders of magnitude lower than for acetylene carbon black, even at high carbon black loading. This strongly suggest that a thin layer of polymer covers the aggregates and separates them in the percolated carbon black networks.

AC measurements are invaluable for resolving small differences in filler micro-dispersion and determining the electrical percolation threshold of cured compounds [222]. It has been reported that contact resistance, temperature and humidity can influence the reproducibility of the measurements. The slope of the conductivity as a function of the frequency is attributed to abnormal diffusion of the charge carriers through the fractal carbon black network that is formed in rubber compounds. The plateau with normal diffusivity of the charge carriers increases with filler loading, whereas the slope that indicates the abnormal diffusivity decreases. The percolation threshold is easily measured by the stepwise increase in conductivity as a function of filler loading [223].

Attempts have been made to monitor mixing of CB in rubbers by means of the time-dependent development of electrical conductivity. Several experiments were based on the concept of placing one electrode in the mixing chamber wall and the other in the flight of a rotor. Changes in conductivity during laboratory mixing of carbon-black-filled compounds have been examined as an online method for measuring carbon black dispersion [179, 224]. A better solution for performing reliable online measurements is described in [179]. The advance is that an insulated ring electrode has been placed in the tip of a rotor flight and the counter electrode on the wall of the mixing chamber. Thus, a high number of signals can be recorded within one revolution of the rotor. This method makes it possible to evaluate a large number of readings and thus ensure a statistically reliable conductivity value. The advantage of the online measurement is that the mixing process can be monitored continuously from the admission of the carbon black to the end of the process.

### 3.8.2 Electrical Conductivity of Carbon Nanotubes

Due to the long length of $sp^2$ hybridization of the carbon atom tubes and the high aspect ratio, the electrical conductivity of the rubber composite is greatly increased in contrast to the carbon black and is shifted to small filler concentrations [225]. When the electrical conductivity of SWCNTs is measured, the results can vary from $10^2$ to $10^6$ S/cm, the exact value depending on the formation of bundles. In the case of MWCNTs the electrical conductivity can vary from $10^3$ to $10^5$ S/cm. The electron mobility, too, is higher in SWCNTs at $10^2$–$10^6$ $cm^2/Vs$. For MWCNTs, the measured electron conductivity is $10^4$ to $10^5$ $cm^2/Vs$ [226]. In the case of multi-walled carbon nanotubes, the electrical conductivity is much higher than for carbon black and depends on the concentration of the almost graphitic planes.

The influence of the mixing process has been studied for multi-walled carbon nanotubes. Increasing the rotor speed from 20 to 60 rpm boosted the surface resistivity from $10^4$ to $10^{11}$ $\Omega/cm^2$. For a mixing time of 7 min, the surface resistivity decreased with increase in rotor speed. The cooling rate, too, changed the surface resistivity [227]. The incorporation of carbon nanotubes into the rubber compounds offers an attractive route to minimize the percolation threshold, as their high aspect ratio and excellent electrical conductivity in turn facilitate the formation of conductive filler networks at very low filler concentrations. For MWCNTs, the percolation threshold is observed at concentrations of 1–1.5 vol.%, depending on the carbon nanotube diameter, the chemical nature of the rubber and the manufacturing technique for preparing the test samples [228]. As the diameter of the multi-walled carbon nanotube increases, the percolation threshold shifts to higher concentrations and the electrical saturation value also decreases. Multi-walled carbon nanotubes increase the electrical conductivity by more than active carbon blacks and electrically active carbon blacks. When a multi-walled carbon nanotube is employed, the difference in electrical percolation threshold inherent in the dispersion in the chemically different rubber and the incorporation can be seen, as shown in Table 3.11 [229].

**Table 3.11** Electrical Percolation Threshold of MWCNTs in Various Rubbers and Processing Consequences [229]

| Rubber | Internal Mixer | Internal Mixer + Two-Roll Mill | Wet-Compounding |
|---|---|---|---|
| NR | 1.2 | 0.7 | 0.4 |
| FKM | 1.2 | 0.9 | 0.7 |
| QM | 1.3 | 1.0 | – |
| NBR (28 ACN) | 1.7 | 1.1 | – |
| NBR (34 ACN) | 1.5 | 0.9 | 0.7 |
| NBR (39 ACN) | 1.3 | 1.0 | – |
| EPDM | 2.0 | 1.3 | – |

The influence of the manufacture of the rubber–multi-walled carbon nanocomposite and the interaction at the phase boundary as well as the polarity of the rubber can be determined by the concentration of the electrical percolation thresholds. The percolation thresholds can be significantly lowered by additional mixing on the roller or, even better, by the wet compounding process, which leads to improved dispersion in the rubbers.

In NBR/N550 (35 phr), the electrical conductivity is very low ($10^{-9}$ S/cm), but adding 5 phr multi-walled carbon nanotubes increases the AC conductivity by eight orders of magnitude to reach $10^{-1}$ S/cm and higher values [229]. In NR/HD silica, the addition of 10 phr carbon nanotubes increases the AC conductivity to $10^{-2}$ S/cm [230]. The use of hybrid fillers formed by a usual filler and a multi-walled carbon nanotube can therefore confer electrical conductivity on precipitated silica compounds.

## 3.9 Surface Activity

The third property that essentially determines filler dispersion and reinforcement in rubber compounds is surface activity, which, however, is not characterized by a specific method in international standards. However, there is no property of a filled elastomer that does not clearly show this interaction between the soft rubber phase and the hard filler phase in the mixed compound or in the vulcanizate. There are a number of properties that point directly to the influence of surface activity, which in comparison clearly show the advantage in phase interaction between the filler and the rubber matrix. The surface activity dominates the rubber-filler interaction as well as the filler-filler interaction. Whereas particle size of primary particles and "structure" of the aggregate are well defined properties with standard industrial definition, surface activity is less well described. This makes it somewhat difficult to define the surface energy. The surface activity is an intensive parameter that dominates the polymer-filler interaction, filler-filler interaction, and even filler-ingredient interaction. There is hardly any physical property of a filled rubber compound that does not show an influence exerted by surface activity and a degree of improvement in reinforcing properties and lifetime of the elastomer. The reinforcing ability of the carbon black or precipitated silica should therefore be the product of the specific surface area and the surface activity.

Surface activity encompasses the structural physical and chemical features of the filler surface that are used to improve the physical-chemical properties of the rubber compound. The surface activity not only affects the rubber-filler interaction, but also has a direct effect on filler-filler interaction, which is important for the dynamic properties of the elastomer. The increase in filler-filler interaction can lead to a decrease in filler dispersion during mechanical mixings, causing a decrease in reinforcement of dynamic properties and the fracture mechanical properties that determine the service life.

All fillers can undergo physisorption within the rubber chains. This process depends on the surface activity of the filler and the chemical structure of the rubber that include specific interaction of segmental units with the surface of the filler, such as preferential interactions of styrene groups with carbon black. In the case of precipitated silica, which shows high interparticle interaction due to the silanol groups on the filler surface, chemisorption with organo-silanes becomes important. Silanization transforms a hygroscopic filler that has strong surface activity into a hydrophobic filler that interacts well with the non-polar rubber and requires good dispersion. Depending on the chemical structure of the filler and the polarity of the polymer, physisorption is different. Due to the different chemical structure, chain length, molecular weight distribution of the rubbers, the interactions with the filler can take different forms that are also important for the surface activity of the filler.

The solubility parameter of the polymer, δ, also plays a decisive role in the evaluation of the interaction with the filler. The thermodynamic interaction parameters, δ, are well described and can be precisely determined. The saturated rubbers have the smallest δ parameter, and the unsaturated rubbers follow with higher δ parameters. In the case of SBR with its styrene groups and the NBR with its acrylonitrile groups, the δ parameter increases in proportion with the content of the more polar structure of the side groups. It can be assumed that the fillers have a significantly higher delta parameter that depend on the polarity of the surface. The functionalization of carbon black is reducing the δ parameter and the difference to the SBR is diminishing. Therefore, the δ parameter of silica is higher than that of carbon black and the silanization of silica reduces the value below that of carbon black. This thermodynamic relationship created by the solubility parameters is shown in Figure 3.26.

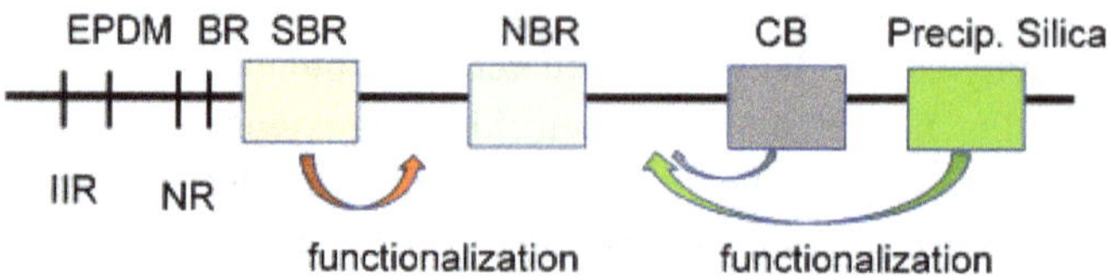


**Figure 3.26** Solubility parameters for rubbers and fillers

The interaction between the rubber and the filler depends on the difference in the δ parameters. If the δ-parameter difference between saturated rubbers and filler is large, there is less intense interaction. The delta-parameter of the unsaturated rubbers is always smaller than that of the saturated filler to the filler. Therefore, the interaction of the unsaturated rubber and the filler is higher and the filler dispension is better in the unsaturated rubber. The silanization of the precipitated silica leads to a reduction in the δ-parameter and the rubber-filler interaction in S-SBR/cis-BR blends yields a high level of filler dispersion. A distinction is made between two influences that describe a good interaction:

- surface roughness
- surface energy

Although the importance of surface activity has been recognized for decades, our knowledge of surface activity is still not complete today. The knowledge is based on the induced dipole-dipole forces and on the orientation effects of the dipoles-dipole interaction, which are described in well-known textbooks [231–233] and in many reports [234–236].

### 3.9.1 Surface Roughness

The morphological similarities of carbon black aggregates with the fractal geometry in nature led to the expression of the surface roughness with the surface fractal dimensions to explain the reinforcement in elastomers in theoretical studies that attribute the influence of surface roughness to a lower rubber-filler interaction [237]. It has been assumed that the random coils of the polymers can better arrange themselves in the voids of the aggregates by reducing the adsorption entropy, but the heat of adsorption remains unchanged. For geometrical reasons, it can be seen that, on the real surface with a certain roughness on a small length scale, fixation of the chains could take place with a small loss of conformational energy. This means that an adsorbed macromolecule adapts better to a rough surface with the same interaction. The chain molecule should "worm its way" into the roughness of the surface without major energy losses, resulting in a higher contribution of interaction energy between the filler and the rubber in the energy balance.

This rather mechanistic view of reinforcement is contradicted by a determination of the surface fractal dimension, $d_s$, according to the yardstick method [236, 239], which is a measure of mechanical roughness, and revealed similar values for all furnace blacks, ranging from $d_s = 1.59$ to 1.64. The very similar surface fractal dimensions could not explain any influence of the surface roughness, from which a difference for the reinforcement of the semi-reinforcing and the reinforcing carbon blacks can be seen. When the corresponding graphitized carbon blacks are measured, the fractal surface dimensions decrease to $d_s = 1.23$ for the graphitized carbon blacks. The surface fractal dimensions for ASTM carbon blacks and the graphitized carbon blacks are shown in Figure 3.27.

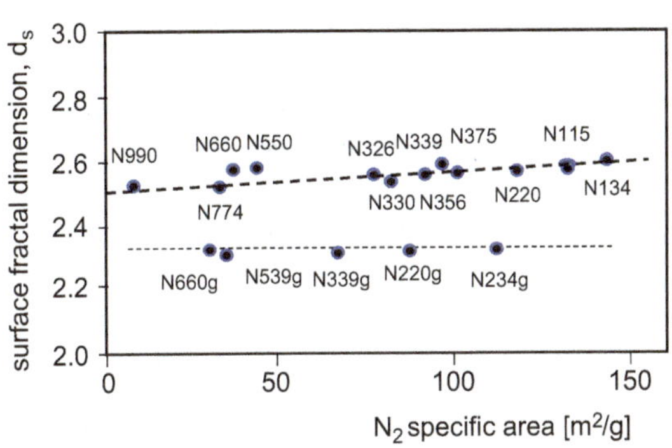


**Figure 3.27**
Comparison of diameter fractal dimensions, $d_s$, for ASTM carbon blacks and the corresponding graphitized carbon blacks [236]

Consequently, statistical gas adsorption measurements carried out at corresponding temperatures have been evaluated. It has been found that, for length scales of 5–6 nm, the furnace carbon blacks exhibit almost the same surface roughness, regardless of their structure and primary particle size. The minor differences are not large enough to explain the theoretically expected reinforcement effects in active carbon blacks.

### 3.9.2 Surface Energy of Fillers

The surface energy of the filler is determined by the chemical structure of the surface of the filler which interacts with the type of rubber. These filler-rubber interactions limit the mobility of the rubber chains in the area of the filler surface. Depending on the intensity of the interactions and the distance from the particle surface, one can distinguish between a firmly bound immobile rubber layer and a less firmly bound rubber shell. As the temperature increases, interaction with the rubber decreases and interaction between the filler particles stays at the same level. While the surface chemistry aspect of the fillers surface activity is today sufficiently characterized, a satisfactory description of the physical chemistry of surface characteristics is still lacking. Present knowledge of the relationships between surface activity and rubber properties is far from complete. The thermodynamic problem has been approached with different methods that pursue different goals. One of the reasons is that there is only a limited number of tools available that can be used to characterize the filler surface energetically. Depending on the type of filler, an interaction occurs between filler and rubber that is based on non-specific dispersive van der Waals interactions and on specific interactions that are generated by dipole-dipole forces. These interactions bind the polymer molecule to the surface of the filler through a physical interaction and have a high strength that determines the detachment of the two phases at a given temperature. Depending on the degree of this interaction, the chains on the filler surface may be considered to be strongly or weakly bound.

A distinction is made between the interactions that differentiate two materials by their surface energy. It is accepted that the nature and strength of molecular interactions vary in different materials, depending on the composition and structure of the molecules. They are determined by the surface free energy, which is determined using the contact angle method. When a drop of liquid is on a flat solid surface, it may remain as a drop of finite area, or it may spread over the surface. The spreading energy is gained by the interaction of the liquid drop and the solid surface [240–242]. The conditions for wetting can also be determined by a drop of liquid placed on a solid surface estimated from the heat of immersion and the nature of the solid surface [243, 244]. The most frequent method used is inverse gas chromatography (IGC), as described in [245]. To an extent depending on the interaction between closely spaced molecules, different types of cohesion forces act between the filler and the rubber matrix.

### 3.9.2.1 Adsorptive Interactions with Solvents

The condition for wetting of the filler with the liquid is that the energy gained in forming unit areas of the filler-liquid interface should be equal or smaller than the required to form unit area of liquid-vapor surface. A graphitic carbon black exhibits a small amount of the energetical heterogeneity responsible for the limited adsorption of water [246]. This method has been used to study the oxidation of carbon black [247, 248]. It was found that isoprene liquids on carbon black N330 increase the immersion heat on the number of carbon atoms [249].

The behavior of silica in rubber is influenced by surface interactions of the silanol groups. To understand the complex nature of these phenomena, a review of adsorption interaction between simple molecules and silica surface is helpful. Heats of adsorption of various organic materials on silica have been studied using a fumed silica with a specific surface of 93 $m^2/g$. The solvents with functional groups of high electronic density interact with the silanol groups via hydrogen bonds, such as methanol, aniline and benzonitrile. The heat of wetting of silanized silica and silica surface was estimated at −130 $mJ/m^2$ and −190 $mJ/m^2$, respectively [250]. The strongest interaction with silica occurred with amines, as a result of acid-base interactions. These adsorptive interactions play a significant role in the preparation and use of silicas. However, the interaction with saturated hydrocarbons, which have low interaction with silica, is different, because of the presence of silanol groups that cannot interact efficiently with non-polar polymers.

#### 3.9.2.2 Sorption Energy

A seemingly simple method for determining filler surface activity is that of inverse gas chromatography (IGC) which delivers a global result on the interaction between the vapors of solvents in infinite dilution. If the surface of the filler is energetically heterogeneous, the retention times show a tail. In these experiments, the retention times, $t_R$, are measured at the peak maximum, which is not entirely correct, due to tailing of the chromatographic peaks that increases with the activity of the fillers. If the gravity centers of the peaks are taken as variables, the slope of the relationship will be higher, which makes this sorption relationship clearer.

The measurement yields the retention time, $t_R$, of solvent vapors, which is a direct measure for the degree of interaction of the solvent with the filler surface. The value of the retention volume indicates the degree of interaction with the filler, but it is not a thermodynamic quantity. The thermodynamic parameters can be derived from the retention time using Henry's law. The retention time, $t_R$, can be converted into the thermodynamic sorption functions. The free energy of sorption, $\Delta G_s$, the heat of sorption, $\Delta H_s$, and the entropy of sorption, $\Delta S_s$, are able to describe the adsorption process on the filler surface. The variation of the standard free energy of one mole of molecular probe from the reference pressure $p_0$ to the adsorbed state at equilibrium with a pressure p of the mobile phase is expressed by the equation:

$$\Delta G_S = -RT \ln (P_0/P) \qquad (3.21)$$

According to the second law of thermodynamics, the sorption enthalpy is described as the derivative of the logarithm of the retention volume to the reciprocal absolute temperature:

$$\Delta H_S = R\, d\,(\ln V)\,/d(1/T) \qquad (3.22)$$

At zero surface coverage, the free energy and the enthalpy are used to describe the sorption process. A linear relationship between the free energy of sorption of homologous n-alkanes and the number of carbon atoms is obtained. The free energy of sorption for one $CH_2$ group can be obtained from the slope of the free energy of sorption $\Delta G_s$ of a homologous series of n-alkanes.

$$\Delta G_{CH2} = -RT \ln V_{(n)} / V_{(n+1)} \quad (3.23)$$

An IGC study of different carbon blacks have demonstrated that, with increase in primary particles, the attraction energy of hydrocarbons on the filler surface decreases. The free energy of sorption affords a way of ranking the filler activity for interaction of carbon blacks with low molecular weight alkanes, as shown in Table 3.12 [251].

**Table 3.12** Energy of Sorption ε and the Surface Energy γ of Furnace Carbon Blacks [251]

| Carbon Black | ε (kJ/mol) | γ mJ/m² |
|---|---|---|
| N110 | 27.2 | 87.6 |
| N234 | 26.3 | 84.8 |
| N330 | 25.4 | 82.0 |
| N762 | 24.8 | 80.4 |

The results indicate a high concentration of active sites on the carbon black surface with small aggregates. For the non-polar and polar hydrocarbons, the dispersive and polar component of the surface energy, $\gamma_s^d$ and $\gamma_s^P$, can be separated:

$$\gamma_s = \gamma_s{}^d + \gamma_s{}^{sp} \quad (3.24)$$

where $\gamma_s$ is the free interaction energy, $\gamma_s^d$ is the dispersive energy component and $\gamma_s^{sp}$ is the specific energy component

The dispersive surface energy, $\gamma_s^d$, is determined via the adsorption increment of a $CH_2$ group from the gradient of the sorption enthalpy of a homologous series of alkanines [252]. This way to present the sorption process is mainly used to demonstrate the difference between carbon blacks, which adsorb primarily non-polar hydrocarbons, and precipitated silica, which can adsorb preferentially polar molecules. The dispersive component of the surface energy ($\gamma_s^d$) for furnace carbon blacks increases linearly with the nitrogen surface area, showing that the main effect is governed by the primary particles of the aggregates [252]. The increase in $\gamma_s^d$ can be understood as increased surface activity that is related to a higher content of amorphous carbon in the primary particle. The reinforcing efficiency is higher when the primary particle size is growing smaller. If the crystal edges become larger, $L_c$, increasing from 1.5 to 1.8 nm, the dispersed component decreases, because the organization is less developed in the smaller primary particle. Even pelletization of the filler causes a difference in dispersive surface energy. The dispersive energy, $\gamma_s^d$, increase with the specific surface area in a linear fashion. Wet-pelletized carbon black show a higher dispersed surface energy than dry-pelletized carbon black. The higher values of dispersive energy for wet pelletized carbon blacks may be associated with high efficiency of the Soxhlet extraction at high temperatures, resulting in more exposed active centers. The polar surface energy $\gamma_s^{sp}$ follows the sorption enthalpy of polar molecules compared with the alkanes having the same molar surface area. In contrast to the dispersive component of the surface energy of silicas, the specific component, $\gamma_s^{sp}$, has much higher surface energy values. Precipitated silica has small dispersive energies, $\gamma_s^{d}$ that reduce the interaction with the polymer, but high specific surface energies that suggest strong aggregate interactions and the development of a filler network.

Precipitated silica shows a preference on the part of the polar silanol groups to interact with polar species. For this reason, the free energy of sorption per $nm^2$ decreases in the sequence: acetonitrile > benzene > heptane [253]. The difference in the polarity of the molecular probes underscores the high filler-filler interaction in precipitated silica, due to the hydrogen bonds between the aggregates, leading to the formation of a strong filler network that becomes effective at much lower concentrations. After silanization of the silica, the polar portion of the surface energy is greatly reduced and the dispersive portion $\gamma_s^{d}$ increases, resulting in the silanized silica acquiring a surface similar to that of the active carbon black. When the precipitated silica is treated with organo-silanes, the specific interaction $\gamma_s^{sp}$ decreases drastically from 16.6 kJ/mol to 10 kJ/mol for benzene at 130 °C. The dispersive energy distribution, $\gamma_s^{d}$, obtained directly from retention measurements with benzene and cyclohexane as molecular probes are increased from VN2 to VN3 at each temperature, as a function of primary particle size. The increase in particle size points towards an attenuation of the surface energy. The difference in $\gamma_s^{d}$ between precipitated and fumed silica may be related to their surface complexity. It seems that the concentration of silanol groups, which have been considered as energy sites is not the dominant factor for the dispersive energy, $\gamma_s^{d}$. On reason for the higher $\gamma_s^{d}$ of fumed silicas may be the flatter surface of the particle. Analysis of the desorption profiles of the gas chromatographic peaks provides adsorption isotherms from which a surface energy distribution is estimated.

For multi-walled carbon nanotubes, the IGC method was applied to describe the dispersive surface energy, $\gamma_s^{d}$, of different producers. For the type produced by Nanocyl and from Arkema, the numbers are 114 $mJ/m^2$ [254]. This method was used for comparative measurements of the free energy of sorption of MWCNTs for the normal alkanes (C5–C8) at 150 °C and carbon black N330 and N550. It was shown that the energy of sorption lies in between that of carbon black 550 and N330 [255], as shown in Figure 3.28.

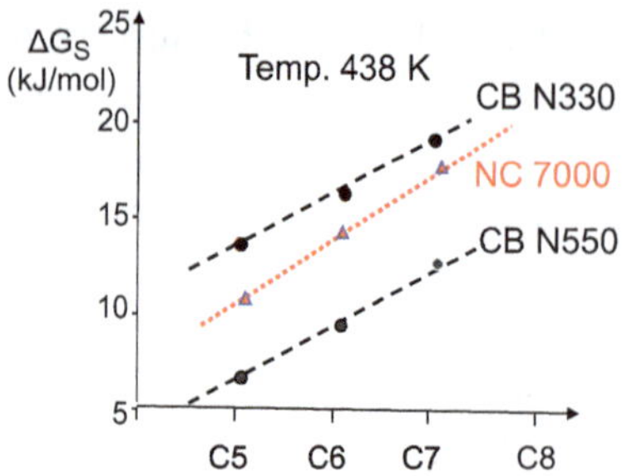


**Figure 3.28**
Comparison of free energy for carbon black and multi-walled carbon nanotubes [242]

With increase in the number of carbon atoms, the free enthalpy of sorption approaches that of carbon black N330, demonstrating the high interaction capacity of the multi-walled carbon nanotubes. In general, $sp^2$ hybridization prefers to form planar σ-frameworks with its neighbors, leaving the orbital perpendicular to the σ-bonds. In CNTs, the $sp^2$-hybridized carbon atom is pyramidalized, meaning that the angle between the orbital is greater than 90°. The pyramidalization angle $\theta_p$ of a carbon atom in the curvature induces local weakening of π-conjugation, which in turn results in the higher chemical reactivity of CNTs. The extent of curvature-induced weakening of π-conjugations in the case of multi-walled carbon nanotubes depends on the diameter of the tube, and the sidewall of the multi-wall carbon nanotubes could be more reactive than a planar graphene sheet and more reactive than graphitized carbon black.

### 3.9.2.3 Surface Energy Distribution

Adsorption studies were conducted as long ago as the 1950s to provide valuable information about the surface activity of carbon blacks. A sensitive variable is the differential heat of adsorption of nitrogen, which shows an energy distribution depending on the degree of occupancy of the carbon black surface [256–259]. At a low occupancy level of 10%, only high-energy adsorption centers are occupied, while a higher occupancy also leads to less active centers, which make a difference between an ASTM carbon black and a graphitized carbon black.

By measuring the heat of adsorption calorimetrically, it has been shown that the interaction with various adsorbates is strong at very low coverages and decreases at coverage of about 0.2 monolayers. Furthermore, it has been found that the adsorption enthalpy of hydrocarbons increases linearly with the number of $CH_2$ groups and is increased by a defined amount due to the influence of double bonds and the influence phenyl rings [260]. The determination of the isosteric adsorption enthalpies at different temperatures is more complex, but the result is more reliable [261]. A distribution of the four different adsorption centers located on the surface of the carbon blacks can be deduced from the adsorption isotherms. Inverse gas chromatography has also been used to determine the energetic heterogeneity of iso-dimensional fillers. A general equation allows the retention data of chromatographic measurements to directly reveal the energetic heterogenity of the absorbent's surface. Differentiating the retention volume presented as a function of pressure determines the energy distribution functions [262–264]:

$$\chi(\varepsilon) = -(p/RT)^2 \, \partial V_n(P,T)/\partial p \quad (3.25)$$

where $\chi(\varepsilon)$ is the adsorption energy distribution, P is the pressure, $V_n$ is the retention volume, and p is the pressure of the molecular probe

Studies have shown that the energy distribution for carbon black and precipitated silica is inhomogeneously distributed according to the surface energies. For carbon black, it been found that the energy distribution of the adsorption of benzene suggests that the peaks show a multimodal curved shape, with the highest value of 30 kcal/mol for N110, N220, N330, N550 and N990 at a temperature of 50 °C [265]. The shape of the distribution curves is a skewed Gaussian curve with a broad tail toward higher energies. The graphitization of carbon black shifts the maximum of the energy distribution to higher values, a fact which cannot be understood easily.

The energy distribution function for the adsorption of precipitated silica with benzene and hexane shows three different peaks as a function of the energy that can be related to the surface chemistry of the silanol groups [260]. It becomes evident that benzene is capable of specific interactions with the silica surface via its localization of a π-electrons. Thus, benzene is able to demonstrate the heterogeneity of the silica surface. For silanized silica, which has no silanol groups on the particle surface, the energy sites distribution measured with the molecular probe shows only a very low peak in the region of 30–37 kJ/mol. Therefore, the silanization lowers the specific component of the surface energy and increases the compatibility of the silanized silica with non-polar rubber.

A more direct measurement of the surface energy distribution can be obtained by volumetric gas adsorption techniques in a complicated method. For the morphological and energetic surface characterization of the filler on an atomic length scale, a theoretical basis developed in 1960 is used to divide the experimental adsorption isotherm into distinct regions to obtain a distribution function [266]. The mathematical expression describing the adsorption isotherm is equal to the integral of the local adsorption isotherms and the distribution function:

$$\Theta(p,T) = \int \Theta(p,T,Q)\, f(Q)\, dQ \quad (3.26)$$

where (p,T) are the experimental adsorption isotherms, (p,T,Q) are the local adsorption isotherms, f(Q) is the energy distribution function and Q is the energy

The experimental adsorption isotherm is measured and the energy distribution is determined from the integral equation using an iteration method [267]. To apply this method a complex apparatus has been constructed at the Institute for German Rubber Technology that is able to work at very low adsorption pressures of the hydrocarbons.

A classic volumetric adsorption apparatus is used, which is equipped with absolute capacitance pressure transducers in the range of $10^{-3}$ mbar to $10^{3}$ mbar. Before adsorption measurements starts, the filler is extracted with toluene and after drying it is degassed overnight at 200 °C at a pressure below $10^{-4}$ mbar. The distribution function is obtained from the 0.001 to 1 monolayer using an analytical numerical iteration to solve the integral of the experimental isotherm. The effect of multi-layer adsorption and lateral interaction between the neighboring adsorbed molecules are taken into account by applying two modifications of Langmuir isotherms:

- multi-layer correction according to the BET concept
- correction due to lateral interactions with neighboring gas molecules

All expressions from the thermodynamic interaction are included to determine the local isotherms [268]. According to the theory it is assumed that different sites with the same interaction energies are arranged in patches which correspond to graphitic micro-crystallites and amorphous carbon in the primary particles. Overall, the sum of the different patches results in an adsorption isotherm from which the distribution of the occupancy can be determined. The integral is normalized to unity, which is important when solid samples with different specific areas are compared.

Static gas adsorption measurements with ethene, ethane, propane, propene, n-butene and 1,3butadiene were performed on a group of different furnace blacks of decreasing specific surface area. The procedure is similar to a BET measurement, but it starts at much lower concentrations of gas pressure [236, 239, 261, 269]. Corresponding to the difference in level and shape of the adsorption isotherms, the amount of highly energetic sites varies significantly at the beginning of the measurement. At low gas pressures, highly active energy sites are covered by the organic vapor. The highest adsorption isotherm has been measured at pressures below 1 mbar for gas carbon blacks, followed by carbon black N220. The number of highly active sites at a given pressure decreases with the specific area from N220 and ends with the graphitized N220 that is composed of $sp^2$ hybridization. The amount of energetic surface sites decrease significantly with the type of carbon black and is a minimum for graphitized carbon black N220. The important result is that all the ASTM blacks demonstrate four different surface energy types. Each energy site corresponds to a topological element on the morphology of a primary carbon black particle, as shown in Figure 3.29 [269].

**Figure 3.29**
Four typical types of surface energies on carbon black grades [269]

The plane of the graphitic micro-crystallites with surface energy measured for ethyne is 15 kJ/mol. This surface site is the most abundant part on the surface of primary particle and increases in inverse proportion with the specific surface area. The next energetic site is the type II, which is formed by “amorphous” carbon with a surface energy of 20–21 kJ/mol. This type II decreases with decrease in specific surface area. “Amorphous” carbon is an important morphological element that provides excellent interactions with rubber chains and leads to good dispersion in rubber compounds. The shorter the formation time for carbon black in the furnace is, the more “amorphous” carbon is formed. The longer the formation time, the higher is the concentration of $sp^2$-hybridized orbitals and that increases the amount of energy sites. It has been shown that this surface energy part disappears in the graphitized carbon black N220. The next-higher surface energy site, type III, is attributed to edges of the graphitic micro-crystallites with a surface energy of 30–31 kJ/mol. The last surface energy site is type IV, which corresponds to slits and cavities between crystallites and a surface energy of 35–36 kJ/mol. Both surface sizes decrease with decrease in specific surface area or increasing primary particle size of furnace carbon black. The contribution of types II–IV increases with the shortening of the reaction time and the disorder present at the surface of the carbon black primary particle. If less time is available for the formation of the carbon black particles, the time for $sp^2$ hybridization of carbon orbitals is significantly shorter and more $sp^3$ hybridizations and smaller micro-crystallites and more gaps between the micro-crystallites are formed. The more carbon atoms present with $sp^3$ hybridization, the better is the dispersion in rubber compounds and the greater is the reinforcement of the elastomer. The highly surface energetic sites (types II to IV) exhibit higher interaction energies, but these are consumed in the formation of filler-filler bonds in the undispersed fillers and cannot compete in the same way as surface energy type II. Gas black, which is produced in small flames, is included in this picture, because the “amorphous” carbon can easily be oxidized during formation. Therefore, the surface energy is reduced and the types III and IV are more active in this carbon black. A summary of the different energy sites on carbon black is provided in Table 3.13, which demonstrates the influence of the manufacturing process and the specific surface area of carbon black [269].

**Table 3.13** Surface Energy Sites in Carbon Blacks [257]

| Carbon Black | I | II | III | IV |
|---|---|---|---|---|
| Gas black | 61 | 2 | 31 | 6 |
| N115 | 69 | 13 | 15 | 3 |
| N220 | 84 | 7 | 7 | 2 |
| N550 | 93 | 6 | 1 | < 1 |
| N220g | 99 | 0 | < 1 | < 1 |

Energy distributions functions, f(Q) for low molecular weight hydrocarbons with saturated and unsaturated molecules of $C_2$–$C_4$ have been measured. The amorphous carbon (type II) is much more pronounced and shifted to higher surface energies. In the case of 1,3-butadiene, the shift is even more pronounced. This shows the specificity of the interactions of the carbon blacks in relation to the degree of unsaturation of the investigated gases, which can be further transferred to the rubber molecules that contain unsaturated polymers. This heterogeneous energy distribution of the adsorption sites has been investigated using special emission FT-IR spectroscopy in the submonolayer range ($\Theta < 0.1$) on carbon black N220 [270]. The shape of the IR adsorption correlates with increase in occupancy with the observed energy distribution. A broad IR adsorption band at approx. 800 $cm^{-1}$ and three bands describing the higher energy places were found. The comparison with the adsorption experiment is obvious, although the IR emission spectrum is somewhat imprecise. It suggests that the reinforcing potential of carbon black is closely related to the energetic sites that can be quantified by the static gas adsorption technique.

The chemical structure of carbon nanotubes is similar to that of carbon blacks, but the surface energy is determined by the energetically more favorable $sp^2$ hybridization, which is lower than that typical of amorphous carbon. Defects in the multi-walled carbon nanotubes due to a 5–7 linkage or a higher proportion of catalytic metals can be assumed. This increases the surface energies and they come a little closer to carbon black.

Static gas adsorption measurements performed on dried precipitated silica show a surface energy site distribution which is somewhat different from that of carbon black. The first challenge is to dry the precipitated silica properly. In addition to the precipitated silica, silanized silica and annealed silica at high temperatures are being investigated [271]. The beginning of the energy spectrum contains the contribution of the siloxane dipoles (Si-O-Si) of surface energy 15 to 20 kJ/mol. The higher energy group with energies greater than 25–37 kJ/mol of 15% is given by the contribution of silanols, but distinct silanol groups cannot be determined from these investigations. When the precipitated silica is heated to 900 °C, the signal of the silanol group disappears and the surface energy is reduced to that of dipolar siloxane groups. Heating of the precipitated silica shows that the high-energy sites disappear and only the signal from the siloxane dipoles remains on the surface. Treatment with monofunctional organo-silanes leads to a similar change in surface energy. The quantitative drop in the surface energy sites falls below that characteristic of the siloxane. This is the effect of shielding of the alkyl chains of the monofunctional silanes on the siloxane surface. This demonstrates an important property of the organo-silanes, one that alters the polar effect of the silicas and creates a filler with a low surface energy. This effect significantly reduces the filler-filler interactions, leading to good dispersion in the rubbers that are needed for the production of passenger vehicle tire treads.

## 3.10 Silanization of Precipitated Silica

A key property of precipitated silica and fumed silica is the reaction of the silanol groups to yield groups that cannot form hydrogen bonds. The generated silanized silica reduces the surface area of the particles, which leads to a reduction in filler-filler contacts and increases the interaction between the filler and the rubber chain. The mixing process assumes the characteristics of a reactor that transforms the silica surface into a less polar one. The silica-silane reaction is a chemical process that occurs during mixing of the silica in the rubber matrix and is essential for lowering the surface tension of the silica. The wide range of organo-silanes provides interactions which are expected to be stronger than that provided by the unmodified silicas.

The chemical process on the surface of silica is crucial for the mixing process. Hence it is crucial to investigate the influence of the chemical reaction in order that production of the rubber compound may be optimized. The silanol groups have to be reduced on the silica surface to ensure both good dispersion with a minimum of organo-silane and, if a bifunctional silane is used, efficient coupling of the silica particles on the non-polar, unsaturated rubber chains [272–275].

The organo-silanes developed in the chemical industry are offered as monofunctionalized silanes or, more importantly, bifunctional silanes. The organo-silanes can reduce the concentration of silanol on the silica surface and provide a chemical link of the silanized silica with the rubber network. Therefore, the use of organo-silanes can significantly reduce the polarity of silica and the viscosity of the rubber compound and increase the physical properties of the elastomer, especially the strength and the abrasion resistance. This combined property of bifunctional silanes was important during the development of "green" tire treads. Bifunctional organo-silanes were the starting point for silanized silica to become an important filler in the passenger car tire industry [276–280].

### 3.10.1 Monofunctional Organo-Silanes

Monofunctional alkyl silanes are used to investigate the silanization reaction of silanols and reduce the concentration of the silanol groups and increase the hydrophobization of silicas. Monofunctional silanes are chemicals that contain a trialkoxy-silyl group linked to an alkyl chain of variable length, as shown in the following scheme:

$(RO)_3\,Si{-}(CH_2)_2{-}CH_3$ $(SiO_3)$
$(RO)_3\,Si{-}(CH_2)_7{-}CH_3$ $(SiO_8)$
$(RO)_3\,Si{-}(CH_2)_{16}{-}CH_3$ $(SiO_2H_{16})$

Depending on the alkyl chain length of the monofunctional organo-silane, the OH groups of the silica surface are diminished and the length of the alkyl chain produces a shielding of the polar surface during the reaction time. The hydrophobization produced by monofunctional silanes, reduces the inter-aggregate interactions of the aggregates and improves the tendency to disperse in the rubber compound. Therefore, the longer the alkyl chains are, the lower the polarity of the silica after the silanization and the better the silanized silica dispersion. This has an immediate influence on the reduction in the viscosity of the rubber compound and a decrease in the shear modulus when the strain amplitude is increased (Payne effect).

Several analytical methods are used to characterize mono- and bifunctional organosilanes in terms of the functions of the silanes in precipitated silicas. The characteristic functional groups are determined by:

- gravimetric measurements to yield the silanization content
- elemental analysis to determine the amount of sulfur and carbon
- FT-IR spectroscopy to identify the functional groups [281, 282]
- differential thermal analysis [10]
- $^{13}C$ and $^{29}Si$ cross-polarization NMR [282, 283]

The extent of hydrophobization of silica through the use of organo-silanes is determined analytically by HPLC and GC [284], which reveals the change in the substitution pattern of the silane and the silanol content in the silica.

For the silanization, the organo-silane must consist of a moiety that can be hydrolyzed and can react with the silanol groups to transform the aggregate into a hydrophobic aggregate. The reaction of the trialkoxysilyl group requires water (about 5%) in the mixture to effect hydroxylation of an alkoxy group. Hydrolysis is a rate-determining step. The reactivity of the organo-silane decreases as the number of alkyl substituents increases [285]:

$$(CH_3O)_3\ Si{-}R > (CH_3CH_2O)_3\ Si{-}R > (CH_3{-}CH_2O)_2\ (CH_3){-}Si{-}R > (C_3H_7O)_3\ Si{-}R$$

Instead of the tri-methoxy group, which triggers the fastest silanization reaction, the producers actually use the tri-ethoxy group, because the eliminated ethyl alcohol does not pose a toxicological hazard and can removed during the reaction.

The chemical process that transforms the silanol groups on the surface of silica into siloxane ether bonds can be studied using monofunctional organo-silanes. The temperature and the reaction times are measured with a rubber process analyzer (RPA). It has been shown that increasing the temperature greatly reduces the response time. The influence of temperature on the degree of silanization is illustrated for triethoxysilane (S203) in Figure 3.30 [286].

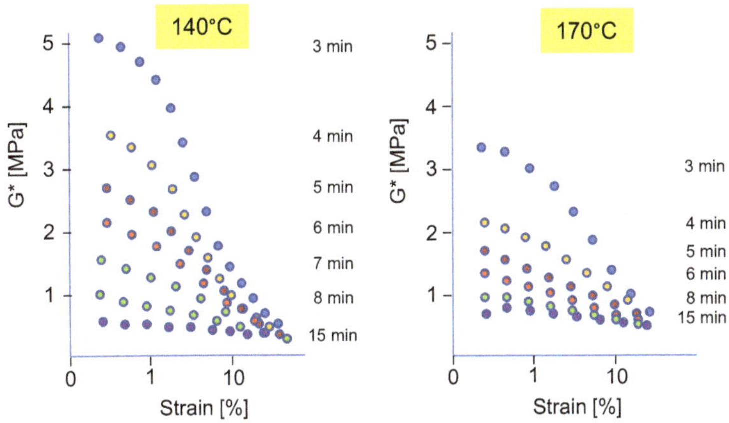


**Figure 3.30** Influence of reaction time and temperature on silanization of silica [286]

At the beginning of the reaction, a pronounced filler network can be observed at low shear rates, which disappears increasingly as a result of hydrophobization with the monofunctional triethoxysilane. The reaction occurs slowly at low mixing temperatures of 140 °C. After 10 minutes, the filler network shows a minimum concentration of residual silanol groups on the surface of the silica. At higher temperatures, 170 °C, the silanization reaction is much faster, but this temperature is not within the processing window of bifunctional silanes. This type of reaction also occurs during mixing in an internal mixer. However, the control of the temperature a time plays a crucial role in the silanization reaction.

Silanization is a multi-step reaction [287]. First, the silane must diffuse to the silanol groups and adsorb on the silica surface; the extent of adsorption depends on the structure of the silane, the compatibility with the rubber matrix and the temperature. The adsorption on the filler surface is influenced by the concentration of silanes and the access of silanol groups to the silica surface. Two possible mechanisms have been proposed:

- direct reaction of silanol groups with alkoxy groups
- hydrolysis of the alkoxy groups prior to the condensation reaction

The fact that the rate of silanization is positively influenced by the moisture content (4–6%) is an important hint that the hydrolysis step is primarily involved at the beginning of silanization. Once this rate-determining reaction is finished, the next reaction step is the formation of Si-O-Si ether bonds. The reaction scheme is shown in Figure 3.31.

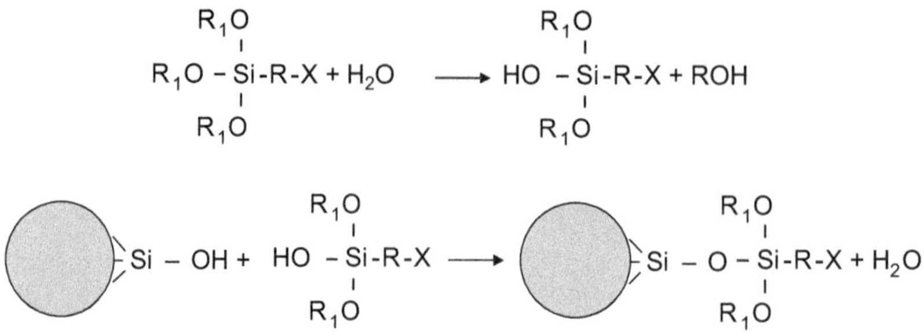

Figure 3.31 The reaction of monofunctional silanes with the silanol group of precipitated silica

The secondary reaction is the sum of several subsequent reactions and is defined as a condensation reaction of silane molecules bound on the silica with further cleavage of ethanol together to form a silanization film on the original silica surface. For this reaction, moisture of 4–6% is essential, as it can provide multilayer coverage of siloxane groups on the surface of the silica.

The silanization is accompanied by alcohol emissions. On the one hand, the alcohol in the compound shifts the reaction equilibrium towards less silanized silica. Therefore, removal of the alcohol from the mixing chamber is necessary for increasing the yield of silanization reaction. On the other hand, alcohol condensation in the mixing chamber result causes the compound to slide along the mixing chamber walls, leading to reduced mixing efficiency and unsatisfactory dispersion as well. The ethanol produced in the reaction should be released from the compound, as otherwise the reaction rate will be slowed down. The emission of ethanol is essential in the mixing chamber to ensure that the secondary reaction during the mixing process goes to completion. The emission of ethanol is significantly reduced by extraction on the internal mixer in order to reduce the toxic effect.

## 3.10.2 Bifunctional Silanes

In case of silica mixing, not only the degree of dispersion of the silica but also the chemical process of linking the silanized silica to the rubber chains plays a decisive role. For the industrial use of precipitated silica, bifunctional silanes are of far greater importance than the monofunctional silanes, which are used more to demonstrate the effect of silanization and the reaction conditions.

With the bifunctional silane, the type of precipitated silica can be brought closer to the use of carbon black and the especially the effect of filler-filler interaction can be reduced more than for carbon black. At the same time, the silanized silica can be connected to the rubber network, which results in better abrasion resistance in the rubber compound. In addition to the triethoxy silyl group, bifunctional silanes possess a reactive group that forms a chemical bond with an unsaturated rubber chain [47, 286]. The chemistry mechanism for the linking of the silanized silica to the rubber chain is the same as that for vulcanization crosslinking. For unsaturated rubbers that are commonly cured with sulfur/accelerator systems, the following types of bifunctional silanes are used:

- sulfane silanes: $[(RO)_3Si–(CH_2)_n]_2–S_x$ with n = 3 and x = 2 or 4
- 3-mercaptosilanes: $(RO)_3Si–(CH_2)_3–SH$
- blocked 3-mercaptosilanes: $(RO)_3Si–(CH_2)_3–S–(CO)–R_1$
- in the case of saturated rubbers cured with peroxides, unsaturated vinyl silanes are used:
  - allyl silanes: $(RO)_3Si–CH{=}CH_2$

To confer the best processing conditions and performance properties on the vulcanizate, it is important to make sure that the bifunctional silanes and the silica are mixed to a very high degree during the compounding process, while ensuring that the reaction with the rubber matrix is prevented and controlled. During mixing, a chemical process occurs that converts the silanol groups on the surface into siloxane ether bonds that can drastically reduce the polarity and improve the interaction with the unsaturated rubber, like S-SBR and cis-BR. The next reaction takes place during the vulcanization step and links the silanized silica to the rubber network.

Initially, geminal Si-OH groups were considered to be the most reactive with organo-silanes. Later, isolated Si-OH groups were considered to have similar reactivity and the highest efficiency when reacting with an organo-silane. By changing the properties of the surface through the use of organo-silanes, the hydrophilic character of the filler aggregate is transformed into a hydrophobic aggregate surface with much improved interactions to non-saturated rubber. This influence of the silanization is expressed by a strong reduction of the solubility parameter and a better interaction to the unsaturated rubber.

The first commercially available bifunctional silane was the 3-mercapto-propyl-trimethyloxy silane for use in non-polar sulfur-cured rubber and 3-methacryloxy-propyl-trimethoxy-silane and vinyl-trimethoxysilane for peroxide-cured rubber compounds [16]. The alkoxysilyl group can react with silanols on the silica surface to form stable siloxane linkage, and the organic functionality can participate in reactions that lead to a linkage with the rubber. Efficiency of coupling increases in a sulfur-cured system with increase in curing agents. This demonstrates that the coupling mechanism is an integral part of the vulcanization mechanism and leads to the interpretation of rubber properties as a function of strong filler-rubber bonding. This coupling system is apparently responsible for the fully reinforcing action of the silanized silica.

Typical of these bifunctional organo-silanes is the short incubation time and the rapid crosslinking reaction to yield unsaturated rubber [280]. At an addition of 1.5 phr, the silica-filled SBR vulcanizate produces nearly equal modulus, a higher tensile strength and lower compression set compared with carbon black N285. For example, 3-mercaptosilane added to a silica-SBR formulation produces a threefold increase in stress at 300%, an increase in tensile strength and a significant reduction in heat build-up. A 3-mercaptosilane with modified alkoxy groups (Si363) has been developed recently that reduces fuel consumption by 8%. [288]. The steric hindrance of the bulk oligomers substituents is overcome by the polar moiety of these substituents. This ensures a high silica affinity, which guarantees rapid adsorption on the silica surface. Furthermore, free silanol groups are shielded by the non-volatile long-chain substituents, which leads to more extensive hydrophobization of the silica. As a result of a modified coupling mechanism, the 3-mercapto group yields nearly 100% coupling. For this reason, only half of the molar amount of silane is needed to achieve superior reinforcement.

Blocked 3-mercaptosilanes, NTX, were invented by GE Silicones as coupling agent, whereby the 3-mercapto group is blocked by esterification with an organic acid, resulting in a thioester [288, 289]. The -SH group of the 3-mercaptosilanes has been replaced by an ester in the grouping -S-(CO)-$R_1$, which has been replaced by the activation residue, which only triggers a crosslinking process after activation. Blocked silanes require a de-blocking agent (proton donors, i. e. trimethylol propane) prior to the start of coupling [290]. Silanization by blocked 3-mercaptosilanes is much faster than with sulfur silanes and shortens the mixing process. However, the process must not be carried out above 145 °C, as some side reactions take place [291, 292].

The development of sulfur silane, which triggers a slower reaction with the rubber than the 3-mercaptosilanes, proved to be advantageous. The next important bifunctional silane is the famous bis-triethoxysilyl-triethyl-tetrasulfane, TESPT (Si69), which is promoted by Degussa as a slower silane and can be used with more flexibility in connection with the use of precipitated silica technology [292–295]. This silanization has led to increased technical use of precipitated silica. The silane is produced in quantities of over 10,000 metric tons per year.

The filler to rubber coupling was demonstrated by the increased reinforcement of the component properties. Depending on the number of sulfur atoms, there are various reaction possibilities. On the one hand, TESPT, which contains on average four sulfur atoms, is considerably less reactive than bis-triethoxysilyl-triethyl-disulfane. On the other hand, TESPD, which contains only two sulfur atoms, requires additional sulfur in the formulation to provide the same effect as TESPT. The difference is the TESPT is crosslinked at lower temperatures (140–145 °C), while TESPD starts to react at higher temperatures (150 °C). It has been shown that silanes with a longer sulfur bridge react faster and lead to a higher coupling efficiency. Nevertheless, the silanes used to bond silica particles on the rubber chain cannot react with all the silanol groups present on the surface [288]. The geminal silanol groups may not react completely, due to steric hindrance. A molecular calculation for Si69 shows that, under ideal conditions, the silica, with its specific surface of 150 $m^2/g$, will be covered only with 8% of the organo-silane to obtain a good interaction with the unsaturated rubber. The kinetics of the silanization of silica with alkoxysilanes have been reported in the literature [293].

It has been shown that the number of sulfur atoms in the organo-silane is important for the reaction rate of the filler-rubber coupling. The length of the sulfur bridge has an impact on the curing rate. The more sulfur atoms the organo-silane contains, the lower the crosslinking temperature and the faster the process takes place. The TESPT contributes some additional sulfur to the compound that others do not. The resulting longer, more flexible sulfur linkages can perform better in dynamic applications. As a large molecule, it requires more time and a given temperature during mixing for adequate reaction with the precipitated silica. However, the mixing temperature should remain below 145 °C, because they can lead to premature breakup of the tetra-sulfide group. In conventional and semi-efficient curing, the organo-silane with a large number of sulfur atoms, such as TESPT, also acts as a sulfur acceptor, because the formation of di- and poly-sulfidic bridges requires more sulfur than the silane contains. The reaction scheme for coupling to the rubber chains is shown in Figure 3.32.

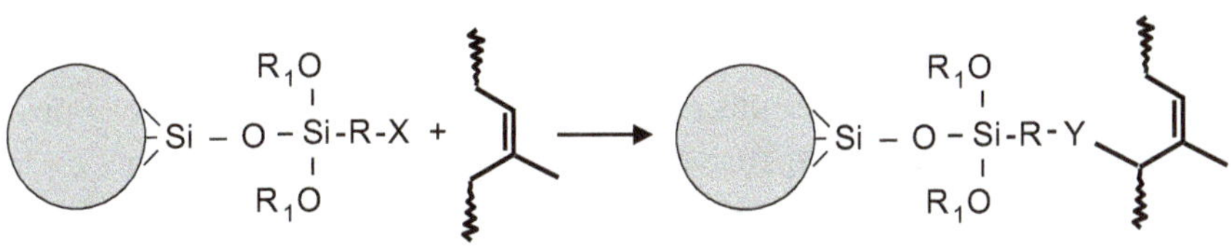


**Figure 3.32** Reaction of silanized silica with an unsaturated rubber

To achieve best processing conditions and performance properties of the vulcanizate, it is important to make sure that silanization during compounding takes place separately from chemical coupling of the silica particles to the polymer and the silanization yield is high. At the same time, the reaction conditions in the internal mixer have to be carefully controlled.

In order to avoid pre-scorch during mixing, the compound temperature has to be monitored carefully and should be below the permissible value for the respective organo-silane, which affects the final temperature in the mixing chamber. Molecular weight differences make the sulfur silanes only half as efficient as the 3-mercaptosilanes. Owing to the fact that silanes react stoichiometrically with silanol groups on the silica surface, they react on a molecular basis and not by phr. Therefore, the molecular weight of the silane has to be taken into account when an optimum silanization of the silica is envisaged. In other words, the type of silane has to be matched to the type of HD silica used. Table 3.14 illustrates the adjustment of TESPD (Si266) and the 3-mercaptosilane (Si363) for HD silicas of different surface area.

**Table 3.14** Comparison of the Dosage of Bifunctional Silanes (phr) Required for Optimal Silanization of HD Silicas

| HD Silica | Si 266 | Si 363 |
|---|---|---|
| 5000 GR | 4.7 | 7.2 |
| 7000 GR | 6.5 | 10.0 |
| 9000 GR | 8.10 | 12.50 |

The mixing process has to comply with are partly contradictory demands. In order to prevent pre-scorch, the mixing temperature should remain low but at the same time be high enough that the ethanol can be emitted within a short time. In most cases, the temperature window for the mixing of silica compounds is relatively narrow, limited by the increasing risk of scorch. The mixing temperature should be in the range of 140–155 °C to accomplish silanization. Targeted temperature management is therefore required for *in situ* dispersion of silica which depends on the mixing tool and the type of silanization. On an industrial scale, several steps are required to mix silica successfully, which lengthen the mixing process, but bring an advantage in utilization of the dynamic properties.

Various methods are used to carry out the reaction with the organosilanes. Wet-process: silane is added to a silica suspension in an organic solvent and the reaction is carried out at 80 °C. Dry-process: silane and the silica acid are intensively mixed and allowed to react at a temperature of 120 °C. *In situ* process: the silanization is carried out *in situ* during the mixing process. The organo-silane is added immediately after the incorporation of the silica in the rubber compound. The silanization takes place during the dispersion phase, and volatile alcohols are formed. Depending on the type of organo-silane, the reaction will take place between 140 and 160 °C.

In order to protect operators from volatile organic carbon (VOC) emissions (ethanol), pre-silanized silicas were brought onto the market. An improvement in the macro dispersion compared with compounds with a bifunctional silane was observed. Due to the different chemical composition of the bifunctional silanes, the measured percentage of ethanol emission is also different. The major differences in composition and VOC emissions between the bifunctional silanes are given in Table 3.15.

**Table 3.15** Compositional Characteristics and Ethanol Emissions of Bifunctional Silanes

| Silane | Sulfur (%) | Silicon (%) | Ethanol (%) |
|---|---|---|---|
| TESPT | 20 | 8.8 | 50 |
| TESPD | 14 | 11.8 | 58 |
| Blocked silane | 8.8 | 7.7 | 37 |

All alkoxysilanes have a positive influence on decreasing filler-filler interaction and lowering the compound viscosities, increasing the modulus and abrasion index through silica-rubber coupling. Unfortunately, the unmodified 3-mercaptosilane carries with it a strong odor which is highly objectionable to factory personnel. The most successful of several attempts to reduce the odor is the use of sulfur silanes.

Coupling of the silica filler to the rubber is one way to improve reinforcement by precipitated silica. However, this procedure requires very lengtly and multi-stage mixing technology and consumes additional mixing energy.

## 3.11 Functionalization of Carbon Black

Over the years, carbon black manufacturers have striven to further improve the efficiency of their products. The main goal has been to increase active sites on the carbon black surface and to increase interaction with the rubber. For this, it was necessary to better understand the activity of the carbon black, to recognize the connection between the reaction time during manufacturing and the activity of the primary particles, and to change the manufacturing process in such a way that these active centers on the surface of the carbon black increased and the proportion of graphitic planes decreased. A start in this direction was better investigation of the graphitization process during production and the disadvantages of the graphitized carbon black in the dispersion and in the rubber compounds. In addition, there were repeated indications that reinforcement of the carbon black can also be achieved by post-treating it.

### 3.11.1 Graphitized Carbon Black

When a conventional carbon black is heated in an inert gas atmosphere at a high temperature of 2700 to 3000 °C, the $sp^2$ hybridization of carbon atoms is fundamentally altered, without change in the volume or the density of the material. The process is characterized by a quantitative change in energetic quality. The entire material assumes the lowest energy value, i. e. $sp^2$ hybridization in the lattice of carbon black, and shows practically no reinforcement of mechanical properties. Graphitized carbon blacks are used as test fillers to show the change in the morphology and to see the role on surface activity and physical adhesion to rubber. The influence of the annealing temperature at 2700–3000 °C is shown in Figure 3.33 [74].

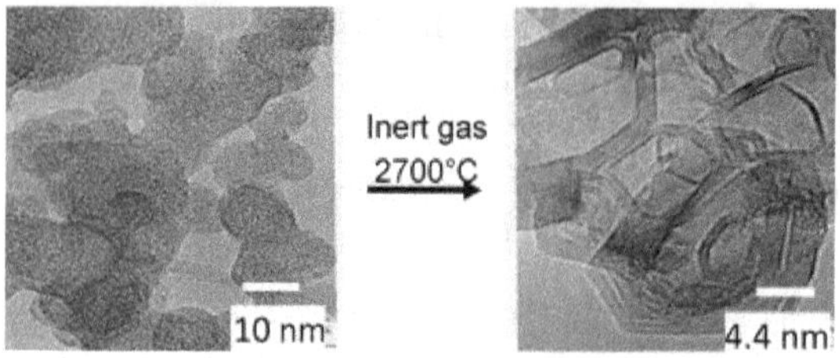


Figure 3.33
Annealing of an ASTM carbon black N220 to graphitized carbon black [74]

As the annealing temperature increases, a change in the structure of the primary particles is observed. They have an extended planar graphitic structure that deforms the primary particle, but does not change the material density. The difference becomes apparent when the specific surface area is determined using an adsorption method, for example the BET measurement. It has been found that the nitrogen specific surface area of graphitized carbon black decreases by 20% compared with the original carbon black [296]. This result indicates that less nitrogen is adsorbed on the surface of the graphitized black, due to the quantitative transition of carbon atoms into $sp^2$ hybridization, which reduces the interaction forces of the nitrogen atoms ono the transformed carbon black surface. The difference in the adsorption capacity of carbon black before and after the graphitization is shown in Figure 3.34.

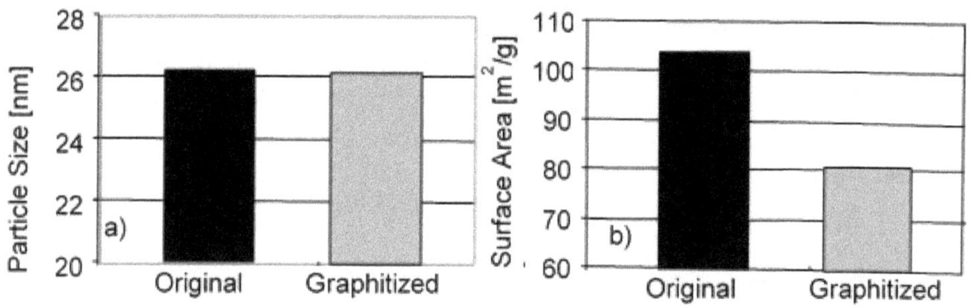


Figure 3.34 Difference in the adsorption behavior of ASTM carbon black and annealed carbon black [296]

Examination of the adsorption properties in unsaturated rubber or the mechanical properties in rubbers reveals that the graphitized carbon black leads to a minimum of interaction to rubber that is not exhibited by non-graphitized carbon blacks, which show significantly higher activity and an excellent reinforcement in vulcanizates. Correspondingly, the interaction with the rubber chains is reduced to a minimum level, which certainly depends on the annealing temperature reached.

## 3.11.2 In-Process Functionalization of Carbon Black

When the carbon blacks were being improved, care was taken to reduce graphitic carbon and to increase the energetic components that interact with the rubber. This has led to functionalized carbon blacks that have improved the physical properties of the rubber vulcanizates. The morphology of the carbon black is changed by the turbulence in the reactor and the velocity of transport through the reactor. One example is "nanostructured carbon blacks" or Ecorax® carbon black produced by Orion (Germany) [297]. Ecorax® has a similar specific area and structure level as the ASTM grades, but a very special surface structure. The micro-crystallites are more disordered and have a higher surface energy, as shown in Figure 3.35 [298].

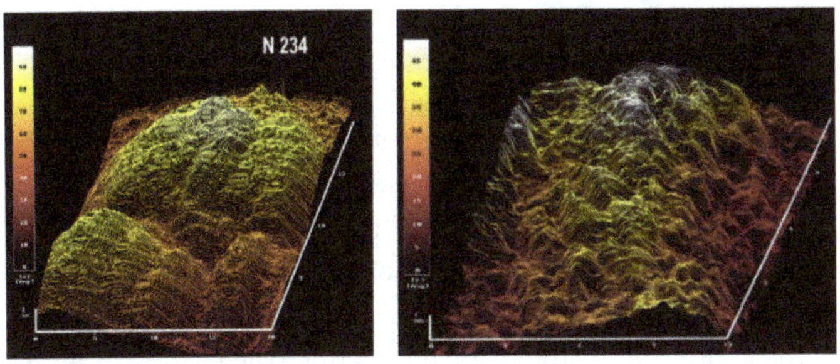


**Figure 3.35** The improved carbon blacks possess high surface roughness and markedly altered surface energy [298]

The regular surface of the primary particles of the ASTM carbon black compared with the unconventional arrangement of the primary particles of Ecorax® carbon black is evident. The nanostructured carbon black has much better filler-rubber interaction, better filler dispersion and enhanced reinforcement of the elastomer. However, the nanostructured carbon blacks are associated with higher production costs, but their use in the manufacture of rubber products is more cost-effective than the silica-silane technology and the wear on the internal mixer and the extrusion is also less pronounced. Ecorax® is a group of new carbon blacks developed for passenger tires, as well as truck tires, because of their combination of very good surface interaction with the non-polar rubber, low Payne effect and low rolling resistance.

Another approach to in-process functionalization is afforded by carbon-silica dual-phase fillers (CSDPF) produced by Cabot (USA) [299–303]. After the aromatic oil has been injected, a silica precursor is introduced into the reactor to produce smaller silica nanoparticles in the primary particles or just decorate their surface. This process reduces the graphitic layers on the primary particle and changes the whole interaction with the rubber. Both types of CSDPF are characterized by reduced filler-filler interactions and higher polymer-filler interactions, which can be shown by strain amplitude dependences of the storage modulus (Payne effect) [298]. The CSDPF are active fillers with a concentration of silicon in the carbon black aggregate of 0.1 to 9.9 wt.% in the CSPDF 2xxx series, and 0.1 to 11.9 wt.% in the CSPDF 4xxx series. The latter group contains a significantly higher content of silica nanoparticles that are located on the surface of the carbon black aggregate. From TEM images, it can be seen that, in CSDPF 2xxx, the silica nanoparticles are distributed throughout the carbon black aggregate, while in CSDPT 4xxx they mainly cover the surface of the carbon black aggregate. This is shown in Figure 3.36 [298].

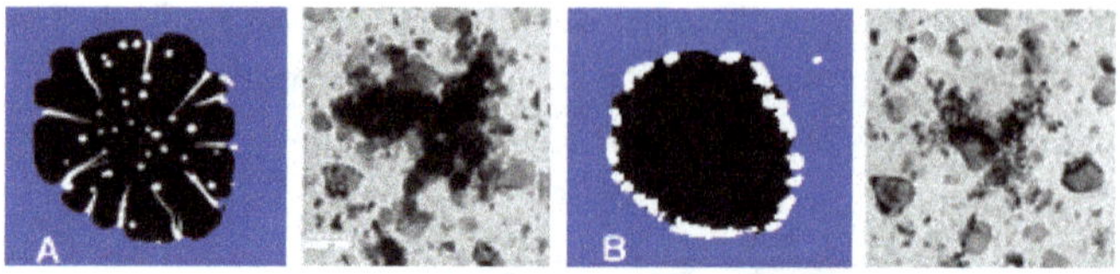


**Figure 3.36** Schematics and TEM images of CSDPF: a) 2xxx and b) 4xxx

In the IGC method, the CSDPF shows a significantly higher contribution to the surface adsorbed energy of the aggregates, indicating the presence of heteroatoms between the carbon atoms. This suggests a high proportion of silica nanoparticles which are incorporated in the graphitic lattice of the primary particles. The resulting CSDPF interact very well with the unsaturated rubbers, such as solution SBR, BR and NR. They are far better dispersed in the rubbers and, above all, they reduce the hysteresis while markedly improving the abrasion resistance of the vulcanizates. The strain amplitude dependence of the storage modulus is smaller than a reduction generated with TESPT in silica-filled S-SBR/BR (75/25). This is a clear sign of a reduction in filler agglomeration and better dispersion in the rubber. The system is less expensive than a coupling silica-silane technology and produces electrically conductive rubber compounds. The use of this advanced filler is less abrasive to the processing equipment compared with the silica filler alone, but CSDPF costs more than traditional carbon black.

A third way to improve the carbon black is functionalization of the surface with oxygen-containing groups during manufacturing. Carbon black treated with a small amount of ozone after production in the furnace can be used in combination with carboxylate S-SBR. The reaction between the two oxygen-containing groups provides for high filler dispersion in the solution rubber and less rolling resistance in the tire tread [304, 305]. These examples show that the carbon blacks can be improved *in situ* during manufacturing.

### 3.11.3 Post-Functionalization of Carbon Black

The more expensive way to functionalize carbon black is post-functionalization with physical or chemical agents that transform the surface of carbon black. Attempts have been made for more than seven decades to improve the adsorption of rubber chains on the surface of carbon black and improve the dispersion and reinforcement of the product. Examples of this are the addition of amines, diazo compounds and polysulfide groups. When 4-aminophenyl disulfide is deposited on the surface of the active carbon black, the dispersive component of the surface energy, $\gamma_s^d$ decreases and better dispersion in rubber is recorded [305]. The influence of chemicals used in vulcanization, when added to the carbon black in "homeopathic" doses, can lead to a significant change in the filler dispersion and an increase in the physical properties of the rubber compounds. With a small amount of ZDMC, a significant improvement in the dispersion of the carbon black (N330) in SBR and NR is produced and an increase in the stress-strain behavior achieved [306]. The influence of a physical plasma on fluffy carbon black has been investigated with different gases (air, oxygen and nitrogen) at atmospheric pressure. The reaction of the plasma gas on the carbon black lasts only a few fractions of a second and yet the surface of the carbon black changes significantly. With air or oxygen, the surface of the carbon black is reshaped so that the carbon black becomes water-absorbent. A plasma generated with nitrogen modifies the surface morphology considerably, by introducing nitrogenous structures into the lattice of the carbon, as can be determined by means of surface analysis. These changes lead to a significant increase in dispersion in natural rubber and SBR and an increase in the stress values and the tensile strength in the stress-strain behavior [307].

### 3.11.4 Recycled Carbon Black

Although the following section applies to the recycling of scrap rubber, recovery of carbon black from these used elastomers is discussed here. Recovered carbon black presents an opportunity to reduce the tire industry's reliance on petrochemicals by replacing a portion of the traditional carbon black with a sustainable and circular alternative, without introducing a performance trade-off. The materials made from waste rubber over the last three decades have a composition that contains silica and quite a lot of ash in addition to the carbon content. The physical properties are in the region of semi-reinforcing black of the category N500 to N700 [308]. In 2020, a European cross-industrial consortium started to improve the production of tires. At a temperature of 400 °C, pyrolysis of rubber shreds yields what is known as recovered carbon black (rCB), which shows some improvements in the semi-reinforcement of rubbers. In some applications, it can partly replace the N700 to N500 series of virgin carbon black. However, if reinforcement is required, rCB may not work on account of the moderate specific surface area and the large content of ash. In addition, approaches are known in which carbon [309], graphitic [310] and other raw materials such as scrap rubber pyrolysis residues are converted into a reinforcing filler. The resulting products attain neither the quality standard nor the reinforcing effect of the carbon blacks made from petroleum and natural gas. The carbon black that is created here has a different composition from the active carbon blacks previously described.

In addition to solid carbon, they contain tar, silica and, above all, a higher proportion of ash. A proportion (up to 20%) of the pyrolysis carbon black can be mixed into a normal rubber compound without any major impairment of physical properties being observed. The product is sold as a semi-active carbon black with similar properties to the carbon black N550–N330 and is offered for non-demanding uses. Pyrolysis oil has been used as a raw material to produce carbon black, called sCB. Three different carbon black grades have been produced with properties according to sN234, sN347 and s550. It can therefore be assumed that sCBs are a simple solution for using the pyrolysis oil, without the need to adjust the properties of rubber compounds [311].

According to a recent report by the European Tyre Manufacturers' Association (ETRMA) some 95% of Europe's end-of-life tires were collected and treated for material recycling and energy recovery during 2019. Two years later, Michelin and Bridgestone outlined a plan to promote and increase the use of recovered carbon black in new tires and other products [312].

### 3.11.5 Functionalized Carbon Nanotubes

Multi-walled carbon nanotubes are difficult to disperse and exhibit a high resistance to "wetting" in a rubber melt. It is known that the performance of multi-walled carbon nanotubes depends critically on the dispersion and interfacial interactions with non-polar and polar rubber. These challenges require the development of effective methods to modify the surface properties by physical or chemical functionalization that can lead to better interaction with the polymer. Carbon nanotubes possess two distinct regions, the end tips and sidewalls that can react with different chemical reactivities. Chemical functionalization is based on the covalent coupling of functional groups onto the carbon scaffold of carbon nanotubes. The reaction can take place at the ends or at the sidewalls. Covalent sidewall functionalization of carbon nanotubes is associated with a change in hybridization from $sp^2$ to $sp^3$ and a simultaneous loss of $\pi$-conjugation on the graphene layer.

Oxidative treatments of multi-walled carbon nanotubes have been reported, which include nitric acid/sulfuric acid, $KMnO_4$ [313] and $H_2O_2$ [314], and oxygen plasma [315]. The resulting oxygen functionalization increases the interaction with the polymer matrix, can provide chemical coupling and leads to better carbon black dispersion. The pentagons present at the end tips of multi-walled carbon nanotubes are more reactive than the hexagons on the side wall. Therefore, functionalization on the sidewall is more difficult to accomplish. Another factor is the curvature of carbon nanotubes for functionalization. The bonding energy on the sidewall tends to increase with decrease in tube diameter. One of the simplest ways to achieve functionalization is to start with fluorination [316, 317]. It has been shown that the fluorination of purified SWCNTs occurs at temperatures of 325 °C and provides a start for many other functionalization. The fluorinated carbon nanotubes have C-F bonds that can enable other reactions, such as reaction with amino, alkyl and hydroxyl groups [318, 319]. Similar reactions have been reported for chlorination and bromination [320], hydrogenation [321, 322], Diels-Alder cyclo-addition [323, 324] and addition of radicals [325, 326]. One of the objectives of functionalizing carbon nanotubes is to improve the dispersion in polymer matrices and to improve the wettability in polymers [327]. However, it is expensive to use functionalized carbon nanotube for industrial application in rubber compounds.

## References

[1] Austin, A. E., Adelson, A., Proc. 5th Conf. on Carbon (1962) pp. 485
[2] Janzen, J., Rubber Chem. Technol., 1982, 55: 669
[3] Kraus, G., Fortschr. Hochpolym. Forsch. 1971, 8: 155
[4] Bruin, W. H., van de Plos, Th., Rev. Gen. Caot. 1964, 41: 453
[5] Given, P. H., Hill, W., Carbon, 1969, 7: 649
[6] Boehm, H. P., Diel, E. Z., Elektrochem., 1962, 66: 642
[7] Studebaker, M. L., Beaty, J. R., in "Science and Technology of Rubber", Eirich, F. R. Ed., Academic Press, New York (1978)
[8] Haws, J. R., Cooper, W. T., Ross, E. F., Rubber Chem. Technol., 1977, 50: 211
[9] Gessler, A. M., Rubber Chem. Technol., 1970, 43: 943
[10] Rivin, D., Rubber Chem. Technol., 1971, 43: 9943
[11] Lewis, J. F., Deviney Jr., M. L., McNabb, C. F., Rubber Chem. Technol., 1970, 43: 449
[12] ASTM D 1619
[13] ASTM D 1618
[14] ASTM D 1506
[15] Loadman, M. J. R., Tidd, B. K., Wake, W. C., in "Analysis of Rubber and Rubber-like Polymers", 3rd ed., (1983) Ch. 12
[16] Wagner, M. P., Rubber Chem. Technol., 1976, 49: 703
[17] ASTM D 1618
[18] ASTM D 1506
[19] ASTM D 8371
[20] Hockey, J. A., Chem. Ind. (London), 1965, 57
[21] de Boer, J. H., Vleeskens, J. M., Proc. K. Ned. Akad. Wat., Ser. B, 1957, 60:45
[22] Heston, W. A., et al., J. Phys. Chem., 1960, 64: 147
[23] Bode, R., Ferch, F., Fratscher, H., Kautsch. Gummi Kunstst., 1967, 20: 578
[24] Fripiat, J. J., Gastuche, M. C., Brichard, R., J. Phys. Chem., 1962, 66: 805
[25] Arminstead, C. G., Tyler, A. J., Hambledon, F. H., Mitchel, S. A., Hockey, J. A., J. Phys. Chem. 73: 3947
[26] Hockey, J. A., Pethica, B. A., Trans. Faraday Soc., 1961, 57: 2247
[27] Bascom, W. D., J. Phys. Chem., 1972, 76: 3188
[28] Hockey, J. A., Chem. Ind. London, 1967, 57
[29] Mueller, R., Kammler, H. K., Wegner, K., Protsinis, S. E. Langmuir 2003, 19: 160
[30] Zhurlaev, L. V., Colloid Surf., A, 2000, 173: 1
[31] Bassett, D. R., Boucher, E. A., Zettlemayer, A. C., J. Colloid Interface Sci., 1968, 27: 649

[32] ISO 7872/2

[33] ISO 3262

[34] ISO 3262/8

[35] ISO 797/9

[36] ISO 5794

[37] Christiades, G. E., EMU Notes in Mineralogy, Vol. 11 (2011) Ch. 6

[38] Schellenberg, M., Schuster, R. H., Report for LANXESS (2011)

[39] Le, H. H., et al., Carbon, 2012, 50: 4543

[40] Zimmermann, H., Schuster, R. H., Kautschuk Gummi, Kunstst., 2011, 63: 40

[41] Donnet, J.-B., et al., Kautsch. Gummi Kunstst., 1986, 39: 510

[42] Gerspacher, M., Wampler, W., in "Basic Elastomer Technology" (Baranwal, K., Stephens, H., Eds.) Rubber Division (2001) Ch. 4

[43] Boller, F., Geisler, H., Schuster, R. H., Kautsch. Gummi Kunstst., 1991, 44: 1166

[44] Medalia, A. I., in "Carbon Black-Polymer Composites", Sichel, E. K., Ed., Marcell Dekker, New York (1976) Ch. 1

[45] Hjelm, R., J. Mater. Res., 1994, 9: 12

[46] Görl, U., in "Handbuch der Kautschuk-Technologie", Hofmann, W., Gupta, H., Dr. Gupta Verlag, Ratingen (2001)

[47] Meon, W., Blume, A., Luginsland, H.-D., Uhrlandt, S., in "Rubber Compounding, Chemistry and Application", Rodgers, B., Ed., Marcel Dekker, Inc. New York, Basel (2004) Ch.7

[48] Guerovich, D., Macosco, C. W., Tirell, M. Rubber Chem. Technol., 2004, 76: 1

[49] Le Meins, J., Modenaers, P., Mewis, J., Ind. Eng. Chem. Res., 2002, 41: 6297

[50] Vermaunt, J., J. Rheol. 2007, 2007, 51: 429

[51] Voit, K., Blume, A., Rubber Fall Colloqu., Hanover (2010)

[52] Wu, Y.-P, Zhang, L.-Q., et al., J. Appl. Polym. Sci., 2001, 82: 2842

[53] Wu, Y.-P, Jia, L.-X., Yu, D. S. Zhang, L.-Q., J. Appl. Polym. Sci., 2003, 89: 3855

[54] Klafke de Azeredo, L., Schuster, R. H., Jacobi, M. M., Kautsch. Gummi Kunstst., 2010, 62: 130

[55] Schellenberg, M., Schuster, R. H., ACS Rubber Division, Cleveland (2013]

[56] Giannelis, E. P., Krishnamorti, R., Manias, E., Adv. Polym. Sci., 1999, 138: 107

[57] Karger-Kocsiz, J., Wu, C.-M., Polym. Eng. Sci., 2004, 44: 1083

[58] Ma, P.-Ch. Kim, J.-K., in "Carbon Nanotubes for Polymer Reinforcement", CRC Press, Taylor & Francis Group, Boca Raton, London, New York (2011]

[59] Graedy, B. P., in "Carbon Nanotubes, Manufacturing, Properties and Applications", J. Wiley & Sons, Inc., Hoboken (2011)

[60] Ebbesen, T. W., Acc. Chem. Res., 1998, 31: 558

[61] Rao, C. N. R., Covinari, A., Acc. Chem. Res., 2002, 35: 998

[62] Schuster, R. H., Wittek, H., Eurofiller, Bratislava (2013)

[63] Warren, B. E., Phys. Rev., 1941, 59: 693

[64] Biscoe, J., Warren, B. E., J. Appl. Phys., 1942, 13: 364

[65] Franklin, R. E., Acta Cryst., 1950, 3: 148

[66] Heckman, F. A., Harling, D. F., Rubber Chem. Technol., 1966, 39: 141

[67] Hess, W. M., Ban, L. L., McDonald, G. C., Rubber Chem. Technol., 1969, 42: 1209

[68] Hess, W. M., Ban, L. L., Norelco Rep. 1966, 13: 102

[69] Hess, W. M., Ban, L. L., Eckert, F. P., Chirico, V., Rubber Chem. Technol., 1966, 41: 1968

[70] Heidenreich, R. D., Hess, W. M., Ban, L. L., J. Appl. Cryst. J., 1968, 1: 1

[156] ISO 6894
[157] Burgess, K. A., Scott, C. E., Hess, W. M., Rubber Chem. Technol., 1971, 44: 233
[158] ASTM D 2414
[159] ISO 4656
[160] DIN 53601
[161] Hess, W. M., Rubber Chem. Technol., 1973, 46: 204
[162] Kraus, G., Fortschr. Hochpolym. Forsch. 1971, 8:155
[163] ASTM D 3493
[164] Medalia, A. I., Sawyer, R. L., Proc. 5th Carbon Conf., New York, Pergamon Press (1963) pp. 563
[165] Mrozowski, S., Chaberski, A., Loebner, E. E., Proc. 3rd Carbon Conf., London, 1959
[166] Dollinger, R. E., Kallenberger, R. H., Studebaker, M. L., Rubber Chem. Technol., 1967, 40: 1311
[167] Voet, A., Whitten, W. N., Rubber World, 1963, 148: 33
[168] Washburn, E. W., Physics, 1921, 7:115
[169] Ritter, R. H., Drake, L. C., Ind. Eng. Chem. Anal. Ed. 1945, 17: 782
[170] Moscou, L., Lub, S., Bussemaker, O. K. F., Rubber Chem. Technol., 1971, 44: 805
[171] Heston, Jr. W. A., Iler, R. K., Sears, G. W., J. Phys. Chem. 1960, 64: 147
[172] Medalia, A. I., J. Colloidal Interface Sci., 1967, 24: 394
[173] Medalia, A. I., Heckman, F. H., Carbon, 1969, 7: 567
[174] Herd, Ch., McDonnald, G. C., Hess, W. M., Rubber Chem. Technol., 1993, 65: 107
[175] Kraus, G., Rubber Chem. Technol., 1971, 44:199
[176] Gerspacher, M., O'Farrell, C. P., Elastomerics 123 (1991)
[177] Gerspacher, M., O'Farrell, C. P., Kautsch. Gummi Kunstst., 1992, 45:97
[178] Ehrburger-Dolle, F., Terence, M., Carbon 1990, 28: 448
[179] Oppermann, B., PhD Thesis, University of Hanover (1993)
[180] Donnet, J.-B., Voet, A., in "Physics, Chemistry and Elastomer Reinforcement" Marcel Dekker, New York (1976)
[181] Wolff, S., Kautsch. Gummi Kunstst., 1981, 34: 280
[182] Herd, Ch., McDonald, G. C., Hess, W. M., Rubber Chem. Technol., 1993, 66: 491
[183] Burgess, K. A., Scott, S. E., Hess, W. M., Rubber Chem. Technol., 1971, 44: 236
[184] Hess, W. M., McDonnald, G. C., Urban, E., RCT 1973, 46: 204
[185] Qiang, D., et al., Am. Soc. Mech. Eng., 2002, 55: 495
[186] Okada, A., et al., Mat. Res. Soc. Symp. Proceedings, 1990, 171: 45
[187] Reinhard, W., Thomas, J. G., Reisinger, A., Polymer, 2003, 44: 7513
[188] Serge, B., et al., J. Polym. Part Sci. Part B Polym. Phys., 2003, 41: 3188
[189] Zammarono, M., et al., ACS Nano 2011, 5: 3391
[190] Schellenberg, M., PhD Thesis University of Hanover (2012)
[191] Krishnamoorti, et al. Chem Mater. 1996, 8: 29
[192] Varghese, S., Karger-Kocsis, J., Gatos, K. G., Polymer, 2003, 44: 3977
[193] Liang, Y., et al., Polym. Testing, 2008, 27: 270
[194] Ganter, G., Gronski, W., Lemke, H., Kautsch. Gummi Kunstst., 2001, 54: 166
[195] Sadu, S., Bohmik, A. K., J. Appl. Polym. Sci., 2004, 92: 689
[196] Mintmire, J. W., Dunlap, B. P., Phys. Rev. Lett. 1992, 68: 631
[197] Tans, J. S., et al., Nature 1997, 386: 474
[198] Hamada, N., et al., Phys. Rev. Lett. Nature 1992, 391: 1579

[199] Grossiord, N., Loss, J., Regev, O., Konig C. E., Chem. Mater., 2006, 18: 1089
[200] Schuster, R. H., ACS Rubber Division, San Antonio 2013;
[201] Zimmermann, H., Chogule, H., Schuster, R. H, Proc. Conf. ICCE-19, Shanghai (2011)
[202] Chogule, H., Schuster, R. H., ACS Rubber Division Meeting, Cleveland (2013)
[203] Wildower, J. W. G., et al., Nature, 1998, 391: 52
[204] Le, H. H., et al., Carbon, 2012, 50: 4543
[205] Desai, A. V., Hague, M. A., Thin-Walled Struct. 2005, 43: 1787
[206] Goldberg, R., Rubber Chem. Technol., 1989, 62: 272
[207] Spain, I. L., in "Chemistry and Physics of Carbon", Vol. 8, Walker, P. L., Jr, Thower, P. A., (Eds.) Marcel Dekker, New York, Basel (1972)
[208] Medalia, A. I., Rubber Chem. Technol., 1986, 59: 432
[209] Kraus, G., J. Appl. Polym. Sci., 1984, 39: 75
[210] Stauffer, D., Aharoni, A., in "Introduction to Percolation Theory", Taylor and Francis, London (1992)
[211] Sichel, E. K., Gittleman, J. L., Sheng, P., Phys. Rev. B 18, 1978, 5712
[212] Probst, N., in "Carbon Black, Science and Technology", Donnet, J.-B., Bansal, R. C., Wang, M.-J., Eds., Marcel Dekker Inc., New York, (1993), Ch. 8
[213] Daire, A., Keithley Instruments, Inc., White paper
[214] Rajeshwar, K., Nikiel, L., Gerspacher, M., O'Farrell, C. P., ACS Rubber Division, Orlando (1999)
[215] Medalia, A. I., Rubber Chem. Technol., 1988, 59: 432
[216] Asada, T., Int. Polym. Sci. Technol., 1987, 14: 25
[217] Carmona, F., Physica A, 1989: 157: 461
[218] Boonstra, B. B., Rubber Chem. Technol., 1979, 50: 194
[219] Voet, A. Rubber Chem. Technol., 1981, 54: 42
[220] Roland, C. M., Peng, K. L., Rubber Chem. Technol., 1991, 64: 790
[221] O'Farrell, C. P., Gerspacher, M., Nikiel, L., Kautsch. Gummi Kunstst., 2000, 53: 701
[222] Lorenz, H. et al., Comp. Sci. Technol., 2009, 69: 2143
[223] Klüppel, M., Adv. Polym. Sci., 2003, 164: 1
[224] Yamada, N., ACS Rubber Division Meeting, Dallas (2000)
[225] Ajayan, P. M., Schadler, L. S., Braun. P. V., in "Nanocomposite Science and Technology", Weinheim-VCH, 2002, pp. 77
[226] Dresselhaus, M. S., et al., in "Carbon Nanotubes, Synthesis, Structure, Properties and Applications", Springer Verlag (2000]
[227] Cho, B.-M., Kim, G.-H., J. Appl. Polym. Sci., 2010, 116: 555
[228] Saito, R., et al., in "Physical Properties of Carbon Nanotubes", Imperial College Press (1998)
[229] Schuster, R. H., 187th ACS Rubber Division, Nashville (2014)
[230] Fritsche, J., PhD Thesis University of Hanover (2009)
[231] Kraus, G., in "Reinforcement of Elastomers", Kraus, G., Ed., Interscience Publishers, New York (1965)
[232] Donnet, J.-B., Bansal, R. C., Wang, M.-J., (Eds.) in "Carbon Black Science and Technology ", Marcel Dekker Inc. New York (1993)
[233] Mark, J. E., Erman, B., Roland, M. C., Eds., in "The Science and Technology of Rubber", Academic Press, Waltham, USA 2013
[234] Avnir, D., Farin, F., Pfeifer, P., New J. Chem. 1990, 14: 197
[235] Kim, S.-J., Reinecke, D. H., Rubber Chem. Technol., 1993, 66: 559
[236] Schröder, A., Klüppel, M., Schuster, R. H., Kautsch. Gummi Kunstst., 1999, 52: 814

[237] Vilgis, T. A., Heinrich, G., Macromolecules, 1994, 27: 784

[238] Huber, G., Vilgis, T. A., Kautsch. Gummi Kunstst., 1999, 52: 102

[239] Schröder, A., Klüppel, M., Schuster, R. H., Macromol. Mater. Eng., 2007, 292: 885

[240] Cassie, A. B. D., Discuss. Faraday Soc., 1948, 3:11

[241] Fowkes, F. M., Wetting Soc. Chem. Ind. Monograph. London, 1967, 25: 3

[242] Schultz, J., Tsutsumi, K., Donnet, J.-B., J. Colloid Interface Sci., 1977, 59: 272

[243] Chessik, J. J., Zettlemoyer, A. C., Adv. Catalysis, 1959, 11: 263

[244] Hill, T. L., J. Chem. Phys., 1949; 17: 520

[245] Conder, J., Young, C. L., "Physicochemical Measurements by Gas Chromatography" (1972)

[246] Jura, G., Hill, T. L., J. Amer. Chem. Soc., 1952, 74: 1598

[247] Young, G. J., Chessick, J. J., Healy, F. H., J. Phys. Chem., 1954, 58: 3137

[248] Busset, H., Martin, J. P., Mendelbaum, H. P., Bull. Soc. Chim. Fr., 1967, 2346

[249] Kraus, G., J. Phys. Chem., 1955, 59, 343

[250] Patrick, W., A., in "The Colloid Chemistry of Silica and Silicates" Iler, P. K., Ed., Cornell Press, Ithaca, New York (1955)

[251] Lansinger, C., PhD Thesis, University Haute Alsace, Mulhouse (1990)

[252] Wang, M.-J., Wolff, S., Donnet, J.-B., Rubber Chem. Technol., 1992, 65: 329

[253] Wang, M.-J., Wolff, S., Donnet, J.-B., Rubber Chem. Technol., 1991, 64: 714

[254] Menzel, R., Lee, A., Bismarck, A., Shaeffer, M. S. P., Langmuir, 2009, 25: 8340

[255] Schuster, R. H., 62nd Canadian Chem. Soc. Rubber Division, (2013)

[256] Beebe, R. A., Biscoe, J., Smith, W. R., Wendell, C. B., J. Am. Chem. Soc., 1947, 69: 2294

[257] Beebe, R. A., Klingston, G. L., Polley, M. M., Smith, W. R., J. Am. Chem. Soc., 1950, 72: 40

[258] Beebe, R. A., Millard, B., Cyanski, J., J. Am. Chem. Soc., 1953, 75: 839

[259] Polley, M. H., Schaefer, W. D., Smith, R. W., J. Am. Chem. Soc., 1951, 73: 2161

[260] Wang, M.-J., Wolff, S., Donnet, J.-B., Kautsch. Gummi Kunstst., 1992, 45:11

[261] Schröder, A., Klüppel, M., Schuster, R. H., Heidberg, J., Kautsch. Gummi, Kunstst., 2001, 54: 260

[262] Rudzinski, W., Waksmunski, A., Leboda, R. J. Chromatogr. 1974, 92: 25

[263] Hobson, J. P., Can. J. Phys. 1965, 43: 1934

[264] Jarowiec, M., Surf. Sci., 1975, 50:553

[265] Wang, M.-J., Wolff, S., Freund, B., Rubber Chem. Technol., 1994, 67: 27

[266] Ross, S., Oliver, J. P., in "On Physical Adsorption", J. Wiley & Sons, London (1964)

[267] Adamson, A. W., Ed., in "Physical Chemistry of Surfaces", J. Wiley & Sons, Inc. New York, 5ed. (1990)

[268] Fowler, R. H., Guggenheim, E. A., in "Statistical Thermodynamics" Cambridge Univ. Press, Cambridge

[269] Schroeder, A., Klüppel, M., Schuster, R. H., Heidberg, J., Carbon, 2002, 40: 207

[270] Schroeder, A., PhD Thesis University Hanover (2000)

[271] Yashkin, S., Schuster, R. H., Proc. ISSHAC-6, Zakopane (Poland), 2006

[272] Stoecker, W., Kolloid. Z. 1956, 149: 39

[273] Boehm, H. P., Schneider, M., Z. Anorg. Allgem. Chem., 1959, 301:326

[274] Wood, D. W., et al., J. Am. Ceram. Soc., 1983, 66: 693

[275] Janik, M., Internat. Report Degussa, 2001, 05701

[276] Wagner, M. P., Rubber World, 1971, 164: 46

[277] Wagner, M. P., Collogue Internationaux, Bischenberg-Obernai, 1973

[278] Pluedemann, E. P., in "Silane Coupling Agents", Plenum Press, 2nd Ed., New York, 1982

[279] Meyer-Simon, E., Thurn, F., Michel, R., German Pat. 214159 (1973)

[280] Thurn, F., Wolff, S., Kautsch. Gummi Kunstst., 1975, 28: 733

[281] Hunsche, A., Görl, U., Müller, A., Knaak, M., Göbel, T., Kautsch. Gummi Kunstst., 1997, 50: 881

[282] Görl, U., Hunsche, A., Müller, A., Koban, H. G., Kautsch. Gummi Kunstst., 1998, 51: 525

[283] Dierkes, W., Nordermeer, J. M. W., Kelting, K.-W., Limper, A., Rubber World, 2004, 229:33

[284] Klockmann, O., Hasse, A., ACS Rubber Division Meeting, San Antonio, May 2005

[285] Beary, F., et al., J. Organometal. Chem., 2001, 625: 210

[286] Klockmann, O., in "Mixing of Rubber Compounds", Limper, A., Ed., Hanser Publisher, Munich (2121) Ch. 4

[287] Görl, U., Parkhouse, A., Kautsch. Gummi Kunstst., 1999, 52: 493

[288] Habekorn, N., et al., W0201680649A1 (2016)

[289] Cruse, R., et al., US Pat. 28673B2 (1997)

[290] Lin, Ch.-J., York, M. W., Code, R., Rubber Chem. Technol., 2017, 90: 126

[291] Dierkes, W., Noordermeer, J. M. W., Kelting, K.-W., Limper, A., Rubber World, 2004, 229: 33

[292] Görl, U., Gummi Asbest Kunstst., 1998, 51: 416

[293] Görl, U., Münzberg, J. ACS Meeting, Annaheim, CA (1997) pp. 38

[294] Luginsland, H.-D., ACS Meeting, Chicago (1999) pp. 74,

[295] Hasse, A., Klockmann, O., Wehmeier, A., Luginsland, H.-D., Kautsch. Gummi Kustst., 2002, 55: 236

[296] Gerspacher, M., Proc. DKT'09, Nuremberg (2009)

[297] Niedermeier, W., Fröhlich, J., Kautsch. Gummi Kunstst., 2003, 56: 519

[298] Göritz, D., Raab, H., Fröhlich, J., Maier, P. G., Rubber Chem. Technol., 1997, 72: 929

[299] Mahmud, K., Wang, M.-J., Francis, R. A., US Pat. 5830930, (1998)

[300] Mahmud, K., Wang, M.-J., Francis, R. A., US Pat 5877238 (1999)

[301] Mahmud, K., Wang, M.-J., US Pat. 6211279 (2000)

[302] Wang, M.-J., Kutsovski, Y., Zhang, P., et al., Kautsch. Gummi Kunstst., 2002, 55: 33

[303] Wang, M.-J., Tu, H., Kutsovski, Y., et al., Rubber Chem. Technol., 2000, 73: 666

[304] Herd, Ch., IOM3 RIEG London (2019)

[305] Tokita, N., Wang, M.-J., Chung, B., Mahmud, K., Nippon Gomu Kokaiashi, 1998, 71: 9

[306] Schuster, R. H., Müller, M., Früh, Th., Hallensleben, Rubber Fall Colloqu. (1998)

[307] Tricas, N., Borros, S., Schuster, R. H., Kautsch. Gummi Kunstst., 2005, 58: 511

[308] Geisler, H., Schuster, R. H., Proc. IRC Lyon (2006)

[309] Baumann, H., Klein, J., Jüntgen, H., Kautsch. Gummi Kunstst., 1982, 34: 843

[310] Charoff, R. P., Erikson, E. H., Müller, D. E., Salyer, I. O., ACS Rubber Division, Los Angeles, 1985

[311] Westenberg, H., Gomez, H., Rechenbach, D., Kautsch. Gummi Kunstst., 2021, 73: 47

[312] Smithers Recovered Carbon Black Conference, Amsterdam (2021)

[313] Eatemadi, A., et al., Nanoscale Research Lett., 2014, 9: 393

[314] Bubert, et al., Diamond Related Mater., 2003, 12: 811

[315] Velasces-Santos, C., Martinez-Hernandez, A. L., Fischer, F. T., Ruoff, R. S., Castano, V. M., Chem. Mater., 2003, 15:4470

[316] Hagemueller, R., Gomans, H. H., Rinzler, A. G., Fischer, J. E., Winey, K. I., Phys. Chem. Lett. 2000, 330: 219

[317] Hamwi, A., et al., Carbon, 1997, 35: 723

[318] Stevens, J. L., Huang, A. Y., Peng, H., Nano Lett. 2003, 3: 331

[319] Zhang, Q. H., Chen, D. J., Science, 1998, 282:95
[320] Unger, E., Graham, A., Kreupl, F., Curr. Appl. Phys. 2002, 2: 107
[321] Chen, Y., et al., J. Mater. Res. 13 (1998) 2423
[322] Chen, J., et al., Science, 1998, 282: 95
[323] Hirsch, A., Angew. Chem. Int. Ed., 2002, 41: 1853
[324] Tasis, D., et al., Adv. Mater., 2009, 21: 1105
[325] Ni, B., et al., Phys. Rev. B, 2001, 61: R16343
[326] Peng, H., et al., Chem. Commun., 2003, 3: 362
[327] Umek, P., et al., Chem. Mater., 2003, 15: 4751

# 4 Mixing Process for Rubber Compounds

The way in which the mixing process takes place determines the degree of filler dispersion and the interactions between the rubber and the filler, which in turn determine the rheological properties and the improvement in physical behavior of the vulcanizates. The final properties of a vulcanizate depend not only on its intrinsic properties, but are also significantly influenced by the way the components are mixed together. The properties of a mixture are determined by the order of addition into the internal mixer, the set mixing parameters and the ejection temperature. The mixing process significantly improves the properties of a rubber compound and can thus ensure that no errors are introduced. The degree of the filler dispersion and the distribution of the filler are very important for the rheological properties, stress-strain behavior, dynamic-mechanical properties and finally the fatigue resistance and crack propagation. The mixing of viscous rubber with solid particles is a complex operation, which may be divided into different operational stages, which are carried out at specific intensities.

When it comes to the mixing process, we now differentiate between traditional dry-mixing, which was previously carried out on a two-roll mill and is now carried out in an internal mixer, and the recently introduced wet-mixing, which has become particularly influential in tire production. The dry-mixing or the melt mixing is the usual procedure for producing filled compounds in an internal mixer or on a two-roll mill. A further development of the mixing process was achieved through wet mixing, which is characterized by simpler production using a rubber latex and a filler slurry to obtain better filler dispersion in the rubber mixture and by this, it allows a higher rubber reinforcement, especially an increase in the fracture mechanical aspects of the filled vulcanizate. The filler dispersion in the wet-mixing is superior to dry-mixing and has a lower energy input, which suggests a high quality of mixing without mechanical breakdown of the polymer during the mixing process. In the following, the different types of mechanical mixing, the strategies pursued during the mixing and the essential steps of the mixing in the internal mixer are presented.

## 4.1 Dry-Mixing

The incorporation of the solid filler into the viscous rubber requires a lot of energy to be applied. Without this energy input, the filler is not or is only insufficiently mixed into the rubber and is not well dispersed in the rubber matrix. Therefore, the mixing process is a multi-disciplinary task that requires consideration of the machine aspects, the relevant processes in a mixing tool, the polymer-specific requirements and the dispersion and distribution of the filler. The result of the dry-mixing process is a multi-phase rubber compound, consisting of a viscoelastic filled rubber phase, which includes a well dispersed filler in the viscous rubber matrix. As the mixing process itself has a severe influence on the end product properties, the relationship between the product quality and the dispersion of the filler are important.

The dry-mixing technology with its technical aspects is the basis for an efficient mixing process. The result of the dry-mixing process is a multi-phase rubber compound, consisting of a viscoelastic rubber phase that includes a dispersed filler that should be distributed uniformly in the volume of the rubber compound. To get a homogeneous compound, the mixing process should support the structural changes in the raw materials that facilitate intimate physical interactions between the compound ingredients and lead to the conversion of these materials into compound quantities that can be easily shaped and molded. To achieve consistency in mixing, it is important to ensure that each component should be uncontaminated by other materials. Each component has to be introduced to the mixer in the correct amount as well as the correct sequence. This is generally summarized under the term “art of mixing” that includes:

- the number of mixing steps
- the mixing procedure
- the operating parameters
- the mixing cycle control

A reproducible compound quality and economical production process requires that the compound ingredients be fed into the mixing tool exactly in quantities indicated in the formulation. A basic prerequisite for this is the metering of all additives in the right order and at the right time intervals. This includes the weighing from tanks that are filled from large silos, computer-controlled hand weighing and precise weighing of small components with a deviation tolerance of at least ± 0.1%. The mixing of rubber is an energy-intensive process in which endeavors are made to achieve maximum dispersion of the solid fillers and good distribution of all formulation ingredients at a beneficial cost to the rubber company.

The process implies the introduction of recipe components, originally supplied in large lumps, bales or granulated form, into microscopic volume elements in the compound that can be processed further in extrusion and in molding steps. With solid fillers, mixing should guarantee that filler pellets and agglomerates are well broken up and form dispersed aggregates that are distributed homogeneously in as short time as possible to reduce the costs of the mixing cycle. The success of such an effort depends on the one hand on the chosen mixing equipment and the selected mixing parameters and on the other hand on the chemical composition of the polymer, its molecular weight and the type and loading of the filler used.

In many cases, the associated question of costs for the production of the mixture is a priority when selecting the filler and the mixing technology used. The speed with which the filler can be incorporated and dispersed determines the economics of the mixing process. It is important to note that the failures made during the mixing process cannot be remedied. Thus, faults can be introduced due to a suboptimal sequence in which the components of the mixture are added, or an unsuitable choice of process parameters that cannot be remedied in the extrusion process. Therefore, the failures made during mixing inevitably lead to quality losses of the rubber compound and these are no longer improved even when the rubber compounds are crosslinked. Therefore, the choice of process parameters as well as the addition of ingredients must therefore be given great attention in the mixing process [1–8]. The mixed compound must have both suitable rheological properties for subsequent downstream processing, improved dispersion of the filler and a morphological structure that will yield the desired or required physical properties of the end product. In most cases this is a difficult task because the formulation ingredients are delivered in forms such as rubber bales, pelletized fillers, liquid oils, hard resins, chips, coated granules and even pastes for low molecular weight resins. Important preconditions for quality mixing are proper storage and handling of the raw materials and precise weighing. But finding a technology to mix the filler safely and without loss in the dispersion on their own machines in the mixing chamber is part of the great skill of the process engineers.

### 4.1.1 Dry-Mixing

The ways of dry-mixing include the addition of the rubber, the filler and the mineral oil in a suitable amount, adding the functional antioxidants and the curing agents to the end of the process [2–8]. Depending on the type and the quantity of the filler used and the chemical saturation or unsaturation and the viscosity of the rubber, the resulting compound may be easy or difficult to mix. The rubber mixing process can be subdivided into three steps:

- feeding the ingredients to the mixer in the correct quantities, at the correct times and at the correct temperatures
- the mixing of the ingredients
- discharge of the mixed compounds
- cooling and shaping

The most important aspect of mixing is consistency in type, quality and quantity of ingredients, as well as in temperature, mixing speed and pressure, mixing stage times and dumping temperatures and times, consistency in work impact, shear rates and shear stress. If one looks at the process from the point of view of the filler, the degree of breakdown of the initial size of the filler when fed into the mixing device and the spatial distribution of the aggregates and the topological structure of the mix are important. From a theoretical point of view, one can distinguish between three kinds of mixing which occur on rubber mixers [7]:

- dispersive mixing
- distributive mixing
- laminar mixing

Depending on the type of mixing, the requirements imposed on the mixing tools are specific. Mixing mechanisms represent borderline cases that can be explained by the characteristic features of the flow processes. Because distributive and dispersive mixing result from different flow fields, the mixing equipment plays a major role in achieving one of these mechanisms as dominant for the rheological behavior of the rubber compound. It is therefore crucial for the quality of a compound whether dispersive or distributive mixing predominates in mixing processes. The dependence on friction in the rubber matrix that arises between the wall of the mixing chamber and the rotors or between the two rotors. Poor dispersive and distributive mixing affect the downstream processing and lead frequently to insufficient manufacturing uniformity, additional energy consumption and waste of raw materials.

#### 4.1.1.1 Dispersive Mixing

Dispersive mixing involves the breakdown of the original size of filler particles into smaller entities tending to be individual aggregates. Once the filler is fed into the rubber matrix, breakdown of the filler pellets starts with the incorporation of the filler agglomerates and subsequent dispersion leads to a progressive increase in the rubber-filler interface that provides phase adhesion between the polymer and the filler. The process of filler dispersion is schematically shown in Figure 4.1.

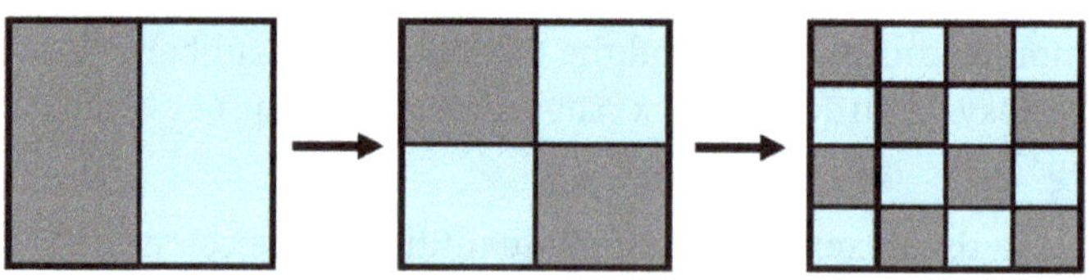

Figure 4.1 Schematic representation of the dispersive mixing

Understandably, the size reduction in filler particles is a kinetic process, which depends on the molecular weight and the chemical nature of the rubber, the filler agglomeration and pelletizing hardness and certainly on the internal mixer and the processing conditions. To substantiate the general statement with practical experience, it should be noted that the dispersion of the filler is promoted by:

- intermeshing rotors which primarily produce elongational flow are better than tangential rotors
- rubbers with a good interaction with particulate fillers
- fillers with a lower structure

The pellets are gradually broken down into agglomerates and then into aggregates depending on the shear rates that can act in the mixing tool. The size reduction in pelletized fillers leads practically to solid objects ranging between several hundred nm and just a few µm. If one considers the carbon black aggregate interaction in an agglomerate as constant, the dispersion becomes more pronounced by increasing the structure of the filler. When the degree of branching of the aggregates is high, the number of neighboring aggregates decreases and therefore the dispersion of the agglomerate requires less mixing energy. In the case of precipitated silica which has a strong inter-aggregate interaction due to the presence of the silanol groups, a poor dispersion of the aggregates occurs [9]. The presence of the silanol groups increases the interparticle interaction and increases the filler-filler interaction. The use of organo-silanes that react with the silanols reduces the filler-filler interaction (NR, SBR, BR). The strong interparticle interaction in layered silicates cannot be broken in a mechanical mixing experiment. The interlayered energy between the platelets is opened in organo-clays, which allow good interaction with polar rubber. The carbon nanotubes can be dispersed quite well in the rubber compounds, even if this is still being debated in the literature.

A wealth of information from mixing chambers indicates that filler dispersion starts only above a minimum energy imparted to the compound. Figure 4.2 illustrates the impact of mixing energy on filler dispersion.

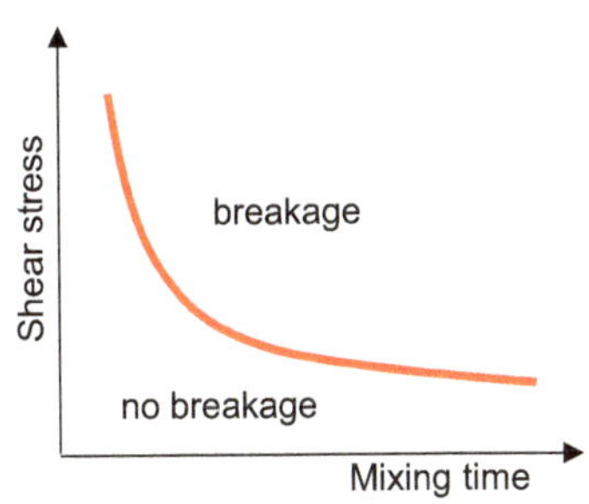


**Figure 4.2**
Impact of mixing energy on filler dispersion

At low shear stresses, filler pellets as such can resist without suffering consistent fragmentation and the pellet fragments are not reduced in size. If the stress is raised above a critical value, breakdown of the filler starts and increases progressively with the energy input.

Concerning the type of flow necessary to sustain dispersive mixing, it has been proven that elongational flow and repeated folding processes increase the efficiency of dispersion more than shear flow [6–8, 10]. Evidence of the effect of the elongational flow has been provided in a test arrangement consisting of four cylinders, two of which each have the opposite direction of rotation. Accordingly, the degree of filler dispersion no longer depends upon stress alone, but more on the total strain imparted to the rubber matrix by the mixing tool. Assuming the total strain or the extensional stress as key variables in dispersive mixing, the question as to how to improve the design of the mixing equipment for increasing the efficiency in dispersive mixing becomes interesting from an engineering point of view. Due to the fact that filler loading is typically high in most of the compounds (> 60 phr) and the mixing time is kept for economic reasons as short as possible, the filler dispersion often does not reach an optimum value. Owing to the mechanical dispersion in mixing tools, however, a part of the filler is always retained in the rubber matrix in the form of pellet fragments and large agglomerates, so an "ideal" dispersion cannot be achieved by using pelletized fillers. The portion of non-dispersed filler is considered to be a trigger for material failure under static or dynamic stress [7, 8, 10]. At the phase boundary of filler agglomerates or large agglomerates there is a detachment from the rubber matrix, which creates cracks that propagate until the sample tears.

In addition, the economics need to shorten the mixing cycle as much as possible, yet that can stop the dispersion at a sub-optimal state. Efforts have been directed at developing suitable mixing tools to disperse the filler at optimum manner. However, compromises have to be made in order to achieve a balance between the good dynamic and ultimate properties of the compound and the cost of mixing and downstream processing.

#### 4.1.1.2 Distributive Mixing

Through distributive mixing a given degree of composition uniformity in the rubber compound is achieved. No time-dependent changes to the size of particles occur by this type of mixing, but instead random distribution of filler particle sizes with a constant size takes place. Thus, distributive mixing can be considered as a statistical process during which the particles do not change their size but their position. Since, in an ideal distributive mixing process, the amount of rubber-filler interface remains constant, the interfacial phenomena which contribute significantly to reinforcement are not influenced in a determining way. Distributive mixing leads to an increase in the homogeneity of the filler in the rubber compound. The homogeneity has to be recorded at length scales much larger than the aggregates size. The randomization process of distributive filler particles is schematically shown in Figure 4.3.

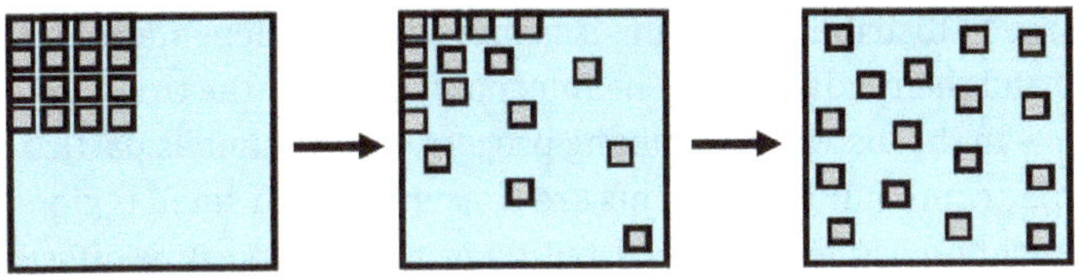

**Figure 4.3** Schematic representation of distributive mixing

The degree of homogenization has a positive effect on downstream processing as well as on the stress-strain behavior, the dynamic-mechanical properties and especially the ultimate properties of the vulcanizate. Accumulations of filler particles in certain parts of a vulcanizate are flaws that can lead to crack initiation if the material is subjected to repeated cyclical loads. Distribution of the filler particles is particularly important when batch feeding is used and minor additives with a particular functionality are concentrated in different regions of the mixer. This applies in particular to the distribution of curing agents which, if they are not homogeneously distributed, form crosslink clusters which can lead to premature failure. Therefore, uniform distribution is important not only for filler particles but also for functional compound ingredients, such as curing agents and anti-oxidants that should be distributed uniformly in the rubber vulcanizate.

However, the degree of homogeneity of a compound depends on the scrutiny to which it is subjected. Since rubber compounds represent multi-phase systems, the decreasing length scale at the sub-microscopic level becomes increasingly heterogeneous, which is also a characteristic of well-dispersed compounds. For practical purposes, the scale of scrutiny should be defined as the minimum size of the regions of segregation that would cause the compound to be imperfect for the intended purpose [11].

In addition to homogenization of the rubber compound, the distributive action of rotors guides the material in the direction of the dispersion zones. As a result of this interaction of the different flows, compounds are usually produced in mixing tools, in which both mixing mechanisms are involved. On two-roll mills, the distributive mixing in the axial direction is not pronounced unless the material is moved across the mill roll. The operator cuts and folds the rubber sheets and improves by this circumferential distributive mixing. A similar effect is achieved when using a stock-blender unit for two-roll mill mixing. On internal mixers the arrangements of the rotor blades provide for repeated splitting of the material flow stream that provides distributive mixing and to some extent laminar mixing. This enables homogeneity to be attained in a short period of time. Driving forces for positional changes of the particles are:

- the rotations of the material within the mixing chamber
- the number of rotations applied to the batch, and
- the extent of material transport in the mixer

In internal mixers, a degree of distributive mixing is achieved which depends on the length of the rotor flights and their axial angle. The internal mixers for the tire industry have a higher efficiency in the distributive mixing process. Distribution is particularly important when minor compound ingredients are concentrated in small regions of a compound. Homogenization achieved by distributive mixing has positive effects on downstream processes as well as on the ultimate properties, lifetime under dynamic loading of the finished products.

#### 4.1.1.3 Laminar Mixing

A third mixing mechanism, namely laminar mixing, is of particular importance in rubber mixing. Here, the goal is to increase the interface between the two layers as much as possible. The increase can be achieved either by the strain shear or by rotational shear. During the dwell time in the mixer, material layers can be stretched by several hundred percent and repetitive straining and stretching provide high laminar mixing effects [12, 13]. The laminar mixing is schematically shown in Figure 4.4.

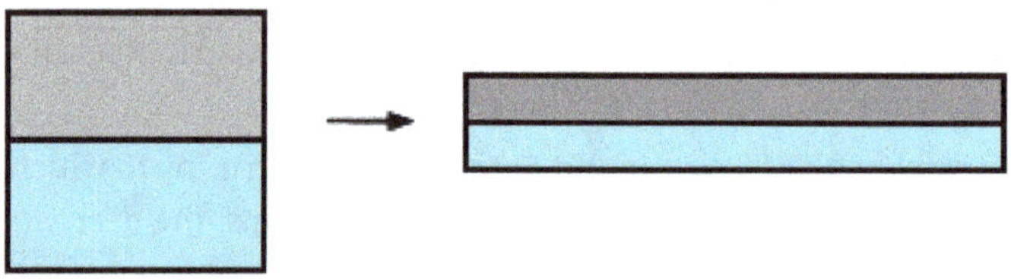

**Figure 4.4** Laminar mixing during strain

Laminar mixing is characterized by the deformations introduced into the rubber compound. They are characterized by the passages through the gaps between the rotor flights and the mixer walls and the passages through the rotor gap area. Typical of laminar mixing of viscoelastic fluids is that the components are deformed in the flow stream, while the thickness and the distance between the flowing layers becomes smaller but at the same time the surface becomes larger [13].

### 4.1.2 Dry-Mixing Strategies

To achieve consistency in mixing, it is important to ensure that each compound ingredient is used within the recommended dosage, the time to be introduced into the mixer and the mixing time. Such a treatment corresponds to the formulation developed by the compounders. In addition, the compound ingredients should be added in the correct amount and in the right sequence in the mixer order to attain the desired properties of the compound [4, 9, 14–16]. In the past, weighing, transporting and dosing the raw materials used to be done mainly manually before mixing started. Technological progress and the demand for consistently high quality of the mixtures have led to a rationalization and automation of the steps upstream and downstream of the mixing process. It is advisable to weigh the ingredients precisely and to convey and charge them automatically to the internal mixer. Microprocessor-controlled systems reduce the range of error in these mixing operations. There are three strategies for mixing rubbers in an internal mixer [10]:

- conventional mixing
- early-oil-addition mixing
- upside-down mixing

The differences between the mixing processes result from the sequence in which the main components (rubber, filler and oil) are introduced into the mixing chamber. Variations of these processes are used to suit the goal of special formulations or the ability of the equipment. In order to avoid undesirable secondary reactions, the sequence of compound ingredients additions and the number of additions should be set.

### 4.1.2.1 Conventional Mixing

Conventional mixing, developed originally for natural rubber, consists of adding rubber first, followed by the filler and finally, when the filler is well dispersed, the liquid ingredients such as mineral oil, antioxidants and curing agents [8]. After the mixing process, the mixture is dumped from the internal mixer as a coherent lump and has to be cooled and formed into strips in the downstream processes, with batch-off systems and strip coolers being used. In the production of strips for cold extruders and injection molding machines, the cooling must be sufficient, the greatest possible lengths are needed and the tray must be cut precisely. Finally, the reactive curing agents are incorporated below the temperature at which the ingredients are able to enter a chemical reaction, like crosslinking. The antioxidants are partly consumed during the mixing process, because of the contact with air in the mixing chamber.

Using the conventional mixing procedure, a good filler dispersion and distribution can be achieved even for compounds that contain fillers with low structure, small particle size and high surface activity [4–6]. The mixing instructions for this mixing strategy can be designed variably, whereby the possibilities of the available mixing unit are exhausted. With more competitiveness, conventional mixing has asserted itself over time, with a become wealth of implementation variants which are characterized by the type of mixture and the implementation potential in the company. Since a longer mixing time is associated with higher energy consumption, conventional mixing is used when higher compound quality and better filler dispersion of the final elastomeric product are necessary. The process has to be optimized for compounds that contain high levels of active fillers. Before starting the mixing process, it has to be decided if the process will be run in a single stage or in two or more stages.

This subdivision essentially considers the incorporation and dispersion of the filler. In addition, also the temperature of the rubber is important. If the temperature is too high, two-stage mixing has to be selected. Therefore, the antioxidants and the curing ingredients are introduced into the next mixing step.

Single-stage mixing includes the techniques in which all components are incorporated and mixed in one stage according to a defined sequence and time of admission to the mixing chamber. The compound ingredients have to be fed into the mixing chamber as quickly as possible. For this reason, the schedule for adding the ingredients is adjusted according to their type and quantity as well as the size of the internal mixer. The mixer has to draw power, or otherwise it is not running at full capacity [17]. There are definite temperature ranges for mixing rubbers that depend on the rotor geometry, the ram pressure and the fill factor of the compound. The effect of preheating the rubber samples in the mixing process was investigated, and it was found that the ram seating time is significantly reduced when the temperature is above 50 °C. Likewise, the filler incorporation time is reduced with increasing the speed of the rotors and the filling volume. Each specific type of rubber has an optimum temperature at which the desired dispersion of compound ingredients in the mix is obtained and the curing system does not react.

The recommended sequence of admission foresees a premix of the rubber to reduce its viscosity to an optimum level, then incorporation of the filler at one or more working points depending on the power consumption in the mixer. The mineral oil is added in small portions or, even better, is directly injected in pulses through injection valves into the mixing chamber. The single-stage mixing methods deliver ready-made rubber compounds and are used when a certain degree of dispersion of the filler has to be achieved. Changes of mixing times can result in changes of compound quality, especially from the degree of filler dispersion or in the case of rubber blends, where a natural distribution of the filler is required. A finishing stage and control of the final temperature of the mixing is necessary to prevent premature vulcanization.

As a matter of fact, when curing agents are incorporated, the temperature of the compound should not exceed the critical level at which the crosslinking process starts. For this, it is necessary to know the scorch time (or incubation time) of the chosen curing system as a function of temperature. Before this critical temperature is reached, the compound has to be discharged. In some cases, curing agents will be added in a second mixing stage, either into the mixer or onto a two-roll mill, to avoid starting the cure process or reducing its efficiency.

As a result of the fact that the curing rate of sulfur-cured rubbers doubles roughly when the compound temperature increases by 10 °C, while for peroxides it trebles over the same temperature interval, very tight tolerances are required for the temperature at which the compound is dumped. For pressure-less vulcanizing extrusion compounds, which require more reactive curing systems, the mixing temperatures should not exceed 80 to 90 °C. For compounds that require molding, the final mixing temperatures can be higher (i. e. 100 to 110 °C) and the use of accelerators with long incubation time (i. e. sulfenamides) is advised. For curing systems with short scorch time, such as sulfur/carbamates, sulfur/thiurams and especially peroxides, it is difficult to maintain the low temperature limit. Here, the principle of the tandem mixer could ensure process safety, because in the greater ramless tandem mixer the rubber compound can be cooled to suitable temperatures that allow addition of curing agents or silanization of the precipitated silica.

The mixed compound has to be removed from the mixer in order to start the next mixing cycle. The hot compound is usually discharged on two-roll mills situated below the drop door of the mixer. Subsequently, the unshaped compound is rolled out to form sheets and is transported on a conveyor belt to a second mill for homogenization and cooling. The sheets are deposited in loops after coating with release agents (i. e. talcum) that lower the surface tackiness of the compound.

Single-stage mixing is popular for productivity reasons, but, as described above, it is feasible only for certain compounds where the mixing time is not limited by the temperature rise. Thus, single-stage mixing is difficult to be achieved with highly filled mixes, due to the high heat input which, despite efficient cooling, cannot be completely removed from the mixture [13].

Two-stage mixing is done by separate production of rubber-filler batches and rubber-filler-oil batches with a different polymer basis, to which crosslinking agents are added at the end when the two ready-made masterbatches are blended in the required proportions. The primary objective of employing this technique is to control the composition of the rubber phases, and in particular the filler loading and the degree of the filler dispersion and for a higher loading of active fillers. As already described, the mixture temperatures reach high values, so that no process safety can be guaranteed for the curing agents, even those with a longer incubation time, are in the mixture. If parts of the mixture are crosslinked, the compound cannot be dimensionally formed and vulcanized and is part of the rubber mix considered as scrap [18].

Using this mixing technique, the rubber is charged to the mixer at once, then the filler is added preferentially in increments to initiate the filler dispersion efficiently. Not all of the contents of the filler are added, but only the amounts that are dispersed within a certain time. Subsequently, the other mixing ingredients are added with the last filler addition. Discharge can be done after a given time or before a critical temperature of the compound is reached. Usually, the compound is sheeted out on a two-roll mill at low friction and stored for 24 h.

Silica compounds only achieve good filler dispersion through efficient silanization. For balanced ultimate properties of the vulcanizates, coupling of the precipitated silica with an organo-silane is required, and only conventional mixing is used [9, 18]. In practice, three to four mixing stages and re-milling are required for sufficient silianization and efficient, good coupling of the filler to the rubber matrix. In order to determine the process conditions for this process, the Mooney viscosity or the Payne effect is determined in preliminary studies. Only when a minimum viscosity or the smallest storage modulus has been determined is the silanization process transferred to the production phase.

Due to the lack of reactive ingredients, the mixing temperature can be increased above the critical temperatures at which crosslinking would start. When the filler content in the discrete phase of a polymer blend has to be controlled, each polymer requires a given filler loading in order to compensate for the effects of a filler transfer that naturally takes place during the final mixing stage. This requires a detailed knowledge of the affinity of the filler for each of the polymers and of the process of filler transfer. The compound should be dumped after this second stage, sheeted out on the two-roll mill and stored for 24 h. The final mix is made by charging first-stage masterbatch into the mixer or on a two-roll mill. After a short time (30 s) the curing agents are added. Discharge has to be done at temperatures below 100 °C. The mix is then sheeted out on the mill and cooled to room temperature.

### 4.1.2.2 Early Oil Addition

This process involves adding the rubber first in the internal mixer and the filler later. After 1–2 minutes of mixing, all the oil is added at once. Dispersion is improved if the addition of the oil is delayed slightly. The method can give a good filler dispersion if the oil is added at the proper time. With early oil addition mixing, it should be borne in mind that diffusion of the oil requires a certain amount of time, which should definitely be observed. This process is used for compounds containing large volumes of plasticizers, but the mixing process can lead to an extended mixing cycle, as the oil has a lubricant effect [4, 10].

### 4.1.2.3 "Upside-Down" Mixing

Upside-down mixing is a simple, rapid process for compounds that contain high levels of inactive fillers and a large volume of plasticizers [10]. Typically, the fillers are added to the mixer first, then the oil and finally the rubber on top. Due to this procedure, dust tends to coat the internal surface of the mixer that can prevent efficient wetting of the metal surface with the polymer. As a result, the duration of the mixing processes can be significantly slowed down. The method is used for special rubbers (i. e. EPDM) when high quantities of fillers are required. If conventional mixing is employed in such cases, crumbling of the initially compact rubber matrix would result and in this state the time necessary to form a compact compound would be quite long.

However, the upside-down method is the fastest and simplest way of mixing. For this reason, this technology is used frequently, regardless of the fact that the final compound quality and the physical properties of the vulcanizate leave something to be desired. In some instances, it may be desirable or necessary to load the mixing ingredients in one shot using the upside-down technology. Under such conditions, the mixer is highly filled at the beginning of the mixing cycle. The upside-down process is recommended for producing highly filled EPDM mixtures. Apart from the crosslinking agents, carbon black, plasticizer and all additives are added and finally the EPDM is incorporated. High filling levels are required due to the high filler and plasticizer content. The crosslinking chemicals are then mixed in on the roller.

The technique is not suitable for compounds containing low structure and active carbon blacks or for compounds that contain high loadings of soft fillers and oil together with highly viscous rubbers. Low structure fillers and active fillers are quite difficult to disperse by this type of mixing. The shear stress provided by the equipment is not enough to achieve good dispersion when the filler loading is high. With very viscous rubbers, upside-down mixing leads to a rapid increase in compound temperature and this results in poor filler dispersion. Upside-down mixing is usually disliked by operators who suspect that dust stop damage and leakage occur more likely than when conventional mixing techniques are used.

### 4.1.3 Mixing Steps

The dry-mixing process consists of consecutive mixing steps that take place in the mixing chamber and are recorded by a mixing diagram. The sequence of the mixing steps is represented by the mixing time of the applied torque [2, 4–8, 10]. An evaluation of the torque curve plotted against mixing time provides all kinds of information about the individual mixing steps. Due to the discontinuous batch mixing, the mixing diagrams recorded for repeated mixing cycles show a relatively strong spread of the measured parameters and a consequence of this spread is that variations in the rheological behavior can be observed and countermeasures initiated during the next mixing process. A conventional mixing procedure can be divided in four distinct stages:

- plasticization
- filler incorporation
- filler dispersion
- filler distribution

Each of these stages contributes in a way to the duration of the mixing process as well as to the final rubber compound properties. For the process engineer, the mixing stages should not last longer than necessary to achieve optimal compound properties in short time and with the lowest energy. As shown in Figure 4.5, the mixing stages can be distinguished by sections in the mixing diagram that differ in the characteristic plot of torque against mixing time [2, 4–7].

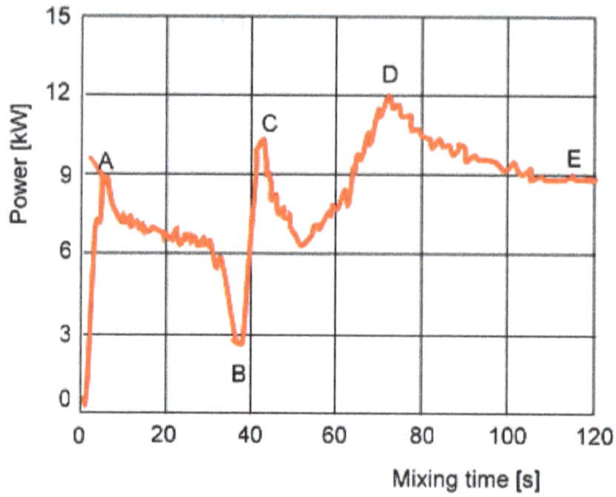


**Figure 4.5** Mixing diagram for a filled rubber compound

The mixing diagram can include the time dependence of other parameters, such as power consumption, ram position, speed of the rotors. They are therefore also referred to as "fingerprints", which provide useful information about the mixing process. The fingerprint of the individual mixing cycle can be compared with an earlier diagram that has been defined as a reference and makes it possible to evaluate fluctuations in the key process parameters and to identify in a mixing cycle a process deviating from the reference mixing process. The occurrence of undesirable influences indicate an irregular mixing quality. Of course, the mixing diagram can take on a different form in many applications.

The influence of the compound composition depends on the type of the polymer, the level and of type of the filler and the amount of low molecular weight ingredients that have an influence on the flowability of the compound [19–22]. Comparison of the mixing diagrams recorded for a mixture over a longer time interval provides the process engineer with information about the stability and reproducibility of the mixing process. An online evaluation of compound stability and for an ejection criterion is discussed in [23]. Unlike the conventional process, torque is no longer represented as a function of time but rather as function of the temperature of the mix. The change in viscosity as a result of the increase in temperature is used to establish a correlation between rotor torque and the reciprocal value of the batch temperature. The two process variables, torque and temperature, are continuously available throughout the mixing operation, so that the plot of the logarithm of the torque versus temperature yields a linear relationship.

#### 4.1.3.1 Plasticization

In the first stage of mixing, namely plasticization, a reduction in rubber viscosity is often a necessary precursor of a mixing cycle. This step is usually required for NR compounds that are introduced into the mixing chamber and have a high viscosity and poor processing performance. In the first stage, the molecular weight and the viscosity of NR should be reduced to a level that are suitable for incorporating an active filler. There are two reasons for plasticizing the non-viscosity-stabilized natural rubber:

- breakdown of the molecular weight of the rubber
- the blending of different rubber

In order to achieve suitable mixing in blends, it is necessary to first mix the two components until a uniform blend morphology is created. With increase in incompatibility of the blend components, the time required to produce a homogeneous rubber blend increases [24].

When the polymer is fed into the mixer, an initial power peak is reached. The height of the peak depends on the viscosity or the molecular weight of the rubber. As NR with high molecular weight exhibits high elasticity, it can produce pressure on the rotor flights. Consequently, the ram pressure has to be increased to squeeze the rubber into the mixing chamber. The power demand in the mastication stage increases by 40% when the feeding temperature is lower than 40 °C. To reduce the power demand, the temperature of the rubber should be increased to 40–50 °C. After this peak is passed, the power demand decreases steadily, due to the temperature rise in the rubber phase and the mechanical breakdown of the rubber chains and entanglements under the shear stress in the mixer. In general, mastication takes about 3–5 minutes, depending on the rotor speed and temperature in order to achieve the desired viscosity.

At moderate temperatures (< 100 °C), mechanical chain degradation takes place, in which the long chains in average break in half, and the average molecular weight decreases and the molecular weight distribution becomes narrower. As the temperature increases above 120 °C, the influence of oxygen on this process becomes noticeable [25]. Degradation of NR follows a reduction of the molecular weight and the molecular weight distribution is strongly increased. The degradation at this high temperature alters the properties of the NR by oxygen functionalities which are incorporated into the chain. It can be seen that mastication of the NR allows an optimum between 110 °C and 120 °C, where the molecular weight does not suffer from oxidative degradation. The influence of oxygen on the decrease in viscosity during mastication can be seen from a comparison with a degradation process in the absence of oxygen. The influence of an inert nitrogen atmosphere on chain degradation does not make the decrease in viscosity when mixing appear as pronounced as when the process takes place in air [26]. The addition of masticating agents increases the degradability of the NR. The desired viscosity is achieved with shorter masticating times and lower mixing energies when mastication chemicals are used. The use of NR grades with constant viscosity (CV grades) shortens the mastication step and contributes to a substantial reduction in mixing times and costs.

### 4.1.3.2 Filler Incorporation

The internal mixer is opened by raising the ram and filler is added into the mixing chamber. The filler incorporation starts with the end of the plasticization stage. The void volume of the filler aggregate is filled by the viscous rubber and the polymer-filler interaction starts. By lowering the ram, the filler is pressed into the rubber matrix and the flow in the mixing chamber follows the direction provided by the rotors. The drag-in process depends on the forces applied by the rotors and the pressure exerted by the ram on the material in the mixing chamber. As a result of the internal friction of the filler particles, the temperature increases and the flowability of the rubber phase gets better, because the viscosity of the compound decreases. The concerted action of these factors leads to the rubber melt being squeezed into the voids of the filler agglomerates and the air of the voids is replaced in the aggregates. The larger the void volume in the filler aggregate, the longer is the incorporation time. The incorporation of filler is associated with an increase in the density of the mixture. The merger of the filler and the rubber realized in the incorporation phase is shown in Figure 4.6.

Figure 4.6 Schematic diagram of carbon black incorporation into the rubber matrix

The filler particles become "wetted" by rubber chains and encapsulated in the rubber melt. At the same time, filler pellets start to break under the effect of the elongational or the shear stress into smaller fragments. In order to efficiently break the pellets, the stress level in the mixing chamber can be raised by increasing the rotor speed and efficient cooling of the chamber wall and the surface of the rotors. In such conditions, the breakage of the pellets is realized in a relatively short time. The higher temperatures allow the rubber chains to flow into the voids of the agglomerates. If the stress level in the mixing tool is low at elevated temperatures, incorporation of the filler into the rubber is delayed and a larger fraction of pellet fragments remains present in the mix, reducing the dispersion of the filler.

The significantly reduced volume of the compound leads to a steady decrease in power consumption. At the end of the incorporation stage, the mix no longer contains "free" filler and no longer emits dust. Because the void volume of the agglomerates is steadily filled with rubber, the entire volume of the compound shrinks and correspondingly the torque values tend to a minimum. After the air in the voids of the filler has been replaced by the rubber, "venting" of the mixer becomes necessary. In practical terms, a higher power peak at the beginning of the mixing cycle leads to smaller pellet fragments in a shorter time. In this stage of mixing, the viscosity of the compound reaches a maximum value as well as the torque or the power consumption.

Repeatedly it has been observed that, at constant filler loading, the incorporation time increases with the average molecular weight of the rubber. The incorporation time for compounds based on high-molecular NR is significantly longer than that for compounds containing NR with medium molecular weight [27]. In the case of synthetic rubbers the selection of grades with optimal viscosity has a similar effect on the incorporation stage. As can be seen from Figure 4.7 the incorporation time of carbon black N330 is strongly dependent on the Mooney viscosity of E-SBR [28].

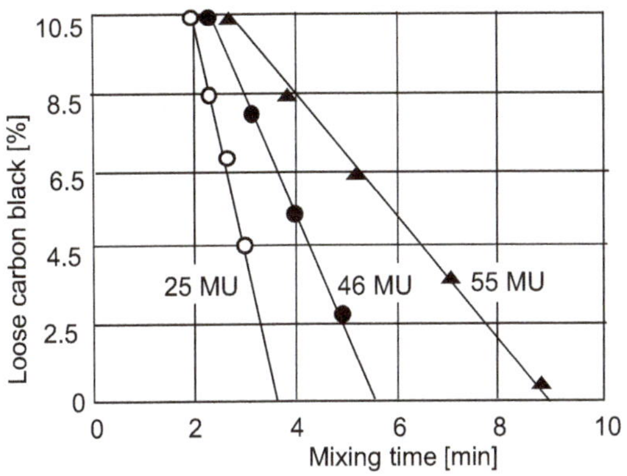


**Figure 4.7** Influence of the Mooney viscosity of E-SBR on the incorporation time of carbon black N330

Doubling of the Mooney viscosity leads to a three-fold increase in the incorporation time for non-oil-extended SBR. Because shortening the incorporation time contributes significantly to reducing the mixing cycle, the use of rubber grades with a viscosity tailored to the mixing target is of particular technical and economic importance. Standing in contradiction to the influence of the chain length of the rubbers on the incorporation is the influence of the structure of carbon blacks. The higher the structure of the filler (DBP-number and OA-number) and consequently the void volume of the carbon black (OAN), the longer is the process by which the rubber chains fill the void volume of the agglomerates. As shown in Figure 4.8, ram displacement increases significantly with the filler structure [29].

As a rule of thumb, the following holds for filled compounds:

- The larger the void volume in the agglomerates, the more mixing is required to fill this space by rubber chain.
- The lower the viscosity of the rubber, the easier the rubber can access the voids.
- The viscosity of the compound increases with the structure of the filler.

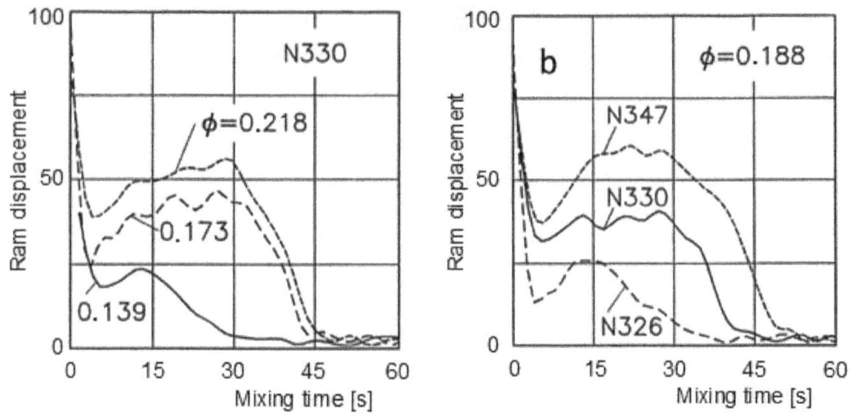


Figure 4.8 Influence of the structure of aggregates on the incorporation time in E-SBR

The incorporation time can be reduced by increasing the ram pressure and the rotor speed and by increasing the temperature of the rotors and mixing chamber as will be shown when the mixing procedure is described. If the ram does not reach its end position in the mixing time, undispersed filler particles will be dropped. It should be noted that the time when the ram reaches its lowest position is the beginning of the mixing process for the filler. The operator should allow for sufficient mixing time after the ram reaches its lowest position to make sure that the filler particles start to be dispersed.

In principle, after dosing of the filler, the ram is lowered fast and the material is pressed into the mixing chamber. By the action of the rotors, the material is transported again in the region below the ram and therefore the pressure pushes the ram back into a higher position. Recording of the ram seating can be done with commercially available tools. Armed with a precise knowledge of the incorporation stage, the operator can adjust the mixing cycle as a function of the type and level of the fillers and the rubber viscosity. However, depending on the degree of mastication and the mobility of the rubber chains, the incorporation stage can vary in duration.

The incorporation time can be determined from the ram displacement and the final ram seating. During incorporation, the ram position shows a repeated up and down movement which progressively lowers to a final position when the incorporation stage comes to an end [17]. As shown in Figure 4.9, the ram seating time increases with the volume fraction of the carbon black in the compound. This result is easy to understand because the volume fraction also increases the total void volume in the mixture.

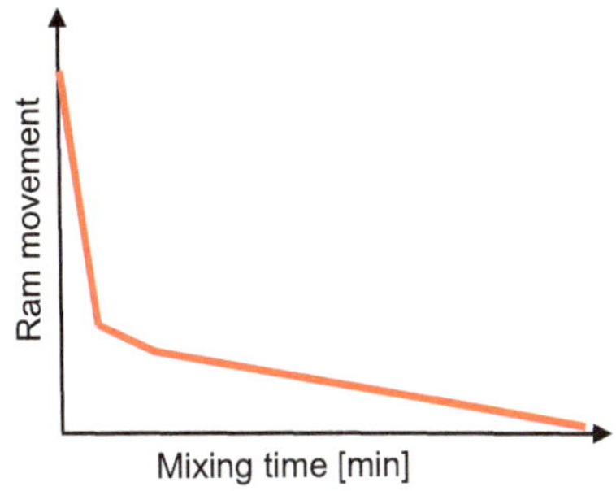


**Figure 4.9**
Ram movement controlled by ram position control system

In the case of precipitated silica, the silanization is carried out in several consecutive steps, with the mixing chamber being like a chemical reactor [30]. The silanization process requires a longer reaction time, which extends to the entire mixing process. With progressive silanization of the silica, the incorporation phase of the silica is significantly improved. To achieve the best processing and performance properties of the vulcanizate, it is important to make sure that, during the mixing process, the coupling agents and the silica are mixed to a very high degree but the reaction with the rubber should be prevented.

The most common mixing technique used in both industry and academia for preparing organo-clay in rubber matrices is melt mixing. The direct addition of layered silicates does not lead to the desired result because the polymer is not able to penetrate into the interlayer. Mixtures are formed which contain unincorporated layered silica. Intercalated or exfoliated organo-clays are easily produced in this way. By adding the ammonium ion, the distance between the layers can be increased and the rubber can penetrate this area [31–34]. The wet-mixing that is discussed at the end of this chapter can be used for the pristine montmorillonite to be intercalated in the rubber phase and provides good rubber-clay interaction. When incorporating MWCNT into a rubber matrix, it is found that the ram seating time is significantly shorter than with active carbon black. The shorter incorporation phase of MWCNT is attributed to the lack of cavities in the aggregates. The difference in ram seating time is shown in Figure 4.10 [35].

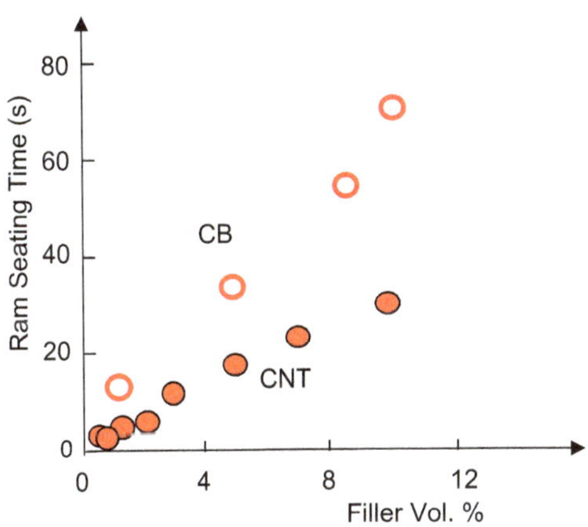


**Figure 4.10**
Ram seating times for carbon black and carbon nanotubes in HNBR compounds

The simplest way of incorporating carbon nanotubes into rubber is melt mixing, whereby the carbon nanotubes should be added at elevated temperatures and shear forces during incorporation [36].

#### 4.1.3.3 Filler Dispersion

The most important step in the dry-mixing process is the dispersion stage or deagglomeration in the mixing diagram. For carbon black and precipitated silica, this stage starts when the pellet fragments are encapsulated by the rubber and a part of the filler surface is "wetted" by the rubber during the incorporation of the filler. If the stress reached in the mixing tool is high enough, the agglomerates are gradually eroded and the size of the agglomerates is exponentially reduced to the aggregate form. As mixing progresses beyond the black incorporation time, the carbon black is deagglomerated and dispersed, reducing the aggregate interaction. Care must be taken that the effective stresses do not fall below the minimum value, due to an increase in the temperature. The dispersion increases with the molecular weight of the rubber and the void volume of the filler as well as the shear rate of the rotors and decreases with the smaller particle size of the filler. This process is visualized in Figure 4.4 by an increase in both torque value and power consumption in the internal mixer. The high viscosity of the compound at the beginning of the dispersion stage enables high elongational stresses and shear stress and elongational stresses to be generated in the efficient mixing zones. The high energy input is necessary in order to break down the filler pellets into the aggregates as quickly and as completely as possible. The pellet strength has a great influence on dispersibility of the filler: any action leading to higher tensile strength of the pellets results in a higher resistance of the pellets against breaking and decreases at the same time the dispersibility and increases the chance that undispersed filler remains present in the compound. The resistance of the pellet fragments to mechanical degradation depends on the cohesive forces between the filler aggregates within the agglomerates as well as the pelletizing technology used during manufacturing. Polymer degradation occurs at rising temperatures and lead to decreasing viscosities and thus to decreasing dispersion forces at the rubber chain. If these forces are of the same order of magnitude as the cohesion forces, the agglomerates can no longer be separated and the dispersion process comes to a standstill.

In order to increase the dispersion and at the same time shorten the duration of the mixing process, it is advisable to increase both the ram pressure and the rotor speed in the internal mixer. By prolonging the mixing process at a high filler concentration, the dispersed particles form three-dimensional filler networks which govern the rheological properties of the compound and the level of the vulcanizate properties. By progressively continuing the dispersion process, the aggregates are surrounded by rubber layers and uniformly distributed in the matrix. This is indicated by an asymptotic drop in the torque value and power consumption. Since temperatures above

130 °C easily occur when working on dispersing highly active fillers, the crosslinking chemicals must be added in a separate operation during final mixing. At the end of mixing process, the dispersion stage and distribution stage of the filler inevitably overlap, so a formal separation of these two mixing steps is not possible. However, it is important that the distribution takes place after the dispersion and that the mixing process is designed for a short time that takes both mixing stages into account.

The desired state of filler dispersion in rubber compounds is dispersion of the filler agglomerates to the level of aggregates (nano-scale dispersion):

- The filler structure must be high.
- The rubber should undergo intensive interaction with the filler.
- The mixing equipment has to be efficient and operated well.

Consistently, the tensile limits for the agglomerates are reached sooner when the surface activity or the filler structure is reduced. Thus, the resistance of the agglomerates to extensional and shear flow, as well as the work required to keep cohesive inter-aggregate forces that keep the agglomerates together, decreases while the structure of the filler increases. In the course of dispersive mixing, interaction between filler and polymer occurs at the interfaces. This interaction is proportional to the surface energy of the fillers and causes the polymer bound to the surface to assume a stiffer consistency.

The influences of the specific surface and the structure as well as the polymer viscosity and the proportion of plasticizer on the incorporation time and the dispersion time are shown qualitatively in Table 4.1.

**Table 4.1** Influence of Incorporation Time and Dispersion Time [5]

| | Incorporation Time | Dispersion Time |
|---|---|---|
| Rubber molecular weight | ⇑ | ⇓ |
| Decreased particle size | ⇓ | ⇑ |
| Increased structure | ⇑ | ⇓ |
| Content of plasticizer | ⇓ | ⇑ |

Various attempts have been made to theoretically describe this important mixing stage. A mechanism that seems to come close to reality is the "onion skin" model [37]. The model assumes that pellets are broken in an initial stage of mixing. Subsequently, aggregates and agglomerates are progressively peeled off from the pellet fragments which reduce in time their size and number in the rubber matrix. The nano-sized aggregates and agglomerates are distributed homogeneously by multiple splitting of the flow stream. The "onion skin" model is shown in Figure 4.11.

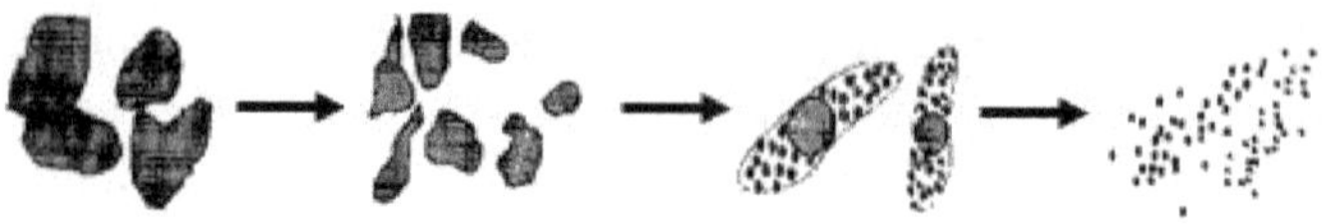

Figure 4.11 The "onion skin" model describes the disintegration of the pellet fragments to form aggregates

It can be understood that the number of small agglomerates and aggregates increases exponentially during the dispersive mixing. The process becomes interesting due to the erosion of the aggregates, when a large number of very small particles enter into a strong interaction with the rubber chains. This interaction confirms good phase bonding in the rubber compound. There is only a slight tendency to detach at the filler-polymer phase boundary, even at high elongation. The process that is emerging here is irreversible. The mistakes made during the process, especially the proportion of poorly dispersed agglomerates, can no longer be eliminated downstream. A number of poorly-dispersed filler fragments, however, remain during dry-mixing in the rubber compound. These undispersed filler fragments can give rise to the appearance of dynamic cuts in the vulcanizate.

#### 4.1.3.4 Filler Distribution

With some delay to the dispersion stage, closely spaced agglomerates and aggregates are randomly transported through the mix. The "distribution" stage is supported by repeated shearing and folding of the mix and continues after the dispersion stage is finished. Due to the randomization, the filler aggregates are separated from each other. And the more homogenized filler distribution is the practical result of this distribution stage that cannot really be separated from the dispersion stage. Differences in local filler concentration are reduced and concentration fluctuations inside the mix are asymptotically reduced and become equal. The quality of the distribution mixing process is evaluated based on the uniform distribution of the components to be mixed. Their concentration in any samples of the mixed material should be the same everywhere if possible. A qualitative measure of mix quality could be the mean of the standard deviations of the concentration of filler particles in the specified sample. Therefore, with increase in homogenization of the filled compound, the flow properties of the compound are standardized, which is an advantage in subsequent shaping processes. However, a certain proportion of undispersed filler (pellet fragments) will remain present after the mixing process. It has been reported that the undispersed filler can act as flaws that can initiate cracks in vulcanizates subjected to cyclical dynamic load. Thus, the quality of the mixing process and the final degree of filler dispersion and distribution play a decisive role in product performance and durability.

The mixing of rubbers and fillers is an energy-intensive process during which endeavors are made to achieve both a good dispersion that ensures a strong filler-rubber phase adhesion and a uniform distribution of the filler that provides easy processing.

Because carbon blacks and precipitated silicas are delivered as pellets with a certain pellet strength, the task of achieving a good mixing quality can be challenging, because a certain amount of the filler is not properly dispersed. The quality of a compound achieved during mixing depends on the one hand on the mixing tool used and the efficiency of the process parameters and on the other hand on the nature of the formulation ingredients, the composition and the order in which the compound ingredients are added. Although dry-mixing is still mostly practiced today, there are a number of disadvantages when working it into the rubber matrix. The disadvantages of dry-mixing techniques are here:

- severe wear on the internal mixer
- poor filler dispersion
- high compound viscosity
- high die swell
- poor processability
- lower abrasion resistance

Although these disadvantages have been known to those skilled in the art of rubber mixing for several decades, changing the mixing technology is a lengthy process and it is dependent on the improvement in the mixing process and on existing market conditions.

## 4.2 Dry-Mixing Technology

Irrespective of the way a mixing chamber is laid out, the internal mixer is the heart of the installation. In the internal mixer, the mixing of the polymer together with fillers, processing oils, antioxidants, curing agents takes place. The production capacity of a mixing line and the quality of the rubber compound produced are determined by the size of the internal mixer and of course by the mixing efficiency. The filled rubber compound has to be produced economically [2, 5, 7]. The dry-mixing technology used in a rubber company can be divided into:

- batch-wise process on internal mixers
- batch-wise processes on two-roll mills
- continuous processes in mixing extruders

The somewhat older technology of the batch-wise process on two-roll mills has now largely been replaced by the batch-wise process in an internal mixer. Most mixing of rubber and filler is carried out in internal mixers, which are manufactured by a number of companies. Mixing of rubber will increase the ability to exert a high localized shear stress in the elastic rubber material during mixing. At the end of the mixing process the rubber compound should be homogenized and a good filler dispersion will have occured. The overall effectiveness of mixing requires both actions, the dispersion and the distribution of the filler in rubber compounds. Mixing devices have to be powerful and must be designed in a way to withstand the forces acting through high viscous and often the presence of solid abrasive compounds. The obtained mix must have suitable rheological properties that allow for optimal downstream operations.

In the following, the mixing tools used to produce rubber compounds in the form of sheets and strips are presented. These semi-finished products are further processed on calenders or extruders in delivery form.

### 4.2.1 Two-Roll Mills

Two-roll mill technology was invented at the beginning of the 19$^{th}$ century for processing NR [4, 5]. Today, two-roll mills are mainly used as downstream machines after the dispersion and distribution has been produced in internal mixers or as preheating units for downstream units. Two-roll mixers consist of two parallel cylinders (rolls) rotating in opposite directions [4, 5]. The working principle of a two roll-mill is shown in Figure 4.12.

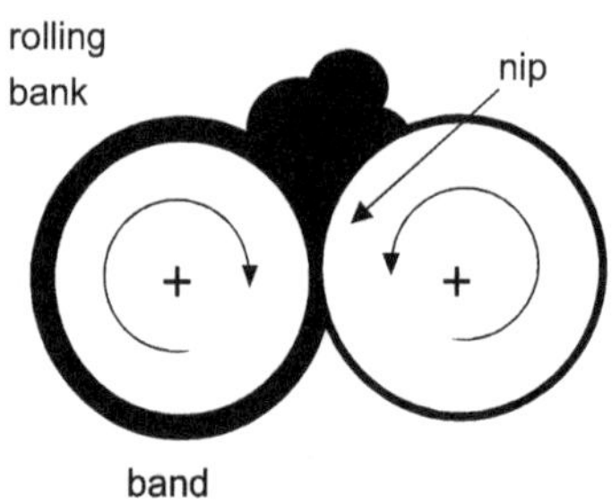


**Figure 4.12**
Schematic representation of a two-roll mill

The rotation of the rolls pulls the compound through the clearance (nip) between the rolls. High strains and stresses are generated in the nip region, whereas homogenization is accomplished by cutting and rolling the sheets. The differential speed of the rolls induces shear ("friction") in the rubber compound between the rolls, which essentially contributes to increasing dispersive mixing of the filler. The higher the friction, the greater are the shear and elongation flow, and thus the intensity of the dispersive mixing increases. The friction also ensures that the mixture head usually runs on the slower-rotating roller [38]. The most important process parameters of a two-roll mill are:

- the throughput through the nip
- various material guide jaws
- cutting devices
- roller temperature control
- stock blender for automation of the dispersive mixing process

The use of hydraulic drives has made it possible to introduce small mills with the ability to drive the rolls either forwards or backwards, allowing high shear to be achieved in the nip between the rolls. Use of such mills appears useful for the mixing of very soft rubbers (i. e. silicone rubber). The specific energy conversions as well as the feed strip temperatures and the mean residence times can be changed. Increasing the friction and decreasing the clearance between the cylinders result in an increase in mixing efficiency through higher shear stresses. From the narrowest point of the nip, a pressure maximum forms, which, in addition to the drag power of the two roll surfaces, also causes a high-pressure flow through the nip. The maximum pressure initiates a flow of material through the roller nip, but also a volume flow against the direction of flow of the roller.

The throughput of the mixed compound is regulated by the clearance and the number of revolutions that correspond to the residence time in the shear stress regime necessary to attain the desired mixing quality. In particular, it is the ability to introduce high specific mixing energies without a significant increase in the compound temperature that makes two-roll mills useful, even in modern rubber manufacturing. Experience shows that often the compound often adheres to one of the cylinders, the one that rotates at a lower speed. This increases the dispersion mixing efficiency in the mixing gap. The mixture sheet runs on the slower-running front roller. However, this is not always the case and can vary from mixture to mixture. The sheet of mixture initially adhering to the slow front roller will switch to the faster rear roller when the gap thickness decreases in line with a certain friction- and polymer-dependent function. An increase in the roller temperature can prevent the transfer of the compound to the roller that rotates with higher angular speed.

In order to facilitate the randomization and the homogenization of the filler distribution, multiple cutting and rolling of the blanket are employed. Useful homogenization of filled compounds is achieved by the "stock blending unit" placed above the two-roll mill. The operating principle of a stock blending unit is shown in Figure 4.13 [5].

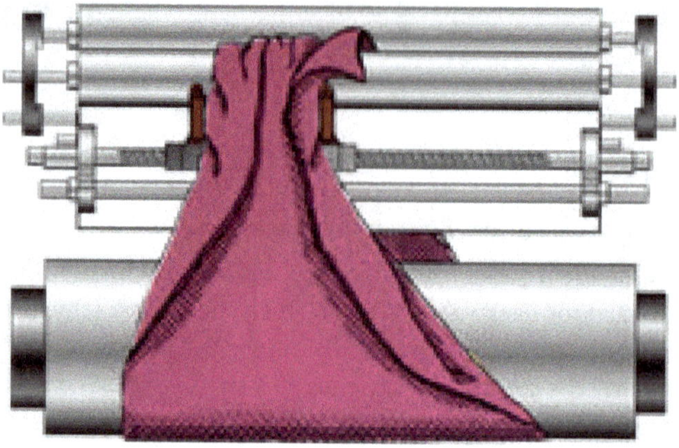

**Figure 4.13** Stock blending on a two-roll mill

To carry out cross-mixing with such a unit, the sheet is cut to form a narrow strip, which is fed manually between guides on the stationary carriage and into the nip of the stock blender rolls. The traversing carriage is then set in motion, and continues to traverse laterally drawing the full width of the sheet into the blender and feeding it back into the mill nip at the diagonally opposite position. The advantages of mixing on two-roll mills are the easy monitoring of the progress made, removal of the compound from the two-roll mill in the form of sheets and the high dispersive mixing action in the nip region and the distribution effect in the roll bead which creates a distribution which is enhanced by cutting the mixture skin or using a stock blender. The active zones for dispersive mixing on a two-roll mill are the rolling bank and the high elongational regions in the nip. However, mixing on a two-roll mill is dependent on operator intervention. Maintaining an active rolling bank is a matter of not overloading the two-roll mill [11]. Typical capacities for a range of machine sizes are indicated in Table 4.2.

**Table 4.2** Capacities (in kg) for Two-roll Mills [11]

| Length (m) | Diameter (m) | Minimum Capacity | Maximum Capacity |
|---|---|---|---|
| 0.75 | 0.35 | 9 | 13.5 |
| 1.5 | 0.55 | 20 | 32 |
| 2.1 | 0.75 | 80 | 136 |

The disadvantages of mixing on two-roll mixes result from the open handling of the filler and other ingredients that can cause house-keeping problems and health risks, especially when hazardous substances have to be incorporated into the compound. Other disadvantages are the prolonged mixing times and reduced possibilities to control and automate the mixing process. Although mixing on two-roll mills has some disadvantages, due to their open structure, this technique leads to compounds with a high degree of filler dispersion, which is why two-roll mills are still used in some production areas.

Often, two-roll mills are used to reduce the viscosity (or the molecular weight) of NR to an optimal value for further processing and filler incorporation. Mill mixing is used for specialty compounds and small batches, and for addition of curing agents in a final stage. Because of the good cooling effect, rubber compounds are produced on two-roll mills in a single stage. More frequently, two-roll mills are used to cool down the mixed compounds dumped from internal mixers and to increase the degree of dispersion of the compounds or to warm up compounds before calendaring and extrusion. However, two-roll mills are still indispensable for mastication of NR and as downstream machines as pre-warming mills for warm-feed extruders, as feeder roll mills for calenders and for the production of reclaim.

### 4.2.2 Internal Mixers

Most mixing of rubber compounds is done in various types of internal mixers. Mixing with internal mixers requires handling the preparation of the mixture in a closed mixing chamber with low dust pollution. This technology enables significantly shorter mixing times and a higher quantity exchange, better reproducibility of the mixing process and a higher degree of automation, harmful ingredients of dust emissions are reduced to a minimum. The supply of rubber in bale form has dictated that internal mixers are mainly batch-operated machines capable of being fed with bales. While there are economic reasons for the operation of large mixers, the investment needed for a mixing chamber is high. Internal mixers of this type have a service life of around 6 years before they need to be overhauled.

A major step forward in mixing technology was the development of a closed internal mixer equipped with a ram to introduce the ingredients into the mixing chamber; it was invented by Fernley H. Banbury in 1916 [39]. The advantages of the internal mixers come from their ability to exert high shear and elongational stress that increase dispersive mixing and the homogenizing effects of a filler distribution in rubber compounds. The production capacity of a mixing line and the quality of the compounds produced are determined by the size of the internal mixer and the mixing efficiency reached in production. The dosing and pre-weighing of the recipe to be mixed takes place in batches, while the downstream processes are operated continuously. Internal mixers consist basically of three sub-assemblies:

- the feed hopper
- the mixing chamber
- the rotor geometry
- the base plate

The structural modules of an internal mixer are shown in Figure 4.14.

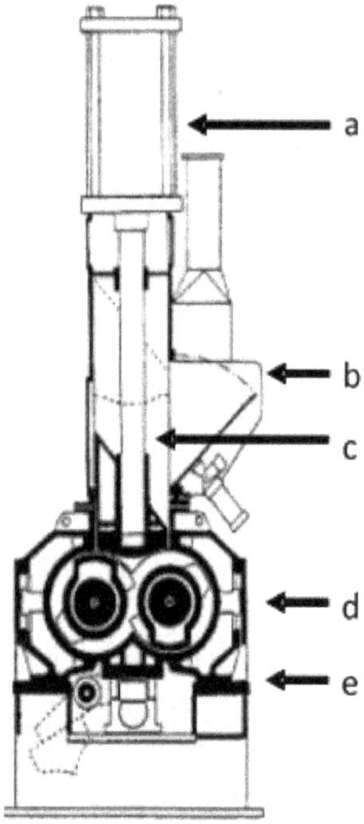


Figure 4.14
Internal mixer: a) feed hoper, b) admission of the rubber and filler, c) ram, d) mixing chamber, e) base plate

Compared with two-roll mills, the advantages of internal mixer lie not only in the closed mixing chamber and low dust emission, but also in shorter mixing times and much larger volume output, the reproducibility of the mixing cycle and automation of the process steps, which makes for increased quality of the rubber compounds.

#### 4.2.2.1 Feed Hopper

The upper part of the mixer consists of the feed hopper, which is an upright housing inside of which is a ram in which the ingredients are supplied to the mixing chamber [7]. The pneumatically or hydraulically operated ram hopper can exert variable pressures between 2 and 12 bar on the mixture. The feed hopper has a charging entry at the front which is opened and closed by a hydraulic cylinder. Through the charging entry, rubber with a viscosity suitable for a good filler incorporation is either charged via a belt weigher in bales or via cuts produced beforehand by a rubber cutter. Low molecular weight formulation ingredients are fed in suitable packages through the charging entry or the openings in the sidewalls of the feed hopper. The ram can be lowered or raised within a vertical shaft. In the raised position of the ram, the formulation ingredients are charged to the mixer. A temperature control inside the ram is realized by a cooling system and serves to prevent heating of the ram surface and sticking of compound ingredients on it. To prevent dust deposition on the top of ram, the surfaces are angled. The dust drops down into the mixing chamber.

Once the compound ingredients have been introduced into the mixer, the ram is pushed down under a constant pressure to seal and close the mixing chamber and to prevent load peaks during mixing. During the mixing process, the ram presses the compound ingredients into the mixing chamber located under the feed hopper so that they become part of the mixing process. The applied ram pressure is 4 to 12 bar. In order to avoid operating problems and to reduce the costs of the compressed air supply and distribution, hydraulic feed hoppers have been developed that replace the old pneumatic feed hoppers. Two different types of hydraulic feed hoppers have been designed: with one cylinder and with two cylinders [17]. The two hopper designs differ with respect to energy consumption. Comparative calculations of operating costs reveal that the single-cylinder feed hopper performs better than the two-cylinder hopper. Other advantages of a hydraulic feeder are:

- reduced noise level
- rapid ram-force setting
- rapid and controllable ram displacement, and
- reduced operating costs

The hydraulic feed hopper ensures both controlled ram displacement and ram pressure. Thus, this module of the internal mixer facilitates repeatable mixing conditions, especially as regards changing the speed of the ram displacement when ingredients are introduced into the mixing chamber. The major advantage of a hydraulic feed hopper is consistency of the mixing operation, independent of a variable air supply in the mixing chamber.

#### 4.2.2.2 Mixing Chamber

The central part of any internal mixer is the mixing chamber, which consists of two overlapping circular halves in which two counter-rotating rotors driven by an electric motor provide the mixing action. Here the different raw materials and functional formulation ingredients are mixed together to form a well-dispersed compound. Due to the high viscosity of the filled compound, considerable forces are required for mixing and, therefore, the mixing chamber must have a solid construction [17].

The walls of the chamber should be hard and provided with wear protection in order to withstand the abrasive action of the solid fillers. The wear protection of the mixing chamber consists of hard steel plates welded on the base steel of the mixing chamber (ram, chamber wall, rotor and drop door). Because the first hard coatings used in internal mixers were made from cobalt alloys that forms cracks during the coating process, they were replaced by alloys with high chromium content that provide a crack-free and corrosion-resistant hard coating. The major advantage of hard coating is the absence of cracks, even after a longer period of use. Crack-free coatings prevent the undermining corrosion which occurs frequently with standard hard coatings and which can give rise to cracks [17]. Understandably, the cost of an internal mixer increases with the quality and hardness of the surface of the mixing chamber.

In order to avoid that fillers and other powdery compound ingredients as well as liquids and pastes can escape between the rotor shafts and the mixer end frames, the mixing chamber needs to be sealed. Depending on the way the surface pressure is applied, a distinction is made between:

- self-sealing dust seals
- spring loaded dust seals, and
- hydraulic dust seals

From the process point of view, hydraulic dust seals show advantages because deposits of materials and contaminations are no longer possible with this new sealing system. Modern mixers are equipped with valves to inject the mineral oils into the mixing chamber. Smaller mixers require one injection valve installed in each mixing chamber half. Large mixers have implemented two valves in each mixing chamber half. This configuration permits not only the injection of larger plasticizer volumes into the compound, but also the admission of different types of plasticizers.

The heat generated during the mixing process can lead to undesirable phenomena, such as degradation of the polymer, consumption of antioxidants and premature crosslinking of the rubber matrix. It is therefore important to remove the heat from the mixing chamber effectively by holes which are drilled near the inner surface of the chamber wall and also from the rotors, by means of a flow of cooling water.

In order to avoid temperatures that impair filler dispersion or facilitate premature scorch, thermocouples are placed in the mixing chambers that record the temperature of the rubber mix. Usually, the thermocouples are positioned on the top of the drop door and in the side wall of the mixing chamber and allow a precise temperature measurement. Good contact between the thermocouple and the rubber mix is a basic requirement for the temperature measurement in the mixing chamber. It has been reported that the temperature measured in the vicinity of the discharge door is 20–30 °C higher than the temperature measured at the side wall. The difference is attributed to compound flow around the thermocouple and more accidental contact of the compound with the chamber wall [17]. The quality of thermocouples has improved significantly in terms of accuracy and speed in recent years. This means that temperatures in the mix are displayed much earlier, enabling better process control.

#### 4.2.2.3 Rotor Geometry

Two rotors with a cylindrical shape are placed in the mixing chamber to ensure dispersive and distributive mixing. Several blades are attached diagonally to the rotor axes and transport the material lengthways and circumferentially and through the gap between the rotors and the mixing chamber wall or between the two rotors.

The rotors are driven by an electric motor while pressure on the mixing chamber is exerted by the ram. Due to the counter-rotating movement of the rotors, the rubber and filler are introduced into in the mixing chamber and carried through the available space between the rotors and the chamber wall. The major functions of the rotors are:

- to impart shear and elongational stress into the compound
- to disperse and distribute the filler particles
- to dissipate the heat generated during mixing

Depending of the rotor design, large strain and concentration randomization can be accomplished between rotors or between the rotors and the wall of the mixing chamber. The material is transported in the longitudinal and circumferential directions through the rotor flights, which are provided with helical openings. The dispersive mixing takes place through the extensional stress during transport through the working gap, while the dispersive mixing takes place through flow divisions and rearrangement throughout the mixing chamber. The extensional and shear flow for effective filler dispersion appears essentially in three regions of the mixing chamber:

- in the nip region, where the rotors rotate in opposite directions
- in the region between the rotors and the wall of the mixing chamber and
- between the rotor flights of a given length and disposed at a defined angle to the direction of rotation

Thus, the rotor geometry controls the degree of dispersive and distributive mixing in an internal mixer. In recent years attempts have been made to numerically simulate stream fields, temperature distribution and transport flows [11]. However, the simplifying basic assumptions regarding the degree of filling, rheological material laws and the resulting deformations in the mixer lead to conclusions that are not yet sufficient to permit an exact calculation of the rotor geometry. Therefore, the rotor geometries are predominantly designed empirically, however, in recent years the rotor geometries and the mixing chamber have been developed on the computer.

The electrical drives that get the rotors moving allow for constant rotor speed, or variable driving mechanisms that permit the speed of the rotors to be controlled in line to meet the need for optimal dispersion and energy savings. In the past, DC motors represented the standard for internal mixers, as AC motors with power of more than 200 kW were too expensive and failed to deliver the high torque in the range of low rotor speed. The efficiency of DC drives has shown to be in the range of 40%. Modern AC drives capable of more than 200 kW have been available since the year 2000 and have become technically the predominant drives. They reach an efficiency of more than 60% [17].

#### 4.2.2.4 Tangential Rotors

The first rotors employed in mixing technology were tangential rotors. Typically, the sum of the two rotor diameters is smaller than the two mixing chamber halves. As a consequence, tangential rotors do not overlap and can rotate independently. Figure 4.15 presents a schematic view of tangential rotors in a mixing chamber. Because the diameters of the rotors do not overlap, each of the tangential rotors can operate at a different speed, providing more shear stress and allowing friction. The speed of the faster-running rotor is 10 to 15% greater.

**Figure 4.15**
Schematic view on a tangential rotor

Assuming the total strain or the elongational flow is a key parameter in dispersive mixing, the design of the rotors is of crucial importance for increasing mixing efficiency. The geometry of the rotor flights, the angular position, the clearance between tangential rotors and the mixing chamber wall, or the width of the rotor flights and the clearance between the intermeshing rotors have been significantly improved over the decades. However, the extent to which dispersive mixing leads to a high degree of dispersion is fundamentally related to the type of filler, the filler loading and the molecular weight of the rubber and the chemical constitution of the chain segments. For example, a blend of S-SBR/cis-BR is used for passenger car tire treads in order to achieve good damping on the one hand and effective abrasion resistance on the other.

The dispersive mixing region of a tangential rotor design comprises the flow zone in the approach to the rotor tip and the zone of small clearance between the rotor tip and the chamber wall. For tangential rotors, the rotor tip is narrow, giving rise to a short flow path and low residence time for the compound in this region. Of particular importance is the axial flow and transverse flow of the mixture in the mixing chamber, the number of mixture redistributions and passages generated by the rotor blades for the rubber mix [4]. The effects of the mixing action of the rotors strongly depend on the type of flow generated by the rotors. This is largely determined by the number and length of the rotor blades, their axial angle, the width and height of the rotor blades. Thus, a complex field of influencing factors have a greater or lesser influence on the dispersion and distribution that depends on the processing conditions and, of course, the compound's composition.

As shown in Figure 4.16, a pressure front is formed in front of the rotor flight, which gives rise to deformation in the compound and, therefore, to high shear and elongational stress. A pressure build-up takes place in the clearance between the mixing chamber wall and the rotor flight, reaching a maximum value in front of the rotor flight. As a result, a mixture of shear stress and elongational stress takes place in the flow front upstream of the rotor.

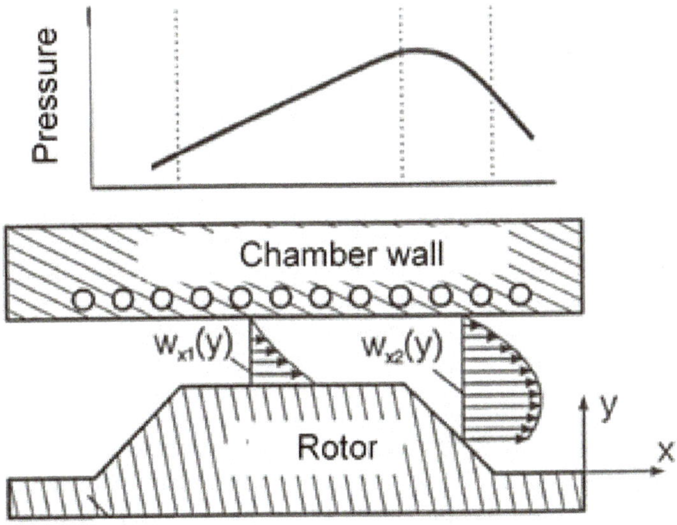


**Figure 4.16** Pressure front is formed in front of the rotor flight

Tangential rotors can be constructed from a single casting that forms the rotor body and the two shafts (one-piece rotor) or from two pieces, where the cast rotor body is made independently of the rotor shaft and secured to it by shrink-fitting (two-piece rotor). The rotors must be robust to withstand the strong forces acting during mixing of filled compounds that exhibit extremely high viscosity. The strength of the one-piece rotors is achieved by the larger rotor body walls, whereas the very stable shaft in the center of the new technology two-piece rotors provides a much higher resistance torque and a reduced risk of being damaged under overload.

In one-piece rotors, cooling occurs through a pipe with several nozzles that sprays the water upward against the inside wall of the rotor body. The cooling is effective only in the top half of the rotor body. Special cooling of the rotors is particularly important. It has been possible to increase the surface area of the cooling ducts, reduce the distance between the cooling ducts and the rotor surface and improve heat transfer, which has almost doubled the cooling of the rotors. In two-piece rotors cooling, also called "super-cooled" rotors, water flows in a spiral channel in the cast rotor body. This cooling system, which is based on forced water flow, provides much more efficient cooling inside the complete rotor body than the spray cooling in the one-piece rotors. As a result of the reduced wall thickness of the two-piece rotors, the cooling efficiency is additionally increased. Consequently, the new technology rotors allow for improved cooling by 50%. This improvement makes it possible to shorten the mixing cycle and at the same time to attain a better mixing quality and better filler dispersion [17]. The mixing efficiency is influenced primarily by the nature of the rotors. Important features of tangential rotors are:

- the number of rotor flights
- the position and length of the blades
- the angular orientation and the profile of the blades, and
- the tip width of the blades

Several zones of dispersive mixing are identified for tangential mixers. The region where the material is transported rotationally provides for efficient folding and mixing streams that are able to generate dispersive mixing. When the mixer is properly filled, the material no longer flows along the rotor flights only, but passes through the clearance between the narrow rotor flights and the mixing chamber wall. Under this regime the mixing action changes from a more distributive mixing to a more dispersive mixing

The different distribution of the rubber compound to be mixed in an internal mixer is determined by the degree of dispersive and distributive mixing in a tangential mixer, in which the distributive mixing is preponderant. [17]. The material to be mixed is positioned in the front of the flight and can follow three flow paths:

- it is transported with the rotor flight
- it can pass over the gap between the blade tip and the chamber wall
- it can be pushed along the flight and pass around the rotor end

Rotational transport with the rotor flight requires a high shear rate and a great power input. The temperature in the flow increases and leads to dispersive mixing. With increasing filling, the flow path along the flights becomes more and more blocked and the material begins to pass over the gap between the flight tip and the chamber wall. Therefore, high shear stresses are generated, such that, with increase in fill factor, the mixer changes its characteristic from distributive to dispersive mixing [7].

When synthetic rubbers that need reinforcing fillers came on the market, the N geometry rotor was developed [40]. The characteristics of N rotors is a long blade and a short blade placed along the rotor axis. The height of the shorter blade is 70% of the longer one. The N rotor type was replaced gradually by the full-four-blade rotor that was originally developed for the purpose of reducing masterbatch mixing times [41]. The rotor consists of elongated flights which force the rubber compound to pass over the mixing tip, thus resulting in more laminar and dispersive mixing, and provide high energy input. This is mainly used for masterbatch mixing of rubber blends and effective re-milling. The two longer blades with a lower angle are on the same end of each rotor and the shorter blades are placed on the opposite side. Due to the long main blade and the low angle and the large flight depth, these rotors permit a rapid dispersion of the rubber compound and provide good discharge properties.

A further development is the synchronous technology rotor (ST®-rotor) developed in 1988 [42]. Characteristic is the presence of long blade on one end of the rotor and short blades at the other end of each rotor. In addition, the helix angles of the two blades on each side are the same, but different from one side to the other. The even speed of the tangential rotors and the angular position of the rotors to each other, which is adapted to the respective mixing task, increases the filling level and reduces the mixing time and mixing temperature.

Therefore, the rotors provide an improved mixing performance. At even rotor speed the synchronous technology allows an increase in the fill factor and a reduction in the compound temperature and thus, a reduction in the mixing time. In order to improve the distribution of the filler and other ingredients and to control the temperature in the mixing chamber, the ZZ2 rotor was developed [43]. The rotor is a four-blade rotor consisting of two long opposing central blades and two short end blades to drag the material from the chamber ends. The ZZ2 rotors typically restrict the flight angles of the main vanes to 40° and shortened their length to more than half of the rotor length, approximately. A second passage is introduced at the side at which the rotor blade started. The rotor is known for its good distributive mixing and cooling abilities. Mixing practice has shown that this rotor concept allows reductions up to 30 °C in discharge temperature when compared with full-four-blade rotors [44]. The application of fluid dynamic studies led to the development of the "new technology tangential" (NTT) rotor [45]. The rotor comprises a variable blade angle along the blades of a four-blade rotor. This has the effect of improving the fill factor and productivity, whilst reducing the variation in rheological properties of the compound. New developments intended to improve the degree of filler dispersion and the thermal performance are the high dispersion supercooled rotor (HDSC) and the maximum dispersion supercooled rotor (MDSC) [46]. The HDSC rotors consist of long blades that are extended to 80% of the rotor length and short blades that form a passage through which the compound has to flow during the mixing operation. The HDSC full four-blade rotor has significantly improved cooling, compared with a normal four-blade rotor. The MDSC rotors are equipped with four long blades; spiral cooling of the rotors guarantees adequate removal of the compound heat and allows for higher compound viscosity and greater shear stress. CAD models of HDSC and the MDSC rotors are shown in Figure 4.17 [46].

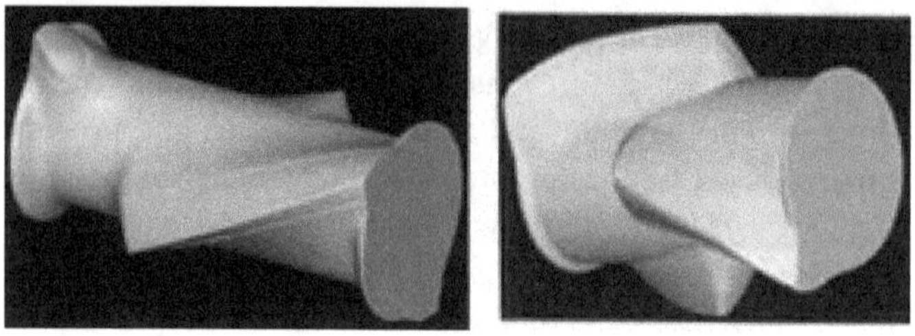

Figure 4.17 New rotors made by Harburg Freudenberger

Tangential mixers are produced with mixing chamber volumes of 250–700 liters and are used for multi-stage mixing in the tire industry. The advantages of tangential rotors are:

- good drag
- greater useful volume
- more rapid emptying of the mixing chamber
- very good machine utilization over the production year

It is reported that shortening the mixing time in large mixers by 5% leads to consistent annual cost savings [47].

### 4.2.2.5 Intermeshing Rotors

Internal mixers with a much different mixing chamber were developed in the first half of the last century [49, 50]. For this type of intermeshing mixer, the sum of the rotor diameters is greater than the two halves of the mixing chamber. The rotors overlap and have to run at the same angular velocity; they do not generate "friction." The dispersing effect of the intermeshing rotors takes place in the area between the rotors. A schematic illustration of an intermeshing mixing chamber is shown in Figure 4.18.

Figure 4.18 Schematic view of an intermeshing mixer

Two main features control the mixing efficiency of intermeshing mixers, namely the narrow clearance between the rotor blades that provide extensional flow and the large surface of the rotors that provide efficient cooling of the rotors and the mixing chamber. The design of intermeshing rotors enables them to act like an extruder screw, pumping material axially but encouraging leakage flow into the clearance between the chamber wall and the rotor for dispersive mixing. As each rotor pumps in the opposite direction, circulation occurs, with forced exchange of material between the rotors at each end of the mixing chamber. Compared with the free space in the mixing chamber, the volume of the rotors is large and so the volume of compound that can be mixed is smaller than in the case of tangential rotors. If the average clearance between the rotors is determined, it can be seen that the tangential rotors have about three times the clearance of intermeshing rotors.

In intermeshing mixers, the material is forced to pass through the gap between the wide counter-rotating rotor blades, where extensional flow is predominant and dispersive mixing is efficient. Due to the narrow gap between the intermeshing rotors, the large surface of the rotor blades and the good cooling of the rotors, the dispersing mixing action is significantly greater than with tangential mixers.

For intermeshing rotors, the approach angle is much steeper and the flow path through the region of small clearance is longer. Consequently, breakdown of the filler agglomerates is accomplished in a shorter time. The geometry of the approach zone promotes at the same time an elastic response in the viscous compound. The clearance between the two blades is the same as that between the blade tips and the mixing chamber. To compensate for the delay caused by the elastic response, the clearance between the rotors in which highly elongational stress occurs is reduced as much as technically possible. The advantages of intermeshing rotors are:

- rapid mixing
- good temperature control
- lower temperatures for dumping
- high energy input and short mixing cycle
- good filler dispersion

Due to the relatively large size of the rotors and the small clearance, the feed time and the take longer for intermeshing mixers. These disadvantages are compensated for, however, by greater mixing efficiency and higher degree of filler dispersion in the final compound. Other advantages of this type of mixer are the thermal performance, as expressed by better cooling and better good co-mastication of rubbers. The incorporation time is somewhat longer in this type of mixer, but mixing efficiency is considerably increased. However, the fill factors recommended for intermeshing mixers have to be smaller than for tangential mixers.

The interlocking rotor is equipped with a long helical blade with a wide blade tip which extends over the full length of the rotor [51, 52]. Two additional short blades are placed on the opposite half of the rotor on each blade end. An interlocking rotor with a long blade of helical geometry and a wide blade tip and two smaller radially spaced blades before and after the long blade have been introduced. The advantage of such a geometry is a reduction in the radial clearance between the core area of one rotor and the tip of the other rotor.

Another concept is that realized in PES rotors in which one long blade is placed in the axial center [53, 54]. Initially during its development, the PES1 rotor had one long blade in the center, with a passage at each end. Subsequently, for the PES3 rotor, the width of the blade tips was enlarged so that more compound could be squeezed between the rotors and the carbon black pellets broken down more rapidly. A pair of PS3 rotors is shown in Figure 4.19 [17].

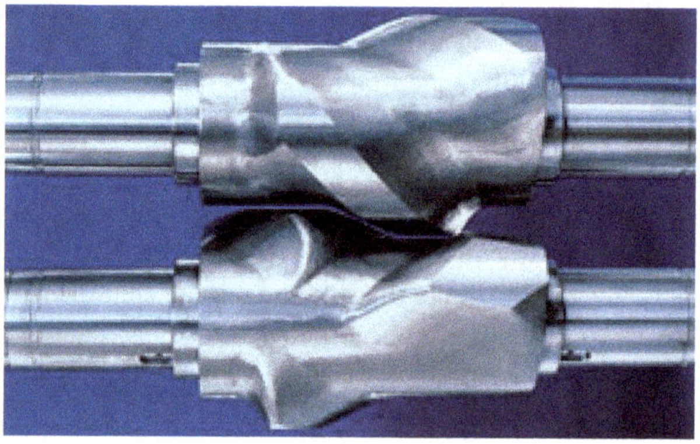

**Figure 4.19** A pair of PES 3 rotors

A relatively long blade carries the compound axially from the center to the ends of the mixing chamber. Two short blades carry the compound back to the center of the mixing chamber. With the new geometry of the PS6 rotors introduced in 2007, an increase in the productivity of 10–15% was achieved [54].

Intermeshing rotors are used to produce compounds for technical rubber goods and are attracting growing interest for tire manufacturing. The narrow area between the rotors leads to a 10–20% higher energy input, compared with tangential rotors [17]. Given the demands for defined and reproducible temperature evolution during mixing of silica/silane compounds, intermeshing mixers are being used increasingly in the tire industry. Intermeshing mixers are produced with free chamber volumes of up to 550 liters.

#### 4.2.2.6 Cooling of Internal Mixers

Cooling depends on the rate at which the heat can be withdrawn from the rubber on the cooling surfaces of a mixer. This is determined primarily by the construction of water-circulation holes around the mixing chamber wall and the cooling system of the rotors. The proximity of water channels to the chamber wall and the rotor surface is essential [7, 55]. The water flow rate and the water temperature determine the cooling efficiency in a mixer. Since heat transfer through rubber is rather slow and the heat transfer depends on the contact of the rubber with the cooling surfaces, intermeshing mixers have an advantage in terms of cooling efficiency, compared with tangential mixers. The benefits of controlling the mixer temperature rather than simply cooling it are nowadays appreciated by the process operators. The benefit of setting the mixer temperature for a particular compound has proved helpful on many occasions. The compound temperature increases with the amount of filler and type of the filler in the formulation [5, 13]. It can be assumed that, in addition to the friction processes that occur during the extension and folding of compound layers, the heat of adsorption released during the wetting of the filler aggregates also contributes to raising the temperature of the compound. The work required to disperse the filler generates heat in the rubber matrix that cannot be transferred rapidly through the surface of the mixing chamber or the rotors. Thus, the compound enthalpy increases in time. The rate of this increase is given by the equation:

$$\Delta H/\Delta t = m \cdot c_v(\Delta T/\Delta t) \tag{4.1}$$

where $\Delta H$ is the change in enthalpy, $\Delta t$ is the time between two temperature measurements, m is the weight of the batch, $c_v$ is the specific heat capacity of the batch and $\Delta T$ is the increase in temperature

The technological progress achieved in cooling technology so far enables efficient cooling of rotors, mixing chamber wall and drop door. Since temperature-controlled parts of the mixer have specific functions, the temperature of the cooling water should not be set at the same level for all. It is common to have the mixing chamber sides at a different temperature than the rotors and the drop door. Thanks to better cooling in the new tangential rotors, the temperature is significantly lowered, in contrast to the case of the old rotors. Comparison of the old rotor cooling and new rotor colling makes it clear how much the temperature can be lowered by bringing the cooling water of the rotors to the wall of the rotors and increasing the circulation on the wall. The efficiency of the better cooling is shown in Figure 4.20 [55].

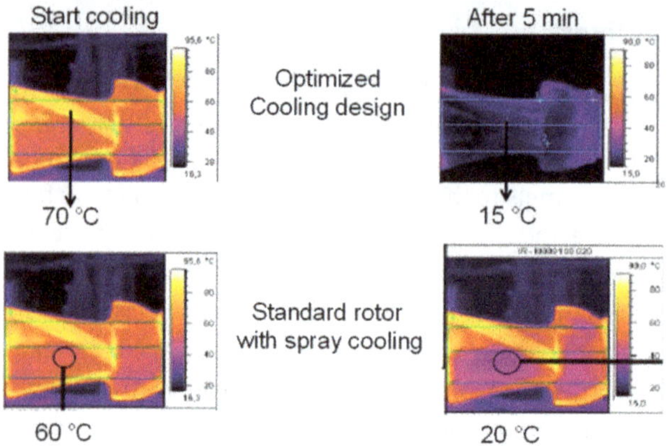


**Figure 4.20** New strategies for cooling the tangential rotors

Optimal cooling requires temperature settings for the different parts of the mixer. Thus, the rotor temperature should be approx. 10 °C lower than that of the chamber wall [56]. In this case, the drag-in and conveying conditions are better fulfilled than for the equal cooling of both parts. In the case of tangential mixers, the cooling water temperature for certain mixtures must not fall below a certain value in order to avoid slipping on the chamber wall. The temperature-controlled surface area in a -liter intermeshing mixer (GK250 E) exceeds the area of a tangential mixer of the same entry volume (GK255N) by approximately 20% [57]. Thus, the intermeshing mixer generally provides better cooling and temperature control than a tangential mixer of the same size.

### 4.2.2.7 Ram Pressure

In standard internal mixers the type of ram, its movement and the ram pressure are adjusted during installation of the machine by the supplier of the internal mixer. Ram movement and ram position have been used to monitor both batch size and progress of mixing [4, 53]. Batch size is assessed by recording the ram position to the end of the mixing cycle. Progress in mixing is assessed via both the ram position and the ram displacement along the mixing cycle. High ram pressure is important primarily during the incorporation stage when a large volume of filler can be added to the masticated rubber.

Since the incorporation of the filler into the rubber matrix implies access of the rubber chains to the voids of the filler agglomerates, the ram reaches its final position only at the end of this penetration process when the volume of the compound is reduced to a minimum value. Monitoring ram seating helps to obtain information about the incorporation time of the filler. With a batch volume matched to the free volume of the internal mixer, the ram displacement initially decreases in a controlled manner and, via an intermediate stage, reaches a stationary final state in which a “dancing” of the ram indicates a property called ram action. With the ram pressure and batch volume correctly dimensioned, the ram should have reached its lowest position after about a third of the mixing time. By recording the ram position, the batch volume (or batch weight) and the fill factor can be correctly determined.

With an optimally set fill factor, the incorporation time can be achieved by quickly lowering the ram in a first step, followed by a short step with intermediate ram speed and then slow lowering of the ram to the end position. This control of the ram speed leads to a significant improvement in the filler dispersion and a shortening of the mixing cycle [13].

At the beginning of filler incorporation, the ram is lowered quickly while, toward the end, the ram speed is gradually lowered. With this type of ram control, the volume of the mix forced on top of the ram can be reduced and the efficiency of mixing increased. With this process, a high pressure is initially generated in the mixing chamber, which promotes breakage of the filler pellets and thus initiates the dispersion process early on. The slow ram movement towards the end enables optimal process management. The power consumption registered during the ram seating time provides useful information. With tangential machines, the power consumption extends over the entire ram seating time.. In the case of intermeshing machines, a sharp increase in power is observed at the beginning of ram seating. After the second power peak has been reached, the ram pressure should be lowered in order to reduce the processes responsible for the temperature increase. Doing this not only avoids undesirable processes but also reduces wear on the mixer. Close control of the ram pressure during the dispersion stage is therefore of special importance to avoid variations in compound quality [5–7]. The ram seating time yields, important information about the sequence of the incorporation stage of the filler and the fill factor of the batch in the internal mixer. There are differences to be noted which are derived from the rotor geometry of the mixer. The speed of ram descent is higher in tangential mixers than in intermeshing mixers. Therefore, in the first case, ram seating time is expected to make up 30–50% of the entire mixing cycle while, with intermeshing mixers, this phase lasts much longer and may amount to 90% of the mixing cycle.

It has been shown that ram displacement can be efficiently used to monitor the incorporation of the filler [13]. The incorporation stage lasts longer if the void volume of the agglomerates (OAN) is increasing and the filler loading is high. The ram seating time increases by more than 50% when the volume fraction of carbon black N330 increases from 0.139 to 0.218. In addition, the ram seating time increases in proportion to the void volume of the filler agglomerates. At a volume fraction of 0.188 that corresponds to a loading of 50 phr, ram displacement during the incorporation time increases in the order N326 < N330 < N339. This influence of the carbon black loading and structure on the incorporation time is shown in Figure 4.8 [29]. Problems arising during mixing have to be observed by the process operator. Dust from the filler can be deposited on top of the ram and can contaminate batches that are mixed subsequently, sticky compound ingredients can adhere to the walls of the hopper and may affect the function of the hopper system.

#### 4.2.2.8 Fill Factor

In order to achieve a high dispersion of the filler in a rubber compound, the volume of the mix must have a certain fill factor, Φ. This process parameter depends not only on the type of mixture but also on the rotor speed and the ram pressure [54]. The degree of filling of an internal mixer can be based on both the mixing chamber volume and the so called "useful" volume. The latter is defined as the chamber volume times the fill factor used in the mixing. The batch weight is defined as the useful volume times the compound density that has to be determined according to the formulation. With the filling degree, flow path along the flights become more and more blocked and the material begins to pass over the gap between the flight tip and the chamber wall [13]. The chamber volume can change due to abrasion when using an internal mixer. This also changes the process parameters needed to achieve an optimal mix quality. Therefore, the chamber volume has to be checked at least once a year.

The fact that the total volume of the raw material shrinks during the incorporation stage requires a knowledge of the compound density at this point in the process. To avoid the undesirable case that the material is dropped from the mixer without having achieved the required mixing quality, the mixing chamber should not be filled completely. Distributive mixing requires multiple folding and transfers of parts of the mixture, a fact which needs a certain free volume in the mixing chamber. If the fill factor is too high, part of the mixture is pushed upwards into the ram shaft, which means that it can no longer participate in the mixing process. The fill factor, Φ, is defined as the ratio of the batch volume to the volume of the mixing chamber:

$$\Phi = (m/\rho)_{comp}/V_{mixer} \tag{4.2}$$

where m is the weight of the batch, ρ is the density of the batch and $V_{mixer}$ is the free volume in the mixing chamber

To avoid any material leaving the mixing chamber unmixed and to achieve the required quality, the mixing chamber must be underfilled. Thus, the free volume of the mixing chamber should be filled at most to 75%. If the fill factor is raised, the throughput of the mixer line could be higher but the degree of dispersion and the quality of the compound would be lower. In the case of an intermeshing mixer, the fill factor should not exceed 65%. In connection with the free volume in the mixer, the contribution of the rotor geometry should also be noted. Tangential four blade rotors allow an approximately 7% higher fill factor than the two-blade rotors that were used at the outset of mixing technology. The discharge temperature of these rotors can also be correspondingly higher.

An easy method by which an operator can determine the optimal fill factor is to follow the evolution of the compound viscosity as a function of increasing fill factor. As compounds with a high macro-dispersion index have a low compound viscosity, a minimum viscosity value is indicated when the compound reaches an optimum degree of dispersion. Depending on the rubber grade and type of filler, the fill factor can vary. For this reason, it is worthwhile to know the fill factor for the internal mixer and compound formulation in order to obtain a quality compound during the mixing process. Studies on an intermeshing mixer with an empty volume of 320 L have shown for mixes of different processability that there is a relationship between the fill factor and the viscosity of the mixture (at 330 $s^{-1}$ and 100 °C), which is expressed by the fill factor, which is inversely proportional to the viscosity of the rubber compound [5]:

$$\Phi = 0.90\text{–}0.12 \cdot 10^{-3}\eta \tag{4.3}$$

where η is the viscosity expressed in Pa s

Another process for determining the "right" fill factor is based on the recommendation to feed the complete formulation ingredient into the mixer and to observe the ram seating time. If the ram does not reach a final position in an acceptable period of time, the batch volume should be reduced until the optimum time for the ram to reach its final position. After adjusting the fill factor, the optimal rotor speed and the ram pressure have to be determined.

#### 4.2.2.9 Base Plate and Drop Door

The base plate of an internal mixer has the function of absorbing the forces generated during the mixing process and to deflecting them into the base of the mixer. Therefore, the base plate must have torsional rigidity, but low weight. At the base plate, a drop door is incorporated to discharge the mixed compound. Figure 4.21 illustrates schematically the base plate and the drop door [17].

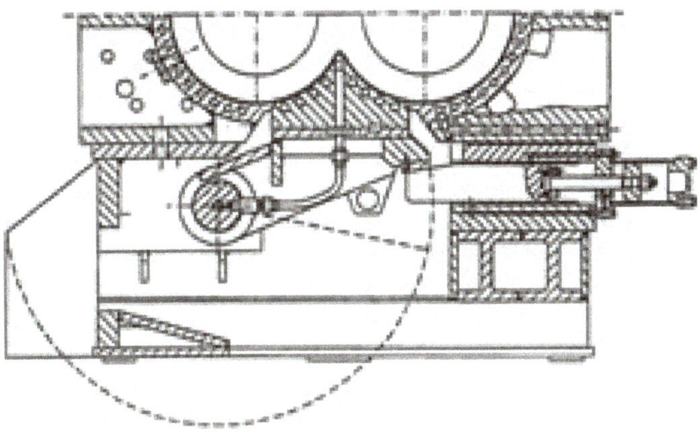

Figure 4.21 Mixer base plate, drop door and latch

A hinged (drop) door, whose flaps can be closed hydraulically by a motor. Like all other components of the mixing chamber coming into contact with the compound, the drop door is hard-coated for high wear resistance. This discharge system makes it possible to shorten the mixing cycle and requires less cleaning during changes of materials [38]. The discharged lumps are different for tangential and intermeshing mixers, because of the way that the batch breaks at the drop door. With tangential mixers a large portion of the batch is located between the door and the rotors, hence a large lump is dumped, followed by material which is carried by the rotors. This process usually takes place in a short period of time, which leads overall to a shorter mixing cycle. With intermeshing mixers in which the door is closer to the rotors, the batch is wound out from between the rotors, and hence a rough sheet is formed. For this reason, the discharge process is somewhat more time-consuming than with a tangential mixer.

### 4.2.3 Comparing Internal Mixers

The number of intermeshing mixers has almost equaled that of the tangential machines [39]. Comparative data for compounds mixed in different types of mixers are relatively scarce, because mixer manufacturers have little interest in making such a comparison. Usually the performance obtained on a laboratory mixer is better than on a production machine [51, 52]. By comparing technological criteria that are achieved by the two types of internal mixers, it has been shown that the energy efficiency and the cooling performance is superior in intermeshing machines, while the feeding and discharge properties that define annual product throughput are superior in tangential mixers [13]. The efficiency of the two internal mixers is shown in Figure 4.22.

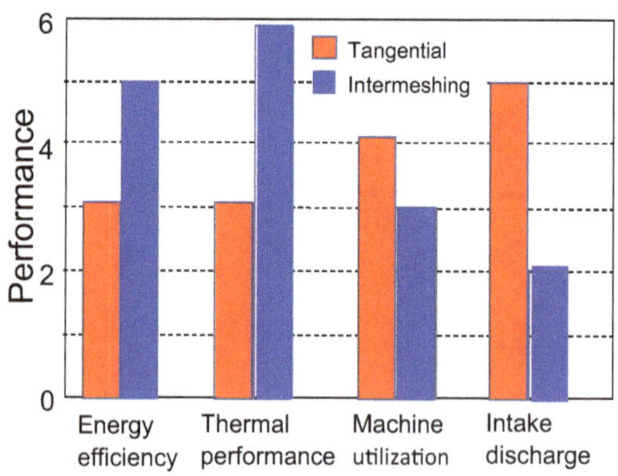


**Figure 4.22** Comparison of the technical advantages of tangential and intermeshing mixers

The results demand the quality and the quantity of the mixed compounds. Mixers with a high volume and short mixing cycles are predominantly used in the tire industry, while good filler dispersion is in the foreground of the production of technical rubber goods. Analysis of the mixing process in recent years has focused on optimization of the mixing process. Computer monitoring of the mixer variables (i. e. power consumption, temperature, mixing time, torque, ram seating) is now offered by practically all manufacturers, regardless of the type of internal mixer. This improvement enables modern mixing equipment to produce quality compounds in large volumes. Adequate process control and an efficient reduction in mixing time are made possible by computer monitoring of the mixing process. From the point of view of cost reduction, however, all factors that influence the process must be taken into account. The cost of rubber mixing is determined by the prices of the raw materials, the machine and labor costs, as well as the overhead costs, which include depreciation and maintenance.

### 4.2.4 Tandem Mixers

Considerable progress in mixing technology has been achieved with the invention of the tandem mixer by Peter, J. (1987). He combined two-stage mixing, mixing the master batch and the final batch consecutively in the same mixer [58]. In the upper internal mixer, the fillers are added to the rubber, the filler is incorporated, and the dispersion process starts. The mix then falls into the much larger ramless intermeshing mixer and is kept at a temperature of 100–120 °C to increase the dispersion of the filler, before the reactive substances are added. The ramless tandem mixer also shows good degradation behavior, which can be influenced in particular by the rotor speed and the mixing time. The concept behind the tandem mixer is shown in Figure 4.23.

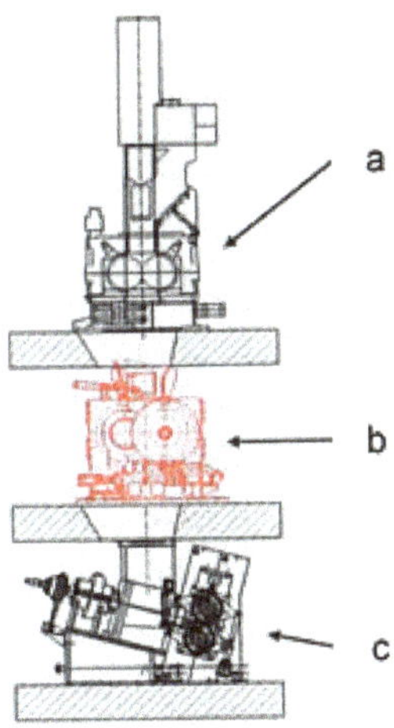


Figure 4.23
Concept behind the tandem mixer: a) upper mixer, b) tandem mixer and c) downstream equipment [58]

After premixing in the upper internal mixer, which may be a tangential or preferably an intermeshing machine, the hot compound is dropped without any intermediate storage into the lower tandem mixer, which is approximately three times larger than the upper mixer. Here, the batch is cooled and the reactive ingredients can be introduced. The lower tandem mixer must be underfilled to ensure efficient cooling of the compound at a low rotor speed (i. e. 5 rpm). The curing agents can be added via a bypass and mixed at an increased rotor speed (i. e. 15 rpm). The important advantage is that the tandem mixer reaches its lowest temperatures at the end of the mixing process. In order to produce final mixes, the curing agents are added at the last second. This splitting of the mixing process into two stages has proved to be a versatile technique for:

- providing good dispersion
- preventing undesired scorching of the compound and
- doubling the throughput of conventional tangential mixers

The advantage of tandem mixing for silica compounds is that in both machines mixing of the compound can be adapted to its unique process requirements. In the upper machine, good filler dispersion can be achieved with high energy input. The ram is used to force the material between the rotors, where it is subjected to high elongation forces. Once the compound reaches a certain temperature, it is dropped into the tandem machine and cooled down to the optimum temperature for silanization. The tandem mixer is equipped with additional venting points to vent the ethanol emitted. Since mixing is performed in parallel in the upper and lower internal mixer, the mixing time for production of silanized silica can be shortened. This technology makes for a significant reduction in mixing time per mass unit of compound for both carbon black and silica, because both mixers simultaneously perform one of the mixing actions.

## 4.3 Wet-Mixing

Despite the various efforts, it has so far not been possible to mix pelletized fillers into a rubber matrix that has a dispersion index of 100%, so that there are remaining large filler agglomerates or undispersed pellet fragments in the mixture that caused reduced physical properties, especially a premature fraction of the vulcanizate under cyclic dynamic load. Consequently, new mixing technology has been used to reduce the amount of non-dispersed filler and to achieve better properties in the rubber compound.

Wet-mixing was introduced to increase the filler dispersion in rubber compounds. However, the longer mixing times and coagulation can affect the productivity of the wet-mixing process. By introducing fast mixing technology provided by the reactor, the filler is well dispersed and the mixing time is reduced to fractions of a second. There exist wet-mixing technologies for carbon blacks and silanized silica and also for fillers with high aspect ratio, such as layered silicates and carbon nanotubes.

### 4.3.1 Carbon Blacks

A great improvement in filler mixing and dispersion has been achieved with wet-mixing of a mixture of a mixed carbon black and NR latex in a hyperbolic jet reactor without any supplementary coagulation agent [59]. Coagulation of the rubber latex starts at a given velocity in the hydraulic jet stream of the rubber-filler slurry, which provides extensional flow and turbulence, sometimes by adding a coagulant. The jet reactor reduces the mixing time and provide a rapid coagulation of the latex that surround the individual aggregates or small agglomerates. Therefore, a high productivity of wet-mixing is obtained by the flow of the NR latex and the not-pelletized carbon black. Coagulation and encapsulation of the filler particles is a rapid process and controllable via the conditions in the jet reactor. Under high turbulent conditions, mixing and coagulation of the latex occur at room temperature in less than a fraction of second. This new technology for mixing rubber compounds was developed by Cabot Corporation that is provided as Cabot elastomer composites (CEC) [60–63]. The advantage is that this wet-mixing offers some benefits over dry-mixing of carbon black into a rubber matrix:

- simplified mixing process
- reduced mixing costs
- elimination of handling free carbon black
- elimination of carbon black dust
- improved filler dispersion
- improved capital efficiency
- improved physical properties

As can be seen from Figure 4.24, wet-mixing of NR and carbon black is a quick process [61].

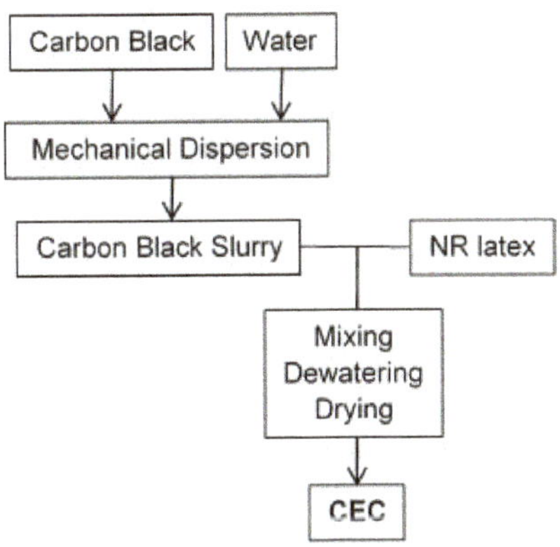


**Figure 4.24**
Schematic representation of the Cabot wet-mixing process

The CEC technology is based on rapid mixing and coagulation and short drying times at high temperatures. The carbon black slurry is prepared by dispersing the filler in water. At high speed, the slurry is mixed with the rubber latex stream. The system is dewatered in an extruder and then dried to further reduce the humidity to below 1%. The dispersion of the carbon black is improved significantly and continuous mixing is feasible. NR-carbon black mixtures obtained by wet-mixing are characterized by an almost ideal carbon black dispersion. Comparison of the macro-dispersion, measured by optical microscopy, of two carbon black N134 vulcanizates reveals no large agglomerates in the test example prepared by wet-mixing. In comparison, a substantial proportion of pellet fragments is still observed after mixing of the compound four times with the same rubber/filler ratio, as shown in Figure 4.25 [62].

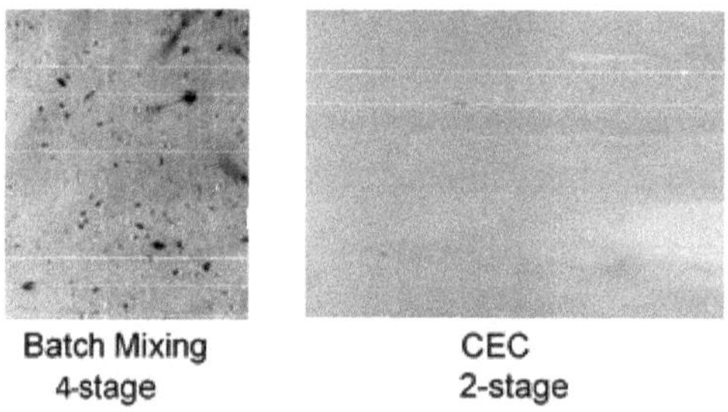


**Figure 4.25**
TEM images of dry- and wet-mixing of NR with carbon black N134

The performance of the rubber compounds obtained by wet-mixing stems from the fact that the polymer-filler interaction is well preserved and polymer degradation is essentially eliminated. The penetration of the rubber into the cavities of the aggregates takes place without a reduction in the molecular weight of the polymer. The dispersion obtained is superior to that of a conventional mixed batch made from the same compound ingredients. When the ingredients of a compound are mixed, they interact with each other to form a multiphase composite. The parameters which impart a significant effect on the composite properties are influenced by the mixing process. After wet-mixing, the proportion of bound rubber is higher than in dry-mixed compounds of the same composition, as was shown for CEC composites. This increase in bound rubber is due to the better access of the rubber chain to the surface of the filler particles. The new technology enables a wide range of carbon blacks to be used. In addition, it is especially helpful when carbon blacks with high specific surface areas (high surface activity) and low structure have to be dispersed in rubber matrices. As a result, the group of active carbon blacks, which are difficult to be dispersed in rubbers, can be greatly enlarged and the ultimate properties of the resulting compounds can be significantly improved above known levels of the reinforcement [61]. The grade of carbon blacks used for wet-mixing can be expanded to high specific area and low void volume of the filler. Significant improvements are achieved, including reduced hysteresis, improved cut-chip resistance, flex-fatigue life and abrasion resistance at high filler loading.

A great effort has been made to produce carbon black master batches by mixing SBR latex with carbon black slurry and then coagulating the mixture by adding an acid. This is a batch process and is carried out in one vessel. In order to obtain the Rubber-Filler Composite (RFC), which was developed by Degussa (Germany), the product is dried in spray-dryers, whereby the outer shell of the pellets is made up of carbon black. The RFC can be mixed in a continuous twin-screw extruder (TSE) or other continuous mixers [64–66]. RFC wet-compounds show much better dispersion in SBR when it is dispersed in a TSE and improved vulcanizate properties, such as tensile strength, reduced Payne effect and better fracture properties. Mixing of the RFC in twin-screw extruders leads to much-improved filler dispersion and very good elastomer properties. The physical properties are improved after the filler dispersion is enhanced in a two-roll mill. This mixing process reduces the dynamic crack growth by three order of magnitudes and increases the life time of the vulcanizate. In this way, two things are secured: a continuous production of rubber compounds and greatly improved physical properties of the elastomers.

## 4.3.2 Precipitated Silicas

A recent development at the EVE Rubber Institute in Quingdao, China, is the wet-mixing of silanized silica in S-SBR/BR blends. The silica was introduced into the solvent and finely dispersed by mechanical means. Both rubber solution and the silica slurry were mixed instantaneously under high turbulent conditions at high pressures. As a result, the silanized silica is fully dispersed in the rubber compounds. The mixture was then sent to a stripping device to remove the solvent at elevated temperatures under low pressure. Compared with silica filled compounds mixed in a dry-mixing, the wet-mixing are characterized by [67]:

- very good filler dispersion
- higher silanization
- strong filler-polymer interaction and weak filler-filler interaction
- improved compound processability
- improved physical properties

The eco-visco-elastomer composite (EVEC) is a synthetic rubber-silica masterbatch produced with a continuous liquid mixing process. Most important, due to the high efficiency of the coupling reaction, the polymer-filler interaction can be considerably enhanced and the filler-filler interaction can be reduced, compared with the dry-mixed process.

### 4.3.3 Layered Silicates

The strong interlayer interactions prevent dispersion of the mechanical two-dimensional platelets of layered silicates. The wet-mixing technology is used to produce intercalated or exfoliated platelets in the rubber mixture. The literature makes references to technologies for wet-mixing. Mixing to disperse pristine layered silica in different rubber lattices is commented on in the reports [68, 69]. A flocculant is added that coagulates the rubber latex and the montmorillonite simultaneously, but the rubber nanocomposites do not exactly intercalate into the galleries between the two-dimensional plates of the layered silicates. As a result, there are non-exfoliated layers in the nanocomposite. Composites containing 20 phr layered silicates are obtained by using NR, SBR, NBR and XNBR in coagulating rubber latex and layered silicates in aqueous solution. In the TEM micrographs, the dark lines represent the intersections of the silicate layers. The micrographs show that there are individual layers and stacked silicate layers with a thickness of 10–30 nm [70, 71]. Therefore, this latex mixing was not a satisfactory technology.

The introduction of continuous dynamic latex mixing (CDLC) developed by the German Rubber Institute in Hanover represented significant progress, with the layered silicates slurry and the rubber latex being combined in a reactor and then coagulated with an acid [72–75]. The CDLC process is shown in Figure 4.26.

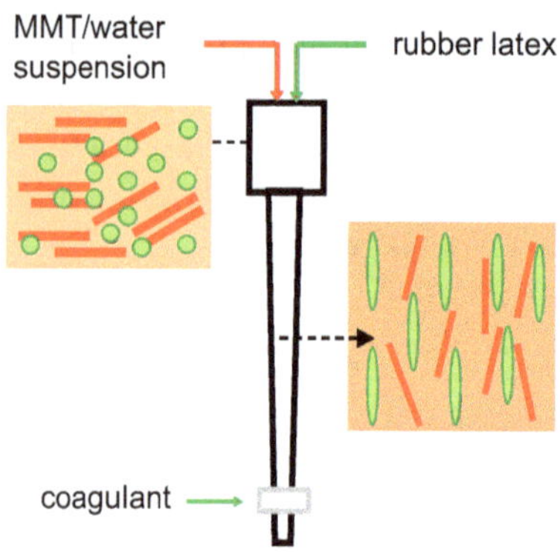


**Figure 4.26**
Continuous dynamic latex mixing of pristine MMT filler

The wet-mixing was performed with pristine $Na^+$-cloisite in NR, NBR, XNBR and SBR latices. It has been proven that high exfoliation (> 85%) of the clay platelets is achieved in fractions of a second. Due to the high exfoliation, the stress-strain properties and the dynamic properties are significantly changed.

### 4.3.4 Carbon Nanotubes

Wet-mixing technology makes it possible to obtain compounds reinforced by multi-wall carbon nanotubes, without damaging the tubes by mechanical mixing [76]. Latex blending was adopted to achieve satisfactory debonding and to preserve as far as possible the original chain length of the multi-wall carbon nanotubes in rubber matrix. The process basically consists in simple mixing of the suspension of multi-walled carbon nanotubes with aqueous latex. The first steps are exfoliation of multi-walled carbon nanotubes entanglements in an aqueous surfactant solution, which is followed by mixing the slurry with the rubber latex to obtain a high dispersion of carbon nanotubes in the rubber [77, 78]. The system is then included in a precipitation process [79]. The whole process is easy, versatile, reproducible and reliable.

#### References

[1] Palmgren, H., Eur. Rubber J., 1974, 147: 30

[2] Funt, J. M., in "Mixing of Rubbers", RAPRA Technol., Shrewsbury (1977)

[3] McKelvey, J. H., in "Polymer Processing", John Wiley & Sons, New York, (1962)

[4] Wood, R. P., in "Rubber Mixing", RAPRA Review Report, Nr. 10, RAPRA Techhnology Ltd., Shrewsbury (1996)

[5] Röthemeier, F., Sommer, F., Eds., in "Kautschuk Technologie", Carl Hanser Verlag, Munich, Vienna (2012)

[6] Hofmann, W., Gupta, H., Eds., in "Handbuch der Kautschuk-Technologie", Dr. Gupta Verlag, Ratingen (2001)

[7] Limper, A., Ed., in "Mixing of Rubber Compounds", Carl Hanser Verlag, Munich (2001)

[8] Mark, J. E., Erman, B., Roland, M. C., Eds., in "The Science and Technology of Rubber", Academic Press, Waltham, USA 2013

[9] Meon, W., Blume, A., Luginsland, H.-D., Uhrland, S, in "Rubber Mixing, Chemistry and Applications", (B. Rodgers, Ed.) Marcell Dekker, Inc. New York, Basel (2004) Ch. 7

[10] Johnson, P. S., in "Basic Elastomer Technology", Baranwal, K. C., Stephens, H. I., Eds., Rubber Division, Akron (2001) Ch. 5

[11] Freakley, P. K., in "Rubber Technology Handbook" (De, S. K., White, J. L., Eds.) RAPRA Technology Ltd. (2001) Ch. 7

[12] Utracki, L. A., Luciani, A., Appl. Rheology, 1992, 10: 1024

[13] Limper, A., in "Mixing of Rubber", Limper, A., Ed., Carl Hanser Verlag, Munich (2001), Ch. 2

[14] Harms, E. G., Eur. Rubber J., 1978, 6: 23

[15] Grossman, R. F., Ed., in "The Mixing of Rubber" Chapman and Hall, London (1997)

[16] Krambock, W., in "Conveying and dosing of rubber, fillers and plasticizers" (German) VDI Verlag, Düsseldorf (1984)

[17] Berkemeier, D., in "Mixing of Rubber Compounds", Limper, A., Ed., Carl Hanser Verlag, Munich (2001), Ch. 1

[18] Nijman, G., in "Current Topics in Elastomer Research", Bhowmick, A. K., Ed., CRC Press Taylor & Francis Group, Boca Raton, London, New York (2008), Ch. 35

[19] Beack, K. C., Comper, L. F., Lowery, V. E., Rubber Age, 1958, 85: 253

[20] Menges, G., Grajewski, F., Kautsch. Gummi Kunstst., 1987, 40: 7

[21] Dizon, E. S., Papazian, L. E., Rubber Chem. Technol., 1977, 50: 765

[22] Cotton, G. R., Rubber Chem. Technol., 1985, 58: 774
[23] Klie, B., Teich, S., Haberstroh, E., Giese, U., Kautsch. Gummi Kunstst., 2015, 67: 31
[24] Schuster, R. H., in "Polymer Science Insights", (de Paoli, M.-A., Ed.) Wiley VCH (2002) Ch. 5
[25] Scholz, G., PhD Thesis University of Hanover (1987)
[26] Abele, M., Kempermann, Th., Kautsch. Gummi Kunstst., 1989, 42: 290
[27] Schuster, R. H., Proc. Workshop Mixing Technology, Freudenberg, (1999)
[28] Cotton, C. R., Rubber Chem. Technol., 1978, 52: 187
[29] Oppermann, B., PhD Thesis University Hanover (1994)
[30] Klockmann, O., in in "Mixing of Rubber Compounds", Limper, A., Ed., Carl Hanser Verlag, Munich (2001) Ch.4
[31] Sui, G., et al., Macromol. Mater Comp., 2007, 292:1020
[32] Shamunharaj, A. M., et al., Comp. Sci. Technol., 2007, 67:1813
[33] Messersmith, P. B., Giannelis, E. P., Chem. of Materials, 1993, 5
[34] Giannelis, E. P., Adv. Materials, 1996, 8:29
[35] Schuster, R. H., Wrana, C., ACS Rubber Division Meeting, San Antonio 2016
[36] Zimmermann, H., Schuster, R. H., Proc. ICCE-19 Shanghai (2011)
[37] Shiga, S., Furuta, M., Rubber Chem. Technol. 1985, 58: 22
[38] Blow, C. M., in "Rubber Technology and Manufacture", Butterwords Sci., London (1982)
[39] Banbury, F. H., US Patent 1,200,070 (1916)
[40] Lasch, A., Frei, K., German Patent 738,787 (1943)
[41] Tyson, D. Z., Comper, L. F., US Patent 3,230,581(1967)
[42] Nortey, N. O., US Patent, 4,744,668 (1988)
[43] Takakura, K., et al., European Patent 1,649,331 (2003)
[44] Valsamis, L. N., Borzenski, F. J., Rubber World, 2008, 225: 38
[45] Kobe Steel, Ltd., Introduction to a new 4 blade rotor, technical information
[46] Limper, A. et al., European Patent 1,649,995 (2008)
[47] Limper, A., Wilson, I., in "Cost Reduction in Rubber Processing", (Graf, H.-J., Ed.) TechnoBiz Communications Co., Ltd. (2014) Ch. 11
[49] Lasch, A., Stromer, E., German Patent 641,685 (1934)
[50] Cooke, R. T., British Patent, 431,012 (1934)
[51] Brennan, A. K., US Patent 3490750 (1970)
[52] Johnson, F., European Patent 0,170,397 B1 (1990)
[53] Wiedemann, W., Schmid, H.-M., German Patent 2836940 C2 (1982)
[54] Wiedemann, W. M., Schmid, H.-M., Kautsch. Gummi Kunstst., 1984, 34
[55] Glasser, P., TechnoBiz Rubber Industry Forum, Pittsburgh (2015)
[56] Berkemeier, D., et al., Rubber World, 2001, 224: 34
[57] Limper, A., Moessinger, J., Internat. Polym. Sci. Technol., 2007, 27: T1–T5
[58] Peter, J., Kautsch. Gummi Kunstst., 1994, 47: 656
[59] Soue, K., Nippon Gomu Kyokaishi, 1989, 6: 308
[60] Marby, M. A., Rumpf, F. H., Podobnik, I. Z., US Pat. 6048923 (2000)
[61] Wang, M.-J., Wang, T., Wang, Y. L., Kautsch. Gummi Kunstst., 2002, 55: 388
[62] Wang, T., Wang, M.-J., McCornell, G. A., WO 2003050182 (2003)
[63] Wang, M.-J., in "Current Topics in Elastomer Research", Bhowmick, A. K., Ed., CRC Press (2008) Ch. 33

[64] Uphus, R., Skibba, O., Schuster, R. H., Görl, U., Kautsch. Gummi Kunstst., 2000, 53: 279

[65] Bogun, M., Abraham, F., Muresan, L., Schuster, R. H., Radusch, H.-J., ACS Rubber Division, Pittsburgh, 2002, p. 107

[66] Luther, S., Bogun, M., Schuster, R. H., Görl, U., Kautsch. Gummi Kunstst., 2006, 59: 473

[67] Wang, M.-J., Song, J. J., Wang, Z., Liu, K., Zhang, H. X., Zhang, X. B., Jia, W. J., Kautsch. Gummi Kunstst., 2020, 73: 45

[68] Zhang, L. Q., Wenig, Y. Z., Wang, Y. Q., J. Appl. Polym. Sci., 2000, 78: 1937

[69] Wu, Y. P., Jia, Q. X., Yu, D. S., Zhang, L. Q., J. Appl. Polym. Sci., 2000, 89: 3855

[70] Wu, Y. P., Wang, Y. Q, Zhang, L. Q., Comp. Sci. Technol., 2005, 65: 1195

[71] Wu, Y. P., Zhang, L. Q., Wang, Y. Q., J. Appl. Polym. Sci., 2001, 82: 2842

[72] Brandt, K., Schuster, R. H., Kautsch. Gummi Kunstst., 2006, 59: 511

[73] Brandt, K., Schuster, R. H., Kautsch. Gummi Kunstst., 2008, 61: 250

[74] Klafke de Azeredo, L., Maldaner-Jacobi, M., Schuster, R. H., Kautsch. Gummi Kunstst., 2010, 63: 8

[75] Schellenberg, M., Früh, Th., Schuster, R. H., ACS Rubber Division Meeting Cleveland (2013)

[76] Schuster, R. H., ACS Rubber Division Meeting, Cincinnati (2012)

[77] Grossiord, N., Loos, J., Regev, O., Koning, C. E., Chem. Mater. 2006, 18: 1089

[78] Georgakilos, V., Gournis, D., Tzitzios, V., J. Mater. Chem., 2007, 17: 2679

[79] Schuster, R. H., Chogule, ACS Rubber Division Meeting Cleveland (2013)

# 5 Filler Dispersion in Rubber

A very important part of the mixing process is how well the pellets break up into their respective agglomerates and aggregates, which starts the long and energy-intensive process of deagglomeration. This deagglomeration is very important in understanding rubber mixing process. Dispersion of the filler in a rubber matrix is a necessary condition for a rubber compound. It is the basis for strong interaction of the filler surface and the chemical structure of the polymer for high reinforcement of the elastomer. In order to achieve filler dispersion in a rubber compound, it is necessary to select the filler that is the most compatible with the rubber. The formulation of the compound is therefore of great importance. For the best formulations, the best mixing process parameters must be selected. The mixing sequence and the incorporation time should not be too short, the dispersion stage should be long enough to achieve maximum breakdown of the pellets such that as many filler particles as possible are present as nanoparticles in the rubber. The physical properties are significantly improved and the ultimate properties, tensile strength, fatigue resistance and abrasion resistance are increased, whereas the dynamic cut propagation is considerably reduced.

A lower dispersion of the fillers in a mixing process can result from on the one hand the type of formulation, where the polymer has not been matched to the filler, or on the other hand the mixing technology, which has scope for improvement, or the post-treatment of the mixture that shows a decrease of the physical properties of the vulcanizate. In the formulation of a rubber compound, the aim is to choose the material parameters optimally and to look for the best interaction between the components. The mixing process should also be carried out under the appropriate process parameters in order to produce a good dispersible rubber mixture. Dispersion involves breakdown of the filler pellets into individual aggregates or small agglomerates. Incomplete breakdown, leaving the filler pellet fragments larger than a few micrometers, is harmful for dynamic cut growth resistance and the abrasion resistance of the elastomer.

## 5.1 Theoretical Aspects

When considering the conditions for a uniform filler dispersion, it can initially be assumed that the filler can be completely dispersed and that there are no pellet fragments in the rubber compound. The final state is a filler surface surrounded by the rubber matrix and no agglomeration of the filler aggregate in the elastomer. For a complete distribution of a filler in a polymer, the contact area per unit volume of filler, A, is given by the following relationship [1]:

$$A = \Phi \rho S \tag{5.1}$$

where $\Phi$ is the volume fraction of the filler, $\rho$ is the density of the filler and S is the specific surface area of the filler

However, this assumption requires 100% filler dispersion and zero pellet fragments or large aggregates in the rubber compound. This situation will not occur and does not represent the real dispersion of the filler aggregate into a rubber compound. The ideal dispersion is a target of any mixing process that involves not only perfect mixing technology but also the right fillers and the most compatible rubbers in the compound. To exemplify the effects of parameters acting in favor of dispersion, consider theoretically an agglomerate that is composed of three aggregates. On the one hand, there is a cohesive force, $F_c$ which keeps the agglomerate resistant against the dispersion process. On the other hand, there is hydrodynamic stress, $F_h$, which tends to separate the aggregates and improves the degree of dispersion. The effect of the cohesive agglomeration force and the hydrodynamic force is schematically shown in Figure 5.1:

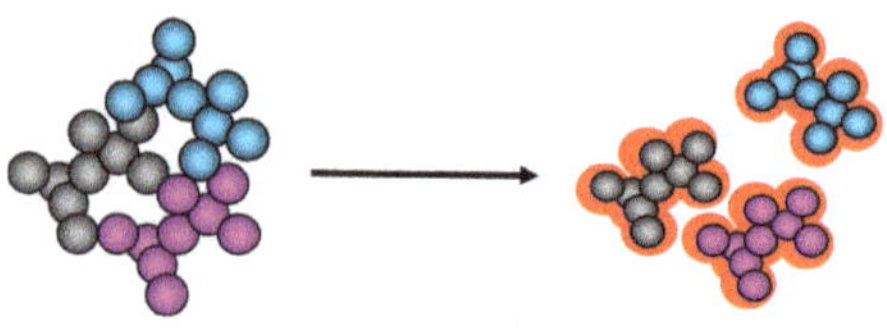

**Figure 5.1** Dispersion of an agglomerate into filler aggregates

Assume that the cohesive force, $F_c$, is defined by the product of the number of average inter-aggregate forces per contact, $\nu_f$, and the interaction energy per inter-aggregate contact H, as described in the equation:

$$F_c = H \cdot \nu_f \tag{5.2}$$

where $F_c$ is the cohesive force, H is the inter-aggregate interaction force and $\nu_f$ is the number of inter-aggregate contacts

The inter-aggregate force depends on the activity of the filler and the number of inter-aggregate contacts in the agglomerate, which is defined by the structure of the aggregate. The time required to attain satisfactory dispersion, once the filler has been incorporated, decreases with increase in structure of the filler, since the aggregates are not tightly packed together. Higher surface area fillers yield stronger agglomerates which are more difficult to disperse. If the aggregates are more branched, the number of interaggregate contacts decreases within the agglomerate. This lowers the cohesion strength of the agglomerate so it can be more easily dispersed. Therefore, the higher the filler structure, the easier the filler dispersion in the rubber compound. If fillers with similar OAN are considered, it is observed that those with larger particle size and lower specific surface area are more efficiently dispersed than those of high specific surface area.

The counteracting hydrodynamic force, $F_h$, which facilitates dispersion in the rubber matrix, comprises several contributions from interfacial tension, $\gamma$, the matrix viscosity, $\eta$, the size of the agglomerates, $R_a$, and the shear stress, $\tau_{xy}$. As a result, the hydrodynamic force $F_h$ can be expressed by the empirical equation:

$$F_h = c \cdot \eta \cdot \tau_{xy} \cdot R_a^2 / \gamma \tag{5.3}$$

where c is the proportionality factor, $\eta$ is the viscosity of the rubber matrix, $\tau_{xy}$ is the shear stress, $R_a$ is the radius of the aggregates and $\gamma$ is the interfacial tension

While the viscosity and the shear stress can be measured quite easily, the interfacial tension has to be approximated. The lower the interfacial tension, $\gamma$, the larger will be the interface between the two components and the stronger the phase adhesion, which enables the aggregates to be separated. The interfacial tension is therefore an adequate parameter for describing and predicting filler dispersion in a given system. The solubility parameter, $\delta$, of the rubber can simply be used to assess the rubber-filler interaction in a sequence of polymers selected for making a product. It is also convenient to use an approximation proposed for rubber blends [2] according to which the interfacial tension can be approximated by the geometric mean of the surface tension of the rubber, $\gamma_r$, and the filler, $\gamma_f$:

$$\gamma \sim (\gamma_r \cdot \gamma_f)^{1/2} \tag{5.4}$$

It should be noted that the adhesion of the polymer on the filler surface is a spontaneous, exothermic process which decreases as the temperature rises. The intensity of the adhesion depends on both the polarizability of the chain segments and the surface energy sites of the filler. In order to disperse the agglomerate into aggregates, the hydrodynamic force $F_h$ has to overcome the cohesive force $F_c$. Only if the filler can be dispersed and the rubber has low interaction tension and a high viscosity can a high degree of dispersion be achieved. From Equation 5.5 is obtained which describes the conditions for rupture of the agglomerate:

$$\eta \cdot \tau_{xz}/\gamma \geq H \cdot \nu_f/c \cdot R_a^2 \quad (5.5)$$

The lower the interfacial tension, the easier it is to overcome the cohesive filler-filler interactions and to obtain a higher dispersion of the filler. The left term of Equation 5.5 contains two temperature dependent values, the viscosity, η, and the interfacial tension, γ, respectively. In order to predict the effect of rising temperatures during mixing, the temperature dependence of both the viscosity and the interfacial tension have to be known. Typically, the temperature gradient of the viscosity is higher than that of the interfacial tension and as a result the dispersion efficiency decreases when mixing is prolonged.

## 5.2 Determination of Filler Dispersion

The extent to which size reduction in agglomerates takes place during mixing depends on the time during which the compound is exposed to extensional and shear stress. Mixing conditions, such as rotor speed, ram pressure, cooling of the rotors and cooling of the mixing chamber wall, play an important role. Better cooling leads to higher viscosity and thus to better transmission of the shear stress to the agglomerate surface. The measured degree of homogeneity depends on the examination to which the compound characterization is subjected. For practical purposes, the scale of checking can be adjusted to the minimum size of objects of segregation that would cause the mixture to be unsuitable for the intended purpose [3]. Thus, a mixture with a statistically random distribution of "sufficiently small" particles can be considered as one with good dispersion. The challenge is to find reliable methods for evaluating the quality of mixed compounds. This requires that experimentally accessible criteria be defined, without which investigations to ascertain the effect of process conditions and material parameters would not provide sufficient information [4].

Direct measurement of filler dispersion is of fundamental importance for achieving information on the efficiency of processing and the quality of the mixed compound. In the past, dispersion of the filler was derived from the vulcanizate properties:

- tensile strength
- elongation at break
- dynamic moduli
- abrasion resistance

However, an assessment based on the criteria does not provide certainty and, above all, in the case of quality fluctuations, which can occur, there are some difficulties attributable to the actual state of filler dispersion. The search for versatile and user-friendly methods that provide information about the morphological organization of filled compounds is faced with a fundamental problem that arises from the fact that the size of the solid objects in filled compound ranges from the level of aggregates in the nanometer range up to pellet fragments in the range of several hundred micrometers or even millimeters [5]. There is no one ruler capable of capturing the different filler aggregates and agglomerates observed in different experiments. The length scale ranges for the different solid objects present in carbon black-filled rubber compounds and some methods used to gain information about them are shown in Figure 5.2.

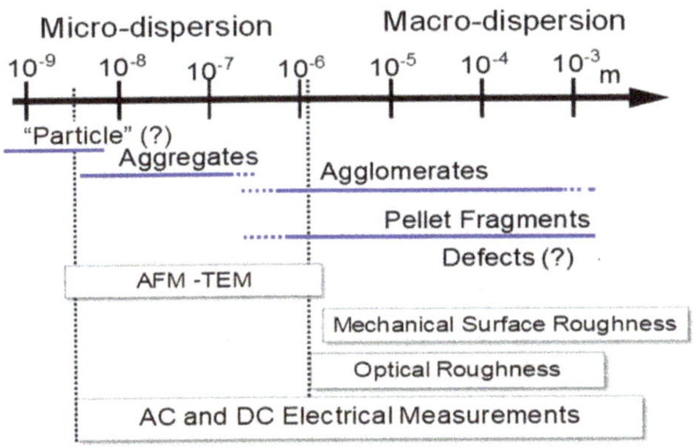


**Figure 5.2** Filler dispersion is a scale-dependent phenomenon

Faced with the limitations in describing the whole range of particle sizes with one ruler, the length scale was arbitrarily divided in two regions. The boundary between the two regimes is not quite set arbitrarily at around 1 µm, but actually results from the resolution of the methods employed. Particles whose size is above 1 µm are deemed to be part of the "macro-dispersions", while particles below this limit are referred to as "micro-dispersions". Basically, this classification according to observation method is made only for practical reasons.

However, the methods for characterizing "macro-dispersions" are straightforward and quite well established, while the micro-dispersion methods do not lend themselves to routine measurements and are still under development. Criteria based on the contribution of filler networks to dynamic compound stiffness have been used to characterize "micro-dispersions". They are useful for establishing a ranking, ranging from poor to good dispersion of rubber compounds, without providing a quantitative assessment in the form of numbers. Nevertheless, direct measurement of filler dispersion is of fundamental importance for achieving more precise information about the quality of the mixed compound in the non-crosslinked state as well as in the crosslinked state. The existing methods of determining micro-dispersions are time-consuming and labor-intensive and are not used in small manufacturing companies. Nevertheless, the results of such methods are very useful for research purposes because they examine the phenomenon of dispersion on a nanoscale.

Other methods exploit the sensitivity of ultimate properties to the occurrence of flaws (i. e. pellet fragments) distributed in the material. A few large objects (> 20 µm) in a compound can substantially decrease tear resistance and increase dynamic cut growth, even for compounds for which a good filler dispersion and uniform filler distribution have been determined. Therefore, the frequency of non-dispersed pellet fragments and large agglomerates should be examined very carefully, especially in the case of rubber compounds, which often show crack propagation in their dynamic behavior. Depending on the frequency of these flaws in a test specimen, it may fail sooner or later. It is therefore of great help to know the susceptibility of a filled rubber compound to the formation of defects that increase the dynamic cut propagation and the abrasion loss.

## 5.2.1 Filler Macro-Dispersion

Basically, microscopic methods are able to determine the size of objects if the wavelength of the radiation used is smaller than the dimension of the objects under consideration. At the beginning of the studies on the dispersion of the carbon black, simple microscopes were used to describe it. For large objects, such as agglomerates and pellet fragments with diameters ranging from a few microns to hundreds of microns, known as undispersed carbon black (UCB), optical microscopy provides accessible information. Even before synthetic rubbers came on the market, filler dispersion – as the most important quality characteristic of filled vulcanizates – was characterized by the use of surface inspection of cut, stretched, and torn surfaces [6, 7]. The breakthrough came with the development of a method for quantifying the amount of carbon black dispersed to agglomerates under a certain size on the basis of optical microscope inspections of cryosections (thickness 2–4 µm) produced by glass knives [8]. Even if the transmission optical microscopy is not suitable for routine tests, it provides a decent basis for evaluating the degree of dispersion from poor to good, based on the frequency of undispersed filler particles and determining the influence of the duration of mixing and the type of filler and rubbers on rheological properties of compounds [9]. A dispersion chart has been successfully used down the years that depicts the differences in carbon black agglomerate size at different percentages of dispersion.

### 5.2.1.1 Optical Surface Roughness Methods

Progress was made when razor blade cuts from a rubber test piece were assessed and examined microscopically. The degree of dispersion was classified on a scale from 0 to 10, where 0 is the poorest dispersion and 10 is the best dispersion [10]. It is convenient to quantify the dispersion of a particulate filler on the basis of the luminous reflectance of razor blade cuts of vulcanized materials. Based on this simple method of sample preparation, qualitative and quantitative characterizations of the degree of filler dispersion can be obtained easily.

Placing the glossy razor cut under a light source that illuminates the cut surface under a given angle enable the undispersed filler agglomerates to be detected. The presence of improperly dispersed filler is shown by irregularities, which usually take the form of circular, convex "bumps" or concave pockmarks on the surface. The light beam is reflected from the flat regions of the surface containing polymer and almost micro-dispersed filler under the incident angle to a detector while curved regions on the surface, originating from large agglomerates and pellet fragments, are reflected in other directions. Consequently, the former areas of the razor cut surface appear as a continuous bright area, whereas the latter areas are seen as dark regions. Thus, the percentage and the size distribution of the undispersed filler particles larger than a few µm can be measured.

A commercially available test device is the Dispergrader [11, 12]. The test specimen is placed in the apparatus and illuminated with a halogen light at an angle of incidence 30° to accentuate surface details. An image is captured with a CCD camera and processed using gray-scale thresholding. Bumps and holes present on the cut surface appear as shadows and are counted to yield frequency/size histograms or percentages of the total area covered by agglomerates and pellet fragments. The percent filler dispersion is calculated based on the image analysis of undispersed agglomerates and the filler volume fraction in the compound. Figure 5.3 illustrates the dispersion quality as measured with the aid of the Dispergrader.

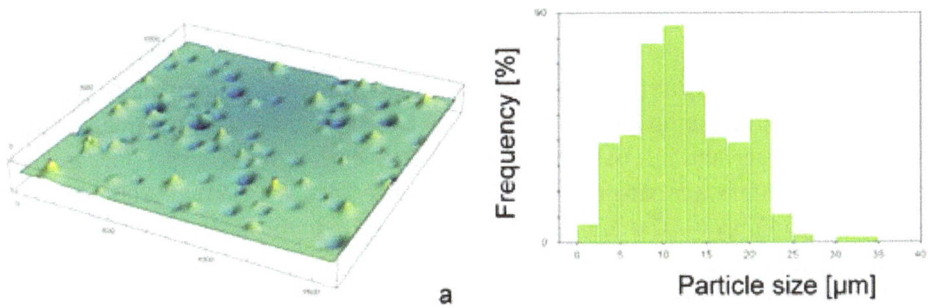


**Figure 5.3** Carbon black dispersion measured by a Dispergrader

The new possibilities for objectively determining filler macro-dispersion are described in articles published by the manufacturers of such devices [13]. The Dispergrader 1000 NT, featuring extended magnification, is claimed to resolve agglomerate sizes of 1.5 µm and greater.

From an optical physics point of view, it is more correct to place the light source insight of the optical microscope and to illuminate the razor blade cut by a vertical light beam [14]. In this case, all the asperities on the surface larger than 2 or 3 times the wavelength of the visual light deflect the reflected light beam at an angle large enough not to be detected by the objective of the microscope. Consequently, the flat regions of the razor-cut surface reflect the light beam back into the lens of the microscope, which results in a bright patch. Conversely, the convex (or concave) regions on the surface that correspond to the pellet fragments and agglomerates larger than the wavelength of the visible light will reflect the light beam and appear as dark spots on the microscope image. The image of the reflected light beam displayed on a computer screen is converted by a charge-coupled device (CCD) camera into a numerical form that can be processed in a computer to finally produce an image that represents the size distribution of solid objects larger than 2–3 µm. The principle behind the "dispersion analysis system" (DIAS) developed at the German Rubber Institute (DIK) is schematically shown on the left side of Figure 5.4.

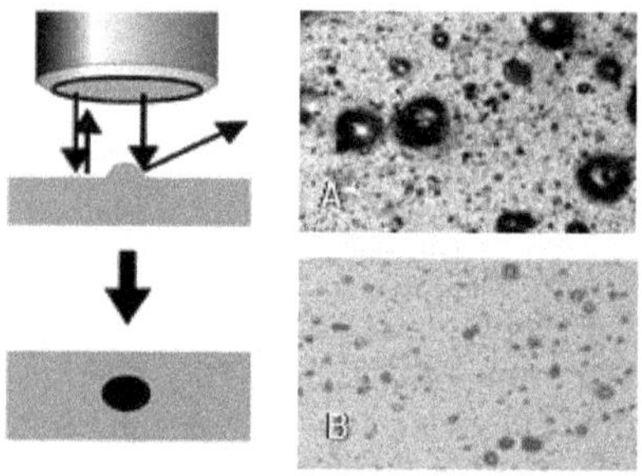


**Figure 5.4**
DIAS system for determining the filler macro-dispersion

A computerized image analysis delivers the size distribution and surface fraction of the filler agglomerates, which can be quantified within several minutes. As can be seen, compound A in the upper image, obtained by upside down mixing, demonstrates poor filler dispersion, characterized by the presence of large agglomerates, while compound B on the bottom prepared by a conventional mixing sequence (first rubber, then filler and oil) has a significant better carbon black dispersion. If compounds with different mixing times are tested, it becomes evident that prolonging the mixing action improves the compound quality. The size of the agglomerates and their total number decrease progressively, due to the mixing properties. The macro-dispersion is expressed as the percentage of dispersed carbon black. A dispersion of 100% means that no undispersed filler particle (pellet fragments and large agglomerates) appears as dark patches on the razor-cut surface. To calculate the dispersion coefficient (or dispersion index), the inspected area containing undispersed filler has to be considered [15]. According to ASTM D2663, the dispersion index (DI) is expressed as follows:

$$DI_{ASTM} = (1 - \Sigma V_{agglom}/V_{filler}) \cdot 100\% \tag{5.6}$$

where $V_{agglom}$ is the volume of the undispersed filler seen under the microscope and $V_{filler}$ is the total volume of the filler

The expression for determining the dispersion index (DI) has to consider the fraction of rubber “occluded” in the voids of the filler agglomerates. Assuming that most of the fillers used possess a structure somewhat higher than the normal one, the average effective solid volume of the agglomerates was considered to be $0.4\,V_{agglom}$. With this correction, the dispersion index (DI) can be calculated from the expression:

$$DI_{ASTM} = (1 - 0.4\Sigma V_{agglom}/V_{filler}) \cdot 100\% \qquad (5.7)$$

The different optical roughness tests are simple, rapid and low-cost processes. With good reproducibility, they provide a dispersion index (DI), which is a quantitative measure of the “goodness” of mixing. However, the same set of optical roughness data can be used to determine the fraction of large pellet fragments present in the compound. This is important in as much as the large pellet fragments (> 30–50 µm) indicate a risk that the compound will fail when subjected to cyclic dynamic deformation. The more such potential flaws that are present in the compound, the higher is the probability that the compound will have a reduced fatigue life.

To exemplify the potential of the DIAS method, Table 5.1 shows the dispersion index (DI), the coefficient of variation of DI and the number of pellet fragments larger than 50 µm for an E-SBR/CB N550 compound mixed in a GK 5 E machine equipped with intermeshing rotors (PES 3) using a fill factor of 70% [16].

**Table 5.1** Dispersion Index (%), Coefficient of Variation and Number of Pellet Fragments Larger than 50 µm as a Function of Mixing Time [16]

| | 45 s | 60 s | 75 s | 90 s | 150 s | 180 s |
|---|---|---|---|---|---|---|
| Disp. index | 74.8 | 86.8 | 94.2 | 93.8 | 94.1 | 95.6 |
| Coeff. of var. | 4.1 | 3.1 | 0.7 | 1.2 | 1.1 | 1.3 |
| Objects > 50 µm | 107 | 35 | 10 | 5 | 7 | 11 |

However, even in the case of the readily dispersible carbon black, the compound contains a number of large, undispersed pellet fragments that will detrimentally affect ultimate properties, such as tensile strength, dynamic cut growth and abrasion resistance. Using the DIAS method, it has been shown that the macro-dispersion of precipitated silicas with similar specific surface area but different concentration of silanol groups (Sears number) on the surface increases primarily as a function of silanol concentration of the polar surface [17]. The results are presented in Figure 5.5.

The greater influence of the silanol group density, which lowers the macro-dispersion much more than the specific surface area of the silica, can be clearly seen. This difference has led the manufacturers of precipitated silica to produce more HD silica with fewer silanol groups, as this can reduce the macro-dispersion.

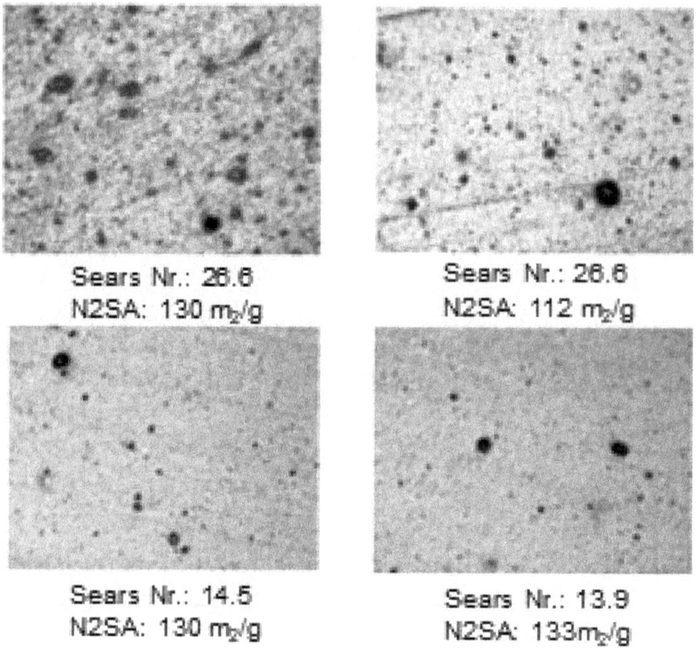


**Figure 5.5** Macro-dispersion of precipitated HD silica in S-SBR [17]

This is particularly important if a compounder wants to find out which changes in the process parameters or the recipe composition leads to a desired improvement in the quality of the mixture. The macro-dispersion of a filled rubber compound can be recorded within a few minutes and the result can be compared with other dispersion indexes from before. There is therefore the possibility of optimizing the mixing parameters and finding the best options for good macro-dispersion, irrespective of whether it is a carbon black or a precipitated silica compound.

### 5.2.1.2 Mechanical Roughness Methods

In contrast to optical roughness, the irregularities caused by filler agglomerates on glossy sections can be determined by means of stylus surface measurements. The method is based on tracking a fine stylus over a cut rubber sample to measure surface roughness caused by the presence of pellet fragments and agglomerates [18]. For the measurements, a surface profilometer fitted with a 2 µm diamond tip stylus is recommended. The sample is prepared by tearing the vulcanized rubber compound to obtain a fresh surface. Artifacts to be avoided are [4]:

- excessive cut marks produced by poor razor blades
- distortions due to compression of the sample during cutting and
- surface dirt that is smeared across the cut surface

The stylus surface determination forms part of industrial standard methods [19]. The most pertinent measures are the number of roughness peaks per unit length and the average peak height. The result is expressed as a roughness parameter, Ra, that is given by the difference of the altitude of a point on the surface and the average altitude of the surface, as shown in Figure 5.6 [20].

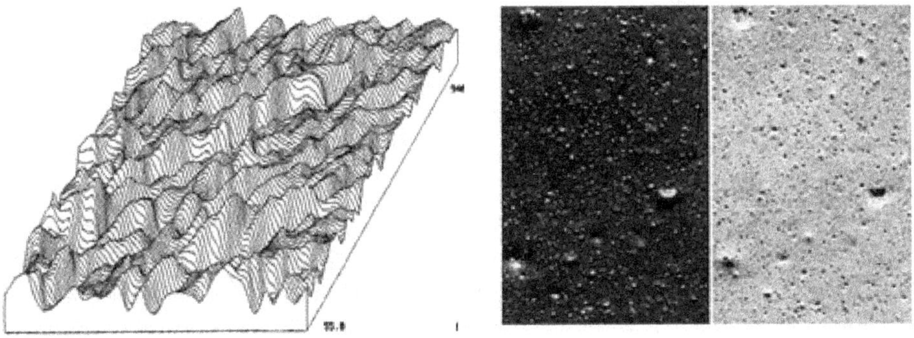

Figure 5.6 Mechanical roughness measurement: a) roughness parameter Ra and b) the macro-dispersion of carbon black N330 in E-SBR [20]

The smaller the mechanical roughness parameter, Ra, the better is the filler dispersion. Interlaboratory tests have attested good reproducibility of the stylus method. Surface roughness measurements are an indirect measure of dispersion, because the stylus does not directly measure the size of the agglomerates. The relationship between the filler dispersion and mechanical roughness provoked by the incorporated filler may vary as a function of polymer type, filler loading and oil content. For this reason, it is helpful to apply the method when working with the same recipe to evaluate the improvements made. For best results, the mechanical roughness method should be calibrated against a more direct method of characterizing percentage agglomeration [4]. The stylus method is often used when examining tire tread compounds. It turns out that, with a prolonged mixing time, the number of pellet fragments decreases and smaller and smaller agglomerates are created, which means that the number of objects that appear in the nano range increases.

### 5.2.2 Filler Micro-Dispersion

"Micro-dispersion" is a nano-scale phenomenon that requires a large magnification when trying to visualize the filler particles. Thus, powerful and sophisticated methods such as electron microscopy (EM) and atomic force microscopy (AFM) at high magnification provide direct images of filler networks and individual agglomerates or aggregates. These methods have been used in the rubber industry in recent decades when the accuracy of determining filler dispersion became very important. However, such methods can only be used by major manufacturers who have a great interest in the effect of the filler dispersion.

#### 5.2.2.1 Transmission Electron Microscopy Method

Due to the great importance of filled elastomers, it is not surprising that transmission electron microscopy (TEM) was used shortly after its invention [21] to analyze structural aspects of carbon black and carbon black micro-dispersion [22–24]. Given suitable preparation techniques, TEM can provide useful two-dimensional images of aggregates or agglomerates which subsequently can be processed to give an aggregate size distribution. TEM offers the highest resolution of any microscopic process and is generally limited to studies of microstructure of primary particles, aggregate structure and is rarely used to determine aggregate size distribution at the nano-scale. The use of TEM requires not only sophisticated instrumentation but also skilled personnel for both the preparation of ultra-thin cuts with an average thickness not larger than 100 nm for correct interpretation of the micrographs.

Conventional TEM projects three-dimensional objects onto a plane; hence the information obtained about the thickness of the objects is only obtained as cumulative information.

Since the transparency of the thin cryosections to the electron beam is limited, precise identification of the individual aggregates become increasingly difficult for highly filled compounds. The distribution can be quantified by an automated image analysis program. However, the conclusions drawn from the inspection of TEM micrographs can be difficult enough to consider such information for designing the mixing process and assess the quality of rubber compounds.

By employing this preparation technique, it has been demonstrated that micro-dispersion of carbon black is improved at the nano-scale, even when macro-dispersion is no longer improved by prolonging the mixing time. Corresponding results are given in Figure 5.7 for SBR compound containing 50 phr normal structure carbon black N330 and mixed in an intermeshing mixer (5.45 L) equipped with PES3 rotors [25]. As can be seen, increasing the mixing time from 125 to 365 s significantly decreases the average aggregate size and the size distribution becomes narrower. For the same content of low-structure carbon black N326, the level of micro-dispersion achieved in the same mixing time is considerably lower. That is shown by a broader aggregate size distribution and a peak value displaced to larger aggregate radius.

Elucidation of the dispersion of nano-fillers in space using only two-dimensional projections remains a major challenge. Specifically, TEM has difficulties in determining where the junctions between aggregates are. Measuring junction width with TEM is difficult, low-precision and time-consuming. Several techniques have been proposed that used rotating sample supports, to determine the dispersion difference, but the range of rotation technically achieved was not great enough to provide reliable results about the spatial arrangement of the primary particles in the aggregates [5]. The preparation of the test specimens takes longer and the image is two-dimensional, which makes accurate representation of the three-dimensional arrangement difficult. TEM investigations provide information about the architecture of filler networks and the change in size distribution of aggregates on the nano-scale as a function of mixing conditions.

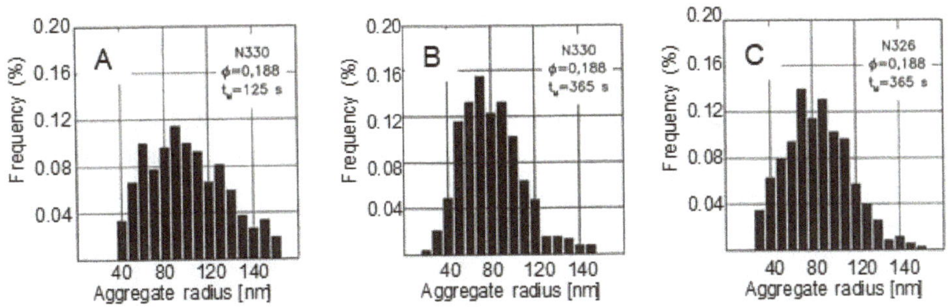


**Figure 5.7** The aggregate radius distribution of CB N330 and N326 in E-SBR as a function of mixing times [25]

This problem was resolved with the advent of 3D-TEM, which combines TEM and computerized cryo-tomography [26]. The images are obtained on 200-nm-thick cryo-sections by the bright-field technique, without staining of the samples. NR compounds filled with carbon black N330 [27] and conventional silica (VN 3) [28] have been investigated with this new 3D-TEM technique. The technique gives an impression of the filler network structure and can estimate the inter-particle distance, which is in fact the inverse of the particle size distribution. The results for NR-carbon black N330 vulcanizates are shown in Figure 5.8 [27].

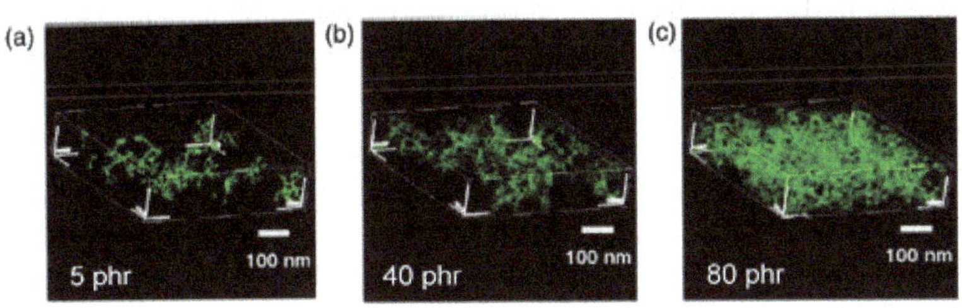


**Figure 5.8** 3D-TEM of carbon black N330 in NR as a function of loading [27]

The formation of the filler-filler network with increase in filler loading can be much better understood when 3D-TEM imaging is used. As the loading of the carbon black N330 in NR increases, the filler network becomes denser, leaving less space for rubber chains in between of the filler particles. The increase in the filler-filler network above the percolation threshold can be clearly seen. This also explains the multiple contacts of individual rubber chains on different carbon black agglomerates. But this method is very complex and only a well-equipped company can afford to examine the micro-dispersion of carbon black or HD silica using the 3D-TEM method. However, 3D-TEM can be expected to deliver new results in the future regarding the micro-dispersion of fillers in rubber compounds.

### 5.2.2.2 Scanning Electron Microscopy Method

Scanning electron microscopy (SEM) has been applied to most of the different aspects of filler dispersion and has proven to be a versatile method. Thus, many dispersion analysis techniques have been complemented by SEM analysis [4]. The image of a microtome cut surface of a rubber with a smaller concentration of carbon black is shown in Figure 5.9. The bumps and depressions caused by deflection of the tear path by the agglomerates can be clearly seen.

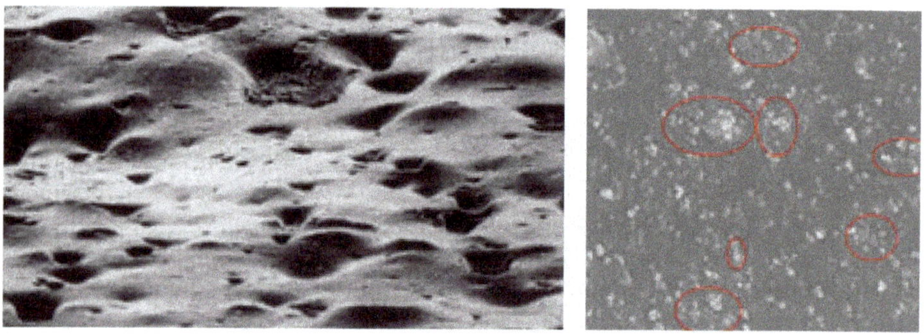

**Figure 5.9** Image of a carbon black micro-dispersion visualized by SEM [4]

The places where there is filler aggregate are clearly highlighted in the SEM image. Another advantage of SEM is the ability to single out individual agglomerates in a sample and provide compositional information from the energy spectrum of the X-rays that are emitted. This can be seen in the image on the right, where the different filler materials are shown.

The ability of SEM to resolve filler micro-dispersion and yield information about the distribution of different fillers can be considered special additional information that is helpful when different types of fillers are used. The effect of the filler loading on carbon black agglomeration in NR/SBR blends has been investigated by SEM. It was shown that up to a filler volume of 12%, the filler particles are uniformly dispersed, while only a small part of the carbon black is agglomerated. In addition, it was shown that carbon black is dispersed more and more evenly with the increase in carbon black content. Above 18%, which corresponds to the percolation threshold, agglomeration and the formation of small clusters appears in most aggregates.

#### 5.2.2.3 Atomic Force Microscopy Method

Since its invention in 1986, atomic force microscopy (AFM) has developed into a powerful technique for imaging materials structure on the nanoscale [29]. The principle of AFM is to place a small tip attached to the end of a thin cantilever in close proximity with the sample surface and to move the tip over the sample surface while recording attractive and repulsive forces through the deflection of the tip. AFM's ability to generate an image contrast based on nano-mechanical properties of the substrate components is a distinctive feature of this type of microscopy. The technique is flexible in many ways that other high-resolution microscopic methods are not, including easy sample preparation and its ability to operate in an ambient climate. Most images are generated in contact mode on an AFM. In this mode, individual agglomerates can be measured and the entire image compared with the other samples. Although the versatility of AFM has made the method a preferred tool for scientific investigations, the subjectivity of visual analysis remains a problem.

AFM offers several advantages over TEM. It enables investigation on a scale of up to a few hundreds of micrometers. The technique does not require any particular preparation of the specimens, except for a smooth surface, which is usually obtained with cryo-microtoming for imaging. It has to be mentioned, however, that the resolution of AFM is lower than that of TEM. Extracting quantitative information about a material's nano-mechanical properties remain more challenging than simply acquiring pictures with a simple contrast [30]. Different scanning modes have been used to record the state of dispersion below and above the µm scale (i. e. non-contact, contact, lateral force, tapping-phase contrast, etc.). Tapping mode is the most frequent and routinely used method for polymer materials characterization.

The potential of AFM for mapping surfaces has been employed successfully to describe filler dispersion on a nanoscale level [31]. A comparison of two EPDM compounds containing 50 phr of the semi-reinforcing and easily dispersing carbon black N762 is shown in Figure 5.10 [32]. The dispersion indices shown in the figures were determined using DIAS. In the compound with lower dispersion index, poor distribution of the aggregates can be seen whereas, in the compound with the higher dispersion index, the aggregates are evenly distributed in the rubber matrix.

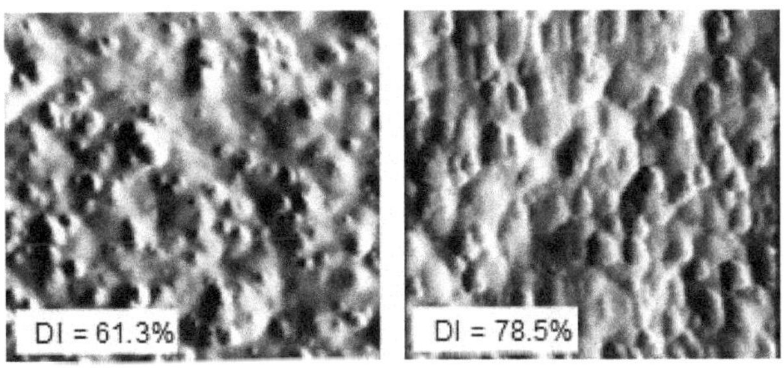


**Figure 5.10** Micro-dispersion of carbon black N762 in EPDM vulcanizate

SBR compounds loaded with silica (5 to 15 vol.%) have been investigated to determine the distribution of the filler and to evaluate the rubber-filler interface [33]. It has been reported that individual aggregates and agglomerates in rubber can be easily observed by AFM. Despite all the obvious advantages that the AFM method has in terms of sample preparation and high resolution, there is still no easy and rapid method for define a degree of dispersion that is valid on the sub-micron scale.

#### 5.2.2.4 Indirect Methods

In addition to the optical and mechanical methods for macro-dispersion, two methods have emerged which are related to the dispersion but do not indicate the size distribution of the agglomerates. They are the storage modulus G′ in the strain-dependent dynamic experiment (Payne effect) and the rheological or electrical percolation threshold of the vulcanizate. Both methods are described elsewhere in this text. However, there is a relationship to the micro-dispersion, but the dispersion is not determined by the size distribution of the aggregates and only applies when comparing different types of the same formulation. However, to gain an insight into the process parameters during mixing or various changes to the mixing technology, indirect methods are often used for the same formulation. The lowering of the storage modulus with increase in the strain amplitude is viewed as an example of increased dispersion of the filler.

## 5.3 Filler Dispersion in Rubber Vulcanizates

The effects of the type of filler on dispersion are closely related to parameters that determine the cohesive strength of the filler agglomerates and the rubber-filler interactions, such as the specific surface area, the branched nature of the aggregates and especially the surface activity. To improve filler dispersion and reduce filler–filler networks in dynamic-mechanical measurements, thermodynamic approaches are used:

- to reduce the surface energy between polymer and filler by filler surface modification and/or polymer modification, or
- increase polymer-filler interaction by filler surface modification, polymer modification or using surface-active and silica coupling agents.

Compounds that must meet high requirements on physical properties are mixed differently than those with high requirements imposed on processability. In the first case, the best possible dispersion has to be envisaged whereas, in the second approach, easy processing with a homogeneous compound of low viscosity and reduced die swell are of major importance. The choice of filler should therefore be tailored to both the intended use of the product and cost-effective processing. The effects achieved during mixing with regard to filler dispersion and compound homogeneity thus depend on the choice of rubber, filler and plasticizers as well as the processing conditions.

## 5.3.1 Dispersion of Carbon Blacks

It is likely that pellet hardness in any filled compound contributes to a certain amount of undispersed filler, which in turn reduces the homogeneity and causes defects in the material that can reduce the fatigue life of the corresponding vulcanizates when subjected to severe dynamic load. The reason for these negative consequences of pellet hardness lies in the wide distribution of hardness, which in turn originates in the complex technology of wet-pelletization. In addition, pellet hardness can differ from one delivery form to the other. In a few reports it has been shown that, with increase in filler specific surface area, the pelletizing process becomes more difficult to handle and the pellet spectrum increases considerably [34, 35].

It has to be noted that the incorporation of hard pellets of carbon blacks requires higher power consumption and, therefore, ram seating times during incorporation and the real mixing cost increase. Shortening the incorporation time increases the annual cost of a rubber mixing process substantially [36]. Therefore, sufficient time should be spent on the incorporation phase of the filler. If this phase is too short, the filler will not be incorporated into the rubber matrix and there will be little dispersion in the rubber matrix. The proportion of pellets with a greater hardness delays the break-up of the pellets, so that the erosion of pellet fragments starts later and at the end of the mixing process the portion of undispersed carbon black in the compound is higher than for filler grades with softer pellets.

### 5.3.1.1 Influence of Specific Surface Area

Several experimental studies have shown that, under a constant set of mixing conditions, the ease of carbon black dispersion increases with increase in size of the primary particles. To achieve dispersion of very small particle size carbon blacks, a high viscosity of the rubber is required which must be maintained as much as possible for a relatively long period of time in the turbulent flow in the mixing chamber. Since these stiff batches generate heat at significant rates, the mixing chamber and the rotors have to be cooled, and such batches dumped before the necessary mechanical work input has been achieved in the internal mixer or the two-roll mill. As shown in Table 5.2, the dispersion index increases progressively as the primary particle size of carbon black increases and the surface activity decreases.

**Table 5.2** Macro-dispersion Index (%) of Carbon Black (50 phr) in S-SBR Compounds at Selected Mixing Times [37]

| Carbon black | 120 s | 240 s |
|---|---|---|
| N115 | 45 | 68 |
| N234 | 56 | 74 |
| N550 | 88 | 92 |

Since the polymer matrix and the mixing conditions remain unchanged, the result can be traced back to the inter-aggregate interactions, which are linked to the primary particle size or specific surface area. The higher the inter-aggregate forces, the stronger the agglomerates will withstand the rupture and erosion mechanisms in the case of extensional and shear flow. The insufficient dispersion index of the large surface area blacks indicate that larger portions of the surface of the active carbon blacks are not wetted by the rubber, which in turn leads to the observation that the bound rubber per unit surface decreases as function of the specific surface area of carbon blacks. Thus, if macro-dispersion is improving for some reason, the micro-dispersion shows a development in the same direction. It can be shown that the filler network, expressed by the storage modulus $G'_0$, decreases with increase in the carbon black primary particle size. The decrease in storage modulus $G'_0$ is attributed solely to the particle density in the compounds. Therefore, at constant filler loading a filler network composed of smaller aggregates contains a higher number of particles of smaller interparticle distance, whereas, with larger aggregates, fewer particles are arranged at a larger interparticle distance. The stiffness of the filler networks is inversely proportional to the inter-aggregate distance. As a result, the Payne effect increases, due to higher aggregate density and the increased filler-filler interactions between the particles. The experimentally observed displacement of the electrical percolation threshold from small filler loads (20 phr) for fine carbon blacks to high filler loads (65 phr) can therefore be explained by the contribution of both the surface energy sites distribution of the aggregates and the geometrical effect of the aggregate size [38]. The mechanism of carbon black dispersion in rubber compounds loaded at practical levels has been examined from three different aspects [39]:

- effect of carbon black loading
- deformation characteristics of raw rubber, and
- rubber chain scission during mixing

It was found that the loading threshold of carbon black, above which efficiency of dispersion is greatly improved, coincides with the percolation threshold of the filler at which filler network, (i. e. Payne effect) of the compound becomes detectable.

### 5.3.1.2 Influence of Void Volume

As a result of the branched nature of carbon black particles, the aggregates are packed less densely into agglomerates and can be easily dispersed in the rubber matrix. This morphological feature reduces the inter-aggregate attractive forces per agglomerate. Low-structure blacks are, therefore, more compact, allowing closer packing and thus greater inter-aggregate attractive forces, which increase the tensile strength of the agglomerates and makes dispersion more difficult. As the degree of branching of the aggregates increases, the number of contacts between the aggregates in agglomerates decreases. This reduces the integral tensile strength of agglomerates and so agglomerates made from branched aggregates cannot withstand high shear or elongational stress. As a result, irrespective of the primary particle size (or specific surface area), the ease of dispersion of the filler increases with the OAN value [40].

A plot of the bound rubber of rubber compounds as a function of the occluded rubber reveals an increasing amount of bound rubber. This indicates that the degree of filler dispersion increases [41]. The occluded layer does not lose configurational freedom or become glassy, as postulated in some publications. Using the "onion skin" model, the evolution of the dispersion index was numerically simulated. The result of the mathematical simulation were proven for carbon blacks of the N300 series with different structure of the aggregates [42]. As shown in Figure 5.11, the dispersion increase in carbon black N347 is high while the low-structured black N326 shows a modest increase in dispersion with mixing time.

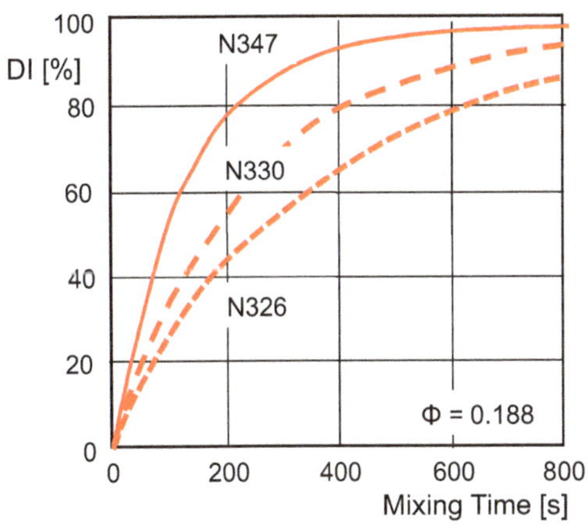


**Figure 5.11** Dispersion increases with the structure of the carbon black aggregates

At the end of the mixing cycle, the degree of dispersion tends to levels that cannot be reached with carbon blacks of a low structure. The use of high-structure blacks saves on mixing time and a high dispersion index is achieved at the end of the mixing cycle. When low-structure blacks are used, incorporation time is shortened, but the dispersion stage is lengthened and ultimately the degree of dispersion in the compound remains low.

The fundamental interrelations between specific surface area, structure and macro-dispersion have been reported for oil-extended SBR 1712 compounds containing 80 phr of carbon black grades varying with respect to both the nitrogen specific surface area and the structure of the aggregates. The compounds were mixed with a constant energy of 1075 MJ/m$^3$. The resulting dispersion indices (measured by the stylus roughness method) are shown as iso-dispersion contour lines in Figure 5.12 [4].

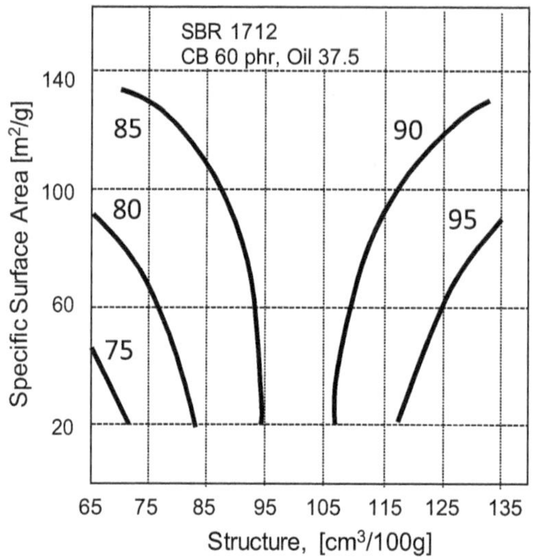


**Figure 5.12** Macro-dispersion of carbon black as a function of the specific surface area and the structure of the aggregates [4]

The expectations relating to the influence of cohesive forces and the number of inter-aggregate contacts in agglomerates are confirmed in this diagram. As far the group of low structure blacks is concerned, it is found that an increase in the specific surface area from approx. 30 m$^2$/g to 140 m$^2$/g results in a moderate improvement in the dispersion index. For the group of high-structure carbon black, the same increase in the specific surface area results in a surprising decrease in the dispersion index from 95 to 90%. Consequently, it is found that carbon blacks of very high specific surface area (> 160 m$^2$/g) and low compressed oil absorption number (COAN) with less than 60 mL/100 g cannot be dispersed by dry-mixing on existing equipment [43].

For upside-down mixed EPDM compounds, in which the polymer is added on top of the filler and oil (100 phr carbon black/ 120 phr oil), it has been shown that dispersion is predominantly related to the carbon black structure. As DBP absorption increases, the influence of the specific surface is on the dispersion index decreases strongly. For high-structure blacks (DBP absorption above 120 cm$^3$/100 g), the values of the dispersion index measured for grades with 80 m$^2$/g or 30 m$^2$/g are within the range of experimental error. If the carbon black is dosed in excess (120 phr carbon black /105 phr oil), the dispersion ratings are lower and increase with decrease in specific surface area.

#### 5.3.1.3 Modification of Surface Activity

The surface activity of particles can be modified while the particle size remains unchanged. Compounds made from carbon blacks with modified surface activity demonstrate better filler dispersion and yield vulcanizates with improved properties.

Essentially, the attempts are directed at improving the dispersion of carbon black compounds for tire applications [44], with the aim of changing the site energy distribution for active carbon blacks to achieve better interactions with the rubber chains, and thus higher phase adhesion. The modification of carbon blacks therefore reduces the portion of graphite crystallites, while increasing the proportion of "amorphous" carbon on the surface. This modification also reduces filler-filler interaction and improves filler dispersibility by enhancing polymer-filler interaction.

For carbon black, the post-processing modifications are used first. Early attempts to modify the surface properties of existing carbon blacks included adsorption of gases [45] and oxidation of channel black [46]. Oxidation of carbon black either by gas phase oxidation or through liquid phase oxidation degrades the carbon black and oxygen functional groups are placed on the edges of the graphitic micro-crystallites. Low-temperature treatment of carbon black N234 with nitrogen oxides leads to surface oxidation and subsequently to better dispersion, as demonstrated by a 14% lower tan $\delta$ of the vulcanizates [47]. Recently, superficial oxidation of furnace blacks has been used for dispersing it in functionalized carboxy S-SBR. The improved dispersion is attributed to strong hydrogen bonding. It has been shown that this leads to reduced filler networks (Payne effect) and a smaller larger interparticle distance, as measured by means of AFM [48]. The effect is attributed to hydrogen bonds between the functional groups of the filler and the functionalized S-SBR.

The reaction of diazonium salts of aromatic amines with active carbon blacks has shown that efficient modification of surface activity is possible [49]. It is assumed that the radicals formed during the decomposition of the diazonium salt react at the edges of the graphitic micro-crystallites and thereby reduce the strong filler-filler interaction without changing the particle size. As a result, the contribution of the high energy sites to the filler-filler interaction decreases, while the contribution of the "amorphous" carbon remains unchanged.

Chemical treatment of ground carbon blacks in suspensions with small amounts quantities of zinc-dithiocarbamate (ZDMC), peroxides and other reactive species provide better dispersion and less hysteresis. The treatment of carbon black N330 with 0.5% ZDMC in aqueous suspension over 24 h at 100 °C leads to a reduction in filler networks by more than 30% [50]. The observed effect of atmospheric plasma on the surface of fluffy black indicates that the surface can be changed in a fraction of a second and therefore the properties of the carbon black and its dispersion in the vulcanizate are improved. Depending on the type of plasma, the lattice of the graphitic carbon black could be changed [51].

Most of the modifications focus on new coupling agents, such as benzofuran [52], N,N′-bis(2-methyl-nitropropyl)-1,6-diaminohexane [53] and p-aminobenzene-sulfonyl azide [54]. When oxidized carbon black is used in combination with bifunctional silanes, the hysteresis and heat generation of dynamically loaded vulcanizates are considerably reduced [55]. Among the bifunctional silanes that can establish a link between the filler and the polymer are several chemicals, including titanate coupling agents [56], zirconate-based coupling agents [57] which have found application polymer composites reinforced with inorganic fillers.

More-interesting experiments, which are useful for large-scale applications, are in-process modifications of the carbon black surface. The main targets of these application-oriented technologies are:

- to reduce the size of the graphitic crystallites
- to increase the concentration of active functional groups that can interact with the polymer

In order to put the advanced technology into practice, changes to the reactor design and processing conditions of carbon black production are necessary. However, such changes are not disclosed in any detail in reports published by carbon black producers. A prominent example is “inversion” carbon black and “nanostructured” carbon black Ecorax® described in Chapter 3. Due to changes in the process parameters [58], the density of graphitic edges is smaller and the heterogeneity of the surface tends not to be as uniform as in ASTM carbon blacks. Accordingly, these types of modified carbon blacks show a narrow agglomerate size distribution. Regardless of the rubber in which these grades are mixed, a shorter mixing time with improved dispersion and less filler networks is reported. The low hysteresis combined with good abrasion resistance qualifies these carbon blacks for use in tire treads [59]. The next example of in-process modified carbon black is that of carbon black-silica dual-phase fillers (CSDPFs) [60–63]. The ease of mixing CSDPF into rubber is demonstrated by lower torque and shorter filler incorporation times and a much better dispersion in the non-polar S-SBR/BR blend. Compared with a vulcanizate that contains carbon black N234, two-fold mixing shows no macroscopic pellets in the compound, whereas the carbon black N234 that is blended in the usual way shows non-dispersed carbon black after four-fold mixing. Evidence for the lower filler-filler interaction of CSDPF is also provided by the storage modulus, which decreases strongly with the strain amplitude. An additional silanization with TESPT enables a further, albeit slight, reduction in the effects of CSDPF. Silanization reduces the Payne effect and increases abrasion resistance. This is not possible with a normal furnace black.

### 5.3.2 Dispersion of Precipitated Silicas

Due to the number of silanol groups on the precipitated silica surface, there is a strong attraction between the aggregates, which gives rise to larger agglomerates in the mixture. Accordingly, there is a high Payne effect that clearly exceeds that of carbon black vulcanizates. As mixing time increases, silica agglomeration in the mixture decreases, as confirmed experimentally by a reduction in pellet fragments [17]. A mixture of precipitated silica in S-SBR shows a modulus of 28 MPa for short mixing times of 120 s, which shrinks to 0.6 MPa after a longer mixing time of 700 s. The dispersion of precipitated silica can be significantly improved by a longer mixing time.

Precipitated silica of various specific surface areas and different concentrations of silanol groups is well dispersed in polar rubber, such as NBR, HNBR and ACM. The dispersion is achieved by acid-base interactions between the acidic silanol groups and the polar segments of the rubber [64–66]. A technological improvement by silica producers is that of the highly dispersible (HD) silicas, which are more porous and can be easily incorporated by the rubber matrix [67, 68]. The time-dependent dispersion process of conventional silica compounds improves with decrease in specific surface area and increase in structure. Compared with conventional silicas, mixing of HD silicas leads to superior dispersion after short processing times.

*In situ* dispersion of silica is a process that circumvents the problem of dispersing silica in rubber [69]. The process begins with swelling crosslinked NR in tetra-ethoxy silane and hydrolysis in which the alkoxy group is replaced by a hydroxyl group. Condensation reactions of silanol groups produce siloxane bonds (Si-O-Si) that ultimately form the spherical silica particles, which are dispersed perfectly in the NR matrix [70]. The radius distribution and number of silica aggregates in a definite volume have been evaluated by the 3D-TEM analysis system. It was found that *in situ* silica nanoparticles had an average radius of 13.3 ± 5.6 nm. Although the method may have potential for industrial applications, such as reactive extrusion processes, there is no industrial-scale application of this approach to date.

Reactive mixing of precipitated silica with organo-silanes facilitates the conversion of hydrophilic silica into a hydrophobic filler, which has a less polar surface and can be easily dispersed. A good reference on the structure, chemistry and application of silanes is provided in [71]. The effects of monofunctional silanes, tri-ethoxypropyl silane (Si203) and tri-ethoxy octylsilane (Si208) on the macro-dispersion and the bound rubber were investigated for HD silica (7000 GR) in NBR and BR compounds [66]. The results are summarized in Table 5.4. It is worth noting that both silanes are assumed to react quantitatively on the filler surface and lead to increased macro-dispersion, but the degree of dispersion in NBR is systematically higher than in BR.

**Table 5.3** Dispersion Index (%) of HD Silica in BR/Silica and NBR/Silica [66]

| Product | NBR | BR |
|---|---|---|
| HD silica | 91.2 | 83.7 |
| HD silica + Si203 | 98.3 | 92.2 |
| HD silica + Si208 | 98.7 | 92.8 |
| HD silica + Si69 | 98.4 | 92.5 |

Due to the lower polarity and the lower solubility parameter δ of BR, the effect of hydrophobization is more pronounced in this rubber. It is conceivable that by increasing the alkyl chain, the area of polar Si-O-Si ether groups is shielded more efficiently and filler dispersion increases to a maximum. The use of monofunctional silanes is beneficial if filler flocculation has to be suppressed and the viscosity of the compound has to be reduced. The bifunctional silanes are of much greater technical importance, since these compounds can provide both hydophobization of the silica particles and chemical bonding of the silica particles to the rubber matrix. The application of these coupling agents greatly improves the processing behavior by reducing the solubility parameter of the precipitated silica. To underline this point, the effect of silanization on the micro-dispersion of precipitated silica in an S-SBR compound is shown in Figure 5.13.

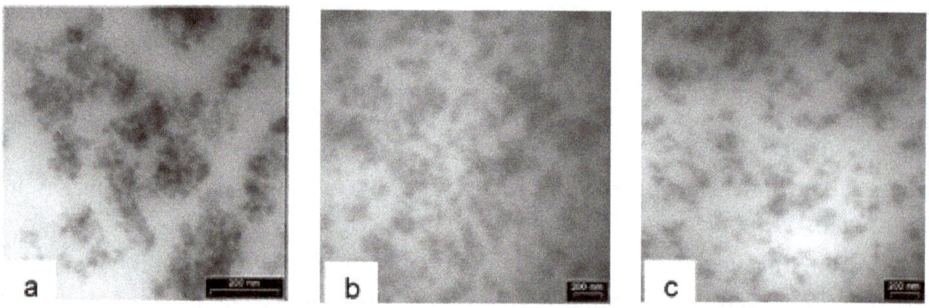


**Figure 5.13** TEM images of S-SBR-containing HD silica: a) not silanized in S-SBR (55V;25 S), silanized silica in S-SBR (25V;25S) and c) silanized silica in S-SBR (55V; 25S)

The non-silanized silica is present in relatively large agglomerates, which appear spatially well separated, due to the low volume fraction in the mixture. As a result of the silanization, the agglomerates in the TEM image are much smaller and better distributed over the area under consideration. The wet-mixing of HD silicas can improve the dispersion in S-SBR. Compared with dry-mixed silica, the wet-mixed compound exhibits only a very low level of undispersed silica, as shown in Figure 5.14 [72].

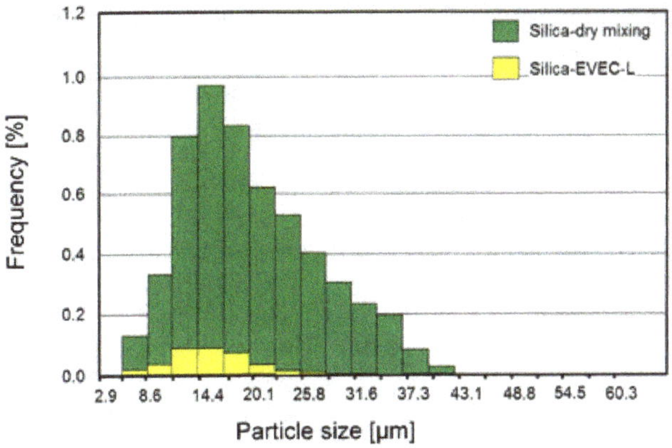


**Figure 5.14** Undispersed silica as a function of the silica compounding: a) dry-mixing (green) and b) wet-mixing (yellow)

The undispersed precipitated silica in the dry-mixed compounds is not only greater in number, but also in size distribution. The maximum diameter of undispersed silica is 14 nm in the case of dry-mixed silica and somewhat lower in the wet-mixed silica, a fact which shows the advantage of this method of manufacturing the compound.

### 5.3.3 Dispersion of Layered Silicates

The dispersion and distribution of pristine layered silicate and organically modified layered silicate depend on the choice of processing technology. A pristine layered silicate can be distributed in rubber latex (NR, SBR or NBR) by wet-mixing and is well dispersed in the rubber. When CDLC technology is used, more than 80% of the platelets are exfoliated in the rubber matrix [73, 74]. At a low volume fraction of the layered silicates, the better dispersion is indirectly observed via the increase in storage modulus in the stress-strain behavior when the pressure during the mixing is increased and via the reduction in the percolation threshold. The incorporation of layered silicates in CDLC processes leads to short distances between the filler particles, such that the properties can be largely modified at the low concentrations of the layered silicates. The intercalation of polymer chains between the MMT platelets is considered to promote clay exfoliation, leading to the ultimate dispersion and alignments of the platelets. The conventional mechanical filler content in rubber is generally 20–60 wt.%, which in turn results in high density and material costs.

In most cases, montmorillonite is transformed into an organo-clay for better dispersion of the filler in the polar polymer. Na-modified montmorillonite modified with ammonia cations exhibits an enlargement of the periodicity of intercalation structure. This is responsible for the different arrangement of the filler platelets in the rubber matrix. The enlargement of the interlayer distance, measured from X-ray diffraction patterns, is taken as evidence of the interaction that has taken place. A mechanism has been proposed to explain nanocomposite formation and exfoliation [75]. If the polymers and the organo-clay have an affinity for one another, the contact between the organo-clay can be increased by peeling the platelets from the tactoids one by one until all the particles are individually dispersed [76]. Organo-clay is well dispersed in engineering polymer matrix [77, 78]. Complete dispersion of layered silicates in a rubber matrix optimizes the number of reinforcing elements available for carrying applied load and deflecting cracks. Exfoliation during dispersion in melt processing has been investigated [79]. The nature of extruder and screw configuration are important to achieve a good dispersion in the polymer [80]. Longer residence time and higher viscosities of the polymer favor a better dispersion.

### 5.3.4 Dispersion of Carbon Nanotubes

The nature of the dispersion of multi-walled carbon nanotubes differs significantly from spherical particles because carbon nanotubes have a small diameter with a high aspect ratio and thus a larger specific surface area. The usual strategy is to maximize dispersion and debundle through good processing, which should not break the length of the carbon nanotubes and inhibit reaggregation. However, achieving good dispersion of the multi-walled carbon nanofiller is a major difficulty. Without proper dispersion, filler aggregates tend to act as defect sites which limit the mechanical performance of the vulcanizates. Because multi-walled carbon nanotubes are anisotropic objects, orientation in the rubber matrix is an important component of dispersion [81]. The goal of dispersion is to homogeneously distribute multi-walled carbon nanotubes throughout the rubber material, such that the applied load can be uniformly distributed in the rubber chains. Dispersion of multi-walled carbon nanotubes has more aspects than size reduction the size or eliminating the powder particles or bundling. Dispersion in rubber is generally influenced by two aspects:

- disentanglement of carbon nanotubes bundles or agglomerates
- uniform distribution of the individual carbon nanotube in the rubber

The ultimate goal of the processing conditions is to separate the individual carbon nanotubes in order to achieve homogeneous dispersion throughout the matrix while adhesion between the nanotubes and the polymer enable effective stress transfer at the polymer-carbon nanotube interface [82]. Insufficient macroscopic mixing can lead to regions of high or low multi-walled carbon nanotube concentrations on large length scales. Stress fields in the direction of processing have a tendency to align the multi-wall carbon nanotubes.

Melt mixing invariably utilizes the shearing forces generated during mixing to disperse the multi-walled carbon nanotubes [83]. Melt mixing leads to strong interfacial bonding, a fact which has been verified by Raman spectroscopy [84, 85]. Multi-walled carbon nanotubes are known to increase the viscosity of the polymer due to their large aspect ratio and high stiffness [86]. A high concentration [~10%] is a common upper limit for uniform dispersion in a melt mixing process. A lower electrical percolation threshold implies better disentanglement of multi-walled carbon nanotubes and better dispersion in the matrix. A disadvantage of using the percolation threshold to characterize dispersion of multi-walled carbon nanotubes is that the microscopy measurements are laborious [87, 88].

In multi-walled carbon nanotubes/rubber systems, dispersion is one of the most critical and important issues because the degree of dispersion determines the reinforcing properties. Characterization of the degree of dispersion is difficult due to the small volume fractions of multi-walled carbon nanotubes. The final dispersion is not only a function of the method used to separate the individual tubes, but also a function of the method used to mix multi-walled carbon nanotubes with the rubber and forming the downstream process and curing.

Measurements of the electrical percolation threshold or rheological percolation threshold have been used to quantitatively determine dispersion quality in rubbers. In general, a lower percolation threshold implies better dispersion and debundling of the filler agglomerates.

The standard way in which AFM is used to characterize dispersion of multi-walled carbon nanotube rubber composites is to image the surface and to infer the characteristics of the bulk dispersion from the characteristics of the surface dispersion. Less common is measurement of the orientation of the multi-walled carbon nanotubes [89]. The use of an electrically charged silicone tip allows the imaging of multi-walled carbon nanotubes not visible on the surface, although the depth is limited [90]. Scattering methods, using either neutrons or X-rays, can measure both nanoscale dispersion and orientation in the small angle regime [91, 92]. High shear forces are responsible for producing well dispersed multi-walled carbon nanotubes on the micro- and nano-scale. However, the dispersion resulting from melt mixing is inferior to that from solution methods which are employed in scientific laboratories. Sonication uses high frequency sound to induce cavitation in low-viscosity solvents in the presence of multi-walled carbon nanotubes, but the sound energy causes the multi-walled carbon nanotubes to break [93]. Using frequencies of 20 kHz has been shown to disperse multi-walled carbon nanotubes well. The efficiency of ultrasonic dispersion depends on the frequency [94, 95]. A low-temperature milling process has been developed to achieve homogeneous dispersion of multi-walled carbon nanotubes in the rubber matrix. The products were able to withstand a 100 °C higher temperature and a 70 MPa higher pressure, compared with O-rings currently in use [96]. The most important factors influencing affecting the quality of dispersion are:

- dispersion energy input into the system
- degree of entanglement of the bulk carbon nanotubes
- interaction of the carbon nanotubes surface with the rubber

A minimum value of approx. 1000 J/mL has been found to effectively disperse a range of multi-walled carbon nanotubes concentration up to 5 vol.%. Higher loadings require higher amounts of energy. Shortening of the multi-walled carbon nanotubes is an unavoidable outcome of dispersion of this product. The reduction in the length of multi-walled carbon nanotubes may negatively impact the reinforcing capacity.

Ways are being sought of combining carbon black or precipitated silica with multi-walled carbon nanotubes in order to achieve better electrical conductivity or better physical properties. In one compound, NR/silanized silica is mixed with a small amount of multi-walled carbon nanotubes to obtain much better electrical conductivity [97]. Similarly, the mechanical properties and the mechanical fracture properties have been significantly improved by mixing a small amount of carbon nanotubes together with carbon black compounds, which increases the good dispersion of the multi-walled carbon nanotubes and improves the physical properties of the elastomer [98].

## 5.4 Influence of Rubber

As the main component of any rubber compound, the chemical composition of the rubber and the molecular weight of the rubber have a direct importance for filler dispersion, which is reflected in the interaction of the rubber chains with the filler surface. As shown from the theoretical aspects, the hydrodynamic force, $F_h$, is related to the interfacial tension between the rubber and the filler surface and the viscosity of polymer. The chemical composition of the chain segments has a similarly graduated influence on filler dispersion as the average molecular weight.

The rubber-filler interaction that determines ultimately the interaction between the polymer and the filler surface depends on whether the rubber chain is saturated or unsaturated. In the latter case, it can contain better-interacting chain segments, e. g. styrene or acrylonitrile, or some functionalities that are introduced into the polymer to enhance polar interaction with the filler surface. The chemical structure of the chain segments determines the affinity of the polymer for a particular filler. Unsaturated non-polar rubbers have a higher affinity for carbon black, whereas saturated non-polar rubbers demonstrate a weak affinity for this type of filler. Polar unsaturated rubbers show good interaction with the filler surface if the filler has polar groups on the surface, such as the acidic silanol groups of precipitated silica.

The average molecular weight and the molecular weight distribution play determining roles in filler dispersion. The higher the molecular weight, the more contact sites the polymer chain has with the filler surface and the lower the chance that adsorbed "train" segments will detach from the filler surface in a desorption process. Rubbers of different molecular weight show a pronounced increase in filler dispersion as the molecular weight increases. The NR with the higher molecular weight leads to an increase in filler dispersion, because interaction with the filler surface depends on the square root of the molecular weight of the NR. Comparison of the filler dispersion in an NR with a molecular weight of nearly 600,000 g/mol shows that it requires nearly double the degree of dispersion of carbon black N330 than NR of only half the molecular weight. This is interpreted as a contribution of the molecular weight or the viscosity of the NR chains.

But a look at the influence of the two different rubbers NR and E-SBR, as shown in Figure 5.20a, reveals that the E-SBR always effects higher dispersion than the NR, although its viscosity is considered to be lower than that of NR [99]. In this case, it is the better interaction of the styrene groups that determines the level of dispersion achieved and not the molecular weight of the rubber. At higher loadings, the interparticle distance decreases progressively. This leads to high molecular weight chains becoming attached to more than one filler aggregate at the same time.

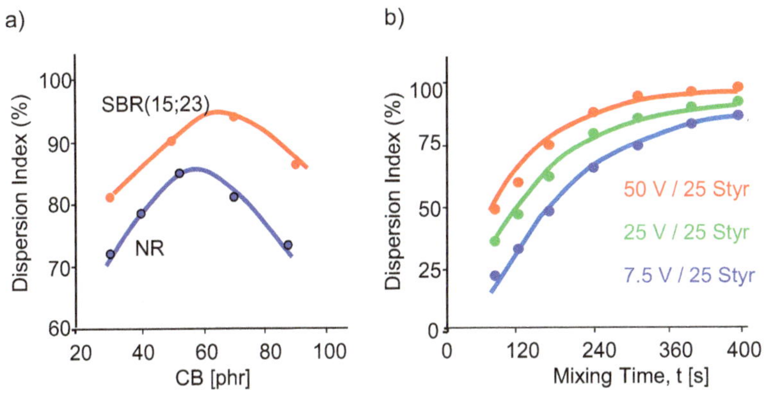


**Figure 5.15** Influence of the rubber on the dispersion of carbon black: a) NR vulcanizate that contents 30phr and 50phr carbon black N234 and b) NR and E-SBR vulcanizates with carbon black N234

When different SBR rubbers that have the same content of styrene groups but an increasing content of vinyl groups are examined, an increase in the degree of dispersion at given mixing times is observed, as shown in Figure 5.15b. The content of vinyl groups increases the viscosity of the S-SBR and, for this reason, better erosion of the pellet fragments will occur at the filler incorporation stage and the filler dispersion stage, which will increase the degree of dispersion.

The unfavorable rubber/filler ratio reduces the overall extent of surface wetting required for efficient dispersion. It becomes evident that the solubility parameter, $\delta$, increases from saturated rubbers (IIR and EPDM), which only interact with the filler surface via weak London forces, to the group of unsaturated rubbers (NR, IR and BR), which can have stronger interactions, such as induced dipole forces and induced dipole-dipole forces, and ends with the copolymer SBR, which contains styrene groups that provide increased interaction with the carbon black. The relationship between the nature of the rubber and the micro-dispersion becomes clear when the electrical percolation threshold is considered [100]. Such a relationship has the form of a linear increase in the micro-dispersion, which was determined via the electrical percolation threshold of the vulcanizates with the solubility parameters of the rubber. This proportionality between the δ-parameter of the rubber and the electrical percolation threshold of the filled rubber underlines the role of the chemical composition of rubber in the micro-dispersion of carbon black. As shown in Figure 5.16, the percolation threshold increases in proportion with the δ parameter of the rubber.

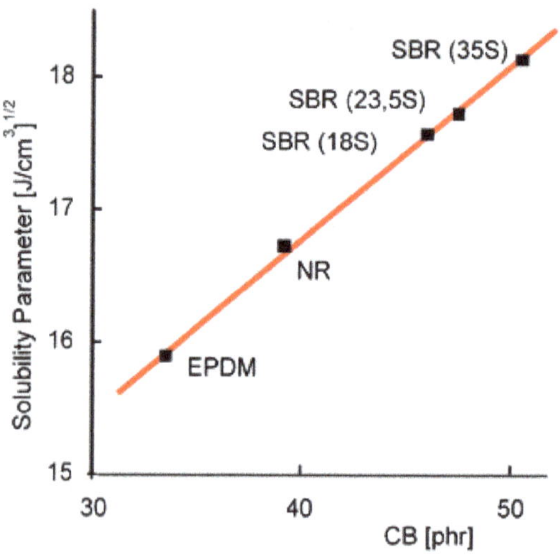


**Figure 5.16** The degree of micro-dispersion depends on the solubility parameter of the polymer

This relationship is of fundamental importance, as it shows that, on the nano-scale, the dispersion process that generates filler aggregates in the compound and leads to the formation of filler networks depends on the interaction potential of the polymer.

Similar behavior has been observed for S-SBR with pre-silanized silica (70 phr) mixed in an intermeshing mixer. Comparison of an S-SBR of narrow molecular weight distribution ($M_w/M_n = 1.42$) with a commercial grade of broader molecular weight distribution ($M_w/M_n = 2.12$) reveals significant advantages in the time-dependent dispersion index of pre-silanized silica [101]. Alongside the macro-dispersion, the influence of the molecular weight of the polymer can be seen in reduced filler networks (Payne effect) and reduced dynamical hysteresis of the compounds, which corresponds to better micro-dispersion of the silanized silica. At constant filler loading, the storage modulus, $G'_0$, of the compound decreases, due to better micro-dispersion, which shifts the percolation threshold toward higher filler concentrations.

### 5.4.1 Influence of Functionalized Rubbers

Efforts have been made to introduce specific interacting groups into rubbers with the aim of achieving a higher degree of dispersion in the compounds and improving the physical properties of filled vulcanizates. The driving force behind these activities is the tire industry with its efforts to improve the rolling resistance and resistance to abrasion and road wear in the tread compounds. Two strategies are generally pursued in the functionalization of rubbers. One consists in modifying existing rubbers to improve the intensity of rubber-filler interaction during the synthesis. The corresponding technique is known as polymer-analogue transformation of rubber. The other strategy pursues the functionalization of rubber by implementing specifically interacting monomers either in emulsion polymerization or in "living" anionic polymerization. Both solutions techniques require synthesis skills, because the molar concentration of the functional groups per chain needs to be reduced to a few percentages in order not to change the dynamics of the rubber. "Economical" functionalization of the non-polar rubber is also necessary so as not to impair the chain dynamics and chain flexibility, glass transition temperature and viscoelastic properties, which are required for good dynamic applications of tires and belts. Chemical groups such as carboxylic groups, heterocyclic amines and thin coupling have been proposed for improving the dispersion of carbon black. For precipitated silicas, the search has concentrated on alkoxysilanes, amine structures, heterocyclic amines, epoxy groups, carboxy groups and hydroxy groups which can enter into a specific interaction with the silanol groups of precipitated silica [102].

Radical polymerization in emulsion and "living" anionic polymerization in solution are preferably used for functionalizing rubbers [103]. In-chain functionalization of E-SBR with hydroxy groups and carboxy groups has been used to improve the mixing behavior and micro-dispersion of carbon blacks in tread compounds [104]. E-SBR with 1.5 to 3% of carboxy groups provides better carbon black dispersion than non-functionalized E-SBR. Emulsion SBR containing 3–5% acrylonitrile displays better rubber-filler interaction and abrasion resistance than unfunctionalized rubber [105]. Terpolymers of E-SBR with hydroxypropyl methacrylate (HPMA) have been studied in silica compounds [106]. The most versatile technique for functionalizing rubbers is that of "living" anionic polymerization. New monomers which contain specifically interacting groups are described for S-SBR in the patent literature. A large number of functional groups that are mainly used in SBRs are mentioned in patents:

- tin-coupled species [107]
- silanol groups [108]
- alkoxy silane [109]
- carboxyl [110]

Two essential pathways have been developed:

- chain-end functionalization
- chain-backbone functionalization

Chain-end functionalization can be realized through both an anionic polar group comprising an anionic polymerization initiator or through a chain-stopping agent. Polymers which are functionalized at both chain ends are produced using bifunctional polymerization initiators [111]. It has been reported that anionically prepared BRs and SBRs can be functionally terminated by active lithium and with polar groups that contain specific interacting groups [112]. Functionalization of the S-SBR leads to a higher dispersion in an S-SBR/cis-BR blend and thus to altered dispersion of the filler in favor of the functionalized rubber. Recently, a polymer producer has published that Nd-BR can be functionalized during synthesis [113]. Accordingly, functionalized cis-BR provides better filler dispersion which is reflected in reduced compound viscosity and dynamic hysteresis of vulcanizates. This achievement is inasmuch important, because cis-BR is used in combination with S-SBR in blends for tire treads. Having both blend partners functionalized makes the filler distribution in the blend phases more equal.

## 5.5 Influence of Mineral Oil

In compound manufacturing, it is necessary to use plasticizing mineral oils, because the highly filled rubber compounds achieve high viscosities and flow limits that render processing more problematic. Plasticizers are commonly used with reinforcing fillers to offset the hardening effect, to impart better processing, and in many cases to reduce the weight of the rubber compounds. If the mineral oil is very compatible with the rubber, it can cause a shift in the glass transition temperature. To ensure good processability, different amounts of mineral oil are proportioned in one formulation. In the compound, the mineral oil has a tendency to compete with the rubber chains for the energetic sites on the filler surface. Due to their lower molecular weight, the mineral oils are more mobile during mixing than the long rubber chains, so that mineral oil constituents can be absorbed on the filler surface in the first phase of the dispersion process, resulting in marked impairment of filler dispersion. The mixing sequence is an influencing parameter that should not be underestimated. If the mineral oil is added together with the filler, the competitive interaction with rubber is very high and dispersion of carbon black decreases. The oil is incorporated quickly, but the maximum torque needed to achieve a maximum carbon black dispersion is relatively low and is not sufficient to effect good dispersion of the carbon black, as shown in Figure 5.17 [114].

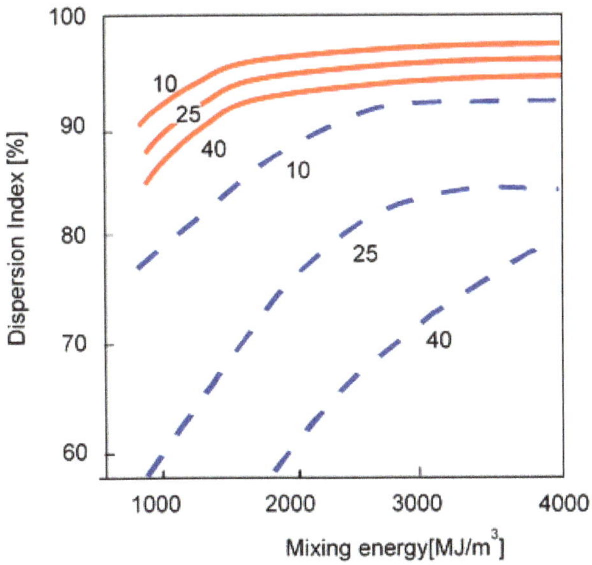


**Figure 5.17** Influence of oil incorporation on the dispersion of carbon black [114]

If the oil is mixed after the filler has been incorporated and is then dispersed, the dispersion index of the carbon black is high, even for the high admixtures of oil. However, it is absolutely necessary to choose the time for the oil injection such that enough carbon black surfaces remain available for oil absorption. If this is not the case, oil incorporation will be delayed. Lubricating films are created in the mixer and it is barely possible to introduce energy into the compound, and this can increase the mixing time. When the processing oil is added at the optimal time during the mixing process, a sudden decrease in the compound viscosity takes place and a drop in mixing temperature will occur.

Differences related to the quantity of mixed filler in the plasticized mixture are nonetheless evident. It should be noted that the influence of the mixing in oil is less pronounced in the case of large-particulate carbon black types. In practice, the admixture of carbon black and oil proceeds sequentially in small portions. Resins and processing agents can enter into the same “competition” for the energy sites on filler surfaces. If larger quantities are added, it is advisable to incorporate and disperse the filler beforehand. Although it is difficult for a machine operator to keep to the all timings for the addition of carbon black and mineral oil and for changes from high to low speeds, it has been clearly shown that missing the crucial timings can lead to considerable losses in manual operation.

## 5.6 Influence of Process Parameters

The process parameters make a significant contribution to filler dispersion in the vulcanizate. This enables the effects of the selected process parameters to be quantified in the mixing process and, if necessary, the mixing process to be optimized with less effort. The associated question of the production costs of mixtures is often in the main consideration when choosing fillers and mixing processes. The speed at which the filler is incorporated into the rubber and dispersed contributes significantly to the economics of the mixing process. However, quality mixing requires optimal use of the individual mixing stages, in terms of the quality of the compounds and the time required.

### 5.6.1 Influence of Mixing Time

The flow generated by the rotors must achieve the dispersive effect through extensional and shear stress in the clearance between and in front of the rotor blade flights, with distributive mixing. The longer this influence of elongational stress can be exerted, the more filler pellet fragments and large agglomerates are broken down and eroded into smaller nano objects. The mixing time is therefore a dominant process variable, because of its direct link with progressive dispersion and distribution in the compound.

Regardless of the type of internal mixer, the dispersion process for the pellet fragments and large agglomerates decreases in accordance with an exponential time law during mixing. The exponential degradation of pellet fragments and large agglomerates during mixing is described by the evolution of macro-dispersion. The evolution of the number of undispersed filler particles and the degree of dispersion as a function of mixing time are shown in Figure 5.18. The speed at which dispersion in the mixture progresses is a function of the mixing time and the composition of the rubber compound.

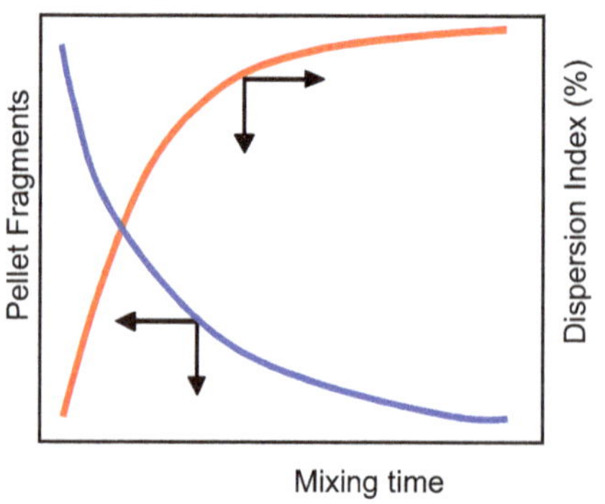


**Figure 5.18**
Influence of the mixing time on the number of pellet fragments and filler dispersion

On the one hand, the temperature of the mixture increases with mixing time, which automatically leads to a decrease in the viscosity of the rubber and has the consequence that the effects of the extensional stress and shear stress on pellet rupture and erosion decrease steadily. On the other hand, polymer degradation takes place, weakening rubber-filler interaction and the phase adhesion forces capable of transmitting external stresses to the filler particle decrease with temperature. A significant step forward has been made with the "onion skin" model. In this model, thin layers are sequentially detached (eroded) from the outer surface of the parent agglomerate by the shear stream of a viscous fluid [115]. The eroded volume fraction is proportional to the surface of the aggregates, to the transmitted tension in the fluid and to the mixing time. The time-dependent relative reduction in agglomerate volume is expressed by the equation:

$$\Delta V_{agglom}/V_{agglom} = c(\tau_{xy}/d_{aggreg}) \cdot t \quad (5.8)$$

where $\tau_{xy}$ is the shear stress, $\Delta V_{agglom}$ is the change in agglomerate volume, $d_{aggreg}$ is the average size of the aggregates, and t is the mixing time

Equation 5.8 describes the dispersion of agglomerates as a function of mixing time. As the mixing time increases, the relative volume peeled off from the agglomerates increases in proportion to the applied shear stress over time. The reduction in agglomerate volume makes it possible to calculate the percentage dispersion index. Using this model, the evolution of macro-dispersion can be calculated for fillers with different structure as a function of mixing time. Results are obtained for the time-dependent evolution of the macro-dispersion by using the theoretical "onion skin" model [42]. The origin of the time axis corresponds to the end of the incorporation stage. As can be seen, the differential dispersion rate (ΔDI/Δt) increases with the degree of branching of the filler (OAN). For low structure filler, the "onion skin" model predicts not only a lower dispersion rate but also a lower final level of dispersion on prolonged mixing times. In contrast, high structure fillers exhibit faster dispersion kinetics and better dispersion on longer mixing. Thus, the observation that dispersion increases for fillers with more branched aggregates is in agreement with numerical modeling.

## 5.6.2 Influence of Rotor Speed

Of the parameters that control dispersive mixing, the geometry of the rotors and the rotor speed at which the rotors are operating are of great importance. This tallies with experience from the mixing chamber and is confirmed by theoretical concepts (see Section 7.8). The movement of the rotors provides opening and folding of layers in the mixed material that contain elongational and/or shear stress [116]. The dispersive and distributive action of the rotors depends on the transport of the material along and across the rotor flights [117]:

- rotational transport of the material with the rotor flight
- flows of the rubber compound between the rotor wings and the mixing chamber
- material transport around the rotor ends

With increase in the number of rotor revolutions per unit of time, breakdown of filler pellets also increases and filler aggregates are better dispersed in the rubber compound. The economic advantages of the shorter mixing times that can be achieved at higher speeds must be carefully weighed against the resulting processing properties. The concerted action of rotor speed and mixing time is shown in Figure 5.19 for an E-SBR compound containing carbon black N330 (50 phr) processed on a tangential laboratory mixer [100]. The response area, which originates from a statistical design of experiments, shows similar graduated influences exerted by the rotor speed and the mixing time on the carbon black macro-dispersion. Even though the response area changes with the filler type and composition of rubber used, the basic tendencies regarding mixing time, rotor speed and degree of filler dispersion realized will still be similar to those shown in the diagram.

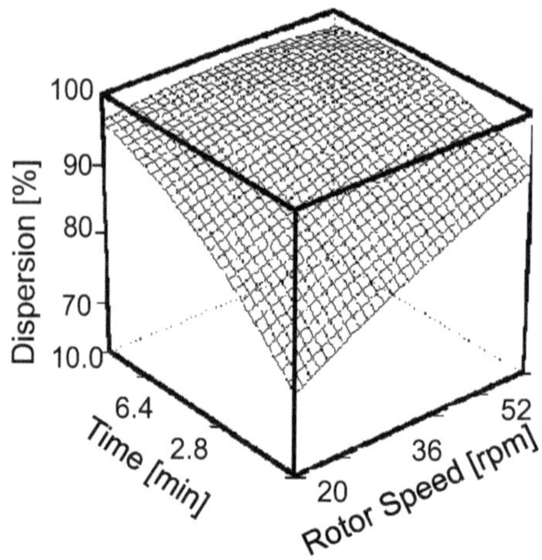


**Figure 5.19**
Influence of the rotor speed and the mixing time on the macro-dispersion of carbon black N330 in E-SBR [100]

The optimum rotor speed depends on the rotor geometry which is designed to achieve good dispersion or uniform distribution of the filler. Thus, with open-blade tangential rotors ZZ2, mixing can be done at higher rotor speeds than with the full four-blade tangential rotors [118]. The ZZ2 rotor is capable of higher rotor speeds at lower specific energy input. The rotor allows more freedom for the compound at the rotor tips, thus limiting the temperature rise in the mixing compound.

A comparison of the rotor speed for internal mixers with a free volume of the mixing chamber of 60 L reveals that a rotor speed of 20–70 rpm is optimal for tangential mixers, while intermeshing mixers are operated at a lower rotor speed of 15–50 rpm. The use of a lower rotor speed for intermeshing mixers is necessary because the temperature of the compound rises due to higher shearing and turbulence flow generated by the rotor shape [119].

Internal mixers equipped with variable rotor speed drives provide more flexibility in finding optimal mixing conditions and running the mixing process at lower costs. The various stages of the mixing cycle can be carried out at different rotor speeds. Such an approach makes it possible, on the one hand, to design the mixing stages in such a way that an optimal dispersion and distribution is obtained and, on the other hand, to reduce the mixing cycle to a minimum, which significantly reduces the costs of compound production. Publications suggest that filler incorporation and the dispersion stage should be performed at high rotor speed. This shortens the time necessary to break the filler pellets into smaller fragments and facilitates an earlier start of the dispersion process. During the dispersing stage, the rotor speed should be kept high in order to promote more intensive size reduction in the pellet fragments per time unit. However, after the second power peak, the rotor speed should be reduced in order to prevent unnecessary heating of the compound and premature scorch processes or premature consumption of antioxidants.

The benefits of variable speed drives cannot be too emphasized enough. The technique was introduced for mixing each compound at its optimum speed. This means that a hard compound (i. e. NBR) can be mixed at a low rotor speed while a soft compound (i. e. EPDM) can be mixed at a higher rotor speed [119]. Keeping other parameters constant, it is always seen that increased shear stress provides a better macro-dispersion. The influence of the shear stress amplitude employed to disperse silica is shown in Figure 5.20 [120].

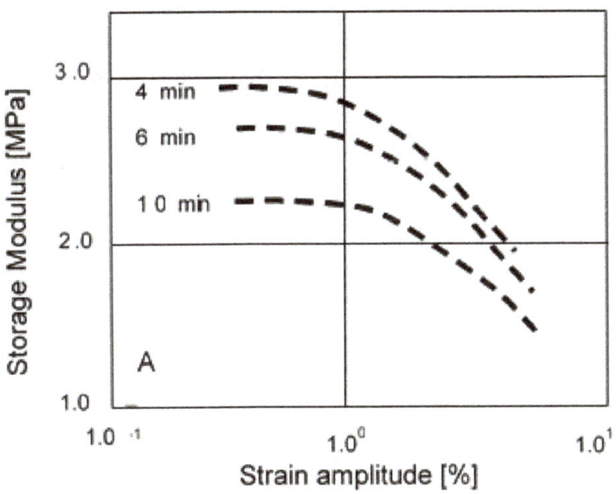


**Figure 5.20** Decrease in the Payne effect with mixing time of carbon black N330 (50 phr) in E-SBR [119]

The macro-dispersion, determined using the mechanical roughness method (stylus method), reveals that conventional silica is dispersed moderately under moderate shear stress (DI = 53%) while high shear stress significantly improve dispersion (DI = 75%). For HD silica, the effect of increasing the shear stress is measurable but not very pronounced, because of the good dispersion of these silica grades. Under higher shear stress, the dispersion index increases from 96% to 97%. A report has shown that the Payne effect, as a measure of filler micro-dispersion, and tensile strength are linearly related to rotor speed and the fill factor [121]. The dispersion index depends on the rotor speed and filler loading. Increasing the rotor speed reduces the effect of mixing time and the Payne effect.

In the case of intermeshing internal mixers, a sharp increase in power is observed at the start of ram seating. It is advisable to keep ram pressures high during the dispersion stage as well, at least until the second power peak is reached. It has been reported that the degree of dispersion increases when a higher ram pressure is used during the incorporation and the dispersion stage [122]. Batch size is an important consideration, both from the economics and a quality point of view. In order to obtain the best filler dispersion and compound properties, an optimum batch volume is required. Experience has shown that an optimum batch volume also yields the greatest throughput, the least batch-to-batch variation, and the best filler dispersion.

The relationship between fill factor, ram pressure and rotor speed with respect to filler macro-dispersion is shown in Figure 5.21. The two diagrams summarize the effects on filler dispersion in terms of general process parameters. They show the basic relationships that were obtained with a tangential laboratory mixer using E-SBR/CB N220 compounds. The y-axis represents the filler macro-dispersion in arbitrary units, while the x-axis plots the fill factor as an independent variable. This value stands for the percentage to which the volume of mixing chamber is occupied by the compound ingredients. A fill factor of 0.5, for example, stands for a mixing chamber that is 50% filled.

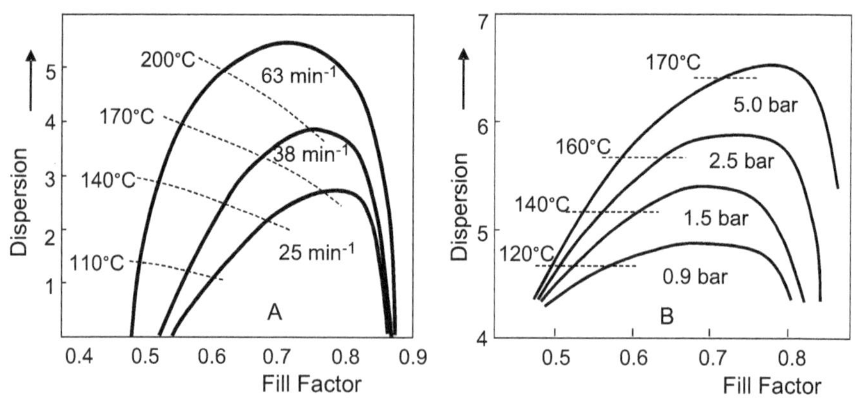


Figure 5.21 Increase in carbon black dispersion as a function of fill factor and a) rotor speed and b) ram pressure [121]

For the diagram presented in Figure 5.21a, all the tests were adjusted such that the laboratory mixer achieved a throughput of 68 kg/h. Each line represents a specific rotor speed. As can be seen in the range of fill factors from 0.6 to 0.8, dispersion increases with increase in rotor speed. Even at sub-optimal fill factors, progress in compound quality, as indicated by an increase in the rotor speed from 25 $min^{-1}$ to 63 $min^{-1}$, is significant. At constant rotor speed, filler dispersion increases with the fill factor until a maximum value is reached. Any increase in the fill factor above this maximum leads to a rapid decrease in the degree of filler dispersion and thus to impaired compound quality. The broken lines in the diagram indicate the drop in temperature of the batch in order to avoid pre-scorch of the compound. If the mixer is poorly filled, passage of the compound along the flights is easy and therefore the material is pushed along the rotor flights toward the edges. In such conditions, distributive mixing primarily occurs. When the fill factor is increased, flow along the rotor flights becomes progressively blocked and the compound passes through the gap between the flight and the chamber wall. The mixing regime therefore changes from distributive to dispersive. However, it must be taken into account that a continuous increase in the fill factor leads to a situation where the flow of the compound through the mixing chamber is impaired. Dead spots in the mixing chamber reduce the mixing efficiency.

As a consequence, it can be assumed that there is an optimum fill factor for every mixing cycle with regard to dispersion quality. Optimization can be achieved by simultaneous decreasing the fill factor and increasing the rotor speed. However, the broken line indicates the temperature in the mixing chamber. Therefore, the compounder can appreciate the maximal temperature before the crosslinking starts. Such an approach can shorten the overall mixing cycle and save mixing energy. Compounds that have to be mixed in a single step should be processed at a slightly reduced fill factor in order to avoid undesirable temperature increase associated with internal friction of the filler.

Figure 5.21b illustrates the influence of the fill factor and the ram pressure on filler dispersion. Basically, the filler dispersion is influenced by both variables in a similar way as in the case discussed above. At constant fill factor, filler dispersion increases with ram pressure. The pressure generated by the ram forces the mixture to flow through the narrow clearance between the rotor flights and the chamber wall. With increase in fill factor, this effect is enhanced, such that, at a constant ram pressure but increasing fill factor, dispersive mixing reaches a maximum. As the ram pressure increases, this optimum shifts to an even higher fill factor. This enables an increasing output of compounds with an increased degree of dispersion. It should be noted that the optimal fill factor depends on the type of rubber and the composition of the compound. However, fill factors greater than the optimum value cause a sharp decrease in dispersion and poor compound quality. It should be emphasized that the ram pressure can be set more precisely with a hydraulic feed hopper than with a pneumatic system.

As the fill factor and rotor speed increase, frictional processes become more pronounced and cause the compound temperature to rise. That has two important consequences. On the one hand, the viscosity of the compound decreases and the shear stress is no more sufficient to disperse the filler agglomerates, and dispersion efficiency decreases. On the other hand, the increased temperature of the compound can be critical for the stability of the rubber when it contains crosslinking ingredients. The premature initiation of crosslinking during final mixing can take place.

The temperature increase in the rubber compound draws heat to the cooling system through the mixing chamber wall and the rotor wall. Cooling of the compound and control of batch temperature are essential for efficient mixing. In this way, heat is increasingly removed from the mixture and the temperature rise is weakened or even converted into a decrease in compound temperature. Preheating the mixer is advantageous because [123]:

- cooling conditions in summer and winter are not the same,
- process temperature optimizations are possible through differential temperature control of the rotors and mixing chamber wall, and
- first-batch effects are reduced.

The time required for heating or cooling the mixing chamber increases with the mixer size, so that the temperature variations during mixing are subject to economic limits.

Better cooling of the mixing chamber causes an increase in the polymer viscosity that results in higher shear stress but a longer flow of the rubber chains into the voids of the agglomerates and, thus, longer incorporation times. The benefit of better cooling is undoubtedly a higher degree of dispersion reached after the incorporation of the filler. This is the impressive outcome of a model experiment performed on a 5.45 L intermeshing mixer, as shown in Figure 5.22 [124].

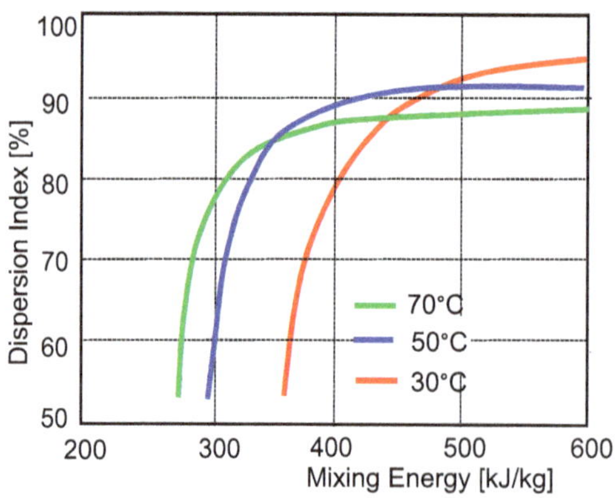


**Figure 5.22**
Influence of cooling the mixing chamber on the development of filler dispersion [124]

With optimal cooling of the mixing chamber, the incorporation time increases significantly, due to the increase in the viscosity of the rubber, but the dispersion rate is at its highest and leads to the highest macro-dispersion. The final improvement in the degree of filler dispersion is an increase of the micro-dispersion that reduces the dynamic cut growth. The choice of water temperature is usually a compromise between conflicting requirements. While low water temperature will delay the incorporation stage but will improve the dispersive mixing, a higher water temperature will shorten the incorporation and implicitly the mixing cycle. In practice, a large range of water temperatures are used in combination with water flow rates to achieve a high heat transfer rate and efficient dispersive mixing.

The improvement in "micro-dispersion" which is the more important effect of efficient cooling is underlined by TEM images showing a decrease in the agglomerate size with better cooling. In addition, the Payne effect measured for compounds obtained under different cooling conditions decreases with better cooling. These results provide an easily accessible information that helps to decide which cooling strategy delivers an optimal quality of the vulcanizates. During basic mixing, temperatures must not exceed or fall below certain limits. For NR compounds intended for dynamically stressed products, a temperature of 140 °C should not be exceeded, due to the risk of chain breakdown. The effect of high-temperature mixing with extended mixing times on tire tread applications has been using low and high structure carbon black and blends of S-SBR/cis-BR and functionalized S-SBR/cis-BR. Raising the temperature increases the mobility of the polymer chains which could raise the diffusion rate of the polymer chains so that they can migrate to positions more favorable for physical adsorption.

To control silanization of silica, the temperature should not fall below 140 °C. To avoid premature coupling of the silanized silica and the rubber, temperatures should not exceed 150–155 °C [125]. The importance of cooling in the intermeshing mixer has been repeatedly highlighted in relation to the silanization reaction. The level of silica silanization reaction and the reduction in Mooney viscosity has been demonstrated for silica-filled tire tread compounds using conventional multi-stage mixing [126, 127].

In order to avoid premature crosslinking reactions during mixing, the temperatures must not exceed 80–120 °C, the exact temperature depending on the curing system used. In any case, the rule is that the chemical reactions take half as long for every temperature rise of 10 °C. Compliance with low temperatures is ensured through the design of the mixers or the choice of process parameters.

The premature crosslinking of portions of the compound as a result of temperature peaks in the internal mixer cannot be verified by measuring the Mooney viscosity or by means of a rheometer curve, since the mixture does not flow during these tests. During downstream processing of the compound, the pre-crosslinked areas behave like an elastically deformable filler and interfere with the shaping and molding process.

It is expected that additional processing on downstream equipment, such as mills and extruders as well as internal mixers for incorporating curing agents, and all operated at relatively low temperatures (< 120 °C), will provide sufficient dispersion and good physical properties. Compounds with the same carbon black loading (50 phr) and very similar specific surface area but different structure have different power consumption in the dispersion stage. The higher the oil adsorption number (OAN), the more energy is required to erode the filler agglomerates and to overcome cohesive filler-filler interactions. In the dispersion stage of the mixing cycle, the values for power consumption converge, because no dispersion work is done here.

The effect of improved rotor design and higher cooling efficiency has been shown for tangential rotors. Compared with full four-blade rotors, the better cooled and geometrically advanced ZZ2 rotors yield significant energy savings in the mixing of tire compounds, as indicated in Table 5.4 [128].

**Table 5.4** Reduction in Mixing Energy for Tire Compounds in MJ [128]

| Compound | Full Four-Blade Rotor | ZZ2 Rotor |
|---|---|---|
| Tread | 84 | 65 |
| Side wall | 95 | 60 |
| Innerliner | 73 | 68 |
| Compound | 82 | 53 |

Since the dispersive mixing effect depends on the level of the induced elongational stress and its duration of action, the mixer that can introduce as much energy as possible into the mixture at a given dump temperature is superior in terms efficiency [128].

The great influence of the degree of filling on the power consumption of the internal mixer before the mixture is ejected has been demonstrated for EPM compounds with 50% filler [129]. A pronounced minimum of power consumption separates insufficient mixing due to insufficient energy input, on the one hand, and excessive viscosity due to underfilling, on the other, from the higher power consumption arising from overfilling. The minimum therefore represents the optimal filling level.

## References

[1] Medalia, A. I., Rubber Chem. Technol., 1972, 45: 1171

[2] Wu, S., in "Polymer Blends", Paul, D. R., Newman, S., Eds., Academic Press, (1978), Ch. 6

[3] Tadmor, Z., Gogos, C. C., in "Principles of Polymer Processing", Wiley, New York (1979)

[4] Hess, W. M., Rubber Chem. Technol., 1991, 64: 386

[5] Gerspacher, M., Proc. DKT'09, Nuremberg (2009)

[6] Ditmar, R., "Analyse des Kautschuks", Hartben, Leipzig and Vienna (1909)

[7] Depew, H. A., Ruby, I. R., Ind. Eng. Chem., 12 (1920) 1156
[8] Leigh-Dugmore, C. H., Rubber Chem. Technol., 1956, 29: 1303
[9] ASTM D2663, Method B
[10] Stumpe, Jr., N. A., Railsback, H. E., Rubber World, 1964, 151: 41
[11] Anderson, L. O., et al., ACS Rubber Division Meeting, Cleveland, (1997) p. 102
[12] Putman, J. B., ACS Rubber Division Meeting, Cleveland, (2001) p. 19
[13] Otto, S., Randl, D., Persson, S., Skroog, L., Tire Technol. Internatl. (1998)
[14] Schuster, R. H., Geisler, H., Boller, F., Gummi Asbest Kunstst., 1996, 96: 816
[15] Medalia, A. I., Rubber Age, 1965, 4: 82
[16] Klie, B., Teich, S., Haberstroh, E., Giese, U., Kautsch. Gummi Kunstst. 2015, 67: 31
[17] Schuster, R. H., Sgodzaj, Proc. Eurofiller 2013, Bratislava, 2013
[18] Hess, W. M., Chirico, V. E., Vegvari, P. C., Elastomerics, 1980, 112: 24
[19] ASTM D 2663 (method C)
[20] Nickiel, L., Gerspacher, M., Yang, H., O'Farrell, C. P. O., ACS Rubber Division Meeting, Dallas (2000)
[21] v. Ardenne, M., Beisher, D., Rubber Chem. Technol. 1941, 14: 15
[22] Ford, F. P., Gessler, Rubber Chem. Technol., 1952, 25: 858
[23] Dannenberg, E. M., Ind. Eng. Chem., 1952, 16: 813
[24] Boonstra, B. B., Dannenberg, E. M., Heckman, F. A., Rubber Chem. Technol., 1974, 47: 1082
[25] Schuster, R. H., Oppermann, B., Schramm, J., ACS Rubber Division, Indianapolis (1998)
[26] Ikeda, Y., et al., Macromol. Rapid Commun., 2004, 25: 1186
[27] Kohjiya, S., et al., Polymer, 2005, 46: 4440
[28] Kohjiya, S., et al., Polymer, 2006, 47: 3298
[29] Magonov, S., in "Encyclopedia of Analytical Chemistry", (R. A. Meyers, Ed.) Wiley, New York (2000)
[30] Gomez, C. J., Garcia, R., Ultramicroscopy, 2010, 110: 262
[31] Zhong, Q., et al., Surf. Sci. Lett., 1993, 290
[32] Alshuth, Th., Schuster, R. H., Krämer, S., Kautsch. Gummi Kunstst., 1994, 47: 703
[33] Trifanova-Van Hoeringen, D., Rubber Chem. Technol., 1999, 72: 862
[34] Keuter, H., et al., Rubber World, 2001, 224: 29
[35] Keuter, H., in "Mixing of Rubber Compounds", (Limper, A., Ed.) Hanser Publishers, Munich (2012), Ch. 5
[36] Limper, A., in "Cost Reduction in Rubber Processing", Graf, H.-J., Ed., TechnoBiz Commun. Co., Ltd., (2014)
[37] Dannenberg, E. M., Rubber Chem. Technol., 1986, 59: 512
[38] Gerspacher, M., Nikiel, L., O'Farrell, C. P., Kautsch. Gummi Kunstst., 2000, 53: 70
[39] Ahagon, A., Miyasaka, K., Nippon Gomu Kyokaishi, 2010, 83: 318
[40] Medalia, A. I., Rubber Chem. Technol., 1972, 45: 117
[41] Robertson, C. G., Bogoslovov, R. B., Roland, C. M., Phys. Rev. E., 2007, 051403
[42] Oppermann, B., PhD Thesis, University of Hanover (1994)
[43] Wang, M.-J., Rubber Chem. Technol., 1998, 71: 520
[44] Wampler, W. A., Carlson, Th. E., Jones, W. R., in "Rubber Compounding, Chemistry and Application", (Rodgers, B., Ed.) Marcel Dekker, New York (2004) Ch. 6
[45] Studebaker, M. L., Rubber Age, 1957, 81: 661
[46] Johnson, P. H., US Patent, 3,318, 720 (1967)
[47] Cataldo, F., Angew. Makromol. Chem., 1999, 270: 81
[48] Tunnicliffe, L. B., Herd, Ch., IOM3 RIEG ATDM, London (2019)

[49] Belmont, J. A., Amici, R. H., Galloway, C. P., US Patent, 5,900,029 (1999)

[50] Müller, M., Schuster, R. H., Proceed. IRC '03, Nuremberg (2003)

[51] Tricas, N., Borros, S., Schuster, R. H., Kautsch. Gummi Kunstst., 2005, 58: 211

[52] Graves, D. F., Rubber Chem. Technol., 1993, 66: 61

[53] Yamaguchi, T., et al., Kautsch. Gummi Kunstst., 1989, 42: 403

[54] Gonzales, L., et al., Rubber Chem. Technol., 1996, 69: 226

[55] Takeshita, M., et al., US 4,820,751 (1981)

[56] Monte, S. J., Sugerman, G., ACS Rubber Division Meeting, Chicago (1982)

[57] Monte, S. J., in "Ken-React Preference Manual-Titanate, Zirconate and Aluminate Coupling Agents", ed. Kenrich Petrochemicals (1993)

[58] Vogel, C., Vogel, K., Niedermeier, W., Freund, B., Messer, P., US Patent 6,056,933 (2000)

[59] Niedermeier, W., Schwaiger, B., Proc. Tyretech. Prague (1999) p. 14

[60] Mahmud, K., Wang, M.-J., Francis, R. A., US Patent 5,830,930 (1998)

[61] Wang, M.-J., Mahmud, K., Murphy, L., Patterson, W., ACS Anaheim, (1997) p. 25

[62] Wang, M.-J., Mahmud, K., Murphy, L., Patterson, W., Kautsch. Gummi Kunstst., 1998, 51: 348

[63] Wang, M.-J., et al., Rubber Chem. Technol., 2002, 75: 247

[64] Killmann, E., Eckert, R., Makromol. Chem., 1971, 144: 45

[65] Wagner, M. D., Rubber Chem. Technol., 1976, 49: 703

[66] Ziegler, J., Schuster, R. H., Kautsch. Gummi Kunstst., 2003, 56: 159

[67] Chevalin, Y., Moravski, J.-C., EP 0157703 B1

[68] Blume, A., Urlandt, S., Rubber World, 2002, 226: 30

[69] Mark, J. E., Pan, S. P., Macromol. Chem. Rapid Commun., 1982, 3: 681

[70] Ikeda, Y., Tanaka, A., Kohjiya, S., J. Mater Chem., 1997, 7: 455

[71] Pluedemann, E. P., in "Silane Coupling Agents", Plenum Press, 2nd ed., (1982)

[72] Wang, M.-J., et al., Kautsch. Gummi Kunstst., 2020, 72(9): 45

[73] Klafke de Azeredo, L., Schuster, R. H., Jacobi, M. M., Kautsch. Gummi Kunstst. 2010, 63: 8

[74] Schellenberg, M., Schuster, R. H., ACS Rubber Division Meeting, Cleveland (2013)

[75] Galimberti, M. Guidice, S., Cipolletti, V., Guerra, G., Polym. Adv. Technol., 2010, 21: 679

[76] Paul, D. R., Robeson, L. M., Polymer, 2008, 49: 3187

[77] Le Baron, P. C., Wang, Zh., Pinnavia, Th., J. Appl. Clay Sci., 1999, 15: 11

[78] Giannelis, E. P., Adv. Mater., 1996, 8: 29

[79] Cho, J. W., Paul, D. R., Polymer, 2001, 42: 1083

[80] Dennis, H. R., Hunter, D. L., Chang, D., Kim, S., White, J. L., Polymer, 2001, 42: 9513

[81] Brady, B. P., in "Carbon Nanotube–Polymer Composites: Manufacture, Properties and Applications", J. Wiley & Sons, Inc., 2011

[82] Bokobza, L., et al., Polymer, 2007, 48: 4007

[83] Goga, R. E., Cohen, R. E., Compos. Part B: Eng. 2004, 35: 173

[84] Ajayan, P. M., Schadler, L. S., Giannari, C., Rubio, A., Adv. Mater., 2000, 12: 750

[85] Hadjiev, V. G., et al., Appl. Phys. Lett., 2001, 78: 3193

[86] Shaeffer, M. S. P., Windle, A. H., Carbon, 1998, 36: 1612

[87] Fakhri, N., Tsouboulski, D. A., Cognet, L., Weisman, R. B., Pasquali, M., Proc. Natl. Acad. Sci., USA, 2009, 106: 14219

[88] Pasquali, R., Phys. Rev. Lett., 2006, 96: 246104

[89] Yonemura, H., et al., Sci. Technol. Adv. Mater, 2008, 9: 024213

[90] Phang, I. Y., et al., Eur. Polym. J., 2007, 43: 4136

[91] Chatterjee, T., et al., J. Am. Chem. Soc., 2008, 130: 6934

[92] Zhou, W., et al., Chem. Phys. Lett., 2004, 384: 6934

[93] Ding, W., et al., Nano Lett., 2003, 3: 1593

[94] Fuchs, F. J., 45$^{th}$ Annual Techn. Conf. Soc. Vacuum Coates, 2002, p. 64

[95] Sharma, S. D., Sing, S., Am. J. Chem., 2013, 3: 96

[96] Deng, F., ACS Nano, 2011, 5: 3858

[97] Fritsche, J., PhD Thesis, University of Hanover (2009)

[98] Wittek, H., Schuster, R. H., Proc. EXPOBOR, Sao Paolo, (2014]

[99] Schuster, R. H., Issel, H.-M., Peterseim, V., Rubber Chem. Technol., 1996, 69: 769

[100] Schuster, R. H., in “Mixing of Rubber Compounds”, (Limper, A., Ed.) Hanser Publ., Munich (2012) Ch.6

[101] Sgodzaj, U., PhD thesis, University of Karlsruhe (2001)

[102] Benko, D., Bates, K., Hennig, S., ITECH, Akron (2002)

[103] Hsieh, H. L., Quirk, R. P., “Anionic Polymerization and Practical Applications, New York (1996)

[104] Thielen, G., Kautschuk Gummi Kunstst., 2008, 61: 377

[105] Senyak, M., Colvin, H., US Pat. 5,310,815 (1994)

[106] Colvin, H., Senyak, M., US Pat. 6,512,053 B1

[107] Ulmer, J. D., Hergenrother, W. L., Lawson, D. F., Rubber Chem. Technol., 1998, 71: 637

[108] Marechal, M., US 8344063 (2013)

[109] Thiele, S., Rulhoff, S., US 8895684 B2 (2014)

[110] Gandon-Pain, S., US 7312264 (2007)

[111] Rempp, P., et al., Adv. Polym. Sci., 1984, 56: 1

[112] Hogen, T. E., Randall, A., Hergenrother, W. L., Rubber World, 2010, 242: 38

[113] Lanxess TSR, Tire Technology Expo, Hanover, (2016)

[114] Hess, W. M., Swor, R. A., Micek E. J., Rubber Chem. Technol., 1984, 57: 959

[115] Shiga, S., Furuta, M., Rubber Chem. Technol., 1985, 58: 1

[116] Freakley, P. R., in “Rubber Technologist’s Handbook”, (White, J. R., De, S. K., Eds.) RAPRA Technology Ltd, (2001) Ch. 7

[117] Limper, A., in “Mixing of Rubber Compounds”, (Limper, A., Ed.) Hanser Publishers, (2012) Ch. 2

[118] Röthemeier, F., in “Kautschuktechnologie” (Röthermeier, F., Sommer, F., Eds.) Carl Hanser, Munich (2001) Ch. 4

[119] Wood, R. P., in “Rubber Mixing”, RAPRA Review Report No. 10, Rapra Technology Ltd., Shawbury, (1996)

[120] Wehmeier, A., “DIK Workshop Mixing” (2003)

[121] Clarke, J., Clarke, B., Freakley, P. K., Rubber Chem. Technol. 2001, 74: 1

[122] Wiedemann, W., Schmid, H.-M., Rubber Chem. Technol., 1982, 55: 363

[123] Berkemeier, D., et al., ACS Rubber Division Meeting, Dallas (2000)

[124] Schuster, R. H., Peterseim, V., Proc. Rubber Fall Colloquium, Hanover (1998)

[125] Welsh, F. E., Richmond, B. R., Keach, C. B., Emerson, R. J., ACS Rubber Division Meeting, Philadelphia (1995)

[126] Gafouri, S. N., Rubber World, 2004, 229: 46

[127] Nijman, G., et al., Proc. Rubber Fall Colloquium., Hanover (2006)

[128] Glasser, P., Proc. TechnoBiz Custom Rubber Mixing Forum, Pittsburgh (2015)

[129] Freakley, P. K., Matthews, B. R., Rubber Chem. Technol., 1987, 60: 618

[130] Leblanc, J. L., Kautsch. Gummi Kunstst., 1990, 43: 888

# 6 Distribution of Filler in Rubber Blends

In addition to filler dispersion mentioned in the Chapter 5, which refers to the homogeneous distribution of the filler in a single rubber phase, the term filler distribution in rubber blends can refer to the practical presence in the different phases of polymer in a blend system, which determines the properties of the rubber blend. It describes the distribution of a filler in the phases of a rubber blend. Most polymer blends are thermodynamically immiscible, having a phase-separated morphology and properties of the rubber blend that vary with the type of the rubber blends, the mixing strategy, the type of filler and the distribution and the dispersion in each blend phase. The filler is distributed in the blend components by selective polymer-filler interactions. The distribution of the filler in a rubber blend is therefore related to the affinity of the filler for the type of polymer, to the specific area, to the structure of the filler and to the mixing technology.

The aim of blending is to combine the desirable features of each blend component in a chemically stable system and to establish the best distribution of the fillers in order to obtain better reinforcement of the rubber blends. An advantage of blending technology is the apparent simplicity of the components used and the lack of new processing steps during blend manufacturing. However, the mixing process depends on the type of polymer and type of fillers added, as well as the mixing technology. Due to their heterogeneous phase morphology, a multi-phase structure blends exhibit stable interphases at both the boundaries of dissimilar rubber domains and the rubber-filler contact surface. It stands to reason, that mechanical properties are related to the thickness of the phase boundary layer and that this depends on the quality of interactions of dissimilar polymer segments and the active sites on the filler surface. The reasons for using rubber blends rather than individual rubbers are:

- reduced compound costs
- improvement in processing

- combinations of rubber properties
- improved product performance

The addition of active fillers to the rubber blends significantly increases the complexity of the mixing process. The filler is distributed in the different phases of the rubber blend in a natural distribution that depends on the rubber-filler interaction in the blend system, or in a forced distribution achieved by masterbatch mixing of compounds to obtain the desired blend distribution. The rheological conditions prevailing during mixing in affect the distribution of the filler, such that, in the case of forced distribution of the filler during production of the mixture, a metastable blending state is reached in which the duration of the mixing process affects the distribution in the individual phases of the rubber blend, which is important for the physical properties of the entire blend. Control of blend morphology, including the domain sizes and interfacial tension, along with partitioning of filler and other ingredients and the uneven crosslinking in blends is necessary for understanding filled rubber blends.

Homogeneity at a fairly fine level is necessary for optimum performance, but some degree of micro-heterogeneity is usually desirable for preserving the individual properties of the respective polymer components. Homogeneity is favored by similar polymer viscosities and solubility parameters, δ, as well as by the presence of polar groups in the polymer composition [1]. Fortunately, miscibility is not a requirement for many applications of rubber blends. A combination of the properties of the original rubbers is required in a rubber blend. Distribution of the filler leads to an additional improvement in the physical properties of the blend component. When the filler is present in the rubber blend, the micro-heterogeneity can be diminished, but can also increase if the filler is mixed selectively into one of the rubber components. The interfacial tension must be low and co-vulcanization is useful in this respect for achieving a good distribution of the filler in both blend constituents [2].

In the following, we will first look at the formation of pure rubber blends, discuss the basics of the thermodynamics of the blends and then move on to the main description of the distribution of fillers in blend phases.

## 6.1 Thermodynamics of Rubber Blends

The most pertinent polymer properties of rubber blends are homogeneity of mixing and cure compatibility. The miscibility of rubbers requires the free energy of mixing to be negative; this can be achieved by exothermic enthalpy of mixing and the entropy of mixing [3–5]. Most rubber blends are not miscible, because mixing is an endothermic process and the entropy contribution is rather low because of the high molecular weights of the blend components. A thermodynamic phase is understood as a space of constant composition and constant physical properties. Therefore, each phase of a polymer blend contributes to the property set of the blend. In materials technology, compatibility is a more general term with a wider variety of meanings than miscibility. In the strict technological sense, compatibility is often used to describe whether a beneficial result occurs when two polymers are mixed together. Compatible polymers are mixtures that show no visible signs of phase separation when blended (e. g. transparency, lower viscosity). Miscibility has to be understood in thermodynamic terms, namely as a consequence of the second law of thermodynamics:

$$\Delta G_m = \Delta H_m - T\Delta S_m \tag{6.1}$$

where $\Delta G_m$ is the free energy of mixing, $\Delta H_m$ is the enthalpy of mixing and $\Delta S_m$ is the entropy of mixing and T is the absolute temperature

If the free energy of mixing has a negative arithmetic sign over the entire composition, the blend is miscible, $\Delta G_m < 0$. Most polymer blends are not miscible systems over the entire composition range, even if the free energy of mixing is negative. In the regions where one polymer is the main phase and the other is present in low concentration, the system forms a single phase and the second derivative of the free energy with respect to the composition is negative:

$$\partial^2 \Delta G/\partial \Phi^2 < 0 \tag{6.2}$$

In the middle of the diagram is a region where the free energy is increasing and a miscibility gap occurs. This region is defined by a negative free energy, and the second derivative of the free energy of mixing with respect to the composition is positive:

$$\partial^2 \Delta G/\partial \Phi^2 > 0 \tag{6.3}$$

The single-phase region and the two-phase region are characterized by the second derivative of the free energy of mixing. The width of the two-phase region decreases as the temperature rises and ends at a critical point where the binodal and the spinodal coincide:

$$\partial^2 \Delta G_m/\partial \Phi^2 = \partial^3 \Delta G_m/\partial \Phi^3 = 0 \tag{6.4}$$

This development is illustrated qualitatively in Figure 6.1, which shows the soluble single-phase region and the miscibility gap.

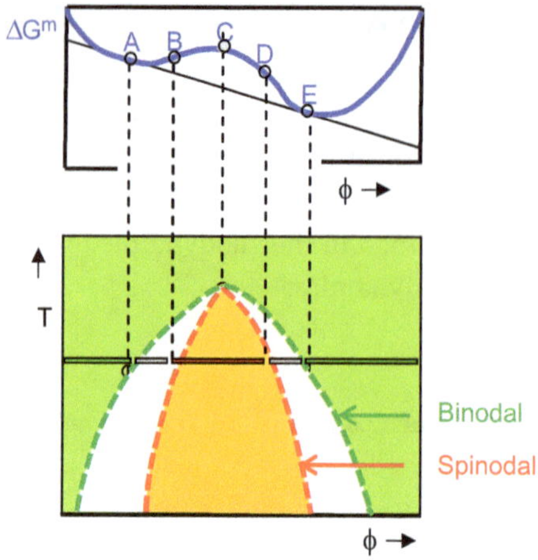


**Figure 6.1**
Phase diagram of a partially soluble polymer blend

Blends of different average molecular weights, $M_w$, have skewed phase diagrams. The critical point is shifted to the region of lower molecular weight [6]. In any mixture of a rubber blend, the boundary between stable and metastable compositions is said to be binodal and the boundary between the metastable and unstable compositions is said to be spinodal. Compositions along the binodal define the phases into which unstable and metastable composition will separate into a two-phase system. Mixtures of rubber blends have endothermic heat of mixing and a low entropy of mixing. They usually exhibit an upper critical solution temperature (UCST), whereas mixtures that have exothermic mixing heats usually exhibit a lower critical solution temperature (LCST). At temperatures above a UCST and below an LCST, the binary mixtures are completely miscible in all compositions. Below the UCST and above LCST, there are large two-phase compositions.

### 6.1.1 Free Energy of Mixing

Attempts to describe the free energy of mixing of polymer-solvent system have been made independently by Flory [7] and Huggins [8]. The entropy of mixing was calculated by using combinatorial algebra to determine the number of arrangements of an assembly of monodisperse polymer chains on the sites of a rigid lattice. Through application of the Boltzmann equation, which relates the number of arrangements to the entropy of the system, an expression for $\Delta S_m$ was obtained. Scott [9] and Tompa [10] were the first to apply the theory of polymer solutions to mixtures of polymers. The combinatorial entropy of mixing can be calculated from the composition and the degree of polymerization of the rubbers. The magnitude of combinatorial entropy decreases inversely in proportion to the degree of polymerization, $x_A$ and $x_B$. The longer the polymer chains, the smaller the combinatorial mixing entropy. There is no provision made for predicting the enthalpy of mixing, $\Delta H_m$, from first principles. The enthalpy of mixing has to be expressed by an empirical polymer-polymer interaction parameter, $\chi$. The Gibbs free energy of mixing for a total volume of two polymers is expressed by [9]:

$$\Delta G_m = (RTV/V_r)\,[(\Phi_A/x_A)\ln\Phi_A + (\Phi_B/x_B)\ln\Phi_B + \chi_{AB}\Phi_A\Phi_B] \qquad (6.5)$$

where $V_r$ is the reference volume which is close to the molar volume, $\Phi_A$ and $\Phi_B$ are the volume fractions of the blend components, $x_A$ and $x_B$ are the degrees of polymerization of polymers A and B, and $\chi_{AB}$ is the polymer-polymer interaction parameter

Miscibility depends on the magnitude of the thermodynamic functions. Miscibility occurs when the entropy term is larger than the enthalpy term. Phase separation occurs when the enthalpy term is larger than the entropy term, which is usually the case for rubber blends. A critical interaction parameter, $\chi_{cr}$, can be calculated from the critical point. All polymer blends with a critical interaction parameter above this value are immiscible [6]:

$$\chi_{cr} = (1/\sqrt{x_A} + 1/\sqrt{x_B})/2 \qquad (6.6)$$

This equation is used to determine the critical molecular weight, $M_{cr}$, which consistently describes the coexistent curve between the single-phase region and the two-phase region as a function of the difference in the δ parameter in a phase separated rubber blend. It is clear that, in most cases, the rubber blends are phase separated even at very low molecular weights, far below the molecular weights of technical rubbers. Only in some special cases can the thermodynamic similarities between the rubbers create the condition for a single-phase system. One such case is the NR/1,2-BR blend system, which forms a single phase for thermodynamic reasons. This relationship between $M_{cr}$ and the square of the δ parameter difference is shown in Figure 6.2.

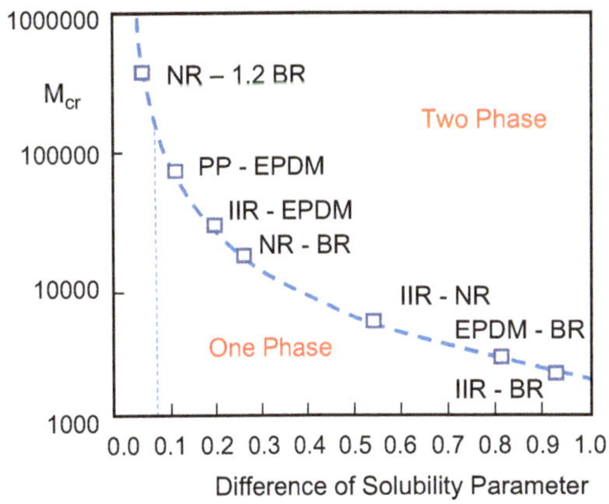


**Figure 6.2** Phase diagram which separates insoluble blends from homogeneous blends as a function of the δ parameter difference in the rubber blend constituents

Binary blends of a cis-BR and various SBRs of increasing styrene content also show a similar curve. Only the cis-BR/E-BR blend results in a critical molar mass in the range of industrially usable rubbers. As the content of styrene groups increases, the critical molar mass decreases substantially and tends toward significantly smaller values as the difference in solubility increases. Without diminishing the importance of the lattice theories, it must be stated that they provide a rough estimate of phase behavior. More complex approaches derived from "equation of state" theories, allow a description of the p-V-T behavior of the blends and use the specific mixing volume and the isothermal compressibility [11]. To test the consequences of the equation-of-state theory, a set of binodal and spinodal curves was computed for a polymer blend by perturbing properties around base case values. The effect of chain length, coefficient of thermal expansion, coefficient of thermal pressure, exchange energy parameter and polydispersity were studied. However, these approaches also require an empirical interaction parameter to fully describe the enthalpy of mixing.

## 6.1.2 Concept of Solubility Parameter

However, the $\chi$ parameter in polymer-solvent systems cannot be predicted from theoretical considerations and has to be determined experimentally. A user-friendly approach which permits a good estimation of phase behavior and interfacial tension is afforded by the solubility parameter $\delta$ introduced by Hildebrand and Scott [12]. To overcome the difficulties in describing the interaction parameter between dissimilar polymer chains, a fundamental assumption from the theory of regular solutions was made [13]. The interaction energy of dissimilar molecules is equal to the geometrical mean of the interactions of similar molecules:

$$\Delta w_{12} = (w_{11} \times w_{22})^{1/2} \tag{6.7}$$

This approximation permits a conversion to the thermodynamic properties of the starting materials. The interaction energy ($w_{ii}$) of low-molecular-weight molecules corresponds to the heat of evaporation that is experimentally accessible from the Van't Hoff law, and the enthalpy of mixing can be written as the difference of the square of the $\delta$ parameters [12]:

$$\chi = (V_r/RT)\,(\delta_1 - \delta_2)^2 \tag{6.8}$$

As the square of the $\delta$-parameter difference is always positive, the condition for miscibility is fulfilled only if $\delta_1$ and $\delta_2$ became similar and the blend system behaves like a thermal solution ($\Delta H_m = 0$) and exhibits entropy-driven miscibility. The solubility parameter can be determined for polymers using the expansion of polymer chains in solution at intrinsic viscosity or the equilibrium swelling of polymer [14], the inverse gas chromatographic method [15] or an empirical approach that uses the incremental contributions of repeating groups (attraction potentials) [16]. All these methods are time-consuming, but deliver suitable results that can provide background information about the compatibility of rubber blends. A comparison of the solubility parameters of different rubbers is shown in Table 6.1 [17].

**Table 6.1** δ Parameter for Rubbers [17]

| Method | EPDM | 1,2-BR (70%) | NR | cis-BR | SBR (15% S) | SBR (23% S) | SBR (40% S) |
|---|---|---|---|---|---|---|---|
| Viscometry | 15.91 | 16.71 | 16.90 | 17.20 | 17.50 | 17.70 | - |
| Swelling | 15.92 | 16.76 | 16.90 | 17.17 | 17.57 | 17.80 | 18.01 |
| IGC | 15.95 | 16.94 | 16.73 | 17.50 | 17.65 | 17.79 | 18.13 |
| Increments | 15.80 | 16.47 | 16.68 | 17.38 | 17.41 | 17.69 | 17.83 |

It has been found that there is a sequence of increasing δ parameters from saturated to unsaturated rubbers. The 1,2-polybutadiene (70% vinyl) has the same polarizability as the natural rubber previously considered in a thermodynamic analysis of this blend. The SBRs exhibit an increase in δ parameter as a function of increase in content of styrene groups.

### 6.1.3 Interfacial Tension

The phase boundary layer in rubber blends is an area in which chain segments of phase A and phase B penetrate each other. In the middle of the boundary layer, the concentration of one phase decreases and the other increases. Phase-separated polymer blends demonstrate interfacial regions between the incompatible domains formed by interdiffusion of dissimilar chain segments. The gain in free energy due to the interpenetration (interdiffusion) of chain segments at the interface has been estimated for non-polar polymers using a lattice theory. The phase boundary of any blend controls the resistance of the blend to external stress [18]. The interfacial tension is defined by the integral of the free interfacial energy. The theories consist in calculating the conformational entropy in the interfacial layer and adopting the known polymer-polymer interaction parameter χ. Application of a lattice model showed that the interfacial tension, γ, is proportional to the square root of the χ parameter:

$$\gamma = \rho_0 RT(\chi \langle R^2 \rangle / N)^{1/2} \tag{6.9}$$

where $\rho_0$ is the average segment density, $\langle R^2 \rangle$ is the squared average radius of gyration, N is the number of statistical chain segments and χ is the polymer-polymer interaction parameter

Instead of the square root of the χ parameter, the difference in the solubility parameters of the two polymers can be used:

$$\gamma \sim \chi^{1/2} \tag{6.10}$$

$$\gamma \sim (\delta_1 - \delta_2) \tag{6.11}$$

If the interaction between the system components deteriorates, the interfacial tension increases and the miscibility gap increases.

The equilibrium thickness of the interface, $a_{IPH}$, is shown to be inversely proportional to the solubility parameter difference. Assuming that the blend components have the same density, the thickness of the boundary layer, $a_{IPH}$, is inversely proportional to the square root of the χ parameter or the difference in the δ parameters of the blend components:

$$a_{IPH} \sim \chi^{-1/2} \tag{6.12}$$

or the reciprocal difference in the solubility parameter:

$$a_{IPH} \sim (\delta_1 - \delta_2)^{-1} \tag{6.13}$$

Therefore, at constant volume fraction of the blend constituents, the amount of the interphases increases with increase in compatibility of the rubber blend [19]. Equation 6.12 and Equation 6.13 can both be used to predict the increase or decrease in interfacial tension as a function of temperature. Thus, the δ parameter difference can be considered as a reliable semi-quantitative criterion for describing both the interfacial tension and the interfacial thickness. As shown in Figure 6.3, the interfacial tension in all blends decreases with increase in temperature [17].

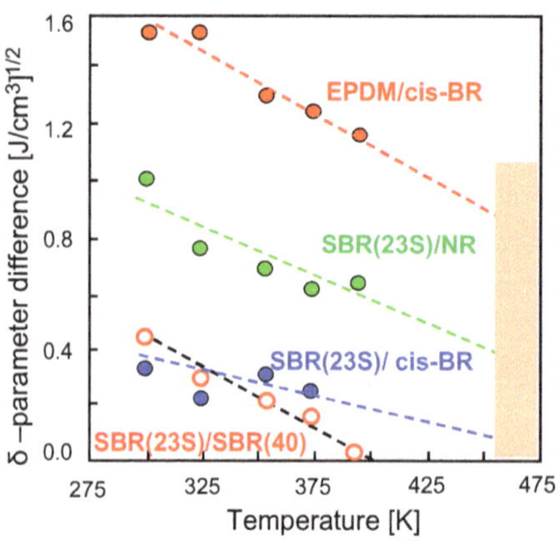


**Figure 6.3** Difference in δ parameters for rubber blends as function of temperature [17]

Incompatible blends exhibit two-phase behavior over the entire temperature range. Rubber blends with lower interfacial tension can exhibit phase transfer into the single-phase region at high temperature. The consistently negative temperature gradient of the interfacial tension shows that, at higher mixing temperatures, the domain size of the dispersed phase and the thickness of the boundary layer increase as a result of decreasing interfacial tension. In this sense, better phase bonding is achieved, which can have a beneficial effect on the physical properties [20]. Attempts to obtain information about the thickness and the volume fraction of the interface have been carried out for the case that one rubber is semi-crystalline, with the assumption that no crystallization of the rubber takes place in the interphase [21]. From such a premise, one can determine the size of the interphase and establish a ranking for different rubber blends. The thickness of the phase boundary layer and the volume fraction in the rubber blend are given for cis-BR in Table 6.2 [22].

**Table 6.2** Thickness of the Interface (nm) and Volume Fraction (%) in the Vulcanizate [22]

| Blend | Thickness $a_I$ (nm) | Volume fraction (%) |
|---|---|---|
| cis-BR/SBR (23 S) | 23 | 84 |
| cis-BR/SBR (29 S) | 11.5 | 54 |
| cis-BR/SBR (40 S) | 5.5 | 28 |
| cis-BR/EPDM | 3 | 15 |

As the δ parameter difference decreases, the boundary layer increases linearly, as predicted by theory. On vulcanized rubber blends, the loss modulus G″ is measured in the glass transition region [23]. It has been shown that the volume fraction of the interphase increases with decrease in δ parameter difference of the blend constituents. Other examples for NR/S-SBR show that the proportion of interphase is up to 25% of the total volume when the S-SBR has more 1,2-butadiene segments [24].

## 6.2 Types of Unfilled Rubber Blends

In a blend consisting of two rubbers, each with different physical characteristics, such as glass transition temperature, coefficient of expansion, dynamic-mechanical modulus, refractive index, optical permeability etc., the presence of phases is indicated by the discrete occurrence of the characteristics of the phase. The number of phases present and the partial solubility of the components are deduced from the assignment of the signals generated during investigation.

### 6.2.1 Development of Phase Morphology

From a technical point of view, it is important to develop the formation of rubber blends under controlled processing conditions. This means that influencing the mixing conditions should lead to a change in the phase morphology. One can see the dependence on the temperature and the composition of the blend phases.

Once this has been found, the same rotor speed, mixing temperature and dumping will then take place under the same mixing conditions. Systems with a dispersed phase morphology have attracted the most scientific interest because these systems can be used to follow morphological change as a function of the process parameters. The essential physical properties, especially the ultimate properties of the vulcanizate, are shown by systems with a co-continuous morphology, which are often implemented in practice [3]. If a rubber blend forms a dispersed phase, this is deformed into fibril-like domains at the beginning of the mixing time which are broken down into smaller and quasi-spherical domains in the course of the mixing process. Such a process is shown in Figure 6.4 for the CIIR/NR (80/20) system [25].

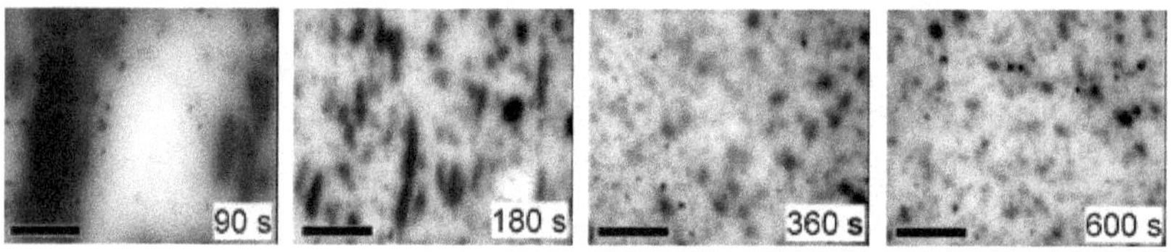


**Figure 6.4** Development of phase morphology in CIIR/NR (80/20) during mixing

A longer mixing time and a higher rotor speed shorten the mixing process. If the factors influencing the size of the dispersed phase are considered, a qualitative picture emerges that is described by the following relationship [26]:

$$\bar{d} = c\,(\gamma/\eta_m) \cdot f\,(\eta_m/\eta_d)\; dV_x/dy \tag{6.14}$$

where $\gamma$ is the interfacial tension, $\eta_m$ is the viscosity of the matrix, $\eta_d$ is the viscosity of the dispersed phase, and $dV_x/dy$ is the shear rate

The relationship between the stationary average domain size at the end of mixing and the interfacial tension have been investigated for a large number of rubber blends of 80/20 composition and under controlled mixing conditions (size of the internal mixer, rotor speed and temperature). The effectiveness at which mixing stress can support particle deformation depends on the interface of the blend constituents. The domain size of the dispersed phase increases linearly with the interfacial tension, as indicated in Figure 6.5 for rubber blends [17].

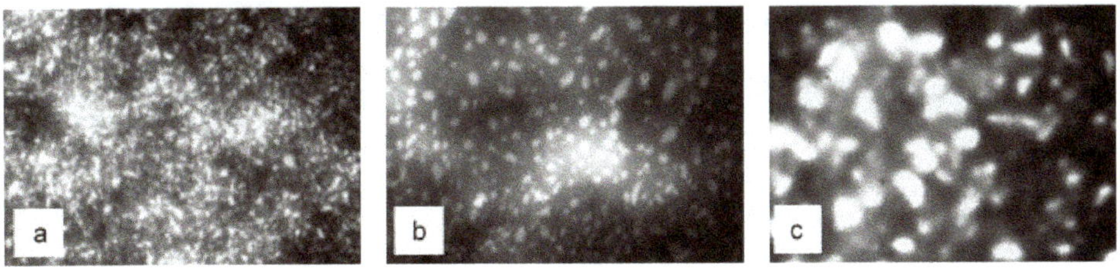


**Figure 6.5** Dispersed size in E-SBR/NR as a function of increasing styrene content. The differences in solubility parameter are: a) 0.43, b) 0.49 and c) 0.73 [17]

The dispersed size of the rubber therefore increases with increase in interfacial tension of the rubber blend. The same proportionality can be observed for a large number of rubber blends. An exception to this rule is the domain size of the strain crystallizing rubber which forms larger domains via the enthalpy of fusion [27]. Compared with older investigations listed in Table 6.3, there are quite large differences between the light microscopy and the new TEM measurements, but the general trend remains valid [28].

**Table 6.3** Average Areas ($\mu m^2$) of Dispersed Phase in 75/25 Pure Gum Blends [28]

| Disp. Phase | NR | CR | BR | SBR | NBR | EPDM | IIR |
|---|---|---|---|---|---|---|---|
| NR | - | 45 | 1.5 | 1.2 | 300 | 1.5 | 2.0 |
| CR | 35 | - | 4.0 | 2.5 | 1.5 | 25 | 15 |
| BR | 0.7 | 4.5 | - | - | 15 | 2.2 | 2.1 |
| SBR | 0.5 | 2.7 | - | - | 20 | 2.1 | 12 |
| NBR | 400 | 1.3 | 17 | 30 | - | 250 | 100 |
| EPDM | 3.5 | 75 | 2.8 | 2.6 | 225 | - | 2.0 |
| IIR | 3.0 | 15 | 3.0 | 4.2 | 75 | 1.0 | - |

The smaller the average dispersed area, the more compatible are the polymers. This generalization is related to the similarities in the solubility parameters, the viscosity of the polymer and rubber polarity. It can be seen that NBR and CR produce the greatest heterogeneity. The smallest domain size is achieved through a higher stress field during the mixing of the rubber blend. The higher the viscosity of the matrix polymer, the smaller is the dispersed domain size. Since the temperature increases during mixing, high-shear blending of polymers of similar viscosities result in the smallest dispersed phase. The domain size is certainly a direct function of the applied rotor speed during mixing. In any blend, a reduction in the stationary domain size is obtained when the rotor speed increases, as can be shown in Table 6.4 [29].

**Table 6.4** Influence of Rotor Speed on Domain Sizes in Selected Blends [29]

| System | Rotor Speed (rpm) | Domain Size (nm) | $\|\delta_2-\delta_3\|$ |
|---|---|---|---|
| NR/SBR (23%) | 65<br>95 | 260<br>120 | 0.45 |
| EPDM/cis-BR | 65<br>95 | 650<br>240 | 0.65 |

As the rotor speed increases, more energy is supplied to the rubber blend, which corresponds to a longer mixing time. With the more compatible blend NR/SBR (23S), a minimum in the domain size is already reached at an energy input of 900 kJ/kg, so that a further increase in the rotor speed only has a minor effect. In contrast, EPDM/cis-BR is far from a state of stationary domain size, so that increasing the rotor speed and the energy input of 1200 kJ/kg only result in a low domain size. If two rubber blends of different δ parameter difference are compared, the strongest effect of the rotor speed is seen in the more incompatible system. That means that the rotor speed has more influence if the rubbers are less compatible and the interphase generated is considerably smaller.

Extensive studies have been performed to establish the mixing conditions that lead to the smallest site on the dispersed rubber [30]. It was found that the domain size has a minimum value at a viscosity ratio, $\eta_m/\eta_d \approx 1$, regardless of the duration of mixing. Using the blend NR/SBR (23 S) (70/30) as an example, the extent to which the viscosity ratio affects the domain size of the dispersed phase after a mixing time of 30 s at a rotor speed of 50 rpm was investigated. The plot of the average domain size at different viscosity ratios shows that the smallest value is reached at a viscosity ratio of 1. This is illustrated in Figure 6.6 [30].

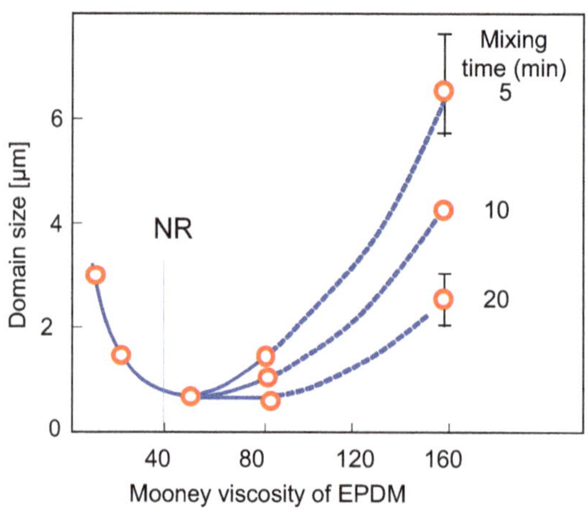


**Figure 6.6**
Influence of the viscosity ratio on domain size in a rubber blend [30]

Extensive mixing studies on EPDM/BR blends of different viscosities have been carried out [31]. It was found that phase inversion occurs at equal viscosities, $\eta_m/\eta_d = 1$, in the concentration region of 0.45–0.55. When the viscosities were changed, phase inversion shifted to the minor component. In this case, the low viscosity rubber is encapsulated by the high viscosity rubber component and then becomes continuous. This is shown in Figure 6.7 [31].

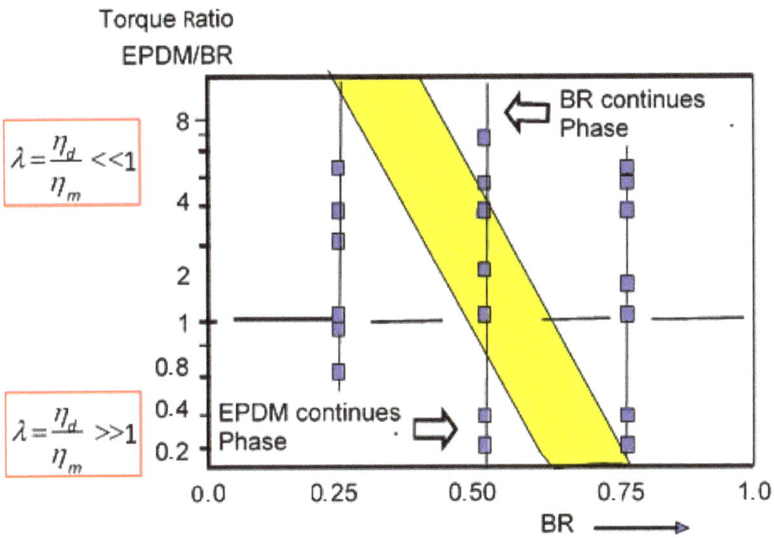


Figure 6.7 The influence of the viscosity ratio, $\eta_d/\eta_m$, on the phase inversion of a rubber blend [31]

If the interfacial tension is increased, the phase inversion region narrows and the properties of the co-continuous morphology are always extended to a smaller region. This dependence of the interfacial tension is particularly important when performing dynamic vulcanization.

If one considers the dispersion work as a function of the mixing energy, one wonders why dispersion does not progress to even smaller domains. With smaller domains, the particles are close together and enlargement of the domain brings an energetic advantage. If the domains become smaller and smaller during the mixing process, a coalescence process occurs, which means that the small domains with an increased particle density per volume also undergo a more pronounced coalescence process. The coalescence process can be observed in rubber compounds that are stored after mixing. The domain size of the dispersed phase of the blend NR/SBR (23 S) observed after three days of storage at 50 °C increases from 120 nm to 150 nm, due to the coalescence. In contrast, EPDM/cis-BR shows an increase in 240 nm at 1500 nm. The different interfacial tension controls the coalescence process. The coalescence process for phase morphology is demonstrated in the TEM micrographs in Figure 6.8.

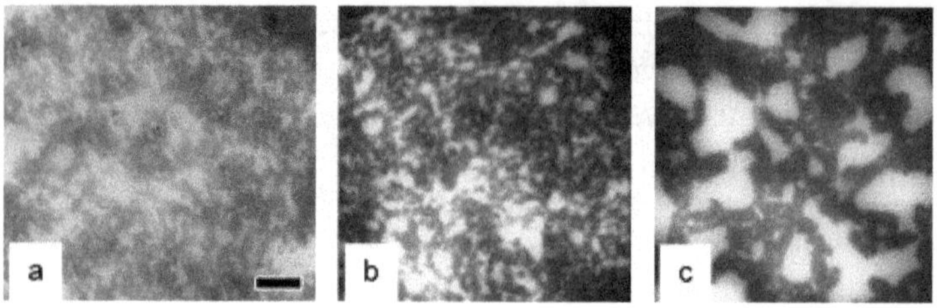


Figure 6.8 Influence of coalescence in NR/SBR (23 S): a) right after mixing, b) 7 d at room temperature and c) 7 d at 80 °C

In other words, the morphology of a rubber blend can change during storage. The coalescence process is greatly increased by the interfacial tension in the rubber blend. The coalescence effect increases with temperature and time until equilibrium is reached under these conditions. If the storage conditions change, the morphology of the blend also changes. The level of interfacial tension is also a decisive factor. The higher the interfacial tension between the two blend components, the greater is the increase in domain size. For blends of SBR/1,2-BR, the change in domain size of the co-continuous blend is much more pronounced. If a filler is incorporated in the rubber blend, the coalescence process is limited to lower levels to an extent depending on the activity of the filler.

### 6.2.2 Crosslinking Density of the Blend Phases

The complexity of the distribution of curing agents increases when blends are used. The distribution of crosslink density between the rubber phases depends on the concentration of the curing agents in each polymer phase. This means that the strongly polar accelerators have a different distribution in the rubber blend than the sulfur. Co-vulcanization has been defined as a network structure consisting of crosslinked macromolecules of both polymers and the interface [32–35]. The nature of the polymer (saturated, unsaturated, polar) determines the distribution of the curing agents in the two blend constituents. Vulcanizates with components having similar curing agents have better properties than those whose components have strong differences in curing rates and the state of cure [36, 37]. Differences in solubility parameters of the rubbers will cause a higher intolerance which can lead to an undesirable influence on the distribution of the crosslinking density, and partition will be further enhanced whenever the possibility of special interactions exists. When phases are small and there is a concentration gradient relative to the concentration dictated by the partition coefficient, the extent to which of migration will be dictated by the diffusion rate [38–41]. For blends of NR, SBR, BR or CR with IIR, CIIR and EPDM, it has been found that rubbers with a low $\delta$ parameter are not crosslinked and can be extracted from the blend [42]. Co-vulcanization in CIIR/BR has therefore been improved by utilizing a combination of different accelerators [43]. Several techniques have been proposed for determining crosslinking densities in vulcanized blends, most of which have proved to lack sensitivity [44–48].

Swollen-state NMR spectroscopy is based on the observation that the signals in NMR spectroscopy of polymers are broader than those of simple molecules and that the signal width increases if the polymer is crosslinked as a consequence of a progressive reduction in chain mobility [49]. The effect is quantified by expressing the crosslink densities measured by equilibrium swelling, having established correlations between peak width and crosslinked densities for elastomers used in a blend. Even if carbon black causes additional broadening, it is possible to use the swollen-state NMR spectroscopy for carbon black-filled blends. Blends differing little in polarity or unsaturation were investigated by the TARC group [50–52]. Of these blends, NR (or IR) and BR received the most attention. Both sulfur/sulfenamides partition slightly in a BR phase that has a higher δ parameter, as shown in Table 6.5 for an optimum cure at 150 °C.

**Table 6.5** Determination of the Sulfur Crosslinks in an NR/BR Blend of Different Blend Ratios and a Concentration of 1.2 phr Sulfur and Accelerators [52]

| NR/BR | Acc. (phr) | XLD in NR | XLD in BR | $S_{NR}$: $S_{BR}$ |
|---|---|---|---|---|
| 70:30 | CBS 1.2 | 1.00 | 1.32 | 0.77 |
| 50:50 | CBS 1.06 | 1.06 | 1.36 | 0.78 |
| 50:50 | MBS 1.14 | 0.98 | 1.36 | 0.72 |
| 50:50 | TBBS 1.08 | 0.98 | 1.42 | 0.65 |

The distribution of crosslinking density is always higher in SBR than in NR. This corresponds to the solubility parameters of the crosslinking agent that is closer to the SBR solubility parameter than that for NR. The highest tensile strength is observed when the crosslinking density in both rubbers is the same. Unequal crosslinking of the blend phases leads to a weakening of the blend resistance, which could lead to a decrease in tensile strength and fatigue properties. Therefore, even blends which are not considered to be problematic and which appear to differ little in either polarity or degree of unsaturation have been shown to undergo uneven crosslinking distribution during sulfur vulcanization.

## 6.3 Methods of Determining Phase Morphology

There are indirect methods that use the value of the glass transition in a rubber blend, but do not provide specific information about the blend morphology. A single $T_g$ is indicative of good compatibility, but does not mean miscibility. Heterogeneous blends clearly exhibit separate $T_g$ values for the individual components. The difference in the individual $T_g$ values must be greater than 15 °C. Dynamic-mechanical spectroscopy is often used to evaluate the rubber blend morphology on cured elastomers. It is less precise than DSC but it is more frequently used in practice [53–59]. Differential thermal analysis (DSC) is a precise method for determining the number of the phases in uncured and cured rubber blends. The $T_g$ of the blend constituents can be greater than 10 °C. Above the glass transition temperature, it shows the crystallization peak of the semi-crystalline polymers. This method is not widely used in companies, and it is only development departments and academic institutions that employ it [58, 60]. Dielectric relaxation spectroscopy is a method that can be used at different frequencies of the alternating current. The samples need to be cleaned and gold sputtered to improve the electrical contact. It can provide information over a wide temperature domain and frequency range [61, 62].

Microscopy methods use the phase contrast between the phases in a blend to show the number and dimensions of the phases. They provide information on the morphology of the rubber blend, the size of the domains, the size distributions and phase inversion. Phase contrast of unfilled rubber blends is studied by optical phase microscopy. The microscope optics produce a phase shift between the diffracted and the transmitted light, resulting in an interference contrast with only small differences in refractive index [28, 42, 63, 64].

Transmission electron microscopy (TEM) is an ideal method because of its high resolution and applicability. It is used for very thin samples (less than 100 nm). The original problem of this method of staining the unsaturated polymer phases with osmium tetroxide ($OsO_4$) has been eliminated by elastically or inelastically scattered electrons penetrating the sample [65–70]. Scanning electron microscopy (SEM) involves much simpler specimen preparation. Smooth cut surfaces are used to determine the backscatter electrons, whereby the emissions increase in the direction of elements of higher atomic number. One advantage is the ability to determine the chemical origin of the individual phases [71–75].

Atomic force microscopy (AFM) has the potential to resolve atoms in rubber blends. It can be used to image both conductive and insulating surfaces at ambient conditions and is suitable for polymers bends, because new types of scanning heads have become available. Potential benefits include high resolution, simplicity of specimen preparation and greater versality in varying the phase contrast [76–78].

Scattering methods describe the size of the phases based on the Bragg relationship. The main size determined by small angle X-ray scattering (SAXS) is close to that determined by TEM measurements. In small angle neutron scattering (SANS), substitution of deuterium for hydrogen in a polymer chain provides the necessary contrast. SAXS and SANS could encounter significant problems in filled rubber blends [79]. In addition to compositional analysis, NMR methods have also been used to characterize the morphology of polymer blends. Segment mobility associated with the rubbery state leads to a widening of the NMR linewidth. Therefore, $^1$H-NMR can be used to assess polymer blend morphologies [80–83].

Generally, rubber blends fall into four different groups at a given mixing temperature:

- Immiscible blend systems that have two phases whose glass transition temperatures are in the same region as those of the blend constituents. The difference in the blend constituents is large and so is the difference in the δ parameters. It can be said that there is no interaction between the phases.
- Partially soluble blend systems still occur most often in rubber blends. In these blend systems, the $T_g$ values of the blend constituents are shifted toward each other. The difference in the δ parameters is smaller. The size of the $T_g$ shift describes the interaction between the two polymers. The mixing temperature always plays a decisive role. The higher the degree of mixing, the lower is the interfacial tension of the blend.
- Miscible blend systems exhibit only one glass transition temperature. Miscible rubber blends can be found when the δ parameters of the individual components are similar. Miscible blends exist for both entropy and enthalpy reasons.
- Metastable blend systems are characterized by a large glass transition between the blend constituents. If two rubber components which have a similar δ parameter are mixed at high temperatures, metastable systems are formed which, when annealed, break down into the individual components and form two phases.

## 6.4 Unfilled Rubber Blends

Phase-separated rubber blends are not expected to mix at the segmental level of the rubber chain, and the morphology is always heterogeneous. The components are segregated into spatially distinct domains and have an interface that depends on the difference of the δ parameter. The dynamics of the components are essentially the same as or very similar to those of the pure materials. Examples of immiscible rubber blends show a large domain size and practically a very small interface between the two rubbers [84]. Saturated hydrocarbon elastomers (IIR, CIIR, EPM and EPDM) are widely used in rubber technology, due to their high strength, low cost, and low dielectric constant. However, they do not have mutually interacting groups that promote strong interactions with other rubbers. All rubber blends with unsaturated rubbers are immiscible blends. There is a different method available for elucidating the phase morphology in unfilled rubber blends, as shown earlier. Thermal analysis by means of DSC and DTG curves are used to investigate NR/BR and NR/E-SBR blends [85]. The changes in blend proportions are in good agreement with polymer ratios in the blend. Pyrolysis gas chromatography has been used to analyze the bound rubber formed in the rubber blend. The results served to determine the rubber blend composition in tire rubber blends [86]. Variations in peak intensity ratio were used to evaluate different rubber proportions in the blend.

TEM analysis can lead to a direct measurement of the domain size. Many strategies for analyzing rubber blends have been developed. One uses differential swelling to analyze EPDM/IIR blends. Figure 6.9 shows the NR/E-SBR blend prepared by two different mixing processes. The blend prepared from solution shows a large domain size of the minor rubber. The result for the same blend prepared in an internal mixer demonstrate the outcome of the strong shearing forces required to blend the two rubbers in the internal mixer. However, in the sample prepared from solution, the additional tension does not exist and the domain size increases.

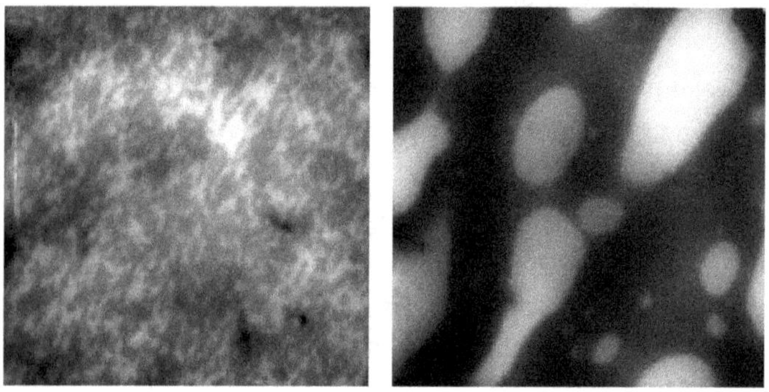

**Figure 6.9** Difference in phase morphology of NR/E-SBR prepared from solution and in an internal mixer

For the cis-BR/EPDM blend, the thickness of the interface was determined to be 2–3 nm using the crystallinity loss method. Atomic force microscopy operated in trapping mode can distinguish between two different polymer domains in blends of NR/synthetic rubber. The images are verified by analyzing the analogue signal histograms. A tapping mode image for NR/EPDM shows poor resolution of the phase boundary. The rubber blends of partial solubility are characterized by a simultaneous shift in the glass transition temperatures. A typical development is illustrated by the distribution of the G″ modulus in NR/S-SBR blend. The shift in the glass transition of the NR phase to higher temperatures and the shift in the glass transition of S-SBR phase to lower temperatures are shown in Figure 6.10 [23].

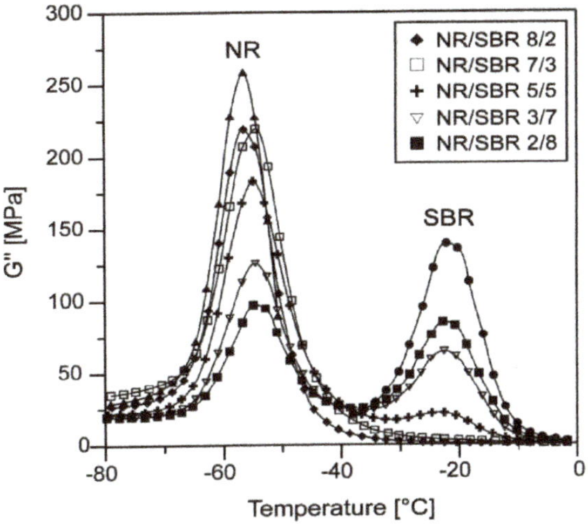


**Figure 6.10**
G″ moduli for both NR/SBR blend constituents at different blend ratios

An important feature is that the S-SBR phase does not exhibit a G″(T) peak at small concentrations of NR, a fact which is due to the phase morphology in this temperature region. At low concentrations, the S-SBR domains are dispersed and act as discrete glassy particles. After the concentration of S-SBR has increased, it has a co-continuous morphology and the G″(T) peak is observed as a peak value. Some aspects are sensitive to the size of the domains as well as the composition and interconnectedness of interfacial regions. The difference in the solubility parameters, δ, indicates how great the differences are and to what extent the blend will exhibit certain properties of the starting components.

Looking at the thickness of the interface, the crystallinity loss in cis-BR blends shows that the interface decreases as the δ parameter difference increases. It can be seen that it is only 2–3 nm for cis-BR/EPDM, 7 nm for cis-BR/E-SBR (40% styrene), 12 nm for cis-BR/E-SBR (29% styrene) and 18 nm for the most compatible cis-BR/E-SBR (23% styrene) [87]. In this way, the difference in the solubility parameter difference, Δδ, can be described with a linear increase in the interfacial thickness. Such an increase also remains for vulcanized samples, although the interfacial thickness has a numerically higher value. On vulcanized samples, the proportion of G″ in the boundary layer can be determined via the thickness of the boundary layer; the result is a slightly larger amount, but the order remains the same [23]. If the proportion of compatibility is increased, the volume of the boundary layer increases up to 25% for NR/S-SBR (70% vinyl; 23% styrene) [88]. Quantitative analysis of E-SBR, BR and NR blends is possible using $^{13}$C-NMR methods [89, 90]. Using magic angle spinning, NMR results were obtained for the vulcanizates of SBR, BR and NR in different proportions [91].

The thermodynamically miscible rubber blends include those whose components have a similar mixing entropy (entropic miscibility) and those exhibiting specific interactions (enthalpic miscibility). The first group includes the blends of 1,2-polybutadiene/1,4-

polyisoprene. This interesting blend has the same δ parameter for NR and for 1,2-BR and a single glass transition, as shown in Figure 6.11.

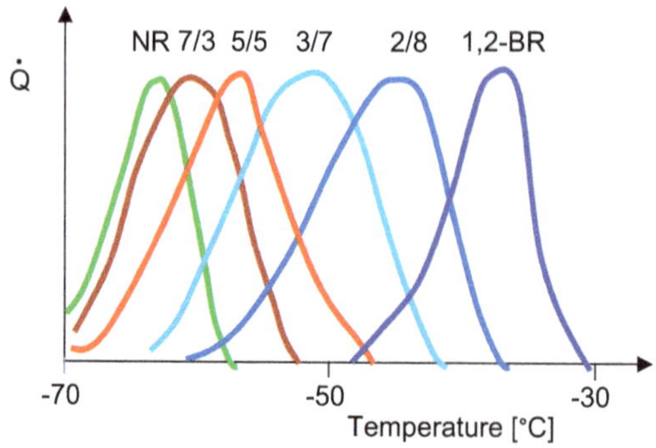


**Figure 6.11**
First derivative of the thermogram of the morphologies for the miscible blend NR/1,2-BR

However, the broad signal at high concentrations (50/50) indicates that a micro-heterogeneity exists in the blend system, but there is no method to resolve this experimentally. Other blends with a single $T_g$ are rubber polyisobutylene/head-to-head polypropene [93], epichlorohydrin/polyvinyl methyl ether [94] and acrylate rubber/fluorocarbon copolymers [95]. Blends of NBR/PVC are among the oldest known enthalpic blends. They are characterized by a single $T_g$, although the TEM resolution indicates some micro-heterogeneity with a very fine dispersion size [96]. Enthalpic miscibility is observed for chlorinated polymers with epoxidized rubber [97], trivial blends of copolymers such as siloxanes [98], various polyolefins [99, 100], nitrile rubbers with different acrylonitrile content [101], and ethylene-propene rubbers [102, 103]. The properties of miscible blends are roughly the average of those of the pure components; the single-phase system can exhibit behaviors not otherwise attainable.

## 6.5 Filler Distribution in Rubber Blends

If the rubber blend is filled, the design requires some degree of compatibility between the immiscible blend constituents and partitioning of the fillers [5, 104]. In addition, filler, dispersion and filler distribution between the distinct phases lead to specific reinforcing effects that improve the properties of elastomers and become attractive for the rubber technology [105, 106]. Various technologies are used to mix rubber blends, including mechanical mixing, which is used in most cases, but also in some cases latex blending and solution blending. The alternative to the use of mechanical mixing is the use of wet-compounding, wchich employs rubber latex and fluffy fillers. This technique is especially used in the production of tires.

The methods of introducing fillers in rubber blends are further divided into ways of distributing the filler that lead to a different filler distribution and different filler reinforcement:

- natural distribution of the filler in a blend
- forced distribution of the filler using masterbatches
- dynamic vulcanization of the filler in one blend phase

The first type of mixing, namely natural distribution of the filler in a blend, consists in adding the filler to the premixed polymer (A + B) in a manner that is similar to mixing a single component. For best results, the polymer components of a blend are usually mixed for a short time prior to addition of the filler.

The second type of mixing is used to control the specific location of the filler and curing agents by forced distribution of the filler in the rubber blends. Separate masterbatches are employed with the target to locate the concentrations of the filler in each rubber phase. The two masterbatches are subsequently blended to produce the final compound that has a special filler distribution. This more difficult method can be applied to produce tread compounds with different filler distributions in each part of the blend.

The third way to immiscible produce a rubber blend is forced distribution of the filler by adding a very small amount of curing agent to a masterbatch to try to keep the dispersion and the distribution of the filler in this phase. With these three types of filler distribution in a rubber blend, it must be ensured that the crosslinking system is optimally distributed throughout the individual phases. Crosslinking of the filled blend is a prerequisite for optimal use of the filled blend system. Due to the distribution of the curing system, the use of the filled blends can be somewhat limited.

A fundamental difficulty of all microscopic methods is the detection of the phase boundary between the polymers with a view to determining a clear distribution of the particles. A phenomenon that occurs in the production of filled rubber blends is the distribution of the filler in the different blend phases depending on the affinity of the filler surface for one of the blend phases. For this reason, the filler can extensively change the rheological behavior of the mixtures and the properties of the filled elastomer. Due to the distribution, the reinforcement of one phase in the presence of the filler change changes the properties of the vulcanizate, which makes analysis of the property behavior of the filled vulcanizates more difficult. This is evident when the filler is preferentially distributed in one of the blend phases, as shown in Figure 6.12.

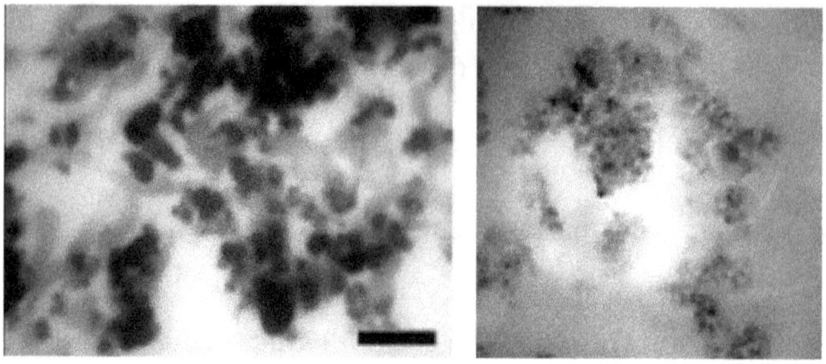

Figure 6.12 Phase morphologies of NR/CIIR/CB and EPDM/NBR/silanized HD silica

Figure 6.12a shows preferred filler distribution in one blend phase, while the second phase hardly contains any filler. The non-uniform distribution causes great differences in the rheological behavior as well as in the mechanical properties of the filled blend. In Figure 6.12b a similar amount of silanized silica is distributed at the interface between the rubber phases. The filler reinforcement is distributed over both phases and shows a reduction in the size of each domain. The distribution of the filler is determined by the nature of the surface of the reinforcing filler and the chemical nature of the rubber, as well as the type of mixing. Fluctuations in the mixing process therefore manifest themselves not only in a different degree of dispersion, but also in a changed filler distribution in rubber blends, which changes the properties of the blend system. Therefore, it is essential for the rheological properties and especially for the properties of the vulcanizate, especially for the ultimate properties, dynamic crack resistance and abrasion resistance.

### 6.5.1 Natural Distribution

Natural filler distribution is a spontaneous process during the mechanical mixing of a filler in a rubber blend, in which the filler particles that come to be in the interphase during mixing can thermodynamically opt for the phase with which the better interactions exist. During incorporation, the filler aggregates are able to select the polymer with the best interactions, such that partitioning in the different blend phases starts in this phase and is continued during the dispersion step and distribution step filler distribution. Of course, the type of mixing, the mixing time and the temperature also play a decisive role in the outcome of the filler distribution [35]. In such a traditional mixing order of addition, the filler is free to go into one of the two rubber phases in accordance with the filler affinity preference.

The thermodynamically motivated filler partition is quantified by a partition coefficient, which indicates the ratio of the filler shares in phases A and B of the blend, $C_A/C_B$. Since neighboring phases of different filler content exhibit strong differences in mechanical response, indicating high stress gradients in their contact region, the effect of fillers depends not only on the filler loading and the affinity of the filler for each rubber blend, but also on the amount of filler and the type of mixing.

However, the main difference in the filler distribution results from the conditions of the filler surface and the chemical structure of the rubber chains.

The distribution of the filler in rubber pre-blends is most often determined by several physico-chemical methods, which are treated in a separate section at the end of the chapter. The compound with the naturally distributed filler also determines the downstream process via a changed viscosity, and a change in die swelling in the extrusion or vulcanization step, marked by filler flocculation in each blend phase, acts as an important parameter that indicates the final distribution of the filler and contributes significantly to explaining the physical properties. Hence the partition coefficient of the filler in each blend phase is an important parameter that indicates the final distribution of the filler and contributes significantly to explaining the physical properties.

A simple distinction is made between rubber blends that have a high interfacial tension with the filler and those that have a low interfacial tension, which presents a thermodynamic challenge for the added filler.

#### 6.5.1.1 Blends with Large δ-Parameter Difference

If a rubber blend is present in which the interaction between the rubber phases is low, i. e. there is a large difference in the solubility parameters, $\Delta\delta$, the natural distribution of the filler in the pre-blend of the rubber can easily be decided in favor for one rubber or the other. The filler is preferentially distributed on the rubber with which the greatest interactions exist. The filler has the opportunity to preferentially distribute itself to the rubber that has the highest affinity for the surface of the filler.

The driving force that controls the filler distribution in blends is the rubber-filler interaction, which is characterized by adsorptive processes of the rubber chains on the active site centers of the filler aggregate. This, of course, depends on the different surface of the filler. The morphology of the primary particles plays a role here, as does the content of silanol groups in the silica, the success of the silanization and the relatively high activity of the carbon nanotubes in a rubber blend. Compared with carbon black, the polar silanol groups of the precipitated silica have significantly different interaction with the rubber chains. The distribution on polar rubbers, such as NBR, is particularly intensive. Silanized silica fundamentally changes the distribution of the silica in the rubber blends. The distribution can vary with the degree of silanization. In the case of layered silicates and carbon nanotubes, mixtures with a distribution of the filler have been found.

Thus, the preferential filler distribution occurs when one of the two rubber components enters into an effective adsorptive polymer-filler interaction, while the other component does not bring about such effective interactions. Small differences in the interaction potential of the different rubbers, which are based on the chemical nature of chain segments, polarity and molecular weight, the filler specific surface area and the mixing procedure, can lead to strong changes in the filler distribution.

The addition of a filler to a rubber blend leads not only to the important filler distribution, but also to phase compatibilization. The presence of a filler-polymer interaction changes the phase morphology of a blend. Whenever a filler is added to a blend, there is a decrease in the domain size of the rubber phase which is readily observed. When a carbon black or silanized silica is added to an IBR/cis-BR blend, the phase morphology becomes more fragmented and causes a strong reduction in the rubber domains because the interaction of the filler with the rubbers has a decisive thermodynamic effect. The morphological consequences on the blend morphology due to the presence of the filler are shown in Figure 6.13 [107].

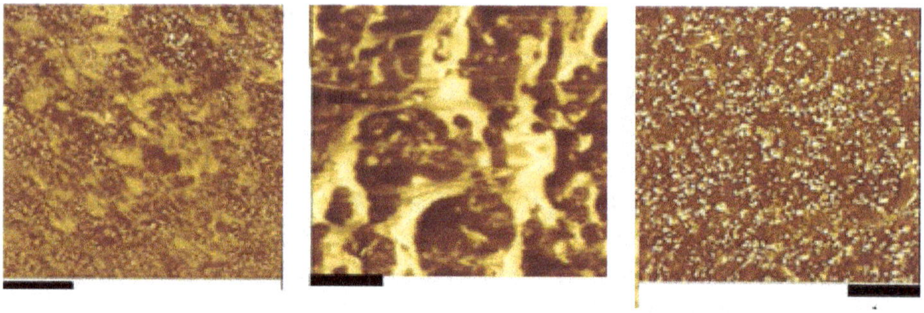

**Figure 6.13** Influence of the presence of filler on IBR/cis-BR blend morphology: a) HD silica on the left image, b) the unfilled blend in the middle image and c) carbon black N234 on the right image

At the beginning of the interest on filler distribution, phase contrast microscopy was used to determine roughly the direction in which the filler is located [63, 108, 109]. The trend-setting success of optical microscopy was transferred rapidly to TEM, which offered a great advantage in the high loading of the filler and the fine dispersion of the rubber. At that time, several different techniques were used, but these are only of historical relevance now [1].

Two phenomena are characteristic of blends possessing a large difference in their polarity or their solubility parameters, δ. During incorporation into the pre-blended rubber, the filler is distributed in the phase where there is the best thermodynamic interaction with the rubber chain segments. If a carbon black is used in a rubber blend made from a saturated rubber (IIR, CIIR, EPDM) and an unsaturated rubber (NR, BR, SBR, CR and NBR), the unsaturated rubber always offers better interaction for the active sites on the filler surface to double bonds of the rubber chain or the functional groups attached to the chain (styrene groups, C-Cl dipoles, acrylonitrile groups), which can give rise to preferential location of the filler in the unsaturated polymer. This is observed for carbon black N330 in CIIR/NR blends.

The driving force for a preferential distribution of the carbon black towards the grey NR phase is the energetically more favorable interaction with the energy sites of the primary particles of the carbon black, while the bright CIIR phase does not contain carbon black. The NR/EPDM blend employed for side walls application with $OsO_4$ shows that carbon black N650 is preferentially distributed in the NR phase. The rubber containing the unsaturated chain always offers the semi-reinforcing carbon black a preferred place in the rubber blend. Without exception, the TEM images show a selective filler distribution in favor of the blend component of higher polarity or degree of unsaturation.

An advantage of dynamic-mechanical characterization of the filler distribution is that a simple technique is used to make a clear statement about the distribution of the filler via the level of the loss modulus in the glass transition region.

The carbon black dispersion is examined for EPDM/cis-BR (40/60). A marked improvement in the carbon black dispersion can be seen with increase in mixing time, because the proportion of the carbon black aggregates shifts preferentially to the cis-BR phase, as shown in Figure 6.14 [108].

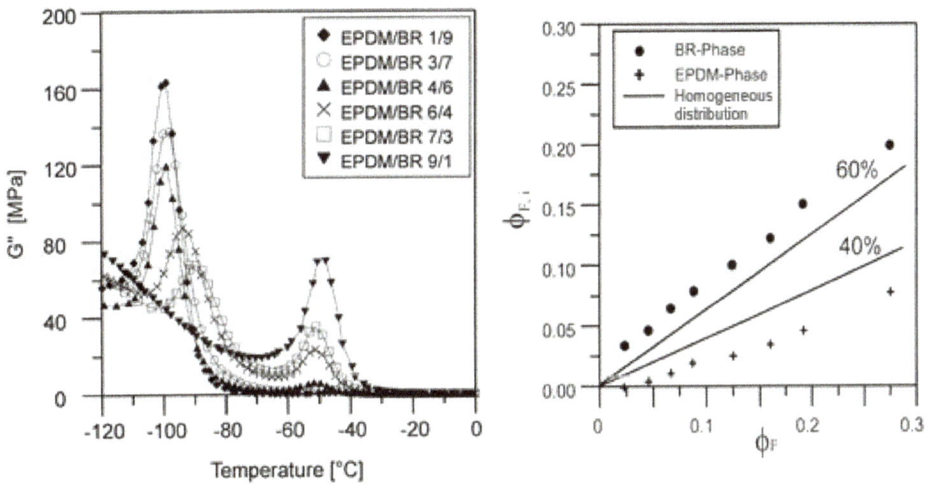


**Figure 6.14** $G''_{max}(T)$ in the EPDM/cis-BR/CB and the carbon black N550 partition in the blend

In EPDM/cis-BR, a specific increase in the $G''_{max}(T)$ is observed in the glass transition temperature range with increase in carbon black N550 content from 5–80 phr. Due to the distribution of the carbon black, the G″ increase is significantly higher in cis-BR than in the non-polar EPDM phase. In contrast to the difficult-to-interpret TEM micrographs, the dynamic-mechanical method shows a clear distribution of the carbon black in the unsaturated cis-BR phase, as shown on the right in Figure 6.14. The partition coefficient is always higher in the cis-BR phase than in the EPDM phase, due to the better interaction with the energy sites of the carbon black [111]. The carbon black partition coefficient, $C_B/C_A$, is 2.6 at higher carbon black concentration, indicating distribution to the unsaturated rubber domain.

Due to the rubber-filler interaction, the size of both domains decreases significantly, which proves the phase compatibilizing effect. The main carbon black content is found not in the non-polar rubber, which can establish only weak London forces, but in the unsaturated rubber, which can establish induced dipole interactions or dipole-dipole interactions with the filler surface. In particular, the functional styrene groups can interact efficiently with the surface of carbon black. Therefore, the saturated rubber in the blend (IIR, CIIR and EPDM) has a minimum of filler aggregates inside the domains. The majority of the filler is located in the unsaturated rubber or is located at the interface between the blend components.

Similar images are obtained using AFM. In an EPDM/NR blend, the preferred distribution of the filler is the NR phase [112]. When comparing the filler distribution in IIR/BR and BIMS/BR, it has been found that there is always a higher filler content in the BIMS phase than in the IIR phase, which indicates the better interaction of the fillers with the styrene units from BIMS [113]. The influence of precipitated silica and silanized silica in EPDM/NBR blends has been investigated, because a system was sought that had both oxidation-resistant and polar properties. Of the two rubbers, NBR always has a stronger interaction with silica [114]. Even at a viscosity ratio, $\eta(NBR)/\eta(EPDM) < 1$, the precipitated silica shows a significantly higher interaction with NBR than with the non-polar EPDM, so that a preferential distribution in favor of NBR occurs. This interaction between the filler and the rubber is attributed to the strong dipole-dipole interaction. If the precipitated silica is now treated with a monofunctional silane, the surface of the silica changes and the distribution of the silanized silica leads to accumulation at the interface of the EPDM. The distribution of precipitated silica (VN3) in NR/CIIR blend is clearly a case of selective partitioning to the unsaturated NR. If the silica is completely introduced first in EPDM and is subsequently mixed with the polar NBR, after a short time the polar rubber phase quantitatively contains the silica [114]. It can be clearly seen that a change in the silica structure leads to a serious change in the distribution. Halobutyl is the polymer of choice for tire inner liner compounds, due to its excellent air and moisture impermeability and fatigue-flex resistance. The BR/BIMS blend has been intensively examined by Exxon Mobile because the modified butyl rubber has a high damping spectrum, which is characterized by an increase in adhesion and slip resistance. If carbon black or silanized silica is added to the BR/BIMS blend, the filler always has a phase-imparting effect. The distribution of the carbon black favors the interface between the two rubbers, whereas silanized silica results in a change in phase morphology with a somewhat more uniform distribution of the reinforcing filler [115]. The preferential distribution of the filler in the thermodynamically more interesting polymer phase has been demonstrated many times.

#### 6.5.1.2 Blends with Small δ-Parameter Difference

Much more important is the distribution of fillers in rubber blends, which have a similar thermodynamic efficiency and where the difference in solubility parameters, Δδ, is relatively small. Determination of filler distribution is much more difficult in blends of quite similar polarity and solubility parameter that are close to each other, such as in the case of NR, BR and SBR used in tread compounds. The filler used is an active carbon black from the N300 to N100 series or a silanized HD silica. However, it is precisely these rubbers and active fillers that are often blended in the tire-makers' mixing chambers in the hope that the properties of the original rubber can be combined successfully. In cases of high filler loading, the phase boundaries of the rubber blend may no longer be discernible. In such blends, optical microscopy cannot accurately discern the distribution of the filler, because of the high loading and the small domain sizes of the filler. However, the content of carbon black and precipitated silica decreases in the order:

NBR > CR > SBR > BR > NR

The functional groups in NBR, CR and SBR are responsible for this good interaction with carbon black and silica. In BR and in NR, the unsaturation is responsible for the good interaction with the filler surface. The swelling technique was used to analyze the carbon black distribution in IR/BR blends at very low loadings of carbon black. However, the BR showed a preference for the carbon black surface. SBR indicated the greatest preference for carbon black, but is quite close to BR, CR and NBR. The NR response to carbon black was distinctly lower [28]. The accumulation of carbon black in the interface was detected in the CR/NR system, indicating a higher activity of carbon black for the CR [116]. TEM images were used to evaluate a series of different carbon blacks in NR/BR blends. The BR phase includes the abundant part of carbon blacks [117]. Furthermore, the carbon black distribution in favor of the BR phase had a significant effect on increase in the rebound resilience and the tensile strength. The fatigue strength increased when the carbon black was preferentially located in the BR phase. An examination of the carbon black N220 distribution in NR/cis-BR at different blend ratios demonstrated that the majority of the filler is in the BR phase. Changing the loading and the blend ratio changes the partition coefficient, as shown in Table 6.6.

**Table 6.6** Carbon Black Partition Coefficient $C_B/C_A$ in NR/cis-BR as a Function of the Blend Ratio and Filler Loading [117]

| Blend Ratio | Filler Loading (phr) | $C_B/C_A$ |
|---|---|---|
| 70/30 | 30 | 1.7 |
| 70/30 | 60 | 1.1 |
| 50/50 | 30 | 2.6 |
| 50/50 | 60 | 1.5 |

As the carbon black loading increases, the distribution in the cis-BR phase decreases. If the tests are done with co-continuous morphology, the partition coefficient increases. This leads to a strong improvement in the reinforcing blend. Pyrolysis-gas chromatography measurements on bound rubber from NR/BR blends of carbon black N330 and N220 show a similar high concentration in the BR phase [118]. The distribution of carbon black N550 in NR/cis-BR blends demonstrate that the crystallization temperature of cis-BR, as measured by DSC, increases with filler loading. For carbon black N550, it has been shown that there is no preference for either NR or cis-BR. When mixed simultaneously with NR/cis-BR pre-blend, the carbon black distributes both rubber domains in an equal form [119]. That is a consequence of the reduced filler activity.

Gas chromatographic analysis of bound rubber was carried out to analyze the filler content in NR/SBR blends. In naturally distributed carbon black of different specific surface area, it was found that the dispersion of the filler is highest in the SBR phase. The preference of carbon black depends on the interaction of the styrene groups with the amorphous part of the carbon black surface. The lower amount of carbon black in the NR phase was attributed to a lower affinity of carbon black for the double bonds in the NR rubber chain, because of the influence of the methyl group attached to the double bond. The much better interaction with E-SBR is determined by the thermodynamic favorability of the mutual interaction with the carbon black surface [120]. In the filler distribution in NR/SBR, it was shown that the molecular weight of the NR can influence the extent of the distribution of carbon black. The higher the molecular weight of the NR, the more carbon black content there is in the bound rubber of the NR/SBR blend. The SBR always contains a maximum level of carbon black in this phase. The results of the carbon black distribution are given in Table 6.7.

**Table 6.7** Impact of Molecular Weight of NR on Carbon Black Distribution in NR/SBR (50:50) [120]

| NR $M_w \times 10^{-3}$ | Bound Rubber (%) | % NR in Bound Rubber | CB in NR (%) | CB in SBR (%) |
|---|---|---|---|---|
| 1203 | 26.6 | 49.6 | 36.1 | 53.9 |
| 1011 | 25.6 | 42.3 | 33.3 | 56.5 |
| 568 | 24.8 | 39.5 | 33.0 | 57.0 |
| 345 | 28.7 | 35.6 | 31.2 | 58.8 |

Even at a high molecular weight of NR, the preferential location of carbon black in the SBR phase enhances reinforcement of the blend. The effect of carbon black morphology has been studied using a series of carbon black varying in specific surface area from 40 to 120 $m^2/g$ and a varying in the structure of the aggregates. The result is shown in Table 6.6 [120].

**Table 6.8** Distribution of Different Carbon Blacks (45 phr) on NR/SBR (50:50)

| Carbon Black | CTAB ($m^2/g$) | CB in NR | CB in SBR |
|---|---|---|---|
| N550 | 43 | 41.2 | 48.8 |
| N326 | 84 | 31.9 | 58.1 |
| N330 | 81 | 34.0 | 56.0 |
| N339 | 95 | 31.8 | 58.9 |
| N347 | 87 | 33.9 | 56.1 |
| N234 | 120 | 30.0 | 60.0 |

In fact, the filler partition is conclusive because the adsorption of a polymer increases with the molecular weight and the high filler attachment to SBR is depicted by the interaction with the styrene groups in SBR. The carbon black distribution in NR/SBR is clearly determined by the specific surface area and the void volume of the aggregate. With increase in activity, the proportion of carbon black in NR decreases continuously, while it increases correspondingly in SBR. The method appears to make sense, but assumes that the bound rubber formation in the blend is the same as it would be in the individual polymers. Analysis of the carbon black content in bound rubber provides useful information about the mixing process. A new measurement technique, the wetting rate, has been introduced. NR shows a higher wetting rate than SBR. The kinetics of carbon black localization in the phase of the rubber blend has been developed. Due to the higher wetting rate of the NR component in the NR/SBR blend, the result show a larger amount of carbon black in the NR phase [121].

The partition coefficient increases with the amount of SBR in the blend. The reliable dynamic-mechanical method developed at the German Rubber Institute leads to the determination of the loss modulus in the glass transition regime to be used to determine the partition coefficient of the carbon blacks N330 in NR/SBR blends [122]. The double bonds of the NR phase seem to have a lower affinity for the carbon black surface than the number of styrene units in the SBR (40S) phase. Probably due to the higher molecular weight, the interaction strength of the carbon black within the NR phase ($M_w$ = 560,000 g/mol) is increased relative to the SBR phase ($M_w$ = 250,000 g/mol). An improvement in the method led to a mass balance of the filler in the rubber blend system. In addition to the content of the filler in the two phases, the boundary layer is also taken into account. This is done using the rubber blend NR/S-SBR (50V, 25S) with an active carbon black. The distribution of a filler in rubber blends that exerts similarly graded interactions to the filler surface has to be evaluated from the loss modulus as a function of temperature. The loss modulus of a NR/S-SBR (50V; 25S) blend and an increased amount of loading shows the highest increase for the interface between the two rubbers, a lesser increase at S-SBR (50V;25S) and a minimum increase in the NR phase. To quantify the filler distribution the G″(T) curves were fitted with spline functions for the pure rubber phases. The region of the interface shows a great contribution in the filled system. Taking into account the stiffness of the filler particles, it can be understood that a large spatial region is involved in strong mechanical coupling of the excluded and the surrounding rubber, leading to a supplementary signal contribution in the vicinity of the interface region. On the one hand, the interfacial signal in filled rubber blends might be overestimated but, on the other, the results indicate larger mechanically active coupling system of the interface and explains the better phase bonding, as shown in Figure 6.15.

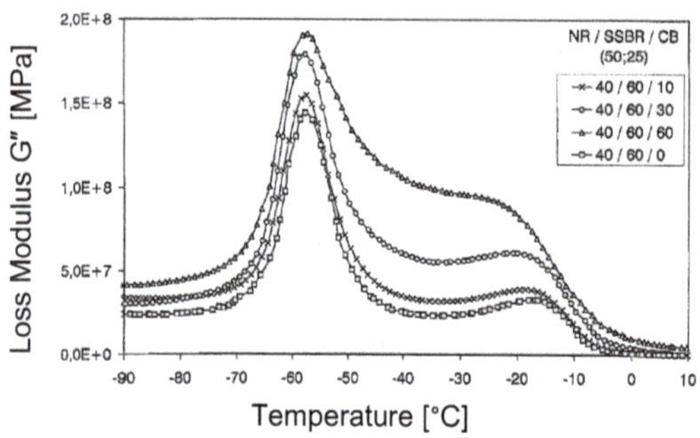


**Figure 6.15**
G″ of N220 in the blend NR/S-SBR (50V; 25S)

A look at the temperature dependence of G″ reveals that the S-SBR phase contains significantly more carbon black than the NR phase. However, in this case, the area between the amplitudes of the filled rubber, the interface, has the highest percentage of carbon black. The present results show that dynamical-mechanical analysis permits quantitative analysis and leads to an estimation of the filler partition. For the 60 phr carbon black, the interface is loaded preferentially with carbon black ($\Phi = 0.66$) while the S-SBR (50V; 25S) phase accommodates a smaller volume fraction ($\Phi = 0.24$) and the rest is present in the NR phase. This shows that, for extended mixing, the contribution of the polymer macrostructure is outweighed by the effect of the filler distribution, which proves to be the driving force. It has been demonstrated that the distribution of the filler between the two discrete polymer phases and the interphase. The location of the carbon black N220 in the interface provides a new functionality in reinforcing the phase boundary, a fact which is shown in Figure 6.16.

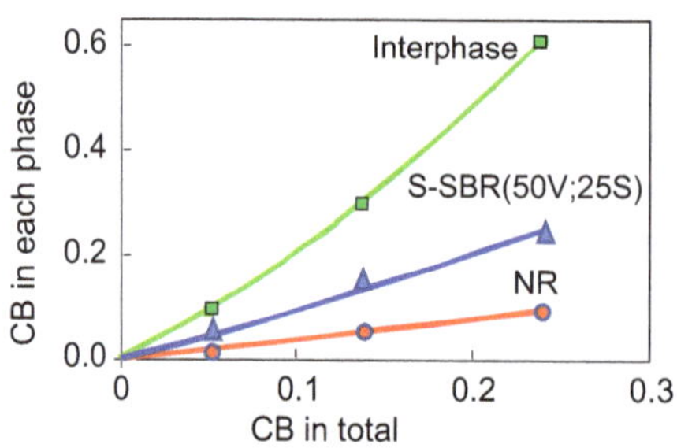


**Figure 6.16**
The partition of carbon black in the interphase between NR and S-SBR

The distribution of carbon black in SBR/cis-BR has been investigated by analyzing the composition of the bound rubber by pyrolytic-gas chromatography [123]. It was shown that the concentration of carbon black increases with the molecular weight of SBR, providing that the second polymer of the blend remain the same. This result was explained by the assumption that SBR and BR form a miscible blend system at the mixing temperature. The distribution of carbon black between the two polymers is not strongly influenced by the specific surface area of the carbon black.

A totally different filler distribution has been found for SBR/cis-BR blends, where the molecular weight of the polymer proved to be an important factor for the location of carbon black [124]. A high molecular weight of the polymer leads to the location of the filler in this phase. Therefore, it is not surprising that carbon black is located in the SBR phase that contains the styrene groups. At equal molecular weight, a slight preference is indicated for SBR. In SBR/BR (80/20) blends, the distribution of N330 was lowest in the natural dispersion in the BR phase.

Raman micro-imaging and small angle neutron scattering have also been used to study the distribution of silica in binary and ternary blends containing cis-BR, NR, SBR and BIMS. The distribution of the filler is shown on contour maps. In general, silica tended to be located near the phase boundaries of BIMS domains [125]. Small angle neutron scattering was used to characterize the size and surface properties of the filler in the elastomer. The BR/BIMS is more sensitive to carbon black than to silica. The interfacial broadening is slowed down by the addition of carbon black [126]. Using tapping-mode AFM, two different polymer domains were distinguished in an unfilled NR/cis-BR blend. To verify the images, analogue signal histograms were analyzed. Silica and carbon black exhibited different behavior in rubber blends. Carbon black is predominantly in the polybutadiene domain, whereas silica tends to prefer the NR domain [77]. From the tan δ values of NR/ENR (50:50) blends, the preferred distribution of the precipitated silica in the ENR phase has been shown [127]. As the concentration of the filler increases, the probability that the silica also partitions into NR increases.

Prior treatment of the silica with an organo-silane dramatically alters filler partitioning. Due to its reduced polarity the silanized silica can be located at the interface between the polar and the unipolar rubber. The distribution of precipitated silica in BR/NBR blends was found to prefer NBR, with a partition ratio of 70/30. Using monofunctional silanes reduces the distribution of the silanized silica in NBR and increases partitioning of the filler in the less polar rubber cis-BR. The changes in filler distribution of the silica is shown in Table 6.9 [114].

**Table 6.9** Distribution of Silica and Silanized Silica in NBR/BR

| Phase | Silica | 1/3 Si 208 | Si 203 | Si 208 |
|---|---|---|---|---|
| NBR | 70 | 40 | 35 | 25 |
| BR | 30 | 60 | 65 | 75 |

Depending on the degree of silanization, the silanized silica has less interparticle interaction and shows better filler partitioning in the BR phase, demonstrating the "carbon-like" behavior in the silanized silica state. The partially and fully silanized silica is distributed in the non-polar BR phase with an increasing partition coefficient. The distribution of the silanized silica can be influenced by the degree of silanization. The non-polar phase can be increased with a certain loading of the filler. This undoubtedly affects the change in the physical properties of the filled rubber blend. In addition, the silanized silica has a strong phase-compatibilizing effect. At the phase boundary, the filler acts as a solid emulsifier and reduces the domain size of the minor blend constituent by a factor of 50.

Another example is a blend of natural rubber and epoxidized natural rubber. NR/ENR(25%) were mixed with two types of HD silica. The introduction of the polar ENR(25%) changed the characteristics of the rubber blend, as it shows a stronger rubber-filler interaction than is the case for NR [128]. FT-IR spectroscopic examination of the bound rubber has been used as a tool for quantitative characterization of the silica distribution in NR/SBR. Silane addition does not affect the wetting of NR but shows a decrease in the wetting of SBR. With increase in temperature in the mixing chamber, the volume of the rubber layer of all mixtures increase, owing to the different rubber-filler bonding. More precipitated silica is found in the NR phase than in the SBR phase [129]. Silica from fly ash particles are used in NR/SBR (50/50) to improve the reinforcement by distributing mainly in SBR phase [130].

With immiscible blends, the first issue to consider is the localization of carbon nanotubes, which in turn will depend on the interfacial energies with the rubber. It is not surprising that carbon nanotubes tend to be concentrated mostly in the blend component with the higher surface energy. Recently, carbon nanotubes have been investigated as fillers in rubber blends to create new functionalities and to improve the properties of tire tread compounds. The reinforcement of SBR by carbon black and MWCNTs has been investigated. A significant improvement in dispersion and tensile strength is observed for samples containing a hybrid filler, composed of carbon black and a small volume of carbon nanotubes [131]. Due to the carbon nanotube activity, the distribution frequently favors the SBR phase in rubber blends NR/SBR [132]. Most of the multi-walled carbon nanotubes are found in the NR/BIIR blends in the NR domains. This result is not due to the activity of the carbon nanotubes, but to the NR which contains phospholipids that interact with the carbon nanotubes surface and so will influence the filler partition coefficient in the rubber blend [133].

While selective filler localization and its effect on the physical properties of carbon nanotubes filled thermoplastic blends have been comprehensively characterized, it is still incomplete in the field of rubber blends. It has been found that, in PP/NR blends ranging from 80/20 to 50/50, there is a reduced percolation threshold from 5 wt.% to 3 wt.%. In predicting the thermodynamically favored location of MWCNTs in this blend, the specific interactions of the phospholipids in NR with the carbon nanotubes was considered quantitatively. Therefore, the carbon nanotubes are selectively localized in the NR phase. Carbon nanotubes in SBR/NBR/NR blends have been characterized by means of the wetting concept and qualitatively supported by TEM and AFM. It was found that carbon nanotubes mainly reside in the polar NBR and nonpolar NR but not in the SBR phase. Such unusual localization of carbon nanotubes in the ternary blend can be explained by taking into consideration the presence of phospholipids in NR that are able to couple with the carbon nanotubes [134].

The dispersion of the layered silicates has been studied by X-ray diffraction and TEM in rubber blends. It was found that these fillers are more compatible and better intercalated by polyurethane rubber than by NR and so they are preferentially located in the more polymer rubber phase [135]. Organo-clays form dispersions in ENR/NR blends, which show that the filler is incorporated in the polar phase ENR and has increased interlayer spacing [136]. The improvement in blend properties has been attributed to organo-clays for the EPDM phase in the EPDM/ CIIR blend by morphological reasons [137]. In this case too, the argument is similar to that for the other fillers. The filler is always distributed in the rubber phase with which the specific surface shows the best physical-chemical interactions. The characterization of morphology development and kinetics of nano-clay distribution in HNBR/NR blends are based on the electrical conductance during the mixing process. The phase localization of nano-clay and the change in the blend morphology have been investigated by online measurable electrical conductance. It became clear that the presence of clay influences the morphology of the blends; a significant change from the co-continuous phase morphology into the dispersed morphology has been observed in HNBR/NR clay composites [138]. The distribution of modified and unmodified nanoclays inside the rubber phases of immiscible rubber blends composed from nonpolar-polar rubbers NR/ENR and NR/cis-BR has been investigated. The distribution of clays at various loadings in the blend was calculated from the viscoelastic properties of the blend. In NR/ENR (50/50), 40% of the Cloisite 30B migrated to the NR phase and 60% went to the more polar ENR phase. However, in the same blend, only 7% Cloisite $Na^+$ was found in the NR phase and 93% in the ENR phase. Again, in the NR/cis-BR (50/50), the NR phase contains 85% Cloisite 30B, whereas 55% Cloisite $Na^+$ migrated to the NR phase. It has been observed that the addition of layered silicates can prevent the coalescence of the polymer blends while retaining improved dispersion after the blending. Exfoliated clay compatibilization has been observed in thermoplastic blends. One explanation is that the nanoparticles concentrate at the interface, preventing coalescence of the particles.

### 6.5.2 Forced Distribution of Filler

An effective technique for controlling the level of filler in the phases of a blend is forced distribution of filler using masterbatches containing a defined concentration of the filler. Phase mixing is a special technique that has been used commercially in the rubber industry to achieve a special property for rubber products such as tires. However, the longer the masterbatches take to mix together, the greater is the likelihood of approaching the natural distribution. Forced distribution of the filler is used primarily in tire blends. Tire rubbers are prepared in internal mixers with tangential rotors and good cooling of the rotors and the mixing walls. The procedure for forced distribution is as follows:

- masterbatch A, which contains rubber A plus a portion of the filler
- masterbatch B, which contains rubber B plus another portion of the filler
- final mix of masterbatch A and masterbatch B
- shaping and vulcanization

Through masterbatch blending, the degree of filling of a blend component is influenced and this is able to change the physical properties of the entire blend. Furthermore, the size of the filled domains is influenced. This includes several influences:

- phase continuity
- domain size
- filler dispersion (agglomeration and networking)
- filler-polymer interaction (bound rubber)

The purpose of this type of masterbatch blending is to use the required filler for rubber A, which needs better reinforcement, while polymer B receives a smaller proportion of the filler. In this way, the physical properties of a blend are controlled when the components are blended. The filler content in mix A and mix B must be chosen such that the final blend contains the correct distribution of the fillers in both rubbers. However, it must be taken into account that, when the preliminary batches are mixed, filler transfer takes place, which requires precise compliance with the mixing procedure in order to achieve the desired forced distribution and avoid fluctuations. In older publications, filler transfer is not taken into account, and so the results from these publications are regarded as a rough guideline that should definitely be corrected. The effect of the incorporation of carbon black on the loss of crystallinity in BR is shown in Table 6.10 [85].

**Table 6.10** Effect of the Incorporation of Carbon Black on the Loss of Crystallinity in of CIIR/cis-BR

| Addition of Carbon Black | Loss of Crystallinity in BR (%) |
|---|---|
| cis-BR preliminary batch | 14.6 |
| CIIR preliminary batch | 12.6 |
| Premixing | 8.8 |
| Both polymers | 13.0 |

In this experiment, the natural distribution is compared with the forced distribution of the filler in the NR/EPDM blend by adding the carbon black and a portion of mineral oil. The results always point to a higher carbon black content in the cis-BR phase, as would be expected from thermodynamics, which determines the kinetics of filler transfer. In the blend added to the CIIR preliminary batch, the cis-BR depends on the extraction kinetics of the filler surface from the CIIR into the cis-BR. However, the filler content shows a slightly lower carbon black content in CIIR.

The role of forced distribution of carbon black in a compound made of NR/EPDM is shown by an AFM experiment presented in Figure 6.17 [139].

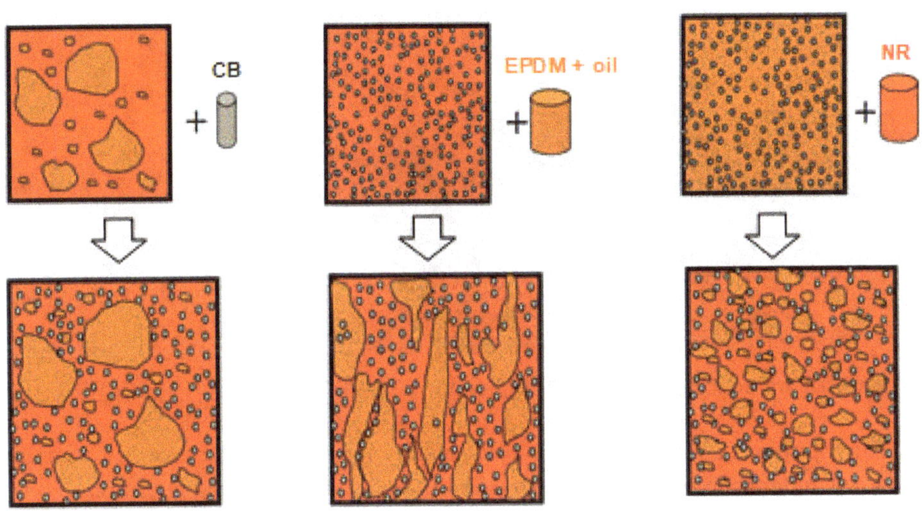


**Figure 6.17** Role of the different types of rubber blending in NR/EPDM [139]

If the two rubbers are mixed and then the filler is added, the natural distribution of the filler shows a preference for the unsaturated NR, and the EPDM phases do not contain the filler. If the filler is mixed first into the NR phase, the subsequently mixed EPDM has no opportunity to extract the filler from the NR phase, because of the low interaction of the saturated EPDM. If the filler is mixed first into the EPDM phase and then the NR is added, an NR/EPDM blend with smaller domains is created and the filler is deposited at the interface between the two polymers. Such a process of forced filler distribution has a greater influence on the desired properties. However, how long it takes to mix depends directly on filler transfer.

Masterbatch mixing of NR/BR (50:50) blends confirms the reduction in domain size induced by carbon black interaction with the different rubber chains [140]. With the increase in the carbon black concentration, the content increases in the BR phase. Since NR has high tensile strength and low crack growth due to strain crystallization, the BR phase requires reinforcement with the filler [141]. In NR/SBR the carbon black has been detected mainly in the SBR phase using masterbatch mixing. In this case too, the duration of mixing can influence the filler transfer, a fact which changes the rheological behavior and, above all, the vulcanizate properties.

A similar picture emerges for forced distribution of carbon black in S-SBR/EPDM, as shown in Figure 6.18 [142]. The TEM image on the left shows the carbon black mainly at the interface between the two rubbers, while the image on the right carbon black only in the SBR phase, because EPDM, with weaker interactions, cannot extract the filler from the SBR phase. The interaction of the carbon black and the saturated rubber EPDM is not strong enough to change this distribution. The interaction between the filler and the rubber therefore has a decisive influence on the outcome of forced distribution of the filler in a blend.

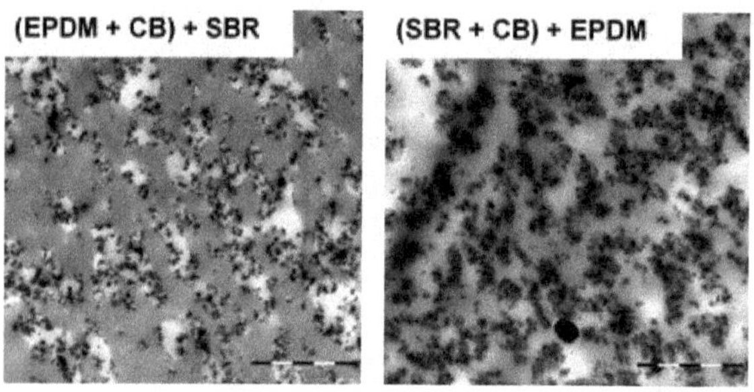


**Figure 6.18** Influence of forced distribution for S-SBR/EPDM

The distribution of the filler rarely corresponds to an equilibrium value, as it is achieved under different mixing process conditions. This is especially true for mixing pre-batches and the filler dispersion realized in this case. If the entire amount of filler is mixed in the rubber of lower polarity and this pre-batch is mixed with a rubber of higher polarity which does not contain any filler, the filler is transferred from the non-polar rubber to the more polar rubber until the entire filler is in this blend phase. Depending on the mixing conditions, especially the mixing time, the rotor speed and the fill factor, the filler partition in the forced filler distribution can change significantly. The extent of the filler transfer and its transfer rate is determined by the balance of the rubber-filler interaction and the mixing conditions.

The mixing of pre-batches of different filler loading can thus be finished after a certain time, which results in a forced distribution that can only be predicted to a certain level because the process engineer can influence the conditions via the mixing time and the rotor speed during the masterbatch mixing and the fill factor in the blend contents. In this way, the distribution of the filler in the polymer phases can be controlled. However, the filler transfer also takes place when mixing pre-blends consisting of rubber of similar polarity, so that the actual filler partition does not correspond to the ratio of the filler content in the pre-batches and has to be determined experimentally. Therefore, the information in the older literature on the filler distribution, which corresponds to the concentration ratio of the pre-batches, should at most be viewed as orientation values. In blends made from SBR/CR, BR/CR, SBR/NBR and BR/NBR, the filler is assigned to the SBR or BR phase [143].

In the case of cis-BR/SBR and cis-BR/NR blends, natural distribution of the carbon black has been examined using the same method [111]. From the resulting partition coefficient, $C_{SBR}/C_{BR} = 1.22$, it was concluded that more carbon black was distributed in the SBR phase. If the entire amount of carbon black is mixed into the SBR in a pre-batch, only a small part of it is transferred into the cis-BR phase when the two pre-batches are mixed, because of the activity of the rubber. Conversely, cis-BR releases about 2/3 of the carbon black loading to SBR. The different partition coefficient is explained by the different time required for filler transfer. However, with these blends, it must be taken into account that some of the crystallinity loss is caused by the interaction in the phase boundary layer.

It can be seen that, with prolonged masterbatch mixing, the partition coefficient of carbon black N550 depends on the mixing process. When the carbon black is filled into the EPDM phase and then the cis-BR phase is added, the filler partition shows a preference for the cis-BR phase, which is similar to that obtained in the natural distribution. The fact that the partition coefficient is altered if carbon black/cis-BR is mixed with EPDM can be attributed to the kinetic aspects of the filler transfer, which takes place during phase mixing. This is shown in Table 6.11 [144].

**Table 6.11** Partition Coefficient $C_B/C_A$ for Forced Distribution of Carbon Black N550 in EPDM/cis-BR (40/60) [144]

| Mixing Process | Partition Coefficient $C_B/C_A$ |
|---|---|
| (EPDM/CB N550) + cis-BR | 3.2 |
| (cis-BR/CB N550) + EPDM | 7.8 |
| EPDM/cis-BR/CB | 3.5 |

In the rubber blends, the distribution of the carbon black in NR/cis-BR and NR/S-SBR blends is determined using the method of G″(T) in the glass transition region. Using a batch mixing process in which the carbon black is preferentially mixed in the NR phase, filler transfer in the final mix increases the cis-BR and the S-SBR phases. The result is an increase in cut growth resistance of the filled blend system [145].

In a ternary blend NR/SBR/NBR, the localization of silica changed with the mixing times. From the masterbatches with a silica content of SBR at 10 phr, and for NR and NBR at 20 phr with final mixing, the silica content increases in NBR and SBR but decreases in NR. This can be explained by the activity of the rubbers for the active sites of the filler surface [146]. The distribution of precipitated silica has been investigated in the NR/ENR blend by using a tan delta in the glass transition regime. A greater change was shown for the more polar ENR relative to NR. The percentage of silica in the ENR phase is reduced by increasing the silica loading. However, regardless of which polymer the silica was added to, it was preferentially located in the ENR phase [127].

Due to the higher wetting rate of the NR component, more silica is found in the NR phase than in the SBR phase in the first stage of mixing of NR/SBR blends. Prolonged mixing transfers the silica from the NR phase to the SBR phase [147]. A combination of scanning transmission electron microscopy and energy dispersive X-ray scattering spectroscopy are used to determine the location of silica in IR/SBR blends. It has been found that the silica is preferentially distributed in the SBR phase, but can be distributed in the IR phase if the volume fraction of this is above 40% [148].

Carbon nanotube/NR masterbatches prepared by pre-dispersing were mixed with NR and subsequently with NBR without filler. The behavior of the filled rubber blend indicates the stronger reinforcing efficiency when the carbon nanotubes are located in the NBR phase [149]. In all cases examined, the forced distribution leads to a distribution that is triggered by the rubber with the highest solubility parameter. The goal here is to achieve a distribution in which the concentration in the rubber of lower solubility parameter $\delta$ is slightly higher in order to increase the strength of this phase.

The last method of making a filled rubber blend has a major advantage. This method achieves a sharp drop in the loss modulus G″ and the rolling resistance of the compound, which is discussed in the chapter on vulcanizate properties. When this third method is employed, one or both masterbatches are treated during mixing with a small amount of crosslinking agent to achieve "step" vulcanization that improves the blend morphology and changes the physical properties in a hitherto unknown way, without changing the blend ratio of the rubbers and the filler content [150]. To prepare the blend, the crosslinking agent is added in small amounts to a masterbatch mixture and then the two rubber mixtures are mixed together, followed by the desired crosslinking agent for the entire blend. In the case of blends of SBR/BR, SBR/NR or NR/BR, the result of such a masterbatch mixture is a significant improvement in the phase morphology and a uniform dispersion and distribution of the filler. This change in phase morphology is shown in Figure 6.19 for SBR/BR and SBR/NR blends.

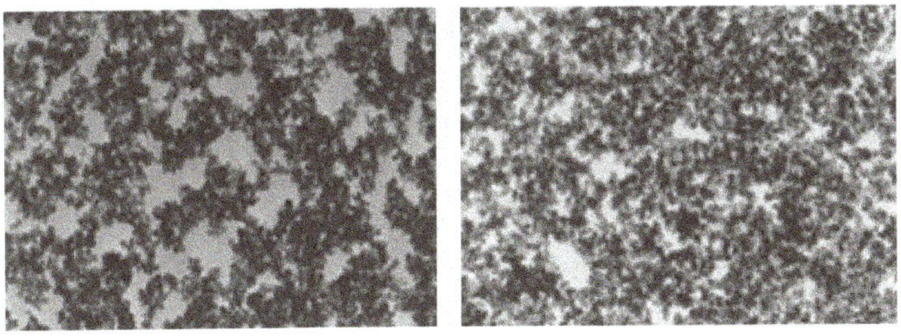

Figure 6.19 Influence of dynamic vulcanization of one blend constituent [150]

The effect of dynamic vulcanization during mixing is represented by the decrease in G″ value as a function of the G′ value. The blend system in which a natural distribution of the filler takes place has the highest G″ value, with high G′ values for SBR/BR/CB. With the transition to masterbatch blending, the system transitions to lower G″ values and through dynamic vulcanization achieves much lower G″ values.

Other physical properties, such as the modulus at 300% elongation and the tensile strength are listed among the physical properties of filled blends. From the tests, a further improvement in the carbon black distribution with a simultaneous increase in the dispersion in the blend is described, which leads to a wide range of mechanical properties with the same composition of the blend.

## 6.6 Filler Transfer during Mixing

Filler transfer is a natural alignment with that polymer phase with which the filler has better interaction at the mixing temperature. Rheological effects have an increased effect on the kinetics of the carbon black distribution and thus on carbon black transfer. Apart from storage and vulcanization, a stable filler distribution transfer can be assumed.

This process occurs with all fillers and is also determined by the polymer phases in the blend. Understandably, in the case of blends with a weakly interacting polymer phase, the interaction is small for one polymer, so that the filler is transferred to that phase which shows better interaction. If the filler is incorporated into the phase with the smaller $\delta$ parameter, after a few minutes of masterbatch mixing, all of the filler will be in the other blend phase, which has a higher delta parameter. The filler transfer takes place when the filler is initially in a phase with which there is little interaction and comes into contact with the good-interacting phase in the flow field of the mixing device. In this way, the filler can go preferentially into the phase where the interaction is highest. The dependencies of the distribution of the filler in connection with mixing conditions were originally answered contradictorily. In some reports, carbon black transfer is not assumed [116, 151] whereas others assume significant filler transfer [152, 153]. This process takes place in all rubber blends. When masterbatch is being mixed, filler transfer leads to a significant change in the properties of the rubber blend.

In EPDM/NR (50/50) blends, the carbon black N650 is preferentially in the NR phase. Unsaturation of the NR phase is the driving force governing the carbon black location. In the case of EPDM/SBR, the carbon black shows filler transfer into the SBR phase. TEM and pyrolysis gas chromatography to measure the distribution of N330 in NR/BR can show filler transfer into BR. The BR phase had a higher affinity for carbon black and the significant transfer took place during the mixing process [28]. The existence of filler transfer has been reported for several blends of NR, BR and SBR [154]. An entirely different trend exists in NR/SBR blends, where carbon black always tends to move away from NR during masterbatch mixing and is located in larger concentrations in SBR, because of the more intensive interaction as shown in Table 6.12 [120].

**Table 6.12** Transfer of Carbon Black During Blending Masterbatches in 50:50 NR/SBR Blends with Carbon Black N339 (45 phr)

| NR masterbatch | 80 | 45 | 30 |
|---|---|---|---|
| In blends | 54 | 35 | 16 |
| SBR masterbatch | 10 | 45 | 60 |
| In blends | 36 | 55 | 72 |

An increase in carbon black of high specific surface area has been observed for SBR blended with NR. It has been shown that the proportion of carbon black in the SBR phase increases with the mixing time, so that, with long mixing times, 60% of this filler is present in the SBR phase. The interaction of carbon black with NR is weak and the carbon black continues to migrate into the SBR on prolonged mixing and during blending of the NR and SBR masterbatches [155]. The natural rubber received the full amount of the carbon black at the start of the trial. After one minute of mixing in the internal mixer, the carbon black aggregates go to the interface between the NR and the SBR and later form smaller SBR phases that coalesce to finally form the morphology. Finally, the SBR domain will contain the higher value of carbon black. The above-mentioned reports confirm that filler transfer is a phenomenon which relates to polymer unsaturation and the role of functional groups like styrene in SBR or acrylonitrile in NBR on the polymer chain and the preferential interaction between the filler and one of the rubbers that effect filler transfer. This specific filler-polymer interaction can be explained by the difference in solubility parameters between polymer and filler.

## 6.7 Determination of Filler Partition

The concentration of filler in the distinct polymer phases of a blend is a great problem. The result of filler partitioning helps to understand the interaction in the blend system, to determine the extent to which a filler distribution is in favor of a polymer, and to understand the direction of filler transfer during the mixing process. In order to solve this problem, various approaches have been pursued, which are discussed in more detail below.

### 6.7.1 Use of Crystallinity Loss

A contribution to this topic was made with blends of CIIR/cis-BR, in which a contribution to the interaction between the two rubbers is made by the change in the melt enthalpy of the cis-BR when a filler is present in the blend [130]. Filled rubber blends have been examined on the basis of the crystallinity loss of cis-BR, on the assumption that crystallinity is reduced in direct proportion to the volume fraction of carbon black and the interaction with the other rubber is weaker to produce a similar graduated reduction. Even if this application only covers a small proportion of rubber blends, it is recognized that the crystalline loss of the crystallizing BR phase has provided a good introduction to this problem [17]. Using this method, it has been possible to estimate the thickness of the interphase and, for filled samples, the distribution of the filler in blends.

### 6.7.2 Composition of Bound Rubber

Since the bound rubber depends on the interaction between rubber and filler, the analysis of the bound rubber in a blend has been used to determine the filler distribution. To reduce uncertainties that arise with the optoelectronic assessment of the carbon black distribution, the composition of bound rubber was determined by the content of the bound rubber from the blends of SBR/ BR, NR/BR and SBR/NR. The bound rubber was analyzed by means of pyrolysis-gas chromatography [124]. The carbon black content in the two blend phases was equated with the proportion of bound rubber of the corresponding polymer markers. It was also assumed that the measured bound rubbers were not strongly influenced by the type of solvent. This technique was used to document carbon black transfer from the NR phase into the SBR or cis-BR phase. Variations in peak intensity ratios were employed to assess different polymer proportions in the blend. Three different pyrolysis processes are listed in the reports:

- quartz pyrolysis tube (500–800 °C)
- electrically heated platinum filaments (800–1200 °C)
- Curie-point apparatus (550–650 °C)

The difficulty with this method lies in the determination of the bound rubber itself. Regardless of the solvent used to extract the non-adsorbed polymer from the mix, one of the two polymers is extracted more extracted than the other. Therefore, the share of bound rubber does not exactly reflect the real composition of the polymeric environment of the filler aggregate. In any case, this method represents a significant advance in the description of the filler partition in the respective blend phases, which is clearer than the microscopic methods.

### 6.7.3 Dynamical Mechanical Analysis

Due to uncertainties in the discussed method relating to the crystallinity loss of cis-BR and the bound rubber distribution in rubber blends at the German Rubber Institute (DIK), two user-friendly dynamic-mechanical methods have been developed. The dynamic-mechanical response of filled rubber compounds follows the general principles of mechanical energy conservation, which state that the total stored or dissipated energy in a dynamically excited blend system results from the energy contribution of the phases that exist in the rubber blend. Since the dissipated energy can be expressed by the loss modulus G″ in the glass transition regime multiplied by the square of the strain, the following relationship is met:

$$G''_{blend} = \Sigma G''_{p,i}\,\Phi_i\,(\gamma_{0,i}/\gamma_0)^2 \tag{6.15}$$

where $G''_{blend}$ is the loss modulus of the blend, $G''_{p,i}$ is the loss modulus of each domain, $\Phi_i$ is the volume fraction of the domains, $\gamma_{0,i}$ is the external strain on the domain, and $\gamma_0$ is the external strain over the whole system

The average strain, $\gamma_{0,I}$, of the i-th blend component in the strained blend is difficult to measure. The simplest way was to calibrate the maximum G″ (T) for an increasing amount of the fillers [156]. The filler concentration was then determined for each polymer. As an indicator of the filler content in each of the blend phases, the damping peak $G''_{max}$ in the glass transition region was chosen. A more complete description that includes the filler distribution not only in the two blend phases but also in the interfacial layer between has been developed by fitting the experimental curve with an equation that can be adjusted:

$$G''_{blend}(\Phi_i, \Phi_{F,i}, T) = \Sigma A_i(\Phi_i, \Phi_{F,i})G''_{P,i}(T) \tag{6.16}$$

This equation considers the two-fold effect of the filler in the i-th phase and has to be solved by spline fitting. The G″ contributions for filled rubber A and filled rubber B are used to adjust the rear of the particular signal and the front of the G″ with the filled rubbers, making shifts on the temperature axis and in filler content. As soon as this is fulfilled, it is observed that an excess contribution occurs in the area between the G″ for the two rubbers, which describes the contribution of the filled interphase of the system. Such measurements have been made for a range of rubber blends and various carbon black and silanized silica in blended vulcanizates that can also contain mineral oil. From the author's point of view, this method is the most complete approach for determining the filler concentration in each phase of the system.

## References

[1] Hess, W. M., Herd, Ch., Vegvari, P. C., Rubber Chem. Technol., 1993, 66: 329

[2] Shereshnev, V. A., Rubber Chem. Technol., 1982, 55: 537

[3] Paul, D. R., Newman, S., in "Polymer Blends", Paul, D. R., Newman, S., Eds., Acad. Press, Inc. (1978), Vol. 1, Ch. 1

[4] Olabisi, O., Robeson, L. M., Shaw, M. T., Eds., in "Polymer-Polymer Miscibility", Acad. Press, Inc. (1979), Ch. 2

[5] Roland, C. M., Rubber Chem. Technol., 1989, 62: 456

[6] Krause, S., in "Polymer Blends", Paul, D. R., Newman, S., Eds., Acad. Press, Inc. (1978), Vol. 1, Ch. 2

[7] Flory, P. J., J. Chem. Phys., 1941, 9: 660

[8] Huggins, M. L., J. Chem. Phys. 9 (1941) 440

[9] Scott, R. L., J. Chem. Phys., 1949, 17: 279

[10] Tompa, H., Trans. Faraday Soc., 1949, 45: 1142

[11] Sanches, I., in "Polymer Blends", Paul, D. R., Newman, S., Eds., Acad. Press, Inc. (1979), Ch. 3

[12] Hildebrand, J. H., Scott, R. L., in "The Solubility of Non-Electrolytes", Dover, New York (1964)

[13] van Laar, J. J., Z. anorg. Chem., 145 (1925) 239

[14] Rehner, Jr., Flory, P. J., J. Chem. Phys., 1943, 11: 521

[15] Guillet, J. E., in "Progress in Gas Chromatography", Purnell, J. E., Ed., Wiley Interscience, New York (1973)

[16] Small, P. A., J. Appl. Chem., 1953, 3: 71

[17] Issel, H.-M., Peterseim, V., Schuster, R. H., Rubber Chem. Technol., 1996, 69: 769

[18] Helfand, E., Tagami, Y., J. Chem. Phys., 1972, 56: 3592

[19] Wu, S., Polymer, 1987, 28:1144

[20] Peterseim, V., Schaper, J., Schaube, J., Schuster, R. H., Rubber Fall Colloquium '94, Hanover

[21] Schuster, R. H., Schaper, J., Hallensleben, M. L., 151st ACS Rubber Division Meeting Anaheim (1997)

[22] Klueppel, M., Schuster, R. H., Schaper, J., Gummi Faser Kunstst., 1998, 51: 508

[23] Meier, J., Klüppel, M., Schuster, R. H., Kautsch. Gummi Kunstst., 2000, 52: 663

[24] Rocha, T. L. A. C., Jacobi, M. M., Schuster, R. H., Meier, J., Samios, A., Int. Macromol. Colloquium, Gramado, Brasil (2001)

[25] Schaube, J., PhD Thesis University of Hanover (2003)

[26] Berger, W., Kammer, H. W., Kummerlöwe, C., Makromol. Chem. Suppl., 1984, 8: 101

[27] Peterseim, V., PhD Thesis University of Hanover (1995)

[28] Callan, J. E., Hess, W. M., Scott, C. E., Rubber Chem. Technol., 1971, 44: 814

[29] Staeber, R., Geisler, H., Schuster, R. H., Report Industrial Association AIF (1996)

[30] Tokita, N., Rubber Chem. Technol., 1977, 50: 292

[31] Avgeropoulos, G. N., Weissert, R. C., Biddison, P. H., Böhm, G. G. A., Rubber Chem. Technol., 1976, 49: 93

[32] Corish, P. J., Powell, B. D. W., Rubber Chem. Technol., 1974, 47: 481

[33] Kerutt, G., Blumel, N., Weber, H., Kautsch. Gummi Kunstst. 1969, 22L, 413

[34] Corish, J., in "Science and Technology in Rubber", Mark, J. E., Erman, B., Eirich, F., Eds., Acad. Press, San Diego (1994) Ch. 12

[35] McDonel, E. T., Baranwal, K. C., Andries, J. C., in "Polymer Blends", Paul, D. R., Newman, S., Eds., Acad. Press, New York (1978) Ch. 19

[36] Amerongen, G. J., Rubber Chem. Technol., 1964, 37: 1065

[37] Rehner, Jr., J., Wei, P. E., Rubber Chem. Technol., 1969, 42: 895

[38] Gardiner, J. B., Rubber Chem. Technol., 1970, 43: 370

[39] Guillaumond, F.-X., Rubber Chem. Technol., 1976, 49: 105

[40] Mueller, G., Frueh, Th., Rubber World, 2000, 222: 28

[41] Roland, C. M., in "Advances in Elastomers" Advance. Structural Materials, Springer Verlag (2013)

[42] Gardiner, J. B., Rubber Chem. Technol., 1969, 42: 1058

[43] Bauer, R. F., Dudley, E. A., Rubber Chem. Technol., 1977, 61: 281

[44] Huson, M. G., et al., J. Polym. Sci., Polym. Lett. Ed., 1983, 22: 143

[45] Huson, M. G., McGill, W. J., Wiggert, R. D., Plast. Rubber Appl., 1985, 5: 319

[46] Honiball, D., Gill, W. G., J. Polym. Sci., Polym. Phys., 1988, 26: 1599

[47] Wang, Y. F., Wang, H. C., Rubber Chem. Technol., 1997, 70: 663

[48] Bauer, R. F., Crossland, A. H., Rubber Chem. Technol., 1991, 64: 585

[49] Tinker, A., Rubber Chem. Technol. 1995, 68: 461

[50] Loadman, M. J. R., Tinker, A., Rubber Chem. Technol. 62 (1989) 234

[51] Brown, P. S., Tinker, A., DKT '94, Stuttgart 1994

[52] Brown, P. S., Cudby, P. E. F., Tinker, A., ACS Rubber Division, Nasville 1992

[53] Bruchdahl, R., Nielsen, L. E., J. Polym. Sci., 1955, 15:1

[54] Moris, M. C., Rubber Chem. Technol., 1967, 40: 497

[55] Bohn, L., Rubber Chem. Technol., 1968, 41: 495

[56] Corish, P. J., Rubber Chem. Technol., 1967, 40: 324

[57] Yoshimura, N., Fujimoto, K., Rubber Chem. Technol., 1969, 42: 1009

[58] Boyer, R. F., J. Polym. Sci., Part C, 1966, 14: 267

[59] Warfield, R. F., Hartman, B., Polymer, 1980, 21: 31

[60] Müller, F., Martin, H., Kolloid Z., 1960, 172: 97

[61] Watson, E. S., O'Neill, M. J., Justin, J., Bremer, N., Anorg. Chem. 1964, 36: 233, 1238

[62] Kremer, F., Schönhals, A. F., Eds., in "Dielectric Broadband Relaxation Spectroscopy", Springer Verlag (2003)

[63] Walters, M. H., Keyte, D. N., Rubber Chem. Technol., 1965, 38: 62

[64] Kruse, J. E., Rubber Chem. Technol., 1973, 46: 653

[65] Heydenreich, R. D., in "Fundamentals of Transmission Electron Microscopy", Interscience Publ., New York (1964)

[66] Glanert, A. M., in "Practical Methods in Electron Microscopy", Vol. 1–4, North Holland Publ. Co., Amsterdam (1972)

[67] Reimer, L., in "Elektronenmikroskopische Untersuchungs- und Präparationsmethoden", 2$^{nd}$ Ed., Springer Verlag, Berlin (1967)

[68] Hess, W. M., Swor, R. A., Vegvari, P. C., Kautsch. Gummi Kunstst., 1985, 38: 1114

[69] Kresge, E., J. Appl. Polym. Sci., Appl. Polym. Symp., 1984, 39: 37

[70] Hess, W. M., Chirico, V. E., Rubber Chem. Technol., 1977, 50: 301
[71] Reimer, L., Pfefferkorn, G., in "Rasterelektronenmikroskopie", Springer Verlag (1977)
[72] Hess, W. M., Rubber Chem. Technol., 1991, 64: 386
[73] Lohmar, J., Proc. Int. Rubber Conf., Stuttgart, Germany, (1985) pp. 91
[74] Trent, J. S., Scheinbein, J. I., Couchman, P. R., Macromolecules, 1983, 16: 589
[75] Röninger, Jr, F. H., Ind. Eng. Anal. Ed., 1979, 5: 251
[76] Bar, G., et al., Langmuir, 2000, 16, 13:5702
[77] Jeon, I. H., Kim, H., Kim. S. S. G., Rubber Chem. Technol., 2003, 76: 1
[78] Johnson, L., Rubber Chem. Technol., 2008, 81: 359
[79] Hashimoto, T., Izumitani, T., Polym. Prepr. Am. Chem. Soc., Div. Polym. Chem., 1985, 26: 66
[80] Carman, C. J., Baranwal, K. C., Rubber Chem. Technol., 1975, 48: 705
[81] Nishi, T., Rubber Chem. Technol., 1978, 51
[82] Cheng, H. N., J. Appl. Polym. Sci., 1988, 35: 1639
[83] Roland, C. M., Rubber Chem. Technol., 1989, 62: 456
[84] Ban, L. L., Doyle, M. J., Smith, G. R., Rubber Chem. Technol., 1986, 59: 176
[85] Sircar, A. K., Lamond, T. G., Rubber Chem. Technol., 1975, 48: 301
[86] Cotton, G., Murphy, L. C., Kautsch. Gummi Kunstst., 1988, 41: 54
[87] Klüppel, M., Schaper, J., Schuster, R. H., Rubber Fall Colloquium '96 (1996)
[88] Rocha, T. L. C. A., Jacobi, M. M., Samios, J. F., Schuster, R. H., Kautsch. Gummi Kunstst. 2004, 57: 377
[89] Komorovski, R. A., Rubber Chem. Technol., 1983, 56: 959
[90] Kinsey, R. A., Rubber Chem. Technol., 1990, 63: 407
[91] Westler, D. D., Rubber Chem. Technol., 1990, 63: 40
[92] Tomlin, D. W., Roland, C. M., Macromolecules, 1992, 25: 2994
[93] Krishnamoori, R., Graessley, R., Fetters, L., Macromolecules, 1995, 28: 1252
[94] Alegria, A., et al., Macromolecules, 1995, 28: 8819
[95] Kader, M. A., Bohmick, A. K., Rubber Chem. Technol., 2000, 73: 889
[96] Matsuo, M., Japan Plastics, 1968, 2: 6
[97] Nagode, G. B., Roland, C. M., Polymer, 1991, 32: 505
[98] Kuo, C. M., Clarson, S. J., Polymer, 2000, 41: 5993
[99] Lohse, D. J., et al., Rubber Chem. Technol., 1999, 72: 569
[100] Reichart, G. C., et al., Macromolecules, 1998, 31: 7886
[101] Bartenev, G. M., Kongarov, G. S., Rubber Chem. Technol., 1963, 36: 668
[102] Dikland, H. G., Van Duin, V., Rubber Chem. Technol., 2003, 76: 495
[103] Ellul, M. D., Gent, A., J. Polym. Sci. Polym. Phys. Ed., 1984, 23: 1953
[104] Schuster, R. H., Angew. Makromol. Chemie, 1992, 202/203: 159
[105] Coran, A. I., Donnet, J.-B., Rubber Chem. Technol., 1992, 65: 1016
[106] Hess, W. M., Swor, R. A., Micek, E. J., Rubber Chem. Technol., 1984, 57: 959
[107] Waddel, Lecture at the DIK Hanover, 1998
[108] Callan, J. E., Topcik, B., Ford, R. F., Rubber World, 1965, 151: 60
[109] Hess, W. M., Marsh, P. E., Eckert, F. G., ACS Rubber Division Meeting, Miami (1965)
[110] Schaper, J., PhD Thesis University of Hanover (1997)
[111] Klueppel, M., Schaper, J., Schuster, R. H., Rubber Chem. Technol., 1999, 72: 91
[112] Alshuth, Th., Schuster, R. H., Krämer, S., Kautsch. Gummi Kunstst., 1994, 47: 695
[113] Rodgers, J. E., Waddell, W. H., Rubber World, 1999, 299: 24

[114] Ziegler, J., Schuster, R. H., Kautsch. Gummi Kunstst., 2003, 56: 293

[115] Tsou, A. H., Waddell, W. M., Kautsch. Gummi Kunstst., 2002, 55: 382

[116] Marsh, P. A., Voet, A., Price, L. M., Mullens, T. J., Rubber Chem. Technol., 1968, 41:344

[117] Hess, W. M., Chirico, V. E., Rubber Chem. Technol., 1977, 50: 301

[118] Lu, J.-G., Zhou, G.-Y., Wu, X.-L., Int. Rubber Conf. Kyoto, Japan, 1985

[119] Massie, J. M., Hirst, R. C., Halasa, A. F., Rubber Chem. Technol., 1993, 66. 276

[120] Cotton, G., Murphy, L. G., Rubber Chem. Technol., 1988, 61: 609

[121] Le, H. H., Illisch, S., Kasaliwal, G. R., Radusch, H.-J., Rubber Chem. Technol., 2008, 81: 767

[122] Schuster, R. H., Meier, J., Klüppel, M., Kautsch. Gummi Kunstst., 2000, 53: 663

[123] Cotton, G. R., Murphy, L. G., Proc. Int. Rubber Conf. Harrogate (England) 1988

[124] Cotton, G., Murphy, L. M., Kautsch. Gummi Kunstst., 1988, 41: 54

[125] Wadell, W. H., Napier, R. C., Poulter, R. P., Kautsch. Gummi Kunstst., 2001, 54: 1

[126] Zhang, Y., et al., Macromolecules, 2001, 34: 7056

[127] Maiti, S., De, S. K., Bhowmick, A. K., Rubber Chem. Technol., 1992, 65: 293

[128] Mamhoran, P., Naskar, K., J. Appl. Polym. Sci., 2016, 133: 43531

[129] Le, H. H., Illisch, S., Radusch, H.-J., Polymer Comp., 2010, 31: 1701

[130] Somotsonpap, N., Wimolmola, E., Markpin, T., J. Appl. Polym. Sci., 2007, 104: 3396

[131] Bokobza, L., Rahmani, M., Belin, C., Brunnel, J.-L., El Bouina, N.-E., J. Polym. Sci., Part B, Polymer Physics, 2008, 46: 1939

[132] Ahmadi, M., Shojaei, A., Polymer Intern., 2005, 64: 1627

[133] Le, H. H., et al, eXPRESS, 2017, 11: 230

[134] Le, H. H., et al., Composite Sci. Technol., 2014, 90: 1809

[135] Varghese, S., Gatos, K. G., Apostolov, A. A., Karger-Kocsis, J., J. Appl. Polym. Sci., 2004, 42: 543

[136] Arroyo, M., Lopez-Manchado, M. A., Valentin, J. L., Carretero, J., Comp. Sci. Technol., 2007, 67: 1330

[137] Ashok, N., Balachandran, M., Lawrence, F., J. Appl. Polym. Sci., 2017, 134: 45195

[138] Ali, Z., Le, H. H., Illisch, S., Thurn-Albrecht, Th., Radusch, H.-J., Polymer, 2010, 51: 4580

[139] Report of Currenta (2012)

[140] Hess, W. M., Scott, C. E., Callan, J. E., Rubber Chem. Technol., 44 (1971) 814

[141] Hess, W. M., Vegvari, P. C., Swor, R. A., Rubber Chem. Technol., 58, 350 (1985)

[142] Schaper, J., Stäber, R., Schuster, R. H., Presentation at the DKG, Fulda (1996)

[143] Callan, J. E., Rubber Chem. Technol., 1971, 44: 814

[145] Wunde, M., Klüppel, M., Rubber Chem. Technol., 2016, 89: 588

[146] Le, H. H., Oswald, K., Illisch, S., Pham, T., Stöckelhuber, K. W., Heinrich, G., Radusch, H.-J., Macromol. Mater. Eng., 2012, 297: 464

[147] Le, H. H., Illisch, S., Heidenreich, D., Wutzler, A., Radusch, H.-J., Polymer Comp., 2010, 31: 1701

[148] Lyu, L., et al., J. Appl. Polym. Sci., 2021, 138: 51443

[149] Kneseng, P., Sae-oui, P., Rattanson, N., Polym. Testing, 2013, 32: 731

[150] Gerspacher, M., Workshop on Reinforcement, DIK (2006)

[151] Marsh, P. A., Mullens, T. J., Price, L. D., Rubber Chem. Technol., 1970, 43:400

[152] Hess, W. M., Scott, C. E., Callan, J. E., Rubber Chem. Technol., 1967, 40:371

[153] Scott, C. E., Callan, J. E., Hess, W. M., J. Rubber Res. Inst. Malaysia, 1969, 22:242

[154] Hess, W. M., Herd, Ch. R., Vegvari, P. C., Rubber Chem. Technol., 1993, 66: 329

[155] Peterseim, V., PhD Thesis University of Hanover (1995)

[156] Klüppel, M., Schuster, R. H., Schaper, J., ACS Rubber Division Meeting, Anaheim (1997)

# 7 Rubber-Filler and Filler-Filler Interactions

Filled rubber compounds can be understood as two-phase systems of rigid particles surrounded by an elastic rubber network which can be crosslinked. The strength of the composite depends upon the rubber-filler interaction and the filler-filler interaction that take place at the nano level. The more pronounced the rubber-filler interaction, the higher the dispersion and the distribution of the filler in the elastomer. The strength of this interaction affects the properties of the filled vulcanizates and leads to a significant strengthening of the elastomer. Both interactions are responsible for the quality of the elastomer. The consequences of the polymer-filler interaction and the filler-filler interaction are the resistance of the elastomer to extreme conditions, which can guarantee safety and comfort in the mobile world.

Each of the four types of filler – carbon black, silica, layered silicates and carbon nanotubes – has a specific interaction with the rubber matrix and leads to different degrees of adhesion with the molecules of the rubber chains. The polymer-filler interface can be considered as an attractive wall that has adhered rubber chains on the surface of the aggregates and the chains have changed their conformational entropy when they are adsorbed by adding small trains of rubber segments that are adsorbed on the filler surface. Between the trains, the rubber chain can form large loops that provide a kind of elasticity and are entangled with the rubber chains in the matrix. The proximal layer is sensitive to the interactions with the filler surface but is also entangled with other chains from the matrix. These rubber-filler interactions shape the rheological behavior and design of the physical properties, especially the ultimate properties that determine the life time of the elastomer. In other words, reinforcement results from an improved polymer-filler interaction and a reduced filler-filler interaction that are responsible for a good filler dispersion, which may be important for certain applications.

In such a system, the presence of the large undispersed pellet fragments should be reduced as much as possible. The undispersed filler fragments and large agglomerates can contribute under dynamic excitations or strong uniaxial elongations to initiate cracks that propagate through the specimen and reduce the life time of the elastomers. If the content of such filler fragments is high, the ultimate properties appear relatively soon.

The interest in rubber-filler interaction has an application-oriented role and is of high scientific interest. The most important questions that arise scientifically very often concern:

- the strength of the rubber-filler interface
- the volume ratio in the filled system

To date, a number of answers have been given to these questions, which are briefly discussed in this chapter. The rubber-filler interactions are based on the physical phenomenon of adsorption of rubber chains on the surface of the filler. Irrespective of the chemical nature of the adsorbates and the types of adsorbents, adsorption is an exothermic process that is favored at lower temperatures. It is of thermodynamic importance and is an important part of what we mean by reinforcement and decreases at higher temperature. The adsorption depends on the activity of the rubber chains, which find a favorable contact point on the filler surface, so that this process takes place with an energy exchange and the chain entropy changes significantly on the filler surface. Although, there is much evidence for the existence of rubber-filler interface, considerable disagreement exists with regard to the thickness of the interface which is related to the early appearance of the ultimate properties. However, this is important in that the mechanical properties are stamped by the type and the volume fraction of the interfacial layer between the filler and the polymer. It became clear that, in rubber-filler systems with high affinity, the amount of the adsorbed chains increases and the thickness of the interface developed can reach the level of the rubber chain gyration diameter. But in addition, there is an important filler-filler interaction, which plays a decisive role, especially in dynamic properties in which the storage modulus $G'_0$ is observed when the strain amplitude increases. Only when this interaction is appreciated the adhesion and energy dissipation in periodic, repetitive processes be described. This filler-filler interaction is the basis for heat development during repeated dynamic excitation and can lead to premature appearance of ultimate cracks.

## 7.1 Rubber-Filler Interactions

The attraction between the rubber chains and the filler surface depends on the chemical composition of the rubber, the molecular weight of the rubber and the nature of the surface of the filler particles. The polarizability of the rubber chains can be improved by the presence double bonds and even more by the groups that can establish specific interaction to the filler surface. This way, the solubility parameter, δ, of a rubber is a useful predictor for the intensity of rubber-filler interactions, the higher the solubility parameter of the rubber, the better the physical adhesion of the chains on the filler surface and the more chains that adhere to the filler surface. The surface of the filler must be pre-determined for interaction with the rubber through certain energy sites that are able to form physical interactions with the rubber chains. The group of "semi-reinforcing" furnace grades, with primary particle sizes in the range of 50–100 nm is considered to be less active than the "reinforcing" furnace blacks with primary particle sizes in the range of 15–40 nm [1]. With carbon black, the $sp^3$-hybridized carbon atoms favor this interaction with the unsaturated double bonds or the presence of styrene or acrylonitrile groups of the rubber.

Through in-process functionalization of the filler surface, a significant change in the surface energies is achieved without significantly changing the specific surface and structure, which significantly increases the rubber-filler interaction. One such improvement is the formation of Ecorax® carbon blacks and the decoration with silica nanoparticles on the surface of carbon-silica dual-phase fillers (CSPDF).

The filler-rubber interaction applies to precipitated silica that interfaces preferentially with the polar segments of the rubber, i. e. acrylonitrile groups. The hydrophilic surface of the precipitated silica is not suitable for the adsorption of non-polar rubbers. Therefore, it requires organo-silanes to reduce the polarity of the silanol groups and create a useful interaction with S-SBR, cis-BR and NR. However, the phase connection must be ensured by using bifunctional silanes, which are able to establish a chemical bond between the filler and the rubber matrix during vulcanization, increasing the resistance to a filler detachment and increasing abrasion resistance [2, 3].

The physical attachment of polymer chains on the filler surface starts with the wetting process during the filler incorporation in the mixing cycle and continues in the dispersion stage and the distribution stage. The final step consists in a rearrangement of the polymer chains to find the high energy adsorption sites on the filler surface that increase the rubber-filler interaction. Therefore, the conditions of the mixing process are essential for the establishing of more rubber-filler interactions associated with improved filler dispersion. There are reasons for stating the performance of filled elastomers has its roots in the properties of the rubber mixing and the downstream process.

One of the problems is the difficulty of dissecting vulcanized compounds in a way to understand the processes that are created during the incorporation stage of the mixing process. The difficulties arise from the question about the nature of rubber-filler bonding, which has been a controversial subject. The interfacial region between the filler and the rubber matrix creates a separate phase, called the interface with its own characteristics that depends on the immobilization of the rubber chains, the morphology of the interface and the physical properties of the less mobile chains. The strength of this interface determines the resistance of the elastomer. It is important to characterize the reinforcement of rubber by the colloidal fillers. The interfacial region is defined as the volume in which the properties deviate from those of the bulk matrix or the filler. In the case of smaller filler particles, it is easy to consider the interparticle distance S between the filler particles:

$$S = d[(\pi/6\Phi)^{1/3} - 1] \quad (7.1)$$

where d is the filler diameter and Φ is the volume fraction of uniformly sized particles on a lattice

This equation shows that, by decreasing the diameter to 10–20 nm, the distance is similar to the radius of gyration of the rubber, and this illustrates that the interfacial distance is in the order of magnitude of the polymer radius or less. The smaller the diameter of the filler, the greater the interactions formed with the surrounding rubber chains and thus the increase in physical properties of the vulcanizate.

### 7.1.1 Adsorption of Polymers in Solution

The advantage of investigating the rubber-filler interaction via adsorption of the rubber from solution is primarily that the rubber chains are not damaged by shear stress. The polymers approach the filler surface by diffusion through the solvent and attaching to the energy sites of the filler surface. First the polymers of lower molecular weight attach to the surface and subsequently they are replaced by those of higher molecular weight, which will be stable in this state. The adsorption of chains occurs if the adsorption enthalpy, $\Delta H_{ads}$, is larger than the sum of the changes in conformational entropy $\Delta S_{conf}$ and the solvation enthalpy, $\Delta H_{solv}$, which results from the polymer solvent interaction. The rather complex polymer adsorption onto solid surfaces can be understood if the polarity of both the polymer and the filler is considered. Given similar polarity of the polymer and the filler surface, adsorption is promoted by physical attractive forces. In the case of a great polarity mismatch (i. e. polarophobic chains and polarophilic fillers) the attachment of the chains to the filler surface is rather modest. This reduces the enthalpy solvation, $\Delta H_{solv}$ which counteracts the enthalpy adsorption $\Delta H_{ads}$, and thus the adsorption process is favored. In order to attach polymer chains from a solution to a solid surface, the adsorption energy must be greater than the solvation energy of the solvent.

According to theoretical concepts, the polymer chains are attached to the filler surface by small sequences of "trains" that are adsorbed on the filler surface. It is assumed that the trains are subjected to dynamic adsorption and desorption, with the statistically constant portion of segments attached over time. Between the trains there are loops that have a reduced entropy and are somewhat immobilized, but local movements are allowed. At the end of the chain, there are free ends that can move more freely. A schematic diagram of the adsorbed rubber chain on the filler is shown in Figure 7.1.

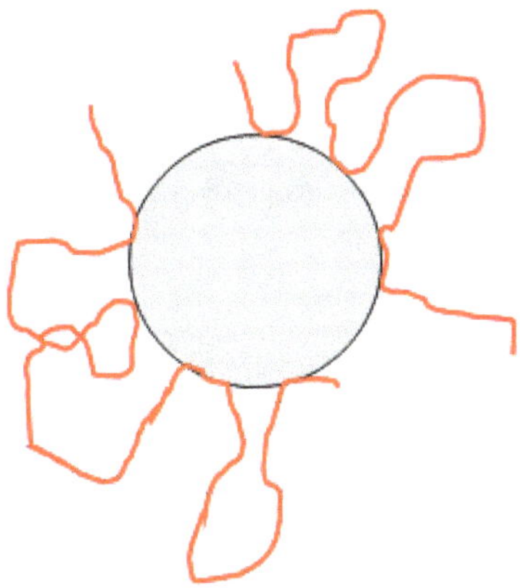

**Figure 7.1**
Schematic representation of an adsorbed rubber chains on a filler surface

The change in entropy in the loops between the short segment trains that are adsorbed on the filler surface can be seen. The loops are wide enough to resemble the statistical tangle and will be connected to the chains from the matrix via entanglements. In this way, an elastic connection to the rubber matrix can be established, although the interface does not have a glassy nature, as proposed in some theoretical studies [9, 10]. The thermodynamic reasons are unlikely that the interface assumes a glassy state, because the interaction to the solid surface is not as strong to confer a glassy layer on the interface. Thus, rate of adsorption, i. e. the mass of polymer adsorbed by a mass unit of filler, increases when the solvent quality decreases. It has been recognized that there are some rules governing the adsorption of polymers on carbon black:

- The shorter chains are adsorbed faster, but after a time they are replaced by the longer chains [4].
- The higher the molar mass of a rubber, the higher is the adsorption rate [4].
- The adsorption of a rubber depends on the specific surface area of the carbon black [5].

The relationship between the adsorption rate of polyisobutylene and the specific surface area of the carbon black is shown in Figure 7.2 [4].

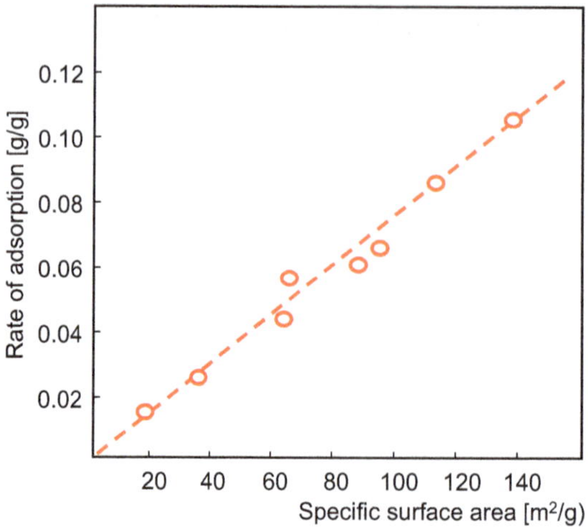


**Figure 7.2**
Correlation of the adsorption rate of polyisobutylene with specific surface area of carbon black [4]

Lastly, temperature also play an important role, whereby lower temperatures are preferred for maximum adsorption. The experiments show a high degree of reliability when the adsorption properties of different fillers are compared in a specific rubber-solvent system [5]. Characteristic of the adsorption of high molecular chains on filler is that only a part of the adsorbed material is taken up again by the solvent during desorption. Once polymer chains are attached physically on the energy sites of the filler surface, they cannot spontaneously desorb from it. Desorption of the rubber chains would require that all physical bonds of the chain with the solid substrate established in "trains" are released at the same time, which is quite improbable. Thus, desorption requires not only supplementary solvent, but also external energy, i.e. heat. The adsorption is governed by the side of the polymer and the filler and certainly the temperature at which the process will take place. It has been shown that the adsorption rate increases with the molecular weight of the rubber. Short-chain rubber is adsorbed faster than long-chain rubber, but after some time, the short-chain rubbers are displaced from the surface by long-chains. Carbon black with high specific surface area and activity takes more time to reach adsorption equilibrium. High structure carbon black adsorbs more rubber than low structure carbon black [6, 7].

Adsorption experiments have proved that the main contributors to the process in solution are the chain length (or molecular weight) and the chemical nature of the chain segments (polarizability). As shown in Figure 7.3, the adsorption of polymer chains from solution follows a strict sequence: always the longest chains present in the solution are attached first [8].

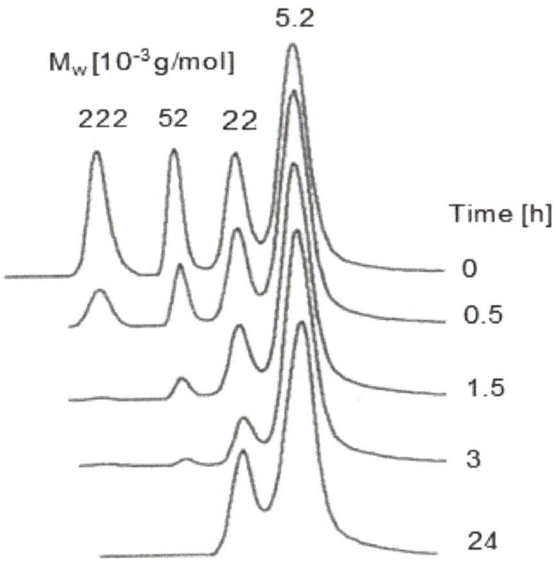


**Figure 7.3**
Adsorption of different polyisoprenes on carbon black N330

This result is in accordance with theoretical considerations that the number of contacts of a polymer chain with the filler surface increases in proportion to the square root of the chain length (or degree of polymerization) and thus, the longer chains have a better chance to become immobilized on the surface [9]:

$$\nu_{Polym} \sim P^{1/2} \tag{7.2}$$

where $\nu_{Polym}$ is the average number of contacts of the polymer chain with the surface and P is the degree of polymerization

The influence of the chemical nature and the polarity of the rubber has been demonstrated by "competitive" adsorption experiments [7, 10]. These are intended to highlight the chemical structure of a polymer on adsorption on the surface of a filler. This is particularly important if the sizes of the polymers are different. As shown in Figure 7.4 in a competitive adsorption experiment, a high molecular weight polymer of low polarity (i. e. polyisoprene) is placed together with a low molecular weight polymer of high polarity (i. e. polystyrene) in a suspension containing ground carbon black pellets [10].

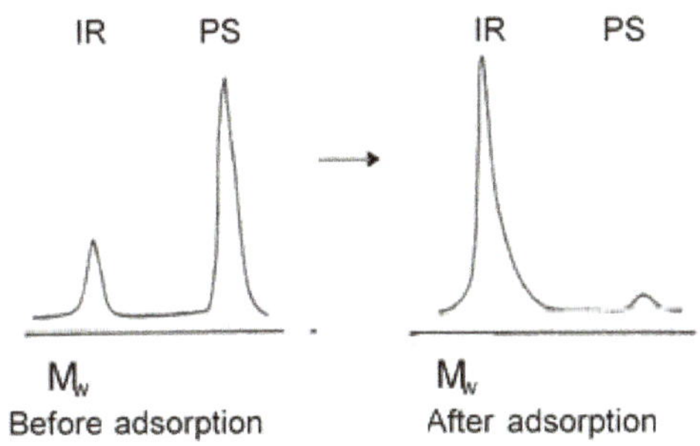


**Figure 7.4**
Competitive adsorption of polyisoprene and polystyrene chains on carbon black N330

As a result, the polymer of higher polarity but significantly smaller chain length is preferentially adsorbed on the carbon black surface, whereas the polymer with the longer chains but less good interactions with the filler surface remain in the solution and can be analytically determined. This result clearly shows that the influence of the interaction of a chain segment is at least equivalent to if not greater than that of entire chain length. The preferential adsorption is attributable to the more efficient interactions of the phenyl groups of the styrene units with high surface energy sites present on the carbon black surface. The existence of these interactions has been proven by FT-IR spectroscopy. A shift of the C-H out-of-plane vibration band (at 699 $cm^{-1}$) in the presence of filler indicates the preferential interaction [10]. In the case of silica filled NBR compounds, a characteristic band (2243 $cm^{-1}$) has been identified using FT-IR spectroscopy which lies next to the band of the nitrile group and indicates the physical interaction between the polymer and the filler. A characteristic IR band indicating the interaction is displayed for the CN group and the silanol groups of the precipitated silica. This method can also be used successfully for practically all physisorption.

The results of selective adsorption of non-polar polymers of similar molecular weight but different chemical nature on carbon black N330 are presented in Table 7.1 [11]. Thermodynamic parameters are responsible for the efficiency of the formation of attached polymer chains on the surface. The higher the solubility parameter, δ, of the polymer, the more polymer can be adsorbed on the filler surface. It can be seen that the partition of the polymer in the adsorbed layer is controlled by the δ parameter the polymers. The larger the δ parameter difference, the greater the amount of the more polar polymer that is attached to the filler surface. The apparently striking result that IR/1.2-BR with only a negligible difference in δ parameter of the polymers shows an unexpected partition towards IR is attributable to the shorter chain length of 1,2-BR.[1)] The intensively discussed question about the type of bonding of the rubber chains has been clearly elucidated for both carbon black and precipitated silica. For carbon black, it has been clearly proved that the attachment of the polymer on the filler surface is a purely physical phenomenon (physisorption).

**Table 7.1** Partition Ratio in Competitive Adsorption Experiments on CB N 330 [43]

| Polymer System | $\delta_1$–$\delta_2$ | Partition Ratio |
|---|---|---|
| PS/IR | 3.31 | 99/1 |
| PS/cis-1,4-BR | 2.72 | 95/5 |
| cis-1,4-BR/1,2-BR | 0.65 | 75/25 |
| cis-1,4-BR/IR | 0.59 | 60/40 |
| IR/1,2-BR | 0.06 | 70/30 |

[1] The 1,2 insertion of butadiene units lead implicitly to a smaller chain length of the polymer.

By heating suspensions of carbon black with polymer chains adsorbed on the surface, the amount of the adsorbed polymer decreases gradually with the temperature, the type of the filler, the polymer constitution and the solvating power of the solvent. Such results are reliable arguments to assess the nature of the bonds between rubber and carbon black as being exclusively physical interactions. The case where chains are linked by coupling agents on the filler surface is considered without any doubt to be chemisorption. The important case within the rubber-filler interaction is precipitated silica linked with bifunctional silanes on unsaturated rubbers. Depending on the type of silanol groups, the interaction with silanes is detected by $^{1}$H-NMR spectroscopy. In recent years, a wide range of sophisticated techniques have been developed for the study of the nature of surface and interfaces which extend and complement the classical techniques, such as surface tension, spectroscopic ellipsometry [12] and surface sensitive FTIR [13]. These methods are rarely used in the investigation of rubber compounds and are beyond the scope of the present chapter.

## 7.1.2 Equilibrium Swelling of Filled Vulcanizates

The consequence of the restricted mobility of the adsorbed rubber phase on the surface of the implemented filler can be observed in an equilibrium swelling test of the filled vulcanizate in a suitable solvent. Numerous reports are devoted to describing the magnitude of the restriction of chain mobility in the filled vulcanizates. This cheap but reliable method provides quantitative information about the level of the interface existing between the filler and the rubber compound for any kind of the filler. Since the rubber chains in the interfacial layer are adsorbed onto the filler surface, their interaction capacity with a solvent is restricted in that the chains are tightly bound to the filler and cannot be solvated by the solvent molecules to the same extend as the free moving chains in the rubber matrix. The degree to which polymer chains from the interface can take up solvent molecules depends on the bonding strength of the polymer on the filler surface and the volume fraction of the dispersed filler. Whereas the crosslinked rubber matrix swells in inverse proportion to the crosslink density, the same solvent has to be used. The polymer chains at the interface swell to a lesser degree, because of the reduced mobility of the chains adhering to the adsorbed layer. The influence of the rubber-filler interface can be qualitatively estimated from the reduction in the equilibrium state if the swelling is carried out in the same solvent and at the same temperature.

The equilibrium swelling of a filled network decreases with the filler loading and the specific surface area of the filler. It can be seen that, with the same filler loading, the increase in the specific surface area of the carbon black leads to a significant reduction in the swelling value [5–8]. Similar experiences are made with the other three fillers. Compared with the swelling values of NBR (34 ACN) of 275%, the addition of

3 vol.% MMT reduces the swelling to 22.5% [14]. Experiments with carbon nanotubes show a greater influence on the swelling of NBR (34% ACN) than with HNBR (34% ACN), since the unsaturated rubber has a better interaction with the $sp^2$-hybridized carbon nanotubes via the double bonds [15]. Experimentally, less swelling was observed in carbon black-filled elastomers, as a result of the additive volume fraction of the filler and the crosslinked rubber [16].

From a thermodynamic point of view, the reduced swelling is the result of the intensive rubber-filler interactions, which were initially interpreted as an increase in crosslinking density due to additional chemical bonds between rubber and the filler [17]. A method that predicts a linear decrease in the normalized degree of swelling as a function of the filler loading ratio has been developed [18]. The normalized degree of swelling is expressed as a linear function of the filler concentration, known as a "Kraus plot":

$$\left(q_f/q_0\right) = c[1 - m_{P-F}\Phi/(1-\Phi)] \qquad (7.3)$$

where $q_f$ is the degree of swelling of the rubber phase in the filled system, $q_0$ is the degree of swelling of the rubber phase in the unfilled system, $m_{P\text{-}F}$ is the rubber-filler interaction parameter and $\Phi$ is the volume fraction of the filler

If the same solvent is used for the equilibrium swelling of a filled vulcanizate, the normalized degree of swelling, $q_f/q_0$, is a measure of the proportion of rubber-filler interaction in the system and at the same time an indicator of the existing rubber-filler interaction. The equation can be used to evaluate the polymer-filler interaction in filled rubber network. It has been shown that the negative polymer-filler interaction parameter $m_{P\text{-}F}$ increases with the specific surface area of the filler, the surface activity and the solubility parameter of the rubber. The slope of the equation, $m_{P\text{-}F}$, is a relative measure of the contribution of the amounts of chains with hindered mobility in the swelling process. The higher the specific surface area, the more polymer chains are adsorbed and reduce the swelling of the vulcanizate. It has been shown that carbon black exhibits a linear polymer-filler interaction, which is determined by the specific surface area and the structure of the aggregates [6]. Furthermore, when comparing the slope of carbon blacks, carbon nanotubes and wet-mixed layered silicates, always those fillers with the highest specific surface area and the highest aspect ratio will show a stronger decrease in negative decay and demonstrate a higher contribution of the immobilized rubber chains at constant filler concentration, which is shown in Figure 7.5. The carbon nanotubes and exfoliated layered silicates provide a much stronger decrease in the equilibrium swelling at the same filler concentration [19].

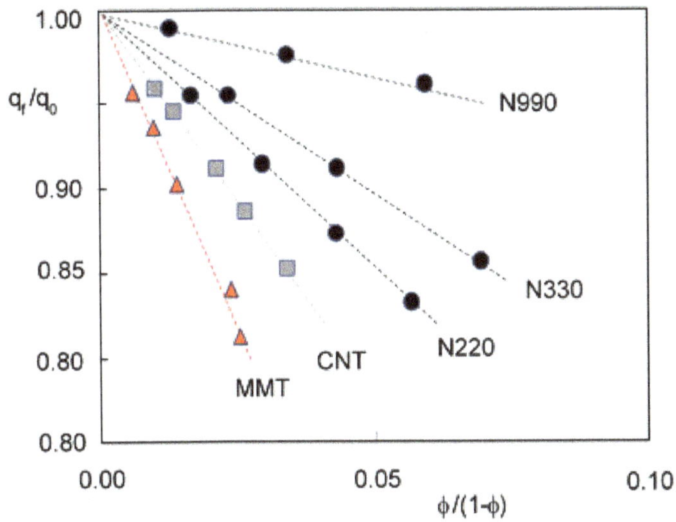


**Figure 7.5** Kraus plot for carbon blacks, carbon nanotubes and wet-mixed MMT in NBR swollen with MEK [19]

When the effect of carbon nanotubes and exfoliated layered silicates is compared with that of carbon blacks, it becomes clear that the higher rubber-filler contact surface realized in the compounds with the two nano-fillers of high specific surface area and the large aspect ratio is responsible for the pronounced reduction in swelling. If a small proportion (5 phr) of an exfoliated MMT is added to an NBR/CB N550, the equilibrium swelling is reduced significantly to lower values, because the high aspect ratio of the individual platelets has a great influence on the interaction with the rubber chains. NBR/HD silica shows a similar effect that can be shown when only 2.25 vol.% exfoliated MMT is added. Due to the high specific surface area and the high aspect ratio, the decrease of the equilibrium swelling has an unusually high effect. This addition of the nanofiller lowers the equilibrium swelling by twice what was achieved with HD silica [19].

Nevertheless, by evaluating the contribution of the immobilized rubber chains from equilibrium swelling, the influence of the δ parameter between solvent and polymer are found to be important for the degree of swelling and this explains why the swelling test with poor solvents causes a greater decrease in the normalized degree of swelling $q_f/q_0$. The investigation of the swelling of filled vulcanizates has the advantage of quantitatively characterizing the degree of interaction between rubber and filler with a certain solvent and providing a better understanding the influence of the filler on the formation of the immobilized rubber chains. However, the result obtained relates to the experimental conditions for which the rubber-filler interaction parameter, $m_{PF}$, applies in the selected swelling agent and the corresponding temperature.

A direct consequence of the reduced chain mobility in the interface should be an increased glass transition temperature. Attempts to confirm this have little success if filled compounds are examined by means of calorimetric measurements or dynamic-mechanical investigations. In contrast, studies on polymer thin films have shown a higher glass transition temperature than the one in the bulk polymer [20]. That a higher glass transition for the interface is not detected in calorimetric investigations of filled rubber compounds can be attributed to the resolution of the methods that cannot detect the small volume fraction of the interface. It has been argued that, depending on the strength of the polymer-filler interaction, physical adsorption of rubber molecules takes place on the filler surface. This interaction leads to an effective immobilization of the chain segments. Depending on the intensity of the interaction and the distance from the filler surface, the mobility of the polymer segments near the interface is lower than in the matrix. This portion of rubber has been identified as an extremely immobilized layer adsorbed on the filler surface that surrounds the filler particle as an outer shell [21]. The amount of rubber in this immobilized state obviously depends on the strength of the polymer-filler interaction. Therefore, the filler surface activity or the surface energy may be considered as a determining factor influencing the effective volume of the filler.

This simple equilibrium swelling of the filled vulcanizates can describe the immobilized rubber layer at the interface qualitatively well, so that an impression is gained of the actual rubber-filler interaction, which means that the interaction between the selected rubber and the filler are relatively well described.

### 7.1.3 Rubber Shell

An important question that rubber science asks concerns the dimensions of the interface between the rubber chains and the filler surface. Such information cannot be derived from the equilibrium swelling of the filled vulcanizates. The formation of an immobilized rubber layer surrounding the filler aggregates in compounds or in vulcanizates has been proved by several methods, but the size of the interface is still hidden. In work on the freezing point depression of benzene in swollen networks, it was concluded that the crosslink density is higher in the vicinity to carbon black. It has been suggested that this occurs at a depth of 35 nm [22]. For SBR, the rubber shell was estimated to be 35–50 nm, while for NR it is 25–30 nm [23]. The interparticle distance of carbon black has been estimated from AFM measurements. It was demonstrated that the distance increases with the particle size, as shown in Figure 7.2 [24].

**Table 7.2** Interparticle Distance of Carbon Black Particles in NR Vulcanizates [24]

| Carbon Black | Primary Particle [nm] | Interparticle Distance [nm] |
|---|---|---|
| N110 | 17±7 | 54±26 |
| N220 | 21±9 | 65±30 |
| N351 | 31±14 | 89±47 |
| N550 | 53±28 | 139±71 |
| N660 | 63±36 | 145±74 |
| N762 | 110±53 | 188±102 |
| N990 | 246±118 | 376±152 |

The measurements can only be used as a guide because the variation in interparticle distance is quite large. A more efficient technique to probe the spatial heterogeneity in filled rubbers with respect to chain mobility is $^{1}$H-NMR relaxation spectroscopy [21, 25–28]. It has been verified by spin-spin relaxation time $T_2$, measured by pulsed nuclear magnetic resonance techniques (Hahn spin-echo) that polymer chains which adhere to the filler surface show a shorter relaxation time, corresponding to a lower chain mobility than the chains in the bulk which are not affected by the interaction with the filler. Although the information is averaged over the sample volume, it results directly from an experimental test without any interference by other factors (i. e. solvent in the equilibrium swelling). Using $^{1}$H-NMR relaxation spectroscopy, the immobilized layer has been investigated in cis-BR and EPDM vulcanizates. Three distinct regions of the rubber were found, characterized by different mobility of the rubber chains:

- unbound rubber of high mobility
- bound rubber in an outer shell, with low immobilization
- tightly bound rubber on the filler surface, with strong immobilization

The results are shown in Figure 7.7 for the spin-spin relaxation measurement system in cis-BR with N110 (50 phr) [21].

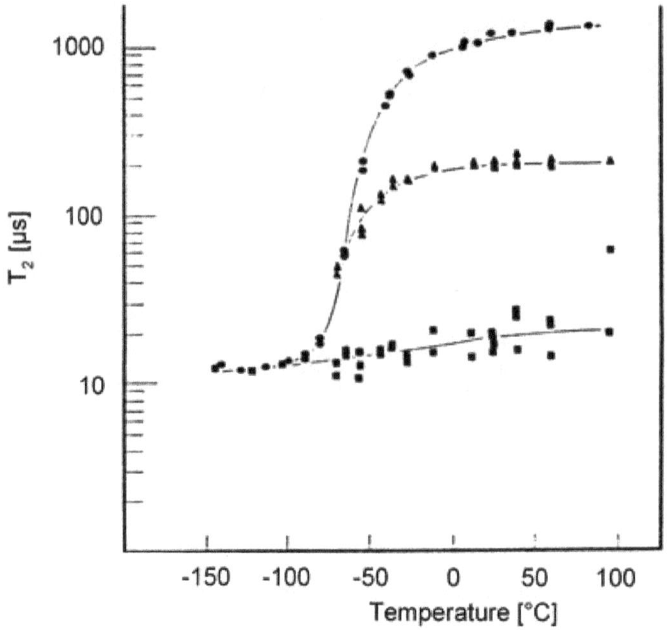


Figure 7.6 Temperature dependence of spin-spin relaxation $T_2$ for a) free BR chains, b) partially bound BR chains and c) strongly bound BR chains

From the results it can be seen that about 2–3% of the total polymer is severely restricted in the segment mobility while 10–20% move in a slightly more flexible but not entirely free state. Although, there is sufficient evidence to show the existence of immobilized rubbers chains on the filler surface, considerable disagreement exists with regards to the thickness of the layers. It is reported that the relaxation time decreases with the degree of macro-dispersion at constant filler volume fraction, which implies better micro-dispersion and, therefore, a larger rubber-filler contact surface [29]. Immobilization of EPDM chains on the surface of carbon black and network structure in the rubber matrix of filled EPDM rubbers have been studied in low-field proton NMR experiments. Advanced $^1$H-NMR relaxation experiments unambiguously show strong immobilization of EPDM chain fragments on the surface of carbon black. The thickness of the immobilized EPDM-carbon black interfacial layer is estimated to be larger than ≥ 0.6 nm [30]. The average number of monomer units per adsorption site is approximately nine to ten, which is interpreted by the authors as preferential chain adsorption on the micro-crystallites boundaries of carbon-black particles. The adsorbed chain fragments form physical adsorption junctions that restrict chain mobility in the rubbery matrix outside of the interface. Representing the normalized relaxation time ($T_{2,f}/T_{2,0}$) as a function of the filler volume, the greater negative slope indicates quantitatively the increasing effect of the rubber-filler interaction [19]. The decrease in normalized relaxation time ($T_{2,f}/T_{2,0}$) is shown in Figure 7.7.

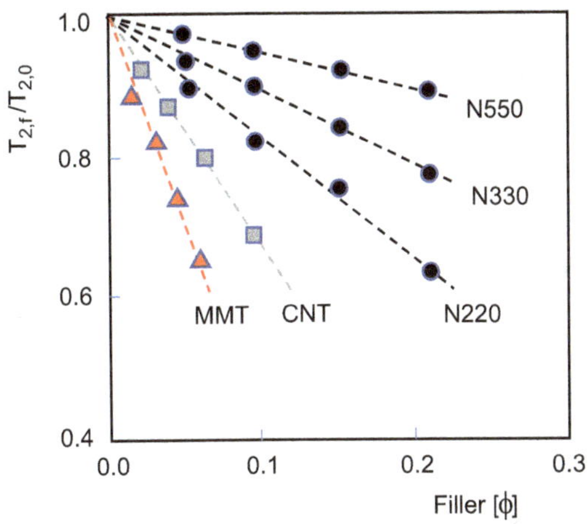


**Figure 7.7**
Representation of $T_{2,f}/T_{2,0}$ for different carbon blacks, carbon nanotubes and wet-compounded pristine MMT [19]

Not surprisingly, the negative slope of the normalized relaxation time $T_f/T_0$ increases with the specific surface area of the carbon blacks, indicating an increasing volume fraction of immobilized rubber in this system. In comparison to the influence of carbon black that results in a medium decrease in the normalized relaxation time, the carbon nanotubes and the wet-mixed exfoliated precipitated silica show a reduction in the mobility that is much higher. For carbon nanotubes and even more for exfoliated MMT, the mobility of the adsorbed rubber layer is a function of the higher specific surface area and the high aspect ratio which provides in this case a larger interfacial region. However, even with this $^1$H-NMR relaxation method, the dimension and extent of the interfacial region can only be derived from comparing the values and not determined directly.

Polyisoprene samples filled with silica generated *in situ* from tetraethoxy-silane by the sol-gel process have been investigated using $^1$H spin-lattice and spin-spin relaxation times that enable insights to be gained into the dynamic properties of both the polymer bulk and the rubber interface and a comparative estimate of the amount of the interface in samples prepared with different compositions and sol-gel reaction times [31]. Three fractions of different mobility have been distinguished by $^1$H $T_2$ measurements and ascribed to immobilized rubber in the interface, rubber bulk, and free chain ends. The immobilized rubber was found to be dependent on sample preparation, with the maximum arising in the sample showing the best dispersion of silica particles in the rubber matrix.

A breakthrough in describing the properties of the rubber-filler interface was achieved by using a special AFM "palpation" technique developed by K. Nakajima at the Institute of Technology in Tokyo [32]. AFM based nano-mechanics seems to be one of the most promising tools for accessing the rubber-filler interface while providing not only structural information but also mechanical property evaluation in the region between the filler and the rubber matrix. An AFM-based static modulus map was used to close in on the understanding of the filler reinforcement effects [33]. With the help of force-distance curves, the Young's modulus in carbon black filled NR compounds was measured at the nano-scale. The regions of high chain mobility in the rubber matrix correspond to a low stiffness or a low Young's modulus. In the rubbery region, a low Young's modulus (approx. 4 MPa) was recorded while in intermediate regions close to the filler particles, Young's modulus was considerably higher (approx. 15 MPa). These results lead to several findings about the state of the interface. The rubber shell does not consist of a glassy layer that surrounds the surface of the filler. Young's modulus increases exponentially from the rubber matrix to the filler surface and reaches the highest value there. On top of filler particles, the modulus was significantly higher than in the intermediate region. The existence of the different micro-mechanical moduli on the interface between the rubber matrix and the carbon black surface is shown schematically in Figure 7.8.

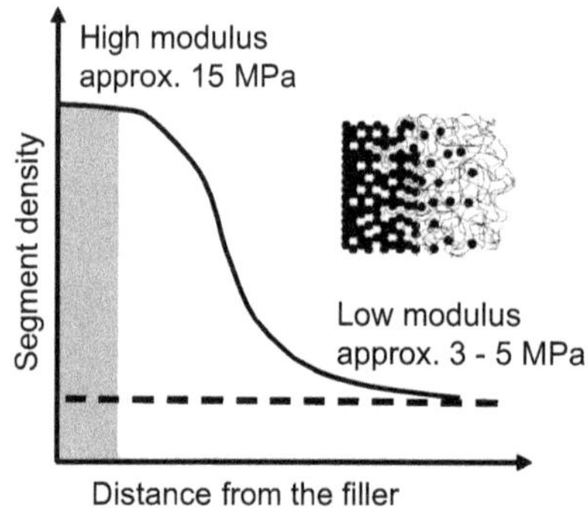


**Figure 7.8**
The elastic properties of the rubber shell on the surface of the filler

In this figure, the limit values of Young's modulus are plotted in the rubber matrix and on the wall of the filler. This proves that there is a very high modulus on the carbon black surface, but it is not as high as a glass-like state. The experimentally verified existence of the intermediate region of higher modulus but significantly lower modulus than that of a glassy state proves that the assumed glassy interface does not exist in filled rubber compounds. The nano-mechanical measurement of the stiffness of the interface illustrates the differences in the physical properties of the boundary layer. In recent finite element calculations, a non-glassy interface has been assumed in order to fit physical properties determined experimentally [34]. The nano-rheological AFM techniques have been expanded also to explain the behavior of carbon black filled vulcanizates in dynamic-mechanical tests [35]. The method has been applied to filled rubbers compounds in order to reveal the different mechanical behavior of the rubber matrix and the rubber-filler interface. The thickness of interfacial layers depends on the nature of the filler, on the filler surface and of course the temperature. With increasing temperature, the thickness of the interface decreases to smaller values, as expected from adsorption theory.

$^1$H-NMR measurements have shown that the thickness of carbon black N220 at room temperature is approx. 2.2 nm. At elevated temperatures, it falls below this value. This small expansion of the rubber-filler interphase results from the existence of the energy sites on the carbon black surface. Because of the amorphous carbon on the filler surface, interaction with the rubber chains is increased and a strong interface is formed. Due to the small interface, electrical conductivity in the vulcanizate filled with carbon black can be achieved via a tunnel effect in a filled system that consists of a fractal system. Even smaller are the rubber interfaces for carbon nanotubes filled rubbers. As carbon nanotubes do not possess stronger interactions than the $sp^2$-hybridized carbon atoms, they cannot adsorb the rubber chains in the same way as carbon blacks [36].

### 7.1.4 Bound Rubber

In a rubber compound, part of the rubber is firmly bound to the incorporated filler through adsorption, which leads to the formation of the bound rubber. The bound rubber is always seen as a sign of a good rubber-filler interaction and is described as a feature of good reinforcement. Since the beginning of scientific investigation of rubber compounds, the portion of bound rubber has been clearly defined as the rubber portion in an uncured filled compound which cannot be extracted by a good solvent, due to the adsorption of the rubber molecules onto the filler surface. The existence of bound rubber was described in 1925 by Twiss [37]. A comprehensive survey of investigations into bound rubber have been published [38–41]. The determined bound rubber value is a useful measure to describe the phenomenon related to the adsorbed rubber-filler interaction. Although, the subject is still a matter of controversy, there are three different arguments concerning the nature of bound rubber:

- the nature of the polymer-filler interaction
- the filler activity
- the rubber polarity

The adsorption of polymer on filler surfaces is a typical phenomenon that appears when the non-crosslinked rubber compound is partly dissolved to remove the free rubber chains. What remains is the bound rubber, a mixture of filler particles and the adhered rubber on its surface. When a filler is admixed to a rubber, strong interactions between both occur in such a manner that a good solvent for the polymer is not able to extract the whole amount of the rubber in a small period of time, leaving a highly swollen bound rubber. After drying, the extracted assembly consists of a certain amount of polymer that adheres to the filler substrate. The bound rubber represents the fraction of rubber (in percent) in the dry rubber-filler gel and is defined by the equation:

$$\text{Bound Rubber}\,(\%) = [(\text{gel} - \text{rubber}) - \text{filler}]\,/\text{rubber} \tag{7.4}$$

Bound rubber is considered to be a macroscopic result of rubber-filler interactions that occur at molecular level. A criterion to quantify the filler surface activity in a certain rubber is the determination of the bound rubber. Without any doubt, the phenomenon of bound rubber is related to the mechanism of filler reinforcement [41, 43]. As far as the filler is concerned, bound rubber is affected not only by the physicochemical characteristics of the filler surface, but also by the shape of the aggregates and the way to mix the rubber compound. With regard to the polymer, both the chemical structure of the molecules (saturated and non-saturated, polar and non-polar) and their molecular weight and the molecular weight distribution influence the level of bound rubber content. However, the bound rubber also shows the dependence of the mixing conditions on the component and the temperature and time during storage of the unvulcanized compound. The proportionality of the bound rubber fraction with the specific surface area of the filler has contributed to an understanding of bound rubber as an indicator for the reinforcing capacity of filler carbon black. Experience has shown that bound rubber is not a specific phenomenon for certain rubber-filler interactions; their amount will depend on the following quantities:

- mixing of the compound
- storage conditions
- type and amount of the filler
- chemical structure of the rubber
- molecular weight and molecular weight distribution of the rubber
- temperature of mixing
- temperature of storage
- time after mixing

However, good comparison possibilities within a test series are possible if the selected test procedure to determine the bound rubber are retained. Bearing in mind that bound rubber is reduced with increase in temperature, it has to be questioned if the extraction time is adequate. In order to understand this, one should consider that the extractable rubber obviously contains chains that are not attached to the filler but are entangled with the tightly bond chains in the rubber-filler interface, so that they will resist the dissolution in the extractant. The chains attached firmly to the filler surface (rubber-filler interface) anchor the entangled chains into an assembly that is not released in short extraction times, which are recommended by test methods. The duration of the extraction is 16 h which is commonly too short. These strongly immobilized chains are not glassy but are closely packed and show a low chain mobility. This layer has a thickness up 3 nm and is likely to behave in a flow field as a supplementary hydrodynamic part of an aggregate. It accounts for 2–4% of the total rubber. The part that consists of rubber chains loosely bond to the interface by entanglements shows a higher mobility than the chain in the interface, but less mobility than the chains in the bulk. This part of bound rubber may have a thickness of 36 nm and accounts for 10–35% of rubber in the compound, depending on the affinity between the rubber and the filler.

### 7.1.4.1 Determination of Bound Rubber

Bound rubber is a parameter which is simple to measure, but there are many factors which may influence the test result. The amount of bound rubber is certainly affected by the conditions under which the bound rubber determination is made (i. e. type of solvent, temperature and extraction time). With regard to rubbers, both the chemical nature (saturated, unsaturated, non-polar, polar) and their macrostructure (molecular weight and molecular weight distribution) can influence the level of the experimentally found bound rubber. However, the bound rubber determination performed on the same compounds in different laboratories show poor accuracy of the values, which is attributable to variations in the experimental conditions.

The choice of solvent is not as critical for bound rubber measurements, provided it is a good solvent for the polymer. The solvents used for NR, SBR and BR compounds are toluene, xylene and benzene. A study of the effect of different solvents on bound rubber in butyl rubber compounds has shown that benzene, cyclohexane, ethylbenzene and xylene gave similar results [44]. The quantity of bound rubber increases with the temperature up to 80–90 °C [45]. In some cases, nitrogen gas has been used to eliminate the effects of oxidation, which can cause chain scission and lead to lower molecular weight of the rubber chains and lower bound rubber. The time of extraction in bound rubber determination is chosen to ensure that equilibrium conditions are achieved. Since the dissolved polymer is in equilibrium with the adsorbed polymer on the filler surface, it is usual to replace the rubber solution with fresh solvents one or more times during the extraction process. The longer the extraction time, the smaller is the bound rubber found in any solvent. It has been published in reports that several days are required for complete extraction of non-polar rubbers, such as NR and SBR. This shows that the bound rubber is a consequence of the physisorption of the rubber on the filler and does not depend on a chemical reaction of the rubber on the filler.

### 7.1.4.2 Influence of the Mixing Process

The formation of bound rubber shows a dependence on the mixing conditions. It is conceivable that bound rubber formation starts during the incorporation stage of the filler as a result of “wetting” of the filler surface. The process continues during the dispersion and the distribution stage. An investigation of bound rubber shows that, in the case of rubbers that are barely mechanically degraded by chain scissions during the admixing of filler, the phenomenon of the bound rubber is equivalent to adsorption from the melt and determines the adsorptive properties of the filler, the polymer and the mixing parameters. The original idea that bound rubber is a gel of filler particles held together by the rubber chains in the form of a three-dimensional network is formed [43, 46, 47]. There is a connection between the mixing of a compound and the proportion of bound rubber. The higher the temperature at which NR and carbon black N330 are mixed on the two-roll mill, the more the bound rubber content decreases, since the influence of the chain modification on the molar mass of the NR is observed [46]. The ratios change during mixing in the internal mixer, since the value of the bound rubber increases significantly with increase in mixing time and the mixing temperature during this process [42]. Figure 7.9 illustrates the development of bound rubber as a function of mixing time [42].

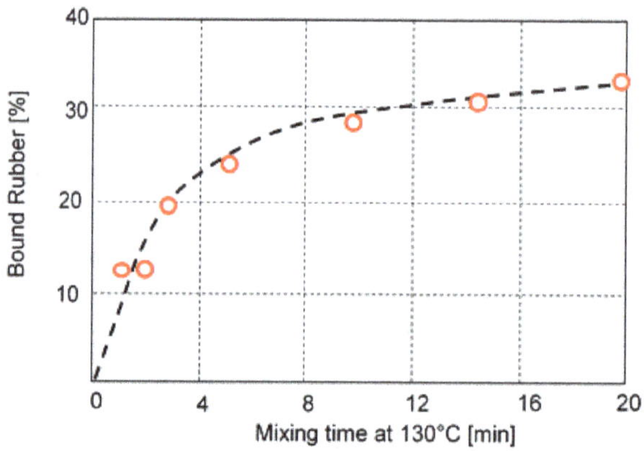


**Figure 7.9** Development of bound rubber during mixing of cis-BR/carbon black N220 (40 phr) at 130 °C

From the qualitative relationship between the torque-time curves and the bound rubber proportion determined at constant time, an interaction between the type of filler and the reinforcement potential has been derived, which increases in the order of the specific surface area and the structure of the aggregates [47, 48]. The adsorption of the rubber chain during mixing involves not only chains that are directly adsorbed, but a large number of chains entangled with them are involved additionally to form the bound rubber. With the uncured mixture, an increase occurs in the bound rubber content of NR compounds, due to the mixing. The more the compounds, are mixed, the higher the content of bound rubber found [49]. In the case of NR/carbon black compounds, the bound rubber formed was explained also by a chemical reaction between the macroradicals of the NR, which form during mixing, and the carbon black [46].

With the storage time of the rubber compound, the value of the bound rubber increases; this shows that, in the absence of elongational and shear stress, the consistency of the rubber-filler gel increases by self-diffusion of rubber chains that have found more efficient energy sites on the surface and are increasing with the contact time of the rubber chains on the surface of the filler. This is observed during the storage time of saturated rubber. After mixing, these adsorb during storage, a process which was previously explained by a chemical adsorption mechanism. It has been postulated that free radicals of carbon black aggregates formed during mixing may be responsible for higher bound rubber when a carbon black with a high oil adsorption number is used in butyl rubber [50]. This mechanism is doubtful for the reason that ozonized carbon blacks do not show any influence on the breakage or the reinforcing effects of butyl rubber [51]. In addition, in the case of butyl rubber, the bound rubber increases significantly with the mixing conditions, especially with the residence time in the internal mixer.

In an experiment, it was shown that the amount of extracted rubber increased with the stirring rate. This is another argument that bound rubber results from physical adsorption. It is not likely that chemically linked chains break down upon stirring. The quantitative desorption of the rubber from the filler surface is an unambiguous argument for the physical nature of the rubber-filler interaction in carbon black filled compounds. However, chemical reactions can occur in carbon black mixes when rubber chains break during mechanical mixing and the resulting macro-radicals are grafted onto the carbon black surface [46].

The rubber-filler compound produced is stored for a while at a certain temperature after the mixture has been produced and the value of the bound rubber can also increase significantly during this time, with the result that a certain change in the viscosity of the compounds occurs. This is illustrated for the increase in bound rubber in compounds that contain different rubbers and different carbon blacks [52].

**Table 7.3** Influence of Storage Temperature on the Bound Rubber in Filled Polymers [52]

| | EPDM/CB 326 | EPDM/N550 | cis-BR/N326 | cis-BR/N550 |
|---|---|---|---|---|
| Initial value | 17.6 | 17.5 | 10.1 | 24.0 |
| 16 h at 80 °C | 25.5 | 22.6 | 31.5 | 47.7 |

The high amount of bound rubber in cis-BR/N550 is due to the better dispersion of this type of the filler and the degree of the filler dispersion. Mostly, the storage time and the storage temperature have a positive influence on the value of bound rubber that is formed. The longer the storage time and the higher the storage temperature, the greater is the amount of rubber chains in the formed bound rubber. This indicates a reorganization of the adsorbed rubber chains in the bound rubber. Experts recognized a connection between the processing behavior and the structure of the aggregates of carbon black at an early stage. A highly structured carbon black always leads to an increase in bound rubber value that increases during compounding and storage in cis-BR or SBR. After a long mixing time, the extrusion shrinkage that occurred was partly explained by the degradation of the rubber and a greater amount of bound rubber [53].

### 7.1.4.3 Influence of Carbon Black

As expected, the amount of bound rubber increases with parameters of carbon black which can be measured independently: filler loading, the specific surface area and the structure of the aggregates. Results of this filler parameter have promoted the idea that the value of bound rubber is a measure of reinforcement. Total bound rubber content increases with the nitrogen-specific surface area. This result is not surprising because, at the same loading, the carbon blacks offer an increased surface area that can be adsorbed by the rubber chains. This is shown in Figure 7.10 for carbon black (50 phr) in E-SBR [5].

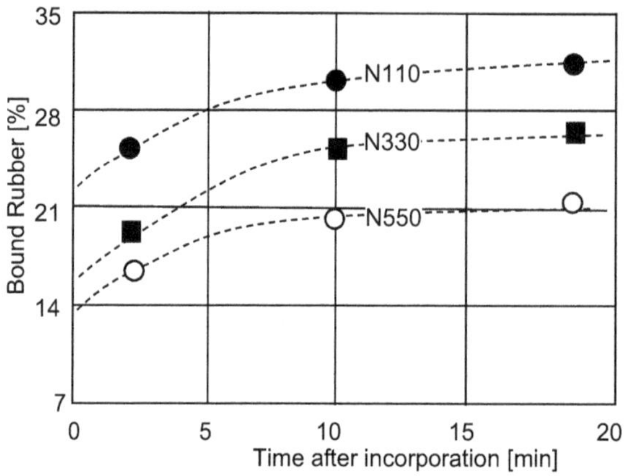


**Figure 7.10** Development of bound rubber during the mixing time for E-SBR and carbon black (50 phr) with increase in specific surface area [5]

The proportionality that occurs between bound rubber content and the specific surface area has often been used as an example in the specialist literature to explain the improvement of physical properties of the elastomer that is known as reinforcement, as a direct influence of the bound rubber on the filler activity. The bound rubber is seen as a direct consequence of the rubber-filler interaction and is associated with the filler activity and the polarizability of the polymer.

For carbon blacks with low surface areas, a coherent bound rubber can be obtained only above a given loading. The reason that bound rubber is formed above the critical loading is the larger interface distance between the particles of the semi-active blacks. For active carbon black aggregates, the rubber chains can be attached to more than one aggregate, which is called multi-attachment. This multi-attachment of the rubber chains to the aggregates at higher loadings can reduce the effectiveness of the filler surface, since a single attachment only would render the molecule unextracted.

Among carbon blacks of similar surface area, high structure blacks generally have a higher bound rubber content because there is more void volume in the aggregates and less ordered graphite micro-crystallites. A comparison of bound rubber measured in compounds filled with carbon blacks of similar specific surface area but different carbon black structure is given in Figure 7.11 [54]. The higher the degree of branching of the aggregate, the more of the rubber chains that can be stored in the void volume, which significantly increases the value of the bound rubber. Therefore, the large range of bound rubber values does not result from different surface activity but from the amount of polymer that exists "occluded" in the voids of agglomerates.

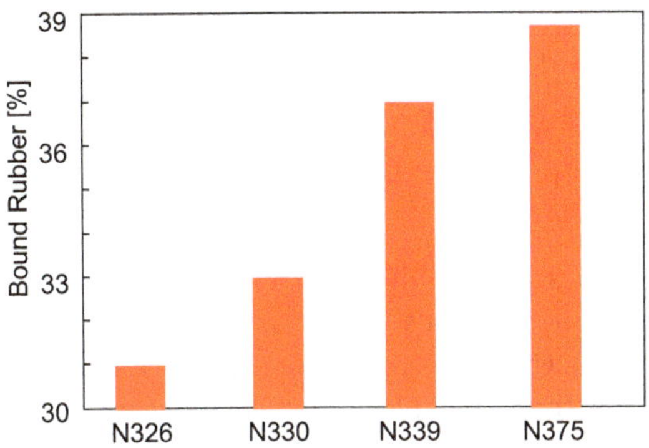


**Figure 7.11** Influence of carbon black structure on the value of bound rubber [54]

Interestingly, high structure blacks show greater bound rubber values than low structure blacks, a fact which is attributable to the increased hydrodynamic volume at the aggregates [55]. The proportion of bound rubber is also positively influenced by the structure of the aggregates. In an experiment, oil-extended cis-BR and SBR 1712 were used to measure the increase in DBP absorption of the carbon black. From the small carbon black structures that allow the formation of small bound rubber, the value increases and reaches a bound rubber value with high DBP absorption, namely 16% for cis-BR and 12.5% for SBR172 [56]. The relationship between the structure of the carbon black and the measured bound rubber is shown in Table 7.3 [54].

**Table 7.4** Correlation of the Bound Rubber with the Aggregate Structure of Carbon Blacks (50 phr) [54]

| Carbon Black | DPB Number ($cm^3/g$) | cis-BR (%) | E-SBR (%) |
|---|---|---|---|
| N220 - LS | 1.24 | 13.7 | 16.8 |
| N220 - HS | 1.68 | 35.5 | 34.3 |
| N330 - LS | 1.15 | 21.5 | 24.4 |
| N330 - HS | 1.70 | 35.9 | 32.0 |

The reason for this behavior is not to be found directly in the rubber-filler interaction, but rather is a result of the branched structure of the aggregates. It has been shown that the bound rubber increases not only with the specific surface area but also with the amount of occluded rubber in the carbon black aggregates. The bound rubber demonstrates a decrease with increasing aggregate diameter and is related to the filler particle morphology. When the diameter increases, the volume of the bound rubber decreases. Therefore, the parallel development of bound rubber and the corresponding tensile strength in the vulcanized system is a result of a mutual connection; this is not surprising because they show the same origins in the rubber-filler interaction. In the case of graphitized carbon blacks, there is a very low tendency to form bound rubber [56, 57]. This lower adsorption effect of the graphitized carbon black is due to the high proportion of graphitic surfaces on the primary particle size of the carbon black, which are unable to bind a large number of the rubber chains on the aggregate surface.

It has been shown repeatedly that, when a bound rubber experiment is carried out at an elevated temperature (> 120 °C), the bound rubber level decreases significantly. By increasing the extraction temperature, the equilibrium of the dynamic adsorption-desorption of the chains is progressively shifted towards desorption. Due to the higher temperature, the adsorption of the rubber chains decreases and they are released from the filler surface. The bound rubber content will continuously decrease as the extraction temperature increases until it is no longer detectable, as shown in Figure 7.12 [46].

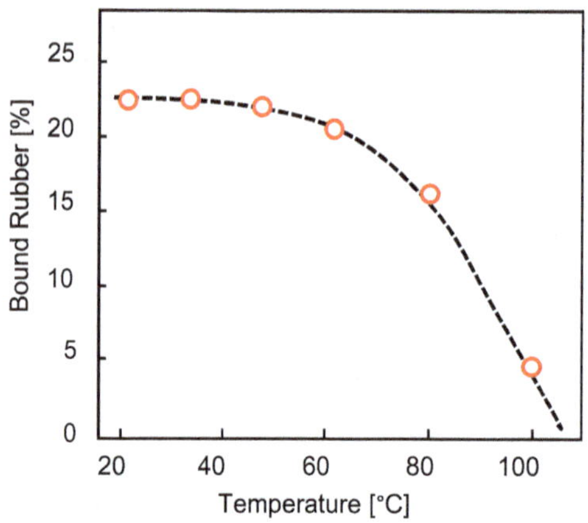


**Figure 7.12**
Reduction in bound rubber with increase in temperature

Since it is unlikely that chemically bonded polymer on the surface can break down over this temperature range, the result demonstrates the physical bonding of the rubber onto the filler surface. As far as the rubber-filler contact surface is concerned, it is interesting to note that, when bound rubber is normalized by the specific surface area of the filler, it has been observed that the bound rubber per unit surface decreases with the increase in specific area of the filler (or the surface activity); this indicates at first glance that low surface area blacks adsorb more polymer chains per surface unit [58]. This striking result, presented in Figure 7.13, does not turn the concepts presented above about rubber-filler interaction on their heads.

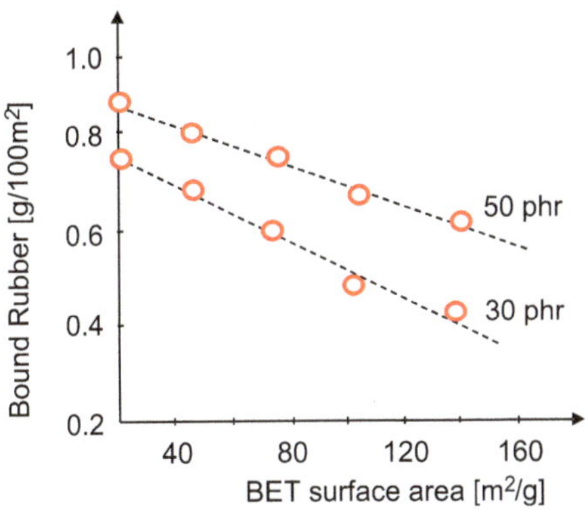


**Figure 7.13**
Bound rubber per square meter of the specific surface area decreases with the specific surface area [54]

This observation demonstrates in fact that the "wetting" realized during mixing does not cover the entire surface of the active fillers. The larger the specific surface area, the more difficult dispersion of the filler during mixing becomes. In other words, the efficiency of overcoming the cohesive forces acting within the agglomerates of fine fillers of small primary particles decreases and the filler-filler interaction increases. Consequently, the polymer-filler contact surface that is effectively "wetted" by the rubber is far less than ideal.

The basic trend is similar at all loadings, but it can be seen that the decrease in bound rubber per surface unit is more pronounced at higher filler loadings. One interpretation for the lower bound rubber per surface unit could be the more pronounced chain scission during mixing, due to greater shear and temperatures in the compound. Another reason may be the multiple segment adsorption that increases with filler loading [59]. Another explanation may be the multiple-segment adsorption which leads to a reduced effectiveness of the filler surface to form bound rubber. Multiple chain adsorption of this kind can take place between neighboring aggregates at higher loadings.

The process of graphitization has little effect on the particle morphology, so that the large effects on bond rubber of graphitized carbon black must be explained in terms of surface deactivation. A number of edges of the graphite crystallites and defects on the surface disappear through the graphitization process and cannot adsorb the rubber chains [60]. Oxidation of carbon black N330 by a small part of ozone leads to a reduction in bound rubber in NR compounds that can be understood as a reduction in the dispersive interaction of the surface and a smaller interaction with isoprene chains [61]. Plasma treatment of N220 increases the filler activity and therefore the amount of bound rubber increases after atmospheric plasma treatment [62]. The increase in activity provides strong evidence that the surface defects on the carbon black surface are responsible for the improved interaction with non-polar rubber.

After adequate mixing, the bound rubber content of wet-compounded carbon black is higher than for the dry-mixed silica of the same composition, as shown by Cabot elastomer composites (CECs) [63]. This is because the bound rubber is formed before the oil has been added to the mixed compound. It has been found that CSDPF compounds having the same composition as compounds with carbon black or silanized silica always have more bound rubber [64]. The greater rubber-filler interaction is triggered by the reduction in the graphitic crystallites on the CSPDF surface. Due to the presence of small silica nanoparticles, the graphite-like structure cannot form and this leads to a higher activity of the modified filler. This significantly reduces the filler-filler interaction and improves the adsorption of the polymer to the surface of the CSDPF. This opens up a number of possibilities to improve the rubber-carbon black interaction, which leads to an increase in rheological behavior. But this also significantly enhances the properties of the vulcanizates filled with carbon black.

### 7.1.4.4 Influence of Precipitated Silica

Depending on the solvent used, similarly high bound-rubber values are measured for silica compounds as for carbon black compounds. In the case of precipitated silica, the silanol groups play a crucial role in the hydrogen bonds and increase the measured bound rubber. The existence of acidic silanol groups on the silica surface that can react with ammonia reduces the bound rubber significantly. However, the measurement of bound rubber in the presence of ammonia reduces the bound rubber in silica compounds from 21% to 1.9%. The results are shown in Table 7.4 [65].

**Table 7.5** Determination of Bound Rubber Treated with Ammonia [61]

| Type of Filler | Ammonia Treatment | Bound Rubber (%) |
|---|---|---|
| Carbon Black N330 | – | 32.4 |
| Carbon Black N330 | Yes | 30.5 |
| Silica VN2 | – | 21.1 |
| Silica VN2 | Yes | 1.9 |
| Silica VN2 – TESPT | – | 41.5 |
| Silica VN2 – TESPT | Yes | 40.0 |

The reduction in bound rubber is due to neutralization of the acidic silanol groups by the basic ammonia. However, silanization with TESPT decreases in the presence of the silanol groups and, therefore, the bound rubber is practically unchanged, because the surface is less polar and shows no reduction in silanol groups during the ammonia treatment. For the same filler loading of the precipitated silica, it can be shown that the bound rubber values in the polar NBR are about four times higher than for the non-polar cis-BR [66]. If the HD silica is treated with monofunctional and bifunctional silanes, the proportion of bound rubber increases in both the NBR and the cis-BR, as shown in Table 7.5.

**Table 7.6** Bound Rubber of Silanized Silica in NBR and cis-BR Compounds [66]

| Silica Modification | NBR | cis-BR |
|---|---|---|
| HD silica | 261 | 61 |
| HD silica + Si203 | 483 | 538 |
| HD silica + Si208 | 667 | 740 |
| HD silica + Si69 | 622 | 641 |

The silanization leads to a significant increase in cis-BR for all three silanes, indicating much better filler-polymer interaction. The use of Si208 and Si69 leads to a greater amount of bound rubber than in polar NBR. This depends on the extent of hydrophobization of the silica surface by the organo-silane used. Thus, the bound rubber content is reduced in S-SBR, BR and NR. Interaction of the polymer chains in the vicinity of the filler interface can be expected to influence the local segmental dynamics [67]. The effect of wet-mixing with silanized silica in solution polymers is different. Compared with dry-mixing that show a bound rubber of 75.3%, wet-mixed silica showed reduced bound rubber of 67.5%. This is related to the increasing interaction between the silica and the rubber as a result of the coupling reaction with the silanes. The lower bound rubber in the wet-mixing may not be caused by the lower extent of the coupling reaction. It may be related to less gelation, because the silane coupling agent is adsorbed on the silica surface and the concentration of the free coupling agent in the polymer is lower.

#### 7.1.4.5 Carbon Nanotubes

Bound rubber measurements have indicated a weaker rubber-filler interaction for NR/multi-walled carbon nanotubes composites composed of NR/carbon black, due to weaker adhesion between the outer layers and the rubber chains [68].

It has been observed that multi-walled carbon nanotubes treated with acid disperse well in the rubber matrix and the bound rubber and the interfacial phase improved in NR composites [69]. Therefore, multi-walled carbon nanotubes lead to a much higher bound rubber content, when compared with carbon black N550 in EPDM, a fact is explained by the higher specific area and the high aspect ratio of the carbon nanotubes.

#### 7.1.4.6 Influence of the Polymer

Bound rubber measurements reveal a significant dependency on the molecular weight of the rubber as well as on the polarity of the rubber [65]. This is in agreement with the results observed for adsorption in solution, which show that the adsorption of polymer chains on the filler increases with the square root of the degree of polymerization. The highest bound rubber values are achieved with rubbers that have an increasing content of double bonds, contain styrene groups or acrylonitrile groups. If oil-extended SBR or cis-BR is present, there is a greater increase in the bound rubber in the first case, because the styrene-containing groups strongly favor adsorption from the melt. In addition to the length of the rubber chain, the proportion of interacting groups also plays a crucial role in the formation of the bound rubber. If a comparison is made between two SBR types of different concentrations and styrene groups, the SBR of higher styrene concentration shows almost double the bound rubber content [10]. However, actually, the big differences in the bound rubber values can be explained by the molecular weight of the rubber. Under the assumption that during mixing the mechanical breakdown of the rubber chains can be neglected, the content of bound rubber increases in proportion to the square root of the molecular weight of the rubber, as shown in Figure 7.14 [70].

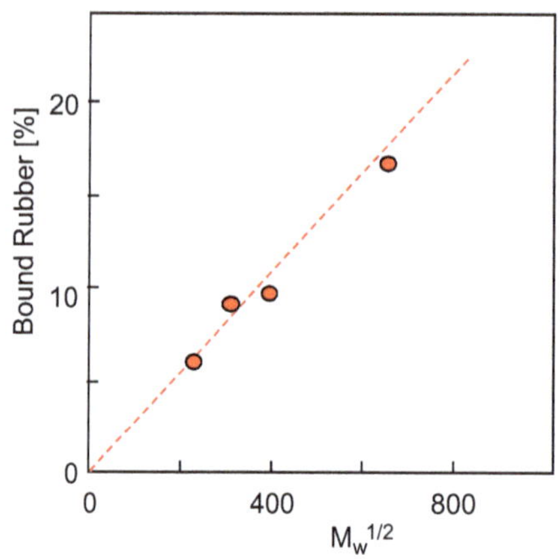


**Figure 7.14**
Increase in bound rubber with the square root of the molecular weight of the polymer in cis-BR/CB N220 [70]

In agreement with the theoretical expectations that long chains are preferentially adsorbed, it has been demonstrated that the average molecular weight of the extracted rubber is lower than that of the originally used rubber when the bound rubber increases [63]. In other words, the long rubber chains are preferably attached to the carbon black particles, which has been described as a model, provided that the rubber is only slightly degraded during the mixing process [71–73]. A similar influence is also observed when rubber chains are adsorbed onto the filler in a suspension. A distribution function of the molecular weight has been derived that can explain the difference in the formation of bound rubber:

$$W_f(M) = \left[1 - \left(AM^n / M^n w\, dM\right)\right] w(M) / (1 - A) \tag{7.5}$$

where A is the total amount of rubber adsorbed on the filler, M is the molar mass of a fraction and n is the molar mass dependence on the adsorption process

During the formation of bound rubber, the higher molecular weight fractions from the melt are selectively adsorbed onto the carbon black. It is clear that the synthetic rubber in the presence of carbon black shows only slight losses in the molecular weight distribution when mixed in the internal mixer. The influence of the polarity of the rubber on the abundance of bound rubber was one of the earliest observations made in this field. For carbon black filled compounds, it has been shown that the fraction of bound rubber increases with the unsaturation of the chain backbone. The content of styrene units in SBR has been identified to increase the bound rubber in carbon black filled compounds, while the content of ACN groups in NBR or HNBR produces similar effects in silica filled compounds [66].

During storage and mixing, the bound rubber value increases depending on the type of rubber, the type and amount of filler and, above all, the time required and the prevailing temperature. With saturated rubbers, such as EPDM, a small increase in bound rubber values to 3% can be observed. In the case of cis-BR compounds, under the same conditions, there is an increase to 8%, which in the case of NR can increase to 12–13% [74].

For SBR compounds filled with the same volume fraction of carbon black, precipitated silica and carbon-silicadual phase-filler (CSPDF), the rubber-filler interactions were investigated via the bound rubber content and the use of solid state $^1$H low field-NMR relaxation spectroscopy. The results indicated that the bound rubber content increased with the specific surface area of the filler. The silica filled compounds showed the highest bound rubber, while the carbon black filled compounds had the strongest rubber-filler interfacial interaction. The dual phase filler compound (CSDPF) presented the lowest Payne effect, characteristic of the best filler micro-dispersion [75].

Besides filler morphology and the polarity of the polymer, adhesion of the polymer on the filler surface depends on the sequence in which the compound ingredients are added to the mixer. For a NR CV50/CB330 (40 phr) compound prepared by the same procedure but containing different formulation ingredients, it has been shown that the adhesion of the polymer on the filler surface (bound rubber) depends on the sequence of admission [74]. The highest bound rubber was found in the compound containing only the carbon black, followed by the compound containing the cure activators (stearic acid and zinc oxide). Even lower bound rubber was registered when antioxidant (IPPD) was added to the compound. This effect was attributed to a radical scavenger effect of the antioxidant that inhibit the gelation of NR macroradicals in the compound. In addition, the more polar ingredients which have poorer compatibility with the less polar rubber, such as NR, will be driven by diffusion to the filler surface where the polarity is higher. This mechanism leads to blocked active sites on the filler surface and contribute to a reduction in bound rubber.

#### 7.1.4.7 Influence of the Rubber Ingredients

The sequence in which the ingredients are mixed is a critical parameter for the formation of bound rubber. The ingredients can affect the adsorption of the polymer on the carbon black surface. This is the case for curing agents, antioxidants and mineral oil. In NR CV50 compounds prepared by the same procedure but containing different ingredients, the highest bound rubber was found for the compound containing only 40 phr carbon black N330, followed in decreasing bound rubber content by carbon black and ZnO and stearic acid. The next one is the compound containing the antioxidant IPPD which is consumed in the mixing stage and reduces the content of bound rubber formation during the mixing procedure [37]. The amount of antioxidants reduces strongly during the mixing cycle.

The lowest bound rubber values are measured for compounds in which plasticizing mineral oil is admitted prior to or together with the carbon black. As has already been shown, the mineral oils have a similar chemical structure to the polymers and occupy part of the active sites of the filler. This blocks the interaction between rubber chains and the filler and leads to a reduction in the bound rubber value. The type of mineral oil also plays a role. Aromatic oils have a similar effect to the non-polar unsaturated polymers, because of the unsaturation of the molecules. A very important role is played by the mineral oil, which reduces the formation of the bound rubber because it can act as radical scavenger [74]. With regard to the addition of oil to S-SBR/BR (75/25)/carbon black N234 (75 phr) and mineral oil (25 phr), the sequence in which the oil is added to the compound has been shown to exert an influence. When the oil is added together with the carbon black the bound rubber was 40%. However, when the carbon black was added first and dispersed in the blend, and then the oil was added, the bound rubber increased to 47%.

### 7.1.5 Phase Adhesion

The important question of cracks forming between the rubber matrix and the filler surface during uniaxial expansion has been attributed to the rubber-filler interaction by the transmission electronic method. This method showed how the cracks propagate between the vulcanized rubber and the filler and how the cracks are deflected by the presence of the filler. This has also deepened our knowledge of the occurrence of cracks in the elastomer. The question of how a filled elastomer reacts when undergoing uniaxial expansion under pressure can be answered by a simple experiment in which the detachment of the rubber-filler interface is viewed as a function of uniaxial deformation. The tests described so far, which prove the existence of a rubber-filler interface of a certain thickness and consistency, say little about the resilience of the interface at high elongations. In order to assess the stress-strain behavior or the fatigue resistance of elastomers, it is important to examine the extent to which the interface can withstand higher stress or cyclic loads.

An interesting aspect is the separation of the polymer from the filler surface. Debonding between the rubber matrix and the filler surface is in fact prohibited by both the magnitude of the interfacial tension and the fact that adhesion of chains on the filler surface is an exothermic process. Thus, it seems evident that the formation of an interfacial layer on the filler surface is a spontaneous process, which strongly depends on the interaction potential of both components. The interaction potential determines the properties of the interface, in particular adhesion to the filler surface. Held by intermolecular attraction forces on the filler surface, the polymer molecules in the interfacial layer may have a restricted mobility but can slip with respect to the filler surface under external stress that deforms the compound. Desorbing from the filler surface and formation of voids can occur at large deformations when rubber-filler interactions reach their limit.

As shown above, phase adhesion is governed by the affinity of the rubber for the filler which in turn is a function of the surface activity of the filler and the interaction potential of the rubber. An electron microscopy method to visualize the rubber-filler adhesion on the nano-scale as a function of the applied stress on rubber compounds has been developed [76]. Microtome sections of filled vulcanizates were fixed on thin PMMA substrates and stretched progressively. The number of detachments and the formation of voids between the rubber matrix and the filler aggregates are recorded as a function of the applied stress. To quantify the easy of detachments the “adhesion index” has been proposed. By definition, this value corresponds to the stress values at which 20% of the filler particles are detached from the rubber phase. It has been shown that the adhesion index increases in proportion to the specific surface area of ASTM blacks, as shown in Figure 7.15.

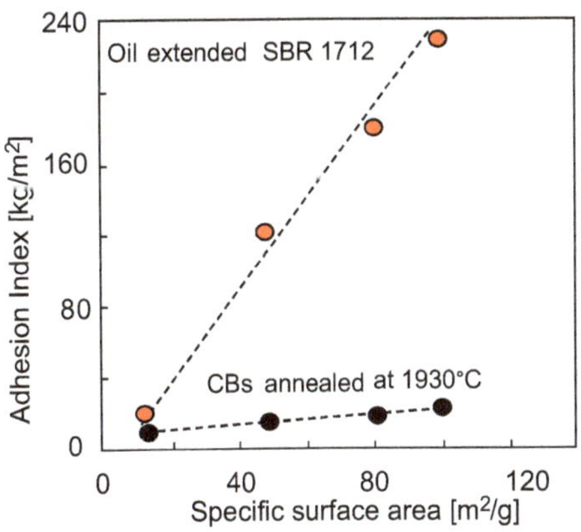


**Figure 7.15**
Increase in adhesion index (%) as a function of the specific surface of ASTM blacks and graphitized blacks [76]

As the primary particle size decreases, the surface energy on the surface of the carbon black increases exponentially, causing the rubber chains to be adsorbed more and more firmly on the surface of the filler. This leads to a decrease in the detachment of the rubber matrix from the filler surface. The activity of the carbon blacks has a decisive influence on the phase adhesion phenomenon. The occurrence of the phase separation is inversely proportional to the specific surface of furnace carbon black. The detachment of the phases can be predicted from the measurements.

In any case, compounds prepared from graphitized carbon blacks show very low adhesion indices that result from the low surface activity of the filler. It has been shown that the detachment is pronounced in compounds containing untreated silica. In addition, it has been observed that the adhesion index increases with the structure of the filler. For example, the adhesion index of carbon blacks in the N300 series increases in the sequence: N326 (120) < N330 (210) < N339 (350). The result is attributable to an influence of the more intensive anchoring of the rubber on branched aggregates. Keeping the carbon black at constant, the adhesion index increases in the sequence of the increasing δ parameter of the rubber.

For compounds that contain precipitated silica, detachment of the polymer from the filler surface occurs at relative low stress values. An important reason for this behavior is the strong interaction of the silanol groups, which determine a high filler-filler interaction. This is a direct consequence of the poor interaction of the polar silica surface with the hydrocarbon rubber. Compounds containing silanized silica that is efficiently coupled to the rubber network do not show detachment from the rubber matrix at the same extension [77]. The TEM micrographs shown in Figure 7.16 demonstrate the resistance of the coupled silicas to the external stress.

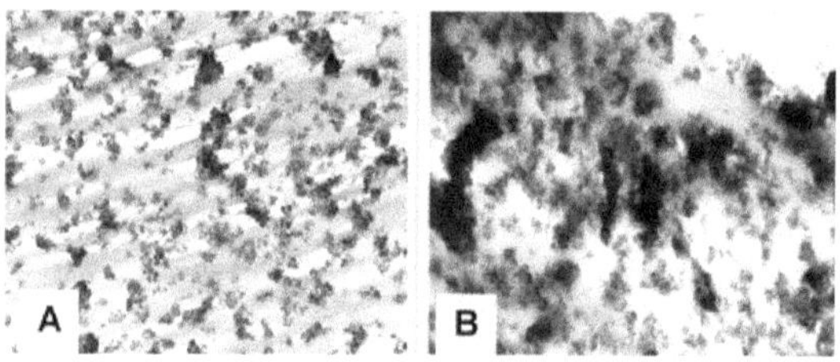


**Figure 7.16**
TEM micrographs show the phase adhesion in a silica compound and the lack of phase adhesion in silanized silica [77]

The improved rubber-filler interaction is documented by a strong increase in stress values in the stress-strain behavior. With the silanized HD silica, the stress values are better than for a carbon black vulcanizate. This makes it easy to explain why the silanization and coupling of precipitated silica increases the abrasion resistance. Even if this way to evaluate the rubber-filler interaction cannot be used routinely, the results demonstrate impressively the importance of the rubber-filler interaction for the mechanical durability of vulcanizates. It has been shown that silanization of silica and coupling with bifunctional silanes results in far less detachment at the same applied stress. The coupling leads to higher stress values and tensile strength.

Few studies on phase adhesion have been carried out with layered silicates and multi-walled carbon nanotubes. But the same physical conditions apply to the phase adhesion of high aspect ratio fillers and the rubber. The greater the rubber-filler interaction, the less the two phases in the vulcanizate will separate.

## 7.2 Filler-Filler Interaction

Filler-filler interaction is just as crucial for the rheological behavior and, above all, the physical properties of a vulcanizate as rubber-filler interaction. Above the percolation threshold, filler-filler interaction leads to filler networks with a fractal structure, which determines the properties of the vulcanizate. This is particularly evident in the dynamic-mechanical properties of the storage modulus and the loss modulus as well as in the fracture properties. The filler-filler interaction is crucial for the effect of the filler network. If a filler network is produced in a rubber compound in the internal mixer, the attraction of the filler particles leads to the formation of a filler-filler network, which particularly determines the dynamic properties of the elastomers. Above a certain filler concentration, a well-dispersed filler forms a three-dimensional filler network that interpenetrates the rubber network and is responsible for the fractal properties of the filler network.

The study of the filler-filler network commenced with the dynamic properties of carbon black filled elastomers. It was observed that the shear modulus, $G'(\gamma)$, decreases in a sigmoidal curve when the strain amplitude increased up to 10%. This is a development that all fillers show in a specific way, which is triggered by the filler-filler interaction. Above this strain amplitude, the filler-filler interaction is no longer important and vanishes to zero. The scientific community uses the "Payne effect" for dealing with the decrease in $G'$ as a function of strain amplitude. The topology of the filler networks has been carefully investigated, since the investigation started by Payne [78]. There have been many studies on the topic in the last six decades, which have significantly improved the understanding of the filler-filler interaction, but this has not yet been completely clarified.

What is certain is that the filler particles do not necessarily have to touch each other. Individual clusters are formed which then lead to a filler network. Due to the strong filler-filler interactions of the Van der Waals type, they are held together. The rheological and the physical properties are dominated by the filler network at small strain amplitudes. The macroscopic stress is no longer transmitted through the rubber matrix alone, but merely through the interactions of the filler particles and the filler clusters in the filler network. This is especially the case for low strain dynamic extensions of filled compounds where the filler-filler interactions account predominantly for the strength of the compound. The appearance of filler networks has been derived from viscoelastic measurements of rheological properties [79, 80], electrical conductivity [81] and dynamic-mechanical properties [82]. Additionally, the existence of filler networks has been proven by TEM images and mechanical tests. In the case of carbon black and carbon nanotubes, electrical conductivity investigations have largely led to a deeper understanding of filler networks, especially in the area of volume resistivity. The aggregates show a tendency to associate in chain-like clusters.

Overall, the filler network has no similarities with three-dimensional rubber networks, which are made up of crosslinked rubber chains. On a small length scale, the filler network is self-similar, but exhibits homogeneous behavior on a large length scale. Nevertheless, it is not clear how the filler network is formed and what type of filler aggregation takes place in the rubber matrix. Due to the overlapping action of the tightly bound polymer chains close to the contact area of filler particles, the mechanical stability of filler-filler bonds increases with increase in the amount of bound rubber. The stabilizing interface can be considered to act as a net encompassing the filler aggregates, implying high flexibility on the part of the filler network.

Theoretical approaches to the formation of filler networks consider either an electrical percolation process [83] or a kinetic cluster aggregation process [84]. While the first concept is based on geometrical considerations, the second assumes a kinetic process that follows from the motion of the filler particles in the rubber matrix during mixing. Fluctuating particles stick together because of their high cohesive forces, $F_c$, and form a continuous assembly of solid particles that interpenetrates the rubber matrix.

All theoretical models that postulate a filler network to describe a reinforcing mechanism assume that the filler-filler interaction leads to mechanically unstable assemblies that can be destroyed by an external force. Direct proof for the correctness of this assumption is that the storage modulus, $G'(\gamma)$, of filled compounds and vulcanizates decreases with the dynamic strain amplitude and the loss modulus, $G''(\gamma)$, is a maximum at a given strain amplitude [85]. This suggests the existence of a physically bonded filler network structure, which governs the small strain dynamic amplitude properties. The research also shows that the smaller the storage modulus attributed to the filler network, the better is the degree of micro-dispersion in the filled system. The concentration of the percolation threshold can be used to determine the formation of the micro-dispersion. Since the filler-filler interaction is discussed in more detail under the dynamic-mechanical properties, only a brief summary of the filler-filler interaction is given below.

### 7.2.1 Carbon Blacks

The experimental results indicate that the effect is primarily determined by the specific surface area and the structure of the carbon black in the elastomer. The rubber seems to act merely as a dispersing medium that does not have a pronounced influence on the overall mechanical behavior of the three-dimensional filler network, but on the entropy elasticity. In dynamic-mechanical measurements, the strain amplitude is systematically increased up to 10%, with a decrease in the storage modulus being observed. The non-linear decrease in the storage modulus $G'$ during the increase in the strain amplitude has been related to a cyclic breakdown of the filler-filler bonds.

Different geometrical arrangements of the filler particles in the particular filler network structure resulting from percolation have been considered in the L-N-B model [86].

Upon stretching of the compound, the filler network is progressively disrupted, which allows substantial energy dissipation resulting in heat generation. When the external stress is removed, the stored energy in the rubber matrix allows the filler particles to reform the filler network, a process termed filler flocculation. In general, higher levels of dispersion at a given loading decrease the total filler-filler interactions by the presence of the weak but elastic polymer interface between the particles. As a result of better dispersion, the loading at which the filler network is formed, known as the percolation threshold, is shifted to higher filler concentrations. Therefore, it may be stated physically that micro-dispersion refers to the formation of filler networks indicated by the electrical percolation threshold. The necessary condition for the formation of a filler network is a small distance between aggregates that allows interaction forces, even if a small polymer film exists between the aggregates. As a consequence, the electrical percolation threshold is a function of the aggregate size and the surface activity of the carbon black. A sequence of NR compounds of increasing carbon black loading has been investigated using 3-D TEM [87]. Figure 7.17 summarizes the major results concerning the inter-aggregate distance and the decrease in the electrical resistivity.

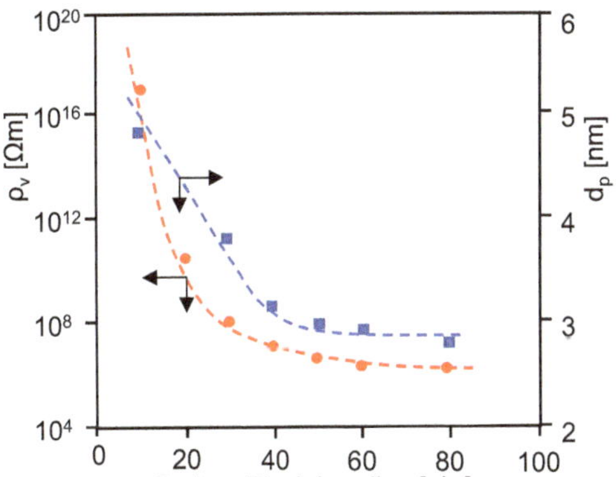


**Figure 7.17**
3D-TEM results show that the inter-aggregate distance is reduced when the filler loading increases [87]

Up to loadings of 40 phr, the inter-aggregate distance decreases from approx. 6 nm to 3 nm. Correspondingly, the electrical resistivity of the compounds decreases from approx. $10^{18}$ to $10^{8}$ Ω cm. Above the critical loading of approx. 40 phr that corresponds to the electrical percolation threshold, the inter-aggregate distance decreases with a less pronounced slope and reaches inter-aggregate distances between 2 and 3 nm. At this distance, electrical charge carriers can be transported through the compound by tunnelling effects [88]. Thus, the compound becomes electrically conductive, a fact which has been proved in several experiments.

Electrical percolation measurements have been carried out on compounds formulated with E-SBR and different grades of carbon black with loading varying from 0 phr to 80 phr [89]. The carbon black loading at which the resistivity starts to decrease is about 10 phr for N115 and as much as 55 phr for N774. Correspondingly, the electrical percolation threshold of the carbon blacks increases with the average aggregate size or with decrease in specific surface area. This sequence corresponds to the decrease in the surface activity as described in the section on surface energy sites. Thus, coarse blacks (N550–N770) are easier to disperse than fine blacks (N330–N110). This is due to the decreased filler activity of the coarse blacks possessing a higher amount of $sp^2$ hybridization on the graphite micro-crystalline layers. It is remarkable that, even at high carbon black loadings, the conductivity is some orders of magnitude smaller than the carbon black conductivity.

When the type of filler is kept constant, it has been seen that the electrical percolation threshold increases linearly with the solubility parameter of the rubber [90]. This proportionality underlines the role of the chemical composition of rubber in the micro-dispersion of carbon black and shows that, as rubber-carbon black interaction increases, the filler network shifts to higher concentrations of carbon black.

For the small-strain modulus of rubber compounds containing a percolated filler network, a power-law behavior for filler loading has been found. Above the percolation threshold, the dynamic shear modulus, $G'_0$, increases exponentially as a function of the filler volume fraction [91]. According to the kinetic cluster aggregation (CCA) model, the exponent $\alpha$ should have the value 3.5. Experimental results on a variety of compounds have shown that the exponent ranges from 2.5 to 3.5 [92]. The exponential increase in the storage modulus, $G'_0$, makes it clear that even small inaccuracies in the filler dosage above the electrical percolation threshold have a major influence on the rheological properties of a compound as well as on the stiffness of the vulcanizates. Fluctuations in the property level have to be avoided through exact filler dosing, which must be ensured before starting the mechanical mixing.

An important feature of filler networks is that the storage modulus, $G'(\gamma)$, decreases with increase in strain amplitude as a result of the filler network breakdown, known as the Payne effect (see chapter 10). As the primary particle size decreases, $G'(\gamma)$ increases exponentially. This increases the Payne effect as a result of the aggregate size and, of course, the carbon black loading. As the electrical percolation threshold is a function of polymer-filler interaction, comparisons of $G'$ for different compounds must be done cautiously. Taking into account that better polymer-filler interaction and filler micro-dispersion shifts the electrical percolation threshold towards higher filler loadings, it becomes clear that, at a given filler loading, the systems with the lower electrical percolation threshold will demonstrate a higher strength on the part of the filler network, $G'_0$, than systems with better micro-dispersion and high percolation threshold.

The temperature dependence of the storage modulus $G_0'$ has been studied intensively. With increase in temperature, an Arrhenius-like decrease in the storage and loss modulus, $G'(\gamma)$, has been found if the deformation amplitude is kept constant. Based on this observation, it has been assumed that the filler-filler contacts can be characterized by an activation energy [93]:

$$N = N_0 \exp(\Delta E_a/R)\,(1/T - 1/T_0) \tag{7.6}$$

where N is the number of filler-filler contacts, $\Delta E_a$ is the activation energy, T is the temperature and $T_0$ is the reference temperature

The activation energy therein, known as the network agglomeration energy, increases with the carbon black loading and the specific surface area. Consequently, the network agglomeration energy decreases with increase in dynamic amplitude as a result of the mechanical breakdown of the filler network during the dynamic deformation.

In the case of carbon black filled compounds, the presence of a filler network in an elastomer can also be proven by means of electrical conductivity. The network agglomeration energy increases sharply when the electrical percolation threshold is exceeded by approx. $10^{-6}$ S/m. The greater the extent of the filler network, the higher the electrical conductivity of the vulcanizates. In this context, it should also be mentioned that, even for the compounds with the highest carbon black loadings, the electrical conductivities are still one order of magnitude lower than typical values for pure carbon blacks ($\approx 10^1$–$10^4$ S/m). This indicates that the detailed character of the filler-filler contacts has to be understood as being more like a filler-polymer-filler contact, due to the rubber interface existing at the surface of the filler [94]. The rubber films between the filler particles reduces the conductivity of the entire system, due to electrical resistivity.

## 7.2.2 Precipitated Silicas

In precipitated silica, which is characterized by polar silanol groups on the surface, the filler-filler contacts through the H-bonds between the silanol groups are of outstanding importance. Due to the silanol groups on the surface of the silica particle, strong, pure filler-filler interaction has to be considered for compounds containing untreated silica. They lead to a high filler-filler interaction and the Payne effect is significantly higher than in those filled with carbon black. In the case of silica-silane systems, the filler-filler interaction is greatly decreased and the Payne effect is lower than in the case of carbon blacks. The corresponding network agglomeration energy has been found to be around 8 kJ/mol [95]. With silanization, the HD silica surface becomes less polar and filler-filler interaction decreases. Increasing the length of the alkyl chains of monofunctional silanes decreases the network agglomeration energies. Through the silanization of the silica, the viscosity of a silica compound can be greatly reduced and remains at a constant low level during storage. Quantitative silanization of the silica surface leads to disruption of the filler-filler interaction and the formation of apparently separated silica aggregates. Silanized silica is usually used in rubber compounds, because the dynamic-mechanical properties of passenger tire treads have lower rolling resistance and higher wet grip.

### 7.2.3 Filler Flocculation

After mixing, the rubber compounds remain, from a thermodynamic point of view, in a transient state. The filler particles will build up again into a filler network, which fundamentally changes the rheological properties as well as the physical properties of the vulcanizate. Aggregates, once dispersed in rubber, may re-form contacts among agglomerates, because of attractive forces and the small distance between themselves. The filler network densifies, resulting in significant changes in rheological and viscoelastic properties. This phenomenon exhibited by the filler network is referred to as filler flocculation, by means of which the filler network increases its stiffness. The process is thermodynamically driven and thus dependent on temperature and the surface activity of the filler and the chemical nature of the polymer chains. The inter-aggregate contacts between the filler particles can be increased after mixing, driven by enthalpic interactions of the filler particles. This is an undesirable effect, because it increases the viscosity of the compounds and the mechanical hysteresis of the elastomer that change the electrical percolation threshold and also the physical properties of the vulcanizate. Re-agglomeration and network formation can be a particular issue in the case of precipitated silica, causing hardening of the rubber compound during the storage period and in the induction stage prior to curing [96].

Similar to observations made on filler suspensions in fluids, it has been found that filler particles agglomerate during the storage of the compounds and counteract, at least in part, the filler micro-dispersion achieved in the stages of mixing [97]. The change in the filler network behavior on annealing is explained by considering the attractive forces between aggregates, even if they are separated by small polymer layers. The lower the viscosity of such separating layers, the easier it is for aggregates to join the chain-link fractal filler clusters that exist on the filler network. The outcome of this re-organization is therefore a strengthening of the filler network. This interaction come into play only when the mean spacing of the dispersed aggregates falls below a critical value, as is the case in an electrical percolated filler network. As the surface energy of the hydrocarbon rubber is lower than that of carbon black or precipitated silica, the driving force behind the flocculation is the energy exchange gained by more inter-aggregate contacts and the reduction in the rubber-filler contact surface that was created by shear stress during mixing.

The first attempts to quantify carbon black flocculation recorded the increase in storage modulus of compounds stored at elevated temperatures and subsequently cured by electron beams at ambient temperatures [98]. It has been shown that annealing at elevated temperatures (150 °C and 175 °C) raises the storage modulus, G′, to values that are 80% higher than for the unannealed reference samples. Further, it has been observed that flocculation occurs when the filler content is above the electrical percolation threshold. Flocculation studies of the small strain mechanical response of non-crosslinked compounds reveal a time-dependent increase in the storage modulus that indicates a strengthening of the filler network during storage time. This reorganization of the filler network is facilitated by the low viscosity of the rubber matrix and higher temperature of the compound. The flocculation rate therefore increases with these parameters.

A deeper understanding of the flocculation process can be achieved by investigations of the temporal development of the small strain storage modulus, $G'_0$, at the smallest possible strain amplitude (i. e. $\gamma = 0.28\%$) using a rubber process analyzer. Such an experiment provides information about the strength of the filler network. Typically, the storage modulus of the freshly mixed compounds increases with time and tends to a plateau value. Figure 7.18 illustrates the temporal development of $G_0'$ of S-SBR compounds filled with carbon black N234 (50 phr) for a step-like increase in the temperature to 160 °C. Depending on the molecular weight of the rubber, a pronounced increase in $G_0'$ is observed in the initial minutes that levels to a plateau at longer annealing times. The most pronounced plateau is observed for the lowest molar mass of the rubber, indicating that the flocculation is facilitated by a low viscosity of the polymer [99].

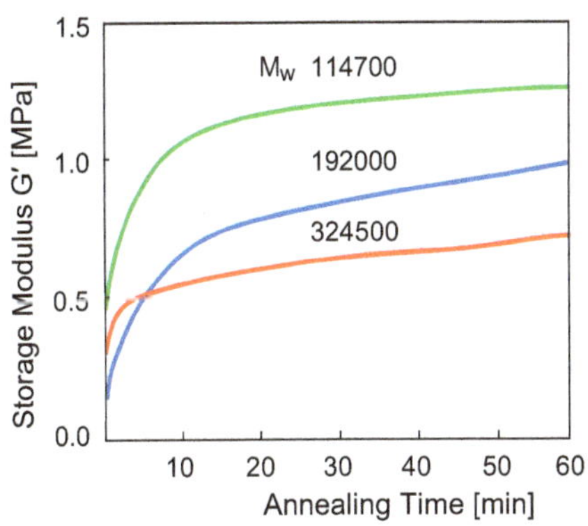


**Figure 7.18**
The increase in filler networks as the molecular weight of the rubber decreases [99]

Such a simple experiment provides information about the storage stability of rubber compounds. Unlike the permanent random placement of aggregates in a lattice, as is assumed by the percolation theory, flocculation provides a means to bridge vacant sites required for the formation of a larger cluster. This flocculation-induced deviation from random aggregate distribution should allow network formation to occur over extended time periods. Taking into account the temperature dependence of the flocculation, it becomes obvious that, with increase in the annealing temperature of the compounds, the process proceeds more rapidly. Thus, it can be concluded that compounds subjected to vulcanization will demonstrate a pronounced flocculation during the induction time, when the exposure to temperature reaches a maximum level. Not surprisingly, the cured compounds demonstrate an increase in $G'_0$ by a factor of 4–5, depending on the curing conditions. Therefore, a prediction of the stiffness of a sample after vulcanization based on the data of the uncrosslinked sample is misleading. In order to determine the influence of flocculation on the stiffness of the vulcanizate, the measurements have to be carried out on it [99]. Nevertheless, variations in the induction time during curing, as a result of changes of curing systems, have to be considered when a constant state of cure of the product is specified.

Measuring the flocculation via the increase in electrical conductivity of carbon black filled compounds has shown that the flocculation intensity increases with the surface activity of the filler [100]. In addition, it has been found that due to filler flocculation the percolation threshold in vulcanizates is shifted to lower filler concentrations. The magnitude of the shift should be known by compounders in order that they may predict the changes of low-strain properties of vulcanizates as a function of the vulcanization conditions. Properly controlled storage conditions of the compound (temperature and time) help to control filler flocculation and prevent undesired variations in compound viscosity.

On account of their polar surface, non-silanized silicas tends to undergo rapid and much more pronounced flocculation than carbon blacks. As can be seen, the precipitated silica aggregates reform their filler network faster because of the stronger hydrogen bonding compared to carbon black filled networks, which reform via the weaker van der Waals interactions. The tendency to reagglomerate stems from the presence of silanol groups that favor the filler-filler interaction via the H-bonds between the aggregates. This tendency can be illustrated during the storage of non-silanized silica compounds. After a certain period in storage, the mixtures must be re-milled in order to regain their original rheological properties. Silanization of the silica greatly weakens the flocculation tendency in non-polar hydrocarbon rubber. The disappearance of the polar silanol groups leads to better rubber-filler interaction and the filler-filler interaction decreases in intensity. The re-agglomeration of fillers has a significant effect on the properties of filled rubber, especially rheological properties of the uncured compounds and viscoelastic properties of the vulcanizates. A comparison of the filler re-agglomeration is much more pronounced in the case of precipitated silica than in carbon black, as shown in Figure 7.19 [101].

The flocculation kinetics are described by various approaches that are presented in projects conducted with tire manufacturers. The influence of the silanization of HD silica using the Payne effect has been examined [102]. In the flocculation process, maintained for 15 min at 170 °C for mixtures without curing agents, a decrease in $G'(\gamma)$ from 6.8 MPa to 2.5 MPa was observed at 60 °C, while a decrease of 1.5 MPa to 0.5 MPa was observed for HD silica silanized with Si208. The result shows that the silanized silica undergoes only a small amount of flocculation, because the interaction with the rubber chains has improved significantly. The difference in filler dispersion that occurs during flocculation can be seen in the corresponding TEM images in Figure 7.20.

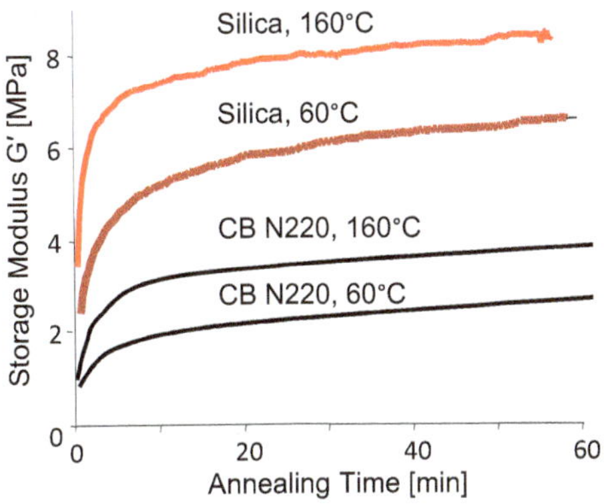


**Figure 7.19** Influence of the annealing temperature on carbon black and precipitated silica in an S-SBR compound [101]

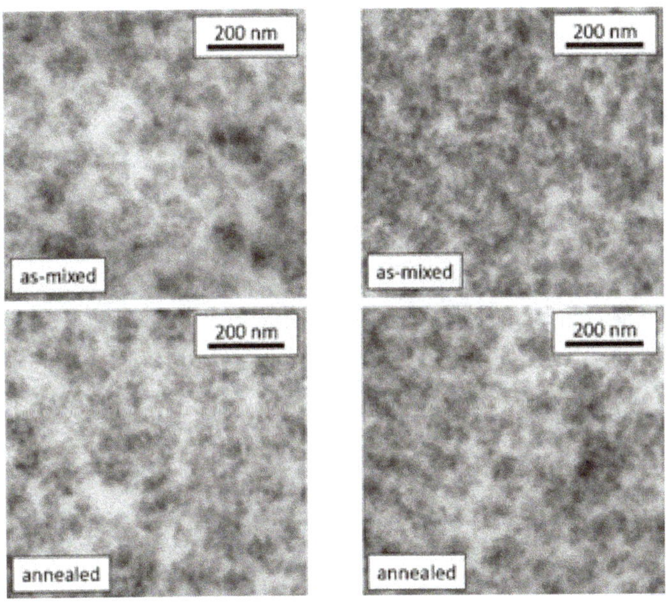


**Figure 7.20** Change in silica dispersion before and after flocculation at 170 °C

Given the process of reorganizing of filler particles during storage, the storage time and temperature must be tracked. However, the main process of flocculation occurs during the scorch time during the vulcanization process, when the viscosity of the rubber compound is very low. This process results in a special filler particle distribution which later has a significant impact on the physical properties. There is a clear increase in filler flocculation, which is explained by a higher $G'_0$. The process is observed for carbon black, for precipitated silica and for silanized silica. In the case of carbon black, a higher electrical conductivity is observed, which shifts the electrical percolation threshold to low concentrations. As for the precipitated silica and the silanized silicas, $G'_0$ is increased by a factor of 2–3. The proportion of filler-filler interaction in the rubber vulcanizate increases significantly in the short period of scorch time, because the viscosity of the rubber chain drops to a minimum and the filler aggregates can form larger clusters. The thin polymer film that lies between the particles is overcome by the interaction energy of the filler and so the filler clusters can widen in the filler network.

## References

[1] Studebaker, L., in "A useful new system for categorizing rubber grade carbon blacks", Phillips Petroleum Co., Technical Bulletin (1971)

[2] Görl, U., in "Handbuch der Kautschuk-Technologie", Hofmann, W., Gupta, H., Eds., Dr. Gupta Verlag, Ratingen, (2001)

[3] Meon, W., Blume, A., Luginsland, H.-D., Uhrland, S., in "Rubber Compounding, Chemistry and Applications", (B. Rodgers, Ed.) Marcel Dekker, Inc., New York, Basel (2004) Ch. 7

[4] Binford, Jr., S., Gessler, A. M., J. Phys. Chem., 1959, 57: 1376

[5] Kraus, G., Dugone, J., Ind. Eng. Chem., 1955, 47: 1809

[6] Kraus, G., Gruver, J. T., J. Polym. Sci., 1971, 8: 571

[7] Dannenberg, E. M., Rubber Chem. Technol., 1976, 48: 410

[8] Schuster, R. H., Bussmann, D., Symp. Carbon Black, Mulhouse, (1992)

[9] Simha, R. H. Frisch, H. L., Eirich, F. R., J. Phys. Chem., 1953, 57: 583

[10] Bussmann, D., PhD Thesis, University of Hanover (1992)

[11] Schuster, R. H., Issel, H.-M., Peterseim, V., Rubber Chem. Technol., 1993, 69: 769

[12] Hilfinker, J. K., et al., Solid State Technology, 1998, 41: 101

[13] Rodenko, K., et al., J. Phys. Chem., 2010, 114: 22566

[14] Klafke de Azeredo, L., Schuster, R. H., ACS Meeting Charlotte (2009)

[15] Chogule, H., Schuster ACS Rubber Meeting Cleveland (2013)

[16] Dannenberg, E. M., Ind. Eng. Chem., 1948, 40: 2199

[17] Bueche, F., J. Polym. Sci., 1956, 135: 97

[18] Kraus, G., J. Appl. Polym. Sci., 1963, 7: 861

[19] Schuster, R. H., Schellenberg, M., ACS Rubber Division, Cleveland (2013)

[20] van Zanten, J. H., et al., Phys. Rev. E 3, 1996, R2053

[21] Kaufman, S., Slichter, W. P., Davies, D. P., J. Polym. Sci., Part A2, 1971, 9: 892

[22] Westlinning, H., Butenuth, G., Makromol. Chem., 1961, 47: 215

[23] Schon, T. G. F., Adler, K., Kautsch. Gummi Kunstst., 1966, 19: 414

[24] Wang, C. C., Donnet, J.-B., 2006, 58: 466

[25] Nishi, T., J. Polym. Sci., Polym. Phys. Ed., 1974, 12: 68

[26] O'Brien, J., Cashell, E., Wardell, G. E., McBrierty, V. J., Rubber Chem. Technol., 1977, 50: 747

[27] Nishi, T., Rubber Chem. Technol., 1978. 51: 1075

[28] Lüchow, H., Breier, E., Gronski, W., Rubber Chem. Technol., 1997, 70: 747

[29] Kuhn, W., Peregi, W., Fei, Z., Giese, U., Schuster, R. H., Proc. Macro UIPAC, Rio de Janiero (2006)
[30] Litvinov, V. M., Orza, R. A., Klüppel, M., van Duin, R. A., Macromolecules, 44 (2011) 4887
[31] Borsacchi, S., et al., Polymers, 10 (2018) 822
[32] Nakajima, K., Nishi, T., in "Current Topics in Elastomer Research", (Bohmick, A., Ed.) CRC Press (2008), Ch. 21
[33] Nakajima, K., et al., Rubber Chem. Technol., 2017, 90: 272
[34] Fukahori, Y., J. Appl. Polym. Sci., 2005, 95: 60
[35] Igarashi, T., et al., Macromolecules, 2013, 46: 1916
[36] Berriot, J., et al., J. Non-crystalline Solids, 2002, 307: 719
[37] Twiss, D. F., J. Soc. Chem. Ind., 1925, 44: 1067
[38] Stickney, P. B., Falb, R. D., Rubber Chem. Technol., 1964, 37: 1299
[39] Kraus, B., Adv. Polym. Sci., 1971, 8: 309
[40] Blow, C. M., Polymer, 1973, 14: 309
[41] Dannenberg, E. M., Rubber Chem. Technol., 1986, 59: 512
[42] Brennan, J. I., Jermyn, T. E., Boonstra, B. B., Gummi, Asbest, Kunstst., 1965, 18: 266
[43] Gessler, A. M., Rubber Age, 1969, 101: 12, 54
[44] Gessler, A. M., Hess, W. M., Medalia, A. I., Plast. Rubber Process, 1978, 3: 141
[45] Dessewffy, O., Rubber Chem. Technol., 1962, 35: 599
[46] Watson, M. W., Ind. Eng. Chem., 1955, 47: 1281
[48] Soutwart, D. W., Inst. Rubber Ind. Conference, Loughborough (1964)
[47] Berry, J. B., Cayre, P. J., J. Appl. Polym. Sci., 1960, 3: 213
[49] Cotton, G. R., Rubber Chem. Technol., 1985, 58:774
[50] Gessler, A. M., Rubber Chem. Technol., 1969, 42: 858
[51] Voet, A., Kautsch. Gummi Kunstst., 1973, 26: 254
[52] Kraus, G., Rubber Chem. Technol., 1965, 38: 1070
[53] Pliskin, I., Rubber Chem. Technol., 1973, 46:1218
[54] Medalia, A. I., Rubber Chem. Technol., 1978, 51: 43
[55] Palmanteer, K. E., Lentz, W. C., Rubber Chem. Technol., 1975: 48: 795
[56] Ban, L. L., Hess, W. M., Papazian, L. A., Rubber Chem. Technol., 1974. 47: 858
[57] Boonstra, B. B., Kautsch. Gummi Kunstst., 1966, 19: 198
[58] Hess, W. M., Chirico, V. E., Burgess, K. A., Kautsch. Gummi Kunstst., 1973, 26: 344
[59] Wang, M.-J., Rubber Chem. Technol., 1998, 71: 520
[60] Wang, M.-J., Wolff, S., Freund, B., Rubber Chem. Technol., 1994, 67: 27
[61] Sheng, E., Sutherland, I. Brandly, R. H., Eur. Polym. J., 1996, 32: 35
[62] Tricas, N., Boros, S., Schuster, R. H., Kautsch. Gummi Kunstst. 2005, 58: 511
[63] Wang, M.-J., Wang, T., Wang, Y. L., Kautschuk, Gummi Kunstst., 2002, 55: 388
[64] Wang, M.-J., Tu, H., Murphy, L. J., Rubber Chem. Technol., 2000, 73: 666
[65] Wang, M.-J., Wolff, S., Tan, E. H., Rubber Chem. Technol., 1993, 66: 163
[66] Ziegler, J., Schuster, R. H., Kautsch. Gummi Kunstst., 2003, 56: 159
[67] Montes, H., Lequeneux, F., Berriot, J., Macromolecules, 2003, 36: 8107
[68] Nah, G., Liu, J. Y., Cho, B. H., Hong, C. K., Gent, A. N., J. Appl. Polym. Sci., 2010, 118: 1574
[69] Sui, G., et al., Macromol. Mater. Eng., 2007, 299: 1020
[70] Cotton, G. R., Rubber Chem. Technol., 1975, 48: 548
[71] Villars, D. F., J. Polym. Sci., 1956, 21: 98

[72] Kraus, G., Gruver, J. T., Rubber Chem. Technol., 1968, 41: 1256
[73] Meissner, B., J. Appl. Polym. Sci., 1974, 18: 2483
[74] Leblanc, J. B., Hardy, P., Kautsch. Gummi Kunststst., 1991,44: 1119
[75] Ma, J., Zhang, L.-Q., Wu, Y.-P., J. Macromolecular Science, Part B, Physics, 2013, 52: 1128
[76] Hess, W. M., Lyon, F. L., Burgess, K. A., Kautsch. Gummi Kunststst., 1967, 20: 135
[77] Fröhlich, J., Educ. Symp., DKT Nuremberg (2016)
[78] Payne, A. R., J. Polym. Sci., 1962, 6: 57
[79] Gerspacher, M., Young, H. H., O'Farrell, C. O. P., Rubber Chem. Technol., 1991, 64: 122
[80] Wolff, S., Wang, M.-J., Rubber Chem. Technol., 1991, 65: 329
[81] Voet, A., Cook, A. F., Rubber Chem. Technol., 1968, 41: 1208
[82] Voet, A., Sircar, A. K., Muller, T. J., Rubber Chem. Technol., 1968, 41: 1243
[83] Staufer, D., Aharoni, A., in "Introduction to Percolation Theory", Taylor and Francis, London (1992)
[84] Meakin, P., Prog. Solid State Chem., 1990, 20: 135
[85] Payne, A. R., J. Appl. Polym. Sci., 1963, 7: 873
[86] Lin, C. R., Lee, Y. D., Macromol. Theory Simul., 1997, 6: 102
[87] Kohjyia, S., Kato, A., Shimanuki, Y., Hasegawa, H., Ikeda, Y., Polymer, 2005, 44: 4440
[88] Balberg, I., Phys. Rev. Lett., 1978, 59: 1305
[89] O'Farrell, C. O. P., Gerspacher, M., Nikiel, L., ACS Rubber Division, Orlando (1999)
[90] Schuster, R. H., in "Mixing of Rubber Compounds", (Limper, A., Ed.) Hanser Publishers, Munich (2012), Ch. 6
[91] Klüppel, M., Heinrich, G., Schuster, R. H., Rubber Chem. Technol., 1996, 70: 243
[92] Schuster, R. H., Proc. "Eurofiller", Bratislava (2014)
[93] Fitzgerald, E. R., Ferry, J. D., Rubber Chem. Technol., 1982, 55: 1569
[94] Pantea, D., et al., Carbon, 2001, 39: 1147
[95] Herrmann, V., Niedermeier, W., Kautsch. Gummi Kunststst., 2010, 62: 559
[96] Mooney, M., J. Colloid Sci., 1946, 1: 195
[97] Mewis, J., J. Rheol., 1978, 22: 381
[98] Böhm, G. G. A., Ngyen, M. N., J. Appl. Polymer Sci., 1995, 55: 1041
[99] Meier, J., Giebeler, E., Klüppel, M., Schuster, R. H., ACS Rubber Division Meeting, Providence (2001)
[100] Gerspacher, M., et al., Kautsch. Gummi Kunststst., 2002, 55: 596
[101] Meier, J., Klüppel, M., Geisler, H., Schuster, R. H., Kautsch. Gummi Kunststst., 2005, 58: 587
[102] Robertson, C. G., et al., Rubber Chem. Technol., 2011, 84: 507

# 8 Flow Behavior of Rubber Compounds

After the mixing is finished, the flow process follows: extrusion, calendering, forming and shaping of the rubber compounds in molds are performed in sequence and on proprietary equipment. The somewhat vague term "processability" comprises all aspects of material behavior which contribute to producing a quality compound that can be shaped and crosslinked into a satisfactory product. Certain compounds are said to have good, others to have poor, processability, depending on whether they are extruded into profiles or pressed into a mold [1, 2]. The rheology for a given rubber compound are significantly different and unique compared to the rheology of thermoplastic materials. Typically, the raw rubber used is an amorphous polymer with low glass transition temperatures (< 0 °C). The behavior of a filler in a rubber compound is easily assessed, via direct comparison with the behavior of unfilled systems. The very small filler aggregates are between 20 to 60 nm in diameter, and form a filler network above of a certain concentration. This highlights the properties of the filler in terms of shear viscosity, reduction in the extruder die and the various flow anomalies. In addition to flowability of the rubber compound, the term processability also covers dimensional stability of extrudates, heat development during extrusion, and tackiness when assembling multi-layered products.

## 8.1 Types of Extruders

Even though the configuration of the extruders serves different processing purposes, they have one thing in common. At the end of the extruder, the compound has to be pressed through an extruder die in order to achieve the desired dimensions of the semi-finished product. An extruder essentially consists of a screw, a cylinder (barrel) and an extruder die. Other essential parts are the machine housing, the motor and feeding device, including the feed roll, as well as heating, cooling, and lubricating systems. Modern extruders are equipped with devices for measuring the bulk pressure and temperature of the compound and with programmable controllers [3–5].

Depending on the purpose in the rubber industry, both warm-feed extruders, an early development, and cold-feed extruders, introduced later, are used. Warm-feed extruders use pre-warmed compound because of their short length. Warm-feed extruders are widely employed in the tire industry and are also used when a high discharge rate is required, e. g., as machines downstream of internal mixers or in combination with cold-feed pin-barrel extruders (duplex extruders) in the tire industry. In cold-feed extruders, the compound is not pre-plasticized in feed or prewarming mills. The extruders are considerably longer (L = 16–24 D) and offer a better temperature control along the screw [2]. Thus, the plasticization of the rubber compound in the extruder is supported by exact temperature control in the barrel wall and the screw over the entire extruder length during extrusion. It is possible to vary the temperature along the barrel with separate temperature-controlled zones. The screws must be self-cleaning. In recent decades, various cold feed extruders have been developed:

- extruders with homogenizing action (i. e. pin-barrel extruder)
- extruders with shearing action
- shearing head extruders

This introduction to extruders should suffice at this point to permit the rheological conditions of the rubber compounds to be discussed in more detail.

## 8.2 Flow Behavior of Unfilled Rubber Melts

An evaluation of the mixed compounds with regard to their processability is therefore desirable from the start of production. The necessary framework is provided by rheology, which describes fluid behavior based on constitutive laws that describe the fluid response of a material to an imposed external stress.

Typically, the energy necessary for deformation is dissipated into heat during the flow of an ideal fluid. Therefore, processability testing has received attention, both to develop the right tests and predict processing behavior, as well as to make quality control precise. The purpose of this chapter is to introduce rheological concepts and to evaluate the flow behavior of polymer, initially of unfilled rubber compounds and subsequently from filled melts. In principle, the deformation of a fluid can take place in shear flow or in extensional flow. The set-ups for measuring in shear or in extension are different and will deliver different parameters. Irrespective of how the external stress is applied, viscosity denotes the resistance of a fluid to transformational deformation at a given rate.

The important property of a rubber melt is the viscosity that can be measured in shear flow and extensional flow. For an ideal fluid, the shear viscosity is described by the proportionality between the shear stress and the shear strain rate, where the proportionality factor is the shear viscosity:

$$\tau = \eta\dot{\gamma} \tag{8.1}$$

where $\tau$ is the shear stress, $\dot{\gamma}$ is the shear rate and $\eta$ is the viscosity

Viscosity can be conceptualized as quantifying the internal frictional force that arises between adjacent layers of a fluid that are in relative motion, as shown in Figure 8.1.

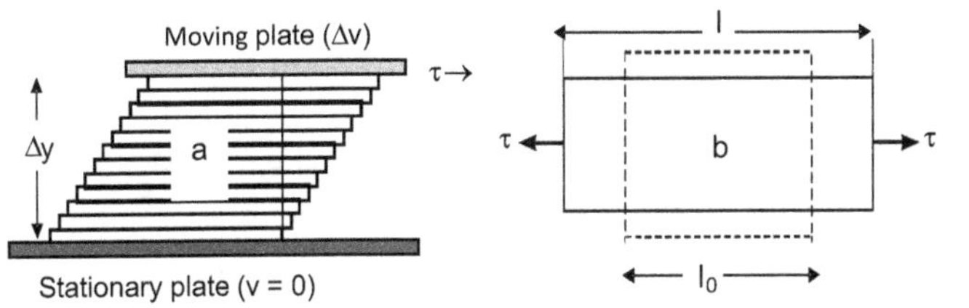


**Figure 8.1** Melt viscosity: a) shear flow and b) extensional flow

In shearing, the variables become clear for the case of a fluid sheared between two parallel plates, with the ground plate considered to be fixed (v = 0) at the end of the wall and the upper plate of the area A, which is adhering on the fluid surface, moving with a velocity $v_{max}$. The shear stress causes a laminar flow of the fluid with a maximum speed at the upper boundary. Therefore, the fluid is sheared along infinitesimally thin layers.

In extensional flow, the viscosity is measured on a test specimen in an extensional manner, as shown in Figure 8.1b. Extensional flow is experienced by rubber samples when the fluid flows through pipes, the diameter of which gradually narrows. Since the volume of the fluid does not change with the applied pressure, an extension takes place in which the speed of the volume elements of the fluid increases continuously. Consequently, the elements of the fluid are stretched in the direction of extension. The extensional stress is proportional to the extensional stretching rate, as expressed by the equation:

$$\tau_E = \eta_E\dot{\varepsilon} \tag{8.2}$$

where $\tau_E$ is the extensional stress, $\eta_E$ is the extensional viscosity and $\dot{\varepsilon}$ is the extensional rate in the direction of flow

The main difference between the shear and the extensional viscosity is that, even at a low deformation rate, the value of the extensional viscosity, $\eta_E$, is three times as large as the value of shear viscosity, $\eta$. Therefore, the parameter viscosity is a complex property that depends on the molecular properties of the non-vulcanized rubber, the temperature and the test conditions to provide the resistance to deformation. The resistance to flow exhibited by ideal fluids results from the intermolecular interactions, as the molecular friction coefficient of the chain segments [6]. In order that polymer chains may undergo translational and rotational movements in a melt, a "free volume" is required in the fluid that increases with the temperature. The larger the "free volume", the greater is the mobility of the polymer chains and the lower the measured viscosity [7].

### 8.2.1 Newtonian and Non-Newtonian Fluids

At low shear rates, a constant zero shear viscosity is observed that depends on the molecular weight, the molecular weight distribution of the polymer and the temperature. The characteristic change in the shear stress at higher shear rates leads to changes in the viscosity. Normally, the rubber chains show a strong decrease in viscosity. Polymer melts that show this behavior are mostly found in common rubber melts. A different curve is shown by polymeric fluids, which increase in viscosity as the shear rate increase. This is known as shear thickening. This different behavior of the polymer melt is a determining factor in the behavior of the viscosity. The viscosity as a function of shear rates is shown in Figure 8.2.

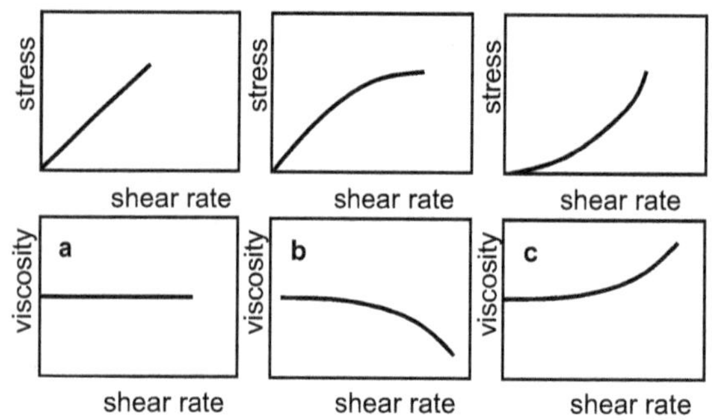


**Figure 8.2** Viscosity of a rubber melt as a function of shear rates: a) Newtonian flow, b) shear thinning and c) shear thickening

In a Newtonian fluid, the shear rate increases in proportion to the applied stress. The viscosity remains constant as the shear rate increase. In the practice of rubber processing, it is absolutely necessary to know this interplay between shear stress and shear rate. Not only is it important to measure the compound viscosity, it is also important to know how rapidly viscosity decreases with increase in shear rate. A distinction is made between fluids in which the shear stress plateaus above a critical shear rate and the viscosity decreases with the shear rate, and those fluids in which shear stress increases with the shear rate, as does the viscosity. The steady-state viscous shear flow of fluids can be described by the Ostwald–de Waele power law [8]:

$$\tau = K\dot{\gamma}^n \quad (8.3)$$

where τ is the shear stress, $\dot{\gamma}$ is the shear rate and K is a consistency constant equal to the viscosity of a shear rate of 1 $s^{-1}$ that corresponds to the operating region of a Mooney viscometer

The arithmetic sign and magnitude of the exponent, n, describe the shape of the flow curve. The power law index, n, is a measure of deviation from simple Newtonian behavior. The value of n indicates the type of flow:

- $n = 1$ corresponds to Newtonian flow
- $n < 1$ corresponds to non-Newtonian flow, in which the viscosity decreases with increase in shear rate (shear thinning)
- $n > 1$ stands for dilatant flow, which means that the viscosity increases with increase in shear rate (shear thickening)

The exponent, n, characterizes the behavior that can be changed by increasing the temperature, as will be shown later. Shear-thinning fluids are characterized by the fact that, at the limit of very low shear rates (or stresses), the viscosity is initially constant, whilst, at the limit of high shear rates (or stresses), the viscosity decreases to a lower level. The occurrence of shear thinning in polymer melts is related to the existence of rubber chain entanglements that provide a sizable internal resistance to flow. However, the flow behavior depends on the time scale over which deformation proceeds. At low shear rates, the entanglements can be released by self-diffusion of chains and the viscosity is independent of the shear rate, known as zero shear viscosity, $\eta_0$ [9, 10]. The critical shear rate, where the viscosity decreases with the shear rate, depends on the molecular weight, the molecular weight distribution and long-chain branching of the polymer. For rubber melts, that exhibit shear thinning flow, the values for the power law index, n, are in the range of 0.1 to 0.3. If the right and left side in Equation 8.3 are divided by the shear rate, a power laws results, which describes the dependence of the viscosity, $\eta_a$, on the shear rate:

$$\eta_a = K\dot{\gamma}^{n-1} \quad (8.4)$$

Thus, for rubber melts that exhibit shear thinning ($n < 1$), the viscosity of the polymer melt decreases by several orders of magnitude as the shear rate increases. At high shear rates, a limiting value of the viscosity should be reached that corresponds to chains fully oriented in the direction of flow. Therefore, shear thinning provides a large spectrum of shear viscosities over the range from low to high shear rates. It is clear that a test method that maps the viscosity at low shear rates, the behavior in the region of high shear rates, cannot be described. In order to gain a realistic picture of the effect of shear thinning, it is advisable not to rely on one method that covers a limited range of shear rates, but to investigate the flow behavior over that range of shear rates. The dependence of the shear flow is shown in Figure 8.3.

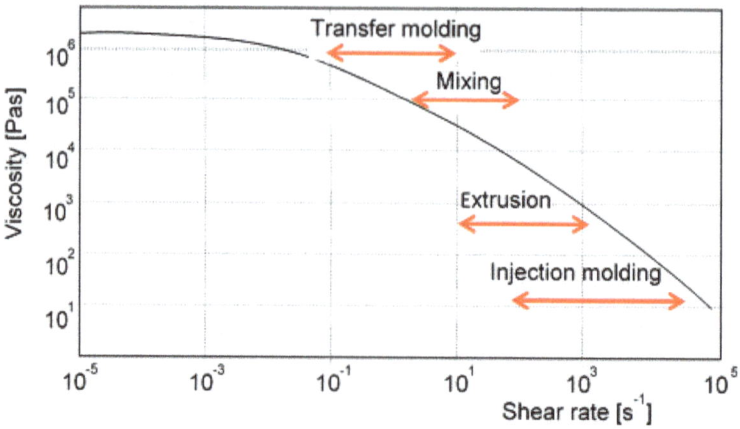


**Figure 8.3** Viscosity of different processes for manufacturing rubber compounds

It is therefore necessary to measure the viscosity in the range of shear rates that are matched to the process. It can be seen that the Mooney viscosity, which can easily be measured, covers an area with low shear rates, which is close to the manufacturing conditions for press molding but is not relevant at all for the processing techniques that require higher shear rates, such as extrusion or injection molding. The capillary rheometer could predict processing in rubber mixing, calendering, extrusion and injection molding. All of these rubber processes were performed at much higher shear rates than a Mooney viscometer can produce. The group of melts that show shear thickening are not as common. When the polymer is sheared, the chains are aligned to a certain extent and can thus form a crystalline super-lattice, which stiffens the melt. At ambient temperatures (20–40 °C), high cis-BR (cis 1,4 units 98–99%) exhibits typical shear thickening [11]. At elevated temperatures, a transition from shear thickening to shear thinning occurs.

### 8.2.2 Influence of Temperature

The viscosity of fluids, low-molecular as well as high-molecular fluids, naturally decreases with increase in temperature, because of the free volume that increases with the temperature and provides more chain mobility. This is a natural phenomenon resulting from decreasing intra- and inter-molecular interaction forces and increasing free space between the molecules that allows more freedom for translational and rotational movements [12–14]. Measurements of the specific volume have shown that the free volume of a polymer has a minimum value at the glass transition temperature that represents the lower end of the rubber elasticity. Above the $T_g$, the free volume increases linearly with the temperature. Consequently, the resistance of the fluid to a deformation process reduces exponentially and follows an Arrhenius type temperature dependence:

$$\eta = \exp(-\Delta E/RT) \tag{8.5}$$

where ΔE is the activation energy, R is the universal gas constant and T is the absolute temperature

The activation energy is related to the type of the rubber chains and the polarity. At low molecular weights, it is also dependent on the average molecular weight.

The influence of temperature on the shear viscosity of LDPE melt at several temperatures is shown in Figure 8.4 [15].

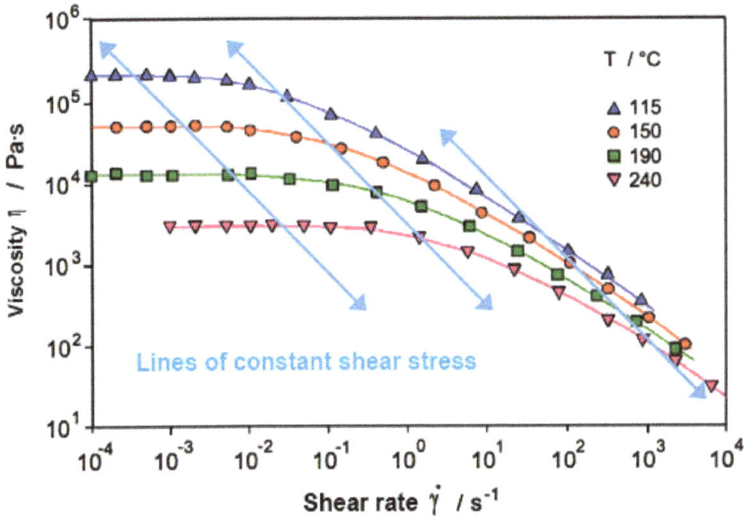


**Figure 8.4** Effect of temperature on the shear viscosity of LDPE [15]

It is clear that the zero shear viscosity, $\eta_0$, decreases roughly by two orders of magnitude when the temperature increases from 115 to 240 °C. The Newtonian plateau decreases progressively with the melt temperature. At high shear rates ($> 10^2\,s^{-1}$), the temperature-dependent decrease in viscosity is less pronounced. The arrows in Figure 8.4 indicate the points of constant shear stress. For the same shear stress, a much higher shear rate can be produced at elevated temperatures that improves the extrusion at higher temperatures. Similar viscosity curves as a function of temperature are obtained for different types of rubber [16].

### 8.2.3 Influence of Molecular Weight and Molecular Weight Distribution

The different influences on the viscosity can be seen more clearly in the case of rubbers that, at low average molecular weight, exhibit a type of Newtonian flow over the shear rates that is known as zero viscosity. It has been shown that zero viscosity is almost proportional to the weight average molecular weight, $M_w$, [17]:

$$\eta_0 \sim M_w \tag{8.6}$$

In contrast to the low-molecular weight area, high-molecular rubber exhibits a zero viscosity, $\eta_0$, that increases with the exponent 3.4 of the molecular weight:

$$\eta_0 \sim {M_w}^{3.4} \tag{8.7}$$

This behavior is dictated by the existence of entanglements in the melt that increase with the chain length. The theoretical framework was modelled by the "tube model" that considers a polymer chain to move with some freedom in an imaginary tube limited by entanglements with neighboring chains [10, 18]. The entanglement spacing is given by the statistical distance between two adjacent entanglements. This value determines the diameter of the tube. The more entanglements exist along the chain, the more the translational movement of the chain (self-diffusion) will be hindered. Any deformation of the entangled chain system that is shorter than the time it takes for the chain to "snake out" or to "reptate" from the imaginary tube encounters the resistance of the transient network and is macroscopically registered as an elastic contribution to the flow behavior. The topological obstructions are well described by the tube model. A schematic representation of the "tube model" is shown in Figure 8.5 [10].

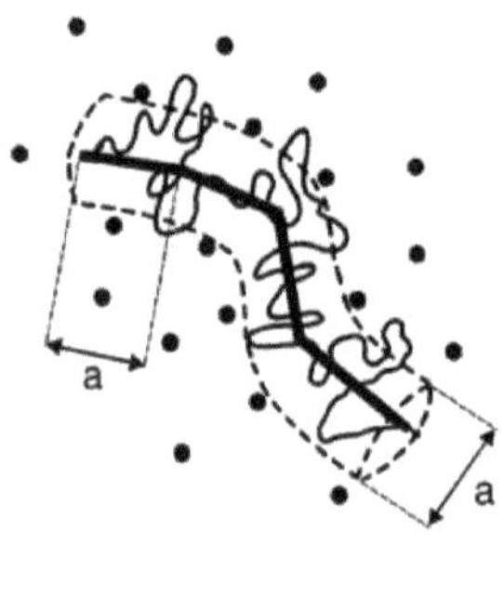


**Figure 8.5**
Schematic representation of the "tube model" [10]

When entanglements become stable over time, they obviously make flow more difficult, hence the higher dependence on the molecular weight. The viscosity behavior of linear S-SBRs of identical microstructure (24 wt.% V, 26 wt.% S), measured by a capillary rheometer, shows a tendency for a small Newtonian regime at shear rates lower than 1 $s^{-1}$ at a molecular weight of $1.5 \times 10^5$ g/mol while, above this molecular weight, a typical non-Newtonian flow behavior with no plateau value is seen, as shown in Figure 8.6 [19].

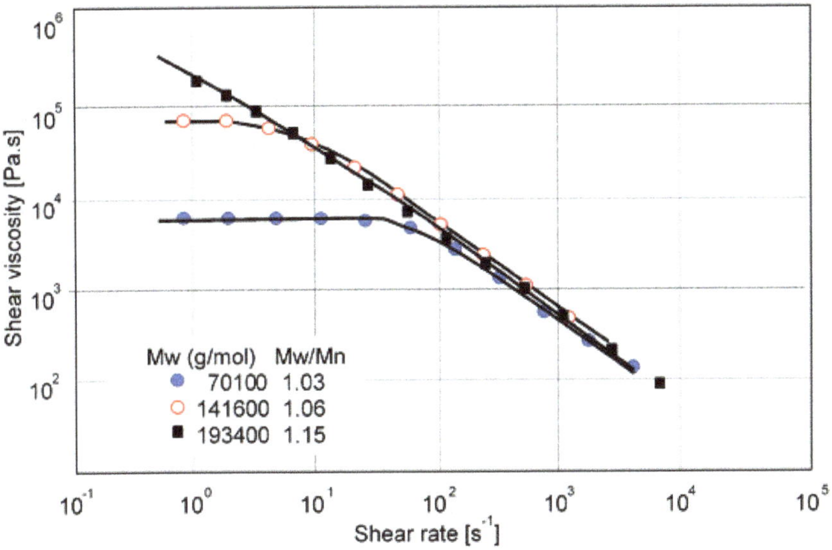


**Figure 8.6** Influence of the molecular weight on the viscous behavior of S-SBR [19]

All else being equal, broadening of the molecular weight distribution results in a melt becoming non-Newtonian at lower values of shear rate, so that the polymer is easier processed. Broadening of the molecular weight distribution results in a decrease in the shear viscosity at constant shear rate, indicating the dominant effect of the fraction of small chains in the rubber melt. Shorter chains in the melt have a shorter reptation time and consequently they release the entanglements in which they are involved much faster than long chains. The influence of molecular weight distribution has been investigated in a model experiment by blending a high molecular weight IR ($M_w = 1.6 \times 10^6$ g/mol) with an IR of lower molecular weight ($M_w = 1.4 \times 10^5$ g/mol) IR [20]. The higher molecular weight component governs the viscosity at small shear stresses, whereas its influence is significantly weakened in the region of high shear rates. The sample with narrow molecular weight distribution ($M_w/M_n = 1.20$) has a consistently higher viscosity at lower shear rates than the sample with a broader molecular weight distribution ($M_w/M_n = 2.50$).

A broad molecular weight distribution for a rubber commonly leads to better flowability and processability than those rubbers with narrow distribution. This applies to rubbers made by anionic polymerization. Although the molecular weight distribution ($M_w/M_n$ = approx. 2) is higher in the technical process than in laboratory samples ($M_w/M_n$ = 1.1 to 1.2), the melt viscosity is still significantly lower than in rubber, emulsion polymerization or coordination catalysis.

Branched polymers are prevented by the entanglements from undergoing translational movements in the melt and offer high resistance to any deformation at low shear rates. Nevertheless, because branched polymers demonstrate a more pronounced shear thinning in almost all cases, the long-chain branching results in a lower viscosity at high shear rates [21].

A comparison of the flow behavior of linear and branched cis-BR of different molecular weight distribution is shown in Figure 8.6. It is shown that, at low shear rates, the linear cis-BR of narrow molecular weight distribution ($M_w/M_n = 2.1$) exhibits a significant lower zero viscosity than the cis-BR with long-chain branching, made by coordination catalysis. This fact is attributable to entanglements of each of the chain branches. When the shear rate increases, the shear-thinning effect leads to a significant change in the viscosity and consequently better processability, as shown in Figure 8.7 [22].

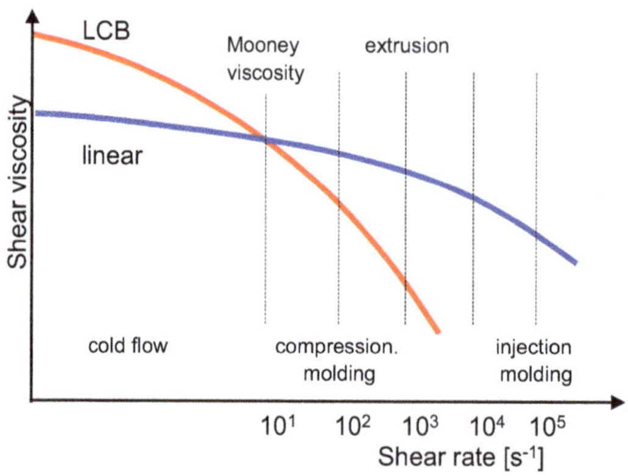


**Figure 8.7** Influence of long-chain branching of cis-BR on shear viscosity over a large range of shear rates

The long-chain branched BR demonstrates the smallest shear viscosity while the linear polymer exhibits the highest viscosity and thus more difficult processability. In this context, it is interesting to note that the flow curves of the different macrostructures cross over in the range of shear rates where the Mooney viscometer is operating and the Mooney viscosity neither distinguishes between the macrostructure of the rubber nor does it allow predictions about rheological behavior at higher shear rates.

For NR/BR blends, it has been observed that the mixing torque is not an additive value of the mixing torques. The blend torque values were higher than for each rubber by itself. It has been shown that, up to 40%, the polymers do not form a single phase [23]. The minimum occurred between the 30 and 40 phr BR. At high shear rates, the flow of the rubber compound is similar to the torque composition [24], even for the more compatible blends, SBR/BR, where the polymers are incompatible at the molecular level in a mixed blend. The development and size of the microheterogeneity affect the flow behavior of the rubber blend, but this is still under debate [25].

### 8.2.4 Extrudate Die Swell

Extrusion has become an important processing technology and many studies have been dedicated to screw and die design. Extruders are conceptually a pump, consisting of a screw to move the material, a barrel around the screw to contain the material, a temperature-controlling system and the extrudate die. The extruder itself has two main functions: to pump the rubber compound through the barrel and to generate enough pressure and to force the rubber compound through a die to give the required cross-sectional profile [26]. Extrudates are produced from mixed rubber compounds using single-screw extruders. In principle, extruders allow the production of storable intermediate products (pellets, strips), uncured finished products, and further processing to give semi-finished products. The geometry of the rubber melt that is extruded depends on:

- the rubber compound
- the geometry of the die
- the length of the die
- the velocity of the rubber melt

It can be seen that the greater the elasticity of the mixture, which has undergone a degradation effect in the extruder, the greater are the geometric deviations. With a little practice, the effort involved in making and running-in a new nozzle should be reduced. With the extruder die swell, which occurs spontaneously, the elasticity affects the entire profile in a relative short time program.

A common phenomenon that occurs with all rubber compounds when forced to pass through a small extruder die is a characteristic change in the extrudate shape when exiting the die. When the rubber exits the die orifice, it possesses elastic "memories" of its dimensions in the barrel so the larger molecular chains try to return to their original position before they entered the die orifice. Extruder die swell represents a spontaneous elastic recovery of the compound elongated in the extruder die. For the practice of extrusion, die swell poses a serious problem, since the simultaneous increase in the diameter of the melt and the shrinkage over the length affects the dimensional accuracy of the extruded profile [27].

The phenomenon of die swell is particularly pronounced with high-molecular polymer melts. Before entering the extruder die, entangled polymer chains in the flow stream occupy a roughly spherical chain conformation that corresponds to a maximum of conformational entropy. By pressing the melt through the narrow die, a sizable amount of energy is forcing the entangled polymer coils in the melt to elongate in the direction of flow and to lose a part of the conformational entropy. This process is associated with an increase in the normal stress. When the time scale of the process is long enough, physical entanglements may relax as well. At the same time, the stretched coils in the flow stream exert a normal stress across the direction of flow, which is interpreted as the elastic memory of the melt. Once the polymer stream leaves the die and the constraint ends, the stretched coils and the remaining physical entanglements regain a portion of the former coil shape and return to the higher conformational entropy. The three-dimensional relaxation of the polymer chains leads to an increase in the cross-section and shrinkage in the direction of flow of the melt. As a result, the diameter of the extrudate increases while the extrudate shrinks in length.

If the extrudate from a capillary viscometer is gauged by an optical light or laser beam sensor, the extrudate diameter can be monitored and compared with the diameter of the die orifice. The common way to evaluate the die swell is to measure the ratio of extrudate cross-section to die cross-section ($d_{comp}/d_{die}$) with a capillary viscometer or a laboratory extruder equipped with a circular die [28]:

$$B_0 = D/D_0 \tag{8.8}$$

where $B_0$ is the die swell and D and $D_0$ is the diameter of the extrudate cross-section

Even if the ratio $D/D_0$ is a relative measure, it affords a practical way to distinguish between grades of polymers (i. e. molecular weight, long-chain branching). Other ways to express die swell utilize the pressure drop along the die or the normal stress difference, both measured in capillary rheometers [29]. Examples of the relationship between the mean molecular weight, $M_w$, and the die swell, $B_0$, have been given [28].

Similar changes are to be expected when using other rubbers (i. e. SBR, NBR and EPDM), provided the molecular weight is above the critical value at which entanglements are formed. Comparative studies have shown that under certain conditions, the values for viscosity, $\eta$, and die swell, $B_0$, determined with the capillary viscometer, provide the best correlation with the processability of rubber compounds [29]. The disentanglement of polymer chains that can occur during flow in the die is a kinetic process. With a longer die and a slower flow rate, less pronounced die swell will be observed. It has been found that the applied shear stress is the decisive factor for this phenomenon [30]. The disentanglement of the chains in the flow stream leads to a reduction in the elastic retracting forces or the normal stress.

There is an ongoing debate about whether long-chain branching increases extrudate shrinkage or not. When the behavior of the melt becomes more elastic because of the presence of long-chain branches, the elastic memory of the melt increases and the extrudate swells as well. However, effects resulting from rubber chain fracture in the die may overlap the memory effects. Therefore, the conditions employed during extrusion can significantly change the extent of die swelling. Increasing temperatures contribute to increasing the flowability of polymer melts and therefore to reducing die swell. The use of a portion of plasticizer which increases the free volume in the melt also acts in the same direction as the temperature increase.

When designing the extruder die properly, it is essential to consider the compound composition, its viscoelasticity, the head pressure provided by the screw and the length of the die. Die swell can help to predict the quality of rubber mixes and to troubleshoot dimensional stability problems in factories that produce products, such as tires, belts and hoses, by extrusion.

### 8.2.5 Viscoelasticity of Polymer Melts

Rubber melts exhibit typical viscoelastic behavior under periodically varying strains and stress. In general, tests are carried out on rubber process analyzers (RPAs) in which the strain amplitude is increased at a constant temperature and deformation frequency. The shear modulus, G′, and the loss modulus, G″, are determined, which provide information about the dynamic behavior. The dynamic behavior is in between that of Hookean elastic solids and purely viscous fluids, and exhibits elements of both deformation types, depending on the time-scale of the deformation. If the deformation acts for a long time, flow and permanent deformation occur, in a manner similar to that in viscous fluids, while rapid deformation induces an elastic response, characteristic of elastic solids. In addition, it is observed that the value of the modulus is time-dependent and reflects the manner in which the experiment is performed. As a result, there is a pronounced dependence on the time and frequency of the cyclical dynamic deformations and the temperature.

When a viscoelastic rubber compound is subjected to a periodically varying stress, a steady state will be reached in which the resulting strain is also sinusoidal, having the same angular frequency but retarded in phase by an angle, δ, which has values larger than 0° and smaller than 90°. The higher the phase angle, δ, the greater is the viscous contribution and the larger the dissipated mechanical energy. For any viscoelastic body, only a part of the energy applied during a deformation cycle is stored; the rest is irreversibly converted into heat [13, 14]. Therefore, the flow behavior of polymer melts comprises viscous and elastic contributions similar to those of the crosslinked vulcanizate. The mechanical parameters that determine the viscoelastic behavior are always dependent on time and frequency and are also determined by the temperature. If the origin along the time axis or the phase angle is selected to coincide with the time at which the strain passes through its maximum, the strain and stress functions can be written as:

$$\gamma = \gamma_0 \sin(\omega t) \tag{8.9}$$

and

$$\sigma = \sigma_0 \sin(\omega t + \delta) \tag{8.10}$$

where $\gamma_0$ and $\sigma_0$ are the maximum amplitudes of strain and stress, ω is the frequency, t is the time and δ is the phase angle

The stress can be subdivided to two components that characterize the storage modulus or modulus of elasticity, G′, which is in phase with the applied strain and the viscosity modulus or loss modulus, G″, which is out of phase:

$$\sigma = \gamma_0 G' \sin(\omega t) + \gamma_0 G'' \cos(\omega t) \tag{8.11}$$

where $G' = (\sigma_0/\gamma_0) \cos \delta$ and $G'' = (\sigma_0/\gamma_0) \sin \delta$

The storage modulus, G′, is proportional to the stress component and is in phase with the applied strain during a deformation cycle. The loss modulus, G″, is proportional to the loss of mechanical energy and is out of phase with the strain. A useful function to describe the ratio of energy dissipated to the maximum energy stored during a cycle is the loss factor, tan δ, defined as

$$\tan \delta = G''/G' \tag{8.12}$$

The term tan δ describes the viscous contribution normalized to the stiffness of the material. For tan δ = 1, the elastic and the viscous contributions to the material behavior are equal.

#### 8.2.5.1 Viscoelasticity of Unfilled Rubber Melts

Basically, the viscoelasticity of polymer melts is attributed to naturally existing entanglements that are formed when the polymer chains exceed a critical length. The entanglements can slide for a certain distance with the strain amplitude, but provide in any case an efficient resistance to shear or extensional deformation. Melts consisting of long rubber chains behave like transient rubber networks and exhibit elastic behavior at short deformations and high deformation frequencies, respectively, but at long deformations and correspondingly small deformation frequencies, the stretched chains can slide over each other and lose the entanglements through self-diffusion. This makes it understandable that viscoelastic melts have a time-dependent or frequency-dependent behavior. The frequency of deformation is a decisive parameter for the extent of elastic contributions. The more entanglements that exist along the chain, the more the translational movement of the chain (self-diffusion) will be hindered. A demanding mathematical treatment of the deformation of an entangled melt describes provides the storage modulus, G′(ω), and the loss modulus, G″(ω), as a function of deformation frequency at small strain amplitudes.

Since available rheometers can only measure in a limited frequency range, it is adequate to track viscoelastic behavior from frequency isotherms that are subsequently shifted along the frequency axis. The frequency axis is restricted to a relatively narrow range. This means that the frequency axis can be shifted according to the temperatures in order to obtain a uniform representation of the moduli G′(γ) and G″(γ) according to the frequency. This yields a frequency master-curve that describes the evolution of the mechanical moduli from very low to very high frequencies, which is entirely relevant for application. The shift factor, $a_T$, on the logarithmic axis is constant for a given polymer. If the temperature is increased, the measuring time is reduced from t to $t/a_T$. On this basis, the curves can be shifted along the time or the frequency axis in accordance with the Williams-Landel-Ferry (WLF) equation [31]:

$$\log a_T = -C_1 (T - T_R) / [C_2 + (T - T_R)] \tag{8.13}$$

where $C_1$ and $C_2$ are polymer-specific constants, T is the measuring temperature and $T_R$ is an arbitrary reference temperature

Regardless of the type of polymer melt, the mechanical moduli, G′(ω) and G″(ω), are described by a characteristic frequency dependence, as shown in Figure 8.10.

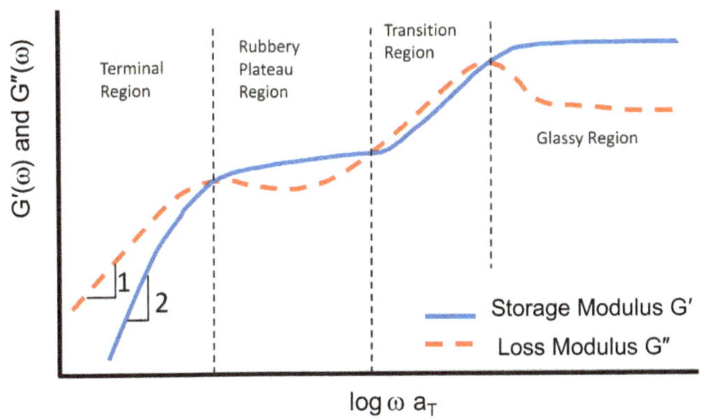


**Figure 8.8** Zones of viscoelastic behavior on the frequency axis

The master-curve describes the viscoelastic behavior of G′(ω) and G″(ω) for four distinct zones:

- terminal zone (or creep zone)
- plateau zone
- transition zone
- dynamic glassy zone

In the terminal zone, the loss modulus, G″(ω), is higher than the storage modulus, G′(ω), at each frequency, indicating the increasing contribution of molecular friction processes to the shorter deformation time [32].

$$\log G'(\omega) \sim 2 \log \omega \tag{8.14}$$

$$\log G''(\omega) \sim 1 \log \omega \tag{8.15}$$

With increase in deformation frequency, the monodisperse polymer melt builds a resistance to oscillatory deformation that is twice as large as the progressing losses produced during cyclical deformation. When polymers of broad molecular weight distributions are investigated, the slope of both functions, log G′(ω) and log G″(ω) as a function of log ω, are reduced to values smaller than 1. This effect can be traced back to the effect of the short polymer chains in the melt that snake out from entanglements in a shorter time and therefore reduce the resistance to dynamic deformation [33]. At the boundary between the creep zone and the plateau zone, the curves of the moduli G′(ω) and G″(ω) cross each other. The intersection between the two modulus curves occurs at the maximum of the G″(ω) curve and represents a reference value that describes the reciprocal value of the characteristic frequency, $\tau_{rep}$, which is required by the monodisperse polymer chain to diffuse out of the imaginary tube. With increase in molecular weight of the polymer, the crossover point is shifted towards lower frequencies or longer relaxation times. Because of the increasing number of entanglements along the chain, the theory predicts an exponential increase in the characteristic self-diffusion time, $\tau_{rep}$, with the molecular weight of the polymer [34].

$$\tau_{rep} \propto {M_W}^{3.4} \tag{8.16}$$

Due to the greater number of entanglements along the chain, the reptation time is shifted to lower values. The cold flow becomes less pronounced. Figure 8.9 illustrates the effect of chain length on the range of the creep zone frequency at which the cross-over of the two functions occurs for G′ and G″ for linear S-SBRs [35].

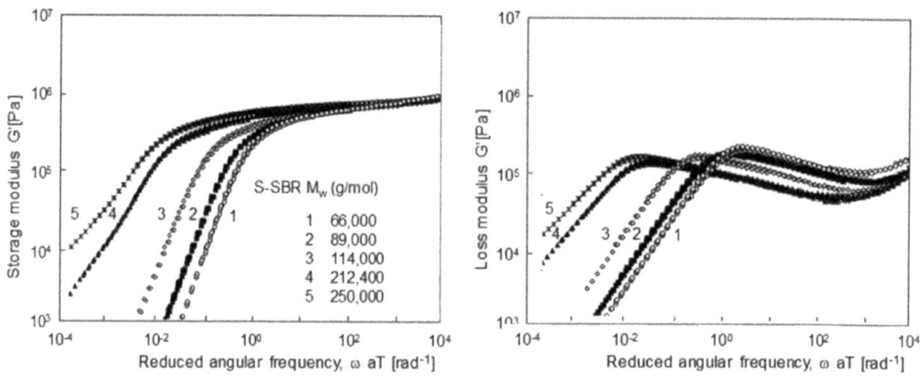


**Figure 8.9** Plot of G′(ω) and G″(ω) as a function of the molecular weight of S-SBR [35]

As shown, the frequency range at which the cross-over of G′(ω) and G″(ω) occurs is shifted by more than two orders of magnitude when the molecular weight of linear S-SBR increases from 66,000 to 250,000 g/mol. The presence of side groups in linear chains increases the chain cross-section and leads inevitably to a decrease in entanglement density. Therefore, at constant molecular weight, an increase in side groups shifts the creep zone and also contributes significantly to reducing the frequency of the cross-over point or the reptation time to significant lower values.

As the deformation frequency increases further, the system enters the plateau zone, where log G′(ω) flattens over a large frequency range, while G″(ω) passes through a minimum value. Due to progressively shorter deformation cycles, the chains are no longer able to diffuse out from the surrounding entanglements of the tube and dissipative phenomena are decreasing. In the plateau, the melt behaves more and more as an elastic solid and flow processes occur to a limited extent. The characteristic value for this regime is the plateau modulus, $G_N$, measured at the frequency where G″(ω) attains its minimum value, as expressed by [34]:

$$G_N = \rho RT/M_e \tag{8.17}$$

where ρ is the density of the melt, R is the gas constant, T is the absolute temperature and $M_e$ is the average molecular weight between entanglements

The plateau modulus, $G_N$, expresses the elastic resistance of the entangled polymer melt to external dynamic deformation. However, it has to be noted that the minimum value of G″(ω) is shifted to lower frequencies when the molecular weight of the polymer increases. Increasing the molecular weight of a polymer does not change the entanglement density and thus the plateau modulus, but leads to an enlargement of the width of the plateau zone in the direction of lower frequencies. With regard to technical polymers of broad molecular weight distribution, a less flat plateau zone is observed. By comparing various polymers, it has been found that the shapes of the curves for G′(ω) and G″(ω) do not depend much on molecular weight, but on the chemical constitution [36, 37]. The magnitude of the plateau modulus, $G_N$, is inversely proportional to the molecular weight of the chain segment between two adjacent entanglements, $M_e$, that is characterized by a rheometric measurement. The higher the entanglement density, the higher is the value of the plateau modulus $G_N$. Because the entanglement density does not depend on the chain length, but on the package of the chains [38] and roughly on the chain cross-section of the polymer, the microstructure of the rubbers has a great influence on the value of $G_N$. The larger the chain cross-section, the smaller is the entanglement density. Therefore, the entanglement density decreases with the size and the amount of side groups attached to the chain backbone. Taking the example of cis-BR with an increased content of 1,2-units, the plateau modulus reaches a maximum value for high cis-BR, prepared by Ziegler-Natta polymerization, while an increase in content of 1,2(vinyl)-units leads to a linear decrease in $G_N$. This is shown in Figure 8.10 [39].

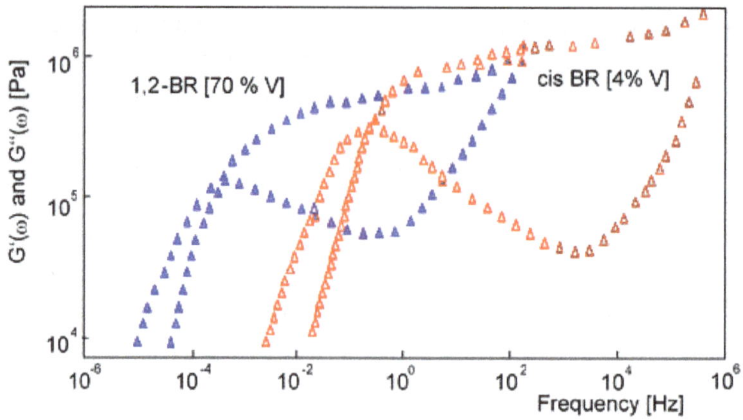


**Figure 8.10** Influence of the vinyl group on the viscoelastic properties of cis-BR

The addition of plasticizers to the rubber melt leads to a decrease in entanglement density, which is attributable to a dilution effect. As a result, the plateau modulus, $G_N$, decreases with the amount of plasticizer. However, the change in the plateau modulus is related to the type and the concentration of the plasticizer [40]. By comparing various polymers, it has been found that the shapes of the curves for $G'(\omega)$ and $G''(\omega)$ do not depend much on the molecular weight, but on the chemical constitution [41]. By making adjustments for the monomeric friction coefficient, it has been shown that butyl rubber enters much earlier into the transition zone, compared with NR and even SBR. This is a direct consequence of the broad glass transition region, which in turn is given by the compact, bulky structure of the isobutylene segments. Energy dissipation at high frequencies become more important since it is known that this process plays a decisive role for tire traction. The glassy zone is not as interesting as the transition or the plateau zone. No configurational rearrangements of chain backbones can occur within a period of oscillation. The stress response for a given small strain is very high and the melt behaves like a glass-like solid.

## 8.3 Test Methods

In order to describe the flow behavior of polymer melts precisely, a number of test methods and the appropriate equipment have been developed over the years [42]. The methods to determine the viscosity of a fluid are based on different experimental approaches. Shear-induced flow in fluids can occur in four different flow modes:

- flow between two parallel plates, one of which moves while the other is stationary
- flow in the annular gap between two concentric cylinders, one of which is stationary while the other rotates

- flow between two parallel plates of between a cone and a plate, with either the plate or the cone rotating or oscillating
- forced flow of a fluid through a capillary due to pressure applied by a piston

In practice, mainly oscillatory rheometers and capillary rheometers are used to measure the viscosity of viscoelastic materials. Chemical reactions in the sample have to be avoided during the test, unless in those cases where such reactions are the goal of the test (i. e. scorch behavior). The first instrument to measure the torque of the material in arbitrary units was introduced by Melvin Mooney in 1936 [43]. The Mooney viscometer was used to measure processability of mixed rubber compounds. The test procedure involves placing a sample on either side of the rotor and filling the cavity by pneumatically lowering the top platen. The platens are electrically heated and controlled at a pre-set temperature (i. e. 100 °C). After closing, a rotor with grooved surface turns at two revolutions per minute, corresponding to a shear rate of approx. 1 $s^{-1}$, while the torque of the test specimen is measured by a force transducer [44]. A schematic diagram of a Mooney viscometer is shown in Figure 8.11.

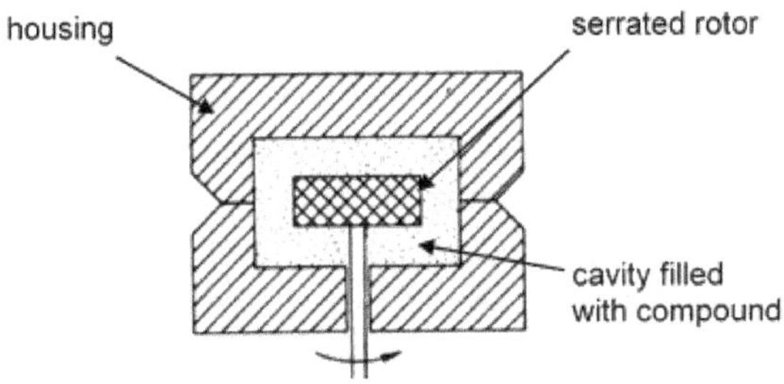


**Figure 8.11**
Scheme of a Mooney viscometer

The viscosity values provided by the Mooney viscometer are given in arbitrary Mooney Units (MU). A value of 100 MU corresponds to a torque of 8.3 Nm. Because of the shear thinning of the rubber samples, the torque values slow down during the first four minutes of the test. The final viscosity corresponds to the lowest value recorded in the last 30 seconds of the test. If sufficient time is given for preheating the test specimen (more than several minutes) and rotation is continued as steady state torque may be observed [45]. The Mooney viscosity measurements of mixed stock give a measure of the amount of mechanical breakdown that has occurred in the mixing process. This crudely relates to the degree of filler dispersion that has occurred in the mixing process. The disadvantage of the Mooney test is the low shear rate that does not allow predictions for the important downstream processes and the temperature indication. It is important to balance the torque measurements without a sample so that the signal coming from the friction of the seal is set to zero while the rotor is moving. Note also that the shear rate of the sample varies depending on the location of the sample. Repeatability, which measures the variations in a single instrument, and reproducibility, which monitors the variation from instrument to instrument, were significantly increased after the introduction of the Mooney viscometer.

After more than 30 years, oscillating rheometers were developed. The measuring principles for torsional rheometers consist in applying a sinusoidal strain in order to assess the resulting complex stress supported by the test specimen. The dynamic viscosity is measured through the application of a sinusoidal strain to an uncured compound in a closed pressurized cavity. Early concepts of cone-plate rheometers [46] and biconical rheometers [47] have been used for low-viscosity fluids and polymer melts. The oscillating disk rheometer (ODR) made process monitoring of rubber compounds quicker and easier than previous quality control testing. This device configuration, the best-known representative of which is the moving die rheometer (MDR), enables easy loading of the test chamber with a sample and a much better thermal contact [48]. The MDR dies have significantly less mass and thus a faster recovery for isothermal tests and higher reproducibility. The test cycle time is reduced significantly if the temperature recovery time is short. The repeatability and reproducibility values are better for MDR than for ODR. The ODR and MDR are still manufactured and provide the backbone for raw rubber and compound evaluation as well as for production release testing.

The rubber process analyzer (RPA), invented in 1992 by Monsanto I&E, introduced a new class of rheometers especially designed for testing rubber in many stages of manufacture. The instrument combines the advantages of the MDR with an advanced technique to vary the deformation amplitude and measuring frequency to allow the measurement of viscoelastic properties as a function of the oscillatory strain amplitude and the shear rate. The RPA can be understood as an advanced moving die rheometer which is capable of performing rheological and dynamic measurements [49]. The relevant parameters are:

- frequency from 0.002 to 5 Hz
- strain amplitude from 0.07 to 125
- temperature from 40 °C to 230 °C

A schematic diagram of the RPA is shown in Figure 8.12.

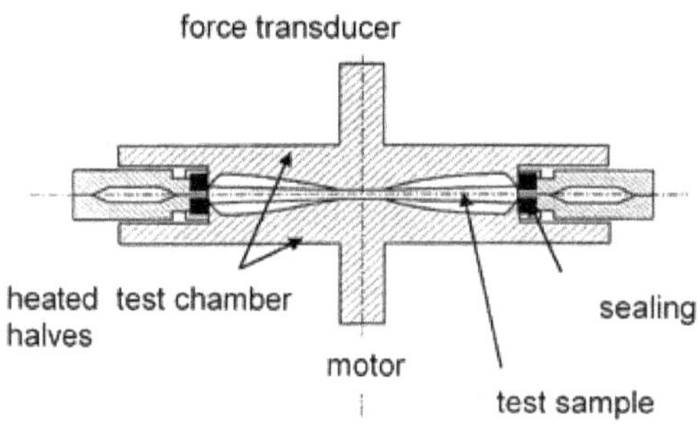


**Figure 8.12**
Schematic diagram of a rubber processing analyzer (RPA)

An RPA consists of a biconical test chamber to prevent slippage. The lower die can be oscillated in torsion at controlled strain and frequency. The use of biconical dies ensures a uniform strain across the entire sample, which is important for filled rubber compounds. The torque measuring device is fixed on the upper die and calibrated with a torsion spring. Like dynamic strain instruments, the RPA measures a complex torque S* that gives the shear modulus, G*, multiplied by a shape factor. The error imparted by the elastomer of the test cavity can be neglected for Newtonian behavior and reaches 10% in the case of a viscoelastic system. The complex torque response, S*, can be divided into an elastic torque, S′, which is in phase with the applied strain, and the viscous torque, S″, which is out of phase with the applied strain. The elastic response, S′, is a function of the amplitude of the applied strain while the viscous torque is a function of the rate of change in the applied strain [50]. The RPA 2000 instrument includes the ability to computer-control other test parameters:

- the oscillation angle
- the oscillation frequency and
- the temperature

Test sequences can be chained together to permit the instrument to assess rheological compound characteristics, cure characteristics and final cured dynamic properties in a single test and with a single sample. It has become the standard instrument for release testing of high-performance raw polymers and tire tread stocks. Since the year 2000, there has been a lot of activity among standard organizations to establish RPA standards to improve quality assurance tests. There are key advantages of using RPA tests for raw rubber or rubber compounds. They provide easy sample preparation, the masses of the dies are low and they allow easy programming of changes in temperature, frequency and strain amplitudes.

Many processes in a rubber plant, such as extrusion, calendering and injection molding have much higher shear rates. Increasing the shear rate causes a given drop in the compound viscosity or the resistance to flow. Capillary rheometers were introduced into the thermoplastic industry after 1950. Capillary rheometers are used to measure the basic rheological properties of polymer melts and to model extrusion processes for compounded rubber at high shear rates. Nearly all capillary rheometer tests start with a warm-up time of 3–6 minutes to allow the sample in the middle of the barrel to reach the barrel temperature. The concept of a capillary rheometer is similar to that of an extruder in that a rubber compound is pressed from a piston (or a small extruder) through a capillary die [51]. A schematic diagram of a capillary rheometer is shown in Figure 8.13.

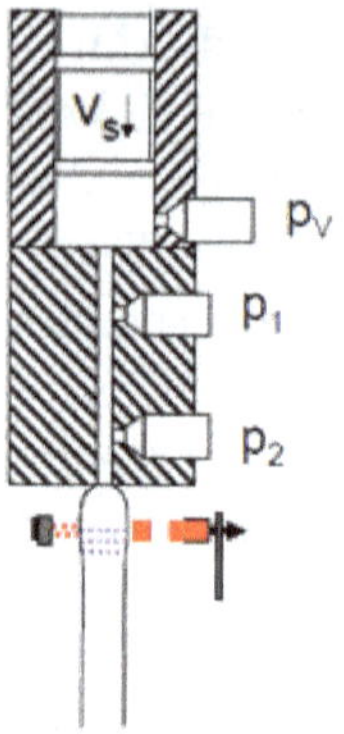


**Figure 8.13**
Schematic diagram of a capillary rheometer

This group of rheometers is mostly used to measure shear viscosity up to very high shear rates (≤ 10,000 $s^{-1}$) at temperatures ranging from ambient to 300 °C. In addition, die swell can be measured very well by optical means. The constant speed of the piston corresponds to the practical conditions of extruding and enables flow anomalies to be detected. The resistance of the compound to flow through the capillary causes a pressure drop. When slit capillaries are used, the pressure drop, $\Delta P = P_1 - P_2$, is measured at different points between the entrance and the exit from the capillary by means of pressure transducers. The viscosity measured by capillary rheometers is expressed by the equation:

$$\eta = \pi R^4 \Delta P / LV \tag{8.18}$$

where R is the radius of the capillary, ΔP is the pressure drop, L is the length of the capillary and V is the volume rate

The viscosity of the melt is a function of the pressure drop in the capillary and the flow rate. The apparent viscosity is determined from the barrel pressure measured by a transducer. If there is a need to measure the "true" shear stress, shear rate and viscosity, then certain correction factors must be applied in a certain way. These corrections are more important when tests are done with orifice dies that have L/D values less than 30. The magnitude of the error increases as the L/D value decreases. Application of the Rabinowitsch correction [52] yields the "true" shear rate and the Bagley correction [53] yields the "true" shear stress while the apparent viscosity is transformed to the "true" viscosity. Dividing the true shear stress by the true shear rate provides the true viscosity. The capillary rheometer is a much better processability tester, but is rarely used in quality control testing because it takes significantly longer to perform tests than the Mooney viscometer.

## 8.4 Rheological Properties of Filled Rubbers

While in the case of unfilled mixtures a deviation from the Newtonian behavior is only observed at high molecular weights of the rubber, filled rubber always exhibits a strong increase in viscosity and a pronounced deviation in the non-Newtonian flow direction. Increasing the solid filler content reduces the volume fraction of the rubber in the compound and increases the physical attachment of the rubber chains to multiple filler particles with higher filler loading. The study of rheological behavior of filled rubbers has both practical importance in terms of the compounds processed and scientific importance as a probe of the compounds' behavior under different test conditions. When dispersing small filler particles in a viscous medium, diffusion process and particle-particle and particle-matrix interaction will play an increasing role as the diameter of the filler drops below 1 µm. It is not only the absolute size of the primary particles and the degree of branching of the aggregates which significantly influence the dispersion process but also the change in the flow behavior of the entire filled melt.

Fillers can form a typical network in a rubber compound, which is reflected in the rheological behavior. The build-up of the filler network leads to an increase in the viscosity of the rubber compound and occurs even at low filler loading. Above a certain deformation speed, this filler network is destroyed, which is expressed by a corresponding decrease in viscosity. Because the filler network can build up again, the viscosity of the mixture can increase again after re-milling. Rheological properties of filled rubbers depend on many factors, such as:

- filler loading
- the size of the primary particles
- the morphology of the aggregates
- the filler surface
- modification of the fillers
- the aspect ratio of the fillers
- filler-filler interaction
- filler-polymer interaction

In proportion to the filler content, the zero viscosity is detected at a low shear rate regime and is transformed in the non-linear regime when filler loading increases to the yield stress. General trends in rheological properties that result by introducing a solid filler depend on the type and the loading, the specific surface area, and the degree of branching of the aggregates (void volume). The extruder die swell decreases with reduction in the elasticity of the compound, which depends on the loading and on the characteristics of the aggregate [54]. At low shear rates, the flow behavior of a filled compound changes from Newtonian to non-Newtonian, with shear thinning behavior at very low filler concentration. At any filler level, they have a higher shear viscosity and require higher shear stress for flow and demonstrate a reduced elasticity. Through adequate filler dispersion, the viscosity of the compound is reduced in rubber compounds.

### 8.4.1 Filler Loading, Specific Surface Area, Structure and Aspect Ratio

How the addition of small solid particles to a fluid affects its viscosity is important to know at the shear rates applied in extruder processing. Many concepts have been developed to describe the flow behavior of rubber compounds containing fillers and, above all, reliable explanations for the mechanism of the rubber-filler interactions in the uncured compound have been provided. The advances in our understanding of the rheological behavior of filled rubber compound provide a basis for tracing the causes of practical problems that arise in production and addressing them by changing the compound composition or the processing parameters. However, a number of practical problems, especially those which indicate changes in the surface activity of the filler, have often still not been answered in a satisfactory manner.

The majority of rheological investigations that deal with filled rubber compounds refer to the changes in shear viscosity provided by particulate fillers relative to unfilled polymer melts. Regardless of the test methods used, the viscosity of a polymer melt increases with increase in volume fraction of the filler. However, the extent of the increase depends on changes in the morphological characteristics, in the filler and especially on the filler structure and surface activity. At this point, we distinguish between the fillers that are iso-dimensional and have a dimension that is not strongly affected by the aspect ratio.

#### 8.4.1.1 Effect of Carbon Black

The flow behavior of carbon black-filled compounds remain relatively simple only at low shear rates; this applies to carbon blacks of large particle size and nearly spherical shape, such as thermal blacks (MT) and flame blacks (FT) of average particle sizes greater than 500 nm. For large spherical carbon blacks with low loading, there is no interaction between the filler particles or the rubber and the filler and the viscosity follow the original Einstein equation, which describes the dependence of viscosity on the filler volume fraction [55]:

$$\eta_f = \eta_0(1 + 2.5\Phi) \tag{8.19}$$

where $\eta_f$ and $\eta_0$ are the viscosity of the filled and the unfilled compound, respectively, and $\Phi$ is the filler volume fraction

The Einstein equation is based on the idea that, since particles suspended in a fluid rotate in response to the application of shear, this rotation disrupts the laminar flow of the fluid and creates resistance to flow. Equation 8.19 predicts that the increase in normalized viscosity caused by the filler is proportional to the volume fraction of the filler. In this equation, the independent variable is the volume fraction. The equation is considered a basic description of "hydrodynamic reinforcement" under conditions in which neither the size nor the morphology of the particles has an influence on the viscosity. Strictly speaking, the equation only applies to ideal viscous fluids. However, as many later studies have shown, it can be transferred to polymer melts with viscoelastic properties.

Attempts to describe the dependence of the viscosity of rubber melts on the volume fraction of carbon black consider the interactions of a pair of spherical particles embedded in the moving fluid. Guth and Gold [56, 57] added a quadratic term to Einstein's equation in order to describe the effect of the mutual disturbance caused by the volume fraction of spherical particles that reduces the volume of the rubber and causes shear amplification in the rubber matrix. The authors introduce factor 14.1 into the quadratic term due to considerations that are not explicitly intelligible from their publications and they formulate the general relationship for hydrodynamic reinforcement as follows:

$$\eta_f = \eta_0(1 + 2.5\Phi + 14.1\Phi^2) \tag{8.20}$$

The Guth-Gold equation describes the viscosity increase arising from the presence of spherical particles and comprises a slight non-linear increase in viscosity with the filler volume fraction that is obeyed up to 50 phr of carbon black MT or FT [58, 59].

Subsequent application-oriented scientific work has led to modifications of the Guth-Gold equation by introducing an amplification factor, f (or shape factor), which, to a first approximation, is intended as a correction for the deviation of the shape of the aggregates from the spherical shape [60, 61]. In light of the knowledge that suspensions of short fibers subjected to shear stress line up to form rod-like chains and therefore offer much higher resistance to flow than spherical particles, the amplification factor was initially interpreted as a function of the aspect ratio of ellipsoidal aggregates. With allowance for inter-rod interaction, the equation for the compound viscosity has been empirically modified as follows:

$$\eta_f/\eta_0 = 1 + 2.5\Phi \times f + 14.1\Phi^2 \times f \ldots \tag{8.21}$$

where $f \gg 1$ is the shape factor

Although the shape factor, f, was initially intended to be related to the anisotropy of an average aggregate of the carbon black, it soon became evident that the aspect ratios determined with the help of TEM images are much too small to even approximate the determined increase in viscosity as a function of the filer volume fraction. The amplification factors, f, that have been in use for a long time are basically nothing more than adjustable parameters without any physical meaning. The shape factor multiplied by the filler volume fraction describes the effective volume fraction that contributes to the increase in the viscosity of the rubber compound.

The effects of carbon black loading on Mooney viscosity have been studied for a series of increasing surface area in NR and E-SBR [62]. The higher the specific surface area of the carbon black, the greater is the increase in Mooney viscosity values, as shown in Figure 8.14.

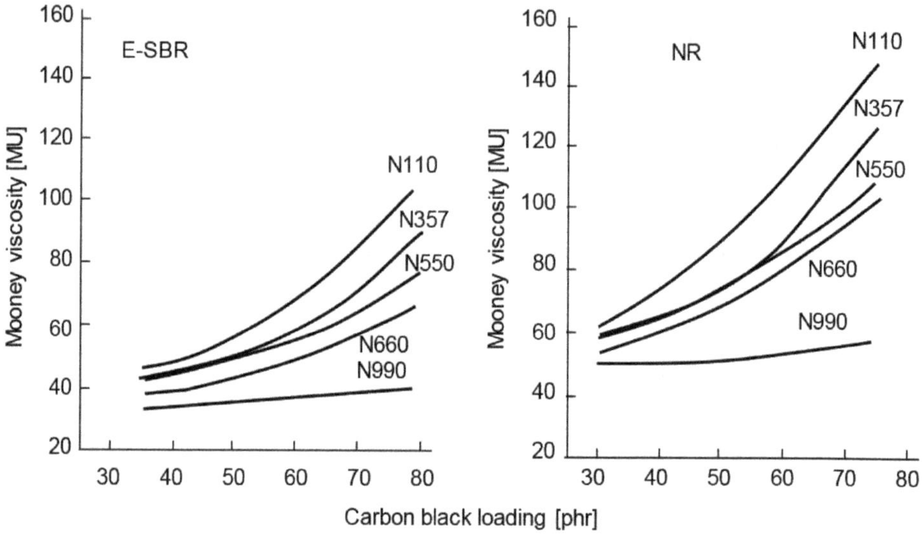


Figure 8.14 Mooney viscosity versus loading of carbon blacks in E-SBR and NR [62]

The shift factor, f, depends on the choice of the carbon black surface activity which increases with the reduction in primary particle size [63]. A plot of Mooney viscosity as a function of fΦ yields a viscosity master-curve which suggests that either increasing the filler loading or increasing the factor f would increase the viscosity. This leads to the conclusion that all furnace blacks are governed by the same mechanism by which the viscosity increases and therefore also the interaction between the rubber chains and the filler at the interface. Consequently, high structure carbon blacks lead in all cases to higher viscosities at the same filler volume fraction.

In E-SBR/carbon black compounds of different specific areas ($14\,m^2/g$ to $164\,m^2/g$), a correlation between the Mooney viscosity and the structure, measured as compressed DBP absorption, has been established. The influence of the branched shape of carbon black aggregates, quantified by the structure (OAN) was considered as a contributor to the high viscosity of rubber compounds. It was realized that the part of the rubber which is "occluded" in the voids of the agglomerates acts as an integral part of the filler. With this part of the rubber volume trapped within the voids of agglomerates and shielded from shear or extensional deformation, the volume fraction of filler increases and correspondingly that of the rubber decreases. As a result, the compound behaves as if it consists of a higher amount of filler. Therefore, there is an equivalence between the filler structure and the filler concentration with respect to the viscosity of the compound. Consistently, the "effective" volume fraction of the filler, $\Phi_{eff}$, has been introduced to represent the filler volume fraction that includes volume of rubber trapped in the voids of the agglomerates. Unlike the empirical amplification factor, f, the effective volume fraction, $\Phi_{eff}$, can be derived from the filler structure.

The effective volume fraction, $\Phi_{eff}$, is always larger than the volume fraction, $\Phi$, by an amount related to the "occluded" rubber and the bound rubber shell. However, occluded rubber should not be confused with the shell of immobilized rubber which is adsorbed on the filler surface. It is the fraction of rubber trapped in the voids of agglomerates and has a low molecular motion.

Equations describe the effective volume fraction as a function of the oil absorption number [60]. The effective volume fraction, expressed as a function of the oil absorption number, is given by the following expression:

$$\Phi_{eff} = \Phi(1 + 0.02139 \times OAN)/1.46 \quad (8.22)$$

The effective inclusion of a rubber portion in the part of the void volume acts to raise the fraction of the filler and increases the effective volume of the carbon black aggregates beyond the mass of the filler. The size of the occluded rubber is responsible for a higher viscosity in the rubber melt [60]. If the compressed oil absorption number (COAN) is preferred as a variable for the effective filler volume, the following expression may be used [64]:

$$\Phi_{eff} = \Phi(1 + 0.0181 \times COAN)/1.59 \quad (8.23)$$

The higher void volume of the carbon black leads to higher compound viscosities than in the case of low structure fillers. The increase in viscosity is more pronounced at low strains, as compared with high strains where the filler network is damaged and the occluded rubber is partly released from its shielding cage by extension of the test specimen. The change in the viscosity is related here to an increase in the OAN or COAN that causes shear strain amplification of the polymer matrix, thus increasing the viscosity at a given loading over the large range of shear rates.

This relationship is known as the principle of structure-concentration equivalence [60]. The higher the structure of the filler, the more viscous is the compound, irrespective of the shear rate. For compounds containing 50 phr of carbon black, it has been found that trebling the OAN doubles the compound viscosity non-linearly. The largest contribution to this increase comes from the high structure blacks (OAN > 80 mL/100 g), which also exhibit a broad distribution of the void volume [65].

However, the described hydrodynamic effects are related to geometrical features of the aggregates and the predicted rheological properties are less strain- and temperature-dependent. Therefore, the non-linear shear-thinning effects observed in filled rubbers cannot be explained by the hydrodynamic effects alone. In addition to the effect of the filler structure, the influence of the specific surface area has to be considered. The firmly bonded layer of immobilized chains has a strong influence on rheological behavior of the compound and should account for reinforcing effects that can be observed in the flow behavior [66].

With a view to the mean molecular weight of the rubbers, it can be seen that the NR compound systematically returns higher viscosity values than E-SBR. Regardless of the molecular adsorption of a polymer chain on the carbon black surface, the effect is sensitive to the entire anchoring of the chain. As large parts of a rubber are adsorbed onto one filler aggregate, it can be adsorbed also onto other filler aggregates, when the loading increases. Such multiple adsorption onto several carbon black aggregates increases the role played by the particle size of the filler [67]. The effective volume fraction was considered to be the total of the filler volume and the bound rubber and was described by the following equation [68]:

$$\eta = \eta_0 (1 - \Phi_{eff})^{-N} \tag{8.24}$$

where $\eta$ and $\eta_0$ are the Mooney torque of the filled and unfilled rubber and $\Phi_{eff}$ is the effective volume fraction, including the bound rubber

The somewhat unusual representation of the shear viscosity as a function of shear stress provides evidence of the influence of the filler level over a wide range of shear stress values [69]. As shown in Figure 8.15, an unfilled NR melt exhibits the expected behavior of a pseudo-Newtonian plateau at low shear stress and hence low shear rate. At high shear stresses ($\geq 10^4$ Pa), the shear-thinning behavior becomes dominant.

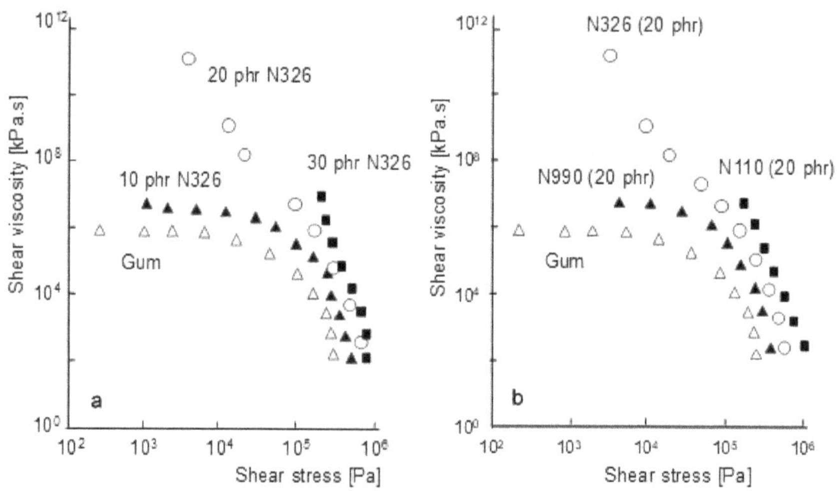


Figure 8.15 Influence of carbon black on shear viscosity of NR compounds: filler loading (left-hand graph) and particle size (right-hand graph) (adapted from [44])

At small filler loading (≤10 phr) of carbon black N326, the compound still shows a Newtonian plateau, which indicates that the continuous NR phase dominates the flow behavior. Above this filler loading, the Newtonian plateau disappears and a transfer to a non-Newtonian flow behavior becomes evident. A suppression of the Newtonian plateau in the shear viscosity function occurs when the filler content is above 12–13 vol.%. This can be seen as the extension in the low-rate range of the non-linear viscoelastic flow that is otherwise observed in pure polymers in the high shear rate range. This non-linearity is induced by the morphology and the filler activity. The system responds more and more as an elastic network and the viscosity tends toward excessive values. The viscosity variation with decreasing shear stress is such that yield stress behavior is suggested. Compounds with high levels of reinforcing fillers exhibit no flow before a stress of 10 Pa is reached.

The relation between the mixing conditions, the morphology of the filler dispersion and the rheological properties has been reported as being a function of the mixing times for SBR/carbon black N330 [70]. A reduction in viscosity during the mixing process was observed concomitantly with the dispersion of carbon black. In the early stage of mixing, the carbon black agglomerates contain immobilized rubber in the space between the aggregates. As mechanical mixing proceeds, agglomerates break down and release a part of the immobilized rubber which causes the effective filler fraction to decrease the viscosity, too.

In addition to filler loading, the degree of branching or the void volume of the filler agglomerates determine the flow behavior of a filled compound. The effect of the structure is shown on a similar type of plot for compounds loaded with 20 phr of carbon blacks of different structure (OAN). As shown in Figure 8.15a, as the filler structure increases (N990 < N326 < N110), there are significant differences in the flow behavior. Whereas a tendency for a Newtonian plateau is shown by carbon black N990, this flow behavior disappears completely when carbon blacks of higher structure are incorporated, and a yield stress behavior is again suggested for the reinforcing carbon blacks N326 and, especially, N110.

Therefore, an equivalence between the effect of filler loading (at constant structure) and the effect of carbon black structure (at constant level) becomes obvious. The same flow functions, $\eta = f(\tau)$, can be produced by using filler concentration or filler structure as variables.

Therefore, fillers with lower structure are used for extrusion, which enable low flow resistance and thus guarantee reduced energy consumption.

The more usual representation of shear viscosity vs. shear rate shows the progressive disappearance of the Newtonian plateau with increasing filler loading and the filler aggregate structure, as was shown in Equation 8.22 and Equation 8.23. Given the same mixing time and different types of carbon blacks, the viscosity of the filled compounds increases with the volume fraction of the carbon black and the shear-thinning flow behavior becomes predominant, but the value of the viscosity becomes more and more similar at high shear rates. An increase in the compound viscosity is intrinsically linked to the specific surface area and the contact surface between filler and rubber. With increase in the loading, the contact surface increases and an increase in compound viscosity is observed, regardless of the polarity differences between rubber and filler and the mixing conditions. Due to this, the fraction of the rubber that adheres on the filler surface increases [71]. This reinforcing phenomenon can be explained by the increase in polymer-filler contact surface per unit volume. A higher contact surface at constant loading enables more polymer chains to be adsorbed on the filler and to form an interface because a multiple adsorption of the chain on several aggregates. This increases the hydrodynamic effective volume of the filler particles and at the same time reduces the proportion of flowable rubber melt. Therefore, the resistance of the filled compound to be deformed under shear or extensional stress increases consistently.

In publications, the bound rubber is often taken as an indicator of filler activity and so it is found that the increase in bound rubber is linked to an increase in viscosity [72]. This approach is quite helpful for estimating the effects of a filler on the viscosity of a compound. In any case, the increase in polymer-filler interaction is accompanied by an increase in adsorbed rubber on carbon black, leading to an increase in shear viscosity. Such an increase in specific surface area and filler activity results in an increase in flow viscosity. It has been shown that in-process modified carbon blacks with a reduced adsorption capacity have a correspondingly lower viscosity. The poor relationship between bound rubber content of filled compounds and the flow behavior is underlined by the fact that re-milling of the compound leads to a rapid decrease in viscosity in filled saturated polymers, such as EPDM, IIR and CIIR. After a storage time, the viscosities can increase substantially [73].

Non-linear effects have been observed in stress-growth experiments performed with oscillatory rheometers of variable rotor speed that allow the shear stress to change when the rotor is suddenly started [69]. With such a device, the change in the torque can be recorded as a function of time.

If the ratio of the maximum viscosity to the viscosity in the steady flow state $\eta_{max}/\eta_{\infty}$ is established, both the influence of the shear rate and the filler type become more evident. The use of a variable-speed Mooney viscometer to perform stress growth experiments on NR and SBR 1500 compounds with various types and levels of CB revealed that the ratio between the "peak" value of the torque measured immediately after starting the rotor and the equilibrium torque after a time of relaxation is related to the activity of the filler in the state of dispersion of the sample. As shown in Figure 8.18, the ratio increases with shear rate for compounds filled with carbon blacks of increasing specific surface area [44].

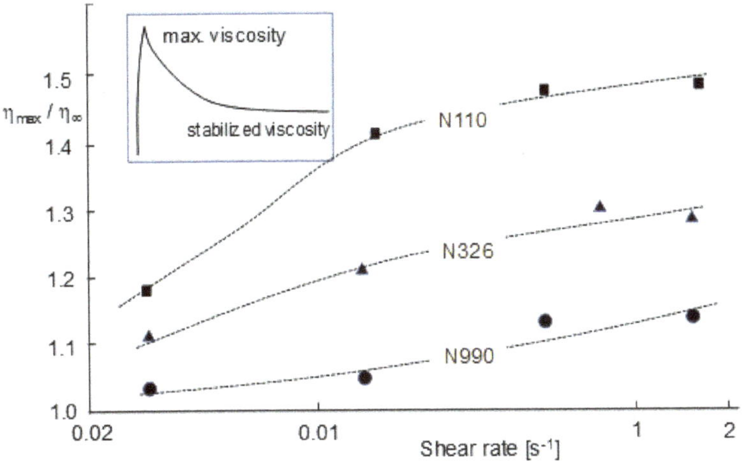


**Figure 8.16** Overshoot experiments on carbon black-filled compounds

In addition to the influence of the filler type and loading, the average molecular weight of the rubber and the degree of filler dispersion also contribute to the level of the "peak" torque value and the stationary torque value. A comparison of the torque values at the beginning of the flow process and in equilibrium makes it possible to estimate the influence of the rotor speed on the very complex shear processes of the compound. It turns out that the effects are greatest for active fillers that are difficult to disperse. Such measurements can provide information about the degradation of the rubber at high filler loading, which is important information for the factory floor. The structure of the filler has an equivalent role to the viscosity with constant loading as the specific surface area. Due to the fact that a considerable part of the rubber matrix is "occluded" in the void volume of the aggregate after the end of incorporation, the ratio of filler volume to rubber volume is shifted, so that the flow behavior in the mixture corresponds to that of a more highly filled compound if the strain amplitude is kept small.

When the strain amplitude in an oscillatory test is increased, the dynamic modulus of the filled compound starts to breaks up the flocculated network structure. The strain-induced breakdown of the filler network is governed by the strain energy. For different concentrations or types of filler, the maximum in the loss tangent, associated with filler network disruption, occurs at the same value of the strain energy (3.5 kJ/m$^3$) [74]. At higher strains, the filler network structure disappears and the main mechanism affecting the flow of filled rubber is the strain amplification in the polymer resulting from the inextensibility of the filler. Carbon black flocculation takes place during storage and leads to an increase in the compound's viscosity [75]. The increase in the viscosity can be reduced by re-milling on a two-roll mill.

The blend SBR/BR and NR/BR was tested with carbon black N330, N351, N299, N110 and N121 in percentages of 25, 50 and 75%. The viscosity was measured with a Monsanto processability tester at a shear rate of 100.7 $s^{-1}$. The viscosity of the NR/BR blends increase slightly with higher BR content, while the SBR/BR compounds show an opposite effect. The influence of the filler structure on the natural distribution of carbon black is shown in Figure 8.17 [76].

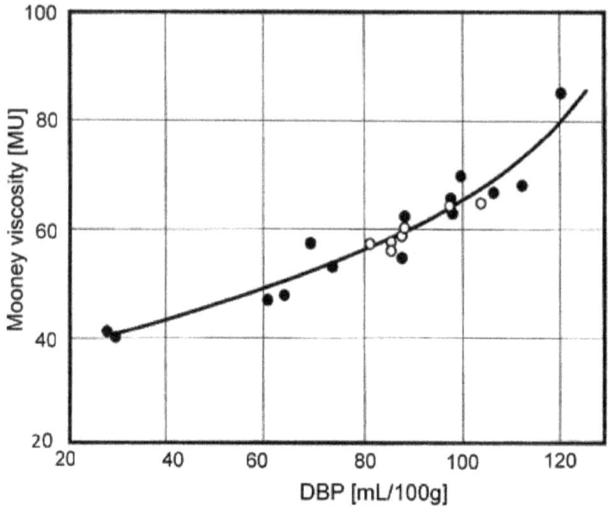


**Figure 8.17** Increase in Mooney viscosity with the structure of carbon black at constant loading (50 phr) (adapted from [48])

All extensional viscosities are associated with great experimental difficulties, which explains some discrepancies. The extensional viscosity as a function of time is superimposed on a master-curve to an almost constant value before rising more rapidly, indicating strain hardening. The extensional viscosity is much higher than the shear viscosity, which remains in the MPa s range and can reach values of 1 GPa s at low elongation rates [77].

In order to validate the performance of the carbon black obtained from the pyrolysis oil compound with NR and S-SBR were produced. The viscosity of sCB N234 according to ML(1+4) at 100 °C was practically the same as the original ASTM carbon blacks. Thus, this new type of carbon black can be introduced into production without any additional effort [78].

During the storage time, the Mooney viscosity of the rubber/carbon black compounds increases, which is attributable to the continuous formation of bound rubber and the simultaneous flocculation of the carbon black aggregates. A re-milling of the compound on a two-roll mill reduces the viscosity of the compound. For the compound obtained by wet-mixing of Cabot elastomer composites (CEC), the increase in viscosity is less pronounced than for the dry-compounded carbon black compounds.

#### 8.4.1.2 Effect of Precipitated Silica/Silanes

Precipitated silica gained a strong influence in 1990, particularly in the production of treads for car tires. However, this filler is characterized by strong filler-filler interaction and low polymer-filler interaction, due to the silanol groups on the silica surface. Adding silicas to rubber tends to build viscosity more rapidly than most fillers. This is because silica particles have silanol groups on their surface and they inter-aggregate with other particles to form strong filler networks that provide high melt viscosities. Precipitated silica has hydrogen bonding holding its agglomerates together, in contrast to with carbon black, which has van der Waals interactions. This difference produces a different shear thinning profile. Using silica filled NR compounds filled with silica as an example, a linear increase in the Mooney viscosity is evident, as shown in Figure 8.18 [79].

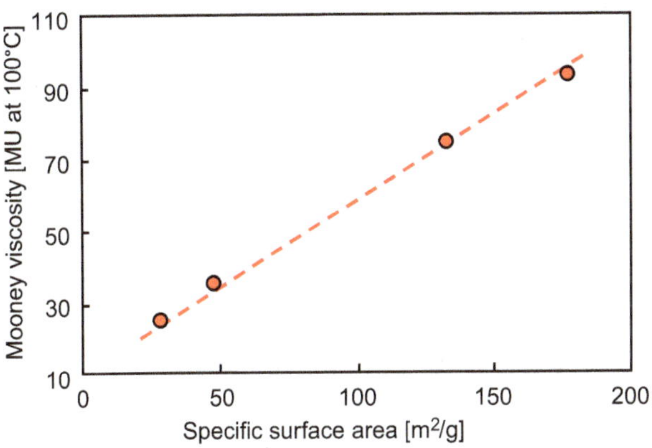


**Figure 8.18**
Influence of silica surface area on the Mooney viscosity at constant filler loading (50 phr) [79]

Even given the incompatibility of the polar surface of precipitated silica with the low polarity of hydrocarbon rubber, there is a clear correlation which can be traced back to the influence of the polymer-filler contact surface between the polar surface and the NR chains; as the molecular weight of NR is high, multiple adsorption can be assumed. If the Mooney viscosity of precipitated silica and carbon black is compared in two rubbers, NR and NBR, the higher viscosity is always in the NBR with the higher solubility parameter, which leads to better dispersion. In order to normalize the effects of both loadings and the influence of the polarity a viscosity difference has been defined:

$$\Delta ML = (ML_{silica} - ML_{carbon\_black}) / ML_0 \tag{8.25}$$

It is clear that the viscosity in the case of NBR does not increase in in the same way as with NBR, due to the strong interactions of the silanol groups with the acrylonitrile groups of the rubber. The effect of the polymer-filler interaction is shown in Figure 8.19.

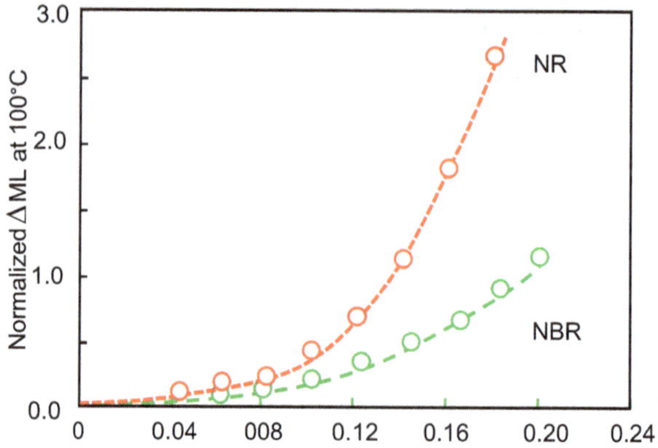


**Figure 8.19** Normalized Mooney viscosity ΔML at 100 °C for silica filled NR and NBR compounds [80]

The Mooney viscosities in NR from silica VN2 and carbon black N110 both increase with filler loading at the temperature of 100 °C, but the increase in the viscosity of precipitated silica is much greater at the filler loading of 0.1%. The difference in Mooney viscosity values is determined by the agglomeration of silica aggregates due to interaction of the silanol groups [80].

A number of chemicals, such as diethylene glycol, triethanolamine at 2 phr, can reduce Mooney viscosity by 30%. The best chemicals are the organo-silanes, which chemically reduce the presence of the silanol groups and therefore the polarity of the aggregates. By silanizing the silica with bifunctional silanes, the filler-filler interaction has been significantly reduced and the silanized silica behaves as a relatively good-natured filler from a rheological point of view, if the difficult task of incorporation is ignored. A distinction is made below between the flow behavior of precipitated silica and silanized silica. Precipitated silica shows a large change in Mooney viscosity during storage. A compound with a less than optimal dispersion exhibits ever-increasing zero viscosity and ever-increasing post extrusion swelling, compared with a material in which the filler has been effectively dispersed.

If the comparison of precipitated silica and carbon black N110 is now repeated for the action of a monofunctional organo-silane (Si208) and TESPT (Si69), the same low viscosity is obtained as was achieved when Si208 served as the silanizing agent. The Mooney viscosity corresponds to that of carbon black N110. However, TESPT drops to half the viscosity of VN2.

The effect of silanization of the precipitated silica depends on the type of silane and the quantity used. Hydrophobization with monofunctional silane, which couples to the silanol groups but cannot react with the rubber, leads to a strong reduction in the silica network and the viscosity of the rubber compound remains low during the storage time. As long as the hydrophobization of the silica is not complete, the formation of a silica network is still possible, but its strength will be lower than that of the unmodified silica [81]. A reduction in viscosity that is brought about by silanes in highly filled compounds is beneficial because silica flocculation reaches a low level in the silanized case. The reduction in Mooney viscosity of S-SBR/BR on HD silica Ultrasil 7000 GR (80 phr) containing more than 4 phr silane is below the level attained with carbon black N234 at 80 phr, as shown in Figure 8.20 [82].

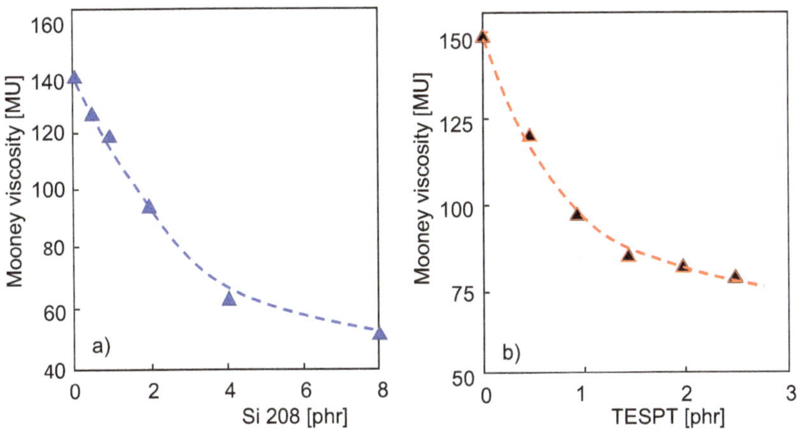


**Figure 8.20** Reduction in Mooney viscosity by silanes: a) HD silica (80 phr)/Si208 in S-SBR and b) VN3 (50 phr)/TESPT in E-SBR [82]

The influencing of the silanization process on the production of S-SBR/BR rubber compounds is a demanding activity, due to the resulting silanization in the internal mixer. The degree of silanization is adjusted according to the mixing temperature and the mixing time using a Mooney or a rubber process analyzer (RPA) in the 2-to-4-stage remilling stages.

As a consequence of the viscoelastic properties of the Payne effect, the linearity limit of $G'(\gamma)$ decreases with the increase in the volume fraction of the silica:

$$\gamma_c = \alpha \Phi^{-\nu} \tag{8.26}$$

With increase in silica loading, the stationary filler network is disturbed earlier and a decrease in $G'(\gamma)$ occurs as the strain amplitude increases. With regard to silica particles, different reports describe a decrease in viscosity [83, 84]. This can be observed for hydrophilic and hydrophobic silica; the values of the exponent are quite different: $\nu = 3.0$ and $\nu = 2.4$ [85, 86].

The difference in Mooney viscosity values depends on the technology used to prepare the rubber compound. For wet-mixed silanized silica (EVEC-L), the viscosity is lower than for the dry-compounded silanized silica and similar to that of carbon black N234 in the same rubber blend S-SBR/cis-BR. It seems that the lower viscosity of wet-mixed silica is related to the adsorption of single rubber chains, while the viscosity in the carbon black system depends on multiple chain adsorption [87]. The dry-compounded silica shows a strong increase when the compound is stored. The hardening of the wet-mixed silica is significantly reduced when compared with the dry-mixed silica. The lower hardening effect is certainly related to the surface modification of silica with the coupling agent.

Like most chemical reactions, silanization is influenced by the temperature, the concentration of the reactants and products, transportation and diffusion processes and the influence of moisture. Good silanization is needed not only to guarantee the best reduction in the viscosity or the Payne effect, but also to reduce compound hardening in storage [88].

#### 8.4.1.3 Effect of Layered Silicates

Because of the platelet structure of layered silicates, the quite high viscosity of the compound filled with clay relative to carbon black or precipitated silica is related to the high specific surface area and the high aspect ratio of the filler. Due to the higher specific surface area and the high aspect ratio, well-dispersed layered silicates are used in pristine formation or as organo-clays at low concentrations. A few phr above the rheological percolation threshold, their range of use is very limited. In addition to the rheological investigations, the type of production is also differentiated, because the intercalation and the exfoliation of the two-dimensional platelets can be quite different. A distinction is made between layered silicates that are introduced via melt blending, solution blending or latex blending. The layered silicates are easily dispersed in water, which acts as swelling agent owing to the hydration of the cations $Na^+$ and $K^+$. The water swelling capacity of layered silicates varies with the type of clay and the cation exchange capacity [89]. Therefore, mixing with rubber latex is a promising route for producing nanocomposites [90]. In NR-MMT and CR-MMT prepared latex compounding the XRD shows a part of the clay is not exfoliated. Latex ARD compounding leads to an exfoliated structure of the clay that improves the properties when it is performed dynamically [91]. Zero shear viscosity of layered silicates is higher than for the unfilled sample. The viscosity of the rubber compounds filled with layered silicates increases with the filler concentration and exhibits strong shear thinning at higher shear rates. The enhancement of the shear viscosity indicates a uniform distribution of the clay particles and a higher extent of exfoliation if the production method is CDLC (continuous dynamic latex compounding).

The Mooney viscosity of NBR and XNBR rubber with exfoliated montmorillonite obtained by the use of the CDLC technique provides higher shear viscosities [92]. This preparation technique provides much better exfoliation of the tactoids at low concentrations of the layered silicates that increase the presence of dispersed exfoliated platelets. The increase in viscosity is triggered by the particle aspect ratio that increases the effective volume of the montmorillonite platelets. The maximum viscosity is obtained at 0.035 $\Phi$ and, above this concentration, the viscosity decreases, due to lower dispersion in the rubber. The resistance to shear deformation is more pronounced when the number of fanned-out tactoids and exfoliated platelets in the rubber compound increases as a result of wet-compounding technique. The mechanical implementation of the montmorillonite shows small increase at low concentrations of montmorillonite, as shown in Figure 8.21 [93].

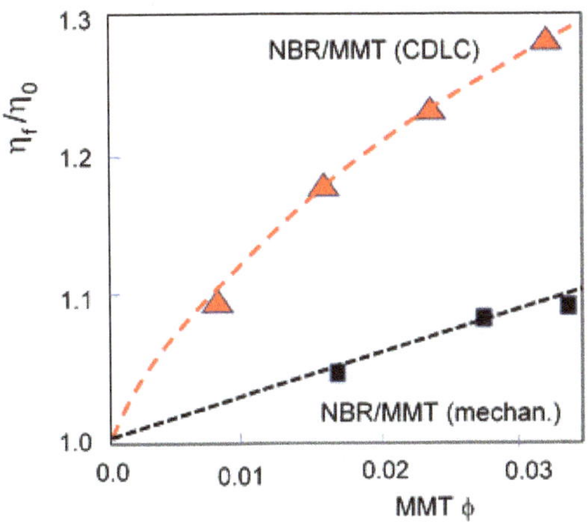


**Figure 8.21**
Relative Mooney viscosity of NBR/MMT wet-mixed and mechanically mixed samples [93]

In non-polar rubber, the viscosity of the compound is less pronounced, because of the smaller filler dispersion and less-exfoliated structure. As the solubility parameter of the rubber increases, the viscosity will change because there is better interaction with the polymer. The Mooney viscosity is higher for NBR/MMT mixtures than for those containing conventional fillers. This result shows the impact of the platelet dispersions on the rheological properties. The resistance to shear deformation is less pronounced for continuous dynamic latex compounding (CDLC) mixed technique. The increase in viscosity is triggered by the particle aspect ratio that can be described by a form factor, f. At a low filler content, the results show a higher form factor for samples prepared by wet-compounding and a decrease in the form factor as the loading increases.

For the unvulcanized layered silicate compounds, the temperature dependency of G′ and G″ are increased when MMT is added to the NBR matrix. At room temperature, G′ increases by a factor of 2.9 and G″ by a factor of 4.4, as a consequence of the large specific surface area and the high aspect ratio. The activation energy of G′ prepared by wet-compounding is larger than that of mechanically mixed platelets, a fact which is attributed to better phase adhesion. The difference between $G'_0$ in the case of dry-compounding and wet-compounding demonstrates that the impact on G′ is higher due to the better exfoliation.

In order to achieve good interaction in an aqueous system across the CDLC mixture, it is necessary to take account of the viscosities of the layered silicate in water. For this, the Brookfield viscosity of some industrial MMT preparations was measured. There was a clear difference in the viscosity, due to the content of the $Ca^{2+}$ and $Mg^{2+}$ ions. The higher this content, the more the Brookfield viscosity increases and thus these layered silicates fail to provide a good material on wet-compounding. The content of these two ions leads to band structures between the platelets, which favor coagulation and do not allow isolation in exfoliated platelets. Of the layered silicates, Nanofil 116 is unsuitable for wet-mixing, because it does not disperse well in water, whereas $Na^+$Cloisite and PGN have good prerequisites for such a mixing process [91].

NR latex mixed with $Na^+$ bentonite varying in concentration from 1 phr to 2.5 phr shows increased viscosity with increase in clay loading [94]. The viscosity measurements can be easily performed at low filler concentrations in rubber matrices, such as IR, ENR, SBR, EPDM [95–97]. The platelets lie parallel to the processing direction. With increase in loading of the modified clay in the rubber compound, the viscosity decreases during shear thinning. For organo-clay, a pronounced shear thinning behavior was determined that increased with the content of organo-clays in BR, SBR and NBR [98], BIMS [99] and fluorinated rubber [100]. The rheological properties of BR, SBR and NBR were investigated by changing the shear rates from 12.2 to 183.7 $s^{-1}$ using a rubber processing analyzer (RPA). The shear thinning and shear viscosity and the die swell decrease with the addition of octadecylamine modified organo-clay. The higher the polarity of the rubber, the more the shear viscosity decreases with the incorporation of organo-clay. This was explained by the unidirectional orientation of the organo-clay in the rubber matrix in the capillary rheometer. It was shown that a compensating cation with a longer alkyl chain leads to a reduction in Mooney viscosity [101]. Organo-clays have the opposite behavior to that of traditional filler. The results depend on the orientation of the organo-clays along the flow direction and on the slippage of the platelets on the polymer chains. Different studies of organo-clays-rubber have been published for NBR [102]. The filler is more intercalated than exfoliated. The influence of the filler concentration on steady shear viscosity is found to be strong at low shear rates and increases with the organo-clay [103]. The reduction in viscosity upon increasing the clay concentration may be due to a breakdown of platelet tactoids.

Due to the high aspect ratio of organo-clay platelets, such fillers increase the melt viscosity [104]. At low strain rates, the elongational viscosity increases with time and reaches a plateau. At high strain rates, the curves exhibit typical strain hardening, and the viscosity goes up more rapidly. In rubber compounds, a percolation threshold for organo-clay appears at low concentrations of $\Phi = 0.01$. In organo-clays, the percolation threshold has been observed to decrease with increase in the dispersion of the tactoids and the increase in exfoliation [105]. In cis-BR/organo-clay, the temperature can be a controlling factor for the exfoliation of the tactoids [105].

The steady shear viscosity measured in the molten state for epichlorohydrin/organo-clay at 200 °C shows an increase with the content of the organo-clay. This increase arises because the interaction and the dispersion of organophilic clay appears to provide resistance to flow. By increasing the content of organo-clay, shear thinning starts at lower temperatures. The zero viscosity and the characteristic times for ECO/organo-clay are given in Table 8.1 [106].

**Table 8.1** Zero Viscosity and the Characteristic Times for ECO/Organo-clay

| Code | $\eta_0$ (Pa s) × $10^{-5}$ | λ (s) |
|---|---|---|
| ECO | 0.44 | 3.27 |
| + organo-clay 2 phr | 0.57 | 2.49 |
| + organo-clay 5 phr | 0.87 | 3.58 |

The viscosity of liquid BR/organo-clay has been investigated, focusing on clay exfoliation and orientation as well as clay interaction with the rubber. Irreversible and reversible viscosity transitions were observed in the temperature range 26 to 136 °C. The transitions were found to depend on the end groups and the molecular weight of the liquid BR [107].

#### 8.4.1.4 Effect of Carbon Nanotubes

The rheological properties of multi-walled carbon nanotubes in rubber compounds depend on the loading of multi-walled carbon nanotubes, aspect ratio and the dispersion in the matrix and also on the type of polymer, the molecular weight and on the interfacial interaction between the multi-walled carbon nanotubes and the rubber. Due to the likewise very high aspect ratio and the high specific surface area, multi-walled carbon nanotubes have a major influence on the viscosity of the rubber compounds. The reason for this is the interactions of the $sp^2$-hybridized orbitals along the length of the nanotubes. At low frequencies, the rubber compound exhibits a typical viscoelastic plateau. As the loading of multi-walled carbon nanotubes increases, the Newtonian viscosity trend drops and a shear-thinning behavior becomes evident. Only relatively low concentrations (< 5 wt.%) of multi-walled carbon nanotubes can be incorporated into rubber compounds, due to the rapid increase in viscosity and subsequent difficulties at high loadings [108, 109].

The multi-walled carbon nanotubes are introduced into the rubber compounds by mechanical treatment, such as milling. After milling, the Mooney viscosity fell from 73 MU to 36 MU and, after further milling up to 60 min, ended at 17 MU [110]. It is clear that, with increase in multi-walled carbon nanotubes content, the Mooney viscosity rises sharply compared with rubber compounds that contains iso-dimensional fillers, as shown in Figure 8.22.

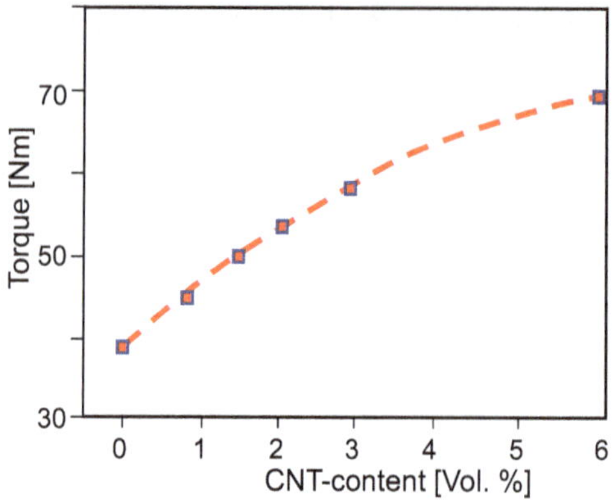


**Figure 8.22**
Torque for carbon nanotube in silicone compounds [110]

Such an increase is seen in silicone rubber where the torque value increases with the content of multi-walled carbon nanotubes. From the mixing torque, it is evident that the viscosities increase in the order:

NR < HNBR < NBR < FKM

In silicone rubber, an exponential increase in the viscosity has been observed in the concentration range up to 7% [111]. In the case of the viscosity of the NBR filled with multi-walled carbon nanotubes, a connection to the concentration of the polar acrylonitrile groups is clearly established. The more of these groups present in the NBR, the higher the Mooney viscosity in the compound. This increases linearly with the concentration of the multi-walled carbon nanotubes in the area where there is high dispersion and an increased shear-thinning viscosity. From the viscosity of the NBR, which is 36 to 39 MU after the concentration of the acrylonitrile groups, this value increases proportionally, due to the mechanical incorporation of the carbon nanotubes. At a concentration of 3 vol.%, it is 49 to 56 Mooney units of the filled system [112]. Such an increase in rubber viscosities is not observed for comparable multi-walled carbon blacks at this concentration level. The difference in rubber melt performance results from the formation of the filler-filler networks that are effective at small deformations. The effects of multi-walled carbon nanotubes on the glass transition temperature depend on the nature of the polymer. An increase in glass transition temperature is caused by a surface that interacts strongly with the polymer, because it undergoes strong adsorption to the polymer [113].

Even in the dilute regime, multi-walled carbon nanotube solutions show clear shear-thinning behavior [114]. A decrease in viscosity at high shear rates occurs because multi-walled carbon nanotubes can become oriented if the shear is sufficient; the shear rate of the transition is a measure of how fast the tubes are able to rotate. The behavior in the high shear region dropped linearly with the logarithm of the shear rate. The zero-shear viscosity increases with the multi-walled carbon nanotubes concentration. A sharp viscosity decrease is observed when the shear is applied to the multi-walled carbon nanotube silicone rubber. Harder to predict is whether the transition from the zero-shear rate viscosity to non-Newtonian viscosity changes with the carbon nanotubes [115, 116].

With an increase in up to 0.05 wt.% of multi-walled carbon nanotubes, the shear-thinning effect increases significantly, because of the high aspect ratio of the carbon nanotubes. Higher shear thinning in multi-walled carbon nanotube-filled systems suggests that the viscosities at high shear stress should overlap with those of the unfilled systems.

At low frequencies, the relaxed polymer chains exhibit a typical Newtonian viscosity plateau. With the addition of the multi-walled carbon nanotubes, the low frequency viscosity increases, whereas the overall viscosity trend decreases with increase in the frequency for a given content of multi-walled carbon nanotubes , indicating that the relaxation of the polymer chains in the nanocomposites is effectively restrained by the presence of carbon nanotubes.

Shear thinning at high shear rates has been found where multi-walled carbon nanotubes are mostly isolated and a power law did not describe the data. A harder to predict response is whether the critical shear rate transition from Newtonian to non-Newtonian flow changes with the addition of multi-walled carbon nanotubes. This transition takes place at lower shear rates than for the iso-dimensional fillers. This behavior is a consequence of the high aspect ratio of the multi-walled carbon nanotubes [117, 118].

The rheology of multi-walled carbon nanotubes-rubber compounds shows a transition from the rheological state when the viscosity changes significantly with increase in filler content to a solid-like behavior in which the viscosity is insensitive to the filler content. The concentration at which this transition occurs is known as the rheological percolation threshold [119]. The rheological percolation threshold of multi-walled carbon nanotubes is strongly dependent on temperature, because the viscous properties are dominated by the network that forms due to the high aspect ratio of the tubes and that can be enhanced at elevated temperatures. Elongational viscosity measurements on multi-walled carbon nanotubes suspensions with different aspect ratios have been performed. The transition to shear thinning is dependent on the length of the aspect ratio [120].

Steady-state oscillatory rheometry gives important clues about the structure and behavior of carbon nanotube-polymer compounds. At low frequencies, the storage modulus decreases with a slope of 2 and the loss modulus with a slope of 1, both as a result of the long-chains nature of the polymer. When multi-walled carbon nanotubes are added to the polar rubber HNBR or HXNBR, the Mooney viscosity increases linearly. The increase in viscosity is more evident from the loading of the multi-walled carbon nanotubes in the more polar HXNBR, which also has carboxylic centers in addition to the acetonitrile groups and provide a stronger interaction. The effect of the multi-walled carbon nanotubes has been compared with the influence of the iso-dimensional carbon black, which does not increase the viscosity at such small loadings. Figure 8.23 shows the effect of multi-walled carbon nanotubes and active furnace carbon black in HNBR and HXNBR [121].

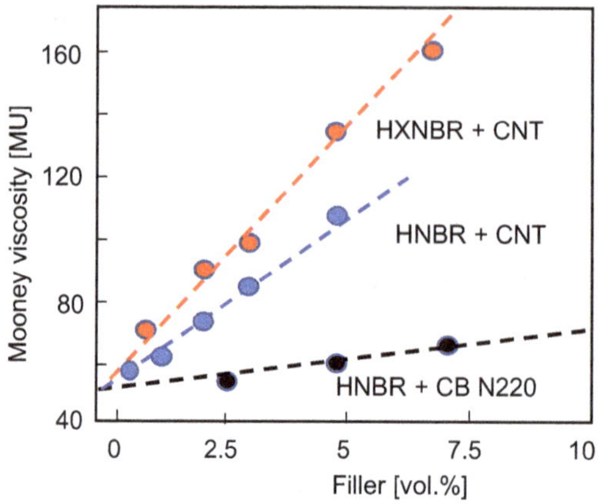


**Figure 8.23** Effect of carbon nanotubes on the viscosity in polar rubbers [121]

In an oscillatory rheometer experiment, multi-walled carbon nanotubes NC 7000 were compared with carbon black N220. The increases in G′(γ) values are shown in the rubber processing analyzer (RPA) test in HNBR in the range up to 10% by volume of the carbon nanotube at a temperature of 80 °C. The exponential increase in the storage modulus, $G'_0$, is very pronounced and leads to values of 30 MPa at 10% by volume fraction. The value of the carbon black N220 of 25% by volume is thus achieved, which represents a reduction in the weight of the compound. As shown in Figure 8.24, the G′ versus filler loading can be plotted in a double logarithmic representation. The rheological percolation threshold for the carbon nanotubes is 1.9 vol.% while for carbon black N220 it is 17 vol.%. Increasing the test temperature changes the rheological percolation in a small manner [121].

In PDMS, an exponential viscosity increase was observed and shows strong shear thinning [122]. In NR, a combination of carbon black N375 with a small loading of carbon nanotubes show a significantly increased Mooney viscosity compared with the level of only carbon black N375 [123]. The shape factor of carbon nanotubes is one order of magnitude higher than the shape factor of the carbon black and this explains the large increase on the viscosity of that system. To describe the effect, an empirical equation has been proposed:

$$\eta_f = \eta_0(1 + 0.67 f_{CB}\Phi_{CB} + 0.67 f_{MWCNT}\Phi_{MWCNT} + 1.62 f_{CB}^2\Phi^2 + 1.62 f_{MWCNT}^2\ \Phi_{MWCNT}^2 \quad (8.27)$$

This effect can be attributed to the high aspect ratio of carbon nanotube, which leads to a high form factor. The influence of the aspect ratio on the effect of the carbon nanotubes can be determined for all physical properties of the rubber compounds.

### 8.4.2 Influence of Filler Dispersion

Dispersion of the filler is realized during mixing process and characterizes the filler-polymer interaction. Therefore, filler dispersion has an effect on the viscosity of a filled compound. Since the Mooney viscosity is easy to measure, it has become common practice to describe the interaction between the polymer-filler in rubber compounds, even if the Mooney viscosity is only valid for a certain shear rate. It has been repeatedly found that, with increase in mixing time, the shear viscosity of a filled compound progressively decreases until it reaches a certain low level at longer mixing times [124, 125]. Therefore, the Mooney viscosity is inversely proportional to the filler dispersion. Due to the progressive erosion of the pellet fragments and large agglomerates, an ever-increasing number of aggregates are transferred to the rubber matrix and encapsulated by a polymer matrix. This increases the local strain amplification of the rubber compound around the filler particle, but, overall, the mixture becomes softer and shows a higher fluidity. During mixing, a progressively higher amount of rubber is released from occlusions in the agglomerate voids and therefore it contributes actively to a better flow [126]. Viscosities of carbon black-filled E-SBR compounds, measured with a capillary rheometer, show an exponential decrease as the mixing time increases, as shown in Figure 8.24 for three types of carbon blacks which are characterized by different aggregate structures [127].

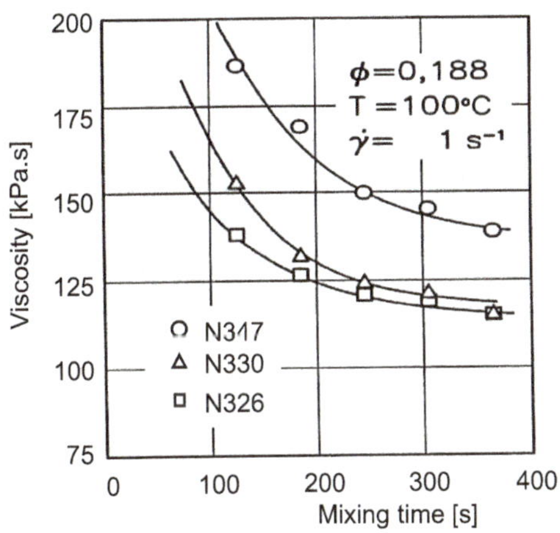


**Figure 8.24**
Influence of the effect of mixing time and the structure of carbon black on the viscosity of E-SBR compounds [127]

The relationship shown between the structure of the aggregates and the viscosity of the mixtures is clearly indicated by the reduction of the viscosity that is equivalent to an increase in filler dispersion. With decrease in the hydrodynamic volume of the carbon black aggregates, the compound viscosity decreases after the incorporation time is finished. The carbon black that has the highest structure leads to the enhanced viscosity in the rubber compound.

For this reason, it makes sense to look at the influence of the mixing time and the rotor speed in the internal mixer on the viscosity obtained in the compound. Monitoring of the effect of mixing time and rotor speed on the Mooney viscosity of SBR compounds mixed in a tangential lab mixer reveals a pronounced decrease in Mooney viscosity, as shown in Figure 8.25 [128].

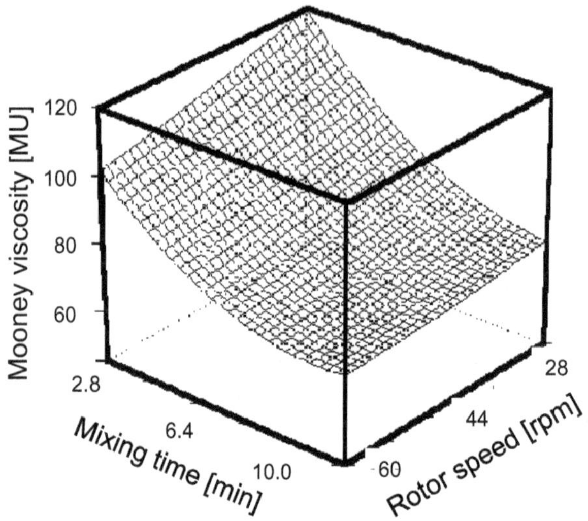


**Figure 8.25** Influence of mixing time and rotor speed on Mooney viscosity of the rubber compound [128]

The increased rotor speed leads to an increase in shear and extensional deformations, which accelerate the breakdown of the pellet fragments and agglomerates. With the increase in the mixing efficiency, more and more filler agglomerates are decomposed and release part of the occluded rubber, which in the distribution stage surrounds the individual filler particles as a film and thus leads to a lower stiffness and lower viscosity of the compound. In any case, the decrease in viscosity does not imply the presence of large agglomerates, which are not degraded and lead to failures in dynamic applications.

The entangled chains can break when the strain imposed by shear exceeds a critical limit. This is more pronounced when filler is present and amplifies the local stress on the polymer chains. It is estimated that shear stress increases by a factor of 2–3, due to the presence of carbon black in the system [129]. The mixing time-dependent degradation of NR in the presence of carbon black N330 (40 phr) has been investigated in a tangential lab mixer. It was shown by GPC analysis that the weight average molecular weight $M_w$ is reduced by half within three minutes of mixing [130]. The process results in unfavorable and insufficient dispersion, which can lead to a higher compound viscosity and a poor surface quality of the extrudates. The use of a 2-stage or a 3-stage mixing process greatly improves dispersion, and leads to greater homogeneity and a superior surface quality of the extrudates.

A comparative investigation on the effect of the rotor design has revealed that the decrease in Mooney viscosity takes place at any fill factor in tangential mixers. A significantly lower Mooney viscosity was obtained in intermeshing mixers at any mixing time. For example, at a fill factor f = 0.65, a viscosity of 105 MU was obtained in the tangential mixer, while the mixture produced in the intermeshing mixer showed a viscosity of only 40 MU [131].

As discussed in the previous chapter, the silanization of precipitated silicas leads to efficient reduction in inter-aggregate interactions of the silanized silica aggregates, resulting in both a general weakening of the silica networks and a constant decrease in compound viscosity to a low level. The drop in viscosity when a monofunctional silane (Si208) is added, which only reduces the filler-filler interaction and increases the rubber–filler interaction, is shown in Figure 8.20. A somewhat smaller effect is evident for the sulfur silane TESPT (Si69), which induces crosslinking with the rubber during vulcanization [132]. With small doses of silane, the viscosity of the compound rapidly decreases to a steady-state value. With the decrease in viscosity, the required number of mixing steps can be determined, thus shortening the real mixing time and minimizing energy consumption [133]. It has been shown that the silanization lasts for a certain time. The influence of this time is determined in the mixing test via the decrease in Mooney viscosity. In this way, the mixing cycle can be reduced to a minimum and a good silica dispersion in the compound can be achieved.

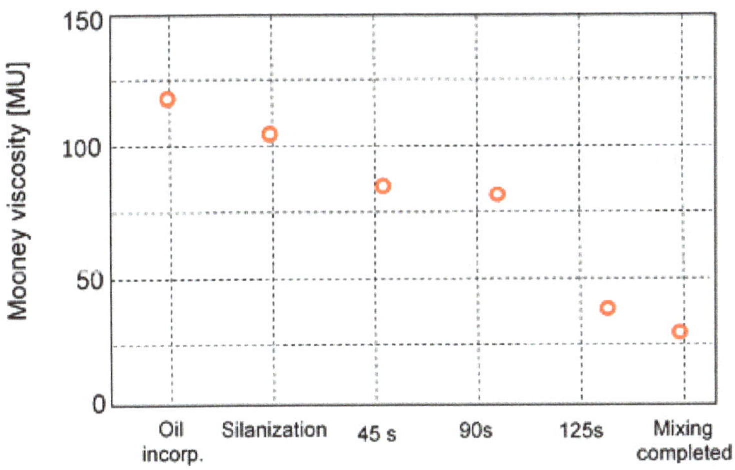


Figure 8.26 Decrease in Mooney viscosity during the silanization reaction of HD silane in an S-SBR compound

Comparison of the effect of different bifunctional silanes has shown that the reduction in the compound viscosity as a function of mixing time depends on the reactivity of the silane. For the same mixing time, the reduction in the Mooney viscosity or the reduction in the Payne effect was found to decrease in the order:

TESPD > TESPT > NTX (blocked silane)

The commonly used TESPT (Si69) has the lowest reduction in Mooney viscosity, because the reaction takes place slowly. In the TESPT with two sulfur atoms, in which sulfur is added, the reduction in the Mooney viscosity is improved. But the somewhat more expensive version of blocked 3-mercaptosilane (NXT) has by far the faster reaction and leads to a significantly greater reduction in Mooney viscosity or the Payne effect. The higher reactivity and the smaller size of the reactive sites increases the silanization efficiency and the greater is the decrease in Mooney viscosity [134].

### 8.4.3 Influence of Plasticizers

The presence of a suitable plasticizer in sufficient quantity lowers the viscosity of filled rubbers and thus improves flowability and formability of the rubber compound and does not cause incompatibility. The use of mineral oils, which make rubber chains more flexible, allows the rubber chain to quickly exit the void volumes of the filler aggregate and diminish the properties of entrapped rubber, reducing the viscosity and favoring filler dispersion. The plasticizing effect is based on the high free volume that the low molecular weight ingredient adds to the free volume of the entire rubber matrix, which increases the mobility of the rubber chains and therefore reduces the viscosity of the compound. A suitable mineral oil is selected that should have a similar solubility parameter as the rubber and will not cause phase separation. Within this range, the rubber-plasticizer system is single-phased and demonstrates a quasi-linear decrease in the viscosity and the glass transition temperature, $T_g$, with the increasing content in plasticizer [135].

Commonly, the influence of plasticizers in a rubber compound is evaluated from its effect on Mooney viscosity. If mineral oils of different chemical composition and viscosities are added to E-SBR, the Mooney viscosity of the rubber compound is sensitive to the viscosity of mineral oils [136]. As for single-phased systems the viscosity of the compound can be estimated from the simple mixing rule:

$$\eta_{comp} = \Phi_{pol}\eta_{pol} + \Phi_{oil}\eta_{oil} \quad (8.28)$$

Statistical design of experiment studies can provide quantitative information on both the influence of reinforcing fillers and the effect of a particular plasticizer. The influence of paraffin oil on carbon black N539 filled EPDM is shown in Figure 8.27 [137].

The contour lines correspond to the compound viscosity as a function of mineral oil content and carbon black loading. In general, the viscosity decreases with increase in oil content and, as already shown, this variable increases with increase in filler loading. It is worthwhile taking the effort to fill such a test plan with data in the appropriate range of shear rates when new raw materials and compounds are used or when new process lines are introduced. The right mineral oil with a similar solubility parameter as the rubber should be selected to obtain the best viscosity effects.

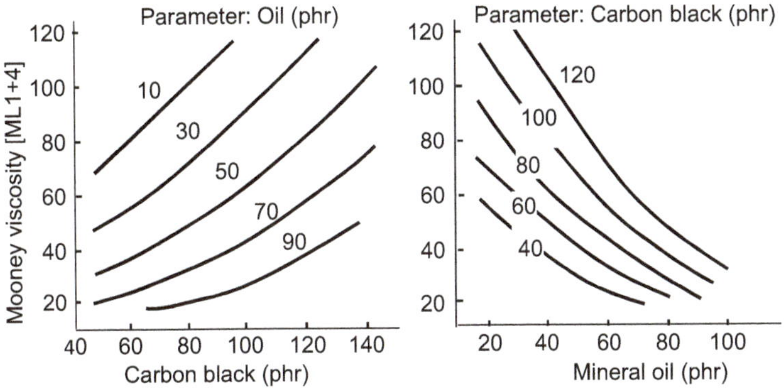


**Figure 8.27** Variation in mixture viscosity as a function of carbon black and mineral oil loading [137]

### 8.4.4 Influence of Storage Conditions

In filled rubber compounds, changes in compound viscosity can be observed after storage periods and as a result of the conditions of storage. The changes are attributed to a reorganization of the filler-filler interaction, known as filler flocculation. During storage, filler flocculation is conveyed to the macroscopic scale by an increase in viscosity as well as the storage modulus, $G'_0$, of the rubber compound [138]. The effect of storage is more pronounced for compounds filled with precipitated silica. The stronger interparticle interactions lead to a pronounced increase in compound viscosity after a short storage time.

With a carbon black and a precipitated silica of comparable surface areas and structures, it has been demonstrated that the increase in Mooney viscosity after mixing is more pronounced and rapid in silica compounds, due to strong inter-aggregate interactions that promote re-agglomeration due to H-bonding of silanol groups [139, 140]. Over a period of 47 days, the storage modulus $G'_0$ of the compound measured with a rubber processing analyzer (RPA) was found to increase by 160% while no change was recorded in the unfilled compound [141].

## 8.5 Die Swell of Filled Rubber Melts

Rubber extrusion basically consists in preheating or softening of the rubber compound, homogenizing and pressurizing the compound in order to push it through a profile die. Profile shaping is the result of a complex flow process. The extruder die is designed to avoid sudden discontinuities as the rubber compound moves through it and often has a contoured head section. In practice, extruder die swell can be quite complex and it might be necessary to modify the die a number of times, before the required profile is achieved. With this concept in view, it is understandable that the decrease in extrudate shrinkage and dimensional stability of the extrudate is promoted by:

- an increased filler loading
- a higher void volume of filler agglomerates
- a higher specific surface area

The condition of the melt flow to be shaped is not completely homogeneous, both locally and with respect to time. Besides the flow properties of the compound, boundary conditions, such as the level of temperature control in the extruder, the throughput and the geometry of the die determine the extrusion process. Furthermore, the adhesion of the compound to the die wall, the multi-dimensional temperature profile of the compound in the extruder and the high gradient in the plastification history of the compound exhibit a transient behavior.

Fillers in viscoelastic polymers can reduce the melt elasticity and hence influence the extruder die swell. The die swell of filled rubber melts correlate with a reduction with the filler loading and the rheological conditions. Therefore, blend compounds in which the unfilled BR is the outer phase will have less die swell than when NR or SBR are the outer layer [142]. Looking at the die swell of filled compounds, it is obvious that the presence of any filler decreases from this undesired effect [143]. As a result of the adsorption of the polymer chains on the surface of the filler particles, both the conformational entropy of polymer coils decreases and the elasticity of the compound decreases. Due to the multiple physical anchoring of long chains, the parts of the chains available for elastic deformations become shorter and shorter. Thus, the elasticity of the rubber matrix decreases for every type of filler with increase in filler volume and so does the above-described "memory" effect of the compound. Lower elastic retractive forces are formed during the forced flow through the extruder die. This limits the storage of elastic energy in the form of conformational entropy in the coils. The relaxation of the filled compound as it exits the die is less pronounced.

The elastic recovery of the extrudate profiles shows the incomplete stress release caused when long-chain molecules are oriented in the extruder die. Sometimes this can occur also through extrudates and melt fracture by the loss of rubber to metal adhesion in the dies at very high shear rates. The higher the temperature and the lower the extrusion rates, the more the elastic recovery of the extrudate profile. The addition of filler leads to a reduction in the elastic component and decrease the relaxation time. High structure carbon blacks will substantially increase the effective volume of the fillers and lead to shorten the relaxation times and are not broken under the shear stress [144].

The semi-automated die swelling tests involves a calculation of die swell by measuring the velocity of the extrudate downstream from the die in a constant volume extrusion process. The proportionality of the die swelling and the velocity assumes that the die swell resulting from compound elastic memory occurs immediately after the compound exits the die [145]. Die swell can help to predict the quality of mixes and to troubleshoot dimensional stability problems in rubber factories that manufacture products by extrusion, such as tires, belts and hoses.

### 8.5.1 Effect of Carbon Blacks

In addition to the concentration of the filler, the effect is primarily dependent on the specific surface area and structure of the aggregates. From a technical point of view, the result indicates that die swell of filled compounds significantly is roughly one order of magnitude below the level observed for unfilled compounds. Similar to unfilled compounds, die swell increases in proportion to with the shear rate but is less evident than in the pure rubber compound. Die swell is reduced by the addition of any type of fillers, due to a reduction in an elastic component [146]. With carbon black-filled compounds, there is a systematic decrease in the extrusion die swell with constant loading when the specific surface area of the filler increases. This effect is undoubtedly determined by the increasing number of active sites on the rubber-filler contact surface.

Due to the principle of structure-concentration equivalence [61], the compounds containing high structure blacks behave as if the filler content were higher and demonstrate better die swelling. The rubber partially immobilized in the voids of the agglomerates is missing from the rubber matrix, which is elastically deformable. This leads to a decrease in extrusion swelling, which is inversely related to the oil adsorption number. Figure 8.28 shows this reduction in die swell for S-SBR/silica VN3 and NR/carbon black N220 [147].

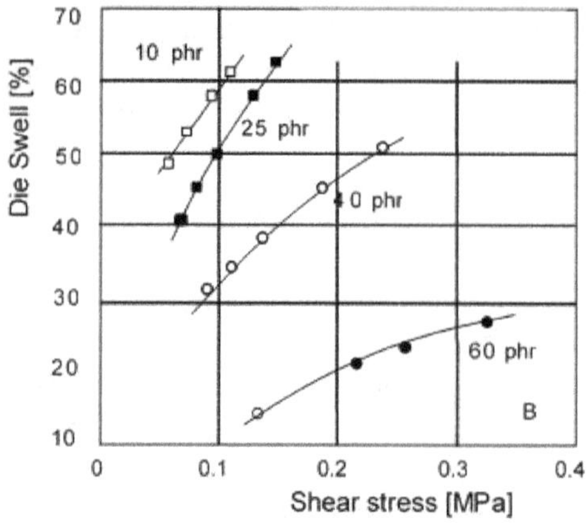


**Figure 8.28**
Influence of carbon black loading and the shear stress on the die swell of NR/CB N220 at 100 °C [147]

The observation that extruder die swell is lower than expected from the corrections for occluded rubber can be attributed to effects of dispersion that cause changes in the "in rubber" morphology. As dispersion progresses, filler agglomerates are broken down and the rubber chains present in the mixture are adsorbed onto a larger filler surface, which leads to a decrease in the elastic properties of the rubber chains. The occluded rubber can act as a non-elastomer filler. For this reason, semi-active carbon blacks that are easy to disperse are used for the extrusion of profiles that do not require abrasion resistance. Proper selection of carbon black may allow smooth extrusion with clear shear edges. The decrease in extrudate swell shown in Figure 8.26 for NR/N220 compounds at 100 °C accompanies the decrease in the filler loading. If the shear stress is higher the die swell increases. In systems filled with reinforcing carbon black above the percolation threshold, extrudate swell decreases to lower values, which simplifies die design and construction. A comparison of carbon black N330 and process oil demonstrates that the concentration of carbon black is more important for reducing die swell than is the presence of processing oil. Comparison of compounds manufactured with different types of carbon blacks have shown that die swell decreases with the DBP adsorption and loading, but not with surface activity [145, 146].

If extrusion shrinkage is considered as a function of mixing time and degree of dispersion of the filler, two opposing influences have to be taken into account. During the incorporation stage, extrudate shrinkage is lowest because of the high amount of occluded rubber. Then, extruder die swell increases during the dispersion stage until the power peak reaches a maximum value. From that characteristic point on, die swell diminishes again due to both polymer fracture which reduces the elastic behavior of the melt and the increasing rubber-filler contact surface created by selective adsorption of the shorter chains on the filler surface [147]. The competition between filler dispersion and chain scission is shown in Figure 8.29.

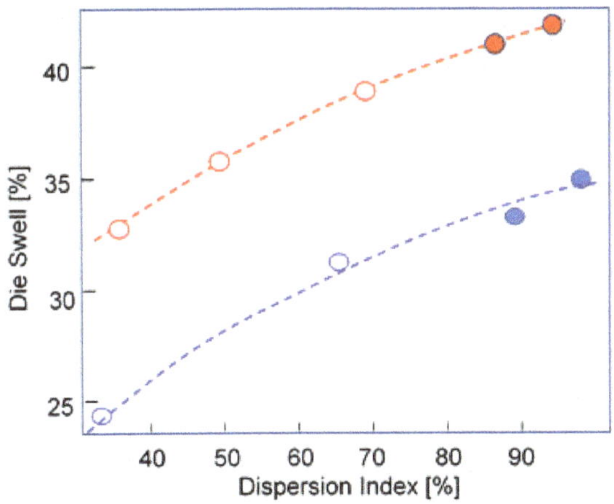


**Figure 8.29**
SBR extrusion die swell as a function of carbon black N326 dispersion at 60/40 and 80/40 carbon black/oil loadings [147]

When the carbon black is introduced first and then the mineral oil, the degree of dispersion is more than 90%. Thus, die swell is reduced progressively with the mixing time and implicitly with the increasing degree of dispersion. The apparent pattern of somewhat higher die swell with increase in oil is attributable to the fact that the black loading remains the same, and increasing the mineral oil reduces the actual percentage of black in the compound. The decrease in die swell with increase in filler loading, as well as the lower severity of extrudate melt fracture, due to turbulent flow, can be assigned to stress dissipation provided by the rubber-filler interaction.

The decrease in extruder die swell as a function of mixing time and energy is in line with the observation that elastic properties of the compound, as measured by the entrance pressure loss in capillary rheometers, decreases with both the filler content and the degree of dispersion [148]. A good correlation has been found between extrudate shrinkage and the storage modulus S′ of unvulcanized compounds measured after large strain that disrupt the filler network [149].

It has been reported found that die swell of a filled compound decreases in inverse proportion to the structure of the filler. The occlusion of the rubber in the voids of the aggregates reduces the elasticity of the compound in a magnitude that is a function of the "occluded" rubber [60]. The observation that die swell is lower than expected from the corrections for occluded rubber can be attributed to effects of dispersion that causes changes in the "in-rubber" morphology. As the dispersion progresses, filler agglomerates are broken down and the rubber chains present in the mixture are adsorbed onto a larger filler surface, which leads to a decrease in the elastic properties of the rubber chains.

It has also been found that the specific surface area of the filler contributes to a reduction in extruder die swell. At lower mixing energies, an effect of diminishing die swell with lower specific surface area has been observed. This has been attributed to the fact that, due to stronger inter-aggregate interaction, the maximum of extruder die swell for fine carbon blacks is shifted to higher mixing energies. A practical consequence of these effects is the better dimensional stability achieved with active fillers on the factory floor.

For E-SBR compounds containing 50 phr of different carbon black grades, two mechanisms die swell recovery have been identified [150]. The fast recovery may be related to the reorganization of the filler network and the slow recovery to the deformation of the entire rubber matrix. The effect of dispersion and oil addition on die swell of SBR compounds has been investigated intensively [151]. The reduction in extruder die swell with the filler loading was confirmed. An increase in 20 phr N326 loading lowers the die swell by roughly 8%. The apparent pattern of higher die swell in presence of oil was attributed to the reduced fraction of the filler in the compound. In addition to the spontaneous swell, the elasticity of the compound affects the shrinkage process over a longer period of time (4 hours). This has an effect on the dimensional accuracy of profiles that are not vulcanized immediately after extrusion.

The presence of plasticizers contributes to reducing die swell and extrusion shrinkage of rubber compounds. The effect of carbon black N330 and naphthenic oil in E-SBR show at a high concentration of both ingredients the best contour diagrams [152]. If the shear rate in the extruder is increased, the boundary fracture region is shifted toward the high concentration of carbon black and the low concentration of the naphthenic oil. The combined effect of filler and plasticizers on extruder die swell can be studied using statistical design of experiments. The contour plots in Figure 8.30 show the interdependence of die swell of an EPDM compound containing carbon black N372 and paraffin oil. The extruder die swell depends on the presence of the plasticizers and the carbon black and is indicated in Figure 8.30 [152].

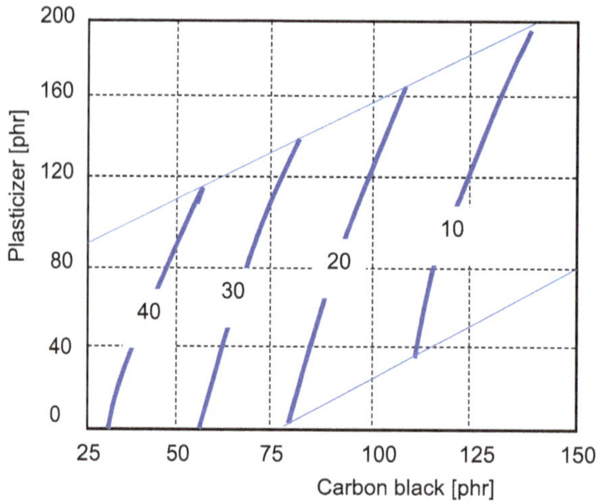


**Figure 8.30** Dependence of die swell on carbon black N472 and paraffin mineral oil loading in EPDM [152]

The extruder die swell of the compound can be kept constant, when the filler loading is accompanied by an increasing content of paraffinic oil. The contour lines indicate the decrease in die swell with carbon black loading at constant concentration of mineral oil. Keeping the filler loading at constant and increasing the amount of oil leads to more die swell. The filler structure has a major effect on the extrusion swell. Due to the structure-concentration equivalence [72], compounds containing high structure blacks behave as if the filler content is higher. The rubber partially immobilized in the voids of the agglomerates is missing from the rubber matrix, which is elastically deformable. This leads to a decrease in extrusion swelling, which is inversely related to the oil adsorption number. Figure 8.31 illustrates this general relationship.

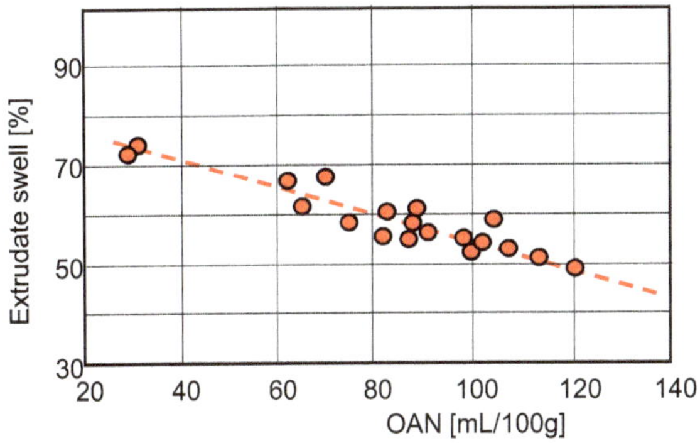


**Figure 8.31** Decrease in extrudate swell with increase in structure of carbon black used to produce the compounds [72]

The observation that extruder die swell is lower than expected from the corrections for occluded rubber can be attributed to effects of dispersion that induce changes in the "in-rubber" morphology. As dispersion progresses, filler agglomerates are broken down and the rubber chains present in the mixture are adsorbed onto a larger filler surface, which leads to a decrease in the elastic properties of the rubber chains.

## 8.5.2 Effect of Precipitated Silicas

Precipitated silica has a very strong interparticle interaction that favors the presence of agglomerates in the compound. Extruder die swell is higher for precipitated silica compounds than for compounds that contain the same loading of carbon black. However, if the loading increases, the extruder die swell decreases strongly, due to the interaction with the rubber chains in the die. The decrease with silica loading and the increase with shear rate are shown in Figure 8.32 [19].

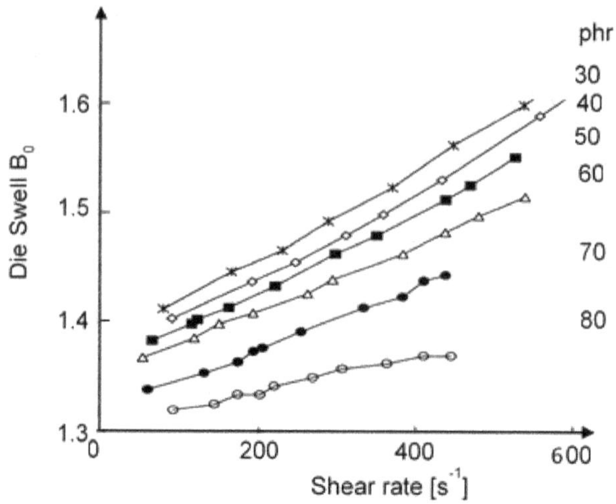


**Figure 8.32** Extruder die swell for the compound S-SBR/silica VN3 with increase in loading and increase in shear rate

The lower die swelling of precipitated silica is expected for the higher filler-filler interaction and the low filler-polymer interaction. If the compound is made with silanized silica, the rubber-filler interaction increases and the influence on the die swell will be much larger than for the virgin silica. The comparison of the extrudate die swell of NR filled with carbon black and silica as a function of the compressed oil adsorption number (COAN) shows that silica causes a greater decrease in the swell [153]. The dimensional accuracy of extruded profiles can thus be better guaranteed. A practical consequence of these effects is the better dimensional stability achieved with active fillers on the factory floor. However, it should be noted that, increased filler loading, structure of the aggregate and specific surface area leads to an increased compound viscosity and thus higher energy consumption during processing.

With silanized silica in S-SBR/BR blends, the extrudate surface has more surface roughness than the non-silanized silica, which is a deficiency related to extrusion. With wet-mixing, the EVEC silica in extrusion process is much lower and the profile surface is smooth, which can be related to less bound rubber and a low flocculation tendency with this filler, because the agglomerates of silica are much reduced, compared with dry-mixed silica compound [87].

## 8.5.3 Effect of Layered Silicates

Higher complex viscosity and storage modulus are observed for layered silicates for SBR compounds when compared with NR, particularly in the low frequency region. But clay platelets have more influence on increasing the viscosity for NR than for SBR, suggesting greater ability of NR for intercalation and exfoliation than SBR. Both $Na^{+}$MMT and ODR-modified organo-clay exhibited lower die swell than that of the unfilled rubber. The die swell always decreased with increase in the filler loading since the addition of the nanofiller lowered the volume fraction of the rubber. The effect is a matter of the rubber polarity. The die swell is reduced more in NBR than in SBR with increase in the concentration of the organo-clay.

## 8.6 Green Strength and Yield Stress

Green strength is the resistance of the uncured rubber to deformation and fracture. A certain amount of green strength is needed in rubber manufacturing for many rubber compounds. It is an important property during the manufacturing of rubber products, such as radial tires, which may thus be more resistant to blow-outs in the second stage of building the product [154]. Also, green strength of a rubber compound is needed in extruding a rubber tube to be used to produce a rubber hose. The natural polyisoprene (NR) has high green strength which is attributed to the strain crystallization [155]. It is accepted that the long sequences of cis-1,4 isoprene units and free fatty acids are responsible for this behavior of NR [156].

Many rubber compounds have a non-negligible yield point. This means that no plastic deformation of the material takes place below a shear stress and this limit has some serious consequences for processing in an extruder. If the shear stress in an extruder falls below the limit locally, flow transport can stagnate. When the filler is added at a sufficient high concentration to establish a filler network, it will produce a response referred to as the yield stress value. The filled rubber compound shows no flow up to the yield stress value, at which it finally starts to flow. This critical stress is given by the stiffness of the flocculated filler network in the compound. At a shear stress above this threshold, gradual degradation of the filler network begins, which enables the plastic deformation. The yield stress behavior of filled compounds could be interpreted as reflecting the response of three-dimensional filler networks to shear stress, which is caused either by contacts between aggregates or, most probably, by rubber-filler interactions and the associated complex morphology.

Simple equations containing the yield stress are useful for viscosity properties, at least over a limited range of stress/rate. In order to describe the phenomenon, the Ostwald-de Waele power law is extended by the yield stress, $\tau_0$:

$$\tau = \tau_0 + k\dot{\gamma}^n \tag{8.29}$$

where $\tau$ is the shear stress, $\eta$ is the viscosity, k is a temperature-dependent factor and $\tau_0$ is the yield stress

It can be seen that two areas are distinguished in flow channel. When $\tau < \tau_0$, there is no flow in the die and, when $\tau > \tau_0$, flow occurs. Equation 8.29 holds for practical reasons, due to the fact that the wall shear rate can no longer be deduced from the volume of flow. The yield stress has to be obtained as a separate quantity. It is therefore important to be able to qualitatively assess the presence of a yield point.

It has been empirically proven that the viscosity and the exponent, n, in Equation 8.29 depends on filler loading. Due to the restricted mobility of the chains tightly attached to the physically bonded interface and the surrounding rubber shell of entangled chains, highly loaded compounds do not flow until the stress exceeds 0.1 MPa. Such compounds require the addition of a certain amount of plasticizer.

### 8.6.1 Effect of Carbon Blacks

Several reports have been published about the yield stress behavior of carbon black-filled compounds, but it is worth underlining here the experimental difficulties in assessing the shear flow behavior of highly viscous materials in the very low shear rate (or stress) range. Any yield stress value that can be derived from such experiments is an extrapolated value, as quoted in the literature on the subject [157]. As the carbon black specific surface area increases, so the yield stress increases. With increase in the carbon black structure, the effective volume of the aggregate would increase and the yield stress is reduced. The higher the surface energies, the larger interfacial area that can contribute to the yield stress [158]. Carbon blacks that are poorly incorporated will give different yield stress values compared to batches which have their carbon blacks better dispersed with fewer filler–filler interactions.

A phenomenon that is not related to yield stress is known as "black scorch". This refers to an increase in the compound viscosity at the end of the mixing process associated with a blockage of the flowability of the mixture. This phenomenon has been repeatedly observed when mixing active carbon black in EPDM. Scorched particles and defects can be found in the compound [159]. It has been shown that the effect is more pronounced for EPDM grades with a high ENB content and high structure blacks. These influencing factors suggest that the "black scorch" is not a kind of yield stress, but a premature scorch, triggered by high temperatures in the final stage of mixing. The appearance of "black scorch" can be suppressed by the addition of small amounts of sulfur (approx. 0.1 phr). The nature of the observed defects has not yet been completely elucidated.

### 8.6.2 Effect of Precipitated Silicas

The precipitated silica shows a higher green strength than carbon black N110 because of the intensive interaction of the silica aggregates [160]. The significantly lower green strength in the case of silanized silica with TESPT and with Si208 leads to a better flow due to the rubber-filler interaction. Silanized silica in S-SBR/BR blends is characterized by a good rubber-filler interaction and is not strongly influenced by the yield stress. The low viscosity is a guarantee of the absence of yield stress at normal silanized silica loadings [161].

### 8.6.3 Effect of Layered Silicates

The high aspect ratio of the layered silicates found in some reports can exhibit higher yield stress values compared with rubber compounds, which can be explained by the intercalation/exfoliation process of these fillers. This is another example for the good interaction with the rubber matrix.

### 8.6.4 Effect of Carbon Nanotubes

The aspect ratio must be high enough to maximize the load transfer between the multi-walled carbon nanotubes and the rubber matrix. A good dispersion not only makes more filler surface area available for bonding with the rubber, but also prevents the aggregate fillers from acting as stress concentrations, which can be detrimental for yield stress. Multi-walled carbon nanotubes can exhibit yield stress if the loading is above the rheological percolation threshold.

## 8.7 Flow Anomalies

The majority of the rubber compounds are processed into endless profiles to produce sealing profiles, hoses, windshield wipers, lining strips for injection molding or strips for lining calendars. For the production of these types of components, the existing production equipment must be selected and the energy consumption evaluated in order to reduce the price without compromising the high quality of the products. However, several requirements have to be taken into account:

- shear viscosity
- reduced extrusion die swell
- homogeneity of the compound
- edge d notching
- restless structure
- porosity
- wall slip
- screw pulsation
- melt fracture

In most fluidic processes, adhesion to the wall is assumed, which means that a melt at an infinitesimal distance from the wall has zero velocity. The aforementioned yield point, which forms a block shear flow, represents a serious problem in extrusion. This means that the volume flow can no longer be deduced from the actual wall velocity and the edge part of the profile is heated by dissipative heating. However, these assumptions are not met when describing the flow behavior of rubber compounds at high pressures and shear rates.

### 8.7.1 Wall Slippage

Experimental studies have shown that rubber compounds exhibit marked slip ratios in the conveyed volume flow as a function of the raw polymers' molecular weight, the fillers and additives, their concentration in the rubber and the processing parameters in the extrusion process. In many cases, when the rubber compound contains a higher amount of filler, the shear rates will show anomalies and the rubber melt will slide over the wall area. The accuracy of the capillary rheometer viscosity depends on the ability of the test material to resist slippage. This is greater at high shear rates. During wall slippage, the pressure of the mixture decreases, and the mixture recovers until the parameters of this undesirable system are re-established. In the subcritical flow range in which the total volume flow conveyed does not exhibit wall slip and above this limit, the compound shows wall slip components. Once critical wall shear slip is attained, polymer melts encounter flow anomalies, such as slip-stick effects and shark-skin effects, pulse-like fluctuations in pressure and problems in ensuring the dimensional stability of the extrudate, which characterize this supercritical flow range.

At higher pressures, flow anomalies occur, which are summarized under the name of wall slippage. From the normal flow the shear rate is greatest between the metal wall and the rubber and lowest at the center. If there is slip between the steel wall and the rubber itself, there will be an error in the viscosity measurements. Slip effects have been shown to occur to varying extents across the entire shear rate in conjunction with severe material stress, which can also lead to failure. Wall slippage describes the behavior of rubber melts that violate the boundary conditions and exhibit slip on the surface of the extruder die. In parallel, the extruded mass flow changes to an extent depending on the extrusion pressure. Experimental studies have shown rubber compounds to exhibit marked slip ratios conveyed as a function of the molecular weight of the rubber, on the type of filler and its loading, on the surface coating of the flow channel [162–165], and on the increase in the dissipative energy at the wall of the flowing system. The presence of unsteady flow and melt fracture during the flow of compounds has significant implications for the processing of rubber compounds. Under wall slip conditions, the total volume flow is divided up into a slip volume, $V_{slip}$, and the shearing volume flow, $V_{shear}$, as shown in Figure 8.33:

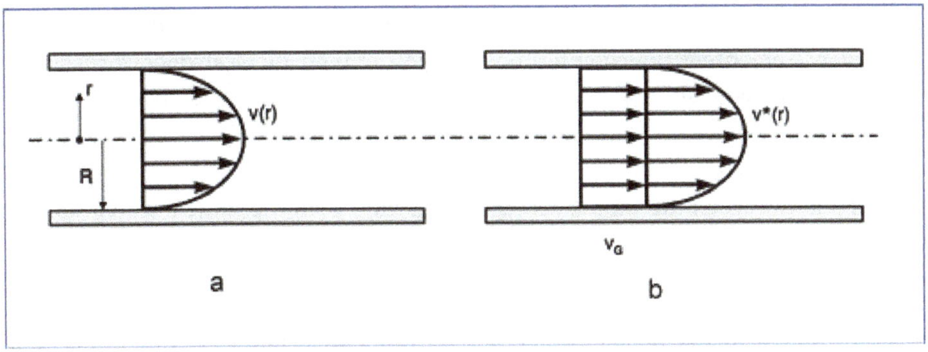


**Figure 8.33** The difference between normal flow and wall slippage

This problem can be solved with the aid of the fundamental equation of motion and the continuity equation with the wall slip as boundary condition. Wall shear stress and shear rate are thus independent of the slip rate, meaning that the total flow is higher under wall slip conditions than under wall adhesion conditions, assuming unchanged pressure loss. Therefore, the relationship should apply:

$$V_{wall\ adhesion} / \Delta p_{wall\ adhesion} < V_{wall\ slip} / \Delta p_{wall\ slip} \tag{8.30}$$

where Δp are the pressure differences

Published reports distinguish between two basic concepts for describing wall slip:

- Coulomb wall slip
- slide film model

The Coulomb wall slip model considers a sliding speed on the wall that corresponds to the ratio of the frictional force to the normal force [166]. The coefficient of sliding friction is lower than the coefficient of static friction. The sliding film model is based on the formation of a low-viscosity boundary layer on the capillary wall, which works like a lubricant. The formation of a sliding layer can be explained by two concepts:

- the dissipative temperature increases on the wall [167]
- the segregation processes at the capillary wall [168]

The flow curves show a plateau area in which the shear stress does not change with increase in shear rate, apart from periodic fluctuations. If a critical shear is exceeded, the flow process becomes unstable with decreasing shear stress, the rubber melt behaves "superfluidly" and detaches from the wall of the extruder. After a while, however, the shear rate falls again below the critical level, which means that wall adhesion starts again. The frequency of this sequence can depend both on the viscoelastic parameters of the polymer and on the geometry factors of the extruder. According to the Coulomb model, the sliding speed depends not only on the friction coefficient, but also on the normal force within the capillaries [169]. Slippage can be caused by a variety of factors. Some rubber compounds have high viscosity values that make slippage likely. Sometimes the chemical composition of the rubber surface is such that there is not enough adhesion between the rubber compound and the metal surface.

At low pressure, the melt glides over the unevenness of the surface. Continuous wall conduction is observed for EPDM and SBR compounds. According to the Coulomb model, wall slip occurs when the wall shear stress is greater than the static friction force. A critical wall shear stress is reported, above which the flow anomaly occurs [170]. Wall slip also depends on the type of filler and the plasticizer used. Mineral oil reduces the internal friction, such that the tendency to wall slip is reduced. Carbon black increases this ratio and promotes the wall slip process. On the other hand, the metal alloy and the surface profile have an influence on wall slip, and so does the geometry of the capillary. For calculation of the slip rate, $v_{slip}$, is given by the equation [171]:

$$V_{slip} = V/WH\text{–}2\,(1 + 2n)\,nH(\tau/k)^{1/n} \tag{8.31}$$

where $\tau = \tau_{wall} = (dp/dL) \times (H/2)$, $V_{slip}$ is the slip rate, $\tau$ is the shear stress, W is the width, H is the height, L is the length of capillary, and n and k are power law parameters under the conditions of wall adhesion

Wall slip can be determined by a simple equation:

$$\text{Slip ratio} = \text{total volume flow } V_{tot}/\text{slip volume flow } V_{slip} \tag{8.32}$$

A rubber compound exhibits a certain slip ratio during the flow process that is independent of the shear rate. With the help of high capillary rheometers, the material and rheological parameters that lead to wall slip have been investigated systematically.

Wall slip can be recorded with the help of flow markers [172] by analyzing the flow pattern at high shear rates. Initial investigations of wall slippage were carried out by M. Mooney at the beginning of the 1930s [173]. By using plate-plate and cone-plate rheometers, it was shown that if the adhesion between the moving plate and the melt is insufficient to transmit the shear stress, the plate slips over the melt, indicating the flow anomaly. In general, it has been found that the critical processing area in which wall slippage occurs can be influenced by the addition of low molecular weight components, such as plasticizers or processing aids [168].

Another factor that causes wall slippage is the increased temperature. High velocities in the flow channel leads to strong dissipative heating. Since the temperature near the wall increases, the adhesion of the melt decreases and the occurrence of wall slippage becomes likely. An interesting consequence of the temperature increase has been observed for high cis-BR [11]. At ambient temperatures, this semi-crystalline polymer shows the characteristics of dilatant flow. As the temperature rises, a transition to shear thinning is observed, and at temperatures above 100 °C, high shear stresses clearly determine wall slippage. The behavior of high cis-BR is shown in Figure 8.34.

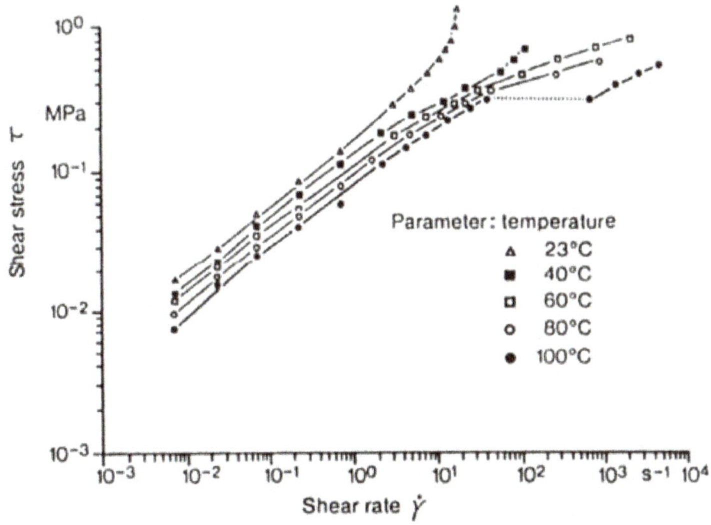


**Figure 8.34** Flow behavior of cis-BR as a function of temperature [11]

Such behavior can be observed with the cis-BR compound on the two-roll mill. At room temperature, the compound rolls smoothly over the roller. At ambient temperature, the compound sheet adheres well to the cylinder wall and the sheets have a smooth surface. As the temperature of the mill cylinder rises, the strength of the compound decreases, enabling partial detachment of the sheet. It is estimated that melt fracture occurs when the entanglement density decreases to about 30% of the original value in the unstressed polymer. The appearance of this imperfection is shown in Figure 8.35 [81].

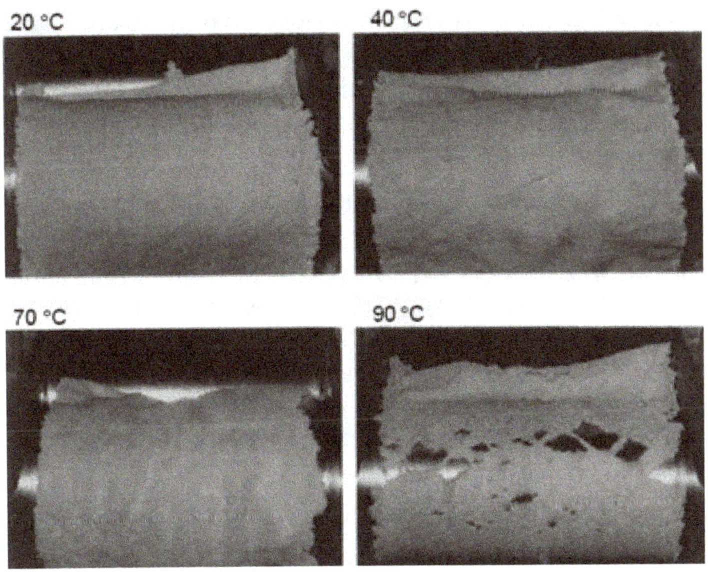


**Figure 8.35** Increasing temperature of cis-BR compounds on the two-roll mill [81]

An apparent slip or lubrication occurs at the wall in the flow of any multi-phase system when the disperse phase moves away from smooth walls. Because this effect is usually confined to a very narrow layer, with a typical thickness of 0.1–10 µm, it resembles the sliding of solids on surfaces.

The wall slip behavior of NR/cis BR blends filled with carbon black and the factors affecting it were investigated at shear rates that varied from 50 to 103 $s^{-1}$ over a temperature range of 90–130 °C. The results show that the measured critical pressures drop when wall slip starts. The wall-slip phenomenon of the test samples happened near the die exit. Above 110 °C, the slip velocity for SBR/carbon black was found to depend on shear stress as a function of the diameter of the extruder die. Sliding appears at high stress whereas the velocity of the flow depends on the shear stress [174]. In the S-SBR/BR system with carbon black N347, the stationary flow range is already pronounced at 80 °C and the wall shear stress is only reached at a shear rate of 40 $s^{-1}$. A critical shear stress appeared, showing that the onset wall slipping was located at 0.1 MPa, which is in good accordance with other publications [175]. Typical examples are compounds containing high levels of mineral oils or synthetic plasticizer. Wall slippage has been shown for NBR melts with a higher level of saturated esters. The effect was attributed to a stress induced change in the plasticizer concentration which leads to an enrichment of the esters in the vicinity of the wall [176].

Similar effects are observed for compounds containing polar ingredients (i. e. Ca stearate, Zn stearate) which exhibit limited solubility. These ingredients can cause phase separation and form lubricating films on the boundary between the compound and the wall [168, 177].

The pressure profiles generated during the capillary rheometer of EPDM at 100 °C show an influence of the molecular weight. For low-viscosity rubbers, it was possible to plot the entire shear rate range free of any adverse effects. In the same “processing window”, a high-molecular EPDM shows extreme fluctuations in the pressure profile already at low shear rates of 1 $s^{-1}$. The higher viscosity evidently leads to more effective processing and results in abandonment of the subcritical flow range and the imposition of the largely subcritical flow behavior, giving rise to difficulties in processing area, as shown in Figure 8.36 [178].

For the low-viscosity EPDM, it is possible to plot the entire shear rate range free of any adverse effect on the pressure profile. An increase in the processing temperature to 140 °C for the same polymer triggers a rise in polymer chain mobility, accompanied by a drop in viscosity during the process. This would have a positive effect on processing behavior. With increase in temperature, the shift in the wall shear stress occurs at higher shear rates. With the polymer viscosity the wall shear stress is significantly shifted to higher values. If the low molecular weight EPDM is given a higher loading of carbon black N550 (20 to 50 phr) and a paraffin oil (20 to 50 phr), wall shear stress increase with both parameters and wall slip takes place at a higher shear rate. As the test temperature rises, the instabilities in the case of low-viscosity EPDM become manifest in shorter cycle times while, for high-molecular EPDM, the subcritical field is shifted to 5–10 s$^{-1}$. The corresponding wall shear rates as a function of shear rate for the two molecular weights of EPDM are shown in Figure 8.37 [178].

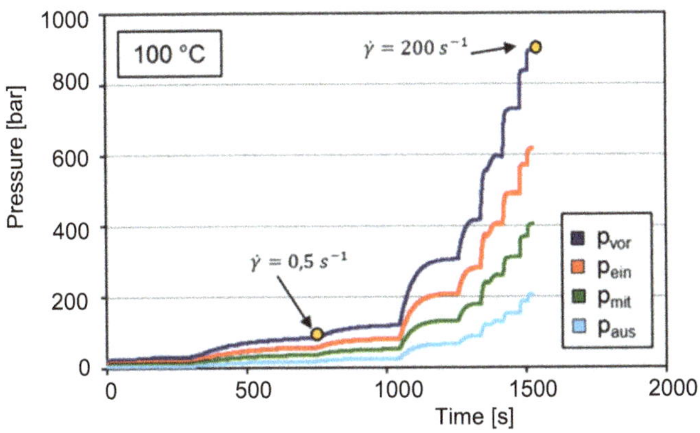


**Figure 8.36** Pressure profile, measured with a capillary rheometer, for low-molecular EPDM at 100 °C in the range 0.1–200 s$^{-1}$ [178]

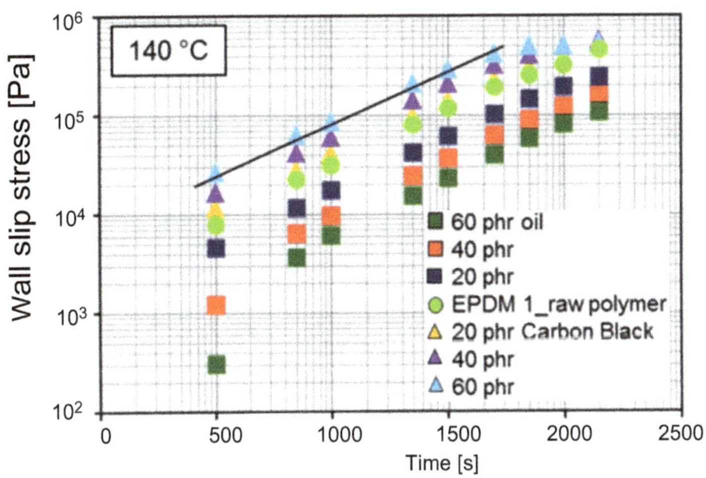


**Figure 8.37** Flow curves as a function of the compound formulation at 140 °C

The filled EPDM compound is more easily processed if the molecular weight is not as high. The flow behavior improves when the processing temperatures is raised. The correlation between slip ratio and shear rates shows that higher processing temperatures result in improved flow behavior and the slip effects are fewer. The rheological behavior of silicone rubber compound as a non-Newtonian fluid has been investigated under wall slippage conditions. The critical shear stress needed to start wall slippage is 0.13 MPa at room temperature, determined by the sudden onset of a pressure drop after this value, which indicates the change in boundary conditions from wall sticking to wall slipping [179].

### 8.7.2 Melt Fracture

The shear rate that produces rough extrudates is often called the "melt fracture" point. Certain combinations of very high molecular weight rubbers with a particularly filler system can produce this type of problem. Slip involving melt fracture can be observed at high shear rates or high extrusion rates. The extrusion rates are dependent on the polymer and on the type and concentration of the filler. The higher the specific surface of the carbon black, the lower is the extrusion rate. This can be seen from the decrease in the extrusion rate as a function of the specific surface area and the structure of the filler. This is shown in Figure 8.38 [153].

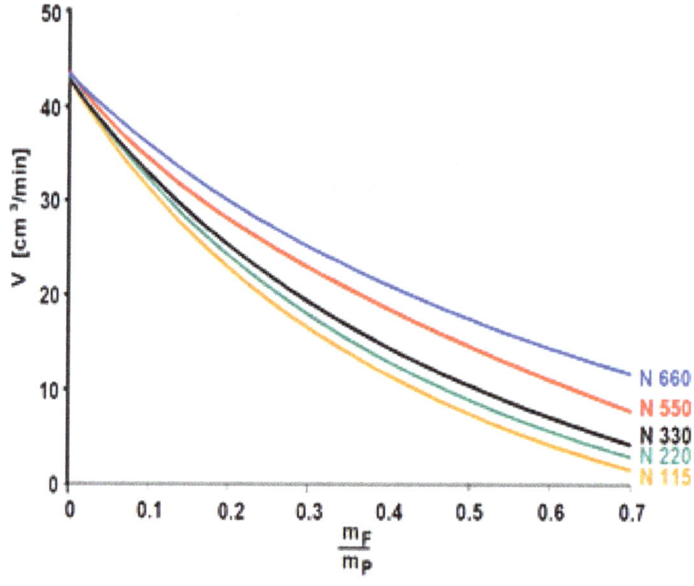


**Figure 8.38** Influence of the carbon black concentration, the role of the specific surface area and the structure on the extrusion speed [153]

High extrusion speeds are over 8 mL/min for N660 and N500. The active fillers can shorten the extrusion process significantly. The extrusion speed of N330 is only 4 mL/min and that of N115 is 1 mL/min to obtain a small surface with the same cross-section and no porosity in the profile. Unfortunately, the extruder speed is often set too high, the material parameters are not properly maintained and melt fracture occurs in the profile. This differs from a perfect extrudate by the appearance of zig-zag edges or irregular shapes, as shown in Figure 8.39.

**Figure 8.39** Comparing a normal profile to that with a zig-zag shape or an irregular shape

This phenomenon leads to a distorted and broken-up extrudate surface. The tendency to melt fracture increases with the average molecular weight, loading with active fillers and certainly with the extrusion process. The onset of melt fracture is indicated by a sudden increase in extruded volume with only a small increase in extrusion head pressure. This can be attributed to a reduction in adhesion between the rubber and the die wall and the appearance of a turbulent flow pattern. It has been reported that melt fracture increases with the molecular weight of the polymer [180]. It has been found that the critical extrusion rate for flow anomalies or melt fracture increases with temperature and die diameter for fixed shear rates [181]. The longer the extruder die at constant diameter, the later the melt fracture appears. Melt fracture leads to irregularities on extrudate profiles and in extreme cases to breaking of the extruded strand. This undesirably phenomenon becomes more and more pronounced with increase in entanglement density in the melt and poorly dispersed fillers during mixing. Using a cavity die the influence of the molecular weight of the polymer and the filler type and loading can be proved rapidly. It has been shown that the content of 1,4-cis-butadiene in cis-BR and the molecular weight of the polymer are responsible for the occurrence of melt fracture [180].

## 8.8 Tack and Adhesion

The ability of an uncured rubber to adhere to itself but not to other rubbers is one of the most important properties for tire fabrication. After adhesion, the resulting system is resistant to separation from each other. This property is known as "tack" or auto-adhesion. Tack is important to the assembly of rubber compounds that need to remain in place prior to vulcanization. A good example is a tire, which requires positioning carcass, belts and tread.

The extrudate layers are held together by their tack or adhesion. If the extruded layers do not have the same composition, this is referred to as adhesion. When the layers are separated, a force is applied to overcome the adhesion. There are three main requirements for tack and adhesion [182]:

- The surfaces must make microscopic contact to permit interpenetration of chains. The process is influenced by the viscosity of the compound, the surface topography and the temperature.
- The interdiffusion must occur quickly in order to fuse the layers and provide adequate auto-adhesion or adhesion.
- The cohesive strength must be sufficient, since it serves as the upper limit of adhesion.

Under normal conditions, the process of interdiffusion causes an increase in tack with contact time. When the diffusion length is on the order of magnitude of the chain coil size, the interfacial region becomes indistinguishable from the bulk and tack becomes constant.

The coil size of a typical rubber molecule is in the range of 10–20 nm. The formation of the contact surface takes longer than the actual interdiffusion process of the chains. It has been observed experimentally that the contact time can be several hours until maximum pull force is set [183, 184].

The roughness of the extrudate surface has a negative effect on tack and adhesion. If the tack is too high, there is a risk that air inclusions between the layers will restrict the interdiffusion of the chains to a smaller surface and thus worsen the tack. If the interdiffusion leads to the disappearance of the phase boundary between the layers, a peel test would damage the surface and a cohesive crack would occur. Overly tacky compounds trap air between the tire components, resulting in poor durability. Insufficient tack makes it impossible to hold the components parts together until the tire is vulcanized. Adequate tack is a must for carcass and belt compounds to maintain a property between plies and belts and between plies and inner liner compounds [185]. The measured peel strength characterizes the green strength of the rubbers. However, as for all viscoelastic properties of rubber compounds, the peel strength depends on the peel rate and temperature [186].

Rubbers with a low $T_g$ and a broad molecular weight distribution, such as NR, cis-BR and SBR, make an important contribution to the measured value of the tack. The low $T_g$ of rubber guarantees high chain mobility and faster interdiffusion. A high molecular weight ensures that the chains are entangled in the interphase. It should be noted that as the chain length increases, interdiffusion of the chains takes place more slowly. NR has excellent tack strength, whereas synthetic rubbers do not show strain crystallization and lead to poor tack [187]. Strain-crystallizing rubbers, such as NR, exhibit superior tack at extensions of more than 300%, due to the dependence of their viscosity and the higher strength on the deformation of the test sample. Only at large strains (≥ 300%) that induce strain crystallization does the material have enough influence to increase auto-adhesion and tackiness of the rubber compound [188]. Mastication helps to increase the low molecular weight fraction in the NR and thus to improve auto-adhesion.

In the case of synthetic rubbers, attempts to increase tackiness by increasing the average molecular weight and long-chain branching does not lead to any significant improvements. The physical network resulting from the entanglements complicates the transport process of individual chains, so that the interdiffusion required for tackiness or adhesion is restricted. Rubbers such as SBR, BR, EPDM and, especially, IIR exhibit reduced tackiness. Blends with NR that show strain crystallization can improve this behavior if the NR content is increased [187].

The addition of active fillers also reduces tackiness and adhesion. Due to the filler-rubber interaction, the extent of which depends on the amount and degree of dispersion of the filler, a number of the polymer chains are adsorptively bound, which ultimately reduces the number of inter-diffusing chains. For tackiness, it is therefore important to increase chain mobility through suitable measures. The use of plasticizers facilitates interdiffusion; however, it entails loss of green strength and thus is not an effective way to improve tack. Tackifying resins can be employed in rubber formulations to take advantage of the rate dependence that the resins impart to the mechanical response [189]. Tackiness of synthetic rubbers is enhanced by adding tackifying resins or even a small quantity of NR. Too much resin becomes detrimental to the curing process. The tackiness, expressed by the peel strength, passes through a maximum in a certain concentration range of the resin. For hydrocarbon resins, which exert moderate effects on auto-adhesion, it has been shown that this happens at higher concentrations [190]. Blending of two or more low-tack polymers does not improve tack. Blends of NR with synthetic polymers increase tack almost linearly as a function of NR loading [191].

## References

[1] Nakajima, N., in "The Science and Practise of Rubber Mixing" RAPRA Technology Ltd., (2010).

[2] Röthemeyer, F., in "Kautschuktechnologie", Röthemeyer, F., Sommer, F., Eds., Hanser, Munich (2001) Ch. 6.

[3] Freakley, P. K., in "Rubber Processing and Production", Plenum Press, New York (1985)

[4] Grossman, R. F., Ed., "The Mixing of Rubber", Chapman and Hall, London (1997)

[5] Lovegrove, J. G. A., "Extrusion of Rubber" RAPRA Review Report No. 34 (1989)

[6] Graessley, W. W., Adv. Polym. Sci., 16 (1974) 3

[7] Boyer, R. F., Simha, R., J. Chem. Phys., 37 (1962) 1003

[8] Schoschke, C., in "A Dictionary of Chemical Engineering", Oxford Univ. Press (2014)

[9] Bueche, F., J. Appl. Polym. Sci., 1960, 4: 107

[10] de Gennes, P. G., J. Chem. Phys. 1971, 55: 572

[11] Eisele, U., in "Introduction to Polymer Physics", Springer Verlag Berlin (1990) Ch. 6.

[12] Bondi, A., J. Phys. Chem., 1954, 58: 929

[13] Vinogradov, G. V., Malkin, A. I., in "Rheology of Polymers", Springer Verlag, Berlin, Heidelberg, New York (1980)

[14] Ferry, J. D., in "Viscoelastic Properties of Polymers", John Wiley & Sons, New York (1980)

[15] Rüllman, M., in "Praktische Rheologie der Kunststoffe und Elastomere", VDI Verlag, Düsseldorf (1995)

[16] White, J. L., in "Science and Technology of Rubber", Mark, J. E., Erman, B., Eirich, F. R., Eds., Elsevier Acad. Press (2005) Ch. 6

[17] Berry, G. C., Fox, T. G., Adv. Polym. Sci., 1968, 5: 261

[18] Doi, M., Edwards, S. F., J. Chem Soc. Faraday Trans., 1979, 75: 38

[19] Sgodzjaj, U., PhD Thesis, University of Fridericiana Karlsruhe (2001)

[20] Kraus, G., Gruver, J. T., J. Polym. Sci., A3, 1965, 105

[21] Hoffmann, M., Krömer, H., Kuhn, R., in "Polymer Analytik", Georg Thieme Verlag, Stuttgart (1977)

[22] Obrecht W., Lecture in the Continuing Education in Rubber Technology, University Hanover (2015)

[23] Folt, V. L., Smith, R. W., Rubber Chem. Technol., 1973, 46: 1193

[24] Graessley, W. W., Adv. Polym. Sci., 1975, 16: 389

[25] Smith, R. W., Andries, J. C., Rubber Chem. Technol., 1974, 47: 64

[26] Johnson, P. S., in "Basic Elastomer Technology", Baranwal, K. C., Stephens, H. L., Eds., Rubber Divison, Akron (2001) Ch 4B

[27] Pliskin, I., Rubber Chem. Technol., 1981, 54: 439

[28] Röthemeyer, F., Kautsch. Gummi Kunstst., 1980, 33: 1011

[29] Freakley, P. K., in "Rubber Technologist's Handbook", De, S. K., White, J. L., Eds., RAPRA Technologies Ltd. (2001)

[30] Bagchi, A. K., Sircar, K. K., J. Appl. Polym. Sci., 1979, 23: 1653

[31] Williams, M. L., Landel, R. F., Ferry, J. D., J. Amer. Chem. Soc., 1955, 77: 3701

[32] Doi, M., Edwards, S. F., J. Chem. Soc. Faraday Trans II, 1978, 74: 1789

[33] Doi, M., J. Polym. Sci., Polym. Lett. Ed., 1981, 19: 265

[34] Graesley, W. W., Fortsch. Hochpolym.-Forsch., 1974, 16: 1

[35] Meier, J., Klüppel, M., Giebeler, E., Schuster, R. H., Kautsch. Gummi Kunstst., 2003, 56: 465

[36] Mancke, R. G., Ferry, J. D., Trans. Soc. Rheol. 12 (1968) 335

[37] Sanders, J. F., Ferry, J. D., Macromolecules, 7 (1974) 681

[38] Witten, T. A., et al., in "Multiphase Macromolecular Systems", Culbertson, B. M., Ed., Plenum Press, New York (1989)

[39] Rocha, T. L. C. A., Jacobi, M., Schuster, R. H., Kautsch. Gummi Kunstst., 2004, 57: 656

[40] Class, B., Chu, C. G., J. Appl. Polym. Sci., 1985, 30: 825

[41] Schuster, R. H., Meier J., Klüppel, M., ACS-Rubber Div. Meeting, Orlando (Fl), May (1999)

[42] Schramm, W. G., in "A Practical Approach to Rheology and Rheometry", Thermo Haake, Karlsruhe (2000)

[43] Mooney, M., Physics, 1936, 7: 413

[44] ASTM D 1646

[45] Dannenberg, E. M., Rubber Chem. Technol., 1975, 48: 410

[46] Porter, R. S., Johnson, J. F., Chem. Rev., 1966, 66: 1

[47] Wise, R., US Pat. 3,387,480 (1968)

[48] Barker, R. I., King, P. D., Pawlowski, H. A., US Pat. 4,552,025 (1985)

[49] ASTM D 6204

[50] Dick, J. S., in "Rubber Technology: Compounding and Testing for Performance", Dick, J. S., Ed., Hanser Publishers, Munich (2009) Ch. 2

[51] DIN 50314

[52] Rabinowitsch, B., Z. Phys. Chem., 1029, 145: 1

[53] Bagley, E. B., J. Appl. Phys., 1957, 28: 624

[54] Patel, A. C., Byers, J. T., Rubber World, 1983, 188: 21

[55] Einstein, A., Ann. Physik, 1906, 19: 289

[56] Guth, E., Simha, R., Gold, O., Kolloid Z., 1936, 74: 266

[57] Guth, E., Gold, O., Phys. Rev., 1938, 53: 322

[58] Kraus, G., Rubber Chem. Technol., 1965, 38: 1070

[59] Bruzzone, M., et al., Chim. Ind. (Milan), 1965, 51: 145

[60] Medalia, A. I., J. Coll. Interface Sci., 1970, 32: 115

[61] Kraus, G., J. Polymer Sci., Part B, 1971, 8: 601

[62] Byers, J., ACS Rubber Division Meeting (1999)

[63] Wolff, S., Wang, M.-J., Kautsch. Gummi Kunstst., 1999, 47: 17

[64] Wang, M.-J., Wolff, S., Tan, E. H., Rubber Chem. Technol., 1993, 66: 178

[65] Hess, W. M., Herd, C. R., Vegvari, P. C., Rubber Chem. Technol., 1933, 66: 329

[66] Kraus, G., Gruver, J. T., J. Polym. Sci., 1971, 8: 571

[67] Wolff, S., Wang, M.-J., Tan, E. H., ACS Meeting of Rubber Division, 1991

[68] Pliskin, I., Tokita, N., J. Appl. Polym. Sci., 1972, 16: 743

[69] Montes, S., White, J. L., Nakajima, N., J. Non-Newtonian Fluid Mech., 1988, 28: 183

[70] Clarke, J., Freakley, P. K., Rubber Chem. Technol., 1994, 67: 700

[71] Wasserman, S. H., Graessley, W. W., J. Rheol., 1992, 36: 543

[72] Leblanc, J. L., in "Filled Polymers, Science and Industrial Applications", CRC Press, Boca Raton (2010), Ch. 5

[73] Wang, M.-J., Wang, T., Wong, Y. L., Kautsch. Gummi Kunstst., 2002, 55: 388

[74] Robertson, C. G., Wang, X., Phys. Rev. Lett., 2005, 95: 0757030

[75] Wang, M.-J., Rubber Chem. Technol., 1998, 71: 520

[76] Hess, W. M., Vegvari, P. C., Swor, R. A., Rubber Chem. Technol., 1985, 58: 350

[77] Cotton, G. R., Thiele, J. L., Rubber Chem. Technol., 1978, 51: 749

[78] Westenberg, H., Gromes, H., Rechenbach, D., Kautsch. Gummi Kunstst., 2021, 73: 47

[79] Görl, U., in "Handbuch der Kautschuktechnologie", Hofmann, W., Gupta, G., Eds., Dr. Gupta Verlag, Ratingen (2001), Ch. 8B

[80] Freund, B., Wolff, S., ACS Rubber Division, Mexico City (1989)

[81] Hasse, A., Luginsland, H.-D., Rubber Chem. '01 Conference, Brussels, 2001

[82] Luginsland, H.-D., Fröhlich, J., Wehmeier, A., ACS Meeting Rubber Division, Providence (2001)

[83] Rueb, C.J., Zukovski, C.F., J. Rheology, 1997, 41: 197

[84] Yziquel, F., Carreau, P.J., Tangui, P.A., Rheolog. Acta, (1999) 38: 14

[85] Zhu, A.J., Sternstein, S.S., Comp. Sci. Technol., 2003, 63: 1113 th

[86] Sternstein, S.S., Zhu, A.J., Macromolecules, 2002, 35: 7262

[87] Wang, M.-J., et al., Kautsch. Gummi Kunstst., 2020

[88] Schaal, S., Coran, A.Y., Rubber Chem. Technol., 2000, 73: 225

[89] Thang, B.K.G., in "Formation and Properties of Clay-Polymer Complexes", Elsevier, Sci., Amsterdam (1979)

[90] Varghese, S., Karger-Kocsis, J., Polymer, 2003, 44: 4921

[91] Schellenberg, M., Schuster, R.H., ACS Rubber Division Meeting, Cleveland (2013)

[92] Azeredo, L.K., Jacobi, M.M., Schuster, R.H., PPS-Americas Meeting Regional, Charleston, (USA) 2008

[93] Azeredo, L.K. Jacobi, M.M., Schuster R.H., Kautsch. Gummi Kunstst. 2010, 63: 8

[94] Stephen, R., et al., Appl. Polym. Sci., 2006, 101: 2355

[95] Austin, J.R., Kontopoulou, M., Polym. Eng. Sci., 2006, 46: 1491

[96] Jeon, H.S., Rameshwaram, J.K., Kim, G., J. Polym. Sci., Part B, Polym. Phys., 2003, 44: 5749

[97] Tian, M., Qu, C., Feng, Y., Zhang, J., J. Mater. Sci., 2003, 38: 4917

[98] Sadhu, S., Bhowmick, A.K. J. Polym. Sci., Part B, 2005, 43: 1854

[99] Maiti, M., Sadhu, S., Bhowmick, A.K., J. Polym. Sci, Part B, 2004, 42: 4489

[100] Kader, M.A., Lyce, M.Y., Nah, C., Compos. Sci. Technol. 2006, 66: 1431

[101] Kim, J.T., Oh, T.S., Lee, D.H., Polym. International, 2003, 52: 1203

[102] Ibarra, L., Rodriguez, A., Mora, I., Europ. Polym. J., 2007, 43: 753

[103] Prasad, R., Pasonovic-Zujo, V., Gupta, R.K., Polymer Eng. Sci., 2005, 44: 1220

[104] Durnuous, A., Kasgaz, A., Makosko, J., Polymer, 2007, 48: 4482

[105] Wang, X., Gao, Y., Mao, K., Macromolecules, 2006, 39: 6653

[106] Lim, S.K., et al., J. Appl. Polym. Sci., 2002, 86: 3735

[107] Wang, X. et al., Macromolecules, 2006, 39: 6653

[108] Liu, C., Zhang, J., He, J., Hu, G., Polymer, 2003, 44: 7529

[109] Fan, Z., Advani, S.G., J. Rheol. 2007, 51: 585

[110] Zimmermann, H., Schuster, R.H., Kautsch. Gummi Kunstst., 2011, 64: 7

[111] Terentiev, E.M., et al., Phys. Rev. Part B, 2006: 73: 422

[112] Chogule, H., Schuster, R.H., ACS Rubber Division Meeting, Cleveland (2011)

[113] Grohens, Y., Brogly, M., Labbe, C., David, M.O., Schultz, J., Langmuir, 1998, 14: 2994

[114] Parro-Vasques, A.N.G., et al. Macromolecules, 2007, 40: 4043

[115] Xu, D.H., Wang, Z.G., Douglas, J.F., Macromolecules, 2008, 41: 815

[116] Song, Y.S., Rheol. Acta, 2006, 46: 231

[117] Feng, Ch., Zhang, H., Zang, Y., Polym. Compos., (2014) 35, 2149

[118] Lee, S.H., Kim, M.W., Kim, S.H., Youn, J.R., Eur. Polym. J., 2008, 44: 1620

[119] Du, F., Scona, R. Z., et al., Macromolecules, 2004, 37: 9048
[120] Tiwani, M. K., Rheol. Acta, 2009, 48: 579
[121] Schuster, R. H., Geisler, A., Chogule, H., Report to Lanxess Company (2014)
[122] Huang, Y. Y., Ahir, S. V., Terentiev, E. M., Phys. Rev. B., 2006, 73: 125422
[123] Cataldo, F., Ursini, O., Angelini, G., Fullerenes, Nanotubes and Carbon Nanostructure, 2009, 17: 38
[124] Boonstra, B. B., Medalia, A. I., Rubber Chem. Technol., 1963, 36: 1190
[125] Touissant, H. E., Unger, W. N., Schäfer, H. O., Kautsch. Gummi Kunstst., 1972, 25: 155
[127] Oppermann, B., PhD Thesis, University of Hanover (1994)
[128] Schuster, R. H., in "Mixing of Rubber Compounds", Limper, A., Ed., Hanser Publishers, Munich (2012)
[129] Ong, L. L., Liu, C. L., J. Rubber Res. Inst. Malaysia, 1982, 30: 81
[130] Issel, H.-M., PhD Thesis, University of Hanover (1995)
[131] Dickstein, R., in "Mixing Technology & Machinery", Technical Course (2003)
[132] Meon, W., Blume, A., Luginsland, H.-D., Uhrlandt, S., in "Rubber Compounding, Chemistry and Application", Rodgers, B., Ed., Marcel Dekker, Inc., New York, Basel (2004) Ch. 7
[133] Reuvenkamp, L. A. E. M. et al., Rubber Chem. Technol., 2002, 75: 187
[134] Dierkes, W., Proc. EU-Project SATPRO Paderborn (2004)
[136] Trimbach, J., Jornadas Latino-Americanas del Technologia del Caucho, Porto Alegre (2012)]
[137] Bunawerke Hüls GmbH, Buna AP Leitfaden für den Praktiker. 1136, (1986)
[138] Böhm, G. G. A., Nguyen, M. N., J. Appl. Polym. Sci., 1995, 55: 1041
[139] Wang, T., Wang, M.-J., Shell, C., Tokita, N., Kautsch. Gummi Kunstst., 2000, 53: 497
[140] Meier, J., Schuster, R. H., ACS Rubber Division Meeting, Rhode Island (2001)
[141] Lin, D. C., Hergenrother, W. L., Hilton, A. S., Rubber Chem. Technol., 2002, 75: 215
[142] Sirkar, A. K., Lamond, T. G., Puiter, P. E., Rubber Chem. Technol., 1974, 47: 48
[143] Monthey, S., 152nd ACS Rubber Division Meeting, Cleveland (1999) pp 35
[144] Cotton, G. R., Rubber Age, 1968, 100: 51
[145] Barker, R. I., Hanna, G. L., Rogers, E. R., Internatl. Rubber Conf. Brighton 1977, pp. 30
[146] Minagawa, N., White, J. L., J. Appl. Polym., Sci., 1976, 20: 501
[147] Hess, W. M., Rubber Chem. Technol., 1991, 64: 37
[148] Robertson, C. G., Roland, C. M., Puskas, J. E., J. Rheol. Sci., 2002, 46: 307
[149] Freakley, P. K., Sirisina, C. J., J. Appl. Polym. Sci., 1997, 65: 305
[150] Pliskin, I., Tokita, N., J. Appl. Polym. Sci., 1972, 16: 178
[151] Cotton, G. R., Rubber Chem. Technol., 1979, 52: 187
[152] Hoper, J. R., Rubber Chem. Technol., 1967, 40: 463
[153] Fröhlich, J., Postgraduate Studies in Rubber Technology, University of Hanover
[154] Buckler, E. J., Briggs, G. J., Dunn, J. R., Rubber Chem. Technol., 1978, 51: 872
[155] Gent, A. N., Rubber Chem. Technol., 1955, 28: 456
[156] Kawahara, S., Nishiyama, N., Matsura, A., Nippon Gomu Kyokaishi, 1999, 72: 288
[157] Barnes, H., J. Non-Newtonian Fluid Mech., 1999, 81: 133
[158] Wang, M.-J., Wolf, S., Donnet, J.-B. Rubber Chem. Technol., 1991, 69: 714
[159] Daniel, C., Pillow, J. G., 99' Rubber Conference Manchester (1999)
[160] Wolf, S., Wang, M.-J., Tan, E. H., Kautsch. Gummi Kunstst. 1994, 47: 873
[161] Wang, M.-J., in "Powders and Fibers" Nardin, M., Papirer, E., Eds., CRC Press-Taylor and Francis Group, Boca Raton (2007)

[162] Tadmore, Z., Gogos, C. C. in "Principles of Polymer Processing", J. Wiley & Sons, New York, (1979)

[163] Han, C. D., Kim, K., Polym. Eng. Sci., 1971, 11: 395

[164] Geiger, K., Kautsch. Gummi Kunstst., 1989, 42: 273

[165] Atwood, B. T., Schowalter, W. R., Rheol. Acta, 1989, 28: 134

[166] Uhland, E., Rheol. Acta, 1978, 18: 1

[167] Fattman, G., PhD Thesis, Univ. Paderborn (2007)

[168] Wiegreffe, S., Kautsch Gummi Kunstst., 1991, 44: 216

[169] Jepsen, C., Räbiger, N., Kautsch Gummi Kunstst, 1988, 41: 342

[170] Bornemann, M., PhD Thesis University of Paderborn (2011)

[171] Osswald, T. A., Rudolph, N., in "Polymer Rheology – Fundamentals and Applications", Carl Hanser Publ., Munich (2014)

[172] Ma, C.-Y., et al., Rubber Chem. Technol., 1985, 58: 815

[173] Mooney, M., J. Rheol., 1931, 2: 210

[174] Malkin, A. Y., Boranov, A. V., Vickulencova, Rheol. Acta, 1993, 32: 150

[175] Crawford, B., et al., J. Non-Newtonial Fluid Mech., 2005, 129: 38

[176] Klüppel, M., Badura, R., Schuster, R. H., Kautsch. Gummi Kunstst., 1992, 45: 614

[177] Menning, G., Rheol. Acta, 1980, 19: 262

[178] Klie, B., Schubert, E., Giese, U., Kautsch. Gummi Kunstst., 2020, 71: 35

[179] Hopmann, Ch., Dach, M., Limantaro, Kautsch. Gummi Kunstst., 2018, 70: 91

[180] Kulbaba, K., Proceed. TechnoBiz Forum on Mixing, Pittsburgh (2014)

[181] Liang, J. Z., J. Mater. Proc. Technol., 1996, 66: 97

[182] Hamed, G., Shieh, C. H., Rubber Chem. Technol., 1986, 59: 883

[183] Bister, E., Borchard, W., Rehage, G., Kautsch. Gummi Kunstst., 1976, 29: 527

[184] Roland, C. M., Macromolecules, 1987, 20: 2557

[185] McDonald, E. T., Baranwal, K. C., Andries, J. C., in "Polymer Blends", Paul, D. R., Newman, S., Eds., Acad. Press, Inc., San Diego, New York, Vol. 2, Ch. 19

[186] Hamed, G. R., Shieh, C. H., J. Polym. Sci., Polym. Phys. Ed., 1983, 21: 1415

[187] Railsback, H. E., Howard, W. S., Stumpfe, Jr, N. A., Rubber Age, 1974, 106: 46

[188] Choi, I. S., Roland, C. M., Rubber Chem. Technol., 1997, 70: 202

[189] Hamed, G. R., Roberts, G. D., J. Adhes., 1994, 47: 95

[190] Wetzel, F. K., Rubber Age, 1957, 82: 291

[191] Morrisey, R. T., Rubber Chem. Technol., 1971, 40: 324

# 9 Physical Properties of Rubber Vulcanizates

Elastomers are entropy-elastic materials whose properties are determined by their viscoelastic behavior. The physical properties of the elastomers are important for use in various areas. Not only the question of what rubber the vulcanizate is made of, but also which filler was used, how the compound was mixed, which dispersion the filler has and how the chemical process of the vulcanization takes place are of crucial importance for the physical behavior of the elastomer. Almost all rubber products are produced to meet a certain specification, which is related to the formulation that is specific for its application to guarantee the product is suitable for the intended use. The resulting vulcanizate can be regarded as a system containing a "rubbery" continuous phase that provides the viscoelastic behavior and a particulate "rigid" dispersed phase that leads to an improvement in the most physical properties and is called the reinforcement of the elastomer. The continuous phase is responsible for the elastic behavior, while the filler plays a specific role in changing the stiffness, stress, tensile strength, viscoelasticity, hysteresis, tear energy, cut growth resistance and abrasion resistance.

Most of the standardized physical properties of elastomers presented in this chapter are measured by standardized test processes and the results can be explained without employing sophisticated approaches. When performing a test, it is also necessary to follow the respective test procedure carefully. This primarily serves to compare the results and to detect errors originating in the composition of the vulcanizates or the manufacturing procedure. However, to fall into the trap of considering testing as being easy and the results as being easy to interpret is unfortunately a frequent occurrence, although the very complex interactions make filled elastomers difficult to understand. The need to elucidate such relationships lies in the causal chain: formulation – processing – curing – product performance.

In the following, the standard physical properties of elastomers the hardness, compression set and stress-strain behavior are reported. In all cases, a comparison of the unfilled and filled elastomers is sought in order to describe the improvement in the reinforcing effect of the filler. This is not only the case for carbon blacks and precipitated silica but also for wet-mixed silicas, the layered silicate and carbon nanotubes show improved properties.

## 9.1 Hardness of Rubber

Hardness is a common and easy way of describing the cured rubber resistance to a localized deformation induced by mechanical indentation. Hardness characterizes the resistance of a vulcanized compound to indentation under a load. The penetration depth of an indenter is measured in the test, as shown in Figure 9.1.

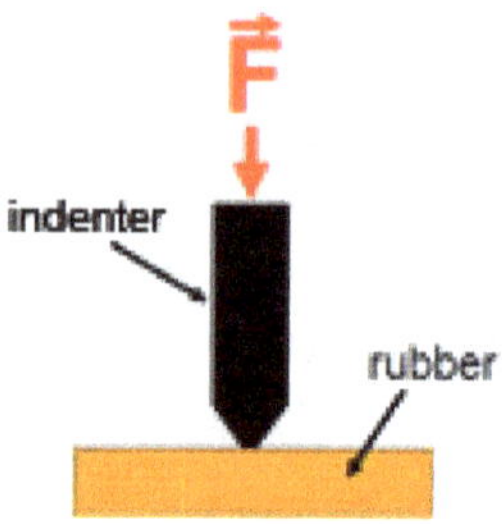


**Figure 9.1**
The principle behind the hardness test

Usually, hardness is expressed as resistance to penetration of an indenter into the surface of a crosslinked rubber on a scale of 0–100. No single device can measure this property across its full range. The test piece must be 6 mm thick and the distance must be 12 mm, so that the stress field is confined to the test piece volume. The hardness Shore A is used for soft rubbers and hardness Shore D is employed for harder rubbers. The Shore hardness is determined using a truncated cone identer. The methods are introduced in standard processes [1, 2]. The results describe a kind of modulus under limited deformation strain deformation paralleled with torsion and compression. Surface hardness is attractive in its simplicity and relative ease of performance, given the availability of pocketsize devices. The international rubber hardness degree (IHRD) uses the indenter as a hemispherical ball [3]. The value of IHRD and Shore hardness are relatively closed. The accuracy and precision of hardness measurements become questionable towards either end of the hardness scale used. Due to the mechanical limits of the test instruments, hardness measurements are rarely expressed more precisely than ± 2.5 points. To help reduce variation in hardness measurements, it is recommended to use a durometer mounted on an operation stand that takes readings automatically after a timeout. Since the measured values depend on the measurement duration due to the viscoelastic properties, the values are read after a defined time. Therefore, hardness tests should not be considered a reliable measure of a design or engineering property, but rather a quick and simple estimate of detecting gross differences in cured compound properties [4, 5].

### 9.1.1 Hardness of Filled Elastomers

It can be concluded that among others, such as crosslink density of the polymer matrix and polymer-filler interaction, the filler agglomeration plays an important role in the hardness of filled vulcanizates. Hardness increases with the crosslinking density of the rubber and the chemical composition of the polymer. However, it has been repeatedly pointed out that hardness values do not correlate with the crosslink density of elastomers [6]. Since two compounds with the same durometer hardness can vary easily by 25% or more with the actual stiffness, such an assumption can mislead the compounder or process engineer.

Since elastomers are mainly used as filled systems, the hardness of the unfilled systems has more academic interest. Tests of hardness are used primarily in engineering fields to evaluate the resistance of filled rubbers. The addition of a solid filler to a soft rubber matrix leads to an increase in hardness because part of the volume of the elastic phase is replaced by solid particles that reduce the viscoelastic properties and increase the hardness by the loading. The hardness of elastomers increases in proportion to the loading of the filler. However, the hardness is less sensitive to morphological variations in the filler particles, but the characteristic parameters such as primary particle size, structure of the filler aggregates, and filler activity are demonstrated. To exemplify this, equivalent loading levels for different fillers that lead to a roughly ten-point increase in hardness Shore A in an SBR compound are listed in Table 9.1. In order to achieve similar hardness values with an active carbon black, the mineral fillers with a significant smaller specific surface require a higher loading in the vulcanizate.

**Table 9.1** Filler Contribution to 10 Hardness Points [7]

| Filler | Loading (phr) |
|---|---|
| Carbon black N330 (80 $m^2$/g) | 23 |
| Ultrasil, VN3 (175 $m^2$/g) | 20 |
| HiSil 532 EP, Ultrasil 360 | 30 |
| Hard clay | 45–65 |
| Calcium carbonate | 80 |

Increasing loading with filler, irrespective of the grade, tends to increase the hardness of the vulcanizate. Therefore, the nature of the filler and its activity can reduce the loading significantly, but the same hardness is retained.

### 9.1.1.1 Influence of Carbon Black

A comparison of different fillers that require the hardness Shore A in NR vulcanizates to increase up to 70 points includes the carbon blacks. For the inactive carbon black N990, 110 phr is necessary while for the active carbon black N330 only 50 phr is necessary [8]. The comparison of the filler loadings is presented in Figure 9.2.

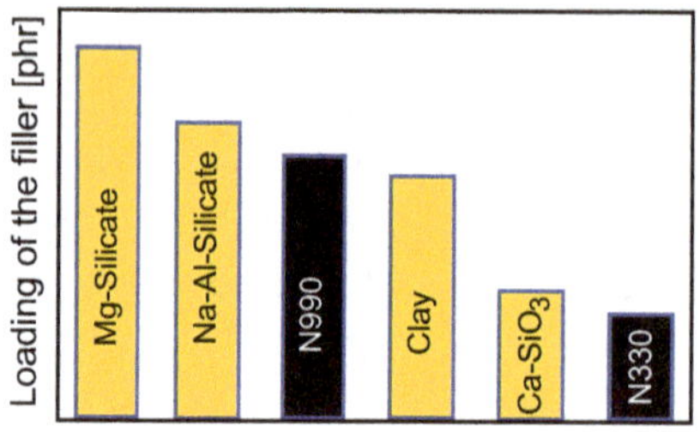


**Figure 9.2**
Amount of different filler required to reach hardness Shore A 70 [8]

Compared with Mg silicates or Al silicates, which require more than 120 phr to reach this hardness, a strong reduction in the amount of active carbon black can be seen. For carbon blacks, it was found that hardness Shore A increases with both the specific surface and the structure of the aggregates (OAN). The relationship between the morphological characteristics of carbon black is shown in Figure 9.3 for E-SBR/carbon black (50 phr) vulcanizates [9].

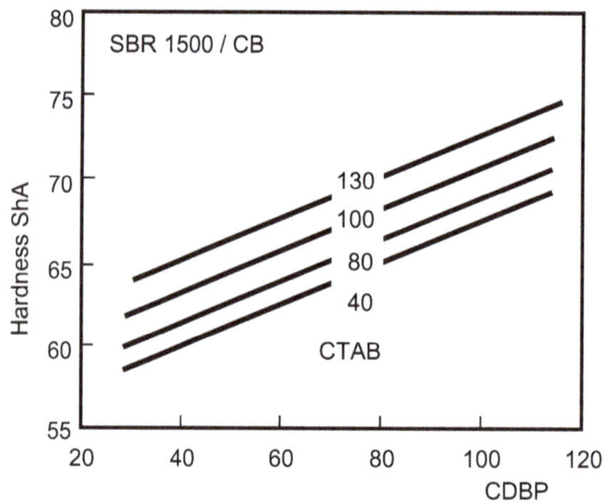


**Figure 9.3**
The increase in the vulcanizate hardness is obtained by increasing the specific surface area or the structure of the aggregates [9]

The proportional increase in hardness Shore A enables a relatively precise indication of the carbon black influence. The specific surface area contributes to higher hardness because the rubber-filler interaction increases the effective filler volume fraction by the interfacial layer. With increase in the specific surface area, the slope of the relationship increases and leads to an increase in vulcanizate hardness. The effect of the structure on hardness is, however, somewhat more pronounced and can be attributed to the effect of the occluded rubber. Due to the incorporation of a portion of the rubber in the voids of the filler aggregates, the apparent filler concentration increases. The increase in the carbon black loading in the polar HNBR increases the Shore A hardness from 63 points while the addition of 25 phr of carbon black N339 increases the value to 76 [10]. Vulcanizates consisting of EPDM are usually made with high proportions of carbon black and correspondingly high proportions of plasticizer. The influence of paraffin oil and carbon black 539 on the hardness Shore A is shown in Table 9.2.

**Table 9.2** Contribution of the Filler and Plasticizer to the Hardness of EPDM Shore A, Expressed by the Relation: Hardness Shore A = EPDM Grade + CB × phr-Oil × phr [11]

| Filler | Paraffin Oil |
|---|---|
| N220 0.35 | 0.33 |
| N330 0.30 | 0.33 |
| N550 0.24 | 0.33 |
| N777 0.22 | 0.33 |
| N990 0.11 | 0.33 |

Regardless of the filler loading, the hardness decreases with an increasing amount of oil in accordance with an almost linear function [12]. Similar additive relationships can be determined with the help of statistical design of experiments as a mathematical function for any number of elastomer systems. If carbon black is obtained from the oil that is produced during the pyrolysis of scrap tire, it shows for N234 the same hardness as ASTM carbon black N234 [13]. Statistically, the CEC gives 1.5 to 2.5 points lower hardness at the practical loading in the rubber vulcanizate. A modified carbon black that is connected to the rubber via a bifunctional coupler can increase the hardness Shore A in some points [14]. This new development of carbon black is able to show a better interaction with the rubber matrix that leads to a higher mechanical hardness.

#### 9.1.1.2 Influence of Precipitated Silica

The difference shown for precipitated silica results from the strong interparticle interactions existing between the silanol groups in the silica networks [9]. Generally, hardness is sensitive to the filler loading, but less to qualitative improvements of filler dispersion. Hardness is more related to the filler volume fraction in the compound than to the integral polymer-filler contact surface, the intensity of rubber-filler interaction and the inter-aggregate distance. The conventional silica requires only 38 phr to show the same hardness Shore A 70, less than carbon black N330 and carbon black N990 [8].

The HD silica shows a smaller hardness than the conventional silica, VN2, VN3. This difference can be explained by the different aggregate size distribution. The effect is shown in Figure 9.4 [9].

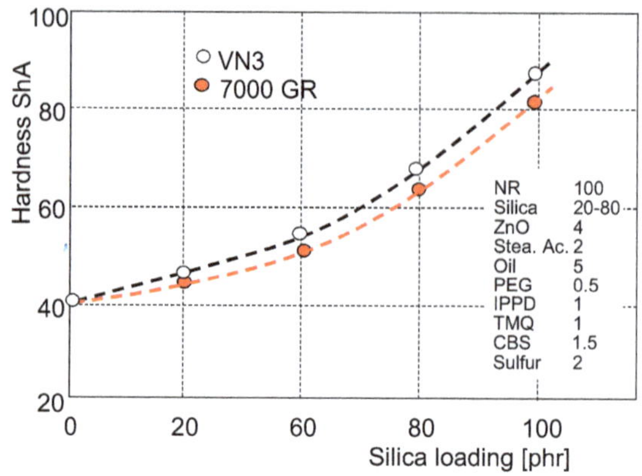


**Figure 9.4**
The influence of the silica grade on the hardness of an NR vulcanizate [9]

In NR, incorporated silica with a particle size of 19 nm has been compared with carbon black N220 with a primary particle size of 22 nm and shows the same hardness Shore A of 64 points [15]. The hardness Shore A of silica-filled vulcanizates shows the expected increase in the hardness with increase in surface areas of the silica. Due to the adsorption of the rubber chains on the filler aggregates, the solid entities can serve as a giant multi-functional crosslink in the crosslinked matrix [16]. The high specific surface area would be expected to have a positive effect on hardness. A comparison of carbon black and precipitated silica made at a level of 50 phr has revealed that hardness increases slightly in the series of carbon blacks, while it demonstrates a higher gradient in the group of silicas with high specific surface area silica (> 140 $m^2/g$). This is shown in Figure 9.5.

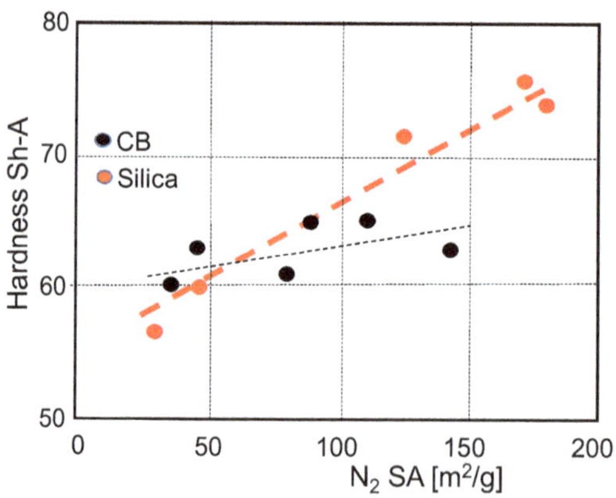


**Figure 9.5**
Comparison of the effect of precipitated silica and carbon black on the hardness of the NR vulcanizate [9]

The dependence of hardness on silica loading is shown by the increase from neat NBR of 57 points to that with a silica content of 25 phr that shows a hardness Shore A of 73 points [17]. Hardness Shore A decreases with the loading of silica in a compound of NR (SMR 209/cis-BR/CBN375/silica (134 $m^2/g$) when the carbon black/silica ratio is 40/20, because the crosslinking density of the blend decreases [18]. An increase in hardness Shore A has been observed for silica filled in compounds NBR/ENR that contains a larger amount of ENR [19]. The addition of ENR improved the hardness because of the rubber-filler interaction that occurs in this rubber.

The hardness Shore A of silanized silica EVEC obtained by wet-mixing in S-SBR/cis-BR is significantly lower than the normal silica obtained by dry mixing. The hardness Shore A of 56 is much lower than that of silica vulcanizate of hardness Shore A 68. As expected, the hardness of the wet-mixed silanized silica increases with loading, but, even with 6 phr more silica, the hardness is still lower compared with dry-mixed silanized silica [20]. The order of hardness Shore A is as follows for the same addition of filler:

N234 > silanized silica dry-mixed > silanized silica wet-mixed

This sequence demonstrates that the hardness is reduced by the type of compounding the mixture. Wet-mixing is able to significantly reduce the hardness.

#### 9.1.1.3 Influence of Layered Silicates

If compared with the mechanically added MMT which does not change the hardness Shore A at low concentration, the wet-mixed pristine MMT leads in NBR (34% ACN) to a hardness of 75 hardness points. Wet-mixing MMT in NR, SBR, and in NBR leads to an improvement in the hardness Shore A, a fact which can be attributed to the high specific area of the clay and the high aspect ratio [21, 22]. The incorporation of MMT into the rubber matrix by wet-mixing results in an increase in the hardness Shore A of 36% for a content of 15 phr ($\Phi = 0.04$), as shown in Figure 9.6 [23].

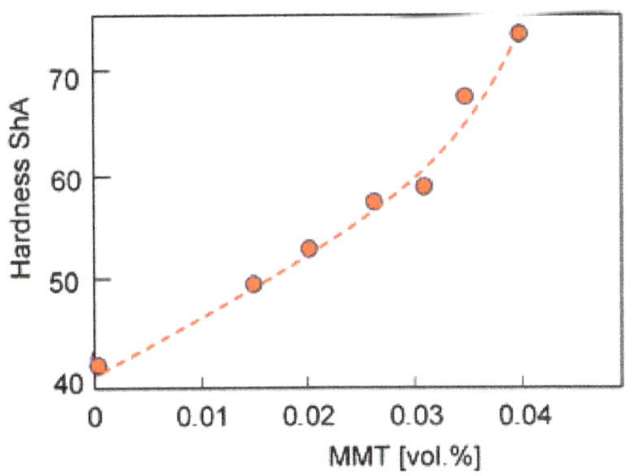


**Figure 9.6**
Increase in the hardness in NBR due to the exfoliated MMT [23]

The hardness Shore A of clays is shown for the wet-mixing obtained by the continuous dynamic latex compounding (CDLC) process and separately for organo-clays obtained by different processes. After an excellent dispersion of MMT (PGN) in NBR latex, the hardness increases with the load of the filler. Above the percolation region of 0.025 vol.%, a stronger increase occurs in the hardness Shore A, due to an outstanding exfoliation in the rubber phase. At 0.04 vol.%, the hardness reaches 70 Shore A points. In the less-polar HNBR (34% ACN), the hardness is lower by 2–4 points [24]. From a geometric standpoint, the difference between random orientation and a parallel alignment in the direction of processing becomes obvious. The hardness is higher for the aligned MMT than for the random orientation.

The addition of a small quantity of wet-mixed MMT to a carbon black N550 filled NBR shows a much higher hardness of the compound. With 3 vol.% of layered silicate introduced by CDLC, the increase in hardness corresponds to 21% of the hardness of carbon black compound. The high gain can be attributed to the degree of dispersion of MMT and the exfoliation in the rubber compound. The incorporation of 1 or 2 vol.% of MMT to carbon black N550 compound leads to a significant increase in the hardness in this compound. Overall, this increase in the 21% shows the reinforcement potential of both fillers. Comparison of the hardness Shore A for an HNBR compound with a hardness of 63 points with an organo-clays (5 phr) reveals an increase in hardness of 72 points, while one with the same amount of carbon black N339 only has a hardness Shore A of 66. If the exfoliation of the MMT platelets does not work, the increase in hardness of the NBR compound is much less. At 10 phr MMT, a hardness Shore A of 62 points is determined, while at 60 phr only 74 points are measured [21].

Organo-clays are intercalated in EPDM and so exfoliation is not possible, due to the poor polarity of the rubber. This problem of poor compatibility is solved in several ways. On the one hand, EPDM is made more polar by the coupling of maleic anhydride and, on the other hand, the phyllosilicate is modified by a suitable modification with octadecyl-ammonium to such an extent that it behaves quite well in EPDM [25]. When crosslinked with peroxide, intercalated structures are formed in EPDM and when crosslinked with sulfur/accelerator, exfoliated structures are formed to a large extent [26]. It has been reported for the example of HNBR/organo-clay that the dispersion of the organo-clays varies with the type of organophilic modifier [27].

### 9.1.1.4 Influence of Carbon Nanotubes

The new group of multi-walled carbon nanotubes are interesting in rubber compounds because their $sp^2$-hybridized carbons can interact with the rubber chains and lead to improved rubber properties. With carbon nanotubes, the high aspect ratio and the specific surface play a special role because the tubes are oriented in the direction of processing and the alignment obtained causes a reinforcing effect. In the case of hardness, the multi-walled carbon nanotubes demonstrate an increase in this value that is dependent on the loading and the diameter of the carbon nanotubes.

The hardness Shore A increases with the multi-walled carbon nanotubes loading which depends on the length of the tubes and the dispersion realized during mixing in the rubber matrix. The addition of cleaned multi-walled carbon nanotubes leads to an increase in hardness of 13 points above the effect achieved by carbon black N326 in NR compounds [28]. Multi-walled carbon nanotubes C150P from Bayer Material Science lead to a higher increase in the hardness than that of carbon black N550 [18]. It turns out that, in NR multi-walled carbon nanotubes with a proportion of 1.5 vol.% lead to an increase in hardness. However, the diameter of the multi-walled carbon nanotube determines the hardness value, as shown in Figure 9.7 [29].

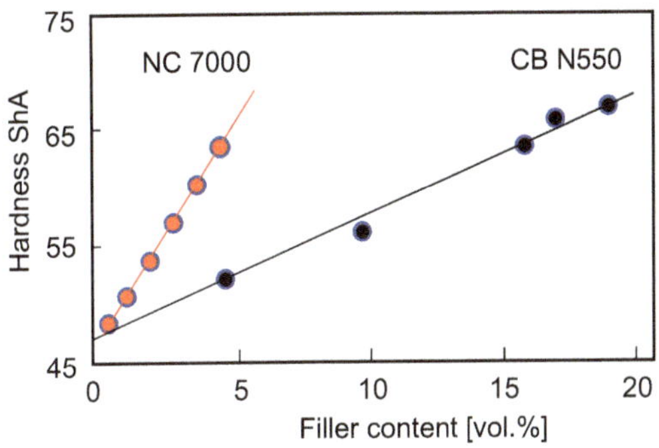


**Figure 9.7**
Hardness Shore A increases at a small filler concentration of multi-walled carbon nanotubes compared with carbon black N550, which increases the hardness at higher concentrations [29]

As the diameter of the multi-walled carbon nanotubes increases, the increase in hardness decreases because the specific surface area and aspect ratio decrease. In addition, a combination of carbon black N550 with 1 vol.% multi-walled carbon nanotubes produce a significant increase in hardness. The related hardness increases in all the rubbers. The IRHD hardness after the addition of 2.5 to 15 phr of multi-walled carbon nanotubes in NR, as compared with the reference carbon black, shows an increase in hardness from 13% to 22% [30]. The hardness Shore A of NBR vulcanizate of unfilled vulcanizates increases from 44 points to 64 points upon adding 5 phr of multi-walled carbon nanotubes [31]. A significant linear increase in hardness Shore A is observed in the region of up to 4 vol.% of multi-walled carbon nanotubes in HNBR. Above this loading, the hardness decreases from linear behavior, because the dispersion in the polar polymer decreases. The more-polar HXNBR shows an increase in hardness Shore A from the unfilled vulcanizate with 43 points to 50 points, obtained by 2 phr of multi-walled carbon nanotubes [33]. In FKM rubber up to a loading of 5 phr, the hardness increases proportional to the loading of multi-walled carbon nanotubes; above this value the unsatisfactory dispersion decreases the hardness in the compound.

Since the precision of the hardness test is not high, small changes in the measured values should not be overestimated. This depends on the accuracy of the hardness measurement and the depth of penetration into the vulcanizate. Nevertheless, compound surface hardness will remain a criterion, which conveys some information about the filled elastomer.

## 9.2 Compression Set

The compression set is used to determine the percentage deformation that remains over a certain period of stress in relation to the initial deformation. When carrying out the test, a deformation of 25% is usually chosen. Compression set (CS) is a widely used test for assessing the recovery of both unfilled and filled elastomers deformed by an external load. Compression set is an important characteristic of elastomeric sealing materials, as spontaneous stress is released during application due to internal pressure or external force variations. Compression set is a simple test to perform and, as such, may not yield the predictive information desired for sealing application [34, 35]. The compression set describes the residual deformation of the test specimen after the release of a static load. The result of the compression set depends on the duration of compression, the temperature at which the test proceeds, the percentage of compression and the crosslinking of the vulcanizate. The test procedure is described in industrial standards [36, 37] and is shown in Figure 9.8.

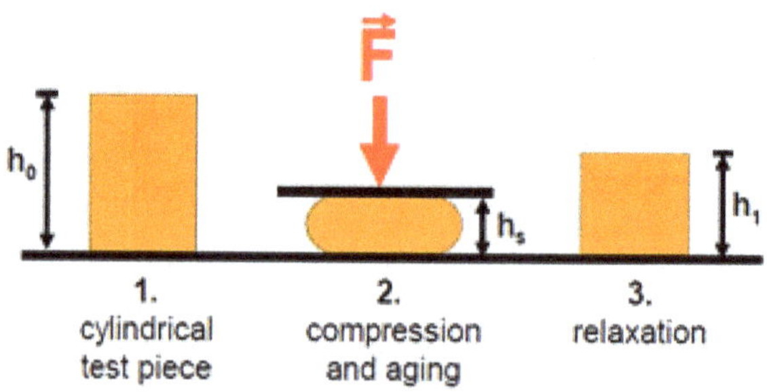


**Figure 9.8** Schematic representation of compression set

The difference between the recovered dimensions of the vulcanizate sample and its original dimensions is defined as the compression set and expressed as a percentage. Standard methods require a compressed sample to be exposed for a fixed time and a defined temperature, and then allowed to recover at room temperature. Compression set is defined as the percentage of original specimen thickness after the specimen has been left in normal conditions for 30 minutes:

$$CS\,(\%) = [(h_0 - h_i)\,/h_0] \times 100 \tag{9.1}$$

where $h_0$ is the initial height of the specimen and $h_i$ is the height after 30 min recovery

The test at low temperature requires the specimen to be kept at low temperatures, with measurement performed after 30 min, still at low temperatures. In several applications, rubber parts are submitted to a constant load or a constant deformation for a long time. Test specimens of different dimensions can produce different test values.

Therefore, compression set is a practical test that describes the deformation strength of the vulcanizate. The compression set is used as a test method to evaluate elastomers for use in sealing elements, damping elements or floor coverings. An engine mount which supports the engine of a car is a good example of the first situation and an O-ring seal is the classic example of the second example. In both cases, it is desirable for the elastomer to fulfill its function over the service time. But because of viscous flow that acts under stress, the retractile function diminishes progressively, leading to changes in the part dimensions. Thus, if the load or deformation is released after a time, the original Shore A of the part will not be regained entirely, due to changes in the internal structure of the material. The value of the compression set is an indicator of how much stress relaxation the vulcanizate underwent while subjected to the load or the deformation.

### 9.2.1 Compression Set of Unfilled Elastomers

Unfilled vulcanizates demonstrate a modest compression set that depends mainly on the type of crosslinked vulcanizates. As was shown, elastomers made from rubbers with a broad molecular weight distribution demonstrate a larger compression set than elastomers prepared from anionic polymerized rubbers with narrow molecular weight distribution. The compression set in this case is attributable to the share of free ends in the networks. Typically, the compression set of unfilled elastomers is lower for peroxide-cured systems than for sulfur-cured systems. In the common group of sulfur-cured rubbers, the compression set increases in the sequence of the sulfur bridges: $S_1 < S_2 < S_x$. This sequence is given by the reactivity of the increasing sulfur/accelerator ratio. While mono-sulfidic crosslinks, $S_1$, are stable and resist the external stress, poly-sulfidic crosslinks, $S_x$, can suffer scission and form new crosslinks that reduce the overall applied stress in the rubber network. The evolution of compression set as function of the crosslink structure is shown in Figure 9.9.

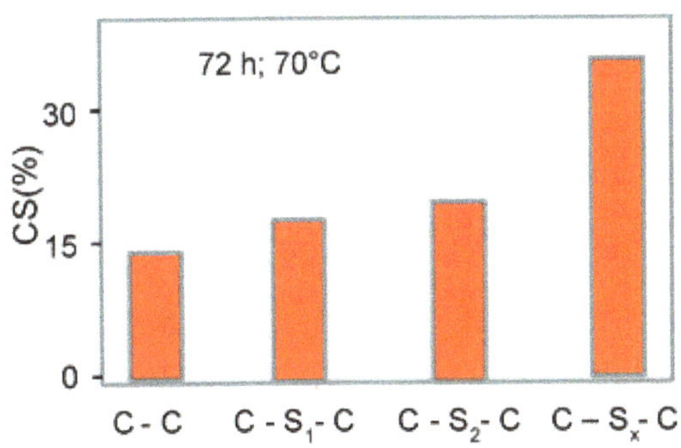


**Figure 9.9** The dependence of the compression set on the crosslink structure

It is accepted that poly-sulfidic crosslinks break under the deformation conditions during the test in order to relieve the stress in the material. The fragments of the poly-sulfidic bridge are able to form new crosslinks that prevent the test specimen from returning to the initial dimensions after the stress has been relieved. High-temperature compression set can be influenced not only by the formation of new crosslinks during the thermal exposure under compression, but also by the conversion of poly-sulfidic crosslinks in reversion-like mechanisms [38].

Elastomers manufactured from polar rubbers, i. e. NBR, show higher compression set values than their non-polar counterparts, a fact which is attributable to the presence of dipole-dipole interactions of the CN groups that form a physical network that overlaps the network produced by chemical crosslinks. The compression set of NBR vulcanizates increases with the ACN content at ambient temperatures and drops to lower values at high temperatures, where the dipole-dipole interaction declines. This is the case in the region of 75–85 °C where the thermal energy cancels the effect of the dipole interaction [39].

## 9.2.2 Compression Set of Filled Elastomers

Unlike the case for unfilled systems, the presence of particulate filler in elastomers leads to more pronounced compression set values because, given significant changes in the compressed state of the elastomers, the fillers can change position and form new interactions with the rubber chains. This effect increases with the filler loading and the activity of the filler and certainly it is more pronounced if the temperature of the test increases. The sliding of chain segments across the rubber-filler interface is a mechanism that contributes to the conversion of elastically stored energy into heat and to partial reduction of excessive stress in localized areas in the nano-region. After releasing the external force, the system cannot return to its initial state because the chains are anchored at energy sites of the filler and cannot leave this energetically beneficial position. Thus, the measured compression set is higher than for the unfilled vulcanizate [40, 41].

### 9.2.2.1 Influence of Carbon Black

The value of the compression set measured under standard conditions increases with the filler loading, the specific surface area and with the test temperature. With increase in the carbon black loading under the extension of the test sample, the polymer chains can be adsorbed on the new surface sites of the carbon black and during the deformation of the test specimen the polymer chains are displaced from their initial energy sites to other active centers, as shown in Figure 9.10.

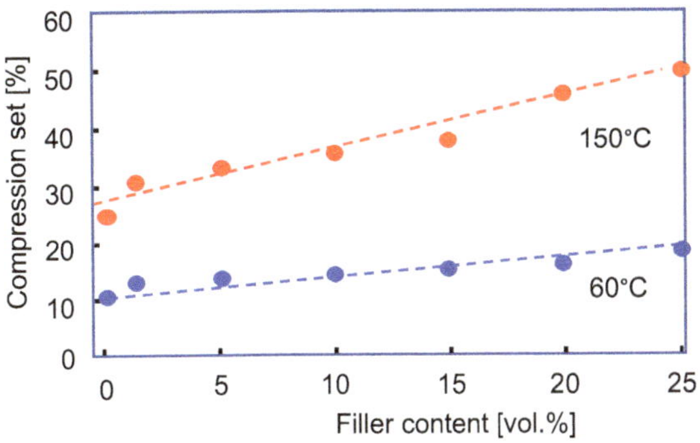


**Figure 9.10**
Influence of filler carbon black loading and temperature on the level of compression set

By releasing the external deformation, the chains are fixed and not able to return in their original state and the prior deformation remains to a certain extent present in the material. The sliding of the polymer chains on the filler surface is facilitated by the increase in temperature on the carbon black surface, so that the extent of the permanent deformation after the test is definitely stronger.

It is observed that, at constant filler, loading the compression set increases with the structure of the filler aggregates. This effect is due to the fact that part of the occluded rubber is included in the deformation of the rubber matrix. With increase in specific surface area of the filler, the compression set increases linearly. In the polar HNBR, the compression set (24 h, 140 °C) is shifted from the original 42.9% to 52.1% with increase in carbon black loading. This effect is attributed to the specific surface area, but above all to the structure of the aggregates. The increase in active energy sites on the filler surface increasingly prevents relaxation of the polymer chains in the initial state. The compression set provides information about the strength under the high loading and other filler-rubber interaction that increases with the temperature.

### 9.2.2.2 Influence of Precipitated Silica

With unsilanized precipitated silica, the slope of the compression set in NR vulcanizates versus the BET specific surface area is higher than for the carbon black [40]. For the same surface area, the precipitated silica shows a higher silanol concentration that increases the inter-aggregate interaction and decreases the dispersion in nonpolar rubber. If the compression set of silica is compared with black as a function of the surface area, the same values for compression set are obtained for a low surface area. But for a higher surface area, the precipitated silica has a much higher compression set than the carbon black compounds, as shown in Figure 9.11 [9].

With silanized silica, which has a better rubber-filler interaction, the values of the compression set are not as pronounced. Therefore, the increase with temperature does not have the high value that non-silanized silica has. The situation changes after silanization with monofunctional or bifunctional silane, which improves the interaction with the silanized silica and rubber and decreases the filler-filler network.

Therefore, the compression set decreases in value. For 50 phr silica in a peroxide-cured NBR hose compound, the compression set at 72 h and 100 °C was 14.5% while the compound containing the 3-mercaptosilane had a compression set of 12.5%.

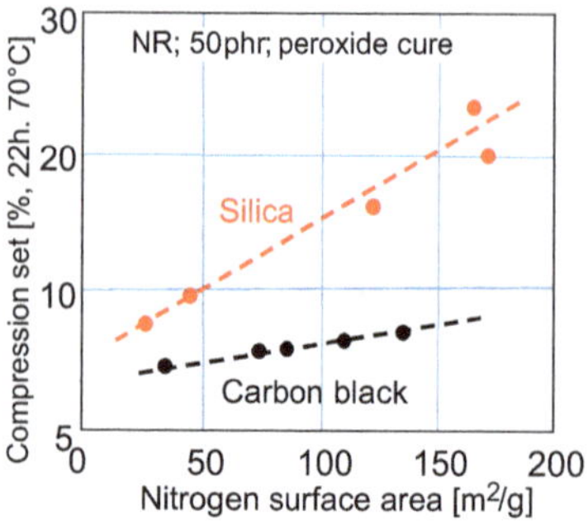


**Figure 9.11**
Compression set of precipitated silica and carbon black as function of the specific surface area [9]

Interpretation of the results is based on the concept that chain slippage on the filler surface takes place under the permanent load and deformation. As a result, the stress peaks in the system are reduced. It can be understood that this mechanism takes place in a larger manner on the polar surface of the silica particles. After releasing the external load, the system does not recover spontaneously to the original topology because the chains are anchored on other energy sites, as at the beginning of the test.

#### 9.2.2.3 Influence of Layered Silicates

The compression set of exfoliated MMT in NBR produced with wet-mixing via continuous dynamic latex compounding (CDLC) leads to a higher compression set than carbon black or precipitated silica. The effect of the higher compression set is given by the high aspect ratio and the high specific surface area. Organo-clays have been introduced into NBR via solution blending, which show a greater increase in compression set than the mechanically blended organo-clays. With the compression set, the organo-clay brings the same increase in HNBR from 42.9% to 49.2% as the structured carbon black N339. The increase in filler content is increased in the compression set with a significant increase in HNBR with a C18 organo-clay (15 phr) to 57.4%. If two organo-clays are used that have different particle sizes (8 µm and 25 µm), different compression set values are observed. There is a 21.8% increase for the larger fillers, while the compression set increases to 36% for the smaller particles [10].

#### 9.2.2.4 Influence of Carbon Nanotubes

This property of the filled vulcanizates has been investigated less frequently in the case of multi-walled carbon nanotubes. There is an increase in the value of the compression set in the direction of the orientation of the tubes, which is more pronounced than with fillers with a small aspect ratio. The compression set of the multi-walled carbon nanotubes has been performed with C150P in EPDM and shows an increase in 26% over the carbon black-filled vulcanizate that depends on the presence of the $sp^2$ hybridization of the carbon atom and especially on the high aspect ratio [18].

## 9.3 Stress-Strain Behavior

Besides hardness testing, tensile testing is one of the most widely used tests for characterizing elastomers, although elastomers are rarely used under high strains. In the tensile test, the test specimens are stretched at a constant strain rate until they crack and the necessary force and change in length are recorded. The temperature has a significant influence on the test results. The tensile test performed on a tensile testing machine is the most commonly used test for an elastomer. The stress-strain behavior of vulcanizates describes the relationship between a gradually applied stress to a test specimen and the resulting elongation at a defined temperature and elongation rate [42–44]. During the tensile test, the test pieces are elongated at a constant rate of traverse until final rupture occurs. When the vulcanized rubber is stretched to high deformations, a characteristic S-shaped stress-strain curve is obtained. The stress-strain test delivers important information on the Young's modulus at very low elongation (< 3%), stress evolution at higher strain and finally at ultimate rupture of the test specimens. The stress value is the tensile force present when a certain elongation is reached, relative to the initial cross-section. In the tensile test, it is advisable to record the complete force-length change curve, as this allows a better assessment of the material's behavior under tensile stress than by specifying the tear strength and elongation at break alone. The area under the stress-strain curve is a sensitive value for assessing the homogeneity of a sample. The stress and strain test can be applied in normal shear, uniaxial mode and in biaxial mode, which is difficult to realize in the laboratory. The common test specimens are dumbbell shaped and cut from a standard molded test sheet (thickness 1 or 2 mm).

Because of the simple sample preparation and the less complicated procedure, the test is often used to obtain information about the crosslink density of vulcanizates, the reinforcing effects of fillers at different elongations and the resistance to uniaxial extension at the rupture of the test specimen. Main results are given at the initial elongations where the Young's modulus is measured, and at the elongation of 100% and 300% and the tensile strength and elongation at break as indicated in Figure 9.12.

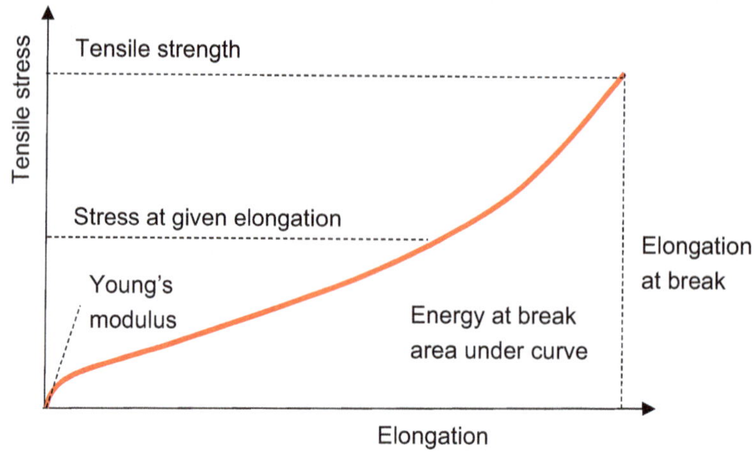


**Figure 9.12** Tensile stress-strain curve for vulcanized rubber

The results depend on the temperature which was previously set. Several effects of filler reinforcement are manifested in this type of measurement that allow quantitative comparison of the type of fillers, but also an estimation of the spatial distribution of undispersed fillers in the vulcanizate. In all, the test provides useful information about the influence of compound ingredient, manufacturing conditions and product quality. The response of vulcanized rubber to a deforming force depends on three variables:

- extension ratio
- rate of strain
- temperature

Tensile strength and elongation at break are affected by mixing, dispersion and distribution of the filler, the presence of contamination, and the cure system, among other factors. Undispersed pellet fragments and particles of different compounding ingredients cause local stress concentrations during the stretching and lead to premature breaks of the test specimen. The stress-strain behavior is limited by such influences, so that at least five repetitions are carried out in order to find an average value. An attempt is made to compensate for the inhomogeneity of the vulcanizate by repeating the stress-strain experiment.

As the strain rate increases and the temperature decreases, the material becomes stiffer and the stress has to be increased to achieve the same elongation. Therefore, the test conditions are important for the magnitude of the measured stress and ultimate tensile and elongation. Although elastomers are rarely stretched in service life to catastrophic failure, tensile tests provide important information about the material response, especially about the:

- type of rubber
- state of cure
- influence of fillers

The important feature in stress-strain behavior of filled vulcanizates is that the stress becomes larger as the extension increases. Generally, tensile values have more meaning in comparisons of compounds, than the values taken by themselves. It is safe to say that if a compound has its average stress value at 100% elongation even 10–15% higher than that of another compound, the first compound is stiffer than the other and possesses a higher modulus. However, the relationship between tensile modulus and the stiffness of the final rubber part is seldom precise. Tensile stress-strain testing of a given compound can provide valuable quality information to detect compounding mistakes in the factory and is useful in compound development because the reinforcing effects of the filler and the influence of the mixing parameters can be observed easily. The tensile test is well suited for estimating the impairment of the elastomer by UV light, ozone and the influence of heat.

### 9.3.1 Stress-Strain Behavior of Unfilled Elastomers

The stress-strain behavior of unfilled elastomers is determined by the chemical nature of the rubber, the molecular weight and the type of the curing agent used. In the region of very small extensions (< 3%), a rubber networks shows a proportional force that leads to proportional deformation. The slope is known as the Young's modulus from which the modulus of elasticity, E, of the rubber can be derived, which is inversely proportional to the molecular weight of the network chains between two adjacent chemical crosslinks, $M_c$, that is demonstrated in theoretical treatments [43]:

$$E = \sigma/\varepsilon = 3\rho RT/M_c \tag{9.2}$$

where ρ is the density of the vulcanizate, $M_c$ is the average molecular weight of network chains, R is the universal gas constant and T is the absolute temperature

The smaller the network chains, the larger the modulus of elasticity, E, of the elastomer. The "affine" network model behind this fundamental equation assumes [43, 44]:

- Each network chain ends in a network point that in turn connects four network chains.
- The network chains have no volume and no interactions with other chains.
- The deformation of the network chains takes place with the same (affine) deformation ratio as the entire network.
- The network is incompressible, has isotropically distributed network chains and no defects (loops or entanglements).

According to statistical mechanics theories, the stress of an affine network is proportional to the number of network chains per unit volume. The relationship between the stress and the deformation ratio, λ, is expressed in molar quantities as follows:

$$\sigma = \rho RT/M_c\left(\lambda - \frac{1}{\lambda^2}\right) \tag{9.3}$$

where σ is the stress value and λ is the deformation ratio (λ = ε + 1)

On the basis of Equation 9.3, the crosslink density of an unfilled elastomer is determined for the limiting case of small extension ratios. It has been repeatedly demonstrated that crosslink density increases linearly when the curing agent (sulfur/accelerator or peroxide) is applied. Under-cured and over-cured elastomers deviate from this linear relationship [45, 46]. Almost all properties of unfilled vulcanizates are related to the crosslink density, as shown in Figure 9.13.

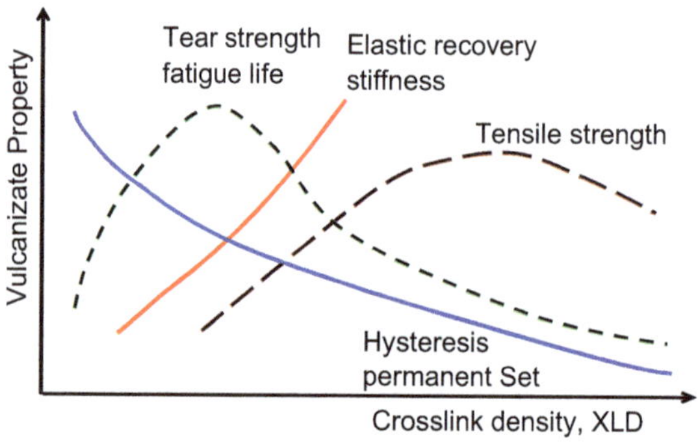


**Figure 9.13** Influence of crosslinking density on vulcanizate properties

The static modulus approximates the behavior of pure elasticity and stiffness of the material and shows a quasi-linear increase, while hysteresis and energy dissipation of the network decreases. In many rubber applications, high elasticity and low hysteresis are desirable. Changes in permanent set and creep often parallel hysteresis. The more complex properties such as tensile strength, toughness and wear resistance reach optimum values at a certain crosslink density. The position of the maxima depends on the nature of the rubber, the curing system used and the crosslink density achieved, so that the difference of the network chains between the chemical crosslink determine its position. As shown in Figure 9.13, the stress-strain curve tends to flatten at medium to high extension ratios. The increase in tensile stress at higher extension ratios is attributed to a non-uniform distribution in the length of network chains, such that short chains are highly oriented and stretched and make a disproportionately large contribution to stress [43, 47]. Contrary to the theory that assumes an affine network has chains of the same length, real networks show a broad distribution of network chains that cannot be avoided in any crosslinking procedure. The non-uniform distribution of network chains reaches its limit of extensibility and implicitly increases the stress values. The higher the inconsistency of the network chains, the higher the increase in the stress value per extension unit. The deviations from ideal networks that cause the limitation of finite extensibility are considered in theoretical frameworks, which are beyond the scope of the present chapter.

In contrast to amorphous rubbers (i. e. SBR or NBR), rubbers that can crystallize under strain (i. e. NR) show a pronounced increase in stress values at extensions above 300% [48]. The increase in stress is caused by formation of strain-induced crystallites in the polymer that act like fillers in strengthening the network that is called the self-reinforcement. The appearance of strain crystallization is shifted to higher extensions with increase in crosslink density. The slope after the upturn of the stress values becomes larger with the crosslink density [49]. This is attributed to the decreasing length of network chains that do not allow the formation of supramolecular organized crystallites [50]. In any case, the effect is more pronounced at low and medium temperatures.

At large elongations, the vulcanized system is no longer able to support the stress concentration and catastrophic rupture occurs. The tensile strength indicates the failure of the material under the selected strain rate and test temperature. Like tensile strength, the elongation at break indicates a limit of material behavior. The integral of the stress values up to the ultimate elongation defines the energy at break, a characteristic value for the elastomer. Tensile strength, $\sigma_b$, of unfilled rubber networks increases with the crosslink density until a maximum, which depends on the curing system, the curing conditions and the deformation rates used in the stress-strain test. There are reasons to state that the non-uniform distribution of the network chains is responsible for the occurrence of the tensile strength maximum shown in Figure 9.13. Depending on the type of crosslinks and the type of rubber, the maximum occurs at different state of cure. Like all processes caused by deformation of elastomers, tensile strength shows a characteristic viscoelastic behavior. The reduced tensile strength, $\sigma_b'$, normalized for the mobility of the rubber matrix ($\sigma_b' = \sigma_b T_g/T$), increases with the deformation rate. Figure 9.14 illustrates the evolution of $\sigma_b'$ as function of the deformation rate, $d\varepsilon/dt$ [54].

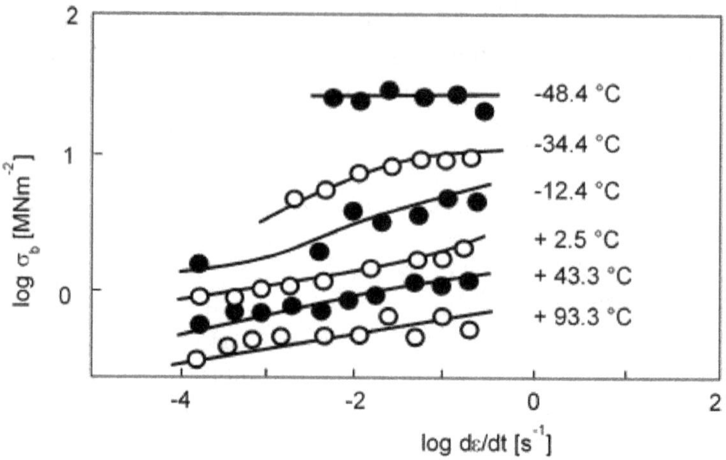


**Figure 9.14** Tensile strength of SBR vulcanizates as a function of deformation rate and temperature

The isotherms shown in Figure 9.14 can be shifted along the abscissa to give a master-curve that describes the material behavior. Using the time-temperature superposition principle, the tensile strength normalized to the glass transition temperature, $T_g$, is described as a function of the strain rate [55]. At high strain rates, the network chains can no longer take part in molecular place exchange processes and, consequently, the tensile strength increases [56]. As a general feature of unfilled elastomers made from amorphous rubbers, tensile strength at constant strain rate increases continuously with decrease in the test temperature [57]. The dependence of tensile strength on amorphous elastomers as a function of temperature and strain rate is shown in Figure 9.15 for an unfilled SBR network [58].

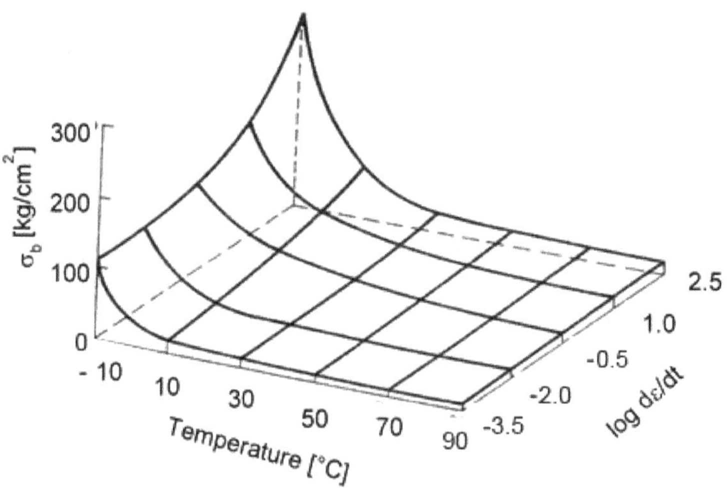


**Figure 9.15** Tensile strength of an unfilled SBR vulcanizate as function of temperature and strain rate

Because tensile strength of an unfilled elastomer increases with the strain rate and decreases with the test temperature, the same tensile strength of a vulcanizate can be set at different temperatures if the strain rate is chosen accordingly. Only at low temperatures, near the glass transition temperature $T_g$ and high strain rates, deviations from the viscoelastic behavior are observed. As a consequence of strain crystallization, the stress concentration in NR vulcanizates is much higher than that in the majority of synthetic rubbers. This is the reason why elastomers based on NR demonstrate high tensile strength (≥ 27 MPa) The increase in stress values with higher deformations is also referred to as self-reinforcement and has an important contribution in reducing cut growth and increasing tear resistance of NR vulcanizates stretched above the onset of strain crystallization [59]. However, the self-reinforcing mechanism is temperature dependent. Above the temperature range at which the strain-induced crystallites are melting, the tensile strength decreases sharply [60, 61]. With increase in crosslink density, the temperature at which the sharp decrease in the ultimate properties occur is shifted towards lower temperatures, due to the smaller degree of crystallinity.

Researchers have attempted to develop mathematical models of how rubber responds to stress conditions. A general treatment of stress-strain relation of rubber like materials was developed for uniaxial elongation, assuming that the material is isotropic in elastic behavior in the unstrained state and incompressible in bulk [62]. In continuum mechanics an elastomer is considered a hyper-elastic material where the strain density function W is a linear combination of two invariants of the Cauchy-Green deformation tensor:

$$W = C_1 (I_1 - 3) + C_2 (I_2 - 3) \tag{9.4}$$

where $C_1$ and $C_2$ are empirically determined material constants and $I_1$ and $I_2$ are the first and the second invariant

The particular form of strain energy is expressed for rather small strains as tension, which is a function of the extension ratio λ of the network [63]:

$$\sigma = 2(C_1 + C_2/\lambda)(\lambda - 1/\lambda^2) \tag{9.5}$$

where σ is the applied stress, $C_1$ and $C_2$ are constants and λ is the extension ratio

Introducing the reduced tension, $\sigma_{red} = \sigma/(\lambda - 1/\lambda^2)$ results in a linear relationship between $\sigma_{red}$ and the reciprocal extension ratio λ.

$$\sigma_{red} = \sigma/(\lambda - 1/\lambda^2) = 2C_1 + 2C_2/\lambda \tag{9.6}$$

The practical meaning of Equation 9.6 becomes more clear by plotting the reduced stress, $\sigma/(\lambda - 1/\lambda^2)$, versus the reciprocal extension 1/λ. Figure 9.16 illustrates such a Mooney-Rivlin plot.

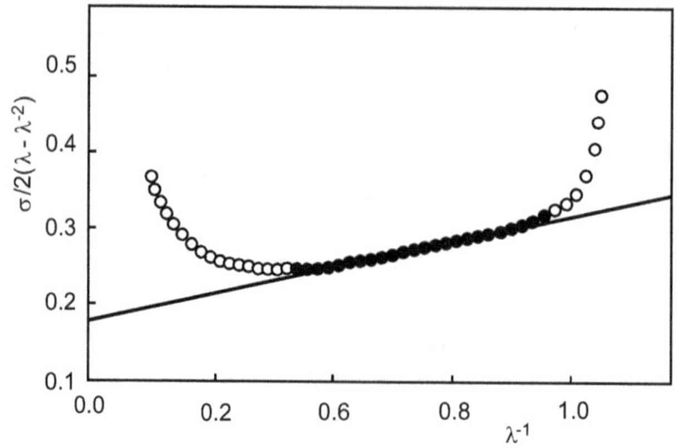


**Figure 9.16** Mooney-Rivlin plot for an NR vulcanizate

At small strains ($1/\lambda$ large) as well as at high strains ($1/\lambda \rightarrow 0$), the reduced tension shows an up-turn. The deviation at small strains might be caused by measurement effects, such as inaccurate clamping of the test specimen. At higher strains the up-turn is attributable to the network chain distribution. The linear part of the plot is used for determining the $C_1$ constant by extrapolation to $1/\lambda = 0$. According to the theory the $C_1$ value is considered as a measure of the crosslink density:

$$C_1 = \rho RT / 2M_c \tag{9.7}$$

where $\rho$ is the density, R is the universal gas constant, T is the absolute temperature and $M_c$ is the average molecular weight of network chains

Whereas $C_1$ is approximately proportional to the number of network chains per unit volume, $C_2$ appears to be rather constant, independent of the degree of crosslinking. Equation 9.6 has been proven to be valid for different curing systems. A number of publications have shown that the crosslink density of unfilled vulcanizates determined by equilibrium swelling is proportional to the constant $C_1$. Although Equation 9.6 is valid at small strains considerable confusion has arisen from its application a larger strain.

To determine the crosslink density of a filled vulcanizate, a specimen is deformed several times up to 80% of its elongation at break. To determine the chemical crosslink density of the filled vulcanizate, the pre-cyclized stress-strain curve is fitted with a material model, whereby the intrinsic elongation is used instead the macroscopic elongation. If the linear behavior is extrapolated after the cyclic pre-deformation the axis of the intersect is independent of the filler content, because the filler-filler interaction is broken down, which means that the remaining stress-strain deformations can be completely described by the concept of the Mooney-Rivlin equations. The determination of the chemical crosslink by means of the cyclic pre-deformation is proved in many cases. This means that at high filler contents, all the filler-filler can be broken up by the cyclical deformation. Because the chemical crosslink density declines when silica is added as a filler, it can be assumed that silica surface can interact with a part of the accelerators and reduce the degree of vulcanization. The remaining contacts result from the chemical crosslinking and show the crosslinking density is always decreasing.

## 9.3.2 Stress-Strain Behavior of Filled Elastomers

The course of a tensile strain curve contains important information about the behavior of the sample with increase in deformation. Implementing active fillers in a rubber vulcanizate the stress increases at any deformation and reaches a maximum at the tensile strength at the elongational break. Iso-dimensional fillers, carbon black and precipitated silica, show an increase by the filler loading, the size of the primary particle or specific surface area, the branching of the aggregates and the surface activity. For nanofillers that have large specific areas and high aspect ratio, they show at very low loadings high increase in the stress values but not as high tensile strength like carbon black. The increase in the stress values is examined for many practical applications.

### 9.3.2.1 Young's Modulus

One of the most important parameters to get from stress-strain behavior is the force-to-strain ratio, which at very small strains has a linear relationship. The higher the force that is required, the higher the modulus of elasticity of the vulcanizate. Depending on the type and quantity of the vulcanization system, the Young's modulus increases significantly. In the presence of a filler, the Young's modulus, which describes the elastic restoring forces in the initial phase of a vulcanized product, changes again. The Young's modulus should be determined using a measuring device that is suitable for detecting small strains.

#### 9.3.2.1.1 Effect of Carbon Blacks

The slope at the origin of the tensile stress-strain behavior is the modulus of elasticity or the Young's modulus. It describes the increases of the stress values at very low deformations that due not destroy the filler-agglomerates in the rubber compound. The occluded rubber does not participate in the deformation, which has the effect of increasing the volume fraction of the filler. Therefore, the Young's modulus increases by the filler loading, the specific surface area and especially by the structure of the aggregates. It is confirmed for any filler used to increase the reinforcement of the elastomer, as shown in Figure 9.17.

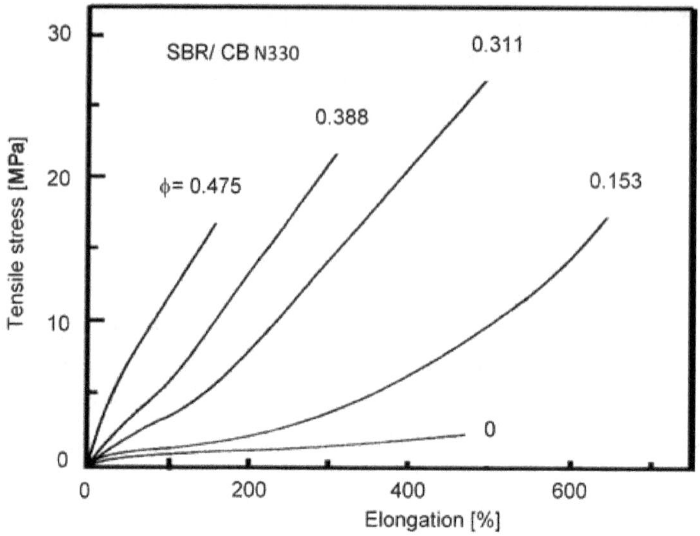


**Figure 9.17** Effect of carbon black loading on stress-strain behavior of SBR vulcanizates

The most important change in the stress-strain behavior of carbon black and of silica is found in the Young's modulus area. Provided that the aggregates are completely enclosed by the rubber matrix, the principle adopted from rheology applies to the modulus of elasticity of the filled vulcanizate [64]:

$$E_f = E_0(1 + 2.5\Phi + 14.1\Phi^2) \tag{9.8}$$

where $E_f$ and $E_0$ are the elastic moduli of the filled and unfilled vulcanizate, and $\Phi$ is the volume fraction of the filler

Several other equations were proposed, but they deviated from each other at higher loadings [65–67]. It is relatively difficult to precisely describe the conditions for the increase in stress at such small deformations, because the aggregates differ from spherical particles and the mobility of the rubber chains is not evenly distributed. A similar equation was proposed for the iso-dimensional fillers [68]:

$$E_f = E_0(1 + 6.7f\Phi + 16.2f^2\Phi^2) \tag{9.9}$$

where f is the shape factor of the filler aggregate

The value of the shape factor f was given different numbers in various reports for carbon black N330. There were small numbers of 1.7–1.9 from TEM studies [69] or 4.7 for SBR/carbon black or 6.7 to explain the strain amplification of NR [70]. The value of the Shape factor was selected according the different application and has no physical meaning. Considering the effect of the filler structure on the Young's modulus the equation was changed by introducing the effective filler volume, $\Phi_{eff}$ [69]:

$$E_f = E_0(1 + 2.5\Phi_{eff} + 14.1\Phi_{eff}^{\ 2}) \tag{9.10}$$

where $\Phi_{eff}$ is the effective filler volume

An effective filler volume with a factor $f_v$ was described in SBR, which converts the filler volume fraction into a filler effective fraction [71]:

$$f_v\Phi = \Phi'_{eff} \tag{9.11}$$

The calculated factor $f_v$ increase linearly with the crushed DBP absorption of carbon black from low values of the semi-active carbon blacks to the active carbon blacks. It was shown that the effective filler volume factor of 0.7 gives a linear relationship to the measured crushed DBP contribution. It was assumed that depending on the strength of the rubber-filler interaction, rubber chains can be adsorbed on the filler surface, which is confirmed by $^1$H-NMR investigations [72, 73].

It was found that the modulus of elasticity of filled vulcanizates increases as the filler particle decrease in size and the structure (DBP or OAN) increases. Two reasons for this behavior are discussed in the literature [74–76]. Firstly, as the particles decreases in size and the particle surface area increases a larger part of the rubber matrix is adsorbed on the filler surface and contributes to form a larger rubber-filler interface and increase the apparent filler volume fraction. Secondly, as particles decrease in size, they tend to form filler networks by self-organization above a critical concentration (percolation threshold). Due to this process the stiffness of the entire compound is increased as long the filler network is not destroyed by the larger strains.

Due to the occlusion of a part of the rubber in the voids of the filler aggregates and agglomerates, the fraction of the elastic matrix is diminished and cannot participate in the elastic deformation. Therefore, the filled vulcanizate behaves like a system with a higher filler content. With most carbon black grades, the individual aggregates occlude rubber to an extent which can be calculated from the OAN or DBP absorption [77]:

$$\Phi_{eff}/\Phi = 0.5\,(1 + 0.02139\,OAN)\,/1.46 \tag{9.12}$$

The determining influence of the structure on the modulus of elasticity makes it possible to increase the modulus of a vulcanizate within certain limits by choosing fillers with a higher structure. Because the effective volume fraction $\Phi_{eff}$ is a variable of both the filler loading and the structure of the filler, the compounder can use one of the two variables to set the desired modulus of elasticity of a vulcanizate. As shown in Figure 9.18 the normalized modulus of elasticity, $E^*_f/E^*_0$, increases linearly with the structure of the filler at constant loading (50 phr) and does not depend on the specific surface area [76].

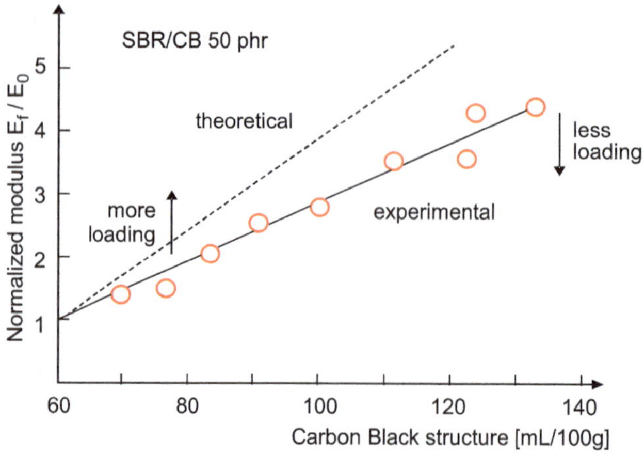


**Figure 9.18** The normalized elasticity modulus $E^*_f/E^*_0$ is a function of carbon black structure

Thus, a specified modulus can be achieved by using a more structured carbon black at lower loading or a less structured carbon black at a higher loading. The application of this relationship, also known as "structure-concentration equivalence principle", can help to reduce weight of rubber parts and at the same time to improve the quality of the product due to the better dispersion provided by the high structure filler [78]. A comparison of theoretical expectations with the experimentally determined increase in Young's modulus concluded that only half of the occluded volume determined in the test contributes to the increase in modulus at small extension ratios [77]. What should be noted is that the theoretical background assumes adhesion between particles and rubber matrix, whereas the increase in modulus indicated by the equation is less likely to be obtained, due to adhesion of the rubber phase to the filler particles.

The difficulty with a quantitative evolution of the stress-strain of filled systems is that the non-linear mechanical properties of the rubber network and the filler-polymer and the filler-filler interaction overlap each other in a complex manner, making a simple separation of the different effects impossible [79].

Using molecular models, it has been demonstrated that the decrease in the filler size as well as the increase in polymer-filler interaction can increase the Young's modulus of filled vulcanizates [80, 81]. Molecular models have confirmed the layer with high chain density near the filler surface that amplify the volume fraction of the filler [82]. Models for rubber that have been used to solve practical problems either analytically or through the use of finite element analysis are reviewed in [83].

#### 9.3.2.1.2 Effect of Precipitated Silicas

In the case of precipitated silica, there is a difference in the stress-strain behavior that depends on the state of the used silica. With precipitated silica, a high Young's modulus is achieved, because the effect of the inter-aggregate interaction of the silanol groups is active and a large filler-filler interaction is present in the vulcanizate. With increasing loading, and the structure of the aggregate, Young's modulus increases in non-polar vulcanizates. In the case of the precipitated silica, there is a difference in the stress-strain behavior of the way silica is used. In accordance with the high degree of aggregation of the non-silanized silica, Young's modulus shows a higher stress in the case of small deformation.

However, if the silica is freed from the silanol groups on the surface by a silane, the process has a significant effect on the stress-strain behavior. Silanization of silica prepared by dry-mixing is characterized by a significant decrease in the inter-aggregate interactions of the silica aggregates and an increase in the rubber-silanized silica interaction. This is seen when the increase in Young's modulus is compared with those of the original silica, at the same loading and comparing with Young's modulus of carbon black. At the filler concentration of 78 phr in S-SBR/cis-BR (70/30) blends, the carbon black N234, dry-mixed silica and wet-mixed silica are compared, as shown in Figure 9.19 [84].

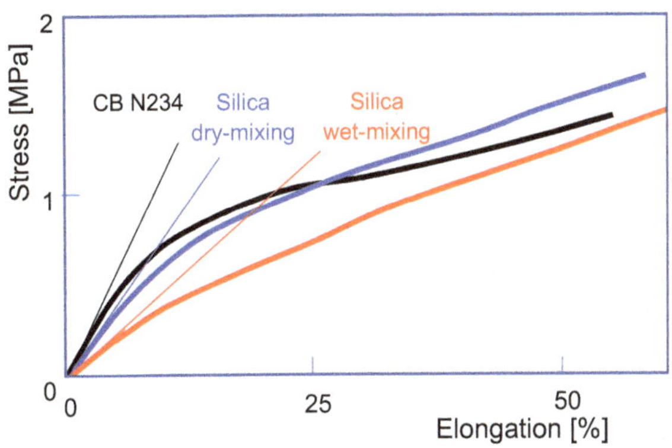


**Figure 9.19**
Influence of mixing process on Young's modulus for silica

The Young's modulus for dry-mixed silica is lower than for carbon black N234, but higher than for wet-mixed silanized silica. The incorporation of the silanized silica in a wet-mixing process that leads to a very small area of undispersed silica shows a lower increase in Young's modulus that is related to an increase in dispersion, decrease in hardness, and an increase in rebound elasticity [84]. The way in which a vulcanizate is produced has a significant influence on the behavior in terms of tensile strain, in dynamic experiments and also in terms of fracture mechanics. If no beaded filler is used, the properties of the vulcanizate can be improved in a certain way.

#### 9.3.2.1.3 Effect of Layered Silicates

With layered silicates, due to their high aspect ratio a strong increase in stress can be expected at small elongations. It is important to know how they were incorporated into the rubber matrix and the form in which they are distributed in the rubber matrix. Due to mechanical mixing, there is no intercalation of the platelets in the rubber matrix and this variant cannot be included.

When a relatively soft matrix is reinforced with layered silicate, the adjacent polymer becomes more restrained mechanically. This enables a significant portion of the load to be carried by the filler, assuming that the bonding between the two phase is adequate [85]. It is clear that the larger the surface of the filler in contact with the rubber, the greater the reinforcement will be. This explains why layered silicates with the high surface area of approx. 800 $m^2/g$ cause dramatic improvements in Young's modulus, even if they are present in small amounts in the polymer. The modulus increases much more sharply.

While the unfilled sample shows a flat, concave stress-strain curve, the exfoliated disks of MMT produced via continuous dynamic latex compounding (CDLC) have a convex shape that undergoes higher stress in the area of small deformations. The stresses achieved are certainly related to the conditions of compound manufacture and cross-linking. The tensile strength increases in proportion to the volume fraction of the exfoliated layered silicate, while the elongation at break decreases continuously. The mechanical incorporated MMT (10 phr) shows a Young's modulus at high filler loadings which is formed from the undispersed clay. Depending on the type of production, a distinction is made between wet-mixing and modified organo-clay, as shown by the results with the same loading.

Compared with the unfilled vulcanizates, the addition of the exfoliated pristine silicates or the intercalated or exfoliated organo-clays results in a significant increase in Young's modulus. The exfoliated MMT obtained by applying wet-mixing (CDLC) demonstrate a pronounced Young's modulus. In NR, the modulus increases from 1.0 MPa to 1.5 MPa at 7 phr, while in SBR at the same rate the increase is 15% and an organo-clay brings an increase in only 10% [86]. In polar NBR, exfoliated MMT platelets above the rheological percolation threshold have a significantly higher Young's modulus, which is 5.2 MPa at a volume fraction of 0.024 [87]. Above the rheological percolation threshold, the increase as a function of the volume concentration is exponential curve with an exponent of 3.4, as shown in Figure 9.20.

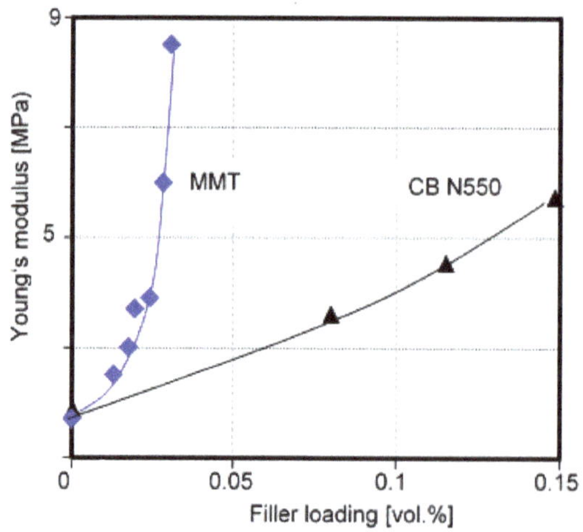


**Figure 9.20**
Young's modulus for exfoliated MMT and carbon black N550 at different loadings [87]

In rubber/clay nanocomposites, it has been shown that a superior reinforcement provided by the exfoliated organo-clay arises primarily from a combination of the high aspect ratio and the surface area. Properly dispersed and aligned organo-clays have proven to be very effective for increasing Young's modulus and higher stress factors in the vulcanizate [88].

#### 9.3.2.1.4 Effect of Carbon Nanotubes

Multi-walled carbon nanotubes exhibit excellent mechanical properties in the pure state, such as a high Young's modulus of 1.2 TPa. The combination of this property along with the density, high aspect ratio and high specific surface area makes multi-walled carbon nanotubes ideal candidates for reinforcing elastomers. However, the mechanical properties depend strongly on the high aspect ratio and on the dispersion state in a rubber vulcanizate, the alignment in the direction of processing flow and the interfacial interaction between the multi-walled carbon nanotubes and the rubber matrix.

The effects are more marked at low multi-walled carbon nanotubes loadings in comparison with NR/carbon black [89]. A linear response of Young's modulus to the composition of multi-walled carbon nanotubes was found to increase at 1 phr from 0.53 MPa to 0.69 MPa, accounting for a 38% increase at 5 phr and a 124% increase at 10 phr multi-walled carbon nanotubes [90]. NR (SMR CV60) show an increase in the region of 7 wt.% to reach 2 MPa. At higher loadings, the Young's modulus decreases because of the poor dispersion in NR. Although multi-walled carbon nanotubes are distributed randomly, the gradual alignments occur during uniaxial stretching. By incorporating multi-walled carbon nanotubes in a SBR in a solution process and subsequent evaporation of the solvent, the dispersed homogeneous fillers show an increase in 10% of Young's modulus [91]. Using 1 to 10 wt.% of multi-walled carbon nanotubes, a 10% increase in Young's modulus was observed [92]. A five-fold increase in Young's modulus is reported for 3 vol.% multi-walled carbon nanotubes in silicone rubber, while this value continues to fall for EVM and EPDM [93]. All the studies reported a good dispersion of multi-walled carbon nanotubes without the need for previous functionalization. It has been suggested that the presence of $\pi$ interaction could play a role in the silicone rubber [94]. In NBR, Young's modulus is significantly increased with the rubber polarity [95]. An increase in Young's modulus of 0.25 MPa is reported for HNBR and of 0.7 MPa for the more polar HXNBR [96]. The comparison of the multi-walled carbon nanotubes samples shows that, with the smaller diameter and the higher specific surface area, a higher Young's modulus is obtained for each rubber matrix. At a concentration of 1.5% by volume of multi-walled carbon nanotubes, the carbon nanotubes with the smallest diameter have the highest Young's modulus [96].

In a combination with carbon black (30 phr), precipitated silica (8 phr) single-walled carbon nanotubes (2 phr) in a blend of NR/BR/IR/SBR (40/20/20/30) showed an increase of 15.4% in Young's modulus [97]. Multi walled carbon nanotubes have been introduced into elastomers in small amounts, with significant differences in dispersion depending on the manufacturing process. A strong increase in the stress values in the initial area of the deformation can be clearly observed. The high increase in stress in the low deformation range is of great importance for many rubber applications. Both the filler loading, the primary particle size, the specific surface area, the degree of branching of the aggregates in the case of iso-dimensional aggregates and the high aspect ratio and specific surface area in the case of layered silicates and carbon nanotubes have a quality effect on the expansion behavior.

### 9.3.2.2 Medium Elongation

At medium elongation, the filler-filler interactions are destroyed continuously, due to the increasing elongation of the test specimen. The filler loading, the specific surface area and the structure of the aggregates play an important role in the increasing the stress in the test specimen. It is not surprising that a medium deformation is usually used to evaluate the performance of vulcanizates, as rubber products are seldom elongated to rupture in service life. It turns out that the particle size and the activity of the surface of the filler play an important role in the interactions, which leads to a significant increase in stress. In addition, the stress in medium elongation are used to illustrate the filler reinforcement and additionally for quality control of the elastomer. In any case, the dispersion of the filler should be considered, because the higher this property achieved in the mixing process, the higher the stress that can be established in the medium elongation.

When the test specimen is stretched up to several hundred percent, the network chains are no longer elongated proportionally to the macroscopic deformation. Some of the shorter network chains can now tear and reduce the stress on the vulcanizate. However, the medium elongation is important for many applications, as the sample exhibits good tensions at the stress level reached and residual elongation after stretching is always at a very low level.

#### 9.3.2.2.1 Effect of Carbon Blacks

Instead of a linear stress-strain behavior achieved for unfilled vulcanizates, the presence of the filler leads to a non-linear increase in the stress value of the vulcanizate. The stress of the elastomer increases as the carbon black loading level increases. The effect increases with increase in the specific surface area and the structure of the aggregates, as shown in Figure 9.21.

Both the type of filler and the filler dosage have a determining influence on the magnitude of the stress values and the shape of the stress-strain curve. As a rule of thumb: for any kind of filler, an increase in loading results in higher stress values. At small filler loadings ($\leq 30$ phr), the increase in the stress is more pronounced at an extension ratio $\lambda \geq 3$ whereas, at high loadings ($> 30$ phr), the reinforcing effect is high over the entire range of extensions. The stress values are meaningful in comparison of compound performance. If the stress values differ in average by 10–15%, it is safe to state that also the stiffness of the compound will differ in this manner.

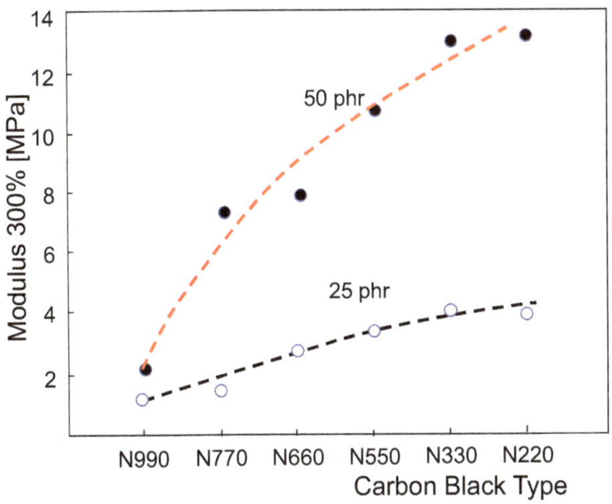


**Figure 9.21**
Increase in the stress values at medium elongation for all types of carbon black

The difficulty in describing the stress-strain behavior of filled vulcanizates lies in the fact that the non-linear mechanical properties of the rubber network and those caused by filler-filler interactions and filler-rubber interactions overlap in a complex manner. A simple separation of both influences is quite challenging. In addition to the observation that the filler clusters are broken down with increase in elongation, it is observed that rubber ribbons that form a super-network in a crosslinked rubber [47]. The super-network is composed of strands created during stretching and polyfunctional junction points, originating from domains of high crosslinking density or a segregation of short chains during deformation. It has been found that the degree of orientation of these ribbons was higher than expected from the external elongation. This is shown in Figure 9.22 [47].

In order to quantify the reinforcing effect of fillers, a number of empirical criteria have been proposed. As a rule, stress values are related to the filler content with a constant extension ratio, or the increase in stress values with different extension ratios. The criteria are used to compare different of types carbon blacks. Commonly, the ratio of the stress values at certain elongations, i. e. $\sigma_{300\%}/\sigma_{100\%}$, is used to quantify the reinforcing effects of the filler. A convenient criterion is the "reinforcing factor", which is the ratio of the stress of the filled system to the stress of the unfilled vulcanizate ($\sigma_f/\sigma_0$) at a given elongation. The "reinforcing factor" defined in this way describes the enhancement of the stress caused by the filler at a given deformation.

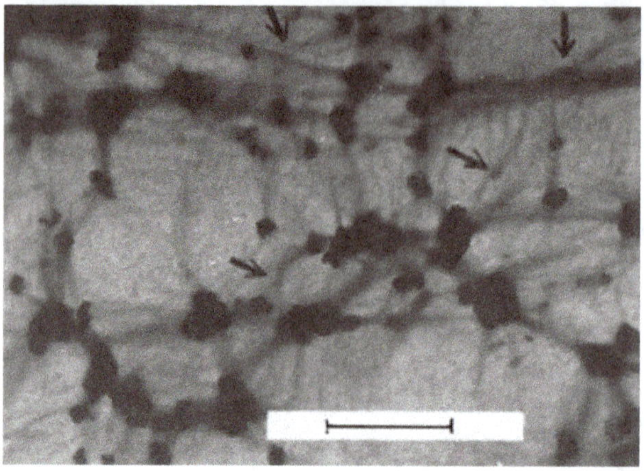

**Figure 9.22** Super-network in carbon black-filled NR vulcanizates

It can be seen that the reinforcing factor reaches higher values ($\sigma_f/\sigma_0 > 3$) above a filler volume fraction of 0.15–0.16, which is close to the percolation threshold of this carbon black type in NR [98]. Thus, the increase in stress values is related to the filler network, although this is mostly damaged by the applied elongation. When vulcanizates are filled with low specific surface area blacks, the stress values and correspondingly the reinforcing factor are reduced. A schematic view of the increase in strength at 300 % elongation is shown in Figure 9.23 as a function of carbon black loading and the specific surface area [99] for the ASTM formulation [100].

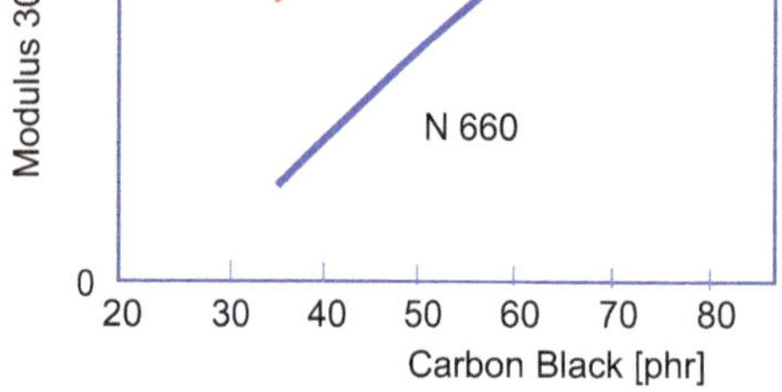


**Figure 9.23** Relationship between carbon black loading and stress at medium elongation in the ASTM SBR formulation

The smaller particle size of N339 produces a significant increase in the strength of the 300% elongation that explains the use of this carbon for tire tread compounds. As shown, the aspect ratio and the high specific surface area of the multi-walled carbon nanotubes govern the reinforcing effect. They provide much higher stress values at far lower volume fractions in the compound than the conventional carbon black. The vulcanizate containing carbon black with high specific surface area and surface activity shows a pronounced increase in the stress value at extension ratios $\lambda > 0.5$ while the vulcanizate filled with the inactive filler demonstrates a moderate increase in stress even at high extension ratios.

A phenomenological experiment was undertaken made that uses the structure-concentration equivalence principle in this strain regime to describe the stress values at 300% extension [76]. At this 300% elongation, the stress value increases much more significantly with the CDBP or OA number. By introducing a shift factor that depends on the structure of the aggregates, the stress at constant elongation can be described by as a function of the COAN of the carbon black [71]. The shift factor is a function of the filler morphology and is not dependent on the polymer. The same stress is achieved by a higher structured black with lower loading or by a lower structured black at higher loadings.

The fact that the stress increases on filling derives from the stress concentration, due primarily to the shape of the carbon black aggregate. This can be illustrated by the stress values of HNBR/CB N339 shown in Table 9.3 [10].

**Table 9.3** Stress for the Elongation of HNBR/Carbon Black N339

| Component | Stress at 50% [MPa] | Stress at 100% | Stress at 300% |
|---|---|---|---|
| HNBR | 1.6 | 2.5 | 10.0 |
| + CB N339 (5 phr) | 1.7 | 2.7 | 11.3 |
| + CB N339 (25 phr) | 2.5 | 4.5 | 20,7 |

Graphitized carbon black with similar specific surface area but low surface activity shows an increasing stress value at high extensions that still remains far below the level provided by the original non-graphitized black. The comparison underlines the role of filler surface activity in the regime of increasing stress values. It can be generalized that stress increases in a moderate manner for fillers with low surface activity but shows a pronounced increase for fillers with high surface activity. At medium to high elongations, the filler activity has a decisive influence on the stress values. Here the rubber-filler interaction is particularly noticeable from the intensity of the attachment of the polymer chains to the filler surface. Compared with an active carbon black, the graphitized black leads to low stress values and even high deformation. Carbon black obtained by superficial oxidation occurs at the active center and introduces oxygen functionalities on the surface area that can increase the stress values in the middle of the stress-strain behavior especially in polar rubber.

At medium extension, debonding of the rubber matrix from the filler surface can occur, especially for semi-reinforcing carbon blacks. The voids produced by this process are a result of the incomplete adsorption of rubber chains on the energy sites of the carbon surface and leads to premature damage of the filled compound. This phenomenon is not as important with active fillers. Here, the rubber-filler interactions are strong enough and dynamic cracking will not occur.

#### 9.3.2.2.2 Effect of Precipitated Silicas

At medium elongation, the precipitated silica leads to high stress values in polymers that possess a low solubility parameter, δ. In NR vulcanizate obtained by peroxide curing, the unsilanized precipitated silica determines the stress values, as shown in Figure 9.24 [101].

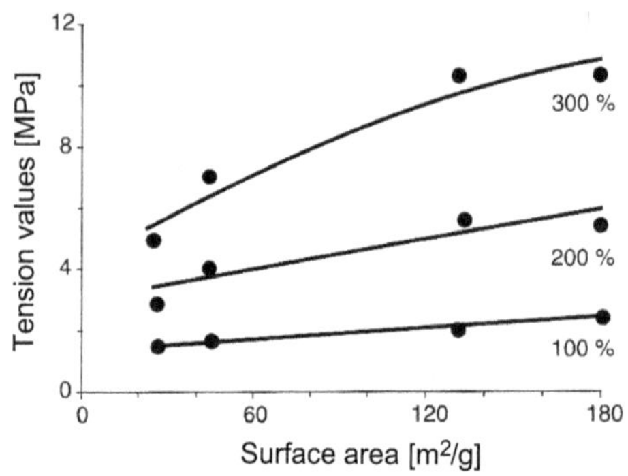


**Figure 9.24** Tension at 100%, 200% and 300% elongation in NR vulcanizates produced by peroxides as a function of the surface area of precipitated silica

Above the low deformation, the loading with precipitated silica leads to a significant increase in the stress values, due to the interaction of the silanol groups between the aggregates. When elongation continue to increase, the high stress can lead to slippage and detachment of the rubber chains from the filler surface. This may result in vacuole formation and elimination of local stress concentration in the vulcanizate [102]. With non-polar rubbers, at a higher deformation rubber-filler debonding is more common, because of the poor interaction between the rubber chains and the presence of silanols on the surface. This leads to rubber-filler detachment and premature failure in the vulcanizate [103].

The addition of 1.2 phr 3-mercaptosilane increases the stress at 300% elongation of non-silanized silica to silanized silica from 67 k/cm$^2$ to 154 kg/cm$^2$ [104]. This increase in tension is accompanied by an increase in the bound rubber from 30.5% to 44.6%. Using bifunctional sulfur silanes TESPT(Si69), the comparison of "green" tire compounds with conventional silica (VN3 GR) and HD silica (Ultrasil 7000 GR) exhibit the higher strength of 300% elongations [105]. A comparison of the effect of different silanes with respect to the increase in stress in silica vulcanizates reveals that monofunctional silanes provide the lowest stress values over the entire range of elongations. The modulus 300% increases with the concentration of TESPT and the amount of sulfur. At a concentration of sulfur (1 phr) and TESPT (4.5 phr), the modulus 300%

has a value of 5.5 MPa. An increase in the TESPT level to 7 phr while retaining the same sulfur content leads to the modulus of 7.9 MPa. If the TESPT is 7 phr and the sulfur content increases to 1.8 phr, the modulus increases to 11.5 MPa. Compared with an untreated silica VN3, the modulus of silanized silica in S-SBR/cis-BR blend with 6.4 phr TESPT increases the modulus 300% from 5.3 MPa to 10.3 MPa.

A four-fold increase in stress has been found for bifunctional 3-mercaptosilanes (Si 263). By coupling the filler, the tensile strength increases by 2 MPa, but the energy at break decreases. A comparison of the stress values at 300% elongation has revealed a stress of 9.6 MPa for silica/TESPT and 11.8 MPa for the HD silica/Si363 filled system in tire tread compounds [106]. In a peroxide-cured EPDM compound with the silane 225, the modulus 300% increases from 2.8 MPa to 4.9 MPa.

The HD silica has a high stress value at deformations of 50% to 200%, which then decreases and is replaced by a wet-mixed silica in the increasing stress region. Thus, the wet-mixing of the silanized silica shows a higher stress value of 300% and more when deformed, which is explained by the better connection to the rubber in accordance with TESPT [84]. Due to the excellent micro-dispersion of the wet-compounding, silanized silica shows higher stress values than the dry-mixed precipitated silica.

At higher elongations, where the agglomerates are totally broken down, the main factor influencing the stress is the polymer-filler interaction.

The effect of rubber/filler interaction on stress-strain behavior for silica filled SBR has been studied in relation to the chain scission of rubber molecules during deformation. The silica surface was modified by several coupling silanes. The combination of electron spin resonance and stress-strain data revealed that, at a given strain, the tensile stress increased with increase in interfacial interactions between rubber molecules and the silica surface, and simultaneously the chain scission was closely related to the mechanical energy applied to the test specimen [107].

#### 9.3.2.2.3 Effect of Layered Silicates

The increase in stress is usually associated with the hydrodynamic effect, which is more marked due to the filler of high aspect ratio and high specific surface area, such as layered silicates. It has been observed that, at a given strain level, a noticeable increase in the stress can only be observed from a filler content above the rheological percolation threshold.

The performance of layered silicates is determined by the manner in which the clay is introduced into the elastomer. It has been found that latex blending at the same volume fraction of clay gives better increase in the stress value than solution blending.

The better filler exfoliation due to wet-mixing leads to a much higher stress over the entire elongation range. Therefore, the reinforcing factor is significantly higher in the test pieces obtained by continuous dynamic latex compounding (CDLC) technology than in mechanical mixing. At a loading of 5 vol.%, the reinforcing factor of wet-mixing increases threefold [23]. While the stress values for unfilled NBR show a flat curve with medium deformations, exfoliated MMMT platelets lead to significantly higher stress values. The reinforcing factor at modulus 200% is 3.8 MPa for MMT at a concentration of 0.04 vol.% and has the same level as for carbon black N550 at 0.2 vol.% strain. Therefore, the exfoliated layered silicates can reinforce the rubber, in the respect of its lower volume fraction and the high aspect ratio.

However, the clay composites made from NBR latex are also dependent on the emulsifier of the rubber, which is reflected in the level of the reinforcement factor [87]. The stress values at 100% elongation are increased by 56% compared with pure NBR.

In NBR, the strength increased with the concentration of the acrylonitrile group. With each deformation, an increase in the stress values according to the polarity of the NBR was observed.

The stress properties of organo-clays with different amine modification are reported for SBR with 15%, 23% and 40% styrene matrix [107]. Even with low organo-clay loading, the organo-modified clay displayed improved stress values that are proportional to the styrene content, and that increase the efficiency in the vulcanizate. The stress values still fall for the solvent used to prepare the incorporation into the rubber matrix in the following order:

$$\text{toluene} > CCl_4 > CHCl_3$$

The effect of processing conditions and the modification to the rubber was investigated in a report that compared EPDM/organo-clay and EPDM-MA/organo-clay at 25 °C and 100 °C [108]. In all cases, the contribution of the organo-clays is significantly higher when the more polar EPDM-MA is used. At 25 °C, the stress value at 300% elongation increases from 2.6 MPa in pure EPDM to 9.9 MPa in EPDM-MA. If the samples are prepared at 100 °C, the stress value increases from 3.0 MPa in EPDM to 12.1 MPa in EPDM-MA. The interaction between the organo-clay and the modified EPDM improves significantly if the organo-clay can be distributed at a higher temperature.

#### 9.3.2.2.4 Effect of Carbon Nanotubes

The suitability of multi-walled carbon nanotubes to greatly increase the stress of a mixture with little loading is a special property of these nanomaterials. With all the rubber vulcanizates used, it is clear that the increase in stress values begins with small deformations. The multi-walled carbon nanotubes are aligned in the vulcanizate, due to good dispersion, and lead to a strong enhancement of the stress values. Below the electrical percolation threshold, the physical impact of multi-walled carbon nanotube is relatively small, but increases very strongly above this micro-dispersion limit [109]. Compared with carbon black, a small amount of multi-walled carbon nanotubes is required to reach the same performance due to a high aspect ratio and high specific surface area.

The diameter of the multi-walled carbon nanotube is crucial for the mechanical effect in a rubber matrix. In NR, the smaller diameter of multi-walled carbon nanotubes leads to an increase in stress at medium strains. For Baytubes 150 at a concentration of 1.5% by volume, a reinforcement factor of 2 is determined, while for Nanocyl NC7000, which has only half the diameter, a reinforcement factor of 3 is achieved [110]. The modulus in the stress-strain behavior increases much more for multi-walled carbon nanotubes than for carbon blacks at an elongation of 100 or 300%, due to the fact that they have a high aspect ratio compared to carbon black. If carbon black and multi-walled carbon nanotubes are compared at the same loading (1 phr to 7 phr) the increase at 300% elongation is from NR matrix at 0.46 MPa to NR/carbon black N220 (7 phr) at 0.76 MPa and finally to NR/multi-walled carbon nanotubes (7 phr) at 1.68 MPa, which represents an increase of 2.4 times [111]. Stress-strain curves of NR/multi-walled carbon nanotubes show a linear stress increase at 100% strain of up to 6 wt.%; above this value the strains decrease and show a smaller modulus since the dispersion is declining [112]. To interpret the variation in stiffness with the multi-walled carbon nanotubes loading, the stress at 100% elongations is fitted to the Halpin-Tsai model [113]. The equation yields for aligned fiber composites in conditions where the modulus of the fiber, $E_f$, is much higher than that of the elastomer:

$$E = E_0(1 + f\Phi)/(1 - \Phi) \tag{9.13}$$

where E and $E_0$ are the moduli of the composite and the unfilled elastomer, f is the aspect ratio and $\Phi$ is the volume fraction of the filler

The experimental data for an aspect ratio of 90 fit the Halpin-Tsai model, because of the well dispersed multi-walled carbon nanotubes in NR. Despite the dispersion detected by TEM, some improvements in the SBR/multi-walled carbon nanotubes have been observed. With the incorporation of 1 phr multi-walled carbon nanotube, the tension increases by 70% [114]. An overview of the stress values at 100% and 200% elongation obtained by carbon nanotubes in FKM, NBR and HNBR is given in Table 9.4 [115].

**Table 9.4** Stress Values (MPa) of Carbon Nanotubes in FKM, NBR and HNBR [115]

| Rubber | CNTs | Stress at 100% Elongation | Stress at 200% Elongation |
|---|---|---|---|
| FKM | 0.00 | 1.4 | 3.7 |
| | 0.01 | 2.8 | 6.3 |
| | 0.02 | 4.4 | 8.0 |
| | 0.03 | 6.5 | 10.5 |

**Table 9.4** Stress Values (MPa) of Carbon Nanotubes in FKM, NBR and HNBR [115] *(continued)*

| Rubber | CNTs | Stress at 100% Elongation | Stress at 200% Elongation |
|---|---|---|---|
| NBR | 0.00 | 1.4 | 1.9 |
| | 0.01 | 2.0 | 3.3 |
| | 0.02 | 2.5 | 4.4 |
| | 0.03 | 4.5 | 7.5 |
| HNBR | 0.00 | 0.8 | 1.1 |
| | 0.01 | 1.9 | 3.2 |
| | 0.03 | 4.4 | 8.0 |
| | 0.04 | 5.5 | 9.2 |

At a higher deformation of 300%, there is an increase from 3 MPa to 9 MPa from the unfilled NBR to multi-walled carbon nanotube-filled NBR [116].

The incorporation of multi-walled carbon nanotubes in HNBR and HXNBR leads to a new material with commercial importance. Stress-strain curves lead to an increase in the stress at medium to high elongations. The samples obtained during the development of the process show a strong increase in the modulus above the electrical percolation threshold, which clearly depends on the direction of the process.

The sample oriented in the process direction always has a significantly higher modulus. With a deformation of 50%, a value of 15 MPa is obtained for the sample in the direction of processing while, perpendicular to it, the stress value will be 6 MPa. With an elongation of 100%, the modulus determined in the direction of processing is 26.5 MPa and perpendicular thereto it is only 13.4 MPa. With the more polar HXNBR, the stresses obtained are somewhat higher because the carboxyl groups also participate in the interaction [117]. With the polar EVM, with a carbon nanotube content of 2.7% by volume, an increase in the modulus of 3.2 is recorded for small deformations of 50%.

An increase in 3.7 in the stress at 100% and 300% deformation is determined for silicone rubber with 3 vol.% multi-walled carbon nanotubes [118]. An addition of multi-walled carbon nanotubes to SBR with 10 phr carbon black leads from a stress without multi-walled carbon nanotubes of 1.13 MPa to 2.9 MPa for 5 phr multi-walled carbon nanotubes. It seems that the combination of carbon black (10 phr) and carbon nanotubes (5 phr) in NBR matrix increase the modulus at 200% from 0.32 MPa to 2.3 MPa, because of the additional contribution of the high aspect ratio of multi-walled carbon nanotubes [119]. The modulus measured at 50% elongation increased by a factor of 2.1–2.6 when multi-walled carbon nanotubes, 5 phr to 10 phr, are added to rubber material filled with carbon black N375. The modulus obtained with 60 phr carbon black N375 increases from 3.3 MPa to 8.4 MPa when 10 phr of carbon nanotubes are additional added [120].

The most frequent functionalization to improve the compatibility and the dispersion with polyurethane rubbers is the partial oxidation of carbon nanotubes. This group would improve the interfacial bonding through the action of an –OH group and –COOH groups of the multi-walled carbon nanotubes with the NCO groups of the isocyanates [121, 122].

In rubber technology, multi-walled carbon nanotubes can be added to the usual filler to make the vulcanizates stronger or to achieve the same tension while reducing the weight of the elastomer. Such an addition of a small amount of the multi-walled carbon nanotubes to carbon black or to a silanized silica can significantly increase the tensile properties, especially in the range of small to medium elongations, and so such a tensile strength is crucial for future applications.

### 9.3.3 Stress Softening (Mullins Effect)

It is assumed that the reinforcement by active fillers is related to the energy dissipation during the deformation of the test sample. The phenomenon can be observed for unfilled rubbers that show strain crystallization and also for the filled vulcanizate. If the dumbbell is stretched twice in succession to an extension ratio, the stress-strain curves are no longer the same, because the initial curve cannot be repeated on subsequent stretching because of material changes which are partly reversible. This is particularly true of compounds containing reinforcing fillers. Such an effect has been termed stress softening. Repeated deformation will minimize stress until an equilibrium is reached where the filler-filler interaction ceases to have any effect. This effect was noted first in 1903 in a technical paper [123]. A half century later, the phenomenon was investigated on vulcanized NR [124, 125] and vulcanizates NR filled with carbon black N330 [126]. The investigations at the same strain rate after a first retraction on a second repetition found that the deflection gives a direct change in stress, which shifts to smaller deformations with the same change with filler content and reaches a value that corresponds to the concentration of the filler. It was observed that if a test specimen is not stretched until rupture, but to a certain elongation, $\varepsilon_1$, and then returned to zero strain, and then stretched again, the stress values for the second elongation are below the ones measured in the first elongation. Above the previous stretching limit, $\varepsilon_1$, the stress value increases again and nearly reaches the values recorded during stretching of the original sample. If pre-stretched dumbbells are allowed to rest for several days or are annealed for several hours at elevated temperatures, preferably *in vacuo* to prevent oxidative processes, the stress recovers partially or fully, depending on the recovery conditions and the type of crosslinks of the rubber. However, because many rubber products are exposed to repetitive stress-strain deformation, the phenomenon can affect end use performance. Not surprisingly, the effect has received great interest in rubber science as a mechanical aging or a damaging process. The uniaxial elongation that define the stress softening, called the Mullins effect, is shown in Figure 9.25.

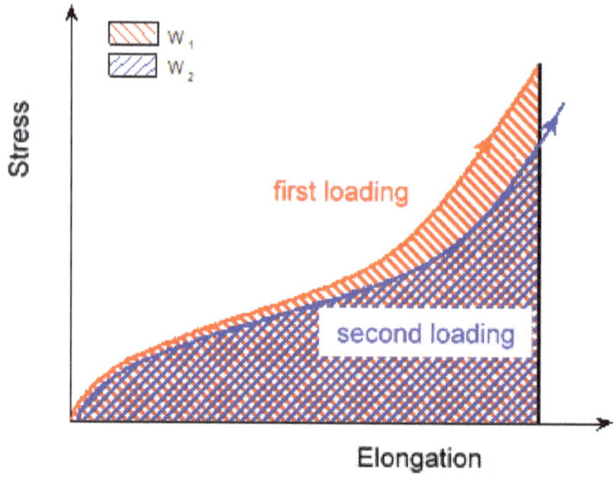


**Figure 9.25** Schematic representation of stress softening effect

To quantify the stress softening effect, the percentage of strain energy loss during two consecutive extensions of a filled vulcanizate is considered representative [126]:

$$\Delta w\,(\%) = (1 - w_2/w_1) \times 100 \tag{9.14}$$

where $w_1$ is the stored energy at first extension and $w_2$ is the stored energy at second extension

The stress softening effect can be understood as a measure of internal energy dissipation that accompanies a decrease in stress peaks and leads to a more uniform load distribution in the direction of the external strain. A residual chain orientation in the direction of strain in the unloaded samples has been detected. An empirical factor X was introduced that is independent of the strain and corresponds to the quadratic hydrodynamic reinforcing term given in Equation 9.13. The strain amplification which relates external strain, ε, of the sample to the internal strain ratio of the rubber matrix is given by the expression [126]:

$$\Lambda = 1 + X\varepsilon \tag{9.15}$$

where ε is the elongation (λ – 1) and X describes the ratio of effective average strain in the rubber phase, λ′

The amplification factor X is assigned to the hydrodynamic reinforcement by carbon black particles [127]:

$$X = \lambda'/\lambda = 1 + 2.5\Phi + 14.1\Phi^2 \tag{9.16}$$

This statement applies if all vulcanizates were initially stretched to the same stress. Tension softening is most noticeable at the second elongation. In subsequent loading and unloading cycles, the effect is only slightly increased until a stationary end value corresponding to the state of use is reached. For reinforcing carbon blacks, the modified equation containing the shape factor X of the asymmetric carbon black particles can be used:

$$X = \lambda'/\lambda = 1 + 0.7 f_s \Phi + 1.62\ f_s \Phi^2 \tag{9.17}$$

The strain amplification factor X can be viewed as a value to assign the modulus increase in filled vulcanizates. Indeed, the result of modeling the stress-strain behavior with the help of the strain amplification factor does not correctly reflect the real deformation behavior of filled vulcanizates, either under small or under large deformation. To explain the impact of the filler on stress-strain behavior, the assumption is made that filler particles are perfectly attached to the rubber matrix. In this way, the external stress can be transmitted through the rubber-filler interface to the filler surface.

The explanations for the stress softening can be grouped in three directions: damage within the rubber matrix by chain scission, appearance of cavities, disentanglements, changes in crosslink density; filler network alteration by rupture of the filler network and loss of percolation; change in the rubber-filler interface by molecular slippage that reduces the applied stress, desorption of chains.

Stress softening of filled vulcanizates can be only partly reversible. This is an indication of an irreversible change in the structure of the filled system. Assuming that material softening is due to damage within the rubber matrix, the average length of network chains, $M_c$, should increase significantly [128]. Several investigations have revealed rather small changes in the degree of equilibrium swelling, indicating that chain scission is not contributing to stress softening. Therefore, the effect could not be attributed either to chain scission during stretching or to vacuoles or detachments on the filler surface [129]. Disentanglements of polymer chains are discussed as an alternative mechanism, which can lead to a lower stress at the second elongation [130]. However, taking into account that, in well crosslinked vulcanizates, the entanglements are trapped between adjacent crosslinks, this mechanism seems to contribute to a lesser extent to stress softening.

In other reports, stress softening on a microscopic level is associated with irreversible damage to the filler-filler interactions. With increase in the pre-strain, a transition takes place from a stiff filler network to a more flexible one with progressively broken physical bonds [77]. Alteration of the filler network by mechanical stretching is supported by the observation that stress softening Δw (%) appears to occur well above the electrical percolation threshold. Nonetheless, while the stress softening is anisotropic, the electrical anisotropy and the change during the deformation are small. The conductivity measurements suggest that filler network alteration is not the cause of the stress softening.

Due to the proven fact that rubber chains are adsorbed on the filler surface, the stress reduction in the second extension was attributed to a molecular slippage mechanism, which is shown schematically in Figure 9.26 [131].

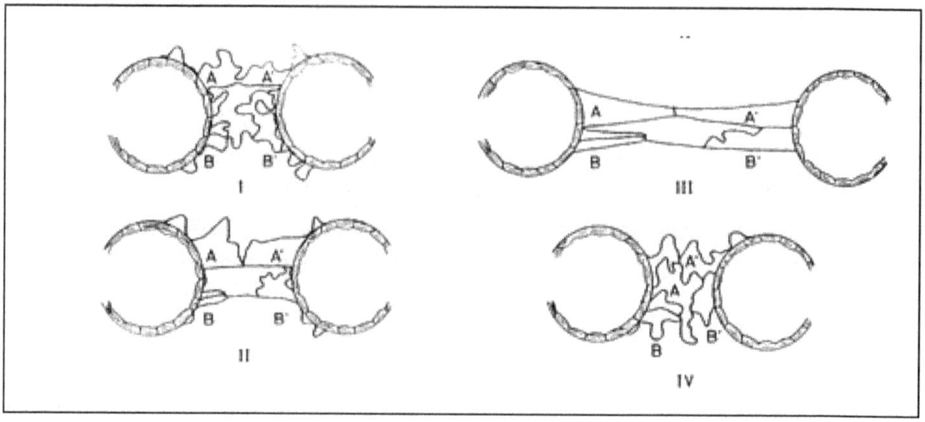


Figure 9.26 Mechanism of stress softening

As the sample is stretched, chains start sliding across the filler surface in the direction of extension. The chain slippage contributes in reducing the applied stress. When the stress that causes the first elongation stops, the rubber strands no longer return to their original position. This results in buckled strands between adjacent filler particles at the beginning of the second extension. Therefore, the second loading to the same elongation requires less stress until the buckled strands will be elongated to the same extent as before. Once the previous extension limit, $\varepsilon_l$, is reached, the buckled strands reach their initial length and thus provide resistance to further stretching, with the result that the overall stress reaches the original stress-strain curve. Heating *in vacuo* increases the molecular mobility without risking aging and thus free chains adsorb on the active sites of the filler surface, restoring the rubber-filler interface to a similar state as before the first stretch. Pre-stretched samples exposed to heat recover completely and become stiffer. Thus, the adhesion of the rubber phase to the filler is of fundamental importance for strain softening. The lower the adhesion, the more pronounced is the sliding of the rubber chains on the filler surface. In most of cases, energy loss is generated by molecular slip and hence the increase in stress associated with motion and rupture in such systems is small. Mechanical hysteresis reduces the efficiency and increases heat build-up.

A further development of the strain amplification concept has been published in connection with the theoretical description of fractal filler networks. For vulcanizates with a filler concentration above the percolation limit, the strain amplification factor considers the overlapping fractal clusters and the fractal dimensions of the clusters [132]:

$$X = 1 + const\ (\xi/a)^d w^{-d} f\Phi\ (2/3 - d_f) \quad (9.18)$$

where ξ is the cluster size, a is the particle size, $d_f$ is the fractal dimension and $d_w$ is the anomalous diffusion exponent of the clusters

Recently, an "interface model" has been proposed to approximate the real stress-strain behavior of carbon black-filled vulcanizates more closely [67, 68]. The model considers the changeable character of bound rubber in relation to the surface site energy distribution of fillers. It distinguishes between rubber strings that are tightly attached to the filler surface and others that adhere loosely to the less active surface sites. Thus, depending on the extension, there is a distribution of soft amorphous chains that can be easily extended and others that are firmly anchored and provide more resistance to deformation. With increase in the surface activity of the filler, the fraction of strongly bonded chains becomes larger and leads to the stress upturn and the high tensile strength shown experimentally. The "interface model" is based on the energy site distribution of carbon black [132, 133]. In contrast to the "strain amplification" concept, which allows an, albeit not exact, estimate of the stress values of filled vulcanizates, the "interface model" serves to explain the phenomenon in many details, but remains almost too abstract or qualitative to be used for predictions of stress-strain effects. Experience shows that the level of stress of filled vulcanizates decreases with increase in temperature. The opposite is true for unfilled vulcanizates, where the stress increases with the temperature, due to the thermo-elastic effect. Taking into account the fact that physisorption of polymer chains on solid surfaces decreases with temperature, the observed decrease in the stress is attributable to a progressively incomplete adhesion of the rubber phase on the filler surface. The reduction in phase adhesion enables chain sliding on the surface during stretching and thus the stress required for deformation decreases. Stress softening is an important property for braking performance and rolling resistance of tires, both related to mechanical energy losses.

#### 9.3.3.1 Stress Softening in Carbon Black Elastomers

Stress softening is actually a consequence of rubber viscoelasticity that can be described by combinations of spring-dashpot models. In addition to viscoelastic softening, strain-induced crystallization can cause large hysteresis during extension-retraction of rubber. The loss of hysteresis energy depends on the filler volume fraction, filler activity and the extension ratio [134, 135]. The increase in both parameters leads to an increase in the hysteresis energy loss. Figure 9.27 shows the change in hysteresis energy loss for SBR/CBN234 vulcanizates as a function of the filler volume fraction and the extension ratio [136].

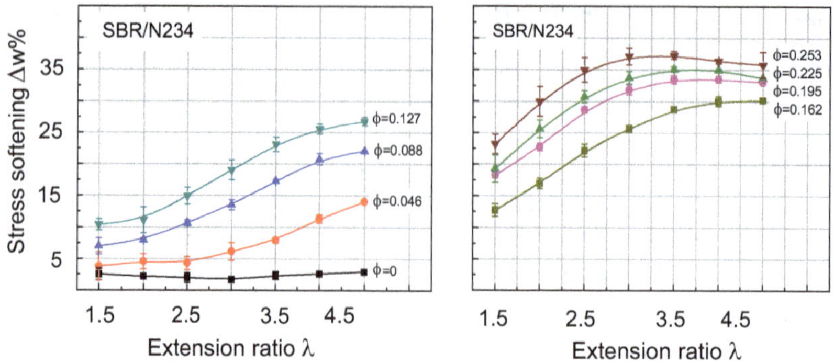


**Figure 9.27** Stress softening as function of filler loading and extension ratio [136]

As can be seen, the curve of the hysteresis energy loss is not linear over a wide range of the extension ratio. Below the percolation threshold of the filler, the strain energy loss only increases above a certain extension ratio, while above this critical concentration the effect increases significantly. At high volume fractions (Figure 9.27, right), the effect is already pronounced, with elongations of 100%. In the case of overloaded vulcanizates, a maximum of hysteresis energy loss is observed for elongations above 300%. The effect of carbon black on the stress softening in NR is an increase in Δw% with specific surface area in the extension up to $\lambda = 4$. The higher energy loss of carbon black with a high surface area may be caused by more intensive agglomeration of the aggregates. At medium to higher elongations, $\lambda > 4$, the contribution of the surface area decreases with the Δw%, especially for the carbon blacks with a high surface area. More interface leads to more polymer slippage.

The carbon black structure leads to higher stress amplification for every elongation of the NR or SBR vulcanizate. The residual elongation that occurs during this process indicates that strain softening is accompanied by damage to the rubber-filler interface. With increase in filler volume fraction, the deformation cycle results in increasing residual elongation. Carbon black N234 with a high dispersive surface activity shows an intensive stress softening in conventional mixing of S-SBR/cis-BR (70/30) blend that reaches a stress value of 16.9 MPa at an elongation larger than $\lambda = 5.5$. The Mullins effect shows an increase to 38% above an elongation of $\lambda = 2$, indicating an increase in rubber chain slippage and detachment from the filler surface.

### 9.3.3.2 Stress Softening in Precipitated Silica Elastomers

Compared with active carbon black, the conventionally blended precipitated silica has a lower stress softening effect that extends over a wide elongation range and the stress. The lower level is achieved because the inter-aggregate interactions of the silanol groups are prominent and hold the silica aggregates together. In an S-SBR/cis-BR (70/30) blend, with a conventionally mixed silica, the stress value of 5 MPa is not exceeded after seven consecutive multi-hysteresis loops at elongation of L = 4. If the silica silanized with TESPT is used in a dry-mixing process, the stress of the vulcanizate will increase to 12.5 MPa. The silanization clearly increases the rubber-filler interaction that can be achieved with the multi-hysteresis. However, if wet-mixing technology is used, the stresses achieved in the same blend have values of 14 MPa and show a higher value than carbon black N234. Due to the silanization reaction and the wet-mixing technology, the energy loss of silanized silica is higher than that of carbon black N234 [137]. The reduced energy of silanized silica increases from small elongations, $\lambda = 2$, with 18% to large elongations, $\lambda = 4$, with 32%. The wet-mixed silica, which shows a much better dispersion and good interaction with the rubber chains, increases from $\lambda = 2$, with $\Delta w\%$ of 15, to $\lambda = 4$ with $\Delta w\%$ of 38, which is higher than for carbon black N234. It can be clearly seen that silanized silica has a major influence on stress softening, which can be exploited in practice.

#### 9.3.3.3 Stress Softening in Carbon Nanotube Elastomers

It has been suggested that slippage of rubber molecules over the filler surface can be a parameter for multi-walled carbon nanotubes that denotes larger values in stress softening. The energy dissipation of NR/multi-walled carbon nanotubes in the first hysteresis cycle is much larger than for NR/carbon black and it decreases in the subsequent cycles. The multi-walled carbon nanotube-filled samples show greater stress softening, suggesting that more debonding and detachment of the rubber chains occurred at the $sp^2$-hybridized carbon atoms compared with the carbon black that possess three stronger adsorption sites on the surface. The difference is shown in Figure 9.28 for HNBR where multi-walled carbon nanotubes are compared with carbon black N220 [138].

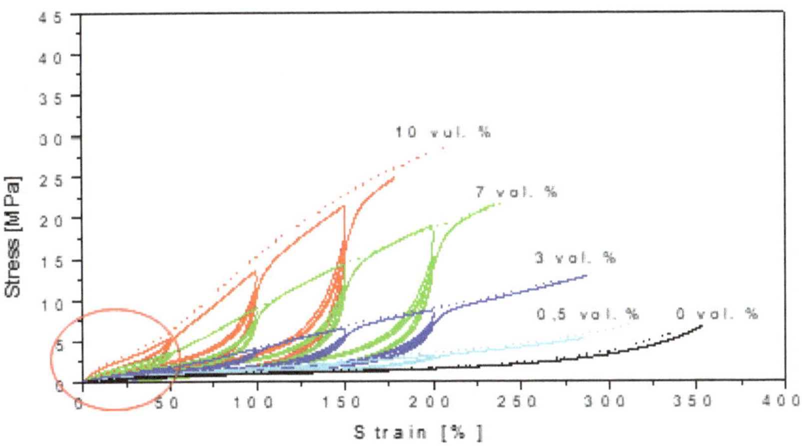


Figure 9.28 The Mullins effect for HNBR/multi-walled carbon nanotubes shows high permanent elongation and a large difference in hysteresis compared with systems that contain carbon blacks

When using 2 phr multi-walled carbon nanotubes, the stress softening is derived mainly from the breakage of the long tubes through interaction with the rubber chains. After resting for 24 h at room temperature, the vulcanizate gains some of its elastic recovery [139]. Stress softening is more pronounced in the samples that contain carbon black or silica and multi-walled carbon nanotubes, which confirms improved interaction with the polymer chains when both fillers are dispersed together [119].

### 9.3.4 Tensile Strength and Elongation at Break

Tensile strength is the maximum stress when the dumbbell specimen breaks during elongation. The energy at break corresponds to the integral of the stress-strain curve up to the ultimate elongation and can be approximated from the product of tensile strength and the elongation at break, $\sigma_b \times \varepsilon_b$. Although the mechanism of tensile failure has not been fully elucidated, it can be regarded as catastrophic uniaxial tearing initiated by accidental defects in the test specimen, such as larger fragments of pellets, micro-wetting or cavitation of the rubber matrix from the filler surface. Impurities, such as dirt, can also cause the dumbbell to break at lower stress. Likewise, volatile compounding ingredients can cause porosity during cure. These voids can also contribute to a lowering of tensile strength. Therefore, tensile strength of a rubber network is capable of dissipating input energies into heat through molecular processes in the bulk of the test specimen. Less elasticity will be available to break the rubber network and increase the fracture energy. Tensile strength has been served as an indicator for filler dispersion and the quality of the efficiency of the vulcanization procedure. The higher the degree of filler dispersion, the higher is the measured tensile strength.

Nevertheless, before methods to determine the filler dispersion were introduced, tensile strength was used to evaluate qualitatively the filler dispersion. Tensile strength and elongation at break of filled vulcanizates are quality control tools to monitor reinforcing effects of nanoscale fillers, inter-batch consistency of compounds and to recognize effects of filler dispersion and processing behavior. However, in order to better determine the tensile strength and the elongation at break, it is necessary to submit several test specimens to this catastrophic test.

#### 9.3.4.1 Tear Strength and Elongation at Break for Unfilled Elastomers

In vulcanizates that do not contain filler, the tensile strength depends on the chemical nature of the rubber, the degree of crosslinking density, the distribution of crosslinks in the network morphology, the type of crosslinks and the conditions of the experimental test. The formation of a structure given by strain crystallization prevents crack and increase the tensile strength by self-reinforcement. In amorphous rubbers, the lack of strain crystallization reduces the reinforcement of the rubber chains and reduces the value of the tensile strength.

#### 9.3.4.2 Tensile Strength and Elongation at Break for Filled Elastomers

By incorporating and dispersing active fillers in vulcanizates with a common cross-link density ($M_c$ = 5000 to 10,000 g/mol), the tensile strength, $\sigma_B$, can be 10 times larger than that for the unfilled vulcanizates if the aggregate size is below 1 µm. The smaller the aggregate size of an active filler, the higher the tensile strength achieved. For inactive fillers with a larger primary particle size and a lower activity on the surface, the tensile strength decreases significantly. A strain-crystallizing polymer, such as an NR or an CR, demonstrates high tensile strength by a self-reinforcing mechanism. Even in the presence of active fillers and a high loading, the tensile strength does not increase as much as with the amorphous rubbers.

Two conditions must be met for the filler to achieve high tensile strength in vulcanizates. Firstly, the system must develop a large stress concentration; secondly, the stress concentration must be supportable by the vulcanized specimen. Because rupture occurs at the weakest point of the system, even if the rubber has complete phase adhesion to the filler particles, the increase in tensile strength is low if the force between the matrix adjacent to the filler is weak. However, without adequate interfacial bonding between filler and rubber, the filler acts as a foreign body at lower strain.

Several mechanisms have been published to explain the tensile strength of filled elastomers. The interpretation of the filler activity, of the interaction between the rubber chains and the filler surfaces, the influence of the specific surface area and the aggregates' structure are included in such explanations. Since the crack cannot develop through the filler particles, it has to propagate around them in the immobilized rubber and this requires more energy to split the system, but prolongs the life time of the filled vulcanizate.

If the test specimen is capable of dissipating input mechanical energy into heat through molecular processes, less elastic energy will be available for the rupture. The incorporation of active fillers is a major source of energy dissipation. The relationship between the tensile strength and energy dissipation during elongation can be formulated. The energy density at break is proportional to the energy dissipated by the rupture and the elongation at break [140]:

$$U_b = c(H_b \varepsilon_b)^{1/2} \tag{9.19}$$

where $U_b$ is the energy density at break, $H_b$ is the energy dissipated by the rupture, $\varepsilon_b$ is the elongation at break and c is a constant that depends on the filler grade and its loading

The alignment of the network and the blunting of the crack tip delays the beginning of the rupture. Since the crack cannot pass through the filler particles, it has to surround them and requires more mechanical energy. Many reports have found that the more dispersed filler in the elastomer, the higher is the tensile strength. If there is a microscopic defect in the structure of the filled vulcanizate, rupture takes place at high stress. Hence, factors such as filler dispersion and rubber-filler adhesion have a significant effect on tensile strength. The ability of filled vulcanizates to support large stress before rupture is promoted by material parameters, such as the specific surface area, surface activity and loading of filler, rubber-filler interaction and a large rubber-filler interface.

However, the non-dispersed filler fragments in the vulcanizate have a limiting effect on tensile strength. The more frequently such defects occur in a vulcanizate, the lower the average tensile strength values recorded in repeated tests. For this reason, the scatter of the tensile test values also shows an inaccuracy about the morphological condition and the spatial distribution of the defects in the sample. The distribution of the tensile test values of a sample provides information about its homogeneity. With iso-dimensional fillers, carbon black and precipitated silica, the tensile strength occurs at high elongation. For fillers of high aspect ratio and high specific surface area, like the layered silicates and the multi-walled carbon nanotubes, the tensile strength decreases to lower values and occurs at a smaller elongation at break.

#### 9.3.4.2.1 Effect of Carbon Blacks

Phase adhesion between the filler and the rubber matrix is dependent on the specific surface area and provides a significant part of the value of the tensile strength. For active carbon blacks, a relationship between the value of tensile strength and the adhesion index on carbon black-filled vulcanizates has been established. When a vulcanizate undergoes stress, the local stress at the tip of the flaw is magnified. Once the local stress at the tip reaches a critical level which depends on the size of the flaws and the rupture energy, cracks will be initiated and new surfaces will be created inside the specimen [141]. It has been theoretically predicted that, if the radius of a micro-void is less than 0.1 µm, its surface energy will exert an additional restraint on expansion. On the other hand, in the case of poor polymer-filler interaction, debonding of the rubber matrix (or de-wetting) from the filler surface will take place and can trigger crack initiation that results in lower tensile strength.

A wealth of observations concerning the influence of furnace blacks on tensile strength of vulcanizates has been summarized. One of the first exhaustive reports on this subject lists the following influences [142]:

- Active carbon blacks increase the tensile strength at low filler loadings, while inactive carbon blacks decrease the value.
- The effect of active carbon blacks on tensile strength passes through a maximum value, which should be used in practical applications.
- The maximum value of the tensile strength decreases with the primary particle size and specific surface area.

- The maximum value of the tensile strength increases with the degree of the filler dispersion.
- Tensile strength decreases with increase in temperature.

At high strain level, the elastomer exhibits a rapid increase in the stress prior to break. This hardening is related to the limited extensibility of some chains between the crosslinks. The influence of filler loading on tensile strength and elongation at break has been recognized as an essential part of filler reinforcement. Generally, tensile strength increases with the increase in filler loading and passes through a maximum, then decreases at higher filler levels, while elongation at break is reduced at low filler loading and decreases systematically at higher filler loading. The tensile strength of NR compounds increases slightly with the type and loading of reinforcing fillers [142]. The main mechanism that causes this difference is the ability to crystallize under strain, which in turn results from the stereoregularity of the polymer which is composed of nearly 100% cis-1,4-isoprene units. The partial transformation under strain from amorphous chain to supramolecular crystalline organization provides a self-reinforcement in the extension direction. Therefore, vulcanizates of self-reinforcing NR usually attain high tensile strength values in the range of 28–32 MPa. It has been found that carbon black does not inhibit the phenomenon of strain-induced crystallization, but its appearance shifts the onset strain of crystallization to smaller extensions. Compared with amorphous rubbers, the crystallizing rubbers lead to higher tensile strength, mainly because the self-reinforcing crystallizing chains occurs.

The dependence of tensile strength on filler type and loading is illustrated for several carbon black types in oil-extended SBR 1712 vulcanizates in Figure 9.29 [142]. Besides the fact that the increase in tensile strength (above a filler loading of 20 phr) increases with the specific surface area, the loading for the maximum tensile strength decreases.

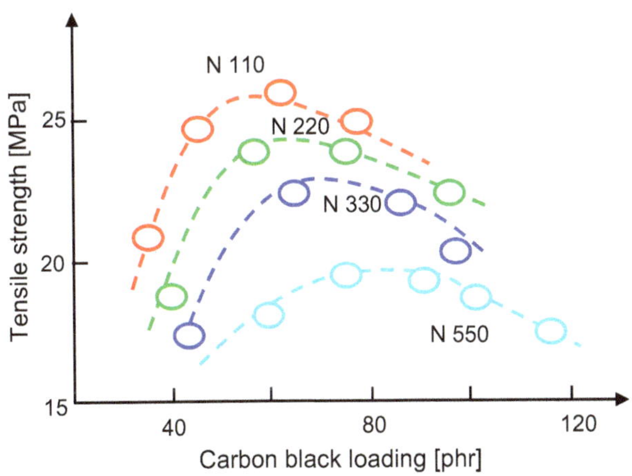


**Figure 9.29** Dependence of tensile strength on the type of carbon black and the filler loading (adapted from [142])

If the primary particle size increases, the maximum of the tensile strength occurs at higher loadings and is less pronounced in the stress at elastomer rupture. The filler concentration at which tensile strength reaches its maximum value depends on the chemical nature of the rubber, the mixing parameters that lead to the specific surface area of the filler, the dispersion of the filler in the rubber and this is certainly related to the type of the internal mixer and the size of the rotor geometries. It is recommended that the filler loading which provides the maximum tensile strength be used in compound formulations designed for elastomers subjected to high stress under severe load conditions. Since the position of the optimal tensile strength depends on the selected test conditions, such as strain rate and temperature, it is advisable to choose the test conditions such that they correspond to the intended use of the elastomer. In contrast to active carbon blacks, the non-reinforcing blacks MT and FT provide considerable smaller strength and do not exceed a maximum value, but decrease with increase in loading. Due to their low surface activity, these coarse fillers are unable to adsorptive bond rubber chains onto the surface. Thus, at larger strains the rubber chains start to slip on the surface and detachments of the rubber phase occur even at moderate strain.

The carbon black structure shows a small increase in the tensile strength at low filler concentrations. Above a maximum, the overloaded vulcanizate shows a decrease in tensile strength. In the overloaded system, the frequency of pellet fragments and large agglomerates increases and can lead to crack-initiating defects. The vulcanizate contains a multi-attachment of the rubber chains to several filler aggregates and therefore the reinforcing effect declines. The elongation at break declines with the filler grade. Higher values are obtained for the carbon blacks of high specific surface areas and the lowest aggregate structure.

Nanoscale fillers with large specific surface areas and high aspect ratio are conducive to processes that reduce crack initiation when well dispersed. It is not surprising that tensile strength increases with these features, while the aggregate shape plays a secondary role.

Fracture mechanics observations which show that the flaw size for initiating tearing increases with the filler concentration support this view [143]. The influence of the particle size of different carbon black grades on the tensile strength was examined in detail. The tensile strength decreases with the mean aggregate area determined in the TEM. For small aggregate surfaces of 50 $\mu m^2$, the tensile strength is as high as 31 MPa and decreases significantly for larger aggregates of 120 $\mu m^2$, resulting in a tensile strength of 26.9 MPa [74]. The tensile strength increases as long as the effective filler content is less than 0.1 and decreases to a low value at higher effective volume when the structure doubles [75]. It has been suggested that the inter-aggregate distance should be large enough to allow the development of a critical crack size. If the initial cracks caused by dewetting are small enough to consume a portion of the required energy, the material will exhibit considerable strength and will be resistant to crack propagation [144].

When it comes to the influence of filler dispersion on tensile strength, it is helpful to reconsider the effects of the major mixing parameters of mixing time and rotor speed. In Figure 9.30, the evolution of tensile strength of E-SBR/CB N330 (50 phr) is shown as a function of mixing time and rotor speed [145].

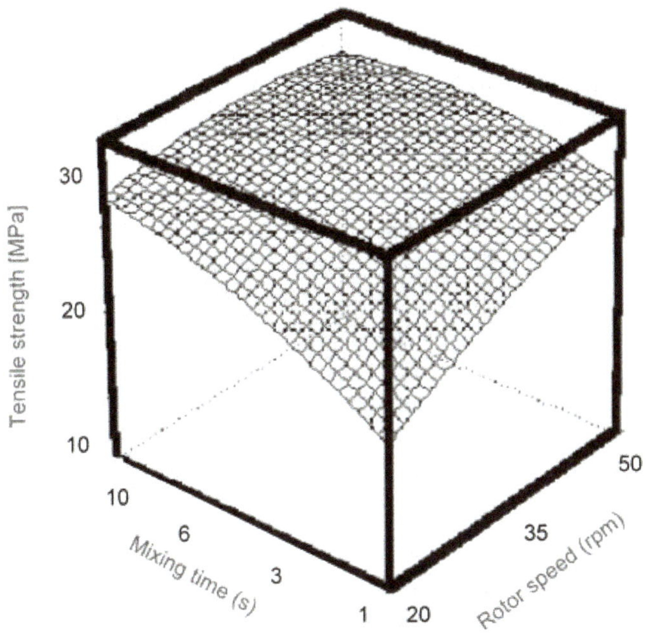


**Figure 9.30** Development of tensile strength in E-SBR/CB N330 as a function of mixing time and rotor speed [145]

The parallel increase in filler dispersion with mixing time and rotor speed leads to an increase in the tensile strength. Laboratory mixed batches usually have higher tensile strength than factory-mixed batches of identical formulation. The reason is that laboratory mixes often exhibit better filler dispersion. Obviously, in factory-mixed batches the frequency of less-well dispersed filler particles and pellet fragments is higher and necessitates appropriate quality control. However, filler dispersion is rarely considered to be a determining quality criterion in factory practice.

The emerging plateau of the tensile strength response surface is attributed to the increased temperature of the compound, which does not allow efficient shear stress in the mixing process, even on prolonged mixing action. If the compound is mixed again after a cooling time, dispersion of the filler can be improved, which leads to an increase in the tensile strength. When using tread grades, small increases in aggregate size cause a pronounced decrease in tensile strength while, with carcass blacks, this influence is significantly weakened, underlining the influence of the filler activity. It has been suggested that fragments of carbon pellets may act as stress sizers by forming nuclei sites, which can increase crack propagation [146]. In the E-SBR/carbon black N234 system, the proportion of carbon black fragments was determined and the decrease in tensile strength was associated with this. A good linear relationship was observed between the two quantities [147].

The question arises as to what extent the tensile strength can be described as a visco-elastic phenomenon and what influence the temperature has at different deformation speeds. The tensile strength of E-SBR/carbon black N330 and E-SBR/carbon black N774 was tested in dosages of up to 100 phr over a temperature range of −50 °C to +200 °C. Application of the temperature-corrected tensile strength, $\sigma/T_g/T$, yields a family of curves, whose shape resembles the modulus of elasticity as a function of temperature and deformation rates [148].

A report published by an EPDM producer describes the effect of paraffin oil on tensile strength for EPDM compounds filled with variable loadings of carbon black 539. It was found that, at any filler loading, the tensile strength decreases by 0.077 MPa for 1 phr of paraffin oil [149]. When the tensile properties of the elastomers containing N234 and the carbon black obtained from the pyrolysis oil sN234 are compared, it was clear that the values of the experimental compound in NR and S-SBR are very similar or only deviate slightly to higher values [150].

The extensibility of the filled network depends on the test conditions, such as the elongation rate and the temperature. Therefore, a failure envelope to characterize the tensile stress has been developed [151]. The stress at break and the elongation at break were measured over a large range of strain rates and temperatures. Such a parabolic curve depends on the carbon black loading, on the type of the polymer and on the crosslink density [152].

#### 9.3.4.2.2 Effect of Precipitated Silicas

Similar to the case for carbon black, the tensile strength of precipitated silica decreases with the primary particle size and increases with the specific surface area in a non-linear direction. For a small specific surface area, there is a strong increase, which declines above the specific surface area of 130 $m^2/g$ for conventional precipitated silica. The elongation at break follows a similar trajectory [101]. The silanization of a conventional silica with 3-mercaptosilane (1,2 phr) can increase the tensile strength from the original 171 $kg/m^2$ to 234 $kg/m^2$, but reduce the elongation at break from 500% to 380% [104].

HD silica shows good interaction with polar rubbers, which is reflected in a high tensile strength of 27.2 MPa, but the elongation at break falls back to 400%. The use of monofunctional silanes reduces the concentration of silanol groups on the aggregate surface and contributes to a lower tensile strength. The tensile strength in cis-BR, which is 17.5 MPa, is was reduced to 14.8 MPa by silanizing HD silica 7000 GR with Si203, while the elongation at break of 440% increases to 750%. When the HD silica is treated with the silane Si208, the tensile strength drops to 24.8 MPa, but the elongation at break increases to 560%. If a longer alkyl chain is used in the silane, the tensile strength decreases further because the Si208 can cover a larger surface of the filler particle. Compared with conventional silica, VN3, the silanized silica with TESPT, decreases from 14.2 MPa to 12.2 MPa. Similar behavior is exhibited by a peroxide-cured EPDM compound that contains silica silanized with Si225 [105]. The tensile strength decreases from 13 MPa to 12.6 MPa. The surface area of the silica surface is larger and the filler interacts better with the rubber [153]. The tensile strength increases with silica loading up to 25% initially, due to the filler-filler interaction [154].

Tensile strength increases to a value of 8 MPa when a silanized HD silica if low specific area is mixed into an S-SBR/cis-BR (70/30) blend. If the silica has a higher specific surface area, it can reach much higher tensile strength. A silanized silica with TESPT (Si 69) has a tensile strength of 18 MPa when it is dry-mixed and, when wet-mixed, the elongation at break decreases, but the tensile strength increases to over 19 MPa. This can be attributed to better dispersion of the silica.

The TEM images indicate much better dispersion of the wet-mixed silica in S-SBR/cis-BR blend, which allows more polymer chains to carry the load and leads to a higher tensile strength [20]. If the carbon black in an NR/cis-BR (75/25) blend is replaced by silanized silica, the tensile strength decreases, because the crosslinking density decreases when carbon black and silanized silica are used [155]. The filled rubber mixes can be obtained by precipitated silica on the wet-compounding technology that shows a better interaction with the unsaturated rubber. Therefore, the dispersibility of wet-compounded silica is significantly higher than with the mixed silica in an internal mixer. Thanks to silane Si747, the well dispersed silica in NR has an excellent tensile strength of 30.8 MPa, which is 17.6 higher than that of latex-compounded NR/silica vulcanizate [156].

#### 9.3.4.2.3 Effect of Layered Silicates

In the case of layered silicates, a distinction is made between the pristine MMT obtained via wet-compounding and the organo-clays modified via quaternary ammonium salts. NR, which exhibits strain crystallization, leads to a significant increase in tensile strength by a factor of 1.2, because of the high value of the strain crystallizing polyisoprene. For pristine clays, an increase in tensile strength can be seen when the loading is lower than 10 phr and the values achieved clearly exceed the value of carbon black N330 in SBR [21, 22]. The effect of MMT on the tensile strength in SBR shows a response to the content of styrene groups in SBR. In SBR (23 S), the tensile strength of the sample containing MMT (7 phr) is increased by a factor of 1.3 [23]. The tensile strength in the wet-mixed NBR/$Na^+$MMT compound reaches 18 MPa and an elongation of 620%, which is significantly higher than the mechanically mixed pristine clay, which only achieves a low tensile strength of 4.2 MPa.

With the MMT derivative PGN (5 vol.%), the tensile strength increases by a factor of 3.4, compared with the reference, while the elongation at break continuously decreases. The tensile strength depends on factors that determine the latex type and the emulgator used and the manufacturing method of the exfoliated platelets, and can also vary in value. After MMT is incorporated and dispersed into the rubber, an increase proportional to the loading of the exfoliated material occurs that increases the tensile strength. Compared with NBR, the course of the stress-strain behavior in HNBR is characterized by a lower increase in stress values, which leads to a tensile strength of 8.5 MPa. The more-polar HXNBR delivers a tensile strength of 27 MPa with MMT at 3 vol.% [24]. If 5 phr of exfoliated MMT is added to an NBR that contains carbon black (40 phr), the stresses increase with small deformations and the tensile strength reaches a significantly higher value.

Tensile strength can be significantly enhanced if the loading of organo-clay is increased [157]. The greatest reinforcing effect has been found for quaternary ammonium salts (methyl-tallow-bis(2-hydroxymethyl) as a modifier for MMT. Exfoliated structures and strongly intercalated structures of the organo-clay are formed. With NR, an organo-clay at 10 phr can achieve a tensile strength equivalent to a normally structured carbon black (Spheron 6400, 40 phr) [158]. In IR/organo-clay, it has been found that the interlayer spacing increases from 2 nm to 14 nm due to penetration of IR chains into the spacing. The longer the amine, the more the tensile strength is increased. All the nanocomposites displayed better polymer-filler interaction and also better ultimate properties in the SBR matrix. The predicted reinforcing factors per number of platelets per tactoids decreases sharply, as shown in Table 9.5.

**Table 9.5** Reinforcing Factor Depends on the Number of Platelets [85]

| Number of Platelets per Stack | Reinforcing Factor in the Halpin-Tsai Equation |
|---|---|
| 1 | 49.2 |
| 2 | 39.7 |
| 3 | 33.5 |
| 4 | 29.0 |
| 5 | 25.6 |

It has been concluded that stacks of platelets reduce the stiffness of nanocomposites due to a reduced aspect ratio and lower tensile stress. With increase in organo-clay loading, the tensile strength of pure NBR increases from 13 MPa to 15.8 MPa when 15 phr of organo-clay is added. The values are significantly lower than for NBR/silica, which achieves a tensile strength of 21.2 MPa for the same amount of silica. If bifunctional silane TESPT is added to NBR/organo-clay, the tensile strength value decreases from 20 MPa to 15.2 MPa at 10 phr organo-clay [159]. The tensile strength increases when an organo-clay with octadecyl-amine in EPDM is added at a rate of up to 10 phr.

The improvement in tensile strength is due to intercalation of the rubber into the galleries of the clay and strong interaction between the filler and the matrix [160].

#### 9.3.4.2.4 Effect of Carbon Nanotubes

In view of the high aspect ratio of the multi-walled carbon nanotubes, it could be assumed that the tensile strength should also assume high values. However, this expectation is not fulfilled, due to the uniform structure of the $sp^2$-hybridized carbon atoms and the absence of a heterogeneous energy distribution. However, the incorporation of multi-walled carbon nanotubes did not lead to an improvement in the tensile strength, such as Young's modulus or the stress values at medium elongation. With increase in elongation, there is a clear displacement of the rubber chains on the surface of the multi-walled carbon nanotubes and, before the end point is reached, the stress-strain behavior takes on a convex shape that indicates decreasing modulus. The tensile strength is always significantly lower than for filled rubber specimens that contain a strain-crystallizing rubber. The tensile strength determined for NR vulcanizates is significantly lower than for active carbon blacks with the same loading. The tensile strengths of multi-walled carbon nanotubes are not very high, a fact which has to do with the phase bonding between the rubber matrix and the $sp^2$ hybridization of the carbon atoms.

The quality yield of multi-walled carbon nanotubes strongly depends on the synthesis and the purification techniques and the specific growth conditions used [162]. This can be observed when comparing multi-walled carbon nanotubes of different diameter in NR vulcanizates. At the same loading (0.015 vol.%), the tensile strength of the multi-walled carbon nanotubes with an average diameter of 9 nm decreases from 10.2 MPa to 7.1 MPa when the diameter of the multi-walled carbon nanotubes doubles [115]. As the diameter increases, the specific surface area of the multi-walled carbon nanotubes decreases and therefore the high tensile strength cannot be achieved. With a purified multi-walled carbon nanotube at high concentrations (25 phr), the tensile strength in NR is 19.4 MPa while it is 23.5 MPa for carbon black N326 [162]. In multi-walled carbon nanotubes, this difference can be attributed to the presence of the $sp^2$-hybridized carbon atoms, which have a lower rubber-filler interaction than the surface energy of carbon black, which has amorphous carbon and the active corners of the micro-graphitic layers. If carbon nanotubes are used in different types of rubber, a strong influence of the rubber matrix can be found in all of them. The higher the concentration of carbon nanotubes, the more the tensile strength in the rubber matrix. A look at multi-walled carbon nanotubes in QM, NBR and HNBR reveals the difference in tensile strength [163]. This is shown in Table 9.6.

**Table 9.6** Tensile Strength of Silicone Rubber, NBR and HNBR Vulcanizates that Contain Carbon Nanotubes NC 7000 [163]

| Rubber | Carbon Nanotubes (vol.%) | Tensile Strength (MPa) |
|---|---|---|
| Silicone rubber | 0.000 | 4.5 |
| | 0.012 | 7.7 |
| | 0.030 | 9.1 |
| NBR (34% ACN) | 0.000 | 3.9 |
| | 0.005 | 4.2 |
| | 0.025 | 6.4 |
| | 0.040 | 9.3 |
| HNBR (34% ACN) | 0.000 | 5.2 |
| | 0.010 | 10.3 |
| | 0.050 | 20.1 |
| | 0.100 | 31.2 |

It is found that the increase in tensile strength in QM is about the same as in polar NBR, which speaks to good interaction in the silicone rubber with the surface of the multi-walled carbon nanotubes. With HNBR, the increase in tensile strength is very high. This is the result of the rubber-carbon nanotubes interaction, which is facilitated by the polar CN groups of HNBR. This strong interaction can have consequences in practical applications.

Multi-walled carbon nanotubes treated with nitric acid to produce -OH and -COOH functionality have higher tensile strength than NBR, because of better bonding of the $sp^2$-hybridized carbon atoms and the CN groups on the rubber chain [164].

The different contents of multi-walled carbon nanotubes in SBR show an increase in tensile strength, compared with carbon black at the same volume fraction. If multi-walled carbon nanotubes are used in a blend of NR/SBR, a systematic decrease in tensile strength with the proportion of NR domains is observed. Preferential distribution of the multi-walled carbon nanotubes in the SBR phase takes place and the increase in tensile strength is greater in the SBR phase. It has been found that the multi-walled carbon nanotubes in NR/BIIR blends have enhanced tensile strength when the blend system is vulcanized with imidazole ions [165]. A similar result has been found for single-walled carbon nanotubes (2 phr) in NR/BR/IR/SBR (40/20/10/30), where the increase was from 16.3 MPa to 17.9 MPa [166].

Diluting expensive multi-walled carbon nanotubes with cheaper reinforcing fillers, without compromising the mechanical properties of the resulting vulcanizate, would have an economic benefit, as the cost of the multi-walled carbon nanotubes would be reduced. In the so-called hybrid systems, the tensile strength increases significantly with the content of carbon nanotubes in combination with silanized precipitated silica. With a proportion of 10 phr carbon nanotubes in the silanized HD silica, the tensile strength increases from 14.2 MPa to 17.1 MPa. The advantage of these hybrid systems is therefore obvious [167]. If EPDM or EVM is filled with carbon black N550 (19 vol.%), the tensile strength is 16.8 MPa and 21.2 MPa. If carbon nanotubes (3 vol.%) are now mixed in, the value rises to 17.9 MPa and 25.8 MPa. Here, too, there is clear evidence of the superiority of a combination of a large proportion of conventional filler and a small proportion of carbon nanotubes [155].

## 9.4 Stress Relaxation and Creep

Elastomers are used in critical applications where long-term performance is essential, such as engine mounts, pipeline sealing rings, antivibration mounting and bridge bearings or earthquake mountings. In such applications elastomers are subjected to static compression over a long period of time

When a constant strain is applied to a rubber specimen, the counterforce exerted by the specimen is not constant but decreases over time. This behavior is termed "stress relaxation". The stress relaxation can be applied to uncured and to cured rubber. In the case of cured rubber, the slow rate of decay offer a rapid deformation can potentially predict better sealing pressures over time or compression set. Stress relaxation is described by the Maxwell model. It builds up a constant stress with a time delay which, if the time is long enough, corresponds to Newtonian yield stress. Conversely, when a constant load is placed on an elastomer, the material generally creeps over time while stress relaxes over time. This behavior is called "creep" and is applied to cured rubber. It can be measured under controlled compression, tension, or shear. A better description of rubber creep is the Kelvin-Voigts model [168]. Both properties depend on the molecular structure of the rubber network, the type and loading of fillers, the temperature and the amount of stress deformation or strain deformation. Test methods at ambient and high temperatures are described in international standards [169, 170].

Stress relaxation is conducted on a displacement-controlled apparatus. The test consists in monitoring the time-dependent stress resulting from a steady strain. In a creep experiment, the evolution of strain is recorded as function of time for a constant stress applied to the test specimen. The test diagrams for stress relaxation and creep are shown in Figure 9.31.

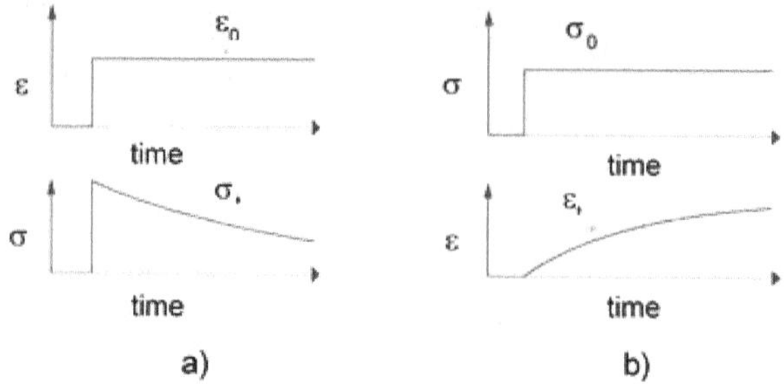


**Figure 9.31**
Scheme for: a) stress relaxation and b) creep

Stress relaxation can be studied in specimens subjected to uniaxial tension, compression or shear. The first loading mode has received much attention in the literature and is considered most appropriate for fundamental studies [171, 172]. Components used for engineering purposes, such as sealing applications, are most frequently subjected to compression [173]. International standards recommend two methods for determining the counterforce exerted by a test piece (cylinder or O-rings) [174]. The first requires compression to be applied and the counterforce to measured continuously at the test temperature. The other specifies that test samples be stored at the test temperature for long periods of time. Stress relaxation measurements are conducted intermittently at room temperature. The reaction forces exerted by the specimen are measured as a function of time.

The relaxation and creep behavior of elastomers is due to both viscoelastic processes and chemical changes in the polymer structure due to thermal and oxidative processes during the deformation period. The processes causing stress relaxation and creep may be physical or chemical in nature. Under normal conditions, both physical and chemical processes occur simultaneously. At low temperatures and over short periods of time, stress relaxation is dominated by physical processes, while at elevated temperatures and longer periods of time, the chemical processes dominate. The key factors which influence the behavior of elastomers under constant stress or constant strain are:

- vulcanization system
- crosslink density
- crosslink structure
- filler type and loading
- thermal-oxidative processes

Physical relaxation is associated with the reorientation of network strands under strain, the disentanglement and rearrangement of entanglements, and the movements of "dangling" chain ends. In filled elastomers, stress relaxation is associated with the breakdown and rearrangement of filler-filler bonds or between polymer chains and filler particles.

The stress relaxation depends on the chain mobility, the loading and type of the filler and is therefore temperature-dependent. The proximity of the operating temperature to the glass transition temperature is the dominant factor in determining physical relaxation rates. Thus, physical relaxation rates are more pronounced when chains become more mobile, as in the case of high cis-BR. The temperature dependence generally corresponds to the WLF equation [175].

The network topology, especially the crosslink density, determines the relaxation rates to a high degree. NR specimens compressed at 100 °C undergo significant changes in crosslink density and crosslink structure. A reduction in crosslink density of 25% and an increase in the mono-sulfidic crosslinks have been detected in a compressed CV-cured sample after 262 h [176]. This may be attributed to the fact that the internal strain during compression increases the rate of rearrangements of sulfur crosslinks. The SEV specimen showed a less pronounced trend.

The incorporation of carbon black into rubber generally results in an increase in relaxation rates. Filled elastomers typically relax faster than unfilled systems. As a rule of thumb, the more active the filler, the shorter is the relaxation time [177]. The effect can be partially explained by the concept of strain amplification [178]. Evidence has been produced for the existence of a distinct relaxation process which occurs within the first 60 s of relaxation at room temperature. This relaxation is attributed to the breakdown of the filler network [179]. However, filler dispersion exerts only a minor influence on stress relaxation and creep.

### 9.4.1 Service-Life Prediction from Stress Relaxation Experiments

Stress relaxation and creep experiments are used to determine the service life of vulcanized rubbers. There are two very different methods for predicting the service life. When viscoelastic phenomena dominate, the time-temperature superposition principle describes the shifting of experimentally determined isotherms along a logarithmic time or frequency axis. When chemical mechanisms dominate, the Arrhenius equation is more often employed.

Figure 9.32 shows the stress relaxation data of NR vulcanizates at different temperatures [179]. The curves show the acceleration of stress relaxation with increase in temperature. The physical relaxation process follows a logarithmic linear regression at lower temperatures and times. The chemical process is seen as an accelerated decrease in the relaxation force at higher test temperatures or extended times.

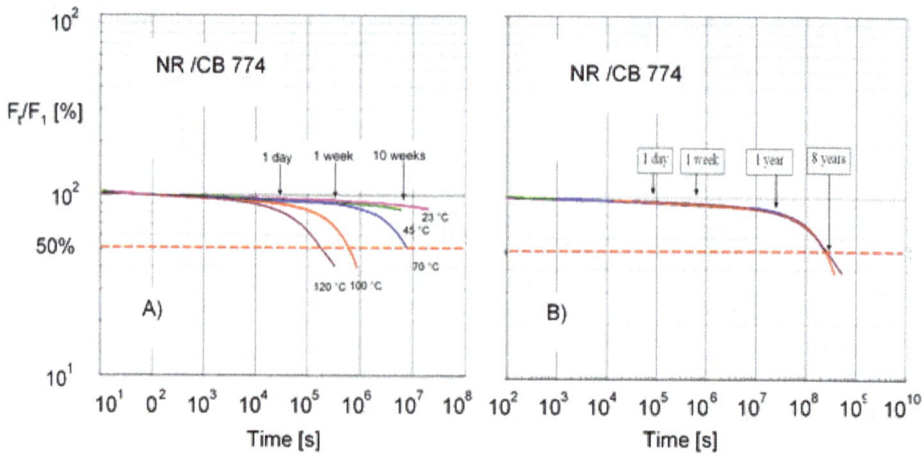


**Figure 9.32** Stress-relaxation experiments for predicting the service life of an NR/CB 774 vulcanizate [179]

The superposition is carried out by shifting each curve horizontally to overlay it on the reference curve at 23 °C, as shown in Figure 9.32. If 25 °C is selected as the reference temperature, each isotherm is shifted horizontally to overlay it with the next and to produce a master-curve. The master-curve should represent the relaxation behavior at 25 °C, leading to predictions over extended time scales. It spans decades and represents the change in the material behavior over time. As can be seen from Figure 9.31, the material suffers a loss of function ($F_t/F_0 \leq 50\%$) over a period of 8 years. The same procedure for sulfur-cured EPDM shows a prolonged service life of 12.4 years. Due to thermal-oxidative processes that occur at elevated temperatures, the service life prediction can be subject to serious errors. The phenomenon of chemiluminescence has shown that, by the start of the accelerated decrease in counterforce $F_t$, the oxidation induction time of the rubber is reached [180].

## 9.5 Barrier Properties

Permeation is the transport of gases and liquids through a crosslinked rubber until a concentration equilibrium is reached. If the amount permeated is plotted against time, a rising curve results which turns into a straight line once the steady state is reached. The incubation time is obtained from the extrapolated intersection with the abscissa, from which the diffusion coefficient is determined. The presence of solid filler helps to reduce the barrier properties of vulcanizate during the transport process of a gas or a fluid through the crosslinked rubber. In the case of iso-dimensional fillers, carbon black and precipitated silica, the barrier properties are reduced only to a small extent and at high concentrations. Barrier properties have long been used in the inner liners of passenger and truck tires, primarily using a butyl rubber grade containing a semi-reinforcing carbon black. Gas adsorption has been tracked in uniaxial deformation of carbon black-filled NR vulcanizate. With the elongation ratio the barrier properties increase with the carbon black N330 loading [181]. However, the barrier properties are reduced at a high carbon black concentration with a relative low effect.

There are disadvantages to using conventional fillers, due to the high loading and the small aspect ratio. Recently, layered silicates of high aspect ratio and high specific surface area were used to improve the barrier properties of rubber vulcanizate and reduce the diffusion through the filled rubber membrane This technique has also been used recently in the inner-liner compounds for car and truck tires. The barrier properties of rubber compounds can be significantly improved by incorporating inorganic platelets of sufficiently high aspect ratio to alter the diffusion path of the penetrating molecules [182].

A significant decrease in gas or fuel permeability is predicted and observed in practice for 2:1 layered silicates. The performance of layered silicates, compared with conventional filler, is clearly demonstrated when the barrier properties are examined. The reduction in the barrier properties in a given rubber matrix is dictated by the platelet structure and the orientation in the vulcanizate. When the platelets are oriented perpendicular to the diffusion direction, the diffusing molecules have to travel around the platelets and the diffusion path increases with the diameter of the platelets and its concentration in the rubber. This leads to a significant reduction in the permeability of the polymer.

To take full advantage of the tortuosity, the platelets can provide an increase in the barrier properties due to their orientation in the processing direction and not curled or curved. The alignment of platelets is affected by the type of processing used to form the test specimen, but also by the concentration of the filler, the molecular weight of the polymer, the crosslinking density, the compatibility between the diffusate and the rubber and not least by the temperature of the diffusion process [85]. The barrier properties of the filled polymer refer to its abilities to impede the transport of diffusant molecules through a given sheet by a diffusion process that is described by Fick's first and second laws. The diffusion process through a sheet of material is shown in Figure 9.33.

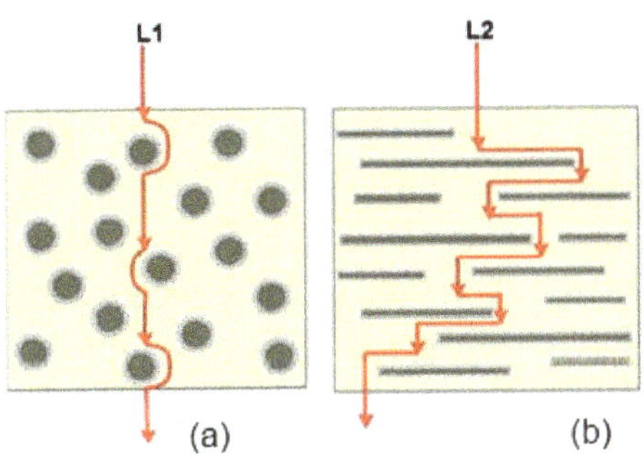


**Figure 9.33**
Schematic representation of the transport of diffusing molecules through a sheet of material: a) iso-dimensional filler and b) two-dimensional platelets

Two parameters are used to describe the permeation process, P, through a polymeric material. The solubility, S, describes the total amount of the diffusant trapped in the sample and the diffusion coefficient, D, of a diffusant molecule in the system:

$$P = DS \tag{9.20}$$

The permeation of small molecules through the rubber is facilitated by the high segmental mobility of the polymer. When a filler is included, the permeability changes with the volume fraction of the filler and the aspect ratio of the filler, which is explained by the concept of tortuous path. According to this approach, the path that a diffusing molecule must travel increases with the presence of oriented layers and an increase in the tortuosity of the diffusing length. The impermeable nature of the filler creates a long path for the diffusing molecules by directing them around the particles.

Various models have been proposed to predict the permeability of a filled rubber. The permeation model uses expressions for quadrilateral platelets and considers the volume fraction of the filler and the aspect ratio, α. Assuming an even arrangement of rectangular platelets of width L and thickness w oriented perpendicularly to the diffusion direction the tortuosity factor becomes:

$$\tau = 1 + L\Phi/2w \tag{9.21}$$

and the reduced permeation coefficient is proportional to the volume fraction of the polymer and the inversely to the tortuosity factor [183]:

$$P_f/P_0 = (1 - \Phi)/[1 + (\alpha/2)\Phi] \tag{9.22}$$

The equation predicts the relative permeation coefficient as a function of the aspect ratio and the filler volume fraction. The framework has been extended by considering the case of regular and random distribution, including disorder and size distribution effects. An expression has been developed that better describes the random distribution [184]:

$$P_f/P_0 = 1/[1 + (\alpha^2\Phi^2)/(1 - \Phi)] \tag{9.23}$$

For very low concentrations of the filler, the denominator becomes $1 + (\alpha^2\Phi^2)$. That is relatively close to another theoretical model that consider the slits between the platelets [185].

The permeation of gases ($N_2$, $O_2$, $CO_2$, $H_2$, Ne) in a variety of elastomers containing carbon black as filler has been measured [186]. Mass transport through filled elastomers is strongly influenced by the type of the filler, the rubber-filler interaction, the aspect ratio and the spatial distribution.

A layered silicate of large aspect ratio and planar orientation can increase the diffusion distance by creating a more tortuous path for diffusing gas or fluids. In recent years, the search for more efficient barrier properties has been geared in particular toward natural or synthetic layered silicates, i. e. montmorillonite, saporite, hectorite [187]. Assuming that the principles are the main reason for the excellent barrier properties of rubber/organo-clay, a more pronounced tortuous path given by an increased aspect ratio in layered silicates would be a decisive clue.

The potential to serve as permeation barriers is inherent in their crystalline structure and orientation during processing. Studies conducted in the German Rubber Institute show that the pristine montmorillonite in polar rubber is able to reduce the permeation of fuel fluid by more than 60% at a concentration of only 3 vol.%. The results are shown in Figure 9.34.

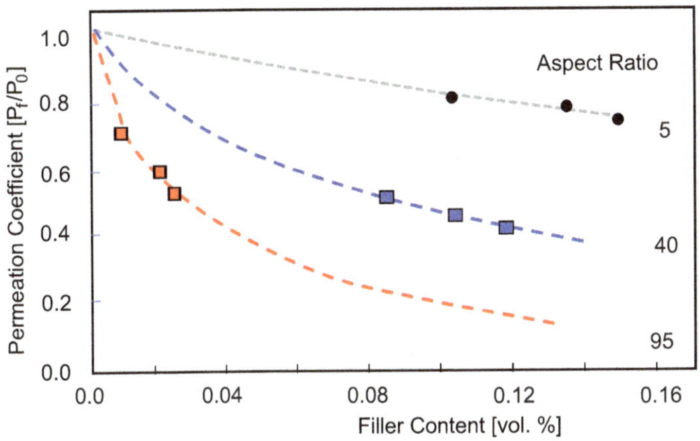


**Figure 9.34** Effect of wet-mixed layered silicates compared with mechanically mixed MMT and carbon black or silica in NBR vulcanizate

It has been observed that gas and vapor permeability of rubbers can be markedly reduced by the incorporation of organo-clays [189–191]. NR/MMT composite prepared by emulsion blending (1, 2, 3 phr) exhibited a 35%, 45% and 50% reduction in oxygen permeability [192]. Organo-clay containing didodecylmethyl-amine resulted in a 50% reduction in oxygen permeability [193]. The inner liner of a tire maintains the air pressure in the tire for an extended period of time when organo-clay is incorporated. The effect of HNBR/organo-clay on the barrier properties is a 40% reduction in the oxygen permeation ratio. The good dispersion and the strong interface for the modified MMT led to the low oxygen permeation [194].

The contribution of the two-dimensional platelets to the reduction of permeation has a significant influence on the amounts of gases and fuel fluids that are transported in vulcanizates. This results in a strong reduction in the transported amount that diffuses through a vulcanizate. Thus, this contribution should provide better environmental compatibility on the part of elastomers filled with layered silicates.

## References


[1] ASTM D 2240

[2] DIN ISO 7619-1

[3] ASTM D 1415

[4] Qui, H. C., et al., Rubber Chem. Technol., 2003, 76: 419

[5] Mix, A. W., Giacomin, A. J., J. Testing & Evaluat., 2011, 39: 1

[6] Walker, J., Proceed. Forum Mixing, TechnoBiz Pittsburgh (2014)

[7] Byers, J. T., in "Basic Elastomer Technology", Baranwal, K. C., Stephens, H. L., Eds., Rubber Division, American Chemical Society, Akron (2001) Ch. 4 B

[8] Magg, H., Lanxess Seminar on Rubber Compounding

[9] Fröhlich, J., Niedermeier, W., Educ. Symp. DKT '09, Nuremberg (2009)

[10] Uhl, C., PhD Thesis, Technical University of Dresden (2007)

[11] Beleen, J., Lanxess Course (2015]

[12] Bunawerke Hüls, Buna AP Leitfaden für die Gummiindustrie, D52/843/851117

[13] Westenberg, H., Gromes, H., Rechenbach, D., Kautsch. Gummi Kunstst., 2021, 73: 47

[14] Frick, A., Borm, M., Muhidhason, V., Kautsch. Gummi Kunstst., 2018, 20: 23

[15] Hewitt, N., in "Compounding Precipitated Silica in Elastomers", W. Andrew Publishers, Norwich, USA, (2007), p .25

[16] Wang, M.-J., Morris, M. D., Kutosvki, Y., Kautsch. Gummi Kunstst., 2008, 60: 107

[17] El-Nasar, D. E., Ward, A. A., Abd El-Massieh, S. R., Kautsch. Gummi Kunstst., 2009, 51: 437

[18] Seppä, T., PhD University of Technology Helsinki (2003)

[19] George, K. M., et al., Kautsch. Gummi Kunstst., 2006, 58: 544

[20] Wang, M.-J., et al., Kautsch. Gummi Kunstst., 2020, 72: 45

[21] Wang, Y. Z., Zhang, L. Q., Wang, Y. Q., J. Appl. Polym. Sci., 2000, 78: 1879

[22] Wu, Y. P., Zhang, L. Q., Wang, Y. Q., J. Appl. Polym. Sci., 2001, 82: 2842

[23] Klafke de Azeredo, L., Schuster, R. H., Jacobi, M. M., Kautsch. Gummi Kunstst., 2010, 62: 30

[24] Schellenberg, M., Schuster, R. H., ACS Rubber Meeting, Cleveland (2013)

[25] Usuki, A., Tukigase, N, Kato, M., Polymer, 2002, 43: 2185

[26] Zhang, H., Peng, Z, Zhang, Y., Polymer Comp. 2004, 12: 147

[27] Gatos, K. G., Sawanis, N. S., Apostolov, A. A., Macromolecule Mat. Eng., 2004, 289: 1079

[28] Zhang, W., et al., Macromol. Mater Eng., 2007, 292: 1020

[29] Zimmermann, H., Schuster, R. H., Project Report for DKG (2011)

[30] Cataldo, F., Ursini, O., Angelini, G., Fullerenes, Nanotube and Carbon Nanostructures, 2009, 17: 18

[31] Kumar, V., et al., Kautsch. Gummi, Kunstst., 2021, 71: 41

[32] Schuster, R. H., Chogule, H., Wrana, C., ACS Meeting Rubber Division, San Antonio, (2014)

[33] Lu, L., Zhai, Y., Zhang, Y., Ong, Ch., Guo, S., Appl. Surf. Sci., 2008, 255: 216

[34] Del Vecchio, R. J., in "Basic Elastomer Technology", Baranwal, K. C., Stephens, H. L., Eds., Rubber Division, American Chemical Society, Akron (2001) Ch. 8

[35] Bateman, L., et al., in "The Chemistry and Physics of Rubber-Like Substances", Bateman, L., Ed., Maclaren & Sons, London, Ch. 15

[36] ASTM D 394

[37] ASTM D 1229

[38] Morrison, N. J., Porter, M., Rubber Chem. Technol., (1984), 57, 63

[39] Obrecht, W., Postgraduate Course in Rubber Technology, University of Hanover

[40] Dannenberg, E. M., Trans. Inst. Rubber Ind., 1966, 42: T26

[41] Edwards, D. C., J. Mater. Sci., 1990, 25: 4175

[42] Flory, P. J., in "Principles of Polymer Chemistry", Cornell Univ. Press, Ithaca, New York, (1953) Ch. 11

[43] Treloar, R. G., in "The Physics of Rubber Elasticity", 3rd ed., Oxford: Oxford University Press, (1975)

[44] Gent, A. N., in "The Science and Technology of Rubber", Mark, J. E., Erman, B., Eirich, F. R., Eds., Elsevier Acad. Press, 3rd ed., (2005) Ch. 1

[45] Coran, A. Y., in "Science and Technology of Rubber", Mark, J. E., Erman, B., Eirich, F. R., Eds., Elsevier Acad. Press (1978) Ch. 7

[46] Class, J. B., Dluzneski, P. R., in "Basic Elastomer Technology", Baranwal, K., Stephens, H. L., Eds., Rubber Division (2001) Ch. 7

[47] Göritz, D., Angew. Makromol. Chem., 1992, 202/203: 309

[48] Katz, J. R., Die Naturwissenschaften, 1925, 13: 410

[49] Ikeda, Y., et al., Macromolecules, 2008, 41: 5876

[50] Dinges, K., in "Polymere Werkstoffe", Batzer, H., Ed., Georg Thieme Verlag, Stuttgart, New York (1985) Vol. 2

[51] Epstein, L. M., Smith, R. P., Trans. Soc. Rheol., 1958, 219

[52] Bueche, F., Dudek, R. J., Rubber Chem. Technol., 1963, 36: 1

[53] Ecker, R., Kautsch Gummi Kunstst., 1968, 21: 304

[54] Smith, T. L., J. Polym. Sci., 1958, 32: 99

[55] Bueche, F. J. Appl. Phys., 1955, 26: 1133

[56] Ferry, J. D., J. Amer. Chem. Soc., 1950, 72: 3746

[57] Sambrook, R. W., Rubber Chem. Technol., 1971, 44: 728

[58] Kraus, G., Fortschr. Hochpolym. Forsch., 1971, 8: 155

[59] Thomas, A. G., Wittle, G. M., Rubber Chem. Technol., 1970, 43: 222

[60] Boonstra, B. B., India Rubber World, 1949, 121: 299

[61] Harwood, I. A. C., Payne, A. R., Wittacker, R. E., J. Appl. Polym. Sci., 1970, 14: 283

[62] Rivlin, R. S., in "Rheology, Theory and Applications", Acad. Press, New York (1956) Vol. 1, Ch. 10

[63] Mooney, M., J. Appl. Phys., 1940, 11: 582

[64] Smallwood, H. M., J. Appl. Physics, 1944, 15: 758

[65] Eilers, H., Kolloid Z., 1941, 97: 313

[66] Brinkmann, H. C., J. Chem. Phys., 1952, 20: 571

[67] Kerner, E. H., Proc. Phys. Soc., Part B., 1956, 60: 802

[68] Guth, E., J. Appl. Phys., 1945, 16: 20

[69] Medalia, A. I., Rubber Chem. Technol., 1972, 45: 1171

[70] Meinecke, E. A., Taftaf, M. I., Rubber Chem. Technol., 1988, 61: 534

[71] Wolff, S., Donnet, J.-B., Rubber Chem. Technol., 1990, 63: 32

[72] O'Brien, J., Cashell, E., Wardell, G. E., Rubber Chem. Technol., 1977, 50: 744

[73] Smith, P. P. A., Van der Vegt, A. K., Kautsch. Gummi Kunstst., 1970, 23: 147

[74] Hess, W. M., Ban, L. L., McDonald, G. C., Rubber Chem. Technol., 1969, 42: 1209

[75] Kraus, G., Rubber Chem. Technol., 1978, 51: 297

[76] Medalia, A. I., J. Colloid Interface Sci., 1970, 32: 115

[77] Kraus, G., J. Polym. Sci. B, 1970, 8: 601

[78] Medalia, A. I., Rubber Chem. Technol., 1974, 47: 411

[79] Wrana, C., in "Introduction to Polymer Physics", Lanxess AG., Leverkusen (2009)

[80] Adnan, A., Sun, C. T., Mahfuz, H., Comp. Sci. Technol., 2007, 67: 348
[81] Yu, S., Yang, S., Cho, M., Polymer, 2009, 50: 945
[82] Ghanbari, A., et al., Macromolecules, 2012, 45: 572
[83] Muhr, A., Rubber Chem. Technol., 2005, 78: 351
[84] Wang, M.-J., et al., Kautsch. Gummi Kunstst., 2020, 72: 45
[85] Pavlidou, S., Papaspyrides, Progr. Polymer Sci., 2008, 33: 1119
[86] Klafke de Azeredo, L., Jacobi, M. M., Schuster, R. H., PPS Charleston
[87] Schellenberg, M., Schuster, R. H., ACS Rubber Division Meeting, Cleveland (2013)
[88] Fornes, T. D., Paul, D. R., Polymer, 2003, 44: 4993
[89] Lopez-Manchado, M. A., et al., J. Appl. Polym. Sci., 2009, 92: 3394
[90] Atieh, M. A., et al., Arabian J. Sci. Eng., 2010, 35: 51
[91] Bokobza, L., Polymer, 2007, 48: 4907
[92] Girun, N., Ahmodem, F. R., Rashid, S. A., Fullerenes, Nanotubes and Carbon Nanostructures, 2007, 15: 207
[93] Zimmermann, H., Schuster, R. H., ICC-19, Shanghai (2011)
[94] Hu, C. H., et al., Appl. Phys. Letters, 2008, 93: 033108
[95] Schuster, R. H., Chogule, H., Zimmermann, H., 180th ACS Rubber Division Meeting, Cleveland, 2013
[96] Schuster, R. H., Geisler. A., Chogule, H., Wrana, C., ACS Rubber Division Meeting, San Antonio 2014
[97] Khumar, V., Lee, D. J., Polymer, 2020, 190: 122221
[98] Schuster, R. H., Wittek, H., Chogule, H., ACS Rubber Division Meeting, Cincinnati (2011)
[99] Gerspacher, M., Wampler, W., in "Basic Elastomer Technology", Baranwall, K. C., Stephens, H. L., Eds., Rubber Division Amer. Chem. Soc., Akron (2001) Ch. 4A
[100] ASTM D 3191
[101] Görl, U., in "Rubber Technology", Hofmann, W., Gupta, H. G., Eds., Gupta Verlag, Ratingen (2001)
[102] Bryant, K. G., Bisset, D. C., 3rd Proc. Rubber Conf., London (1954)
[103] Wehmeier, A., DIK Workshop, Hanover (1998)
[104] Dannenberg, E. M., Cotton, G. R., Rev. Gen. Caoutch. Plast., 1974, 5: 343
[105] Meon, W., Blume, A., Luginsland, H.-D., Uhrland, S., in "Rubber Compounding, Chemistry and Applications", Rodgers, B., Ed., Marcel Dekker, Inc., New York, Basel (2004) Ch. 7
[106] Fröhlich, J., Lecture at Further Education in Rubber Technology, Hanover (2014)
[107] Suzuki, N., Ito, M., Yatsuyanagi, F., Polymer, 2005, 46: 193:
[108] Varghese, S., Karger-Kocsis, J., Gatos, K. D., Polymer, 2003, 44: 3977
[109] Endo, M., et. al., Am. Chem. Soc. Nano, 2011, 5: 3858
[110] Zimmermann, H., Schuster, R. H., Meeting of DKG, Hanover, (2011)
[111] Nah, Ch., Lim, J. Y., Cho, B. H., Hong, Ch. K., Gent, A. N., J. Appl. Polym. Sci., 2010, 118: 1574
[112] Kolodziej, M. Bokobza, L., Bruneel, J.-L., Comp. Interface, 2007, 14: 215
[113] Halpin, J. C., J. Comp. Mater., 1969, 3: 732
[114] Chen, X.-H., Song, H.-H., New Carbon Mater., 2004, 19: 214
[115] Chogule, H., Wittek, H., Schuster, R. H., PMA, Bratislava (2013)
[116] Cho, B. M., Kim, G.-H., J. Appl. Polym. Sci., 2020, 116: 555
[117] Schuster, R. H., Wrana, C., ACS Rubber Division Meeting San Antonio (2016)
[118] Zimmermann, H., Schuster, R. H., Kautsch. Gummi Kunstst., 2011, 63: 40
[119] Bokobza, L., Rahmani, M., Belin, C., Brunell, J.-L., el Bounia, N. E., J. Polym. Sci., Part B, Polym. Phys., 2008, 46: 1939

[120] Cataldo, F., Orsini, O., Angelini, G., Fullerenes and Carbon Nanotubes, 2009, 17: 18
[121] Sen, R., et al., Nano Letters, 2004, 4: 459
[122] Cho, J. W., et al., Macromolecules Rapid Comm., 2005, 26: 412
[123] Bounasse, H., Corriere, Z., Ann. Fac. Sci., Toulouse, 1903, 5: 257
[124] Mullins, L., J. Rubber Res. 1947, 16: 275
[125] Mullins, L., J. Phys. Colloid Chem., 1950, 54: 239
[126] Haarwood, J. A. C., Mullins, L., Payne, A. R., J. Appl. Polym. Sci., 1965, 9: 2993
[127] Mullins, L., Rubber Chem. Technol., 1969, 42: 339
[128] Boonstra, B. B., J. Appl. Polym. Sci., 1967, 11: 389
[129] Fukahori, Y., in "Current Topics in Elastomer Research", CRC Press (2008) Ch. 18
[130] Roland, C. M., Rubber Chem. Technol., 2006, 79: 429
[131] Klüppel, M., Adv. Polym. Sci., 2003, 164: 1
[132] Fukahori, Y., Rubber Chem. Technol., 2003, 76: 548
[133] Fukahori, Y., J. Appl. Polym. Sci., 2005, 95: 60
[134] Mullins, L., Tobin, N. R., Trans. Inst. Rubber Ind., 1966, 33: 1
[135] Dannenberg, E. M., Brennan, J. J., Rubber Chem. Technol., 1966, 39: 592
[136] Li, F., Wang, M.-J., Schuster, R. H., Proc. Intern. Seminar on Elastomers, (ISE), Quingdao (2016)
[137] Wang, M.-J., et al., Kautsch. Gummi Kunstst., 2020, 11/12, 72: 49
[138] Schuster, R. H., Fall Rubber Colloquium Hanover (2018)
[139] Lu, L., et al., Appl. Surf. Sci., 2008, 255: 2162
[140] Haarwood, J. A. C., Payne, A. R., J. Appl. Polym. Sci., 1968, 12: 889
[141] Hamed, G. R., Hatfield, S., RCT., 1989, 62: 143
[142] Studebaker, M. L., in "Reinforcement of Elastomers", Kraus, G., Ed., Interscience Publ. (1965) Ch. 2
[143] Gent, A., in "The Science and Technology of Rubber", Mark, J. E., Erman, B., Eirich, F. R., Eds., Elsevier Acad. Press, 3rd ed., (2005) Ch. 1
[144] Eirich, F. R., Colloques Internationaux du CNRS (1975)
[145] Schuster, R. H., in "Mixing of Rubber Compounds", (A. Limper, Ed.) Hanser Publ. (2012) Ch. 6
[146] Gent, A. N., Pulford, C. T. P., J. Mater. Sci., 1984, 19: 3612
[147] Peterseim, V., Schaper, J., Schaube, J., Schuster, R. H., Fall Rubber Colloqu. (1994)
[148] Ecker, R., Kautsch. Gummi Kunstst., 1968, 21: 99
[149] Bunawerke Hüls GmbH. Broshure 1136 (1986)
[150] Westenberg, H., Gromes, H., Rechenbach, P., Kautsch. Gummi Kunstst., 2021, 73: 47
[151] Smith, T. L., J. Polym. Sci., A1, 1963, 35: 27
[152] Smith, T. L., Polym. Eng. Sci., 1977, 17: 129
[153] Ziegler, J., Schuster, R. H., Kautsch. Gummi Kunstst., 2003, 56: 159
[154] El-Naskar, Ward, A. A., Abd El-Massieh, S. L., Kautsch. Gummi Kunstst., 2009, 51: 434
[155] Zaformehrashian, R., et al., Eur. J. Chem., (2012) 9, 1102
[156] Sun, Y., Cheng, Z., Zhang, L.- Q, Liang, H., Li., Ch., J. Appl. Polym. Sci., 2020, 137: 49526
[157] Finnigan, B., et al., Int. J. Nanotechnol., 2007, 4: 496
[158] Arroyo, M., Lopez-Manchado, M. A., Herrero, B., Polymer, 2003, 44: 2447
[159] Fritsche, J., Klüppel, M., J. Phys. Condens. Mater., 2011, 79: 429
[160] Gatos, K. G., Karger-Kocsis, J., Polymer, 46: 3069
[161] Shaeffer, M. S. P., Sandler, J., in "Processing and Properties of Nanocomposites", Advani, S., Ed., World Science Ltd., Ch. 1

[162] Zhang, W., et al., Macromol. Mater. Eng., 2007, 292: 1020

[163] Schuster, R. H., Report to InnoCNT (2012)

[164] Perez, L. D., et al., Polym. Eng. Sci., 2009, 49: 866

[165] Le, H. H., et al., eXPRESS Polym. Letters, 2017, 11: 230

[166] Bacosova, D., Bakosova, A., Polymers, 2022, 14: 3039

[167] Fritsche, J., PhD Thesis, University of Hanover (2009)

[168] Tobolski, A. V., Rubber Chem. Technol., 1957, 30: 427

[169] ISO 815

[170] ISO 1653

[171] Dunn, J. R., et al., Rubber Chem. Technol., 1960, 33: 423

[172] Mercurio, A., Tobolski, A. V., Rubber Chem. Technol., 1960, 33: 72

[173] ISO 3384

[174] Williams, M. L., Landel, R. F., Ferry, J. D., J. Amer. Chem. Soc., 1955, 77: 3071

[175] Ronan, S., PhD Thesis Dublin Institute of Technology (2009)

[176] Bornstein, D., Pazur, R. J., Polymer Testing, (2020), 88, 106566

[177] Mullins, L., Tobin, N. R., Proceed. 3rd Rubb. Technol. Conf. London (1954)

[178] Andrews, E. H., Rubber Chem. Technol., 1963, 36: 777

[179] McKenzie, C. I., Scanlan, J., Polymer, 1984, 25: 559

[180] Ronan, S., Santoso, M., Schuster, R. H., Proceed. Intern. Seminar on Elastomers, Freiburg (2007)

[181] Reichert, W. F., Göritz, D., Rubber Chem. Technol., 1993, 66: 1

[182] Paul, D., Robeson, L. M., Polymer, 2008, 49: 3187

[183] Nielsen, L. E., J. Macromol. Sci., A1, 1967: 92

[184] Frederickson, G. H., Bicerano, J., J. Chem. Phys., 1999, 110: 2181

[185] Cussler, E. L., et al., J. Membr. Sci., 1988, 38: 161

[186] Solomon, G., van Amerongen, G. J., J. Polym. Sci., 1947, 2: 355

[187] Okamoto, M, in "Polymer/Layered Silicate Nanocomposites", Rapra Review Report, B 163, Rapra Technology (2003)

[188] Yang, X., Giese, U., Schuster, R. H., Kautsch. Gummi Kunstst. 2008, 60: 38

[189] Sinha Ray, S., Okamoto, M., Progr. Polym. Sci., 2003, 28: 1539

[190] Takahashi, S., et al., Polymer, 2006, 47: 3083

[191] Paul, D. R., Robeson, L. M., Polym. Nanotech.: Nanocomp. Polym. 2008, 49: 3187

[192] Donnet, J.-B., Custodero, E., in "The Science and Technology of Rubber", Mark, J. E., Erman, B., Eirich, F. R., Eds., Academic Press, 3rd ed. (2005), p. 367

[193] Wu, Y., Jia, Q., Yu, D., Zhang, L.-Q., Polym. Test., 2004, 23: 903

[194] Gatos, K. G., et al., Macromolecular Rapid Communication, 2005, 26: 915

# 10 Dynamic-Mechanical Properties

A vast array of rubber parts is subjected to periodic cycles of loading and unloading, such as tires, belts, engine mounts and others. Dynamic deformation demands high stability and long-lasting functionality from the elastomers. For the industrial use of elastomers, evaluations of the dynamic-mechanical properties are important for understanding end-use performance. Dynamic tests are used to characterize the deformation behavior of elastomers under periodically changing loads. The question of how elastomers respond to repeated cyclic stress has received much attention in the scientific literature. The term dynamic-mechanical properties refers to the filled elastomers subjected to periodically acting forces. The forces cause the material to deform, but do not cause any significant damage (e. g. cracks).

The cyclical deformations of elastomers frequently occurring in practice are unordered sequences of pulses of different deformation amplitude, followed by relaxation times of varying duration. Thus, elastomers are rarely exposed to uniform cyclical stress. However, both types of periodic loading are subsumed under dynamic-mechanical properties. Harmonic sinusoidal deformations of the material are most often used for test purposes. This procedure is based on a simpler experimental design, but has the disadvantage that such sinusoidal deformations rarely occur in reality, which means that the transfer of laboratory tests to real conditions requires a good knowledge of physics. The parts of the dynamic tests that are carried out under pulses with different slopes and times are much closer to the practical application and require a much higher measurement effort.

Reinforcing fillers change significantly the dynamic-mechanical properties of elastomers. The special viscoelasticity of percolated filler networks introduces additional complexities into the dynamic behavior of rubber vulcanizates, so that a qualitative limit of the properties can be observed. In the multiphase system, filler loading, particle size and structure, filler surface activity and, in particular, the degree of dispersion are controlled. In this context, the type of the rubber and its state of cure have an additional influence. A special feature of filled elastomers is the pronounced strain-amplitude dependence, which is of no significance for unfilled elastomers. However, a detailed description of the viscoelastic behavior of unfilled systems under dynamic loading provides a basis for quantifying the influence of reinforcing fillers in elastomers and for a better understanding of reinforcing mechanisms. Theoretical considerations deal on a mathematically demanding level with the influence of a periodically applied sinusoidal stress on the deformation that occurs as a function of time or frequency of use for crosslinked polymers of different compositions. If the following presentation of the theory on the dynamic properties is not detailed enough, the reader should turn to textbooks on this topic [1–5].

## 10.1 Viscoelasticity of Elastomers

The dynamic-mechanical properties are dependent on temperature, strain amplitude and frequency. The presence of a filler has a considerable influence on the storage modulus and loss modulus. The dynamic properties of a viscoelastic elastomer lie between a Hookean elastic solid and a purely viscous fluid. These materials exhibit elements of both types of response, depending on the time scale of the deformation. If the deformation acts for a prolonged time, flow and permanent deformation similar to those in viscous fluids occur while rapid deformation elicits an elastic response, characteristic of elastic solids. The course of a time- and frequency-dependent viscoelastic function of stress and strain is shown in Figure 10.1.

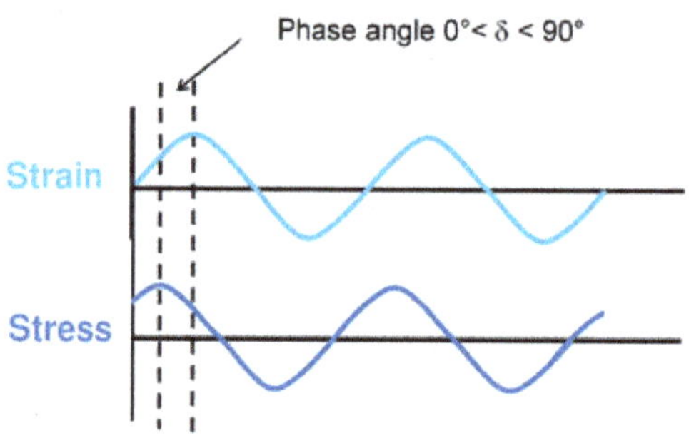


**Figure 10.1** Applied stress and resulting strain for a viscoelastic material

Repeating the deformation may lead to a change in the dynamic properties due to a rearrangement between rubber and filler. To understand the dynamic behavior of rubber, it is useful to first examine the behavior of an ideal elastic material and an ideal viscous fluid. An ideal elastic material obeys Hooke's law, according to which the stress applied to the body corresponds to a proportional deformation. A sinusoidal shear stress, $\tau(t)$, corresponds to a simultaneous sinusoidal shear deformation, $\varepsilon(t)$:

$$\tau(t) = G\,\varepsilon(t) \tag{10.1}$$

Thus, ideal solids react instantaneously to stress, such that there is no phase difference between the sinusoidal excitation and the deformation. The shear modulus, G, is defined by the ratio of the amplitude of stress and deformation:

$$G = \tau(t)\,/\varepsilon(t) \tag{10.2}$$

The mechanical energy stored in the first half-cycle is completely released when the load is relieved in the second half-cycle. The energy stored during the deformation is completely converted into kinetic energy when the ideal solid is relieved. The energy dissipated during a deformation cycle is therefore zero. The ratio of deformation to force is constant. For an ideal fluid, which behaves like a dashpot, the proportionality between shear stress and shear rate is given by:

$$\tau(t) = \eta\,\dot{\varepsilon}(t) \tag{10.3}$$

When the shear stress reaches its maximum value, viscous deformation begins and, when the shear stress is zero, the deformation reaches its maximum value. Thus, between stress and deformation is a phase difference is 90°. The energy input is completely and irreversibly converted into heat by internal friction during the entire deformation cycle.

When a viscoelastic material is subjected to a simultaneously varying stress, a steady state is reached in which the resulting strain is also sinusoidal and has the same angular frequency, but is retarded in phase by an angle, δ, which has values between 0° and 90°. For any viscoelastic body, only part of the energy applied during a deformation cycle is stored, the rest is irreversibly converted into heat.

While the mechanisms involved in periodic deformation are complex, it is useful to describe the behavior of a viscoelastic system using a combination of elastic springs and viscous dampers [1–3]. A summary of this topic is given below. The deformation strain and the stress are functions of frequency and time:

$$\gamma = \gamma_0 \sin(\omega t) \tag{10.4}$$

and

$$\sigma = \sigma_0 \sin(\omega t + \delta) \tag{10.5}$$

where $\gamma_0$ and $\sigma_0$ is the maximum amplitude of the strain and the stress, t is the time, and δ is the phase angle between the stress and strain

Stress can be divided further into two components, one in phase and the other out of phase:

$$\sigma = \sigma_0 \sin(\omega t)\cos\delta + \sigma_0 \cos(\omega t)\sin\delta \tag{10.6}$$

Therefore, the dynamic stress-strain behavior of an elastomer can be expressed by the storage modulus, G′, which is in phase with the strain and by the loss modulus, G″, which is out of phase:

$$\sigma = \gamma_0 G' \sin(\omega t) + \gamma_0 G'' \cos(\omega t) \tag{10.7}$$

where $G' = (\sigma_0/\gamma_0) \cos \delta$ and $G'' = (\sigma_0/\gamma_0) \sin \delta$

The storage modulus, G′, and the loss modulus, G″, define the complex modulus, G*:

$$G^* = (\sigma_0/\gamma_0) = G' + iG'' \tag{10.8}$$

The complex modulus, G*, is represented as a two-dimensional vector in a right-angled triangle.

Vulcanizates of different stiffness will have smaller or larger complex moduli and vulcanizates with different levels of energy storage or, conversely, energy dissipation will have different phase angles, δ. Consequently, the storage modulus, G′, and the loss modulus, G″, are expressed as follows:

$$G' = G^* \cos \delta \text{ and } G'' = G^* \sin \delta \tag{10.9}$$

The storage modulus, G′, is proportional to the stress component and in phase with the strain applied during a deformation cycle. The loss modulus, G″, is proportional to the loss of mechanical energy and is out of phase with the strain. A useful function to describe the ratio of energy dissipated to the maximum energy stored during a cycle is the loss factor, tan δ, which is defined as:

$$\tan \delta = G''/G' \tag{10.10}$$

The meaning of tan δ is to describe the viscous contribution normalized to the stiffness of the material. The energy loss during one cycle at strain is given by the expression:

$$\Delta E = \int \sigma d\gamma = \pi \gamma^2 G'' = \pi \sigma_0 \gamma_0 \sin \delta \frac{-b \pm \sqrt{b^2 - 4ac}}{2a} \tag{10.11}$$

One can use the loss compliance, J″, which is defined as $G''/G^{*2}$ or $G''/G''^2 + G'^2$. In this case, the energy loss during one cycle of strain is given by:

$$\Delta E = \omega \sigma_0{}^2 G''/G^{*2} = \omega \sigma_0{}^2 G''/\left(G''^2 + G'^2\right) \tag{10.12}$$

Depending on whether $\gamma_0$, $\sigma_0$ or $\gamma_0\sigma_0$ is kept constant during the deformation, the energy loss is proportional to G″, J″ or tan δ, respectively. It has been recognized that, depending upon the molecular motion of the vulcanized rubber, the dynamic hysteresis of unfilled rubber shows a strong temperature- time or temperature-frequency dependence.

Dynamic-mechanical properties describe the viscoelastic response of crosslinked rubbers to periodic or cyclical, often sinusoidal, deformations at small strain amplitudes ($\varepsilon = 0.01$ to 10%) over a large range of deformation frequencies ($\omega$) and temperatures. At small strain amplitudes provided by the dynamic tests, the moduli exhibit linear-viscoelastic behavior, i. e. the applied stress is proportional to the strain.

The simplest model is Maxwell's dashpot [6], which consists of a single spring attached in series to a damping cylinder. If the time in which the mechanical load is applied is much shorter than the relaxation time, then the elastic properties of the spring dominate the Maxwell element. Conversely, if the mechanical load is much longer than the relaxation time, the Maxwell element behaves like an ideal fluid. Both elastic and viscous behavior can be described between these two extremes. The Kelvin-Voigt model is based on a parallel connection of spring and dashpot. The total stress consists of the sum of the stresses acting on the elements, whereas the deformations are the same for each element and correspond to the total deformation. If the duration of the loading is much longer than the relaxation time, then the spring dominates the mechanical response [7]. Both mathematical models are shown in Figure 10.2.

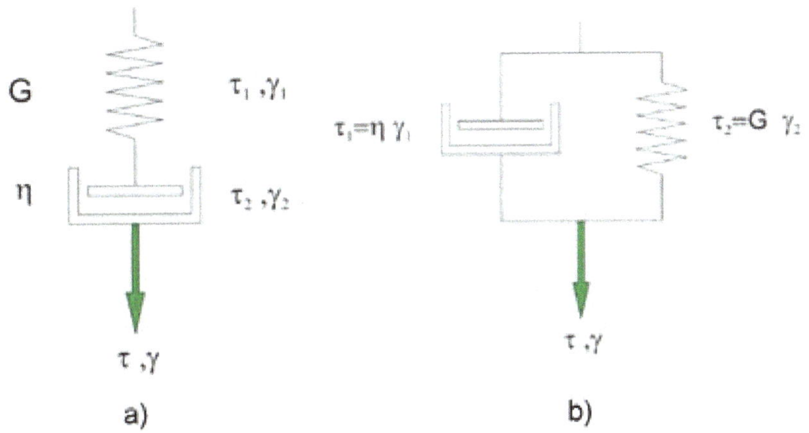


**Figure 10.2** Mathematical models for describing viscoelastic behavior: a) Maxwell model and b) Kelvin-Voigt model

For a quantitative description of the frequency-dependent dynamic-mechanical behavior of real systems, several elements or combinations of modified elements must be used. A response more closely resembling that of a real polymer is obtained by adding a second spring in parallel with a Maxwell unit. This model is known as the standard linear solid [8].

Most dynamic-mechanical test devices operate on the basis of forced, non-resonant oscillations in different deformation modes, such as torsion, extension, compression, simple shear and double shear. The cyclical deformation modes are usually modeled by a sinusoidal curve. Forced oscillation methods are preferred when temperature dependence, strain-amplitude dependence and frequency dependence are to be investigated. To investigate the influence of one of these variables, the other two have to be kept constant. Recalling that the response of elastomers to a deforming force is a function of the level of strain, the temperature and the rate at which the force is applied, it follows that $G^*$, $G'$ and $G''$ will have different values under different conditions.

In dynamic-mechanical tests, the complex modulus, $G^*$, and the time delay between the applied force and the strain the phase angle, $\delta$, are measured. Using fast Fourier transformation, the dynamic moduli, $G'$ and $G''$, are calculated. In addition to advances made in the precise execution of the movement modes, electronic data acquisition and data processing are contributing to the measuring quality of modern test machines. The inertia of instrument parts limits the frequency range of the instrument. Servo-hydraulic drives are used to enlarge the frequency range up to 1 kHz. However, the strain amplitude must then be reduced correspondingly. In addition, the response time of force transducers is also a limiting factor for the use of high frequencies.

Measurements with varying temperatures are carried out at constant strain amplitude and constant frequency with a small deformation amplitude and a constant temperature gradient (1 K/min or 2 K/min) under precise temperature control of the specimen. Any conduction of heat toward or away from the sample through metal parts (clamps, mounts, etc.) has to be reduced as much as possible. The larger the sample, the smaller the temperature gradient the heating rate should be. Heating up too quickly leads to inhomogeneous heating of the specimen and thus to an erroneous measurement result. Experimentally, consideration must be given to temperature accuracy and thermal control of the instrument. A dynamic-mechanical spectrometer needs to be calibrated a few times a year.

In contrast to unfilled elastomers, in filled elastomers the two moduli depend on the deformation amplitude. Strain-amplitude sweeps are recorded at constant temperature and constant low deformation frequency (1 to 10 Hz) of the specimen. The applied strain-amplitude sweeps are small in order to keep the measurement within the linear viscoelastic region. While the deformation amplitude plays a subordinate role in the testing of unfilled elastomers, there is a practical requirement to increase the deformation amplitude to more than 10% for filled vulcanizates. The test is important for characterizing the energy dissipation or the dynamic hysteresis under progressively increasing dynamic strain. The test provides information about filler microdispersion.

Frequency and temperature have an opposing effect on the viscoelastic properties. An increase in temperature produces the same change as a decrease in frequency. The frequency dependence is tested at constant temperature and constant low strain amplitude. In most commercial instruments, the frequency ranges between $10^{-2}$ Hz and 50 Hz. To achieve frequencies of up to $10^3$ Hz, machines with servo-hydraulic drives are required. The frequency isotherms measured at increasing temperatures are manipulated to extend the range of analysis to master-curves that cover a wide frequency range and allow estimation of rolling resistance in the plateau region and skid properties at high frequencies.

In order to better utilize the experimental parameters for a dynamic-mechanical investigation, it is necessary to distinguish between carrying out the measurement under constant deformation, constant stress or constant energy. Depending on the choice of test, it is possible to select the important characteristics of applied force, elongation of the material or energy loss. The differences between the test methods are shown in Figure 10.3 [4].

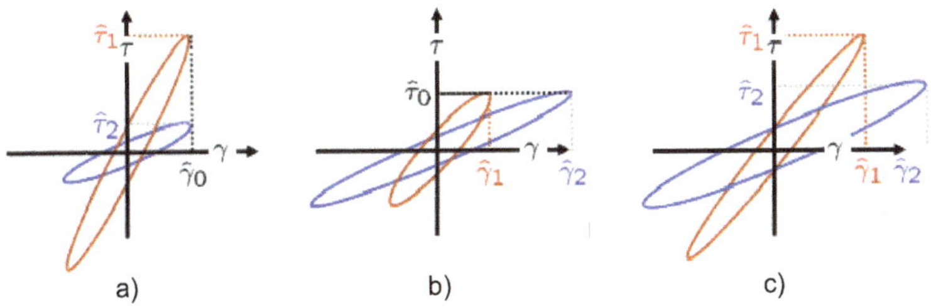


**Figure 10.3** Three routes to choosing the experimental parameters for dynamic testing: a) constant deformation, b) constant stress and c) constant energy [4]

To select the right dynamic test, the test at constant deformation that gives the ratio of the loss moduli of the two materials, $G''_1/G''_2$, is chosen. The result can be used to select the right vulcanizate for a rubber belt. Controlled stress situations use an alternate form of the modulus called compliance. This alternate form allows the data to follow a more linear trend. In the case of constant stress, the result must use the ratio of the loss compliance, $J''_1/J''_2$, where $J'' = G''/(G''^2 + G'^2)$. This test result indicates the best material for engine mounts. At constant energy, the ratio of tan δ is has the highest elastic energy after the test. This type of measurement is required for comparing the elastic energy in a rebound resilience test of two different vulcanizates.

It can be seen that various dynamic-mechanical test processes lead to a given application and the corresponding test parameters that are suitable for an application. It should be noted that the measured energy loss depends not only on the mechanical properties of the vulcanizate, but also on the mode of deformation. The energy loss for an experiment performed at constant stress and constant strain is given by the following equations [9]:

$$\text{constant stress: } E_{loss} = \pi\sigma_0{}^2 \left[G''/\left(G'^2 + G''^2\right)\right] \tag{10.13}$$

$$\text{constant strain: } E_{loss} = \pi\varepsilon_0{}^2 G'' \tag{10.14}$$

$$\text{constant energy: } E_{loss} = \pi\sigma_0{}^2 G''/G' \tag{10.15}$$

where $\sigma_0$ and $\varepsilon_0$ are the respective amplitudes of the stress and strain

As a consequence, elastomers of different stiffness deliver different energy losses when deformed dynamically at constant stress and constant strain.

Measurements of the ultrasound velocity and attenuation factor, α, provide information at high frequencies for specimens in the order of 10 mm. Excitation frequencies up to 0.5 MHz can be achieved with ultrasonic waves. Commercial instruments that use ultrasound wave propagation for performing dynamic-mechanical tests are not available on the market. An ultrasonic spectrometer uses two similar transducers facing each other. Between the transducers is a specimen-holder wheel, driven by a stepper motor working on the revolver principle, and providing space for six specimens of the same material. The specimens have different thicknesses to reduce transmission errors. One space in the holder wheel is left empty to serve as a reference for the calculation of the attenuation factor, α. The position of the holder wheel ensures that the ultrasonic wave propagates perpendicularly to the specimen. Dynamic-mechanical analysis (DMA) is an important method for characterizing the viscoelastic properties of rubber vulcanizates. General descriptions of DMA methods can be found in international standards [10–12].

## 10.2 Rebound Resilience

A useful method for determining the mechanical energy that remains in an elastomer utilizes the rebound resilience. The simplest and yet a very useful dynamic way of evaluating the magnitude of the energy dissipation or the loss of elasticity during a cyclical deformation is the rebound resilience test. Rebound resilience is defined as the ratio of recovered energy to expended energy. There are two methods for determining the rebound resilience:

- falling metal bodies as indenters
- hemispherical indenters attached to a pendulum

The rebound resilience is calculated from the rebound height for vertically swinging pendulums and from the rebound angle for horizontally swinging pendulums. A simple pendulum test involves a falling ball to measure the energy from initial potential energy. When the pendulum method is used, the rebound resilience corresponds to the ratio of rebound height to the initial height, as shown in Figure 10.4.

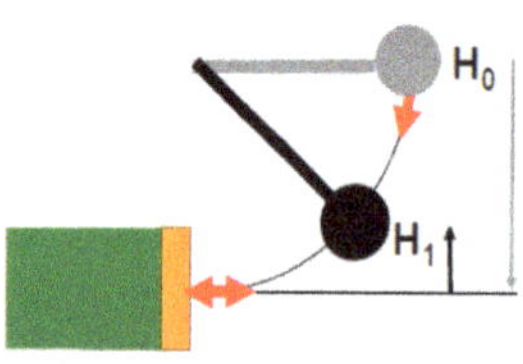


**Figure 10.4**
Schematic representation of the rebound resilience test

The method is standardized [13–15]. A good correlation between the logarithm of rebound resilience and tan δ has been reported [16]. The measured rebound resilience, R, is inversely proportional to the hysteresis loss of an elastomer measured at a frequency of approx. 120 Hz:

$$\tan\delta = -\ln R/\pi \tag{10.16}$$

where R is the measured rebound

The test allows the energy loss to be measured at constant energy. The rebound resilience test is the only one that provides direct information about the mechanical hysteresis factor, tan δ, measured at constant energy. Unfortunately, the test delivers this information only for a given frequency and strain amplitude if the impact is regarded as one half-cycle of dynamic steady oscillation. The measurement is usually performed at ambient temperature but, with some effort, rebound tests can be performed over a large range of temperatures [17].

## 10.2.1 Rebound Resilience in Unfilled Elastomers

The elasticity of unfilled elastomers depends on the mobility of the polymer chains, which is given by the chemical constitution and is also influenced by the crosslink density, the crosslink structure and the temperature at which the test is conducted.

Highly elastic compounds measured at room temperature exhibit rebound elasticity of 60 to 75% or higher, intermediate elastic compounds have values of 40 to 60% and low elasticity compounds that enter into stronger intermolecular interactions have a lower rebound elasticity of 10 to 40% [18]. An overview of rubber elasticity as a function of test temperature is given in Figure 10.5 [19].

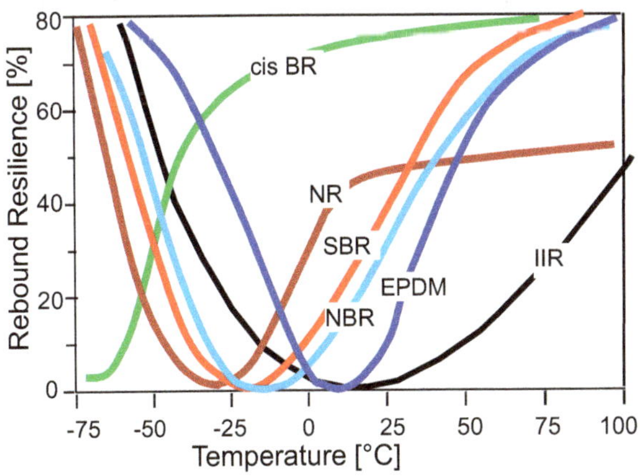


**Figure 10.5** Rebound resilience for different rubbers as function of temperature [9]

As the glass transition is approached, the rebound values decrease progressively and reach the lowest elasticity and the highest energy loss at the glass transition. The position of the glass transition on the temperature axis depends on the chemical nature of the polymer, the amount of side groups in the chain backbone, the existence of polar chain segments and the crosslinking density.

With increase in temperature, the elasticity of rubber increases steadily and the losses due to molecular friction processes decrease as a result of the increasing free volume in the system [20]. The highest elasticity in the group of non-polar rubbers is demonstrated by cis-BR. The higher the content of cis-1,4-butadiene units, the lower is the glass transition temperature and the lower is the energy loss measured in the rebound test. An increased content in 1,2-butadiene units results in increased intermolecular friction and therefore a proportional increase in $T_g$. The elasticity measured at constant temperature decreases linearly with the content of 1,2-units. NR and IR possess in each chain segment a methyl side-group that provides intermolecular friction and therefore reduces chain mobility. More pronounced is this structure-property relationship in the SBRs. With increase in styrene and vinyl content, which can be adjusted via the polymerization parameters during anionic polymerization, the elasticity of the vulcanizate progressively decreases and its ability to dissipate mechanical energy increases. Unlike the other rubbers, IIR exhibit a broad $T_g$ range, with very low elasticity and very large energy dissipation. This behavior is attributed to the special molecular structure of IIR, which is responsible for the low free volume [21]. The broad damping curve can be imagined as the result of a mixture of rubber chains with different glass transition temperatures. The IIR chains can obviously assume a variety of different motion states. The elastic behavior of the diene rubbers is only achieved at temperatures above 80 °C. At elevated temperatures, where NR starts to undergo thermo-oxidative oxidation, the elasticity of IIR is superior to that of NR. Due to strong intermolecular interactions, polar rubbers, NBR, HNBR and ACM exhibit considerable hindered chain mobility that gives rise to a higher $T_g$ value and a larger contribution by dissipative processes in the rebound test. The temperature range over which the elastomer is in service has to be known precisely and the test conditions should be selected to closely match the conditions of use. Tests performed at room temperature have a limited meaning if the vulcanizate is put into service at elevated temperatures.

### 10.2.2 Rebound Resilience in Filled Elastomers

The interpretation of the rebound resilience for carbon black vulcanizates considers the intensity of the rubber filler-interaction at the interface between the carbon black surface and the rubber matrix. Due to the increasing number of favorable adsorption sites on the carbon black surface, the freely movable chain segments are not as much, which limits the elasticity of filled rubber compounds. In addition, due to the rather large deformation amplitude, the rubber-filler interface is stressed, which can lead to molecular slippage and thus to a further dissipative process. Consistently, measured at constant energy the elasticity decreases because the rubber-filler interaction and the dissipated energy increase with filler loading.

#### 10.2.2.1 Effect of Carbon Black

If E-SBR is mixed with a larger number of carbon blacks, the vulcanizates show an increase in tan $\delta$ with the specific surface area in rebound tests, which decreases somewhat with the specific surface area [22]. When applied to filled vulcanizates at constant temperature, the rebound resilience test reveals that elasticity decreases linearly with the filler volume fraction or with the hardness of the elastomer. Figure 10.6, left, shows the decrease in the elasticity as a function of carbon black content and specific surface area and Figure 10.6, right, shows the influence of carbon black structure [23].

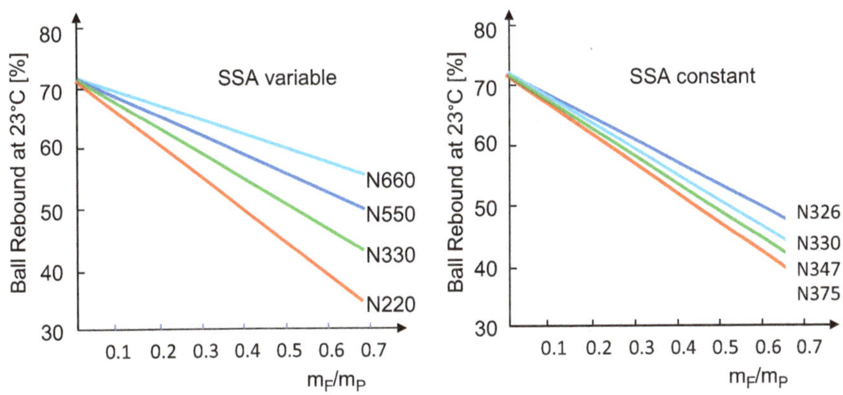


**Figure 10.6** Influence of filler loading, surface area and structure of aggregates on ball rebound of rubber vulcanizates (according to [23])

In the case of a carbon black-filled SBR vulcanizate, it has been found that the elasticity of the filled network increases when the inter-aggregate distance of the aggregates increases [24]. When SBR is filled with different furnace blacks, an increase in effective volume of the carbon black was shown to lead to a rebound resilience mastercurve [25].

A plot of rebound resilience vs. dispersion index of a SBR1712/BR tread compound filled with 50 phr carbon black N330s, shown in Figure 10.7, shows that the elasticity of the vulcanizate increases non-linearly with the macro-dispersion index [26].

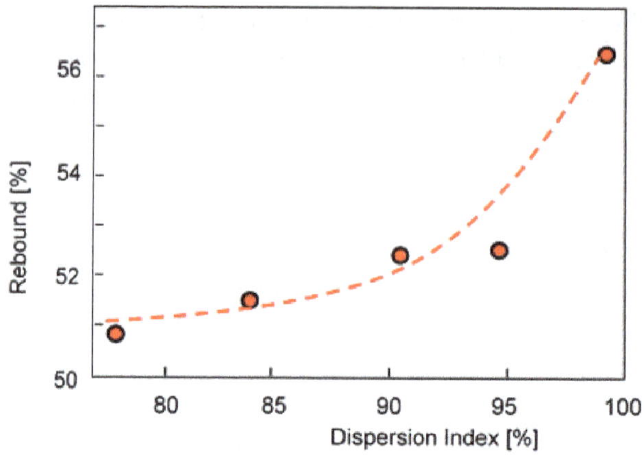


Figure 10.7 Influence of filler macro-dispersion on rebound of SBR 1712/BR vulcanizates filled with 50 phr of carbon black N330 (according to [26])

The sharp increase at the macro-dispersion index of 95% indicates the importance of filler micro-dispersion for the elasticity of vulcanizates. The fact that filler aggregates are surrounded by thin polymer films reduces friction in the composite and increases the elastic response of the material. Thus, the damping properties of vulcanizates are reduced by efficient mixing and better micro-dispersion of the filler is achieved in the vulcanizate. The rebound resilience of NR/cis-BR (50/50) blends containing different carbon black grades varies more widely. Carbon black N330 increases with the dispersion index of the compound. The lower structured carbon black N326 has the lowest elasticity, due to the very low dispersion kinetics, and poor dispersion, due to the higher number of neighbors in the agglomerate. The special case of carbon black N110 shows that a maximum of elasticity can be attributed to the agglomeration of the aggregates and the state of percolation in the vulcanizate at low loadings [26]. The addition of plasticizer in filled compounds has an influence on the elasticity. For EPDM vulcanizates filled with variable loadings of carbon black N539, the elasticity at room temperature decreases slightly with the content of paraffin oil. At high filler loadings (> 100 phr), the effect of the oil is small [27]. Therefore, the elasticity is much lower and increases less with temperature at higher filler loading.

### 10.2.2.2 Effect of Precipitated Silica

In contrast to carbon black, the elastomers filled with precipitated silica show a higher elasticity at a higher specific surface area, which is characterized by low interactions with the rubber chains and promotes the agglomeration of the aggregates. Interaction with the rubber is greatly improved by the silanization, so that the rebound resilience returns higher values [28]. Rebound resilience testing on S-SBR containing silanized silica shows higher resilience than the untreated silica, because of better rubber-filler interaction that increase the elasticity of the vulcanizate [29]. The good interaction between epoxidized NR and the silica shows a significant increase in elasticity when HD silica is used in NR/ENR 25, even without the addition of a silane [30]. The silanized HD silica will always show better interaction and higher elasticity than the non-silanized precipitated silica, as the interaction between the silanol groups has been destroyed by the silanes and the interaction with the rubber has increased significantly. Therefore, the remaining elasticity of the rubber show higher values after the rebound resilience test [31].

#### 10.2.2.3 Effect of Layered Silicates

When MMT is introduced into rubber by wet-compounding, the outcome is predominantly exfoliated platelets that interact well with polar rubbers. This leads to a strong interaction with the polar rubber chains, so that the rebound resilience is low, which is mainly due to the specific surface and the high aspect ratio. Due to the platelet shape, dispersed layered silicates do not lead to an increase in rebound resilience, but to a significant drop that increases with loading. If the rebound resilience is plotted as a function of hardness, the MMT exfoliated in wet-mixing by continuous dynamic latex compounding (CDLC) has significantly lower elasticity than a rubber/carbon black N220 vulcanizate, as shown in Figure 10.8 [31].

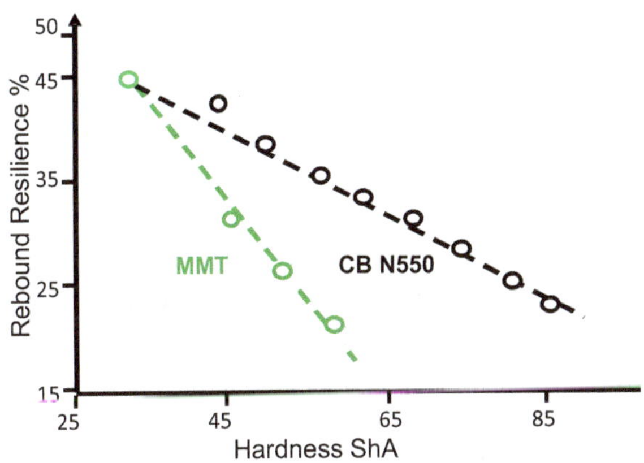


**Figure 10.8**
At constant hardness, MMTs exhibit higher hysteresis and a reduced rebound resilience than is the case for CB networks [31]

If intercalated or exfoliated organo-clay platelets are used, the rebound resilience is also reduced and the mechanical energy is converted into heat. The dispersion of SBR/silica (53 phr)/organo-clay (7 phr) has been compared with SBR/Silica (60 phr) and again it shows that the organo-clays reduced the elasticity of the elastomer [32]. Exfoliated layered silicates, in their original form or after modification as organo-clays, exert a small influence when it comes to increasing the elastic properties of the vulcanizate. They are oriented by the processing direction and always return a high tan $\delta$ when the rebound resilience is measured.

### 10.2.3 Effect of Carbon Nanotubes

If the rebound resilience is determined in the vulcanizates filled with multi-walled carbon nanotubes, there is an increase in the residual energy in the vulcanizate. This increase in elasticity is due to the low volume fraction and the high aspect ratio in the vulcanizate and the high specific surface area. As a result of the higher elastic energy, such a system can be used for the development of spring elements. The low filler concentration of multi-walled carbon nanotubes and their high aspect ratio help to increase the rebound resilience. When comparing the rebound resilience of HNBR/multi-walled carbon nanotubes NC7000 with that of HNBR/carbon black N220, it makes sense to use the hardness of the vulcanizate as a reference. As the hardness increases, the elasticity of the vulcanizates in the HNBR filled with carbon nanotubes is significantly increased, as shown in Figure 10.9 [33].

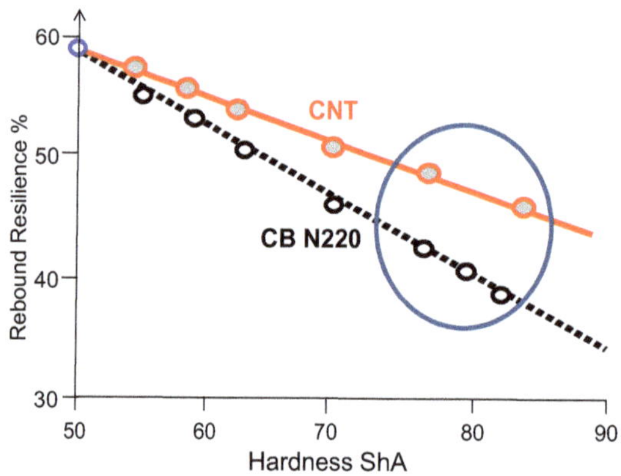


**Figure 10.9** At constant hardness, carbon nanotubes lead to higher rebound than the active carbon black N220 in HNBR vulcanizate [31]

With a hardness Shore A of 80, the increase in elasticity is 7.5% higher for elastomers filled with multi-walled carbon nanotubes than in the case of carbon black. Due to the high aspect ratio, carbon nanotubes in the specimen are aligned and do not hinder the increase in elasticity by overlapping the carbon nanotubes in the compound. On this basis, better elasticity of the vulcanizate, rubber compounds have been produced in practice.

## 10.3 Temperature Dependence of G′(T) and G″(T)

As most rubber products are used under dynamic load, it is useful to know the dynamic properties of the unfilled and filled vulcanizates over a large range of temperature, strain amplitude and deformation frequency. Only by having this background knowledge can the reinforcing effects of the filler be evaluated properly and the reinforcing mechanism better understood. In many cases, improvements in the rubber material are achieved by using the best rubber-filler combination. The simplest dynamic experiment that can be performed with a rubber vulcanizate is to study how G′ and G″ evolve over a wide temperature range and how the filler affects both. The temperature profile of the shear modulus is a property that distinguishes elastomers from thermoplastics and thermoplastic elastomers. It distinguishes between the glass-like range, the elastic range of use and the thermal decomposition of the rubber. The temperature profile is divided into four ranges in which the polymer has characteristic properties. In the energy-elastic range below the glass transition temperature, the polymer chains are frozen in a glass-like manner. The transition from the glassy state to the rubbery state is important over one order of magnitude of the drop. In this area the temperature dependence is particularly pronounced. The increase in G″ in the glassy state and the sharp drop in the rubbery state can also be of strategic advantage for the development of certain rubber/filler vulcanizates. The entropy-elastic range is characterized by a low shear modulus and a low loss modulus. The last range corresponds to the thermal decomposition of the rubber at higher temperatures. The consequences of incorporating particulate filler in a rubber matrix are seen in the glass transition region, but are even more pronounced at 30–40 °C above $T_g$ in the rubbery plateau. Here, the enhancement of G′(T) and G″(T) is affected by the type of the filler, the filler loading, the specific surface area and structure of the aggregates and, last but not least, the degree of filler dispersion achieved during the mixing process.

### 10.3.1 Temperature Dependence of G′(T) and G″(T) in Unfilled Elastomers

Temperature-dependent measurements of G′(T) and G″(T) performed at constant frequency and low strain amplitude yield information about the position of the glass transition on the temperature axis at service temperatures. Below the glass transition temperature, the free volume in the polymer is very small and the movements of the polymer segments and adjustment of their position cannot take place in the time scale required for normal dynamic testing. This leads to lower dynamic energy dissipation, because the polymer appears as a rigid material with a high modulus of elasticity based on enthalpy elasticity [34]. As an unfilled glassy elastomer, its free volume increases, such that initially side-chain movements (γ-transitions) are activated and subsequently, as the temperature continues to increase, chain segments composed of 4–7 monomer units begin to move and the glass transition temperature, $T_g$, is reached (β-transitions). The thermal energy becomes comparable to the energy barriers to segment rotations and the molecular adjustments become quick enough to follow the dynamic strain. Long-range contour changes of polymer chains take place, associated with entropic elasticity and low resistance to strain. The material enters the “rubbery” region characterized by low modulus and low energy dissipation.

The glass transition temperature, $T_g$, marks the natural limit for entropy-elasticity of the network and thus it depicts the limit of low-temperature application. The lower the $T_g$ of the rubber, the greater the flexibility of network chains and the less risk there is that rapid external deformation will lead to a scission of network chains. Therefore, to improve abrasion resistance of tires, rubbers such as high cis-BR with very low $T_g$ (−100 °C) have to be used in combination with well dispersed fillers. Low $T_g$ rubbers exhibit low molecular friction during deformation and therefore have low energy dissipation during deformation. Even though fillers significantly increase the dynamic hysteresis of vulcanizates, a small background contribution from the rubber may add value to the product. Between the glassy and the rubbery regions, there is a transition zone where the modulus of elasticity drops exponentially by several orders of magnitude and the loss modulus passes through a pronounced maximum.

Temperature sweeps recorded with a dynamic-mechanical spectrometer reveal three curves G′(T), G″(T) and tan δ (T) that contain a specific information about the glass transition temperature and the maximum loss of energy at the boundary from the glassy to the rubbery state. After passing the glass transition, the storage modulus decreases by roughly three decades along a sigmoidal curve. At the inflection point of the G′(T) curve, the glass transition temperature is recorded that corresponds to the maximum of the loss modulus G″(T). As tan δ is defined as the ratio G″/G′, its maximum value is shifted to higher temperatures than the inflection point recorded in the G′(T) curve or the maximum of G″(T). The glass transition temperature, $T_g$, of an elastomer depends primarily on the chemical constitution of the rubber, the molecular weight and the crosslink density [35]. The presence of side groups along the chain backbone increases the rotational potential of the chains and increases the glass transition temperature that defines the role of the rubber in different applications.

The increase resulting from non-polar side groups (i. e. vinyl groups or styrene groups) is illustrated for SBR in Figure 10.10, left, and the effect of polar groups (i. e. nitrile) is shown for NBR in Figure 10.10, right. The effect of the molar concentration of side groups decreases in the sequence: nitrile > styrene > vinyl [36].

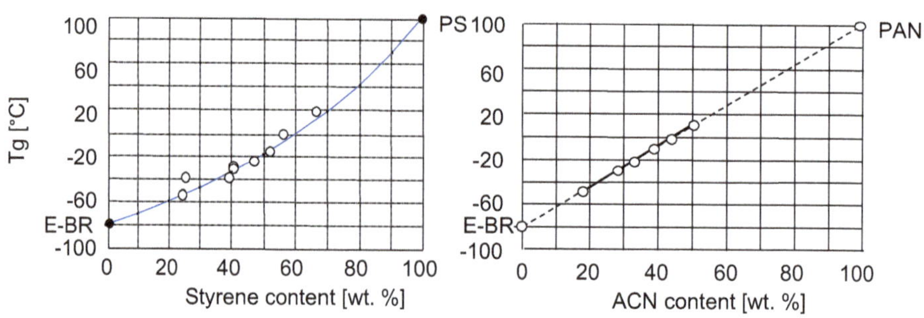


**Figure 10.10** Increase in glass transition temperature with co-monomer; left: styrene content in SBR; right: acrylonitrile content in NBR

Thus, by controlling the co-monomer ratio in the copolymer, the elasticity and energy loss can be precisely tailored via the monomer ratio and polymerization conditions.

Empirical rules describe the increase in $T_g$ in co-monomers as a function of the co-monomer composition [37, 38]. This strategy has been pursued by the polymer producers for S-SBR. It was shown that a range of S-SBRs having the same $T_g$ is accessible by changing the ratio of vinyl and styrene units. This influence of the co-monomers in respect of changing the $T_g$ of the rubber finally allows tire properties to be fine-tuned [39].

Theoretical concepts hypothesize that the chain mobility at $T_g$ is equal for all polymers. Consequently, for a temperature difference, $T\text{-}T_g$, the chain mobility is similar in different rubbers, whatever the composition of the rubber chains [40]. In this way, a reference state for which the dynamic properties of the polymer and the relaxation times are similar is found. The recorded glass transition temperature, $T_g$, of unfilled vulcanizates depends on:

- the state of cure
- the presence of plasticizer
- the frequency at which the test is performed

The process of cross-linking leads to a reduction in co-operative movements in the elastomer. A higher amount of thermal energy is needed to bring the network into a similar same state of mobility as in the melt. This leads to a slight increase in the glass transition temperature. As a consequence, the glass transition temperature increases by 3–8 °C with the crosslink density, irrespective of the curing system. The addition of plasticizer improves the low-temperature flexibility of vulcanizates, which is reflected in a decrease in $T_g$ that characterizes the efficiency of the plasticizer [41]. Empirically, plasticizing efficiency is defined as the change in $T_g$ produced by the molar concentration of plasticizer, $\Delta T_g/\Delta c$. Plasticizers interact with rubber chains on a molar basis. This means that each individual molecule of plasticizer contributes to the free volume of the entire system. At the same dosage (i. e. phr), products with a lower average molecular weight are therefore more effective at lowering the glass transition temperature. For long-term use, products of medium molecular weight and lower volatility are preferred in practice. The plasticizing effect is exerted only at concentrations at which the plasticizer is soluble in the rubber matrix. Above the solubility limit, the low-temperature flexibility of vulcanizates remains constant and cannot be improved by adding more plasticizer. On the contrary, the plasticizer forms a separate phase and tends to diffuse to the surface of the vulcanizate, causing undesirable blooming. Depending on the type of plasticizer in paraffin oil, naphthenic oil, aromatic oil, highly aromatic oil and the type of rubber used, plasticizing effect varies greatly [42].

If the measuring technique senses molecular movements, the frequency of the cyclical deformation contributes to the increase in the glass transition temperature. This is evident in the low frequency range (1–50 Hz) and is even more pronounced when the test is performed in the high frequency range. Ultrasonic spectroscopy is used to record an attenuation peak whose maximum corresponds to the glass transition process at high frequencies [43]. The storage modulus G′(T) typically decreases by a factor of $10^3$ on passing from the glassy region to the rubbery plateau zone. In the temperature interval of the rubbery plateau, the segment displacements are much faster than the time scale of the modulus measurements and the energy losses described by G″(T) gradually decrease. This corresponds to an increase in material elasticity, as shown by the rebound resilience experiments. The rubbery plateau is characterized by a slight increase in the complex modulus G* (T), as theoretically required. The level of the complex modulus increases with the state of cure and decreases with the concentration of plasticizing oil.

In ultrasound measurements, the glass transition is measured using the maximum of the attenuation factor, α [44]. At 0.5 MHz, the glass transition for cis-BR, NR and IIR lies at temperatures of −48 °C, −5 °C and +18 °C. This increase in the glass transition, $T_g$, is a function of the higher testing frequency of the ultrasonic measurements. For S-SBR, the influence of vinyl groups and styrene groups has been investigated by keeping the other segment groups constant. In all cases, the loading of SBR and NBR by side groups increases the energy dissipation, due to the movement of these groups in the vulcanizate. The energy dissipation measured is $\alpha_{max}$/mole fraction. For S-SBR, the energy dissipation is: styrene groups = 0.05, vinyl groups = 0.08. For NBR, the ACN groups show the highest energy dissipation of 0.12, which explains the lower elasticity at ambient temperatures [45]. The ultrasound test makes it possible to better understand the influence of the side groups of a rubber and to better estimate it on the desired properties.

### 10.3.2 Temperature Dependence of G′(T) and G″(T) in Filled Elastomers

Although many reports focus on unfilled systems, the bulk of investigations into dynamic properties are devoted to determining the influence of the filler on the temperature dependence of vulcanizates. The temperature dependence is the easiest to design for the development of G′ and G″ and represents an introduction to the dynamic properties of filled vulcanizates. Under dynamic stress conditions, a filler-reinforced elastomer exhibits a higher storage modulus, G′(T), and a higher loss modulus, G″(T), than its unfilled counterpart over the full range of temperatures. The incorporation of active fillers affects the dynamic-mechanical moduli in the glassy state and rubbery states, as the measured values describe the reinforcement by the fillers at each temperature. Generally, a moderate 3–5 °C shift in glass transition temperature is caused by the filler, irrespective of whether the polymer is physically attached to the filler surface or anchored by coupling agents, as in the case of precipitated silica. It can be seen how much G″ can increase in the glass state as a function of the loading of the filler or how high G′ and G″ can increase in the rubber level to clearly show the effect of the filler. These values describe the activity of the filler in the rubber matrix and can demonstrate the benefits of fillers in selected rubbers.

#### 10.3.2.1 Effect of Carbon Blacks

In order to investigate the dynamic-mechanical influence of the carbon black, the glass transition temperature is increased by increasing the temperature and measuring the shear modulus, G′(T), and the loss modulus, G″(T). The dynamic-mechanical behavior has been determined for E-SBR and carbon black N330 in torsional vibration tests conducted at temperatures ranging from −60 °C to +160 °C. An increase in G′ and G″ can be seen in the glassy state. However, the transition to the rubbery state is characterized by a decrease of several orders of magnitude for both moduli, with G′ and G″ increasing significantly with loading [46]. As advances have been made in dynamic-mechanical instruments, the measurement results have become more precise and so measurements have been extended to a variety of carbon blacks and various rubber vulcanizates. For carbon black N234, which has established itself in the tread compound of tires, the temperature-dependent curve was measured from −55 °C to +100 °C, under a deformation of 5% and at a frequency of 10 Hz [47]. The results are shown in Figure 10.11.

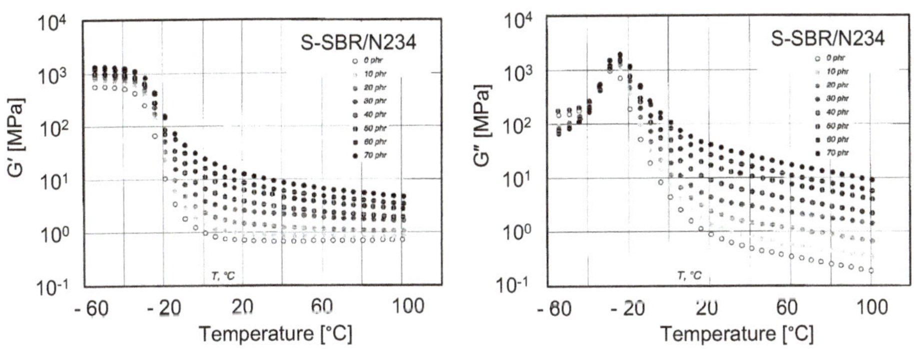


**Figure 10.11** Temperature dependence of storage modulus G′(T) and loss modulus G″(T) for S-SBR/N234 [47]

From the first two figures, it is evident that the influence of increasing the loading of carbon black N234 in the glass transition regime leads to a shift of $T_g$ to higher temperatures, which is shown by G′, and to an increase in G″ to higher values. The most important fact is the decrease in both moduli into the rubbery region, which occurs 30 °C above the $T_g$. Therefore, reinforcing carbon blacks provide a strong increase in G′(T) and G″(T) at a given temperature. However, at any temperature, the storage modulus G′(T) is higher than that of the loss modulus G″(T). At a given temperature, the factor $G'_f/G'_0$ increases more than factor $G''_f/G''_0$. The result for the measured tan δ is shown in a semi-logarithmic plot in Figure 10.11. Tan δ at temperatures in the glass transition region is related to the wet grip of tires on the road, which must be high to ensure grip on the road and must have a lower value at high temperatures (> 70 °C), where tan δ is an indicator of the rolling resistance of tires.

The filled system shows the maximum value in the $T_g$ region and the lowest tan δ at high temperatures. With increase in filler content tan δ, the grip values increase in the glass transition region, but decrease at high temperatures.

A consistency analysis shows that the increase in G″(T) in the glass transition maximum leads to the description of the polymer-filler interaction parameter [48]:

$$G''_{max,f}/G''_{max,0} = 1 + \alpha\Phi \tag{10.17}$$

where $G''_{max,f}$ and $G''_{max,0}$ are the maximum loss modulus of the filled and the unfilled vulcanizate, and α is the rubber-filler interaction parameter

The proportionality factor, α, increases with decrease in the primary particle size and with the structure of the filler aggregates. This increase can thus be determined as a function of the filler concentration and the degree of rubber-filler interaction. The transition from the glassy to the rubbery state requires an additional amount of energy to increase the chain mobility in the immobilized rubber layer at the filler surface. At constant loading, both the storage modulus G′(T) and the loss modulus G″(T) increase with the specific surface area or surface activity. Several reports point out that semi-reinforcing carbon blacks provide less-high G′(T) values as, and lower G″(T) values than, reinforcing carbon blacks, which can increase the moduli over a wide range [49].

Interestingly the values of G′(T) and G″(T) decrease with temperature in carbon black systems, while they increase in unfilled elastomers in accordance with the theory of ideal rubber elasticity. The presence of the active filler influences the entire network topology and so the behavior of the filled elastomers no longer obeys the fundamental law of ideal rubber networks. In Figure 10.11, when the temperature in S-SBR/carbon black N234 (70 phr) is increased, the storage modulus G′(T) decreases from 11 MPa at 20 °C to 7 MPa at 100 °C while the unfilled S-SBR show an increase in G′(T). The loss modulus, G″(T), goes from 7.5 MPa at 20 °C to 1 MPa at 100 °C. This increase can thus be determined as a function of the filler concentration and the degree of rubber-filler interaction [50]. In contrast to unfilled vulcanizates, which demonstrate a slight increase in storage modulus, $dG'(T)/dT > 0$, and a decrease in energy losses, $dG''(T)/dT < 0$, the storage modulus and the loss modulus of filled vulcanizates decrease with test temperature. The temperature gradient of the two mechanical moduli, G′(T) and G″(T), is negative. Since the extent of the temperature-related changes in the mechanical moduli depends on the type and amount of filler used, it is advisable to determine these changes by precise measurements [51]. The discrepancy in the temperature gradient of the dynamic-mechanical moduli should be taken into account when designing physical properties, especially for elastomers that are used at elevated temperatures.

As a general rule, it can be stated that parameters which lead to a higher degree of dispersion bring about a decrease in energy loss, expressed as G″(T) or tan δ(T). This holds for parameters that govern the mixing process as well as functionalization of the polymer and filler or the use of dispersants. However, the importance of surface characteristics to dynamic properties has not been emphasized as much as it should have been. For example, graphitization of carbon black leads to an increase in hysteresis that is related to the reduced percolation threshold of the carbon black and poor dispersion in the rubber matrix.

With regard to tire applications, it has been established that repeated straining of the compound through skidding can be approximated as a process of constant energy input involving different temperatures and frequencies [48]. Rolling resistance refers to movement of the whole tire corresponding to deformation at temperatures of 50–80 °C or low frequencies of 10–100 Hz. In the case of skidding or wet grip, deformation due to asperities in the road surface is approximated by the loss at 0–15 °C or a frequency in the range $10^4$–$10^7$ Hz. From a viscoelastic point of view, the necessary information can be obtained from the temperature dependence of tan δ. Recording the energy dissipation, expressed by tan δ, over a large temperature range provides practical information about the magnitude of both processes. Thus, an ideal tread compound should have a high tan δ value at temperatures in the range of −20 to 0 °C for the purpose of high traction and wet-skid resistance, and low tan δ values at 50–80 °C for low rolling resistance [50]. This relationship is shown in Figure 10.12.

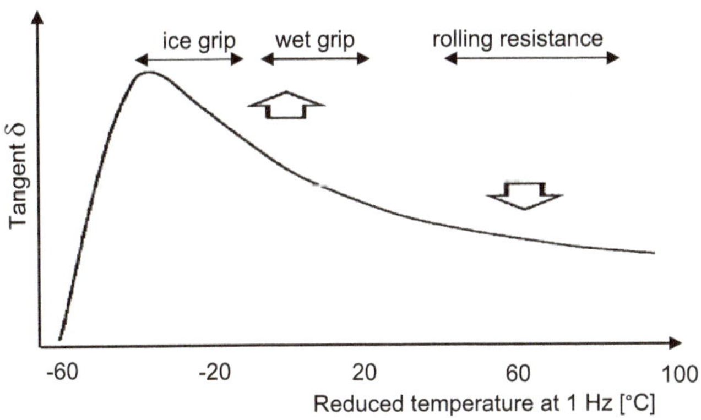


**Figure 10.12** Representative ranges for grip and rolling resistance indicated by the temperature dependence of tan δ

Wet grip describes the stress generated by the resistance of the road surface and the movement of the tire tread on the surface at a frequency of 104–107 Hz at room temperature [52]. The demand for compounds that offer higher wet grip and traction on the one hand and lower rolling resistance on the other hand has been met by polymer developments and improved fillers. A major change in dynamic properties, both

G′ and G″, can be achieved by functionalizing polymers and in-process functionalizing of the filler surface [53] and more readily dispersible carbon black grades [54].

Carbon black manufacturers are tackling this challenge by further developing the performance of carbon blacks. A representative example of the surface modification of fillers is the group of carbon-silica dual-phase fillers (CSDPF) with a hybrid surface. When incorporated into S-SBR/BR blends (70/30), these fillers provide good filler dispersion at short mixing times. As shown in Figure 10.13, the temperature dependence of G″(T) and tan δ(T) for CSDPF approaches that of the silica/TESPT "green tire" formulation. The loss modulus and tan δ at 0 °C are significantly higher, indicating better grip properties. At 50–80 °C, the tan δ curve is even much lower than that of the "green tire" formulation [55].

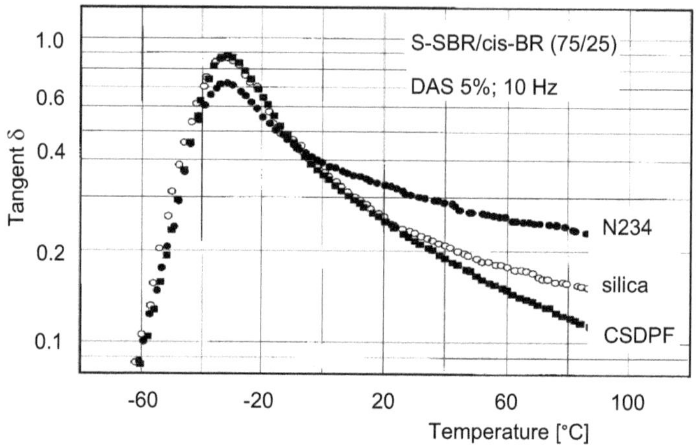


**Figure 10.13** Temperature dependence of tan δ for S-BR/cis-BR tire tread compounds filled with carbon black N234, silanized HD silica and CSDPF (according to [55])

The lower filler-filler interaction and the resulting weaker agglomerates of CSDPF manifest themselves as high hysteresis at low temperatures and low hysteresis at high temperatures on a curve of tan δ vs. temperature. The TESPT modification of CSDPF, which reduces tan δ at high temperatures and the rolling resistance, can be seen as a further improvement.

Ecorax® carbon black developed by Orion Engineered Carbon shows a higher increase in G′ and G″ in the rubbery area, compared with the corresponding ASTM carbon black, which is related to a higher surface energy modification, but a similar specific surface area and structure of the aggregates. Compared with a carbon black N121 that has a tan δ of 0.198 at 100 °C, experimental black EB172 shows a reduced tan δ of only 1.178 under the same conditions. This improvement tends to reduce the rolling resistance and increase the abrasion resistance [56]. Better interaction with the S-SBR/cis-BR blend is inferred from this. A further development is the use of new rubbers obtained from the oil of scrap rubber pyrolysis. At 60 °C, the dynamic stiffness E* and the loss angle tan δ are the same for NR/carbon black sN234 as for NR/carbon black N234. Therefore, sidewall quality will not be affected, compared with tires made from the oil of pyrolyzed rubber [57].

An excursion into the field of ultrasonic excitation of vulcanized rubbers leads to investigated rubber-carbon black compounds. Carbon black N 347 In NR and NBR has been investigated and the results are shown in Figure 10.14 [58].

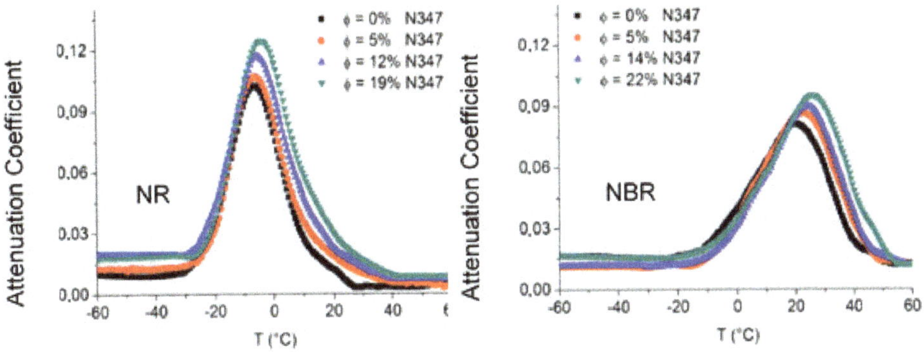


**Figure 10.14** The attenuation coefficient in the system NR/CB 347 and in NBR/CB 347 [58]

The attenuation coefficient, α, in NR and in NBR is shifted to much higher temperatures, a fact which indicates an influence of the higher measuring frequency. They show an increase in the attenuation maximum of 13% in NR and 23% in NBR for a maximum of 22 vol.% filler. That means that dispersion is much better in the case of NBR. In SBR, the concentration of styrene units and vinyl units could be varied independently of each other. The largest increase in the attenuation coefficient was observed when the vinyl concentration was varied. Therefore, the ultrasonic method is more sensitive to the interface which forms between the carbon black surface and the surrounding rubber.

When an active carbon black N220, N330 and N550 are used in S-SBR, the attenuation peaks do not appear to be influenced by the different grades. However, after the maximum attenuation, the carbon black of smallest primary particle size exhibits the highest attenuation. Such a shift on the temperature axis is allowed, as shifts to higher values are associated with higher deformation frequency. The result has revealed an inverse proportionality to the rubber-carbon black interaction. If this interaction is not very pronounced, the reduced sound velocity is assigned a higher value. S-SBR with the lowest styrene content shows the highest increase in reduced sound velocity, whereas the highest styrene content and good interaction with the carbon black shows the lowest increase in reduced sound velocity, such as NBR/carbon black 347. Clearly, therefore ultrasound spectroscopy reveals properties at high frequencies, which is not only important for tire applications.

### 10.3.2.2 Effect of Precipitated Silicas

When precipitated silica is used as an active filler, a distinction must be made as to whether it is untreated or functionalized with silane. The condition of the silica surface determines the viscoelastic properties that are triggered by this filler. Precipitated silica exhibits the same complex modulus G*(T) at room temperatures as carbon black of the same specific surface area. As the specific surface area increases above 100 $m^2/g$, the G*(T) rises to higher values that are not reached by the most active carbon blacks. The complex modulus G*(T) for precipitated silica with a surface area of 180 $m^2/g$ reaches values of 16.5 MPa. Therefore, the storage modulus G′(T) and the loss modulus G″(T) demonstrate an increase with specific surface area, a fact which can be explained by the strong silanol interaction between the silica aggregates. Looking at tan δ, the values for precipitated silica are lower than for carbon blacks having the same surface area. This is a consequence of the higher storage modulus G′(T). Due to the differences in the behavior of the complex modulus and tan δ, replacing the furnace carbon with the precipitated silica can lead to dynamic properties which cannot be obtained with furnace carbon [28]. However, such a change encounters considerable difficulties for vulcanizates with sulfur.

The use of silanes solves this problem. This can already be seen by observing the increase in the G″(T) in the glass transition region. The larger the rubber-silica contact surface and the more intensive the rubber-silica interaction, the higher is the value of the interaction parameter, α, from Equation 10.15. In silica-filled cis-BR vulcanizates, α increases linearly with the degree of silanization, which indicates good interaction between the silanized silica and the rubber matrix and leads to an improvement in rubber properties. In contrast, silica silanization in NBR leads to a decrease in α, indicating that the good interaction between silanol groups and acrylonitrile is destroyed [48]. This difference in $G''_{max}(T)$ in cis-BR and NBR compounds is shown in Figure 10.15.

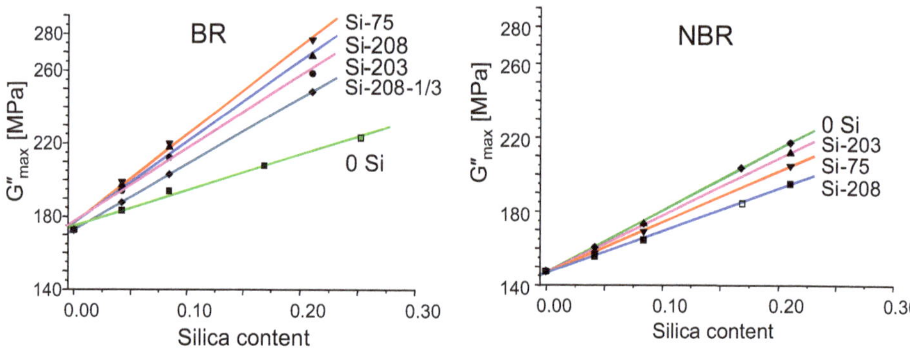


**Figure 10.15** Change in loss modulus as function of silanization of silica in cis-BR and NBR vulcanizates [48]

The reinforcing effect is due to filler loading, implying a rise in storage modulus G′(T) with silica concentration. The sample without silane displays a strongly increased loss modulus, G″(T), as compared with silanized silica, a fact which is attributed to fewer flexible filler-filler contacts in the silane free system, due to strong dipolar interactions that imply filler-filler bonds [59]. A 3-mercaptothiol is more active than a disulfided silane, because it delivers a compound that has twice the silica-silane functionalization [60]. Chain-end functionalization and in-chain functionalization of S-SBR lead to better interaction with precipitated silica. Compared with non-functionalized S-SBR, the tan δ value of the vulcanizate is increased by 5% at 0 °C while a decrease of 38% is observed at 60 °C [61]. Functionalization of E-SBR with 3% hydroxylated methyl methacrylate (HPMA) in E-SBR/BR blends filled with silica/silane increases tan δ in the wet skid temperature region (−10 °C) [62].

Silanization of precipitated silica leads to a decrease in tan δ. Comparison of the effect of different bifunctional silanes in green “tire” silica compounds reveals that the more effective hydrophobization of HD silica provided by mercapto silanes leads to a reduction in tan δ of almost 60% at 60 °C [63]. The value of tan δ decreases as the loading of silica increases and the height of the amplitude and the glass transition temperature shifts by more than 5 °C. The value of tan δ decreases as the loading of silica increases and the height of the amplitude and $T_g$ shift is more than 5 °C [64]. In the absence of specific polymer interactions, *in situ* synthesized silica nanoparticles (diameter 10 nm) show that the polymer segments within a few nanometers of the filler particles exhibit relaxation times up to 2–3 orders of magnitude slower and a reduced increment in heat capacity, which is reflected in the loss modulus [65].

Relative to that of carbon black vulcanizates, the very low tan δ of silanized silica vulcanizates in the range from −30 °C to +40 °C is certainly due to the stronger silica aggregates which contain occluded rubber unable to participate in energy dissipation [50]. Taking the carbon black-filled oil extended S-SBR/BR (70/30) tread compounds as a reference, the wet skid resistance increases in the order:

carbon black < CSDPF < precipitated silica

The other important tire property, rolling resistance, increases in the order:

carbon black < silica < CSDPF

This shows a serious advantage of the silanized HD silica and the CSDPF in reducing the tan δ and thus, implicitly, oil consumption. It has been recognized that the mechanism involved in wet skid resistance is more complicated than expected, as lubrication effects and adhesion friction play a role at the micro-scale. However, the loss modulus and tan δ at low temperatures are considered to be qualitative predictors of these contact mechanical phenomena.

Ultrasonic spectroscopy has shown that the addition of precipitated silica to SBR and NBR compounds leads to a stronger attenuation factor, α, and shifts the temperature at which attenuation takes place, as indicated in Figure 10.16 [66].

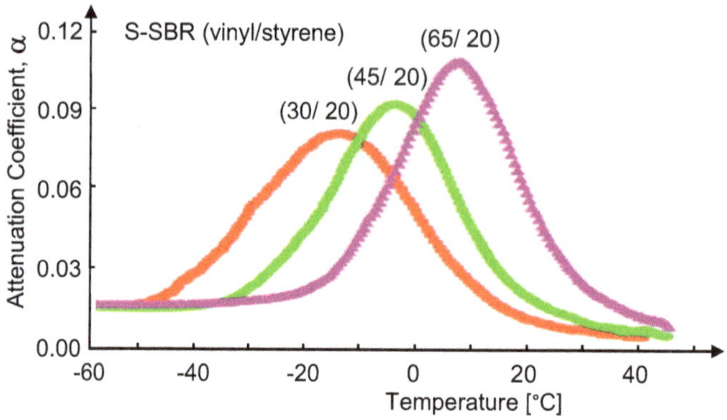


**Figure 10.16** Attenuation coefficient, α, for S-SBR of variable vinyl content [66]

The addition of silica induces an increase in the attenuation factor, α, and leads to broader peaks in both polymers. The asymmetry of the ultrasonic attenuation hints at a dependence on the glass transition temperature. In S-SBR where the styrene content was varied, it can be seen that an increase in polarity ascribed to a higher styrene content induces a small variation that can be ascribed to the surface treatment of the silica. In NBR, silanization leads to a reduction in the interactions between the ACN groups and the treated silane. The reduced sound velocity shows a slight increase in NBR and a nearly threefold increase in SBR of low styrene content, which should be ascribed to the stronger silica-silica bonds that prevent better filler dispersion. The silanization of silica is reflected in a reduction in the attenuation area. However, the extension of the dynamic properties to the high frequency range provides direct values of the attenuation coefficient for the compound.

#### 10.3.2.3 Effect of Layered Silicates

Dynamic-mechanical analysis is often used to study the rubber-layered silicates composites. The addition of $Na^{+}$MMT to the NBR matrix results in a remarkable increase in the storage modulus G′(T). This means that 10phr layered silica in NBR has the effect of carbon black N550 of 50 phr. That increase is related to the fact that layered silicates are exfoliated in the polar rubber. The exfoliation of layered silicates not only favors better dispersion, but also improves the storage modulus G′(T) of the rubber compound. One objective has been to achieve the same mechanical reinforcement of rubber/carbon black with much less MMT or organically modified clay. In terms of dynamic properties, the storage modulus of nanocomposites is markedly higher than that of pure rubber, due to the high aspect ratio and the high specific surface area. The influence of the exfoliated layered silicates produced by wet-mixing with continuous dynamic latex compounding (CDLC) shows a shift in the glass transition temperature for small volume fractions and an increase in the G′(T) and G″(T) behavior in the rubber range, which is shown for NBR/PGN in Figure 10.17 [32].

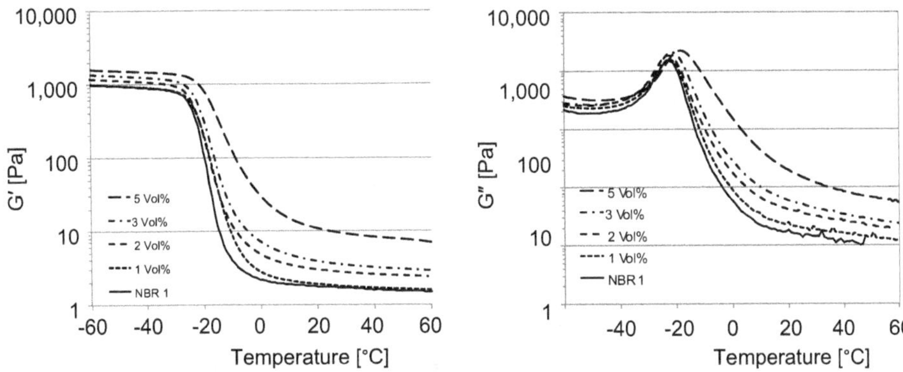


**Figure 10.17** G′(T) and G″(T) for exfoliated NBR/pristine montmorillonite [32]

The differences from the NBR vulcanizate can be seen in a double logarithmic plot of log ($G'_f/G'_0$) vs. log Φ. Above the rheological percolation threshold (Φ = 2.3 vol.%) there is an exponential increase to 3.8, which indicates the strong interaction demonstrated by the polar rubber. The loss modulus shows an exponential increase to 4.5, but with smaller values. The reason for the strong exponential increase in $G''_f/G''_0$ is the high aspect ratio of these platelets and the good interaction between the physisorbed polymer chains on the surface of the MMT.

The NBR/MMT vulcanizates yielded by the CDLC process have the highest reinforcement in the glass transition region. Due to the CDLC technology, the degree of dispersion is high and the aspect ratio measured by TEM in the range of 90–100 [69]. When the dynamic measurements are made at high frequencies (0.5 MHz), the $T_g$ of the vulcanizate changes to high values. If the attenuation coefficient of the dispersed NBR/MMT compound is considered, it can be seen that the role of the nanofillers leads to a higher ratio to the filler concentration than carbon black [66].

Hybrid systems consisting of carbon black and a small loading of MMT (5.3 phr) show a clear increase in the normalized $G'_f/G'_0$ at 20 °C before increasing to 2.5 times in the logarithmic plot, which reflects the good interaction of the exfoliated MMT with the dispersed carbon black and the polar HNBR. The use of organo-clay (6 phr) in NR increases the storage modulus from originally 0.6 MPa to 4.2 MPa [70].

The treatment of layered silicates with octadecyltrimethyl-ammonium leads to an increase of the storage modulus G′(T) in EPDM that is higher than that obtained by mechanically introduced pristine clays. Therefore, the increase in the dynamical properties is considerable. In EPDM-MA, however, the identical organo-clay at 10 phr has a storage modulus of 13 MPa at room temperature [71]. The shear-alignment of organo-clays in IR vulcanizate exhibits an increase in storage modulus when the volume fraction exceeds the rheological percolation threshold. The storage modulus is markedly higher than that of pure NBR or HNBR filled by mechanically mixing with the organo-clay. HNBR/organo-clay shows an increased influence on the storage modulus and decreased loss factor, tan δ. When the clay is modified with methyl-tallow-bis(2-hydroxy-methyl) quaternary ammonium, the best increase in the G′(T) value is achieved, due to the interaction of the -OH groups of the organo-clay and the acrylonitrile group of HNBR [72]. Hybrid fillers of carbon blacks and organo-clays in NR, IR [73] and in NR/SBR blends [74] have been used to improve the storage modulus in the elastomer. The remarkable effect of the dynamic-mechanical properties led to the proposal of a synergistic effect between the two different fillers and the rubber.

#### 10.3.2.4 Effect of Carbon Nanotubes

Investigation of the dynamic-mechanical properties yielded some evidence that well dispersed multi-walled carbon nanotubes impact the mobility of the polymer chains themselves, causing changes in the glass transition temperature and increasing the dynamic-mechanical moduli G′(T) and G″(T) [75]. In general, elastomers that contain multi-walled carbon nanotubes show better dynamic-mechanical properties than those filled with carbon black, suggesting that the reinforcing mechanism is related to the high aspect ratio and the increased specific surface area [76]. While alignment is necessary for maximizing the storage modulus in a particular direction, it is not always beneficial, as the aligned multi-walled carbon nanotubes tend to have highly anisotropic properties, and the storage modulus varies with the orientation in the elastomer.

Damping describes the trend of materials to reduce the amplitude of oscillations in an oscillating system. The combination of extremely large surface area and weak interfacial bonding with polymer along with low mass densities of carbon nanotubes means that the frictional sliding of the nanotubes with the polymer matrix can cause significant energy dissipation and provide damping capability [77]. There are several factors governing the damping properties of carbon nanotubes vulcanizates. The effect of carbon nanotube dispersion can enhance the damping properties of the elastomer.

The temperature dependence of G′(T) and G″(T) increases the $T_g$ and affects G′ in the rubbery region. The peak value of G″(T) also increases after the addition of multi-walled carbon nanotubes [78]. It has been shown that the storage modulus G′(T) and the loss modulus G″(T) change with the amplitude of the oscillations, with multi-walled carbon nanotube loading and with the diameter of the nanotubes. The addition of carbon nanotubes has been found to enhance the thermal stability of the elastomer [79]. In dynamic-mechanical analysis, a temperature sweep demonstrates an exponential loading that leads to a storage modulus of 23.4 MPa at 5.4 wt.% of carbon nanotubes in NR [80]. An increase in the storage modulus and the loss modulus G″ has been recorded for SBR/carbon nanotubes [81]. The normalized loss modulus $G''_f/G''_0$ is on the same curve for NR and FKM and above the values obtained for carbon black. Significant differences in terms of dynamic properties are measured on the basis of the diameter of the carbon nanotubes. The larger the diameter of the carbon nanotubes, the more $G''_f/G''_0$ shifts to higher concentrations of the carbon nanotubes and the smaller is the respective increase, a fact which can be attributed to the decrease in the specific surface area of the carbon nanotubes.

## 10.4 Strain-Amplitude Dependence of G′(γ) and G″(γ)

An excellent method for determining the influence of the filler-filler interaction in rubber vulcanizates uses the increase in G′(γ) in the strain amplitude at constant temperature and the stress frequency. As the strain amplitude increases, the storage modulus G′(γ) decreases in a sigmoidal curve and the loss modulus G″(γ) passes through a maximum. This behavior was first observed in 1952 [81] and studied in detail 10 years later by Payne, after whom the strain-amplitude dependence is often called [82–85]. At high strain amplitudes, the storage modulus G′(γ) and the loss modulus G″(γ) show a strong reduction to the values of the hydrodynamic reinforcement of the filler and the effect of the crosslinking. Therefore, the dependence on strain amplitude can illustrate the filler-filler interaction and the mechanism for breaking down the filler cluster aggregates. Inter-aggregate interaction has been classified as being the cause of the complex nature of filler attraction [49]. In an amplitude-dependent dynamic-mechanical experiment of the storage modulus G′(γ) and loss modulus G″(γ) under increasing deformation, amplitude is an indication of the non-linearity of the stress-strain relationship [7]. The storage modulus G′(γ) decreases according to a sigmoidal curve, while the loss modulus G″(γ) passes through a maximum.

At small deformations (< 0.1%), the values above the filler percolation threshold increase exponentially with the filler loading, as observed by Payne. The exponential increase in the storage modulus G′(γ) leads to the formation of the filler-filler network, which is characterized by G′(γ) at the smallest double strain amplitudes. The increase of the double strain amplitude leads to a continuous breakdown of the filler network that has the sigmoidal form. After passing the inflection point, the filler network is broken into individual filler clusters. The maximum of the filler breakdown occurs at the strain amplitude of 2–3%, where G″(γ) shows maximum energy dissipation, and then it declines to lower values. The extent of the filler-filler interaction in the filler network, which is influenced by the presence of the polymer, leads to differences in the behavior of the filler network. Therefore, the strain-amplitude dependence of G′(γ) and G″(γ) describes the breakdown of the filler network as the strain amplitude is increasing and delivers important information about the energy dissipation of materials undergoing dynamic-mechanical testing. It can be seen that the shear modulus decreases with increasing oscillation period and the viscoelastic body becomes softer. The loss modulus takes on small values both when the strain amplitude decreases and when it increases. The position of the maximum in between is determined by the relaxation time. The shape of the amplitude-dependent storage modulus G′(γ) and the loss modulus G″(γ) in a carbon black-filled SBR is shown in Figure 10.18.

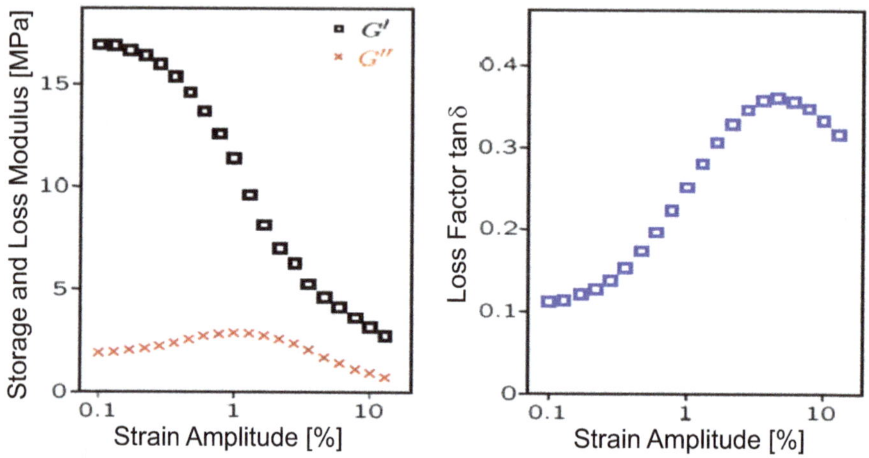


**Figure 10.18** Increasing the strain amplitude decreases G′(γ) and leads to a maximum for G″(γ)

The first tests on NR vulcanizates filled with carbon black were carried out on channel black, gas black and thermal black as early as 1941. Dynamic testing with forced oscillations showed that the complex modulus increases with the specific surface area and loading of the filler with the same cross-linking [86]. A similar behavior by the fillers has been observed for the synthetic rubbers SBR, CR, NBR and IIR that were available at the time. The storage modulus, G′(γ) decreased with the deformation amplitude [87]. The non-linear strain-amplitude dependence of G′ and G″ was first observed for carbon black-filled NR vulcanizates [82]. Payne, who used a self-built dynamic-mechanical tester that allows precise measurements, published a fundamental framework on the strain-amplitude dependence of the dynamic moduli and hysteresis. The interest in the strain-amplitude dependence of G′(γ) and G″(γ) was triggered by the effort to improve tire performance, particularly by the attempts to reduce the rolling resistance of tire treads, which is related to dynamic hysteresis up to 10–12% strain amplitude. The proportion of energy loss due to tire tread is estimated at 40% of the energy losses by the whole tire. Energy loss during the increase in strain amplitude is important not only for tire treads but also for vibration mounts. Therefore, the strain-amplitude dependence of the storage modulus G″(γ) and the loss modulus G″(γ) have become an important quantity for testing the application in practice.

### 10.4.1 Effect of Carbon Black

When a carbon black-filled rubber vulcanizate is subjected to oscillatory deformation with a progressively increasing strain amplitude, the stress necessary to deform the filled vulcanizate decreases non-linearly with increase in strain amplitude, and the loss modulus pass through a maximum, as illustrated for S-SBR/N234 vulcanizates in Figure 10.19 [50].

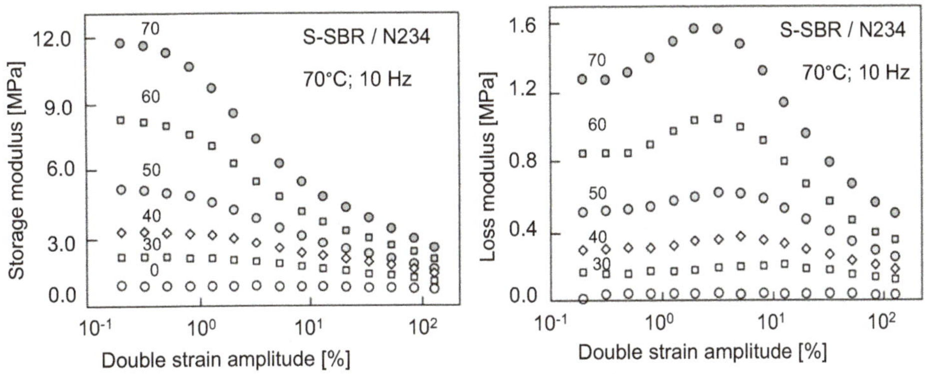


**Figure 10.19** Double strain-amplitude dependence of G′(γ) and G″(γ) as a function of carbon back N234 for S-SBR/N234 vulcanizates (according to [50])

Consequently the storage modulus G′(γ) of the system increases exponentially with the carbon black loading and as the deformation amplitude becomes larger, the value of G′(γ) decreases according to the sigmoidal curve and finally reaches the points small values corresponding to the modulus given by the hydrodynamic effect of the fillers based on the structure of the aggregates. The decrease in the storage modulus G′(γ) from an initially high value, $G'_0$, given mainly, if not only, by the stiffness of the filler-filler network, leads to a limiting value at high strain amplitudes, $G'_\infty$, that corresponds to sub-clusters and agglomerates. The loss modulus G″(γ) that describes the mechanical energy loss transformed into heat shows, from very low values, an increasing value that passes through a maximum that lies at the inflection point of G′(γ). After passing through the loss maximum, the loss modulus G″(γ) decreases to smaller values. The development from a three-dimensional filler network to a system that contains more and more individual clusters is shown in Figure 10.20.

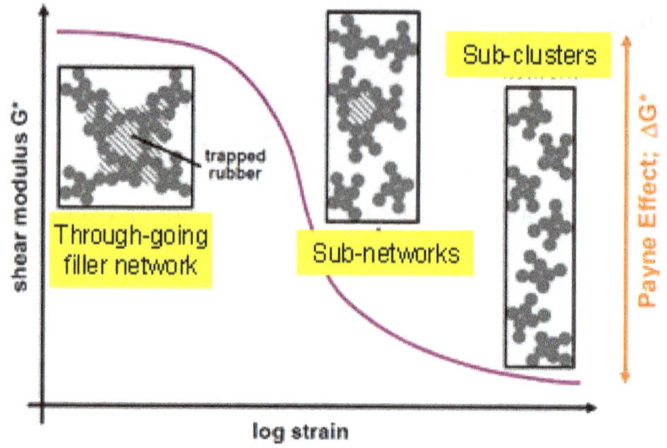


**Figure 10.20** When the strain amplitude increases, the through-going filler network is continuously destroyed

However, even at high strain amplitudes, the filler network is not completely broken down at the amplitude range of 10% or larger. The limiting value, $G'_\infty$, depends on the strain amplitude realized with the test method. The difference between the measured storage moduli, $G'_0 - G'_\infty$, represents the mechanical stiffness provided by the filler network. Naturally, the breakdown of the filler network is also associated with an increase in the dissipative response of the filled vulcanizate. The effect is largely reversible once the strain is released and can be annealed at a higher temperature.

G′ of carbon black N330 (110 phr) was measured for comparison with unfilled rubber [85]. The results show a strong influence exerted by the type of the rubber and dispersion achieved in the overloaded system, as shown in Table 10.1.

**Table 10.1** Influence on Storage Modulus, $G'_0(\gamma)$, of Carbon Black N330

| Rubber | Storage Modulus, G′ [MPa] |
|---|---|
| NR | 30.5 |
| CR | 23.5 |
| NBR | 55.1 |

The strain-amplitude dependence of storage modulus $G'_0(\gamma)$ and $G''_0(\gamma)$ for NR with various loadings of carbon black N110 increases with the loading. At all concentrations of the carbon black, the storage modulus is independent of the strain amplitude when the latter is less than $10^{-4}$%. The loss modulus $G''(\gamma)$ could be measured only at strain amplitudes above of $10^{-5}$% and was constant prior to an increase at deformations approaching the level at which the storage modulus markedly decreases [88].

As with the dynamic-mechanical properties, the strain-amplitude dependence becomes more pronounced with high filler loading above the electrical percolation threshold, irrespective of the filler particle size and the structure. However, when the filler reaches the percolation threshold, $G'(\gamma)$ increases with a high exponent, as the expression shows [89, 90]:

$$G'(\gamma) = \Phi^{\alpha} \tag{10.18}$$

The background of this scaling law is attributed to a kinetically cluster aggregation (CCA) model of fractal filler networks. The exponent, α, was expected to be 3.5 in theory, but in practice it varies between 2.5 and 3.5, depending on the type of filler and the vulcanized rubber [91]. Due to the exponential increase in $G'_0$, the vulcanizate properties react sensitively to changes in filler loading and variations in the curing conditions.

To avoid changes in the stiffness of filled vulcanizates, it is necessary to add the filler to the compound in the exact dosage and to follow the mixing procedure and the downstream process in the same way. In addition, a comparison of the Payne effect, $\Delta G'(\gamma) = G'_0 - G'_\infty$, at an arbitrary loading, i. e. 50 phr, risks misinterpreting the real contribution of the Payne effect. With the loading selected for comparison, the filler that percolates at a lower concentration has inevitably the highest Payne effect and the filler that is better dispersed and has a higher percolation threshold will have a low Payne effect. Since fillers with high activity and strong filler-filler interaction percolate in any rubber matrix at smaller concentrations, they demonstrate a higher Payne effect at any chosen loading [92]. A comparison of the Payne effect in different systems therefore only provides qualitative information that does not explain the existing differences in the filled network, which depend on the filler activity and its interaction with the rubber chains.

In contrast to $G'(\gamma)$, which decreases progressively with increase in strain, the loss modulus $G''(\gamma)$ passes through a maximum at moderate strain amplitudes for any loading. After passing through this maximum, the loss modulus G″ decreases rapidly with further increase in strain amplitude. As a result of the increasing density of the filler network, the maximum of the loss modulus above the percolation threshold shifts steadily to smaller-double strain amplitudes. Qualitatively the same observation is made for other filled systems, regardless of the chemical nature of the rubber, its microstructure and the grade of carbon black used.

Based on the assumption that the energy loss of the filled rubber during dynamic strain is controlled by the breakdown and re-formation of the filler network, the following expression has been derived for the loss modulus, $G''(\gamma)$ [93]:

$$G''(\gamma) = G''_\infty + c'\varepsilon^n{}_c(G'_0 - G'_\infty) \tag{10.19}$$

where $\varepsilon^n{}_c$ and $c'$ are constants, $G'_0$ and $G'_\infty$ are the leveling values of G′ at low and high strain amplitudes, and $G''_\infty$ is the loss modulus after total destruction of the filler network

This equation takes into account the filler-filler interaction in the vulcanizate. The higher the filler-filler interaction, the more noticeable is the loss modulus $G''(\gamma)$. The changes in the filler-filler interaction also change the energy loss when the strain-amplitude dependence is increased.

#### 10.4.1.1 Influence of the Specific Surface Area

At constant carbon black loading, the storage modulus $G'(\gamma)$ increases linearly at low strain amplitudes as the specific surface area increases [94]. The increase in $G'_0(\gamma)$ with the specific surface area can be understood on the basis of the increase of the filler-filler interaction in the filler network. The influence of $G'_0$ is shown in Figure 10.21.

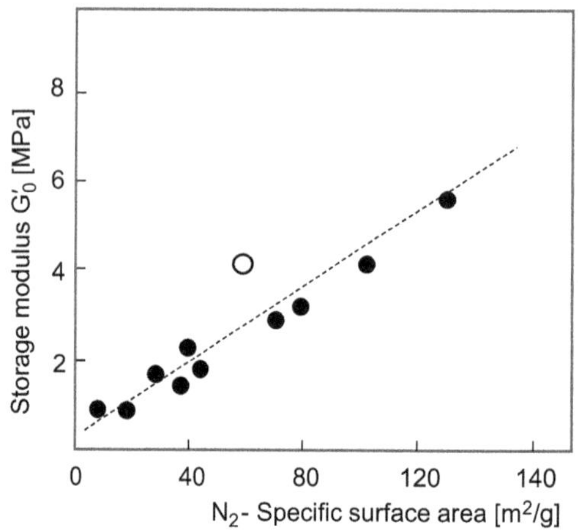


**Figure 10.21** Storage modulus $G'_0(\gamma)$ of carbon blacks with increase in specific surface area

The $G'_0(\gamma)$ values increase linearly with the specific surface area of the carbon blacks. An exception to this rule is acetylene black, which has a high aggregate structure and is formed by thermal decomposition of acetylene. The increasing specific surface area of the carbon blacks is established in the complex modulus $G^*$ and the $\tan\delta$ as a function of double the strain amplitude [23].

The strain-amplitude dependence of $\tan\delta$ at 70 °C for S-SBR with carbon black N234 shows an increase similar to $G''(\gamma)$ which is can be compared with $G''(\gamma)$ at higher strain amplitude (approx. 10%). When the strain-amplitude measurements are repeated at low temperatures (0 °C), the picture is quite different. Over a certain range of strain amplitudes, $\tan\delta$ decreases with increase in filler loading. For highly loaded vulcanizates, $\tan\delta$ is even higher than the unfilled counterpart at high strain amplitudes. Under these conditions, cyclical disruption and re-formation of the filler-filler network would also lead to additional energy dissipation. As the two processes increase with the strain amplitudes, $\tan\delta$ would increase steeply in the strain experiment [24].

Besides the inter-aggregate distance, the carbon black activity has to be considered [95]. The carbon black surface has high energy that increases with the specific surface area and certainly increases the strong inter-aggregate interactions in the filler network. With increase in the primary particle size, the active sites decline and lose their activity. Many reports document the influence of $G'_0(\gamma)$ in double strain-amplitude experiments in this direction [96]. Specifically, the strain-amplitude dependence at tan δ has been compared using S-SBR filled with carbon black (50 phr). With increase in surface area the storage modulus $G'(\gamma)$, the Payne effect increases and shows a strong tendency toward agglomeration in the case of small aggregates [97]. For cis-BR containing furnace blacks, the strain-amplitude dependence is exhibited over the entire amplitude. As the specific surface area increases, the amplitude dependence is represented by higher storage moduli $G'(\gamma)$ in the order N660, N339, N220 and N110. The horizontal range of $G'_0(\gamma)$ at small deformation amplitudes increases with decline in the specific surface area, a fact which can be attributed to a strongly decreasing interaction between the filler-filler particle, as shown in Figure 10.22 [98].

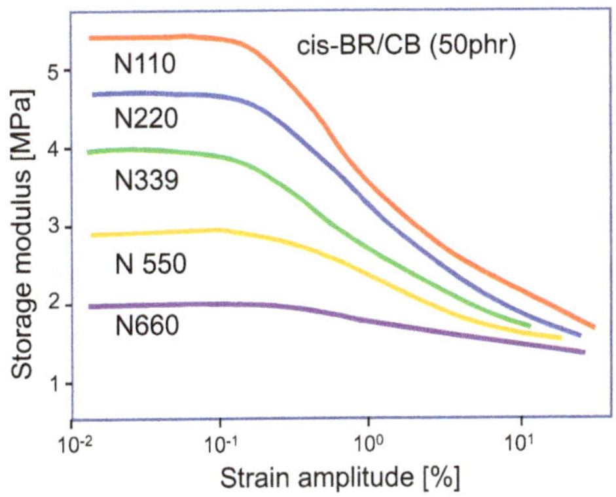


**Figure 10.22**
Comparison of Payne in cis-BR vulcanizates with increase in the specific surface area of carbon black (according to [98])

Comparison of the Payne effect reveals that the carbon blacks of high specific surface area form stiffer filler networks and demonstrate higher energy losses than do low surface area carbon blacks. In line with these results, it has been found that an increase in carbon black loading and a decrease in particle size the raises rolling resistance of tires. Dynamic hysteresis, as expressed by tan δ, increases with specific surface area [96, 99]. Broadening the aggregate size distribution decreases the hysteresis and the rolling resistance at constant specific surface area and structure [100]. It can be generally observed that, compared with conventional products, carbon blacks with broad aggregate size distribution, obtained either from the furnace or by blending of grades with different aggregate size, yield lower tan δ values at temperatures in the rubber state [101, 102]. This effect is attributed to the increase in the inter-aggregate distance, due to broadening of the aggregate size distribution, which leads to a less dense filler network.

### 10.4.1.2 Influence of Carbon Black Structure

Studies have been made of the influence of structure of aggregates on the $G'(\gamma)$ and $G''(\gamma)$ on double the strain-amplitude dependence [103]. Comparison of twelve carbon blacks, which differed in their DBP number and specific surface area, revealed that $G'(\gamma)$ increases in the strain-amplitude measurements, the experiment being based on the effective volume fraction:

$$G'_f = G'_0(1 + 2.5\Phi_{eff} + 14.1\Phi_{eff}{}^2) \tag{10.20}$$

The portion of rubber trapped or caged within the void volume of the filler loses at least some of its identity as part of the rubber matrix and behaves like part of the filler in terms of strain-dependent dynamic properties [104]. Therefore, the effective volume of the filler increases, such that the increased modulus at low deformation is governed primarily by the apparent filler concentration. The breakdown of the filler network with increase in strain amplitude helps to release trapped rubber, with the result that the effective filler volume fraction and hence the modulus decrease. Comparison of compounds obtained from three carbon blacks (N326, N330 and N347) of similar surface area but differing in structure has revealed that the effect of structure is crucial for the dynamic hysteresis. The carbon black with the higher structure yields a higher effective volume fraction, due to more rubber being occluded in the voids of the agglomerates. As a result of the increased storage modulus $G'(\gamma)$, the tan $\delta$ is merely reduced. From a practical standpoint, it is significant that the storage modulus $G'(\gamma)$ can be approximated with reasonable fair accuracy from the rubber modulus and the oil adsorption number. For a series of carbon blacks (14–164 $m^2/g$) in NR vulcanizates, a dependence of the $G'(\gamma)$ values on the DBP number has been found [104], which was confirmed for oil extended SBR 1712 [106]. Thus, the importance of the structure of the aggregates in a carbon black compound has been determined in the case of filler-filler interaction, which plays a particularly important role in dynamically stressed vulcanizates. The storage modulus at high strains correlates well with the structure of carbon black used in SBR test recipes, with an inverse correlation with the primary particle size.

### 10.4.1.3 Influence of Carbon Black Activity

Any change in surface activity of the filler that reduces the cohesive inter-aggregate forces in agglomerates will result in a shift of the percolation threshold and therefore in the magnitude of the $G^*_0$ and $G'_0$ or the $\Delta G' = G'_0 - G'_\infty$, measured at constant loading and constant frequency. The carbon black clusters break down and form sub-clusters as the strain amplitude increases. The filler activity controls the level of $G'_0(\gamma)$ as well as the chemical composition of the rubber. Therefore, there are parameters to reduce the strain-amplitude dependence of $G'(\gamma)$ and $G''(\gamma)$. With regard to the dynamic-strain amplitude and mechanical hysteresis, it has been found that the Payne effect $\Delta G'$ and the loss modulus $G''$ of graphitized carbon blacks are higher than those of the non-graphitized counterparts, as shown in Figure 10.23 for carbon black N234 and carbon black N300 [23].

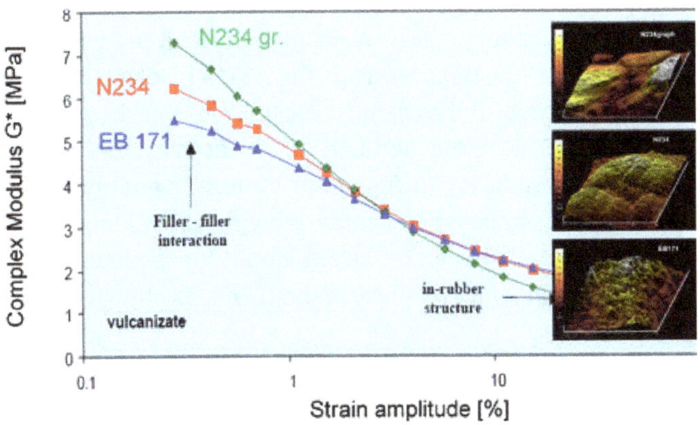


**Figure 10.23** Comparison of carbon black activity: a) carbon black N234, b) experimental black EB 171 and c) graphitized N234 and the corresponding STEM images

At very low deformation amplitudes, the $G'_0(\gamma)$ of graphitized N234 increases above the level of ASTM black. A similar trend is observed for a carbon black of lower surface activity. The higher structure in N347 shows a shift to higher $G^*_0(\gamma)$ value than for carbon black N326. However, the experiments show a strong influence of the graphitized carbon blacks to deliver high $G'_0(\gamma)$. The Payne effect becomes more pronounced, because the filler-filler interaction exceeds the rubber-filler interaction. The effect of graphitization of carbon blacks on the Payne effect is attributed to more developed agglomerates, although the question arises as to why such agglomerates are formed.

The graphitization of carbon blacks has been investigated on filled NR vulcanizates with carbon black N330 (70 phr) that was annealed at 850 °C, 1050 °C, 1800 °C and 2700 °C. The graphitized carbon black in this sequence shows an increase in $G'_0$ due to stronger filler-filler interaction and continuously reduced filler-polymer interaction as $sp^2$ hybridization in the graphitized carbon blacks increases [107]. It was known that the influence of graphitized carbon black increases with the loading [108]. However, compared with the polymer, the surface energy of graphitized carbon black is still high, indicating a tendency to form larger agglomerates, as the adsorptive interactions with rubber chains are on a lower level.

The change in carbon black activity through *in situ* functionalization during the manufacturing process will have a significant effect on the reduction of the $G'_0(\gamma)$ level at low strain amplitudes, due to a much better rubber-filler interaction. Such a property is demonstrated by the experimental carbon black shown in Figure 10.24 in which the morphology is the same as for ASTM black N234, but the distribution of the "nanostructured" carbon black is totally different and makes it possible to reduce $G'_0$ [109].

In combination with superior dispersion, these carbon blacks have a lower Payne effect and lower tan δ values at 60 °C than carbon black N234 of similar specific surface area. Testing of truck tires carried out by TÜV Bayern show less power consumption at higher speed with Ecorax® but similar tread wear [110]. Tan δ measured at 100 °C, which is a proxy for rolling resistance, shows a strong decline in the value for the nanostructured carbon black EB 172 compared with the ASTM black N121. One example of the successful modification of the carbon black surface is carbon black-silica dual-phase fillers (CSDPFs) [55]. Due to the modified surface activity, these types of fillers show not only significantly reduced stiffness and hysteretic energy losses (tan δ) but also a higher abrasion resistance when compared with ASTM blacks. It is interesting to make a direct comparison between the strain-amplitude dependence of tan δ for vulcanizates filled with CSDPF and HD silica/silane (Si69), as shown in Figure 10.24.

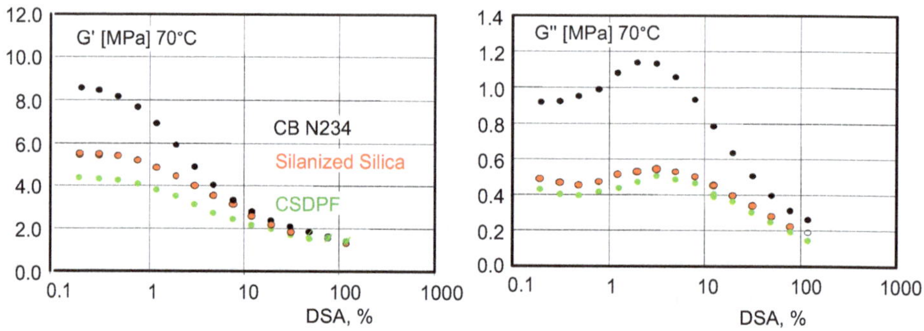


**Figure 10.24** Strain-amplitude dependence of G′(γ) and G″(γ) at 70 °C for S-SBR compounds that contain carbon black N234, silanized HD silica and CSPDF [55]

The surface modification reduces the inter-aggregate interactions and results in better filler dispersion and softer filler networks with lower dynamic hysteresis than in the case of the silica/silane filled vulcanizate. It has to be realized that this effect is a consequence of the superior filler dispersion and does not depend on coupling of the filler to the rubber matrix. It has been found that CSDPF doped with 3.55% silicon on the surface gives the lowest Payne effect, with even lower values than silanized silica. This improvement stems from the significant reduction in interaction between the graphite-like micro-crystallites by the silica nanoparticles that occupy the surface of the carbon black. This creates a better rubber-filler interaction, which results in better dispersion of the CSDPF. In addition, the hysteresis of the compound at low temperatures is improved by the use of the CSDPF [111].

The functionalization of the S-SBR and the carbon black can reduce the rolling resistance to a very low level. The S-SBR is functionalized by adding -COOH groups to the rubber chain. The carbon black N234 is oxidized on the surface and also contains

-COOH groups. The resulting H-bonding between the functionalized rubber and the functionalized carbon black leads to a reduction in the filler-filler interaction and an increase in the rubber-filler interaction. The combination of both groups leads to a modified S-SBR/modified N234 elastomer that exhibits a decrease in $G'_0$ from 11.4 MPa to 4.2 MPa and a clear reduction in the Payne effect in this modified elastomer [112].

Recently, attempts have been made to chemically couple the carbon black aggregates firmly to the rubber network, whereby a shift of the loss modulus G″(γ) to higher deformation amplitudes can be observed, similar to the silanization of silica with bi-functional silanes [113].

#### 10.4.1.4 Influence of Temperature

At constant frequency, G′(γ) and G″(γ) exhibit a decrease as a function of the test temperature. The decrease in the strain-amplitude dependence is related to the improvement in the interaction between the filler surface and the rubber chains. The influence of temperature is shown for S-SBR/CB N121 in Figure 10.25 [7].

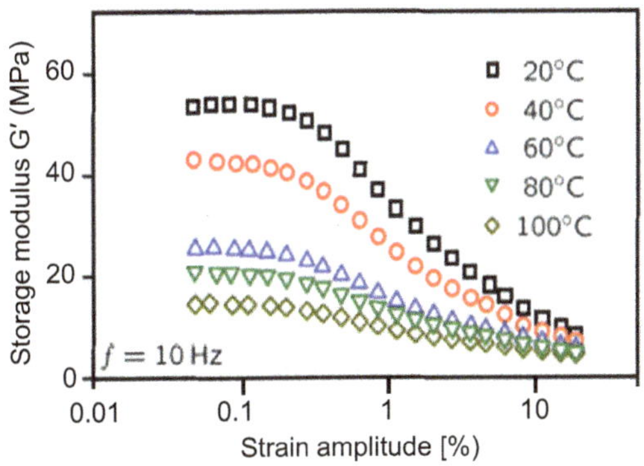


**Figure 10.25**
Decrease in G′(γ) and G″(γ) with test temperature for S-SBR/CB 121 (70 phr) [7]

Between 20 °C and 100 °C, the storage modulus, G′(γ), decrease strongly from 58 MPa to 15 MPa and the loss modulus G″(γ) undergoes the maximum decrease from 12 MPa to 3 MPa and a shift of the maximum to lower strain amplitudes. In the case of S-SBR/CB 234 (70 phr), the storage modulus $G'_0$ at 0 °C is relatively high at 48 MPa and decreases at 70 °C to only 14 MPa. The maximum G″(ω) at 0 °C is 13.2 MPa, decreasing at 70 °C to 2.5 MPa.

In both cases, the physical description of the temperature dependence can be attributed to the influence of temperature on the viscosity of the rubber chains. This makes it possible to reduce the filler-filler interaction and increase the filler-polymer interaction. Therefore, the filler networks break more easily at high temperatures, leading to less-high dynamic moduli. Raising the test temperature would weaken inter-aggregate filler-filler interactions and diminish the modulus of the rubber by reversing chain immobilization. The temperature dependence of G′(γ) and G″(γ) has been discussed for different carbon black grades in S-SBR and NR. The decrease in the dynamic moduli $G'(\gamma)$ and $G''(\gamma)$ with increase in temperature is related first to the free volume of the rubber phase and second to the dissipation of the carbon black grade. It is recommended to match the test conditions for a material under consideration as far as possible to those prevailing in the real use case of the filled elastomer and not to rely on test results obtained at temperatures and frequencies that differ from those in application. This improves the hysteresis behavior and is used to optimize winter treads. The higher loss modulus $G''(T)$ at low temperatures results in good wet grip, while the low loss modulus $G''(T)$ at high temperatures results in good rolling resistance.

#### 10.4.1.5 Influence of Carbon Black Dispersion

With constant carbon black loading, the Payne effect is significantly influenced by the intensity of mixing in the internal mixer. Mixing conditions that are designed to improve filler dispersion, increase the inter-aggregate distance, reduce the Payne effect and the dynamic hysteresis, and so result in low rolling resistance do not have adverse effects on skid resistance and sliding, which is a high frequency property. However, the Payne effect concerns both the filler network and the rubber-filler interface, as will discussed later. The longer mixing process can provoke a reduction in the overall Payne effect and the $G'(\gamma)$ obtained are significantly smaller. In SBR 1712/carbon black N220 (69 phr), the Payne effect has been reduced threefold by increasing the mixing time from 1.5 min to 8 min [114]. It has been found that the viscosity of the rubber is not responsible for this reduction, but that the dispersion of the carbon black led to it [103].

In any mixing trial, an increase in the dispersion phase leads to a significant reduction in the storage modulus, which is measured as a function of the strain amplitude. Attempts to increase the mixing time have led to a significant reduction in the storage modulus of each rubber vulcanizate [115]. The rotor speed makes for better erosion of the carbon black agglomerates and lowers the strain-amplitude storage modulus $G'_0(\gamma)$. For example, the storage modulus measured over a certain deformation amplitude can be used as an indirect criterion for dispersion. Therefore, $\Delta G'$ vs. $\Delta\gamma$, is used to qualitatively monitor the degree of dispersion of a mixture if the rubber compound has the same formulation. Such investigations are often carried out using the rubber process analyzer (RPA). If the filler dispersion is compared over a long period of time with a mixture of the same formulation, a difference in $G'_0(\gamma)$ can clearly be seen which is due to the differences in the weight of the components and also in the way the mixing is carried out. An increase in mineral oil concentration will cause a decrease in $G'(\gamma)$ and $G''(\gamma)$, but the storage modulus decreases more than the loss modulus.

 

### 10.4.1.6 G-Plot

When the strain-amplitude dependence of G′(γ) and G″(γ) is being considered, it proves to be somewhat complicated to assess the differences in both moduli between different vulcanizates. Above all, it is important to identify the vulcanizate that has the lower energy dissipation at the same storage modulus. The relationship between G″(γ) and G′(γ) can be converted to similar representations of mechanical compliance through a Cole-Cole diagram [116]. Since both values, G′(γ) and G″(γ), are a function of double the strain amplitude, γ, it is mathematically correct to represent G″(γ) as a function of G′(γ) [117]:

$$G'(\gamma) = f_1(\gamma) \text{ and } G''(\gamma) = f_2(\gamma) \tag{10.21}$$

Therefore:

$$G''(\gamma) = f_3[G'(\gamma)] \tag{10.22}$$

In a G-plot, G″ and G′ values recorded at the same strain amplitude are plotted in accordance with Equation 10.21. The simple possibilities of this model allow the amount of the loss modulus for a storage modulus to be estimated, as shown in Figure 10.26 [118].

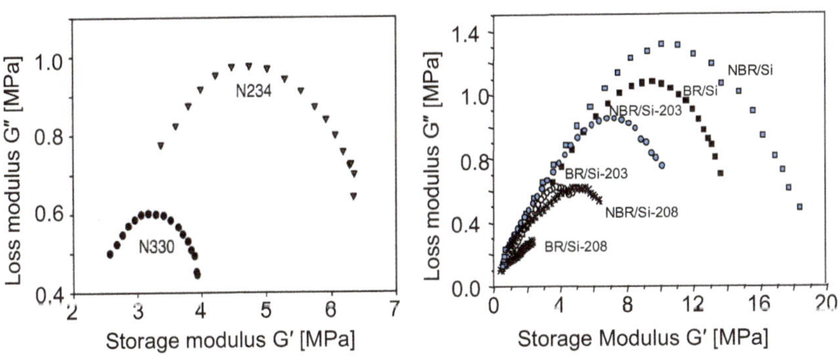


**Figure 10.26** Comparison of filler effects on dynamic moduli using a G-plot

The values measured at the lowest double strain are on the right of the diagram. With increase in double strain amplitudes, the values move to the upper left of the plot. The G-plot permits easy evaluation of the loss modulus at constant storage modulus. Figure 10.26 (left) compares the strain-amplitude characteristics of two carbon black-filled vulcanizates. The carbon black of smaller surface area exhibits lower hysteresis at any deformation amplitude. Figure 10.26 (right) shows the influence of rubber polarity and the effect of monofunctional silanes on the loss modulus of silica filled vulcanizates.

As the dispersion of the filler increases, the maximum of G″(γ) decreases and shifts to smaller G′(γ) values. The G-plot makes it easier to see the benefits of the formulation and also to compare different types of mixes.

#### 10.4.1.7 Models of the Strain-amplitude Dependence of G′(γ) and G″(γ)

The non-linear strain-amplitude dependence of G′(γ) and G″(γ) of filled vulcanizates is the subject of several theoretical models seeking to explain the filler-filler breakdown. Three models look at this phenomenon in different ways. The first model considers only filler-filler agglomeration, the other approach sees a solution for this important phenomenon in the interaction between filler and rubber, while the third is a combination of the other two models in which the interface that develops between the two phases describes the problem.

The "dynamic cluster network" model uses the destruction and re-formation of filler-filler agglomeration in a kinetic approach [119]. However, such a model is unable to explain the aspects of dynamic strain softening. An extension of this model has been proposed which describes the fractal situation of the filled rubbers [120, 121], especially the modulus recovery that is observed when a previously strained material is left to rest for a sufficient period at the appropriate temperature. The second approach is the "dynamic adhesion model" and defines a filled network by the intensity of the polymer-filler interactions and recognizes the importance of the rubber-filler interaction [122, 123]. The model that considers the interaction of the interface during the strain-amplitude experience is called the "immobilized layer" model [124].

Filler network breakdown is the background assumption underlying the "dynamic network model" which assumes that physical bonds between filler aggregates are continuously broken and reformed in a kinetic process during dynamic deformation [119]. It is recognized that the loss modulus depends on energy consumption arising from the disruption and re-formation of filler agglomerates. Less breakdown of agglomerates at small strains will result in less energy loss and a lower G″(γ). With increase in strain amplitude, more aggregates will be broken and re-formed, leading to an increase in G′(γ). When the strain amplitude increases further, the disruption of the agglomerates is more severe and no re-formation takes place. Therefore, G″(γ) will decrease due to the decline in energy dissipation. It is assumed that the part of the filler network destroyed per deformation cycle, $R_b$, depends on the number of filler-filler contacts and the deformation amplitude, γ, or on a deformation amplitude, $f_b$:

$$R_b = k_b N f_b \tag{10.23}$$

The fraction of the new formed filler network per deformation cycle is proportional to the number of filler-filler contacts that are destroyed:

$$R_M = k_m (N_0 - N) f_m \tag{10.24}$$

where $N_0$ is the total number of filler contacts, $k_m$ is a constant and $f_m$ is a function that depends on the deformation

In the equilibrium state, the number of destroyed filler networks is equal to the newly formed filler network, $R_b = R_m$. Therefore, the number of stable filler-filler contacts at a given amplitude is represented by:

$$N = N_0/(1 + k_b f_b/k_m f_m) \tag{10.25}$$

Assuming that the storage modulus, G′(γ), is proportional to the number of filler-filler contacts, the equation of the dynamic moduli becomes:

$$G'(\gamma) = G'_\infty + (G'_0 - G'_\infty)/[1 + (\gamma/\gamma_c)^{2m}] \tag{10.26}$$

where $G'_0(\gamma)$ is the storage modulus at low deformation, $G'_\infty(\gamma)$ is the limit value at high deformation, and $\gamma_c$ is a characteristic deformation amplitude that depends on $k_b$ and $k_m$

Accordingly, the loss modulus is calculated by considering only the separation of the fillers:

$$G''(\gamma) = 2G'' + (G'_0 - G'')(\gamma/\gamma_c)^m/[1 + (\gamma/\gamma_c)^{2m}] \tag{10.27}$$

The derived model has been modified several times to render it suitable for practical measurements. As a result, this model is qualitative and cannot be used for a quantitative description of the G′(γ) modulus without some empirical modifications [125]. The modifications to the theory lead to good agreement of the G″(γ) and G″(γ) values with the experimental results, as shown in Figure 10.27.

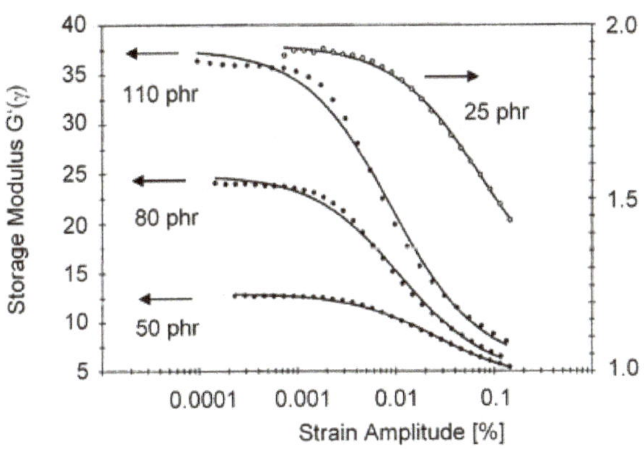


**Figure 10.27** Matching results for G′(γ) using the dynamic cluster network model

An extension of the dynamic filler model is the "cluster-cluster aggregation model" (CCA) [121, 122]. A fractal network structure that refers to a space-filling configuration of kinetically aggregated filler clusters was considered. The model is based on the assumption that the particles can fluctuate around their mean position in the rubber matrix. Flocculation of the filler particles leads to a space-filling configuration of fractal clusters. Each strain amplitude corresponds to a dynamic equilibrium, which is characterized by the size of the cluster. The theory has been applied to NR/N326, NR/330 and S-SBR/N326. The CCA model provides a power law between $G'_0$ and the exponential filler volume fraction, as shown in the following equation:

$$G'_0(\gamma) = \Phi^\alpha \tag{10.28}$$

where $\Phi$ is the volume fraction of the filler and $\alpha$ is the exponent

According to the original work, the exponent is $\alpha = 3.5$. However, for other carbon blacks and other fillers, an exponent with a smaller number has been found [126, 127]. Nevertheless, the CCA theory describes the interaction between the fractal filler particles to form the fractal clusters of the filler network. Although the strain associated with the disruption of the filler network varies with the filler type and deformation modes and frequencies, the maximum loss occurs at the same value of the product of the stress and strain amplitudes, and is constant for a given polymer. For carbon black and silica, this critical strain energy is in the range of 2–4 kJ/$m^3$ [128].

The dynamic adhesion model assumes that the rubber chains are physically adsorbed on the filler surface. The idea is that there are high-energy sites and low-energy on the surface of carbon black. On the carbon black surface, there is a reduced number of sites of strong adsorption and a majority of sites of weak adsorption [122]. A quantitative adhesion model that assumes an equilibrium of the rubber chains based on adsorption-desorption of chain segments with two types of surface site energies of the filler has been developed [123]. The model describes the surface of carbon black in a rubber compound as locally consisting of either free sites or sites occupied by stably adsorbed or unstably adsorbed polymer chains. For the derivation of $G'(\gamma)$ and $G''(\gamma)$, filler-filler contacts are replaced by filler-polymer contacts of different strengths. The result is a relationship that assigns the proportion of stable filler-polymer contacts to an amplitude-independent contribution and uses the proportion of mechanically unstable filler-polymer contacts to describe the amplitude-dependent decrease in $G'$. The storage modulus $G'(\gamma)$ and the loss modulus $G''(\gamma)$ are described by the following relationships:

$$G'(\gamma) = G'_s + G'_i/(1 + c\gamma) \tag{10.29}$$

and

$$G''(\gamma) = G''_s + G'_i c/(1 + (c\gamma)^2) \tag{10.30}$$

where $G'_s$ and $G''_s$ are the moduli on the strong adsorption sites and $G'_i$ and $G''_i$ are the moduli on the weak adsorption sites

The model describes the decrease in $G'(\gamma)$ and passage through the maximum of $G''(\gamma)$ quite well, as shown in Figure 10.28.

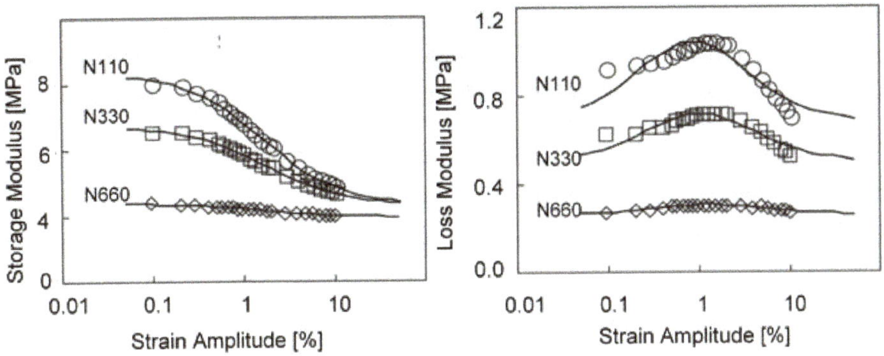


**Figure 10.28** Modelling of the strain-amplitude dependence of G′(γ) and G″(γ)

The model satisfies the dynamic stress softening of G′(γ) reasonably well, but fails to exactly match the asymmetric shape of the G″(γ) versus strain curves [123]. This adhesion model uses the knowledge that there are different adsorption sites on the surface of a carbon black which decisively influence the dynamic properties G′(γ) and G″(γ).

The last of the proposed models is the "immobilized layer", which offers an explanation of the deformation-amplitude behavior under multi-modal excitation [124]. It assumes that a filler cluster can participate in both filler-filler and filler-polymer contacts. The model of the "immobilized layer" is shown in Figure 10.28 [124].

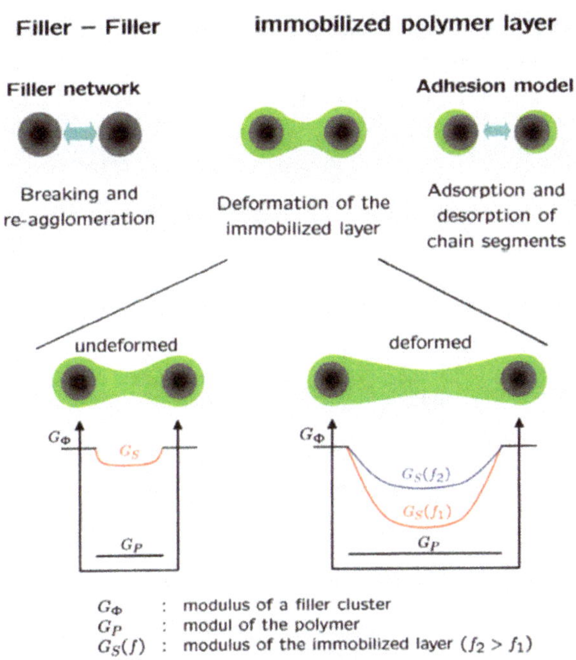


**Figure 10.29** The immobilized layer model tries to explain both filler-filler interaction and filler-polymer interaction [124]

It can be seen that the rubber-filler interface participates strongly in the strain amplitudes of filled elastomers. During deformation, the filler cluster can be broken into several smaller sub-clusters. When the material is stretched, the distance between the fillers increases. This leads to a reduction in the modulus in the intermediate layer and a softening of the system as a whole. This model explains the frequency- and temperature-dependence of the amplitude-dependent behavior in more detailed way. It takes account of the deformation of the entire carbon black cluster and the rubber-filler interface.

## 10.4.2 Effect of Precipitated Silicas

It is important for the dependence of the stress amplitude to indicate the form in which the precipitated silica has been mixed into the rubber [28, 31]. In the simplest case, untreated precipitated silica is used, which is able to build up a strong filler-filler network via the silanol groups. This filler agglomerate returns a high modulus $G'_0(\gamma)$ and a high maximum of the storage modulus in a strain-amplitude dynamic-mechanical test. Both lead to a pronounced Payne effect, $\Delta G'(\gamma) = G'_0 - G'_\infty$. Even during the mixing process, it has been observed in an RPA study that the filler-filler interaction decreases as the mixing time increases. If the mixture is crosslinked after a short period of storage, the storage modulus $G'_0(\gamma)$ of a silica vulcanizate may be influenced [129]. In addition, it should be mentioned that the complex modulus, $G^*(\gamma)$, in the uncured compound is low, but shifted to much higher values after vulcanization. This is shown in Figure 10.30 for HD silica with a specific surface area of 167 $m^2/g$ and silanization with a monofunctional silane, Si216 (8 phr), and with a bifunctional silane, TESPT (10 phr) [31].

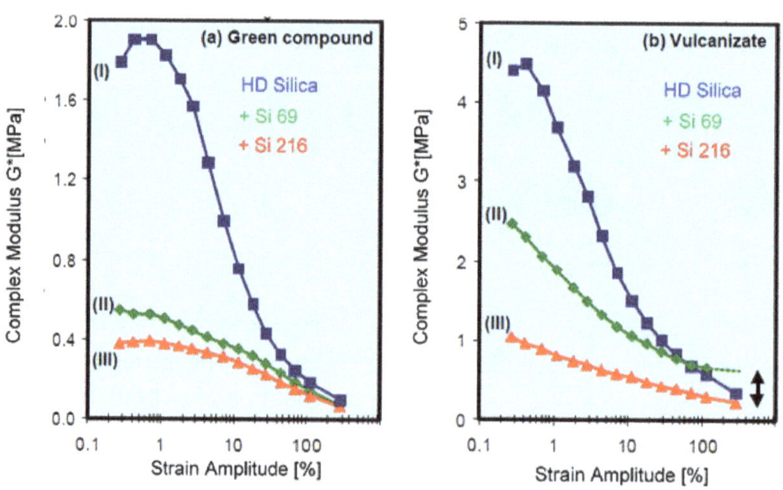


**Figure 10.30** During vulcanization, the strain amplitude increases the complex modulus of the silica-containing vulcanizates and the silanes decrease the Payne effect [31]

In the unvulcanized compound, the complex modulus of the HD silica is 2 MPa, while it is significantly lower in the silanized HD silica, where silanizing with TESPT yields a value of 0.6 MPa and silanization with Si216 leads to an even smaller value of 0.4 MPa. After vulcanization, the complex modulus of HD silica shifts to 4.5 MPa while the modification with TESPT moves to 2.6 MPa, which constitutes a four-fold increase and a doubling of the silanization effect with Si216 after vulcanization. The increase in the complex modulus, G*(γ), after vulcanization is attributable to the flocculation process that takes place during the induction time before the start of the crosslinking process. At this high temperature, the viscosity of the rubber is at its lowest and the reorganization of the filler aggregates can take place relatively easy as long as crosslinking has not developed. This process also shifts the rheological percolation threshold to lower volume fractions of the precipitated silica. For this reason, the stiffness and hysteresis of vulcanizates cannot be estimated from strain amplitude sweeps of the unvulcanized compounds, as measured using RPA.

The specimen without silane displays a strongly increased storage modulus as compared with carbon black elastomers, a fact which is attributable to less-flexible filler-filler contacts in the silane free system, due to the strong dipolar interactions of the silanol groups between the aggregates. As the proportion of silica is increased, the Payne effect increases exponentially over the deformation amplitude and the $G'_0$ increases exponentially with the filler fraction. The breakdown of the HD silica takes place at higher strain amplitudes than is the case for carbon black-filled systems. In the vulcanized systems, the specific surface area of the precipitated silica increases and the value of the storage modulus increases linearly with the BET specific surface area. The shift to lower storage moduli occur at higher strain amplitudes when the particle size of the precipitated silica is smaller. For the loss modulus, G″(γ), the maximum shifts to higher strain amplitudes. This development of primary particle size is shown in Figure 10.31 [28].

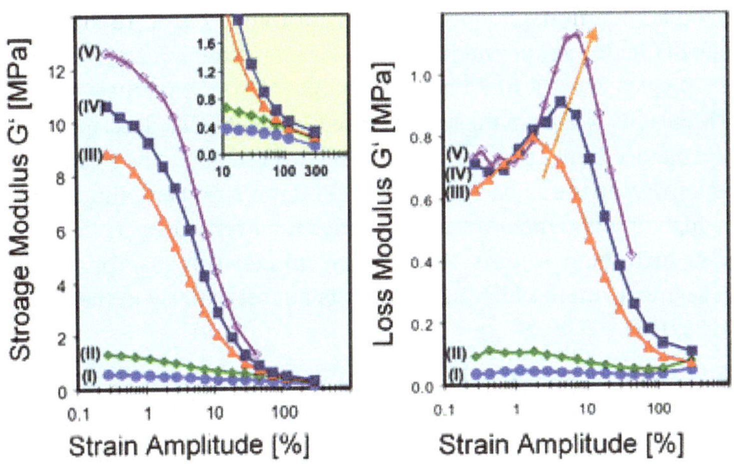


**Figure 10.31** The influence of the surface area on the strain-amplitude dependence of G′(γ) and G″(γ) in S-SBR vulcanizate

The significant increase in the two moduli is governed by the interaction of the silanol groups on the silica surface which increases the silica agglomerates. As the specific surface area increases, the inflection of the G′(γ) and the maximum of G″(γ) shift to higher strain amplitudes. Comparison of conventional silica, VN2, with an $N_2SA$ of 130 $m^2/g$ and highly dispersed silica, HD silica 7000 GR with an $N_2SA$ of 170 $m^2/g$, on one hand, and carbon black N234 with an N2SA of 130 $m^2/g$, on the other, reveals that the precipitated silicas have a higher storage modulus G′(γ), a fact which explains the higher proportion of silica agglomerate. The breakdown of the filler network occurs at higher double strain amplitudes in the case of HD silica, and this is attributable to the presence of polar interparticle interactions that strengthen the filler network. For this reason, the maximum of the energy dissipation G″(γ) shifts to higher double strain amplitudes. A significant shift is shown by the maximum of the loss modulus, G″(γ), which shifts from very small deformation amplitudes to high ones, depending on the specific surface area. Like carbon black N234, precipitated silica is influenced by the structure of the agglomerates. The higher the structure of the silica, the more extensive is the agglomeration into larger particles; and the Payne effect increases. As the specific surface area increases, the Payne effect becomes more pronounced.

Compared with carbon black N234, HD silica yields a high storage moduli, G′(γ) and G″(γ), in respect of strain-amplitude dependence. At 0 °C, $G'_0$ is 18.5 MPa for silica compounds and 14.9 MPa for carbon black, while at 70 °C the relative difference between the filler system increases. The silica system reaches 9.3 MPa while the carbon black system only reaches 5.4 MPa. This fact indicates that the filler-filler interaction is better developed and the effect is stronger in the silica-filled vulcanizates [131]. Newly developed HD silica (Z Premium 200 MP, Rhodia Group) yields lower values for tan δ than other silica grades of similar surface area. This silica has a relatively small diameter of particles and a relatively large aggregate diameter. Compared with a HD silica of the same CTAB the tan δ at 60 °C is lowered from 0.170 to 0.145 [132].

Elastomers filled with precipitated silica show, similar to the case for furnace blacks, a decrease in G′(γ) and a maximum G″(γ) as the temperature increases. To illustrate this with an example, in CR, the compound filled with VN3 has a modulus of 37.7 MPa at 0 °C whereas the HD silica 7000GR has a modulus of 32.7 MPa, so that these values are reduced to 10.4 MPa at 60 °C and in the second case to 9.4 MPa. The description of the temperature dependence of the storage modulus G′(γ) is related to the formation and rupture of a bond between the silica and the rubber. At low strain amplitudes, the activation energy is high at low temperatures and drops to lower values as the temperature increases. At high temperatures, the equilibrium between adsorption and desorption of chain segments on the filler surface shifts almost entirely in the direction of desorption [133, 134].

However, the polarity of the silanol groups is able to control the filler-filler interaction. HD silica mixed with S-SBR of the same styrene content and a variable vinyl content shows better filler dispersion and a reduced Payne effect for the S-SBR of higher vinyl content. Effects similar to those obtained for silica/silane can be achieved by using partially epoxidized SBR or cis-BR. When the amount of epoxy groups is small ( ≤ 7%), a reduction in $G'_0$ occurs, which shows that the Payne effect is less pronounced than in the case of silanized silica (VN3/Si69) by approximately 1 MPa. The Payne effect for the small epoxidation is shown in comparison to the silanization of precipitated silica in Figure 10.32 [135].

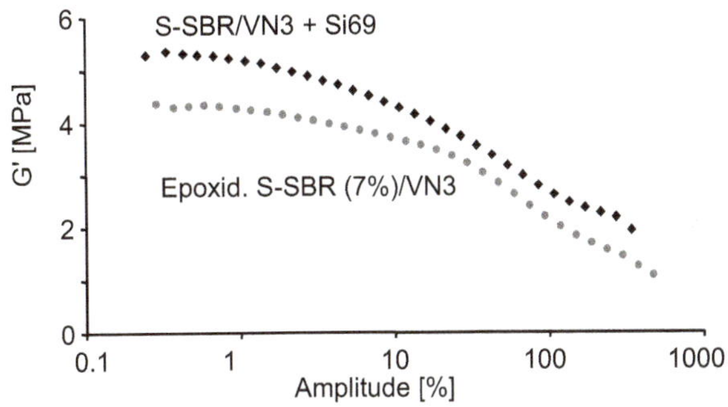


Figure 10.32 The rubber-filler interaction can be increased by slight epoxidizing of S-SBR without silanizing the HD silica [153]

The epoxidation of the S-SBR and cis BR is relatively minor, because the viscoelastic properties have not been changed. An even simpler procedure is to partially epoxidize BR oligomers, which are subsequently blended with the base S-SBR at a certain ratio that does not violate the compatibility rules. With the help of the S-SBR/epoxidized BR oligomer, it is easy to achieve a lower Payne effect in silica-filled vulcanizates than in the more difficult silica/TESPT vulcanizate of the "green" passenger tire. The new compound containing epoxidized oligomer/S-SBR/silica compounds is much cheaper than the silanized S-SBR and is shown in Figure 10.33 [136].

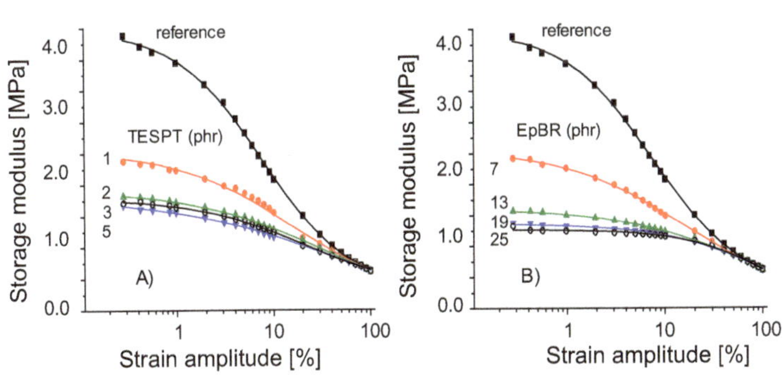


Figure 10.33 Partial epoxidation of cis-BR oligomer blended with S-SBR without silanization of HD silica has a similar effect to silanized silica in the S-SBR vulcanizate [136]

It becomes clear that the same strain-amplitude dependence of the storage modulus, as in the "green tire" formulation, can be achieved with a slightly epoxidized liquid rubber blended with a non-functionalized S-SBR and a non-silanized silica. The functionalized liquid oligomer has a sufficiently low molecular weight to dissolve in the S-SBR and not show phase separation. This additive to the compound formulation eliminates the technically complex, multi-stage silanization of the precipitated silica and is always a cost-effective solution. When NR is transformed to a partially epoxidized NR, there is a much better interaction of the silanol groups with the epoxy rings of the rubber and, because of this interaction, the Payne effect is lower than is the case for silanized silica. It has been found that increasing the epoxy group from 4 to 14 mol.% reduces the Payne effect from 88% to 91% [137].

#### 10.4.2.1 Advantage of Silanized Silica

The silanization of precipitated silica by mono- or bifunctional silanes provides a much better interaction with the rubber that increases the filler dispersion and leads to a systematic reduction in the strain-amplitude dependence [31, 138, 139]. The exponential increase in $G'(\gamma)$ is significantly reduced to lower values when the silane transforms the polar silica into a non-polar silanizate silica. Silanization diminishes the $G'_0(\gamma)$ values, lowers the maximum of $G''(\gamma)$ and shifts the maximum to other energy dissipations. The effect of silanization depends on the quantitative reaction of the silane with the silanol groups. This is illustrated in Figure 10.30, which shows the decrease in the strain-amplitude dependence during the silanization reaction in cis-BR. As silanization progresses, the rubber-filler interaction increases, which improves the development of the dispersion index for the filler.

In the case of monofunctional silanes, the length of the alkyl chain determines the reduction in the polarity of the silica and promotes less silica agglomeration and better contact with the rubber chains. This can be observed when Si203, Si208 and Si216 are used to silanize the silica. The length of the alkyl chain in Si216 leads to a strong improvement in reducing the storage modulus. It has been found that grafting Si208 or Si216 onto silica provides the non-polar energetic surface for increasing the compatibility with the unsaturated polymer. When such a silanized silica is incorporated into NR or S-SBR, a low-filler network with a much lower Payne effect is formed.

Quantitative silanization of the silica surface with triethoxypropyl silane (Si203) reduces the stiffness of the filler network, $G'_0(\gamma)$ by 40%, whereas, with triethoxyoctyl silane (Si208) which possesses a longer alkyl moiety, there is a reduction of 80% [140, 141]. This demonstrates that the shielding of the silica surface by the silane is plays a determining role in the strain-amplitude dependence of $G'(\gamma)$ and $G''(\gamma)$. The viscoelastic properties of S-SBR/HD silica (80 phr) measured at 50 °C with increase in loading of Si208 show that the strain dependence is reduced by the amount of added shielding silane [143]. The reduction in $G'_0(\gamma)$ as a function of the shielding silane is used to illustrate the efficiency of the reaction, which is shown in Figure 10.34. The decrease in the storage modulus to a final level can be determined from this reduction in the value of $G'_0$ (γ).

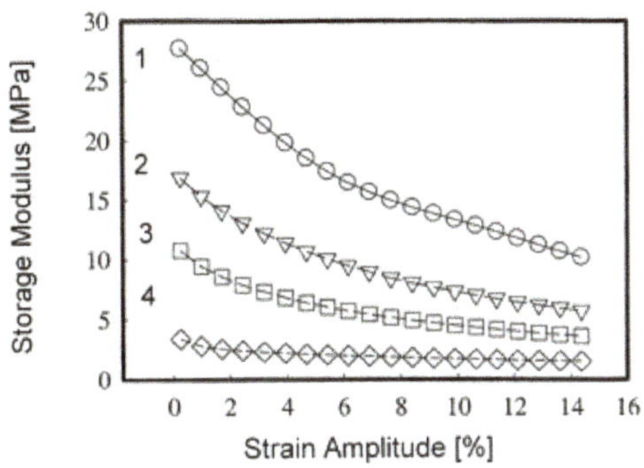


**Figure 10.34** Silane loading reduces the agglomeration of HD silica in S-SBR [143]

A comparison of the efficiency at reducing the Payne effect and hysteresis of bifunctional silanes has shown that bifunctional silanes causes a reduction in G′(γ) and G″(γ) and links the silica to the rubber matrix via sulfur linkages [144]. Also, a comparison of TESPT with Si363 has shown that, at a constant silica loading, the vulcanizates containing 3-mercaptosilane exhibit only half the Payne effect and a lower tan δ, irrespective of the mixing time employed for preparing the compounds. The difference is due to the coupling efficiency of the 3-mercaptosilane, which is reduced by 45% [145].

The question of efficiency also arises when using TESPT and the blocked 3-mercaptosilane (NXT). At different reaction times, the RPA test returns a significantly lower $G'_0$(γ) for silane NXT, which is significantly lower than TESPT, as shown in Figure 10.35 [146].

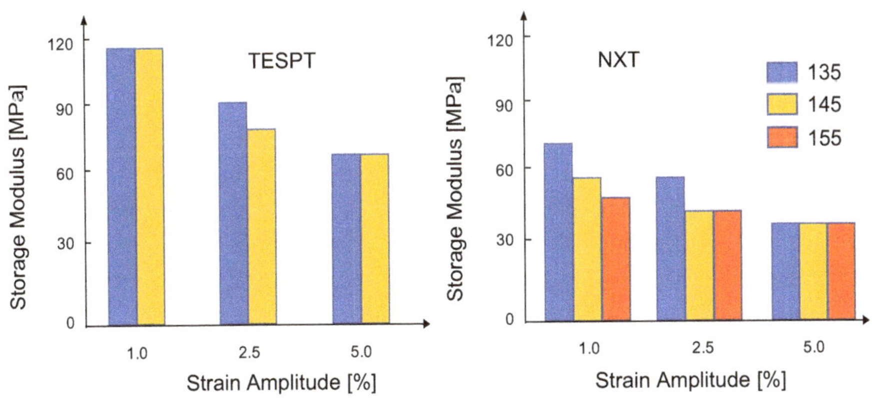


**Figure 10.35** Influence exerted on the storage modulus G′(γ) by the different silanes that can react with HD silica [146]

If the silica is treated with TESPT, the Payne effect in an NR vulcanizate is aided by the presence of proteins that can further reduce the drop in storage modulus and result in a smaller Payne effect [147]. If HD silica in an NR compound is silanized by TESPT, the Payne effect is greatly reduced. The storage modulus $G'_0(\gamma)$ has a value of 7 MPa, which is below that of the carbon black N110 vulcanizate measured at 9.8 MPa and considerably lower than the value for unsilanized silica of 26.5 MPa, as shown in Figure 10.36 [148].

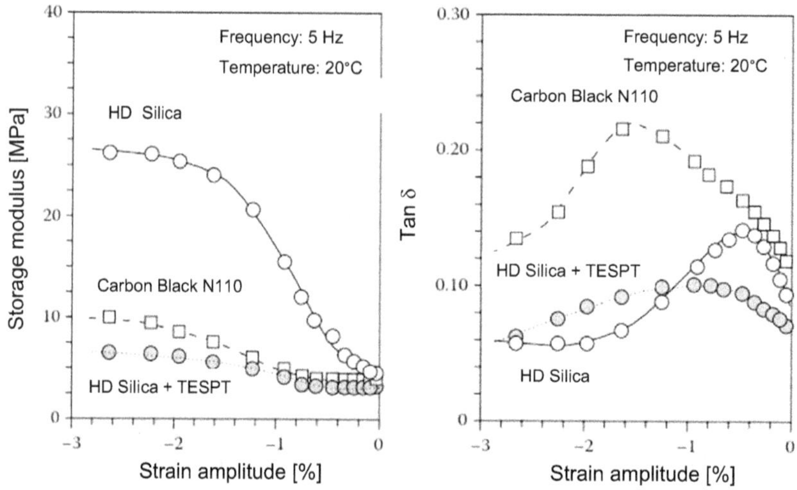


**Figure 10.36** Comparison of HD silica, silanized HD silica and carbon black N110 on the strain-amplitude dependence of $G'(\gamma)$ and $\tan\delta$ for an NR vulcanizate

The plot of $\tan\delta$ as a function of deformation amplitude makes it clear that, in the case of the silanized silica, the maximum shift to smaller double strain amplitudes is closer to the maximum shown by carbon black N110. However, the broad maximum confers an advantage on rolling resistance.

A major difference can be observed on the strain-amplitude dependence of silica prepared by dry-mixing and wet-mixing. Compared with the carbon black N234-filled vulcanizate, the silica compounds exhibit a lower Payne effect, due to the coupling reaction of the silica with the rubber, which reduces the difference in surface energy between the silanized silica and the S-SBR/cis-BR(70/30) blends and increases the polymer-filler interaction. The greater the amount of modified silica, the lower the storage modulus of the silica in the wet compounding. Therefore, the filler-filler interaction is smaller in the wet-mixing precipitated silica, which is reflected also in increased filler dispersion. A comparison of the storage modulus $G'(\gamma)$ is shown in Figure 10.37 [149].

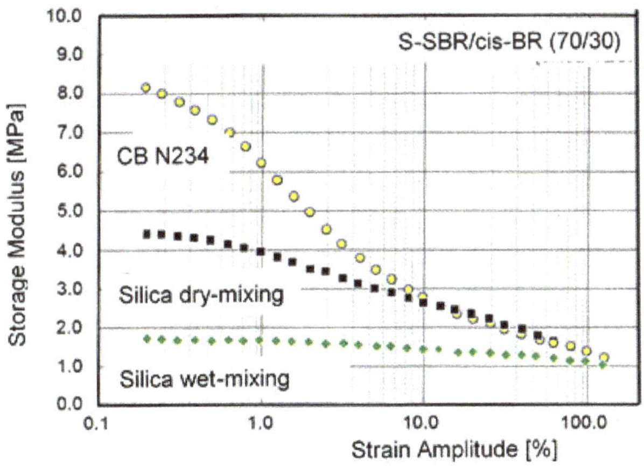


**Figure 10.37** Comparison of the strain-amplitude storage modulus for wet-mixing silica with dry-mixing and carbon black N234 [149]

The strain amplitude of S-SBR/cis-BR is given to exemplify the difference between the different compounding procedures, which affect also the dynamic-mechanical properties.

When two fillers having different surface characteristics are blended, the interaction between the aggregates of different fillers would influence filler networks in the polymer matrix. The question is whether filler networking in a blended filler is more pronounced. The mixing ratio of carbon black N234 (specific area: 119 $m^2/g$) and silanized precipitated silica (specific area: 170 $m^2/g$) reduces the tan $\delta$ values at temperatures of 70–100 °C and effects an improvement of the rolling resistance [23]. With increase in the amount of silica in the blend, the rolling resistance improves from 21% to 48% when the silanized silica is the only filler in the compound. In terms of wet grip, the improvement was compared with carbon black N234, which achieves 100%. A silica content of 90 phr led to an improvement of 9%, as measured with a Grosch LAT tester.

## 10.4.3 Effect of Layered Silicates

The efficiency of a high aspect ratio is dependent on how the intercalated and exfoliated platelets are produced in a rubber matrix. For exfoliated MMT prepared by the continuous dynamic latex compounding (CDLC) technique, a high $G'(\gamma)$ is obtained at low volume fractions. It can be seen that loading of MMT is accompanied by an exponential increase in $G'_0$, with the polarity of the rubber and the emulsifier used in the rubber latex playing a decisive role, as shown in Figure 10.38 [150].

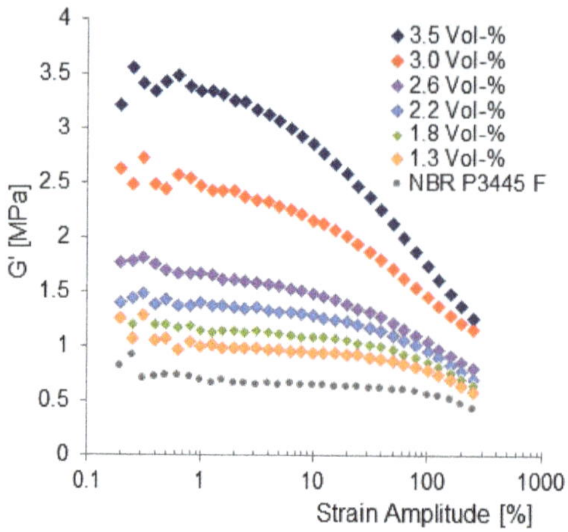


**Figure 10.38** The exponential increase in the Payne effect of exfoliated MMT in NBR vulcanizate [150]

This shows that the low loading of the exfoliated layered silicate can lead to a very high storage modulus, due to the high specific surface area and the high aspect ratio. For 3 vol.% (8 phr) of exfoliated MMT, a $G'_0$ of 2.8 MPa is obtained that is equal to 80 phr of mechanically incorporated non-exfoliated MMT and 50 phr of carbon black N550. From the rheological percolation threshold, which is only 2.3 vol.%, strain amplitude storage modulus $G'_0(\gamma)$ increases significantly to high values. This leads to the conclusion that 1 phr of MMT is approximately equivalent to the effect of 13 phr of carbon black N550. An addition of exfoliated MMT in an amount just below the percolation threshold increases the storage modulus of carbon black N550 (60 phr) from 8.2 MPa to 20 MPa. If there are 80 phr of carbon black, 28 MPa can be achieved with this amount. These amplifying effects are the outcome of a synergistic interaction between MMT and carbon black. A doubling of the storage modulus occurs with 2 vol.% exfoliated MMT. Thus, this material is quite capable of significantly influencing the Payne effect.

### 10.4.4 Effect of Carbon Nanotubes

In the case of multi-walled carbon nanotubes, which are well dispersed in non-polar rubbers, the dynamic strain-amplitude moduli are primarily given by the aspect ratio. Due to alignment in the parallel and perpendicular direction of the manufacturing process, an growing increase in the strain-amplitude dependence of $G'(\gamma)$ and $G''(\gamma)$ can be observed. In NR, multi-walled carbon nanotubes NC7000, an exponential increase in the storage modulus has been determined at room temperature. At a loading of 10 phr, a $G'_0(\gamma)$ of 5 MPa is achieved. As the temperature increases, $G'_0(\gamma)$ decreases. Carbon nanotubes at 3 phr in NR lead to a value of 1.4 MPa at 22 °C, while the same filler only yields 0.83 MPa at 80 °C. This change in the Payne effect with temperature must be taken into account when carbon nanotubes are used [151].

A pronounced exponential increase in the storage modulus G′(γ) up to volume fractions of 10% can be seen, which exceeds the influence of an active carbon black N220. The storage modulus returned by NC7000 in HNBR is shown in Figure 10.39 [152].

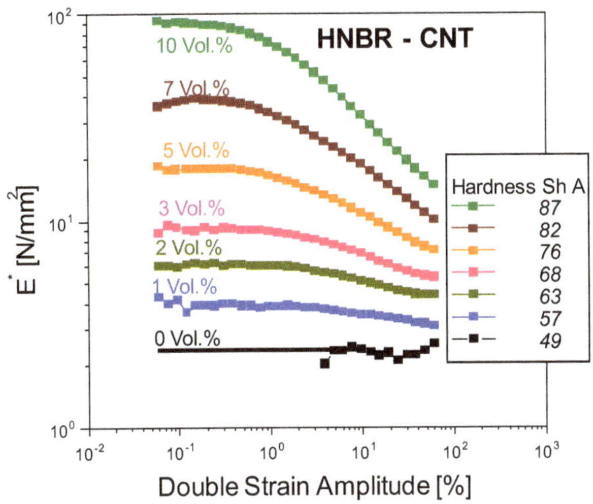


**Figure 10.39** Payne effect increases exponentially with the loading of carbon nanotubes NC7000 [151]

It can be seen that a quantity of 3 vol.% of carbon nanotubes produces a $G'_0(\gamma)$ that corresponds to the volume fraction obtained with CB N220 of 16 vol.%. The value obtained with 10 vol.% cannot be obtained by a carbon black loading. The particularly high Payne effect certainly has an influence when the carbon nanotubes are incorporated into a carbon black compound or a silica compound at small loadings. If wet-mixing of carbon nanotubes is used with HNBR, a reduction of half a percent in the electrical percolation threshold is obtained. The curve of log G′ vs. log Φ decreases compared with carbon nanotubes introduced by dry-mixing. In NBR, the percolation threshold decreases with the amount of polar acrylonitrile group:

$$\text{NBR (28\%ACN) } \Phi = 1.4; \text{ NBR (34\%ACN) } \Phi = 1.2; \text{ NBR (39\%ACN) } \Phi = 0.87$$

The performance of multi-walled carbon nanotubes in SBR was better than that of carbon black N330. This suggests their use in tires of low rolling resistance and high fatigue loss [153]. Carbon nanotubes have been incorporated into car tire treads to improve the performance, especially in wet traction. The influence of the carbon nanotubes on tire performance was investigated on the basis of the dynamic properties. Replacement of 50 phr carbon black N234 by 20 phr carbon nanotubes was able to maintain the hardness, but increased the dynamic moduli, which was beneficial for the tire performance. The multi-walled carbon nanotubes improve traction by 1.5% and traction on a wet road by 6.5% that is related to an improvement in the hysteresis and a decrease in storage modulus in the high frequency domain ($10^4$ to $10^8$ Hz). The trade-off was a 7% higher tire rolling resistance coefficient and 33% worse wear resistance for the tires containing carbon nanotubes [154].

Multi-walled carbon nanotubes can be combined with carbon black or with precipitated silica to increase the electrical conductivity, but also to increase the Payne effect. A combination of carbon black N220 and carbon nanotubes NC 7000 in HNBR shows an increase in $G'_0(\gamma)$ by a factor of 6 and a reduction in the electrical percolation threshold, as shown in Figure 10.40 [152].

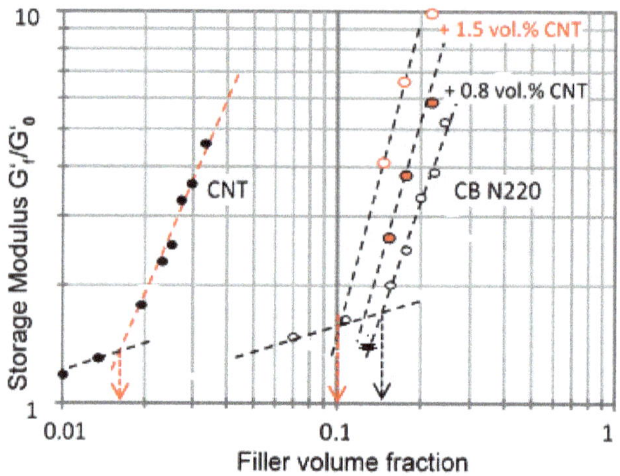


**Figure 10.40** Storage modulus obtained by combining carbon black with carbon nanotubes (0.8 and 1.5 vol.%) in HNBR [152]

A combination of a conventional amount of active carbon black and a small dose of carbon nanotubes increases the strain-amplitude storage modulus by several orders of magnitude. Consideration could certainly be given to saving weight in the material by omitting a considerable proportion of the carbon black and obtaining the same storage modulus with small amounts of carbon nanotubes. In addition, the elastic properties of the vulcanizate are increased significantly by multi-walled carbon nanotubes, reproducing the effect on high rebound elasticity of the compound.

## 10.5 Frequency Dependence of G′(ω) and G″(ω)

Elastomers can exhibit dynamic-mechanical behavior at different frequencies and temperatures. It is assumed that the wet grip of a tire tread is related to a dynamic strain that is generated between the rubber and the road surface at high frequencies and that the rolling resistance is related to the movement of a tire tread at low frequencies [53]. The frequency dependence of the dynamic-mechanical moduli contains valuable information about the viscoelastic properties, ranging from constrained flow to properties governed by entanglements and high frequency phenomena such as skid, friction and braking [49]. In the case of skidding or wet grip, the stress is generated by the asperities of the road surface and the frequency of these deformations is in the range of $10^4$ to $10^8$ Hz at room temperature and depends on the roughness of the road surface, the relative velocity of the sliding tire tread and the dynamic damping properties of the filled elastomer. Timing belts and other products are also exposed to similarly high deformation frequencies and require a viscoelastic frequency procedure.

### 10.5.1 Time-Temperature Superposition

In practice, the testing of such high frequency dynamic-mechanical properties is often limited by the frequency of the test instruments, which can be increased only in a small range at a given temperature. The fundamentals for understanding the frequency dependence of G′(ω) and G″(ω) are presented here for unfilled systems and are based on the temperature-time superposition principle, the WLF equation, which was presented in the chapter on the flow behavior of rubber networks. The time-frequency superposition was proposed in 1955 by Williams, Landel and Ferry [40]. The relationship is derived from the theory of free volume and its temperature dependence. The material exhibiting linear viscoelastic behavior obtained at different temperatures in a range of increasing temperature in a range of increasing frequencies can be superimposed on a master-curve by a horizontal shift along the frequency axis. The shift factor $a_T$ depends on the composition of the vulcanizate. More precise measurements of the frequency dependence of G′(ω) and G″(ω) take account of the changes in density with temperature by applying an additional vertical shift factor, $T_R\, \rho_R/T\rho$. Assuming a constant free volume for amorphous polymers at the glass transition temperature, the WLF relationship is obtained for the viscoelastic processes. The equation for the shift factor is known as the Williams-Landel-Ferry relationship [40]:

$$a_T = \exp[-C_1\,(T - T_R)\,/\,(C_2 + T - T_R)] \tag{10.31}$$

where $T_R$ is the reference temperature, and $C_1$ and $C_2$ are constants which depend on the nature of the rubber

The parameters $C_1$ and $C_2$ vary with the choice of $T_R$. It has been shown that the values can change from one rubber to another [155]. The WLF shift equation has an important ability to predict results at difficult or impossible test conditions using data from conditions that are readily possible. In frequency-dependent dynamic measurements, both the temperature and the strain-amplitude are kept constant and only the frequency of the deformation is varied. The frequency range depends on the type of measuring device used. The frequency range covered by the measuring device at a given temperature usually extends from 0.1 to 50 or a maximum of 100 Hz. As the available dynamic-mechanical test apparatus can only measure over a limited frequency range, the frequency dependence is measured at a number of temperatures and the isotherms obtained are shifted horizontally by a shift factor that is characteristic of the rubber under consideration and in accordance with the WLF method. However, the tests on vulcanized rubber are easier to perform than similar tests on rubber melts, where the material can flow.

## 10.5.2 Frequency Dependence of G′(ω) and G″(ω) in Unfilled Elastomers

Valuable information about the influence of the deformation frequency on viscoelastic moduli can be derived from frequency master-curves of G′(ω) and G″(ω) over several decades of frequencies. The fundamental difference from rubber melts is the fact that polymer chains can no longer undergo large transition movements in the creep zone. Therefore, an exponential increase in the storage modulus is not observed at very low frequencies. Another feature is that entanglements are trapped between chemical crosslinks and the plateau modulus comprises the contribution both of the entanglements and of chemical crosslinks. Some of the useful characteristics that can be deduced with frequency sweeps include the estimates in raw rubber concerning the molecular weight and the molecular weight distribution. However, conclusions about the molecular nature of the rubbers taken from the frequency sweeps should be applied with care. The WLF equation has been successfully applied to individual rubbers. The WLF parameters $C_1$, $C_2$ and $T_R$ for calculating the shift factor $a_T$ according to Equation 10.31 are provided by way of example for NR, SBR and NBR in Table 10.2. The reference temperatures $T_R$ corresponds to the glass transition temperatures measured at 1 Hz [140].

**Table 10.2** WLF Parameters for Shift Factors $a_T$ of Rubber [156]

| Rubber | $C_1$ | $C_2$ | $T_R$ [°C] |
|---|---|---|---|
| NR | 11.8 | 38.5 | -57.2 |
| SBR | 12.9 | 42.9 | −39.4 |
| NBR | 10.9 | 31.3 | −18.8 |

It should be noted that the shift factor $a_T$ resulting from the WLF relationship is valid only at temperatures above $T_g$. The molecular interpretation of the time-temperature equivalence is based on the transition state theory of rate processes [157]. This assumes that a simple interrelation exists between the time scale and the temperature. Thus, the viscoelastic behavior of a system at a given temperature can be related to that at another temperature by changing the frequency scale. The WLF equation works well for many linear rubbers before they are used in a filled rubber compound. As a direct consequence of crosslinking, which does not allow translational movements of network chains, the frequency dependence of G′(ω) and G″(ω) differs significantly from of rubber melts, as shown in Figure 10.41 for an unfilled NR vulcanizate.

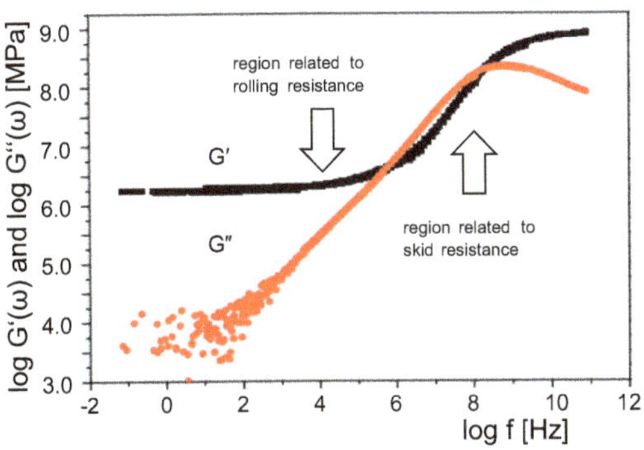


**Figure 10.41**
Master-curve of frequency- and temperature-dependent measurements on unfilled NR vulcanizate

From low frequencies to the plateau zone, G′(ω) increases steadily with a small gradient in the double-logarithmic plot, but does not show the pronounced increase found in the case of rubber melts. In this frequency range, G″(ω) increases significantly and reaches a maximum in the plateau zone. The height of the plateau is specified in terms of both the molecular weight-independent plateau modulus $G_N$ and the crosslink density or the state of cure of the vulcanizate. Taking into account that crosslinking traps the rubber matrix entanglements, it becomes clear why the plateau modulus, $G_N$, is important for the evaluation of dynamic properties. The G″(ω) values in the plateau zone are considered to provide information about energy dissipation processes, which are essential for the rolling resistance of tires, whereas the transition zone at higher frequencies delivers information concerning traction, sliding and grip properties [158]. On approaching the transition zone, the situation is much the same for polymer melts and crosslinked polymers. Both viscoelastic functions, the storage modulus G′(ω) and the loss modulus G″ (ω), strongly increase with the frequency, because chain mobility is unable to follow the oscillatory movement over time. At very high frequencies, where the polymer chains cannot follow the deformation within the period of oscillation, the material becomes hard, although its physical state does not change. Most of the energy is dissipated during oscillation and therefore the following holds:

$$G''(\omega) > G'(\omega) \tag{10.32}$$

The relevant range that needs expansion of the frequency scale is around $10^5$ to $10^6$ Hz. Such a high frequency cannot be produced by dynamical instruments except the acoustic spectrometer. The loss modulus G″(ω) of the vulcanizate in this frequency range determines the contact mechanical behavior in addition to the surface roughness and the velocity of the sliding vulcanizate on the asperities of the road [159, 160]. On the basis of G″(ω) values determined from master-curves, the dynamic damping of vulcanizates can be specifically designed in order to achieve optimal values for dry traction, wet traction, ice traction and shortening the braking distance of vehicles.

However, conventional dynamic-mechanical measurements cannot access the high-frequency range, and so, based on the frequency-temperature superposition principle, tan δ at 1 Hz is often used as a predictor of traction for tire tread compounds. Detailed superposition needs to be conducted for each distinct compound, because tan δ at 0 °C and a frequency of 1 Hz do not constitute a universal predictor [161].

The great importance of the principle of time and temperature equivalence lies in its physical consistency, which enables dynamic material properties to be predicted in temperature and frequency ranges that are not accessible by commercial test devices. The advantage of the method is that material developments in components can be carried out more efficiently by characterizing the properties of laboratory samples. In addition to the temperature and frequency, however, it is important that the conditions to which the product is exposed in service be reproduced as exactly as possible in the dynamic test. The WLF principle has also been applied to rubber blends [162]. For the unfilled blend S-SBR/cis-BR (75/25) at a reference temperature of 25 °C, a discontinuity is found in the master-curve of tan δ at high frequencies that might be caused by the difference in the solubility parameters of the original rubbers.

The use of ultrasound spectrometers that measure wave propagation through the elastomers allows direct measurement of an attenuation coefficient, α, without the need to resort to the WLF shifting technique. With the help of this technique, the high frequency dynamic properties for unfilled and filled vulcanizates become accessible. Results have been published in several reports [44, 45, 58]. These show the effect of frequency by comparing the loss signals G″(T) at low frequencies and the attenuation coefficient, α, at 0.5 MHz, as measured by ultrasonic spectroscopy [58]. Compared with the low frequency regime, the attenuation peak, α, is significantly shifted to higher temperatures. Interestingly, it has been found that, at high frequencies, the vinyl groups in SBR contribute more to the dissipation process than styrene units.

### 10.5.3 Frequency Dependence of G′(ω) and G″(ω) in Filled Elastomers

The time-temperature master-curve can only be produced for a filled vulcanizate if, in addition to the frequency-dependent horizontal shift, a vertical shift is also performed, which should take into account the filler-filler interaction. The constructing of a smooth G′(ω) and G″(ω) master-curves from isothermal frequency sweeps of filled vulcanizates is not simple. A horizontal shift using the $a_T$ determined for the rubber does not yield smooth master-curves for either G′(ω) or G″(ω).

This is shown in Figure 10.42.

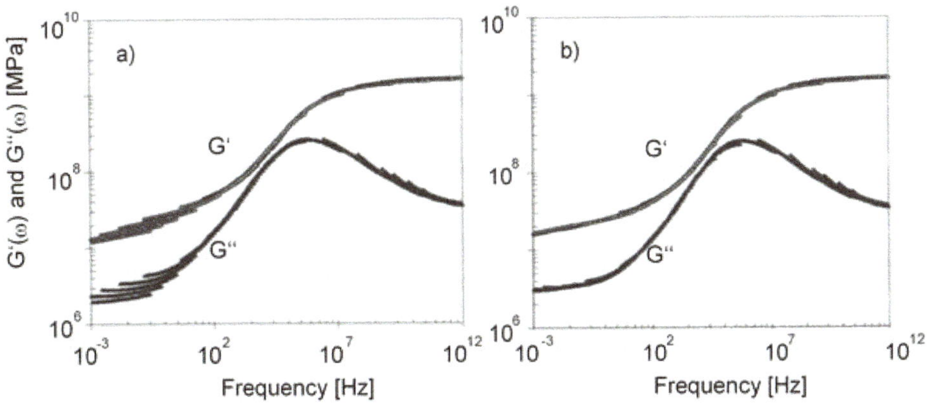


**Figure 10.42** Frequency master-curves using only the horizontal shift (left) and the vertical shift (right)

The shift leads to a feather-like family of curves that does not allow any statement to be made about the value of the dynamic moduli, G′(ω) and G″(ω), in a certain frequency range. It should be noted, that when active nano-particle fillers are present in the polymer matrix, the temperature-dependent viscoelasticity of filler-filler networks necessitates the application of some vertical shifting that compensates for the reinforcing effects of the filler. Once a vertical adjustment of the shifted isotherms is applied, a smooth master-curve is obtained.

A perfectly elastic tire would not dissipate sufficient energy to stop the vehicle during braking. To overcome the contradictory requirements of low rolling resistance but adequate and safe braking performance, tread compounds need to dissipate energy during the higher dynamic deformation encountered during sliding. This requires a sufficiently high glass transition temperature on the part of the polymer matrix (> −45 °C), so that the dynamics of the polymer segments are in the order of $10^5$ Hz or higher at ambient temperature. When tire tread rubber is sliding at speeds in the m/s range over the asperities on a road surface, the contact surface is exposed to high-frequency deformation in the MHz range [164].

Only with this method can the time-temperature superposition be reasonably applied to filled vulcanizates. It has been observed that the vertical shift factors increase with the specific surface area of the filler at constant loading. This indicates that the vertical shift factors are related to surface activity and correspondingly to the increase in the filler-filler interaction. Due to the polymer-filler interaction, the properties depend on the individual components and their interaction. For example, in the tire industry there is a good correlation between passenger tire wet traction and dynamic properties measurements at normal operating temperatures and a frequency of 200 Hz. A different layer of immobilized rubber shell will form the surrounding of the filler aggregate. Using S-SBR/cis-BR (75/25) compounds filled with carbon black, silanized precipitated HD silica and carbon-silica dual-phase fillers (CSPDF) and aromatic oil the master-curve are obtained in the temperature range of −60 to +60 °C and the frequency range of 0.032 to 32 Hz [165]. The log $\omega a_T$ varies over several decades. In the master-curve of log $G'(\omega)$ versus frequency, an increase at low frequency to high log $\omega a_T$ values rises from 0.5 MPa to 3 MPa at high frequency. Similarly, the log $G''(\omega)$ varies over the same ranger of frequencies from 0.20 MPa to 2 MPa, where it passes through a maximum. A comparison with silanized silica or CSDPF demonstrates that log $G'(\omega)$ values at low frequencies are smaller when CSDPF and silanized silica are present. Both $G'(\omega)$ and $G''(\omega)$ of silanized silica and CSPDF compounds have lower values at low frequencies and higher values at high frequencies [166].

The dynamic moduli, $G'(\omega)$ and $G''(\omega)$, are useful for predicting dynamic properties in the high-frequency region where no values are available from direct measurements. The frequency range for skidding and optimum braking depends on both the speed of the vehicle and the roughness of the road. While the first factor is usually known, the second has to be measured or approximated. Much effort has been expended on establishing equations that describe braking as a function of loss modulus and the parameters mentioned above. Consequently, tire tread compounds have been developed to reduce the braking distance of cars.

An ultrasonic spectrometer is used to measure the energy dissipation in the glass transition region at 0.5 MHz using the attenuation coefficient, $\alpha$. A comparison of the effect of carbon black N347 on the attenuation coefficient has revealed an increase in the amplitude of the signal and the area of the damping signal, as illustrated for NR and NBR [58].

The broadening of the attenuation signal towards higher temperatures can be attributed to polymer-filler interaction that governs chain attachment to the filler surface [167, 168]. The dissipation maxima increase with the filler content in both vulcanizates. In the NR vulcanizate, the filler loading increases the damping maximum proportionally and causes a small shift in overall signal toward higher temperatures while, in the NBR vulcanizate, the damping maximum and the signal area are shifted toward higher temperatures. It has additionally been found that chain units have positive effects on damping at high frequencies. For S-SBR, it has been found that $\alpha_{max}$ increase in proportion to the styrene content and the filler volume fraction. For silica filled S-SBR, it has been shown that the attenuation peak increases linearly with the silica loading. The effect of vinyl units on high frequency damping has been found to be greater than that of styrene units [169].

Through the use of ultrasonic spectroscopy, a good correlation between the wet-traction ranking of seven industrial tread compounds and the attenuation coefficient, $\alpha$, corrected for the frequency-dependent $T_g$ shift, has been found in a multi-client project performed at the German Rubber Institute [170]. The correlation is shown in Figure 10.43.

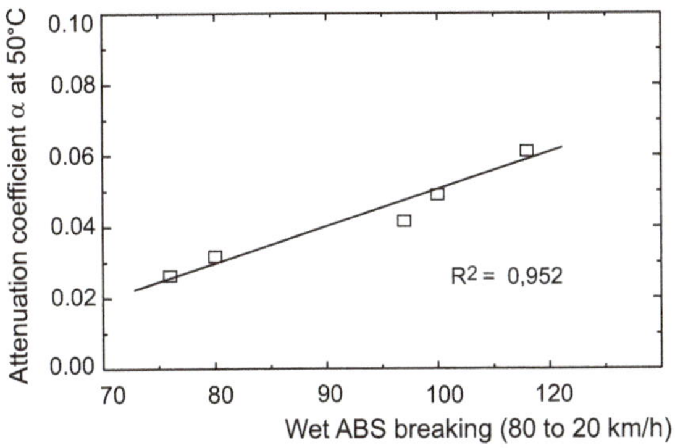


**Figure 10.43**
Correlation of the attenuation coefficient, α, with wet ABS braking [170]

The compounds with the highest attenuation exhibit the best wet/traction values. This result underlines the importance of methods that can measure high frequency dynamic properties directly. Unfortunately, this method, which would also be important for other products exposed to high deformation frequencies is not available commercially.

For rubber friction studies, for which the high-frequency viscoelastic properties are an important input, the strain can be very large and it is important to account for non-linearity in the viscoelastic response. However, rigorous accounting of non-linearity is difficult. For rubber friction on road surfaces, a direct measurement of viscoelastic modulus in response to an oscillating strain is considered a good approach, because the pulsating deformation acting on a rubber surface sliding on a hard, rough road surface is similar to the oscillatory deformation generated in a DMA experiment. This is because road asperities slide along the rubber surface with only little pre-strain. The shear modulus obtained without pre-strain is probably better than the modulus obtained in tension with pre-strain [153].

## References


[1] Treloar, R. G., in "The Physics of Rubber Elasticity", 3rd ed., Oxford University Press, Oxford (1975)

[2] Ferry, J. D., in "Viscoelastic Properties of Polymers", Wiley & Sons, New York (1980)

[3] Elias, H. G., in "Macromolecules", Hüthig Verlag, Heidelberg (1981)

[4] Eisele, U., in "Introduction to Polymer Physics", Springer Verlag, Berlin (1990)

[5] Strobl, G., in "The Physics of Polymers", 3rd ed., Springer Verlag, Berlin, Heidelberg (2007)

[6] Maxwell, B., J. Polym. Sci., 1956, 20: 551

[7] Wrana, C., in "Polymer Physik", Springer Spectrum, Berlin, Heidelberg (2014)

[8] Zener, C., in "Elasticity and Anelasticity of Metals", Chicago Univ. Press, Chicago (1948)

[9] Roland, C. M., Rubber Chem. Technol., 2006, 79: 429

[10] ASTM D 5992

[11] DIN 53 513

[12] ISO 4664

[13] ASTM D 1054–66

[14] DIN 53512

[15] ISO 4662

[16] Gehman, S. D., Rubber Chem. Technol., 1957, 30: 1202

[17] Meyer, R., Kautsch. Gummi Kunstst., 1981, 34: 750

[18] Rehage, G., Kautsch. Gummi Kunstst., 1981, 34: 715

[19] Bayer AG-KA, Butyl and Halobutyl Compounding Guide for Non-Tyre Applications, 12/ 1992

[20] Boyer, R. F., J. Polym. Sci. Symp., 1975, 50: 189

[21] Fusco, J. V., in "Basic Elastomer Technology", Baranwal, K. C., Stephens, H. L., Eds., Rubber Division, American Chemical Society, Akron (2001) Ch. 9E

[22] Kraus, G., Angew. Makromol. Chem., 1977, 60/61, 2015

[23] Fröhlich, J., Niedermeier, W., Educ. Symp. DKT '09, Nuremberg (2009)

[24] Wang, M.-J., Wolff, S., Tan, E.-H., Rubber Chem. Technol., 1993, 66: 178

[25] Wolff, S., Wang, M.-J., Kautsch. Gummi Kunstst., 1994, 47: 17

[26] Hess, W. M., Rubber Chem. Technol., 1991, 64: 386

[27] Gessler, A. M., Hess, W. M., Medalia, A. I., Plastics and Rubber, Processing, June (1978)

[28] Görl, U., in "Handbuch der Kautschuk-Technologie", Hofmann, W., Gupta, H., Eds., Dr Gupta Verlag, Ratingen (2001)

[29] Vaikunartam, S. R., et al., Rubber Chem., 2017, 90: 467

[30] Manoharan, P., Naskar, K., J. Appl. Polym. Sci., 2016, 133: 43531

[31] Meon, W., Blume, A., Luginsland, H.-D., Uhrland, S., in "Rubber Compounding: Chemistry and Applications", Rodgers, B., Ed., Marcel Dekker, (2004) Ch. 7

[32] Schellenberg, M., Schuster, R. H., ACS Rubber Division Meeting, Cleveland (2013)

[33] Schuster, R. H., Wrana, C., ACS Rubber Division Meeting, San Antonio (2016)

[34] Flory, P. J., "Principles of Polymer Chemistry", Cornell Univ. Press, Ithaca, New York, (1953) Ch. 11

[35] Nordsiek, K. H., Kautsch. Gummi Kunstst., 1986, 39: 599

[36] Obrecht, W., Lecture at Advanced Training Course in Rubber Technology, University Hanover (2015)

[37] Gordon, M., Taylor, G. S., J. Appl. Chem. 1952, 2: 493

[38] Fox, T. G., Bull. Am. Phys. Soc., 1956, 1: 123

[39] Mouri, H., Hall, J. E., 146th ACS Rubber Division (Pittsburgh)

[40] Williams, M. L., Landel, R. F., Ferry, J. D., J. Amer. Chem. Soc., 1955, 77: 3701

[41] Geisler, H., Scholz, G., Schuster, R. H., Kautsch. Gummi Kunstst., 2002, 55: 653

[42] Wommelsdorff, R., Hamann, W., Kautsch. Gummi Kunstst., 1992, 45: 24

[43] Gerspacher, M., Nikiel, L., ACS-Rubber Division Meeting, Dallas (TX) April (2000)

[44] Kroll, J., Alshuth, Th., Schuster, R. H., Kautsch. Gummi Kunstst., 2006, 59: 33

[45] Oprisoni, C., Alshuth, Th., Schuster, R. H., Kautsch. Gummi Kunstst., 2008, 61: 301

[46] Ecker, E., Kautsch. Gummi Kunstst., 1968, 21: 304

[47] Wang, M.-J., Workshop "Theory of Reinforcement", Hanover, (1996)

[48] Ziegler, J., Schuster, R. H., Kautsch. Gummi Kunstst., 2003, 56: 159

[49] Medalia, A. I., Rubber Chem. Technol., 1978, 51: 437

[50] Wang, M.-J., Rubber Chem. Technol., 1998, 71: 520

[51] Schuster, R. H., Decenial Congr. Polymer Competence Center, Leoben (Austria) 2012

[52] Saito, Y., Kautsch. Gummi Kunstst., 1986, 39: 30

[53] Nordsiek, K. H., Kautsch. Gummi Kunstst., 1985, 38: 178

[54] Wolff, S., Wang, M.-J., in "Carbon Black, Science and Technology", Donnet, J.-B., Bansal, R. C., Wang, M.-J., Eds., Marcel Dekker, Inc., New York (1993) Ch. 9

[55] Wang, M.-J., Mahmud, K., Murphy, L. J., Kautsch. Gummi Kunstst., 1998, 51: 348

[56] Niedermeier, W., Fröhlich, J., Kautsch. Gummi Kunstst., 2003, 56: 519

[57] Westenberg, H., Gomez, H., Rechenbach, D., Kautsch. Gummi Kunstst., 2021, 73: 47

[58] Schuster, R. H., Oprisoni, C., ACS Rubber Division Meeting, Milwaukee (2011)

[59] Klüppel, M., Meier, J., Heinrich, G., in "Constitutive Models in Rubber", Busfield, J., Mure, A., Eds. (2003)

[60] Ahn, B., Kim, D., Kim, K., Comp. Interfaces, 2019, 26: 585

[61] Thiele, S., Belgart, D., Holzleg, M., Kautsch. Gummi Kunstst., 2008, 70: 244

[62] Colvin, H., Senyak, M., US Pat. 6,512,053 B1

[63] Klockmann, O., Proc. DKT `06, Nuremberg, (2006)

[64] Zaformehrabian, R., et al., E-J. Chem. 2012, 9: 1102

[65] Fragiadakis, D., Bokobza, L., Pissis, P., Polymer, 2011, 52: 3175

[66] Oprisoni, C., PhD Thesis University of Hanover (2011)

[67] Kader, M. A., Kim, K., Lee, Y.-S. Nah., C., J. Mater. Sci., 2006, 41: 7341

[68] Klafke de Azeredo, L., Schuster, R. H., Jacobi, M. M., Kautsch. Gummi Kunstst., 64: 7

[69] Gatos, K. G., Karger-Kocsis, J., Polymer, 2005, 46: 3065

[70] Abdollali, M., et al., Iranian Polym. J., 2008, 17: 519

[71] Gatos, K. G., Karger-Kocsis, J., in "Rubber Nanocomposites", Thomas, S., Stephens, R., Eds., John Wiley&Sons (Asia) Pte Ltd, Singapore (2010)

[73] Galimberti, M., et al., Compos, Sci., Technol., 2009, 69: 1206

[74] Cataldo, F., Macromol. Symp. 2007, 247: 67

[75] Ra, D., et al., Amer. Soc. Mech. Eng., 2002, 55: 495

[76] Nah, Ch., Liun, J. Y., Cho, B. H., Hong, Ch. K., Gent, A. N., J. Appl. Polym. Sci., 2010, 118: 1574

[78] Lu, L., Zai, Y., Zhang, Y., Guo, G., Appl. Surf. Sci., 2008, 255: 2162

[79] Ismail, R., Ibraham, A., Rusop, M., Adnan, A., Nano-Sci. Tech., 2019, 2051

[80] Zimmermann, H., Schuster, R. H., Kautsch. Gummi Kunstst., 2011, 64: 7

[81] Fletcher, W. P., Gent, A. N., Inst. Rubber Ind., 1953, 29: 266

[82] Payne, A. R., J. Appl. Polym. Sci., 1962, 6: 1

[83] Payne, A. R., J. Appl. Polym. Sci., 1963, 7: 57

[84] Payne, A. R., J. Appl. Polym. Sci., 1963, 7: 432

[85] Payne, A. R., J. Appl. Polym. Sci., 1964, 8: 2661

[86] Gehman, S. D., Woodford, D. E., Stambaugh, R. B., Ind. Eng. Chem., 1941, 33: 1032

[87] Dillon, J. H., Prettman, J. B., Hall, G. L., J. Appl. Physics 1944, 15: 309

[88] Roland, C. M., Lee, G. F., Rubber Chem. Technol., 1990, 63: 555

[89] Klüppel, M., Heinrich, G., Rubber Chem. Technol., 1995, 68: 623

[90] Klüppel, M., Schuster, R. H., Heinrich, G., Rubber Chem. Technol., 1996, 70: 243

[91] Schuster, R. H., Proc. Mater. Res. Soc., Boston (2003)

[92] Schuster, R. H., Proc. Eurofller, Bratislava (2013)

[93] Kraus, G., J. Appl. Polym. Sci., J. Polym. Symp., 1984, 39: 75
[94] Payne, A. R., PhD Thesis, Durham University (1965)
[95] Schroeder, A., Klüppel, M., Schuster, R. H., Heidberg, J., Carbon, 2001, 30: 401
[96] Hess, W. M., Vegvari, P. C., Swor, R. A., Rubber Chem. Technol., 1985, 58: 350
[97] Ayala, J. A., Hess, W. M., Dotoson, A. O., Rubber Chem. Technol., 1990, 63: 747
[98] Robertson, C. G., Rubber Chem. Technol., 2011, 84: 507
[99] Patel, A. C., Lee, K. W., Elastomerics, 1990, 122: 22
[100] Hess, W. M., Klamp, K., Rubber Chem. Technol., 1983, 56: 390
[101] Kraus, G., Proc. Int. Rubber Conf. Brighton, UK (1977)
[102] Stacy, C. J., Chirico, V. E., Kraus, G., Rubber Chem. Technol., 1975, 48: 538
[103] Medalia, A. I., Rubber Chem. Technol., 1973, 46: 877
[104] Medalia, A. I., J. Colloid. Interface Sci., 1970, 32: 115
[105] Kraus, G., Angew. Makromol. Chem., 1977, 60/61: 215
[106] Ulmer, J. D., Chirico, V. E., Dizon, E. S., Rubber Chem. Technol., 1975; 48: 592
[107] Payne, A. R., J. Appl. Polym. Sci., 1965, 9: 3245
[108] Medalia, A. I., Laube, S. G., Rubber Chem. Technol., 1978, 51: 89
[109] Niedermeier, W., Freund, B., ACS Rubber Division, Nashville (1998, p. 18
[110] Fröhlich, J., Niedermeier, W., Proceed. DKT '16, Nuremberg (2016)
[111] Wang, M.-J., ACS Rubber Division Meeting, Anaheim (1997)
[112] Herd, Ch., IOM 3 RIEG ATDM (2009)
[113] Frick, A., Borm, V., Murolidkoran, V., Kautsch. Gummi Kunstst., 2018, 70: 2
[114] Boonstra, B. B., Medalia, A. I., Rubber Age, 1963, 46: 892
[115] Geisler, H., Staeber, R., Schuster, R. H., Report to AiF (1996)
[116] Cole, K. S., Cole, R. H., J. Chem. Phys., 1941, 9: 341
[117] Gerspacher, M., in "Carbon Black Science and Technology", Donnet, J.-B., Bansal, R. C., Wang, M.-J., Eds., Marcel Dekker, New York, (1993)
[118] Schuster, R. H., Meier, J., Klüppel, M., Kautsch. Gummi Kunstst., 2000, 53: 563
[119] Kraus, G., J. Appl. Polym. Symp., 1984, 39: 75
[120] Klüppel, M., Heinrich, G., Rubber Chem. Technol., 1995, 68: 623
[121] Klüppel, M., Schuster, R. H., Heinrich, Rubber Chem. Technol., 1997, 70: 243
[122] Funt, J. M., Rubber Chem. Technol., 1987, 61: 842
[123] Maier, G., Goeritz, D., Kautsch. Gummi Kunstst., 1996, 49: 18
[124] Sternstein, S., Zhu, A., Macromolecules, 2000, 67: 7262
[125] Ulmer, J. D., Rubber Chem. Technol., 1996, 69: 15
[126] Pian, J. M., Dorget, M., Palierne, J. F., Pouchelon, M., J. Rheology, 1999, 43: 305
[127] Schuster, R. H., Eurofiller Bratislava (2012)
[128] Robertson, C. G., Wang, X., Phys. Rev. Letters, 2005, 95: 075703
[129] Sgodzaj, U., PhD Thesis, University of Karlsruhe (2001)
[130] Müller, M., Klüppel, M., Schuster, R. H., World Amazon Rubber Conference, Manaus, (1999)
[131] Wang, M.-J., Rubber Chem. Technol., 1999, 72: 430
[132] Guy, L., et al., Kautsch. Gummi Kunstst., 2009, 51: 383
[133] Wrana, C., in "Introduction to Poylymer Physics", Lanxess (2009)
[134] Clement, F., Bokobza, L., Monerie, L., Rubber Chem. Technol., 2005, 78: 2127
[135] Rocha, T. L. A. C., Schuster, R. H., Jacobi, M. M., Kautsch, Gummi Kunstst., 2004, 57: 656

[136] Jacobi, M. M., Braum, M., Rocha, T. L. C. A., Schuster, R. H., J. Appl. Polym. Sci., 2007, 104: 2377
[137] Nuinu, P., Sirisinha, C., Suchiva, K., Daniel, P., Phinocheep, P., J. Mater. Res. Technol. 2023, 24: 2155
[138] Wagner, M. P., Rubber Chem. Technol., 1976, 49: 703
[139] Thurn, F., Wolff, S. Kautsch. Gummi Kunstst., 1971, 28: 733
[140] Ziegler, J., Schuster, R. H., Kautsch. Gummi Kunstst., 2003, 56: 159
[142] Schuster, R. H., Müller, M., Früh, Th., Proc. Rubber Fall Colloquium, Hanover (1998)
[143] Scuratti, A., Lin, Ch., Rubber Chem. Technol. 2006, 79: 170
[144] Robertson, C. G., Lin, Ch., Rachaitis, M., Roland, C. M., Macromolecules, 2008, 41: 2727
[145] Niedermeier, W., Schweiger, G., ITECH Akkron (2006)
[146] Dierkes, W., PhD Thesis, University of Twente (2006)
[147] Sarikaro, S., Dierkes, W., Noordermeer, J. W. M., Rubber World, 2012, Nov. 26
[148] Wolff, S., RCT 1996
[149] Wang, M.-J., et al., Kautsch. Gummi Kunstst., 2020, 74(9): 45
[150] Schellenberg, M., Schuster, R. H., ACS Rubber Division Meeting, Cleveland (2013)
[152] Bokobza, L., eXPRESS Polym. Lett. 2012, 6: 213
[153] Chen, X.-H., Song, H.-H., New Carbon Mater., 2004, 19: 214
[154] Shoo, H.-Q., Wei, H., He, J. H., Rubber Chem. Technol., 2018, 91: 609
[155] Ferry, J. D., in "Viscoelastic Properties of Polymers", Wiley & Sons, New York (1980) Ch. 11
[156] Fritsche, J., Klüppel, M., J. Phys. Condens. Mater, 2011, 23: 035104
[157] Glasstone, S., "Textbook of Physical Chemistry", 2nd ed., Macmillan, London (1953)
[158] Hall, D. E., Moreland, J. C., ACS Rubber Division Meeting, Dallas (2000)
[159] Heinrich, G., Klüppel, M., Rubber Chem. Technol., 2000, 73: 578
[160] Persson, B., J. Chem. Phys., 2001, 115: 3840
[161] Mouri, H., Akutagawa, K., Rubber Chem. Technol., 1999, 72: 960
[162] Roland, C. M., Ngai, K. L., Macromolecules, 1992, 25: 363
[163] Banzuin, C. G., Esenberg, A., J. Polym. Sci., Part B, Polym. Phys., 1986, 24: 1021
[164] Roland, C. M., Rubber Chem. Technol., 2006, 79: 429
[165] Wang, M.-J., Lu, X., Mahmud, K., J. Polym. Sci., Part B. Polym. Phys., 2000, 38: 1240
[166] Murphy, L., Wang, M.-J., Mahmud, K., Rubber Chem. Technol., 2000, 73: 55
[167] Landry, C. J. T., et al., Macromolecules, 1993, 26: 3702
[168] Fragiadakis, D., et al., Polymer, 2005, 46: 6001
[169] Oprisoni, C., PhD Thesis University of Hanover (2011)
[170] Schuster, R. H., Alshuth, Th., Industrial DIK Project (2005)

# 11 Fatigue and Fracture of Elastomers

Mechanical fatigue and crack resistance of filled elastomers is understood as a progressive weakening of physical properties due to repeated stress and strain to such an extent that the product loses its functional values in use. Fracture toughness is an important property, because it reflects the ability of the elastomer to resist crack propagation and fracture. The progressively decreasing functionality of the vulcanizates results from structural changes in the material. Gradual deterioration of a vulcanizate during cyclic deformation is followed by crack initiation and the concomitant loss of stiffness and functionality of the vulcanizate. The crack propagates and ends with complete fracture of the vulcanizate. Internal damage to the material, such as crack initiation and propagation, cavitation and debonding of the filler from the polymer, contribute to the multiple phenomena that are summarized under the term fatigue. However, fatigue life is not only the result of dynamic treatment, but is also caused by other circumstances. The factors that affect the fatigue life depend on the vulcanizate, the mechanical loading history, the different dissipative mechanisms in the material and the environmental effects, such as the low and high the temperature, humidity and the presence of ozone in the atmosphere. However, the degree of filler dispersion achieved by efficient mixing will result in an improvement in the fatigue life and durability of the elastomer.

Macroscopic fracture of vulcanized rubber is a process in which network chains are broken and new surfaces are created as a result of loading. Loading may continuously increase up to fracture or it may be applied periodically, typically at lower levels, until fatigue fracture occurs. For example, a vulcanizate which breaks at 25 MPa under uniaxial tensile loading may fracture at just 5 MPa if loaded repeatedly. Considering the viscoelastic nature of elastomers, it is not surprising that fracture mechanics-properties are dependent on the stress rate, strain amplitude and test temperature and are controlled by the viscoelastic properties of the elastomer. This suggests that processes of energy dissipation are involved in rubber fracture. The appearance of cracks in elastomeric materials depend on their composition, the types of filler, the degree of filler dispersion and on the dynamic load, time and temperature to which they are exposed during a test or in service. The addition of fillers enhances energy dissipation through several mechanisms and improves the fracture properties of the vulcanizate. For a better understanding of the facts, it is helpful to clarify questions such as:

- where and under which conditions cracks are initiated,
- what governs crack propagation once cracking is initiated,
- which are the main influencing factors,
- which general laws describe the evolution of the cracks, and
- how can fracture be inhibited by compounding and processing measures.

In this context, microscopic processes that contribute to a reduction in the failure properties must be understood and taken into account, such as [1]:

- strain energy dissipation from filler-polymer detachment,
- mitigation of stress concentration by load transfer through the particle when the chain breaks, and
- crack growth deviation around filler particles.

Such processes are discussed in the following chapter. In view of the fact that abrasion resistance was first identified as a reinforcing property of carbon black, this chapter concludes with a brief summary on the abrasion and wear resistance of filled elastomers. Complex problems that arise in anisotropic elastomer products, such as tires or timing belts, where there is failure of adhesion at the interfaces in addition to failure of the filled elastomer phases are beyond the scope of this chapter.

## 11.1 Material Flaws

There are a number of material flaws that indicate an increase in crack propagation. Macro defects in the vulcanizate can shorten the service life of the vulcanizate by causing flow lines and weld lines that can initiate cracks, but the micro defects generated by the porosity of the elastomer and, the presence of dust can also influence the functionality of vulcanizate. Contrary to their macroscopic appearance, elastomers are not homogeneous materials on a microscopic scale. Heterogeneities inherent in the material may serve as stress raisers and precursors to cracks when subjected to high stresses and repeated cyclic deformation at high strain. In the case of unfilled elastomers, heterogeneities may be caused by:

- a broad network chain distribution,
- crosslink clusters, or
- insufficient co-crosslinking of the phase boundary in blends.

The energy required to create a new surface area is considerably greater than the intrinsic fracture energy necessary to rupture bonds in this area, which means that other processes of energy dissipation are involved in rubber fracture. The intrinsic inhomogeneities in filled elastomers result from the type of filler and the mixing efficiency:

- large filler agglomerates and pellet fragments,
- detachments of the rubber matrix from the filler surface

Some extrinsic features like scratches, dirt particles, sharp inclusions may also act as stress raisers that can initiate cracks under dynamic load. The existence of such flaws leads to stress concentrations in certain regions of the material where if the average applied stress exceeds a critical level, the crack will be initiated and start to grow [2]. Although the formation of superficial scratches was avoided by vulcanizing compounds in polished glass molds, the breaking strength was not found to increase significantly [3]. This experiment demonstrates that intrinsic sources of flaws, such as large agglomerates, are the main failures in filled vulcanizates. The failure is driven by intrinsic flaws related to geometric singularities or inclusions of a different nature [4]. The damage mechanisms and the evolution of defects under fatigue loading have been investigated using optical microscopy, X-ray diffraction and X-ray tomography [5–7].

Effective flaw sizes in the range of 20–50 µm were estimated from studies on dynamic crack growth, which covered different polymer types, various fillers, different curing systems and other compounding variables [8]. Flaw size depends to some extent on crosslink density, filler type and certainly the degree of dispersion of compound ingredients [9]. The scale of some flaws is summarized in Table 11.1.

**Table 11.1** Typical Flaw Size in Rubber Vulcanizates

| Flaw | Size (µm) |
|---|---|
| Pellet fragment | 200 |
| Smallest visible flaw | 100 |
| Carbon black agglomerate | 20 |
| Coarse carbon black particle | 0.5 |

More recent views on the problem assume that no intrinsic flaws pre-exist in the elastomer [12]. It appears that up to now there has no clear distinction been made between crack nucleation and crack initiation in these theoretical approaches.

The question of stress distribution at the tip of a spreading crack and the location of a flaw has occupied fracture mechanics experts from an early stage. In order to simulate the size of an intrinsic defect on a test specimen, a flat incision or notch of length

$l_0$ in the specimen is considered. The increase in stress at the tip of this flaw, $\tau_t$, in relation to the applied stress, $\tau$, is given by the expression [13]:

$$\tau_t/\tau = 1 + 2\,(l_0/r)^{1/2} \tag{11.1}$$

where $l_0$ is the depth of the edge flaw and r is the radius of the tip in the unstressed state

The defects trigger crack growth when a critical stress value is exceeded. Cracking occurs under cyclic loading of elastomers even when the stress is far below the ultimate tensile strength. Accordingly, longer and thinner edge flaws create more stress and are more effective at initiating cracks than enclosed flaws. After crack starts, due to an initial incision or notch, a characteristic radius is formed on the crack tip that does not depend on the original radius [14]. The specimen only breaks if the product of stress and radius exceeds a critical size at the crack tip [15]. According to this mechanism, the rubber does not immediately detach from the filler surface, but experiences a rupture interval near the filler surface, nucleated by the formation of small voids. The voids expand under the stress generated and when the expansion reaches the breaking extension of the matrix, the vulcanizate cracks at an applied critical stress, $\sigma_{cr}$. The critical stress is dependent on the filler-rubber interaction, which defines the modulus for $\sigma_{cr}$ [16]. If the radius of the voids is less than 0.1 µm, their surface energy significantly limits further expansion and the system is resistant to a crack initiation [17]. If the dewetting stress is lower, the failure stress is small and cavitation will occur at the surface between the filler surface and the rubber matrix [18, 19]. Based on the energy balance, the equation proposed for the critical stress of detachment is:

$$\sigma_c = (2\pi G_{if} E/3d_p)^{1/2} \tag{11.2}$$

where $G_{if}$ is the interfacial fracture energy, E is Young's modulus and $d_p$ is the diameter of the filler particle

The equation indicates that $\sigma_c$ not only depends on the interfacial energy between the filler and the rubber, but also increases with decrease in particle size of the filler. The more the particle size of the filler is reduced, the higher is the critical stress and the vulcanizate will not show rubber-filler phase detachment [20, 21]. Intrinsic and randomly distributed flaws decrease the reproducibility of the tests, which generally show an impressive scatter of the fracture data. Such behavior can already be observed for tensile strength, where the inhomogeneities of the flaws that occur are clearly visible. Even for fractures of filled elastomers, a high number of test specimens is required to demonstrate the reproducibility of the material. To improve reproducibility, a prescribed cut is made in the test specimen to cause stress concentration at the tip of the cut. In most cases, the methods used have changed little over the years, although there have been refinements and considerable advances made in instrumentation.

## 11.2 Mechanical Fatigue

Fatigue can be understood as any change in material properties after prolonged exposure to deformation. Repeated cyclic loading of elastomers causes a reduction in strength, which can lead to cracking and fracture. The type of failure depends on the geometry of the product, the loads and the environmental conditions. Mechanical fatigue is the result of dynamic processes that lead to a progressive weakening of the physical properties of the vulcanizates. The loss of properties experienced by the part before its functionality is no longer takes protracted periods of repeated uniaxial tension, compression, torsion, bending etc. and is even more pronounced under multiaxial deformation. Flex fatigue resistance of an elastomer is measured by the time or cycles required for a test specimen to degrade crack, or fail from repeated cyclical deformation. It is therefore important to know how long an elastomer can withstand the repeated deformation cycles with large amplitudes at relatively low frequency without losing its specified functionality. The number of cycles required to break a specimen into two pieces at a particular stress for a stress-controlled or strain for a strain-controlled test is called its fatigue life. The precision of the test is determined by test methods that are run under defined conditions and require a high number of test specimens.

The methods and processes used for this cannot be denied a certain amount of empiricism. Fatigue tests are proposed and carried out all kind of variants and, although the physical significance of the tests is not clear in many cases, the decrease in the quality of the material is always evident. Some of the methods presented below show the failure of a filled rubber based on properties that can be described as a reduction in quality. In these methods, failure is determined by the quality of the rubber, the amount and type of filler and oil added and, of course, the crosslinking agent. It can be seen that the flaws in the material significantly accelerate the failure. The less intensively a vulcanizate is mixed, the more distributed pellet fragments it contains and the more likely is the systematic occurrence of a crack in the elastomer.

### 11.2.1 Fatigue Testing

Over the last 100 years, empirical methods with variable loading modes and different severities of test conditions have been invented. Various methods have been developed that subject the material to repeated uniaxial, biaxial or triaxial deformation. However, the fatigue tests require a large number of test specimens in order to characterize the influence of material defects. In principle, it is not possible to identify the material parameter that is causing the failure but it is possible to narrow down the responsible parameters and explain the failure of the material. All the important fatigue methods have very poor repeatability and reproducibility. To show significant differences, one should run a large number of test specimens with the same formulation. Due to the importance of the fatigue behavior of rubber products, a variety of flex cracking and cut growth tests have been developed for the purpose of generating data on fatigue life. By using large enough specimen sizes for the tests and closely monitoring them as they begin to deteriorate, test precision can be improved. The test devices work according to one of four different deformation modes [22]:

- periodic loading between fixed stress limits (in tension or compression at constant load limits)
- periodic loading between fixed strain limits (in tension or compression at constant strain limits)
- reversed shear stress generated by torsional deformation
- reversed bending stress in one or two dimensions

Due to the importance of material fatigue phenomena, a large number of fatigue test devices have been developed over the last century. In order to keep the demands on the material as high as possible, complex cyclic deformation was preferred. In the tests, the wear behavior under strong dynamic loads is assessed. Compression and simple shear are deformations of great practical importance in engineering components. Earlier investigations had pointed to failures of rubber cylinders under repeated compression and shear.

A simple but compelling method consists in subjecting the material to repeated cyclic compression and measuring the number of deformations until the material displays a variant in the mechanical behavior that corresponds to fatigue [23]. The two variants of the mechanical test are shown in Figure 11.1.

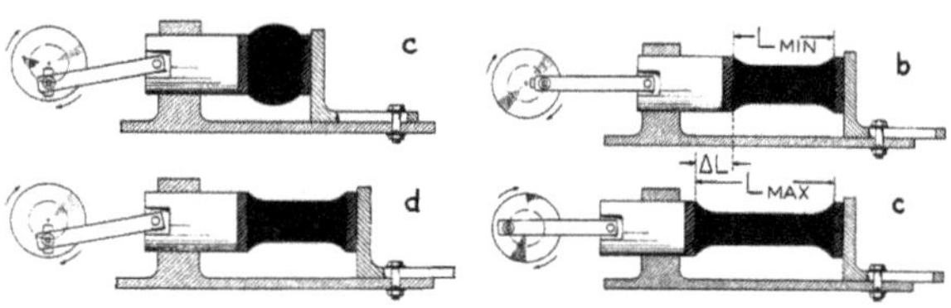


**Figure 11.1** Compression and elongation of a pre-stressed rubber specimen [11]

In the first variant, the vulcanized material is compressed and elongated under defined conditions. During compression, the vulcanizate exhibits bi-axial elongation in the middle of the specimen that causes dynamic damage to it. In the second variant the pre-stressed vulcanizate is repeatedly elongated to higher deformationto show the importance of the applied extension of the test specimen. These two methods were used to make predictions about the service life of the filled vulcanizates, although this was not widely known.

Other methods can change the shape of the material in three dimensions and have a much higher functionality. The Goodrich Compression Flexometer, introduced in the

1930s, characterizes the durability of filled elastomers [24, 25]. Cylindrical test specimens are subjected to cyclic compression at a frequency of 30 Hz and 1 MPa static load, followed by dynamic strain until porosities occur in the material. The measured properties are:

- compression fatigue
- permanent set resistance
- heat build-up
- blow-out time
- resistance time to damage, $N = t \times n$

The test delivers the change in temperature over time and the strain of the specimen under compression over time. If the load is increased to 2 MPa, the repeated deformation leads to a catastrophic deterioration (blow-out) of the material. Modern test devices are equipped with sensors to gauge dynamic load and strain. The internal damage of large test specimens in a compression flexometer test is shown in Figure 11.2.

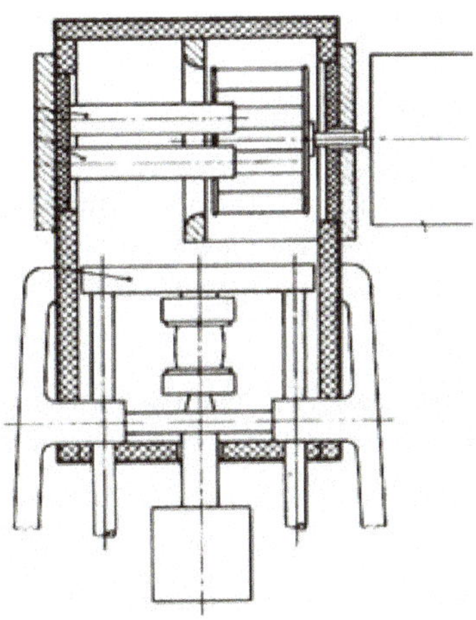

**Figure 11.2**
Goodrich compression flexometer

The occurrence of porosity and eventually large cavities in the interior of the material can induce increased stress through the repeated bi-axial stretching in the middle of the specimen, leading to increasing heat build-up in the material during repeated compression. The Goodrich Flexometer also has an alternate and more severe test condition that can result in a "blow-out". This method has more sensitivity to subtle differences in dynamic properties produced by formulation differences.

The De Mattia Fatigue Rotational Flexometer is the most commonly used for bending tests with different degrees of severity [26]. The tests can be conducted under the following conditions:

- resistance to fatigue
- heat build-up at the end of the test
- residual deformation
- rotational speed of 875 or 1500 rpm

The specimen is clamped at one end in a stationary grip and the other end in a reciprocating grip. The degree of bending can be adjusted by the operator. Undamaged strip-shaped test specimens with a notch on the side are subjected to a periodic bending movement at a predetermined frequency. The specimen is inspected until the first crack is detected. For crack growth, the test is interrupted at certain intervals to determine the crack length and size. The load cycles required to reach a certain level of crack length are assessed. The function of the De Mattia fatigue tester is shown in Figure 11.3.

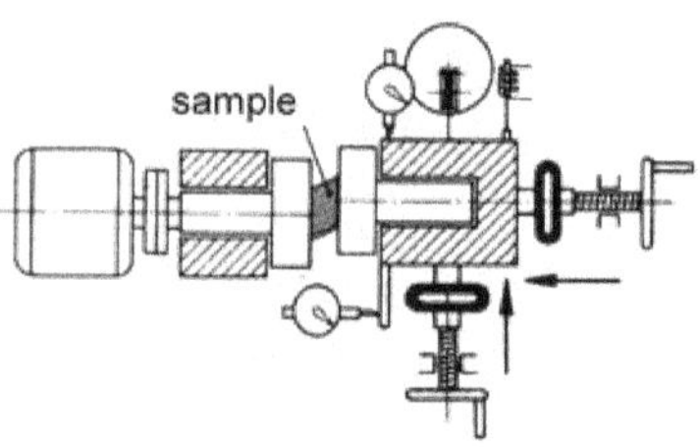


**Figure 11.3**
Illustration of a De Mattia fatigue tester

The onset of crack formation is dependent on environmental conditions (temperature, humidity, ozone level). In the fatigue test, the cracks produced are only formed above a critical strain [18]. Crack formation is especially pronounced under the influence of ozone.

The Du Pont fatigue tester utilizes constant load amplitudes. This device is used for testing the fatigue failure of a rubber vulcanizate under alternating stress. Based on Wöhler's experimental design, it employs a motor to rotate a circular cantilever specimen that is loaded at its free end. The evaluation of the test is based on the number of cracks to be examined in the specimen. The device can be used to determine the service life of a rubber material under different stress conditions.

The Monsanto fatigue to failure tester is used to stretch and release specimens via a continuously rotating cam. The rubber specimen is shown in Figure 11.4.

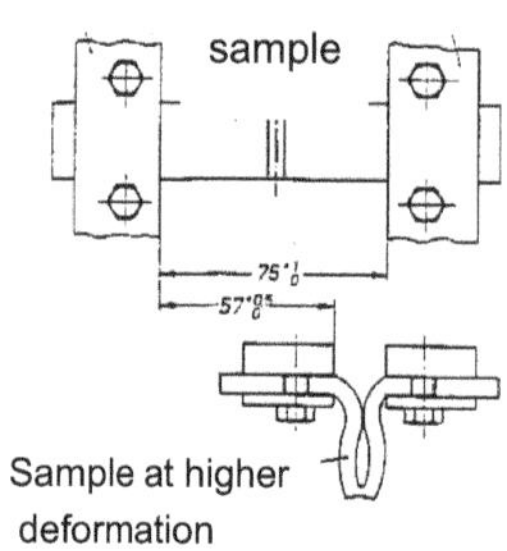


**Figure 11.4**
Test specimen for the Monsanto fatigue to failure tester

The tester records the number of cycles applied to a specimen until it is destroyed by fatigue. The device can measure:

- the number of flexes needed to produce a cut of defined size
- the number of flex cycles during which no cuts are formed

By applying cyclic loading conditions, a large number of specimens can provide fatigue results. A versatile life-cycle property can be established for the filled vulcanizates.

For all tests, it has to be taken into account that a viscoelastic response forms the basis of the repeated deformation. The test temperature should be kept as constant as possible, because fatigue can vary with temperature by a factor of 4 [7]. The advantage of these tests lies in their simplicity and the possibility of establishing an empirical ranking of the materials. This means that changes in the recipe or in the mixing process or the downstream flow can be more easily identified as the cause of flaws. The disadvantages are the subjective influence of the operator and control over the deflection of the specimen.

Another problem encountered with many flex fatigue tests is heat build-up, amplifies the effects of repeated stress and strain. Inside a specimen, the temperature rises quickly due to the bi-axial deformation and this leads to a chemical reaction in the elastomer. The main problem with this type of damage is the difficulty in tracing the failure of the material back to a specific cause and of quantifying it. Therefore, the evaluations of a material on the basis of fracture mechanics tests do not necessarily agree with the results from practice. Despite the low specificity of classical fatigue tests, general rules for the fatigue behavior of elastomers have been established [3].

### 11.2.2 Fatigue of Elastomers

Fundamental work on the fatigue of elastomers has shown that the deformation of a test specimen can occur under compression or under elongation deformation. It has been shown that compression delivers a more rapid fracture because of the bi-axial elongation in the middle of the specimen. Fracture mechanics studies conducted on bonded rubber cylinders subjected to uniaxial compression suggest that the locus of crack growth defines a parabolic surface which essentially covers the area of the specimen biaxially stretched during the cyclic test [27]. A comparison of NR and E-SBR vulcanizates has shown that the fatigue resistance of NR is significantly higher than that of amorphous E-SBR because it can crystallize under strain. The additional physical network nodes between the NR chains lead to self-reinforcement, which causes the storage modulus to increase sharply [23]. The crack growth in NR and E-SBR increases linearly as a function of the number of cycles. As in the compression test, crack propagation in the De Mattia flexometer test returns a significantly higher value for E-SBR than for NR in the unfilled state. When the rubber is filled by carbon black N330 the crack propagation in E-SBR is significantly lower and only slightly higher than in filled NR, as can be seen in Figure 11.5 [28].

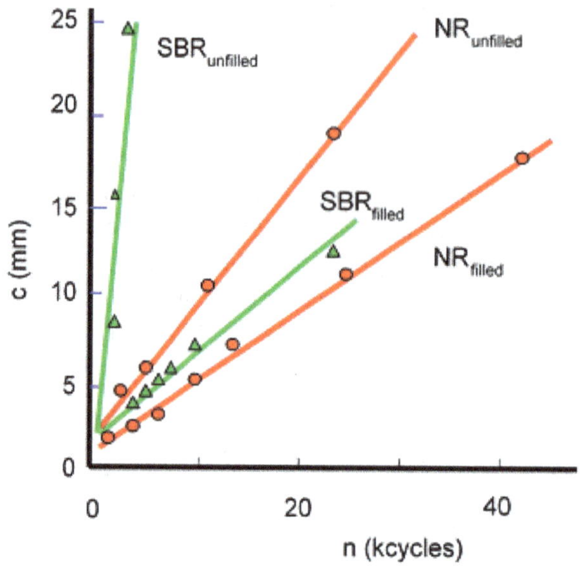


**Figure 11.5** Increase in cut growth in the De Mattia flexometer for NR and E-SBR specimens

Fatigue resistance of vulcanizates is primarily related to the flexibility of the network chains. The less energy dissipation that takes place during cyclic deformation, the lower is the fatigue of the elastomer. This becomes evident for cis-BR grades obtained by different coordination catalysts in Ziegler-Natta polymerization. With increase in the cis-BR content of the rubber, the De Mattia crack growth is reduced at the same crosslinking density. The results are confirmed by Monsanto fatigue-to-failure tests that reveal an impact of the catalyst used in the coordination polymerization process [29]. In a Ziegler-Natta polymerization, the content of cis-butadiene units increases in the order:

$$\text{Ti}-\text{BR} < \text{Ni}-\text{BR} < \text{Co}-\text{BR} < \text{Nd}-\text{BR} \quad (11.3)$$

Due to the increased crystallization of the cis-polybutadiene, the fatigue properties increase linearly as a result of the kcycles that are survived. This shows the importance of the high cis content of BR, which significantly improves the fatigue properties. With Nd-BR, which provides the highest cis content in BR, the highest fatigue resistance and the lowest abrasion are achieved in a car tire vulcanizate.

Changing the chemical structure of the CR by modifying it with xanthate or mercaptan brings about a drastic change in the fatigue resistance in the normal state and in the state after aging. The xanthate-modified CR always has a significantly longer fatigue resistance of 220 kcycles at 55% deformation amplitude than the CR modified with mercaptan, which achieves 150 kcycles. Even after 7 days at 100 °C, the fatigue resistance has decreased, but the order remains the same [30].

With increase in crosslink density, vulcanizates demonstrate a progressively increasing fatigue resistance until a maximum is reached. The maximum depends on the cure system and the rubber matrix on the crosslinking density. Slightly under-cured vulcanizates exhibit the highest fatigue resistance and a better dynamic behavior than over-cured vulcanizates, which show a rapid decrease in the fatigue resistance. Rubber vulcanizates based on higher ratios of accelerator to sulfur possess more mono-sulfidic crosslinks resulting in better heat aging and poorer flex fatigue resistance. The hardness of the cured rubber has a great influence on the fatigue resistance as well. Regarding the crosslink structure, it has been found that poly-sulfidic crosslinks provided by CV sulfur-curing systems provide a better De Mattia fatigue resistance than EV and sulfurless curing systems in NR and E-SBR compounds [31]. For EPDM vulcanizates, it has been found that sulfur crosslinks increase the number of cycles to failure by an order of magnitude over the level attained by peroxide curing. This observation demonstrates the advantage of conventional sulfur vulcanization that is conferred by the length of the sulfur bridges of the crosslinks. The use of co-agents in combination with peroxides leads to an improvement of 40% compared with sulfur curing [32]. Anti-degradants, which protect particularly unsaturated elastomers from oxidative processes, promote the fatigue resistance of vulcanizate. In carbon black-filled NR/BR blends, it has been found that the kilocycles to failure at large strains increased substantially by 6PPD (2 phr) and even more so by the mixture 6PPD/TMQ (2 phr/1 phr) [33]. The effect of the antioxidants is shown in Figure 11.6.

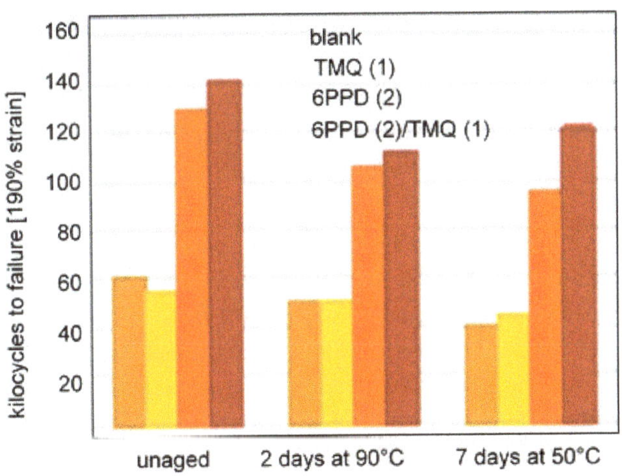


**Figure 11.6** Effect of antioxidants in NR/cis-BR/ CB N550 after aging [33]

The anti-degradants that contain p-phenylenediamine are basically equipped to reduce oxidative attack by oxygen and ozone. Anti-degradant agents guarantee good fatigue resistance to NR, but very little to SBR vulcanizates, which depends on the mechanism of protection [34].

### 11.2.3 Heat Build-Up

If the test specimen is compressed repeatedly in a sinusoidal or in pulse mode, some of this deformation energy is dissipated and converted into heat. This effect, known as heat build-up, occurs in the middle of a large-volume vulcanizate and triggers chemical processes that attack the rubber chain and lead to the formation of large spaces. The temperature rise is measured at the bottom of the test specimen in a Goodrich flexometer test. During deformation-constant tests in the Goodrich flexometer, heat development is observed, which increases with carbon black loading [35]. The temperature rise in cis-BR and carbon black N234 is shown in Figure 11.7 [36].

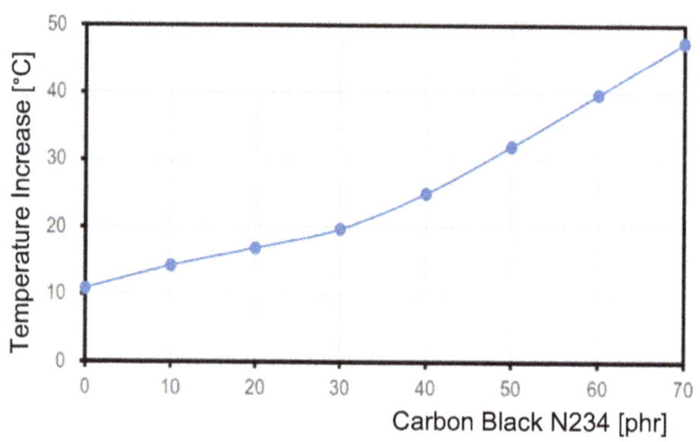


**Figure 11.7** Influence of carbon black N234 on heat build-up in cis-BR [36]

If the proportion of carbon black is kept constant and dispersion occurs to the maximum extent, the temperature increases with the specific surface area of the filler. With active carbon blacks, this increase is much more pronounced than with semi-active carbon blacks, which is an unintended consequence of filler activity. For NR/CB N330, there is a 15 °C rise for every 10 phr. With semi-reinforcing carbon blacks, this increase is much smaller.

If the increase in temperature is a cumulative value for heat generation, it can be correlated with tan δ, which that describes the heat increase in one deformation cycle. Thus, the heat build-up correlates with the measured tan δ in the dynamic temperature dependence [37]. A study of the heat build-up in SBR vulcanizates filled with 17 carbon blacks of specific area ranging from 30–139 $m^2/g$ and crushed DBP ranging from 57–107 mL/100 g and a loading up to 50 phr has shown that the relative temperature, $\Delta T_f/(\Delta T0 - 1)$, leads to a master-curve for heat build-up for SBR vulcanizates [38].

If the internal appearance of the specimen is examined after a different load duration, a number of small cavities appear first in the middle of the specimen, where the bi-axial deformation is greatest. If the test is continued, large cracks form which no longer correspond to the original function of the vulcanizate. Such a progression is shown in Figure 11.8.

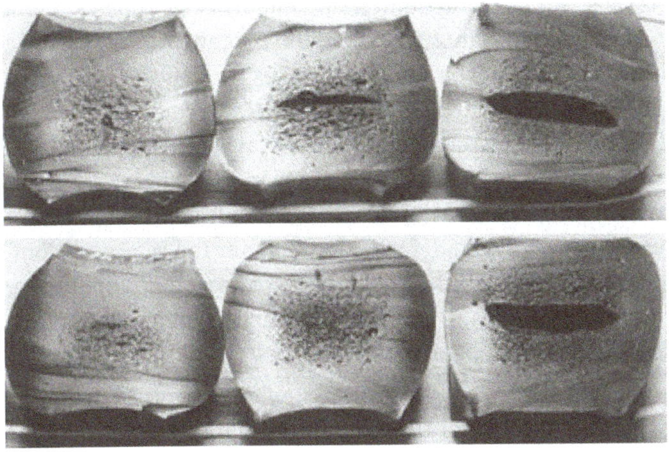

**Figure 11.8** Internal damage of a specimen in the Goodrich flexometer

Thus, the flexometer can provide information about the degree of material damage. The formation of bubbles can be attributed to different contributions, such as poor choice of type of filler or use of the wrong mineral oil; in this case, the temperature at which the specimen was loaded was too high.

Looking at the carbon blacks, the group of Ecorax® carbon blacks has a higher energy site distribution on the amorphous carbon atoms that leads to better rubber-filler interaction and to a significant reduction in filler agglomeration. The temperature development of the flexometer test shows a lower heat build-up than for the ASTM carbon blacks. These results were published by Orion Engineered Carbon and are shown in Figure 11.9 [39].

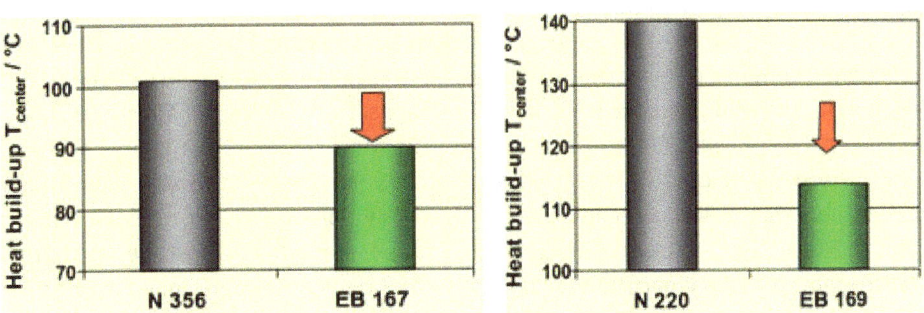


**Figure 11.9** Comparison of the ASTM carbon blacks with the Ecorax® carbon blacks on the Goodrich flexometer test [39]

The comparison was performed on carbon blacks of similar specific areas and DBP structure, but the heat build-up in the case of the Ecorax® carbon blacks was much lower because of the increased interactions of the amorphous carbon atoms with the rubber chains and a better rubber-filler interaction. Increasing the surface energy of the Ecorax® carbon blacks brings about a reduction in the temperature. The reduction in temperature rise is 12% for EB 167 and 25% for EB 169.

In the Goodrich flexometer test, an NR compound that contains carbon black and a small amount of multi-walled carbon nanotubes exhibits a significant increase in temperature by the end of the test. The heat build-up increases from the value of 64 °C for NR/CB N375 to the 129 °C shown by the combination of CB375 and multi-walled carbon nanotubes (15 phr) [40]. The low proportion of multi-walled carbon nanotubes which have a high aspect ratio generates a disproportionate increase in temperature development in the flexometer test. Permanent set after the Goodrich flexometer test reveals much more pronounced permanent deformation than for the reference compound extensively filled with carbon black. The phenomenon of heat build-up is more intense for specimens with high loadings of multi-walled carbon nanotubes, which show better performance at elevated temperatures.

If non-silanized silicas in NR or in SBR are examined in the flexometer test, it can be seen that the heat development in the case of HD silica is considerably lower than for conventional silica VN3 [41]. At a loading of 60 phr, VN3 shows a heat build-up of 155 °C whereas the HD silica has a temperature increase of only 110 °C. This difference is a sign that a compounder should use the HD silica, which has different porosity distribution, which in turn leads to less energy dissipation and improved dispersion. If tan $\delta$ is used as a reference, the precipitated silica always generates higher heat build-up than the carbon blacks, due to the silanol groups, which are able to provide a higher filler network. With increase in the structure of the filler aggregates, the difference in heat build-up becomes larger.

## 11.2.4 Woehler Curves

The most important phenomenological fatigue relationship for predicting fatigue life is the "Woehler" curve, which relates a parameter describing the cyclic stress-strain state to the number of loads endured until failure. This engineering concept was developed in England in the 1870s to estimate the µservice life of railway wheels [42]. However, the concept allows to evaluate also the service life of filler rubber vulcanizates. The Woehler curve describes the applied dynamic stress in a stress-controlled test or the strain in a strain-controlled test as a function of the deformation cycles to failure. Normally, the stress range (or strain range) is plotted against the number of cycles to failure on a logarithmic scale. A typical Woehler curve for material used for engine mounts is shown in Figure 11.10 [43].

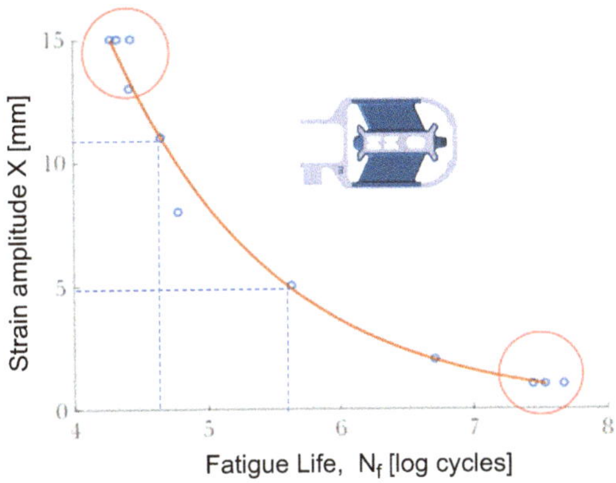


**Figure 11.10**
Woehler curve for a carbon black-filled NR compound intended for an engine mount [43]

The fatigue life, N, can be expressed as an exponential function of either stress or strain. Importantly, the fatigue life increases asymptotically as the dynamic stress or strain is reduced, indicating a lower limit for fatigue. Usually the curve is constructed on stress or displacement levels with one test piece for each level. Based on the variation of a functional property (i. e. stiffness) an end-of-life criterion can be defined. The test parameters are chosen so that the temperature in the vulcanizate does not rise too high and thus damage the material, and the frequency of cyclic loading roughly corresponds to that in practice. Fatigue behavior is mostly driven by early crack initiation and subsequent propagation. Experiments have shown that cracks initiate before 20% of the fatigue life and grow slowly with the fatigue cycles, except at the highest stresses or strains, respectively. On the assumption that dissipated energy accumulates along the fatigue cycles, a criterion for predicting fatigue properties based on mechanical and thermal properties has been proposed recently [44]. Woehler curves show the combined effects of several non-linear deformation processes active during the experiment. They depend heavily on crack precursors present in the material (i. e. defects, inclusions, interface failures, cavities etc.) which vary from specimen to specimen, batch-to-batch and from lab mixed compounds to factory mixed ones. Consequently, the tests require many specimens of identical composition. They are lengthy and affected by a wide range of experimental errors. As a consequence, large scatter of the measured fatigue life is found for test specimens without a defined pre-existing cut. Despite the experimental effort entailed in determining Woehler curves, the prediction of fatigue life is of great importance for the application of filled elastomers in practice.

This challenging procedure has been employed to determine the fatigue life of NR-metal laminated bearings for earthquake protection [45]. In the virgin specimen, the cut is more efficient at the interface between the metal face and the rubber material and reduces the fatigue life with increase in shear strain. At a shear strain of 200%, the number of cycles to final degradation exceeds $10^4$, but at a shear strain of 600% the cycles do not exceed $5 \times 10^7$ in number. For a specimen aged for 70 years, the number of cycles shifts lower as the shear strain increases. The results are summarized in Figure 11.11 [45].

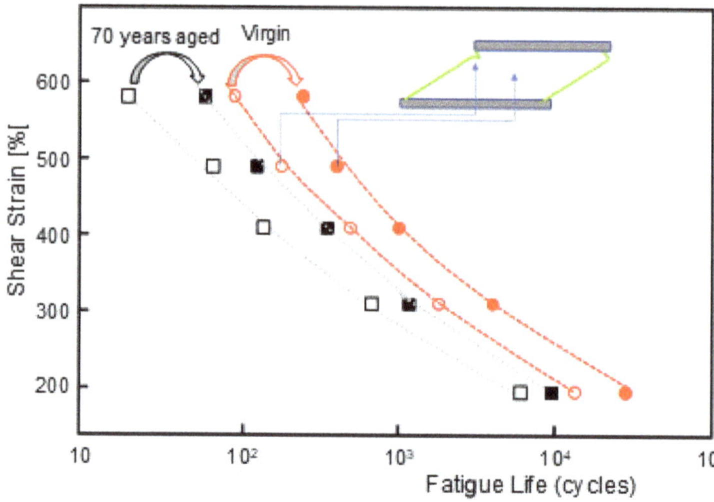


**Figure 11.11** Woehler curves for an earthquake protection device

The results can be misleading, as the fatigue life is sensitive to both the magnitude and the type of applied stress. This explains why results obtained in the laboratory, rarely correlate with service performance. Moreover, most products are subjected to more complex deformation patterns in service than in a laboratory test. For this reason, the tests have to be repeated in order that the statistics of the test may be satisfied. The test conditions are tailored to the loading modes for a particular product and are considered to be functional tests. The results of these tests are used to assess the suitability of a product. Such tests are not designed to establish which material parameter causes the failure, but rather to gain an understanding of the end-of-life conditions.

A number of phenomenological criteria for evaluating fatigue life can be found in the literature. Some of these criteria remain fully macroscopic and do not provide any explanation for the complex process [12, 46, 47]. More recent approaches are seeking to use macroscopic fatigue criteria as a basis to understand better the crack initiation [48] and crack propagation [49]. An understanding of fatigue behavior and failure requires a knowledge of both the thermo-mechanical response of the material and the degree of filler dispersion. The development of filled elastomers resistant to fatigue remains a difficult task, because failure is strongly related to the heterogeneity of the material, especially to the existence of flaws generated by insufficient filler dispersion.

### 11.2.5 Tear Strength

The major mode of failure of an elastomer, unfilled or filled, is some kind of tearing. Tearing is understood as the act of breaking apart the elastomer by force, without the aid of a cutting tool. The tear strength of a rubber material is a measure of how well a material can resist the effects of tearing when it is subjected to protracted stretching. Tear strength is an indication of endurance and toughness and leads to a longer service life.

Tearing can be triggered by several imperfections and defects in the elastomer that act as incipient failure sites. When the material is subjected to high strain or repeated cyclic deformation (even at moderate strain), a microscopic crack develops that can progressively propagate through the material until catastrophic failure of the piece. The stress at the tip of the crack irreversibly ruptures rubber chains as the crack propagates. Any process at the molecular level that helps reduce stress at the tip of the crack or to increase resistance to chain rupture helps to reduce crack propagation and increase tear strength. The force required to initiate a tear is quite different from the force necessary to propagate the tear.

Tear strength is particularly evident when elastomers containing poorly dispersed reinforcing fillers are torn. Since the product life of many elastomers depends on their tear behavior, there is great interest in test methods for gathering reliable information about tear strength and cut growth under conditions that approximating those faced by the product in service life. As tearing occurs in a wide range of products and is also involved in abrasion and fatigue processes, it is not surprising that a number of concepts for tear tests have been developed over the years. A widely used empirical measure of an elastomer's resistance to tear initiation and tear propagation under uniaxial tension is the tear resistance (tear strength). Tear resistance is determined on notched tensile test specimens that have a defined thickness, h. By definition, tear resistance is the ratio of the averaged force peaks to the initial thickness of the test specimen:

$$\text{Tear resistance} = F \times h \ \ (\text{N/mm}) \tag{11.4}$$

where F is the mean force (in N) and h is the thickness of the test specimen (in mm)

Two types of tearing are observed in the trouser tests:

- stable tearing
- unstable tearing

Stable tearing is when the force fluctuations are slight and the rate of tear propagation is uniform. Unstable tearing or slip-stick tearing occurs when the tear growth is arrested and reinitiated at fairly regular intervals. This type of tearing leads to surface irregularities with periodic "knots" that depend on the tearing rate and the temperature. Figure 11.12 provides an overview of the crack formation that has taken place.

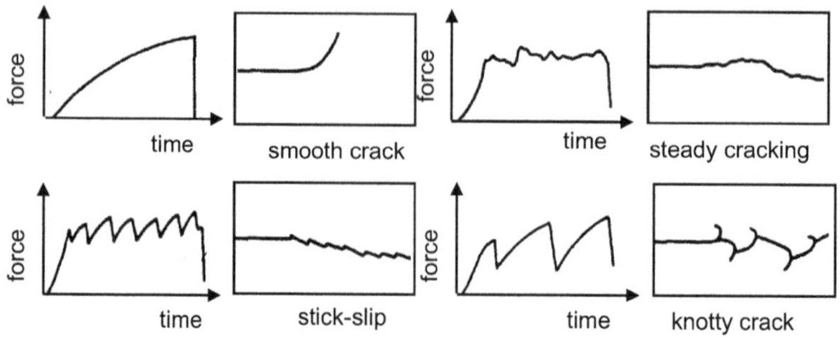


Figure 11.12 The different shapes of crack formation

The result of the test depends on the geometry of the test species and, of course, on the test conditions. A large number of test geometries have been proposed for measuring tear resistance [14]. ASTM defines three different test specimens, A, B and C [50] while DIN recommends one test piece with different geometries [51]. The specimens employed to measure tear resistance have an incision on the edge. Generally, these specimens do not yield the same tear resistance values, as is seen in the ASTM and the DIN standards. Depending on how tear resistance is measured, specimens of the same composition will show different results. The application of the test method is therefore somewhat arbitrary.

Usually, a tensile specimen with an incision at an edge or a trouser specimen is placed between the two holders of a universal stress-strain tester and a uniform tensile force is applied until the crack propagates through the specimen. As a result of the tensile force, a spectrum of force peaks is recorded as function of time. The mean force taken over a selected period of time is taken for evaluation. Figure 11.13 illustrates the change in tensile force over time.

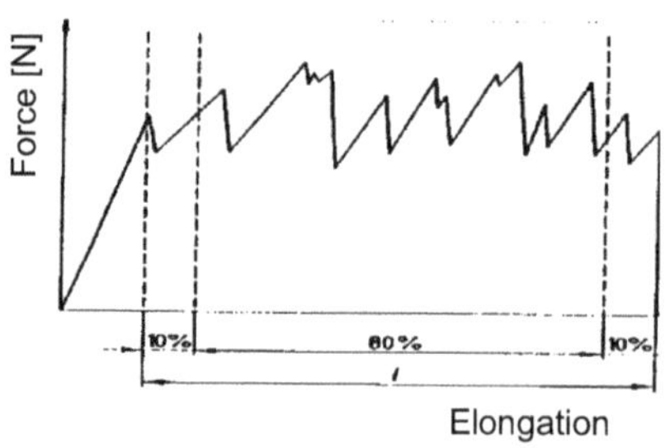


Figure 11.13
Force versus elongation for a tear process

A number of other test geometries have been proposed for this test. Today "constrained trouser" specimen and the single edge notched specimen (SEN) or pure shear specimen are predominantly used to determine tear resistance. For notched specimen, the measured force comprises the force required to propagate the tear and the force needed to elongate the specimen. However, the proportion of these forces depends on the geometry of the test specimen. Therefore, a test performed on different test specimens may not yield the same tear resistance value [14]. Table 11.2 illustrates the influence of the test specimen on the tear resistance of different vulcanizates.

**Table 11.2** Influence of Test Specimens on the Tear Strength of Vulcanizates [14]

| System | Trouser | SEN |
|---|---|---|
| NR | 12.5 | 16.4 |
| SBR | 4.0 | 2.4 |
| BR | 1.5 | 0.9 |
| NR/CB | 25.1 | 21.6 |
| SBR/CB | 18.5 | 17.1 |

Between the trouser measurements and the results from the SEN measurements are large differences. However, the strain-crystallizing NR leads to higher values than the amorphous SBR. The large difference in tear strength is still a matter of discussion. However, for filled vulcanizates, the difference between the measured values is smaller than for unfilled vulcanizates. Tear resistance measurements on unfilled and filled vulcanizates show differences that can be attributed to the state of cure of the rubber as well as to the rubber-filler interaction and the degree of filler dispersion.

#### 11.2.5.1 Tear Strength of Unfilled Vulcanizates

Generally, the tear resistance test does not relate to rubber product performance directly, but, if performed on the same test specimen under constant conditions, it provides a ranking of reinforcement which can be useful for judging material properties and to choosing the optimal formulation or the most efficient processing technology. The tear resistance of unfilled vulcanizates is fairly low, with the exception of those elastomers made from strain-crystallizing rubber, such as NR or CR. Polymers that tend to strain crystallization result in high tear resistance. The amorphous rubbers, such as SBR, NBR, ACM, CSM, EPDM, BIIR and CIIR, have a much lower tear strength, while FKM, BR and especially silicone rubber Q, have a modest tear strength. Particularly at high temperatures, EPDM vulcanizates yield tear strength values that are comparable to those of NR vulcanizates [52]. A higher tear strength can be achieved with the use of selected curing systems. In general, sulfur cure provides higher tear strength than peroxide cure. Thus, the length of the sulfur bridge, which exhibits an elastic resistance before the specimen is ruptured, is compared with the short and inelastic C-C connections that do not have such a property.

With any curing system, the tear resistance increases with the crosslink density until a maximum value is reached [53]. The type of rubber and the curing system has a strong influence on the value of the tear resistance. The tear strength depends on the state of cure and the length of the sulfur crosslinks, as shown in Figure 11.14 for E-SBR, NR and EPDM vulcanizates.

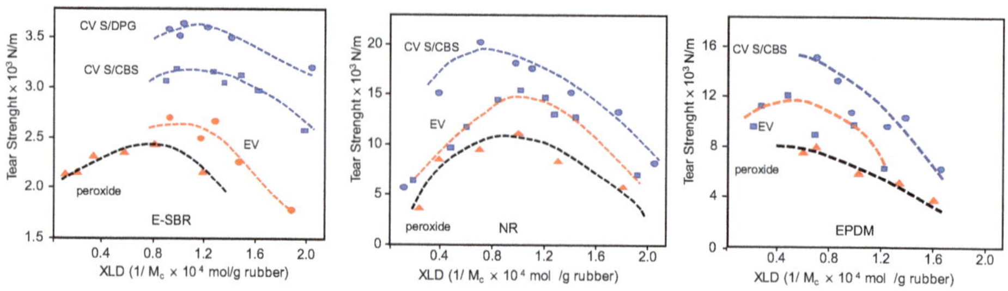


**Figure 11.14** Tear strength of differently crosslinked E-SBR, NR and EPDM [53]

Regardless of the rubber matrix, the maximum value of tear resistance increases in the following order of crosslink structure:

$$C - C < S_1 < S_x$$

In the case of E-SBR, different CV curing systems result in specifically increased tear resistance. The increasing tear strength suggests that both the flexibility of the crosslinks and the reactivity of the moieties formed by the scission of sulfur crosslinks that contribute to the tear resistance. The crosslink rupture at the crack tip associated with the tear leads in the case of poly-sulfidic crosslink structures to species that can create new crosslinks and thus weaken crack propagation.

In addition to the crosslink density and crosslink structure, the level of tear resistance depends on the ability of the polymer to crystallize under strain. Rapid strain crystallization builds up the resistance to the propagation of the crack. NR and CR vulcanizates that are able to show strain crystallization demonstrate significantly higher tear resistance than amorphous vulcanizates that do not possess this self-reinforcement. The molecular mechanism of this self-reinforcement is the same as that discussed for tensile strength.

If tear resistance is recorded as a function of temperature, the influence of strain crystallization becomes more evident. Regardless of the type of crosslinking, the tear resistance of NR vulcanizates decreases with temperature, because the crystallites are progressively broken down at temperatures above 100 °C. At about 110 °C, strain-induced crystallites are no longer formed, so that the tear resistance remains at a low level above this limit. Fracture in strain-crystallizing rubbers has been studied by conducting simulations at the continuum level [54–56]. Simulation of fracture in polymeric materials have been performed at the molecular level using coarse-grained models [57–60]. The results of such numerical simulations support the experience gained from the tear tests.

#### 11.2.5.2 Tear Strength of Filled Vulcanizates

The presence of a dispersed filler in a vulcanizate leads to a strong increase in the tear strength. It can be seen that the dependence of the tear strength on carbon black loading is similar to the increase in tensile strength. In stable tearing the tear force fluctuates only slightly and the rate of tear propagation is continuous. Unstable tearing is referred to as "stick-slip" tearing. The tear growth is arrested and reinitiates at fairly regular intervals. Therefore, the tear force varies widely, from a minimum at the tear onset to a maximum at tear initiation. This type of tear is named knotty tearing. The tear strength values measured for filled vulcanizates are considerably higher than for their unfilled counterparts. If the loading of carbon black is increased, the tear strength increases significantly and reaches a maximum at which the lower specific surface area shifts to higher loadings and then decreases again. For the production of high-performance compounds, this is the maximum to aim for. Only the non-active carbon blacks show a moderate increase without a maximum. The behavior here is similar to that of the tensile strength in uniaxial stress-strain behavior.

Compared with NR, SBR and BR polymers, the filled vulcanizates used in car tire treads inherently have a much higher tear strength due to the rubber-filler interaction and the good dispersion. When stress is applied to a crack, an orientation of the strain-crystallization of NR or an alignment of amorphous rubber chains in SBR and BR occurs in the region of the tear tip, which is associated with local deformation and increases the tear strength in the filler vulcanizates.

For any type of active carbon black, the tear resistance increases with loading until a maximum value is reached. The force maxima in cracking tests on filled vulcanizates can be taken into account when optimizing the crack properties of the vulcanizates. However, the qualitative tendency of tensile strength must be considered. Inactive carbon blacks, i. e. flame black FT and thermal black MT show a steady increase in tear resistance, albeit at a very modest level. This is shown in Figure 11.15.

For SBR vulcanizates, it has been shown that, at constant carbon black loading and constant COAN, the tear resistance increases with the specific surface area. A linear decrease in tear resistance is observed for carbon blacks of constant surface area when the COAN increases. The decrease is less pronounced for low surface area blacks [7]. The tensile strength is related more to the rubber-filler surface of the dispersed aggregates and less to the hydrodynamic volume of the filler aggregates. The details for specific area and crushed OAN are shown in Figure 11.16.

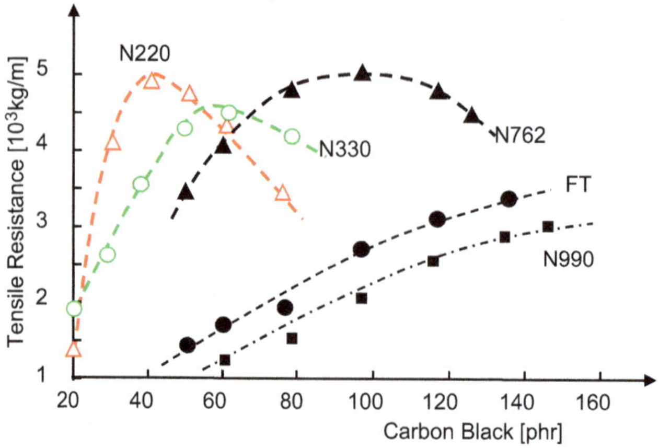


**Figure 11.15** Tensile resistance of different carbon blacks in E-SBR

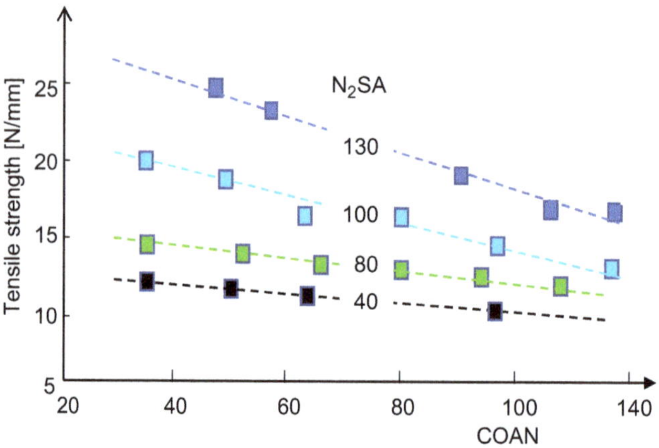


**Figure 11.16** Influence of the specific area and the structure of carbon black on the tensile strength of E-SBR

The tensile strength of precipitated silica is always higher than that of carbon black at the same loading and surface area, because of the existing silanol groups. The stronger interaction of the silica particle stems from the silanol groups. However, large differences between the tear stress measured by different test specimens have been found [61]. In the case of silica-silane coupling vulcanizates, the polymer-filler interaction mainly stems from covalent bonds that fix the polymer chains on the filler surface, leading to a lower tear strength. The effect is more developed for the vulcanizates that contain wet-mixed silanized silica. Compared with S-SBR/cis-BR (75/25) filled with carbon black N234 or with dry-mixing silanized silica having a tear strength of 40 N/mm or 46 N/mm, wet-compounded silanized silica exhibits a low tensile stress of just 36 N/mm [62].

Tear strength is improved in all layered silicate loadings in comparison with other carbon black or silica composites and could improve resistance to crack propagation [63, 64]. Failure of rubber/organo-clay upon tensile loading and tensile stress starts with crack initiation. As dispersion of the platelets allows the creation of voids in their vicinity, the amount of dissipated energy is high enough to withstand higher values than before. Furthermore, the length of the crack path is increased, and the crack travels along a "zig-zag" route that can be considered as another mechanism of energy dissipation [65]. For an EPDM containing both a compatibilizer of EPDM-maleic anhydride and an organo-clay modified with octadecyltrimethyl-ammonium, the ultimate properties are markedly improved at concentrations up to 15 phr [66]. The crack growth path at high deformation of rubber containing the same amounts of layered silicates shows a reduction in crack growth. The shortening of the crack growth distance can lead to greater resistance to crack propagation [67].

The tear strength of rubber and multi-walled carbon nanotubes vulcanizates has a similar effect to that of tension. It increases under carbon nanotube loading and decreases when the ratio of the rubber increases. When a blend of carbon black or silanized silica and multi-walled carbon nanotubes is used. the tensile strength is greatly improved, because of the high aspect ratio of the carbon nanotubes. An increase of 40% with carbon black/multi-walled carbon nanotubes in the ratio of 10/5 phr has been reported [68]. The tear strength of carbon nanotubes at 10 phr fails to show a large difference between carbon black N550 and carbon nanotubes C150P in EPDM. One explanation for this could be the lower degree of dispersion of the multi-walled carbon nanotubes in EPDM compound [69].

## 11.3 Tearing Energy

A concept developed in the context of fracture mechanics to characterize tearing processes is that of tearing energy, T. This characteristic value is truly independent of pre-existing notches and the geometry of the test specimen and operates on the thermodynamic bases. This engineering criterion, which describes the cracking behavior of materials, was formulated by Griffith in 1924 for solid materials, e. g. glass [70]. It is obvious that tearing energy is dissipated irreversibly in the region around the growing tip of the crack, which is heated. A failure in a material under stress deformation grows into a fracture or crack if the energy stored in the material is greater than the free energy of the fracture or crack surface generated. The energy required to create a unit surface by the propagation of a crack through a material is defined as the tearing energy.

There are inherent strength properties, such as tensile strength and fatigue, which are triggered by flaws in the rubber vulcanizate. In some recent publications, the term energy release rate is used as an equivalent for tearing energy. Tearing energy is understood as the rate at which energy is transformed as a material undergoes fracture and is expressed in energy per unit area (J/m$^2$). The tearing energy is not dependent on geometric parameters of the specimen, but on the composition of the material, such as the type of rubber composition, the tendency to crystallize under strain, the filler activity and the dispersion attained during mixing. There are several viscoelastic parameters that influence the tearing energy, such as the chain mobility of the rubbers and the influence of temperature. It makes big difference whether crack development takes place close to the glass transition region or at a much higher temperature.

Since it is very difficult to calculate the local stress at the crack tip of an elastomer, filled or not, it has been proposed that an engineering definition for tearing energy, T, be adopted. Accordingly, tearing energy is defined as the energy required to create a surface by increasing the crack in vulcanized rubber [71]:

$$-(dW/dA) = T \tag{11.5}$$

where dW is the stored energy, dA is the fracture area formed and T is the energy per unit area

The concept of tearing energy has helped to classify the vulcanized rubber compounds behavior better. It can be helpful for classifying the different types of tearing without, but without providing an explanation of the actual mechanism of the process. This must be sought in the molecular mechanisms of orientation and relaxation of the polymer chains and, implicitly, in the rubber-filler interaction.

The tearing energy is much greater than required to create a new surface. It is thus obvious that the energy is dissipated irreversibly in a highly localized region at the tip of a growing tear. The concept of tearing energy is valid for assessing mechanical fatigue cracking, as a simple plot of crack propagation rate, dc/dn, is a function of tearing energy. The magnitude of the tearing energy is understood as a fracture mechanics parameter of the material that is independent of the test method. It has been found that, during cyclic pulsation loading, the crack rate is determined by tearing energy or the maximal energy release rate [72].

Tearing energy can be either measured experimentally or calculated from model considerations [73]. A correlation exists between tearing energy and both the strain distribution around the tip of the crack and the strain rate applied during the test [74]. Crack propagation in real elastomers is influenced by the irreversible energy dissipation processes that occur at the crack tip. Although the concept and importance of tearing energy testing has been long recognized, several reports refer to arbitrary tearing stress. The usefulness of the tearing energy concept for material characterization will be underlined in the following section.

### 11.3.1 Processes for Testing Tearing Energy

Tearing energy is measured on standard test specimens subjected to uniaxial strain or to cyclic deformation. A number of test specimens have been developed for which analytical expressions for tearing energy exist. The three most common tests are the:

- constrained trouser test
- pure shear test
- single edge notched test

The specimens used in these test methods are shown in Figure 11.17.

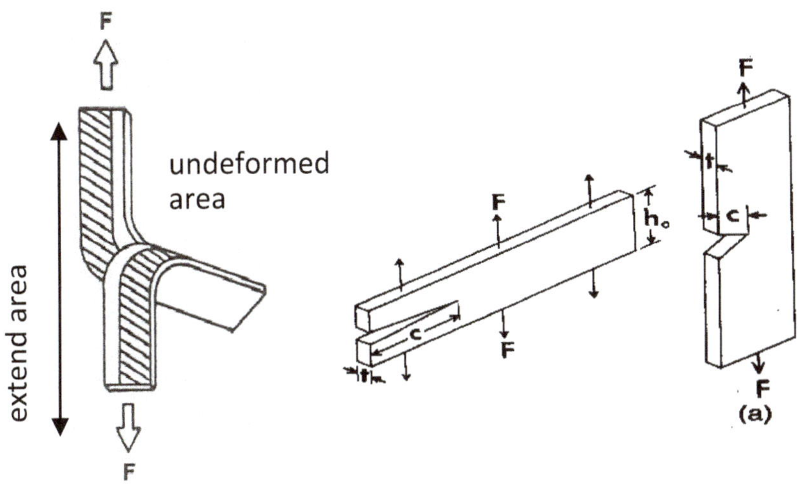


**Figure 11.17** Test specimen for testing tearing energy

The trouser crack growth test is a frequently performed test for determining the tearing energy of elastomers or the tear resistance described above. The two strips of the trouser test pieces are fastened in the holder of a stress-strain machine and pulled apart at defined strain rates; this is also known as out-of-plane shear mode of crack propagation. In order to avoid crack deviation, the two strips of the constrained trouser specimen are fitted with metallic strips; hence the term "constrained". The energy criterion for the tearing test performed with the constrained trouser specimen is:

$$T = 2f/h \tag{11.6}$$

where F is the force and h is the specimen thickness

Therefore, the tearing energy depends on the composition of the vulcanizate and the diameter of the specimen, which must be determined. The constrained trouser test thus affords an easy way to determine the tearing energy.

Tearing energy can be measured on pure shear tests specimens fixed between long parallel clamps. The deformation responsible for cracking corresponds to the applied shear. With this specimen geometry, the length of the incision has no influence on the value of the tearing energy. The test can be performed under uniaxial tension or cyclic tension. The tearing energy for this test geometry is given [75]:

$$T = Wl_0 \tag{11.7}$$

where W is the elastic energy density and $l_0$ is the length of the specimen

Alternatively, crack growth tests can be performed on single edge notched tensile specimens. The test can be performed under uniaxial tension or cyclic tension. The energy criterion corresponding to this specimen geometry is given by:

$$T = 2kc_0W \tag{11.8}$$

where $c_0$ is the initial length of the edge notch, W is the elastically stored energy and k is an extension-dependent function which takes into account the excess stress in the area of the crack tip

The size of the initial flaw $c_0$ and the elastically stored energy determine the magnitude of the tearing energy. For larger non-linear extensions, it is necessary to consider that the function k decreases with the square root of the extension ratio ($k = \pi\ \lambda^{-1/2}$) [14, 75]. The stored mechanical energy can be obtained from the uniaxial stress-strain experiment. In this case, the tearing energy no longer depends on the applied length of the crack but does depend on the energy density and the extension ratio. It is worth noting that all test geometries other than single edge notched tensile specimens are independent of the crack length.

In order for an initiated crack to propagate in an unfilled vulcanizate, the energy density acting at the crack tip must be higher than the energy needed for rupturing of the polymer chains in the plane in which the crack is propagating. Consequently, tearing energy, T, is related to the average strain energy density, $W_t$, at the tip of the crack of the diameter, d [76]:

$$T = W_t d \tag{11.9}$$

The energy required for tearing vulcanizates is partly used for the effective rupture of the polymer chains at the crack tip and partly dissipated through relaxation processes at the crack temperature.

Due to viscoelastic relaxation processes, the tearing process is determined by the temperature and the tearing rate (m/s) at which a test is carried out. These influences on tearing energy result from the mobility of the network chains. Since the tearing energy is being determined on a viscoelastic compound, the speed of deformation will play a decisive role. As a result, the tearing energy of vulcanizates increases with increase in tearing rate and decrease in temperature.

## 11.3.2 Tearing Energy of Filled Elastomers

Tear propagation energy is a material-specific value that is determined by hysteresis effects during crack propagation. Due to hysteresis, part of the energy introduced is dissipated and is no longer available for crack growth. The values for tear propagation energy are significantly increased by active fillers or the strain crystallization of the rubber. The reinforcement of elastomers by particulate fillers is a particularly prominent effect in increasing tearing energy. Although the small particle size carbon blacks have a favorable effect on tear energy, particularly with regard to the initiation of the crack, increases of the aggregate structure are slightly detrimental because the high modulus will reduce the effective radius of the tear tip. The rubber-filler interaction contributes to the resistance to cyclic periodical stretches and leads to an increase in tearing energy. A pronounced rubber-filler interface, good phase adhesion by active fillers, high bound rubber and a very good filler dispersion make the formation of crack surfaces much more questionable. The resistance of a filled elastomer is attributable to an intensive rubber-filler interaction, a high degree of filler dispersion and, of course, the chemical nature of the rubber.

Adding fine particle fillers, such as carbon black or silica, to rubber compounds causes an increase in tearing energy. For carbon black filled rubber, the alignment of polymer chains can be facilitated in the tear tip by slippage and detachment of the polymer chains from the filler surface, due to the relative low polymer-filler interaction. This oriented structure is formed parallel to the direction of applied stress, and perpendicular to the direction of tear growth, leading to higher tear energy. Table 11.3 provides a comparison of the influence of carbon black N330 (50 phr) on the tensile strength and tearing energy of NR and SBR vulcanizates [77].

**Table 11.3** Increase in Tearing Energy and Tensile Strength Due to Carbon Black

| Rubber | Filler | TS (MPa) | T (kJ/m$^2$) Pure Shear | T (kJ/m$^2$) Trouser |
|---|---|---|---|---|
| NR | – | 28.3 | 6 | 12 |
| NR | N330 | 28.5 | 12 | 50 |
| SBR | – | 2.3 | 1.4 | 3.2 |
| SBR | N330 | 28.78 | 7 | 6 |

For tensile strength measurements the SBR samples show the influence of the carbon black for the filled vulcanizate. The number of filled SBR is ten times higher than in the unfilled sample. For NR and SBR, the tear energy shows a strong increase due to the carbon black filler. However, the type of the measurement influences the result of the tear energy, as can be seen for the pure shear and the trouser measurements. Only above 100 °C does the number of elongations to failure for NR decrease significantly, as strain crystallization is no longer effective. Depending on the test specimen, the tensile strength and the tearing energy increase in NR because of the appearance of strain crystallization, which is not the case with SBR vulcanizates. At the same loading, fillers having small aggregate size and high surface activity confer greater resistance to crack propagation.

With carbon black of constant structure, the tearing energy increases significantly with decrease in particle size or increase in specific surface area. The marked enhancement of tearing energy is associated with a change in the character of the tearing process, from relatively smooth tearing to discontinuous stick-slip tearing (knotty tearing), where the tear deviates from a straight path.

The increase in tearing energy increases with the mixing time of the compound. Dispersion of the filler exhibits proportionality, which means that the degree of the filler particles is a determining factor in for the value of the tearing energy. This beneficial effect of mixing efficiency is shown as a function of filler dispersion in Figure 11.18 [78].

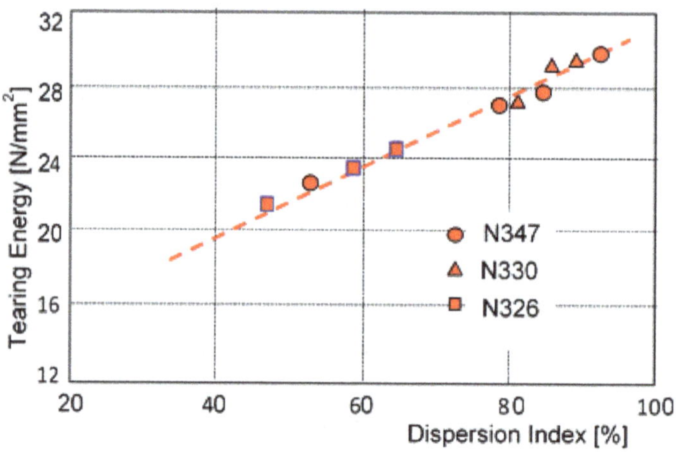


**Figure 11.18** Increase in tearing energy with the dispersion coefficient in E-SBR vulcanizate [78]

This relationship illustrates the advantages of dispersive mixing for gaining high performance vulcanizates. If fillers having the same specific surface area but different structure of the aggregates mixed into the rubber are selected, those fillers with the high DBP number will attain the highest tearing energy values, which is a sign of aggregate branching. The better the filler is dispersed, the more crack propagation is hindered. Since the crack cannot propagate through the filler particles, it has to move around them, thus the path of the crack is changed many times when the density of the filler particles in the system increases.

Electron microscope examinations of the cracks produced have shown that the path on which the crack propagates through filled vulcanizates runs from one filler particle to the other. In the case of vulcanizates containing carbon blacks of low specific surface area, it has been found that the rubber phase separates from the aggregates and the crack runs along the rubber-filler interface [77]. If the adhesive forces between the filler and rubber are stronger than the cohesive forces in the bound rubber layer, the crack takes place in the rubber matrix. It is assumed that the crack runs in the direction of the highest stress concentration in the material and that the filler leads to an extension of the crack path. However, at high filler loadings, there are always more filler particles that are not surrounded by a rubber layer and that touch each other, with the result that the crack can propagate unchecked between the particles. The overloading of the vulcanizates leads to a decrease in tear strength. The crack can propagate also between the two neighboring filler aggregates.

Looking at the influence of the experimental parameters, namely the strain rate and temperature, an interesting tear is seen in the E-SBR/carbon black N330 in the range of 0 °C when the material comes close to the glass transition region [80].

At higher stretching rates, the tearing energy increases several times over, compared with the unfilled counterpart. Some authors hold the test procedure responsible for the irregularities of the T-surface [81]. This strengthening action is, however, restricted to a specific range of tear rates and test temperatures that depend on both the type of the filler and the rubber [82]. Compared with that of unfilled vulcanizates, the fatigue resistance is increased substantially in the presence of well dispersed active fillers, as shown in Figure 11.19 for NR and E-SBR vulcanizates [17].

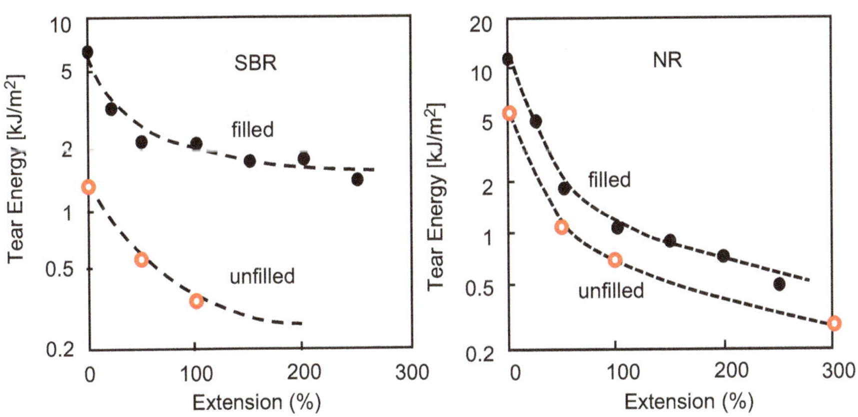


**Figure 11.19** Comparison of tearing energy in unfilled and filled SBR and NR vulcanizates

The linear plots of the cut length, c(mm), versus the number of flex-cycles, n, show propagation of the cut. The steeper the slope, the more rapidly the crack propagates.

As illustrated, cut growth is lower in unfilled NR vulcanizates than in unfilled SBR vulcanizates; this is a consequence of the strain crystallization in NR. There is a prevailing concept that failure properties of strain-crystallizing rubbers are less affected by fillers. This concept is based on the idea that at the crack front, where the stresses are highest and reinforcement is most required, crystallites blunt the crack tip and toughen the material locally, thus reducing the need for filler reinforcement. The reduced cut propagation in filled SBR is more pronounced than in filled NR, indicating that rubbers which undergo strain crystallization do not require the same reinforcement. This shows greatly enhanced tearing energies over a wide range of stretching rates and temperatures (−20 to 80 °C) at which strain-crystallization occurs at the crack tip. It appears that the effects of strain-crystallization outweigh the viscoelastic effects [75].

Even at low elongation, the tearing energy in the SBR vulcanizate increases fourfold, while in the NR vulcanizates it doubles. This big difference between the two vulcanizates shows the improvement in the filler in the amorphous vulcanizate. However, with increase in severity of the cyclic elongation, the tearing energy decreases in both systems. But the difference from the unfilled system is much more pronounced in the amorphous SBR than in the strain-crystallizing NR. At high temperatures (> 80 °C), strain crystallization declines and the tearing energy drops sharply to values shown by amorphous vulcanizates. At high temperatures, strain crystallization has lost its significance. At low temperatures and high tear rates, molecular organization into crystallites cannot take place in short period of stretching, as the crack tip advances and tearing energy approaches again lower values. For this reason, the tearing energy increases at low temperatures, which is particularly the case with vulcanizates made of amorphous rubber.

In ternary rubber vulcanizates NR/SBR/BR (50/20/30), increase in carbon black N330 loading is accompanied by an increase in the compression stress. The compression stress value is important for applications such as dampers or seals. At 60 °C, the value of compression stress increases by 350% compared with the unfilled ternary rubber, which remains at 0.53 MPa.

For the vulcanizates that contain silanized silica, the polymer-filler interaction is mainly from covalent bonds that fix the polymer chains on the filler surface. This would limit the orientation of polymer chains in the tip area, leading to a lower tear energy.

## 11.4 Dynamic Crack Growth

An important area is the determination of crack growth propagation as a function of the deformation rate, the tearing energy and the chemical composition of the rubber. The scientific basis for the selection of materials for fatigue life optimization is the determination of crack growth over a wide range of tearing energies. When small stresses are applied periodically to a test specimen, a crack will grow incrementally by one length per cycle, dc/dn. Usually the cyclic deformation is applied in a sinusoidal form to test specimens of different geometries in uniaxial tension or compression mode [12]:

- single edge notch
- pure shear
- simple shear
- dumbbell

In many cases, the crack propagation is observed under sinusoidal deformation. The measured crack propagation depends on the range within stress or strain is varied. However, test methods that follow pulsed deformations with defined strain amplitude and time between pulses are more application oriented [83]. There are many variables to adapt the test to practical applications. It is possible to change the loads severity or set the load sequence, vary the R-ratio or introduce different pulses with a variable dwell period between the pulses. One method is to use the Tear Analyzer to investigate dynamic cut growth, adapting the strain amplitude and the time between successive pulses to the practical application. Therefore, this method is the one that is closest to that which is obeyed in real situations. These tests are very time-consuming, but they are able to vary the conditions during the test so that the load cycles are as close as possible to those in practice [12]. Figure 11.20 shows several load cycles that can be applied to crack deformation.

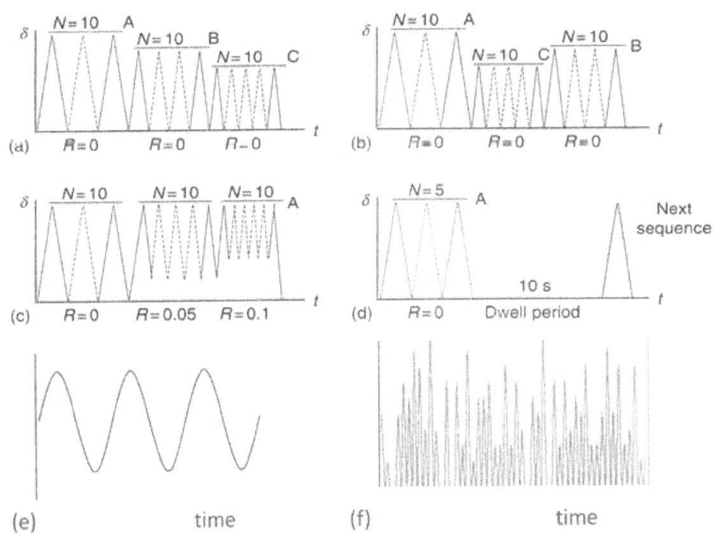


**Figure 11.20** Amplitude test signals applied to pure shear specimens: a) load severity, b) load sequence, c) R-ratio, d) dwell time, e) constant amplitude and f) variable amplitude signal

Each of these stresses prompts the material to fail in different ways and contributes to the understanding of the processes involved in deformation. The severity of the test is expressed by the so-called R-ratio, which is defined as the ratio of the minimum parameter (stress or strain) to the maximum parameter. An R-ratio of zero corresponds to fully relaxing conditions, whereas decreasing R-ratio tests describe more severe test conditions. At constant tearing energy, cut propagation, dc/dn, increases with the R-ratio [84]. The influence of the R-ratio will be discussed below.

### 11.4.1 Crack Growth in Unfilled Elastomers

Crack growth in unfilled elastomers is important for determining the influence of polymer morphology, chain dynamics and crosslinking. Four distinct regimes of dynamic crack propagation have been identified, based on the tearing energy for $R = 0$ for unfilled vulcanizates. The increase in the crack growth regime, dc/dn, as function of tearing energy, T, is shown for NR and SBR vulcanizates in Figure 11.21 [85].

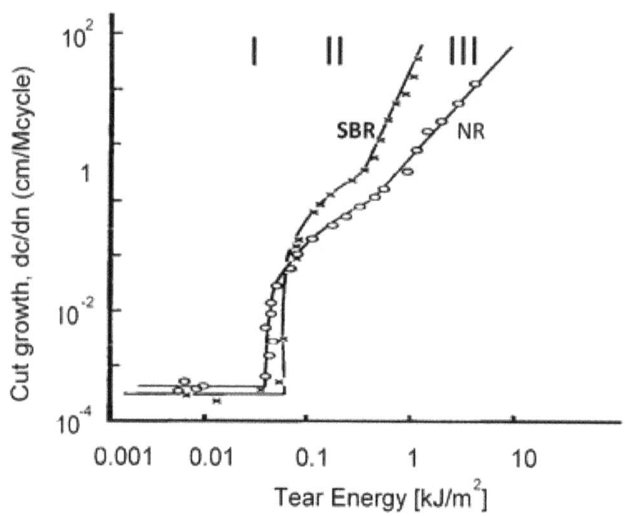


Figure 11.21 Crack growth regimes of SBR and NR vulcanizates as a function of tearing energy

As long as the tearing energy remains below a threshold, $T_0$, no mechanically induced crack growth occurs. Crack growth proceeds exclusively due to environmental attack at a very low level. The speed of tear growth in unfilled elastomers is essentially constant and limited by the concentration of ozone. *In vacuo* no crack will grow below $T_0$. A transition zone with an increase of several orders of magnitude follow the initial regime. The crack growth rate is quite independent of tearing energy, with a somewhat linear form. Strain-crystallizing NR shows a reduced crack growth when compared with amorphous SBR. After this transition zone, there is a range over which the dynamic crack propagation can be expressed by a power law for the tearing energy:

$$dc/dn = BT^{\beta} \tag{11.10}$$

where n is the number of deformation cycles, and B and β are elastomer-specific material constants

The equation can be integrated if the tearing energy, T, is known as an analytical expression, and the exponent β and the pre-exponential factor B are determined empirically. The crack grows until a size is reached at which there are relatively few cycles left until failure. The crack growth accelerates and enters an unstable crack regime in which rapid tear growth occurs, which is also called a catastrophic tear. This limiting energy decreases with the crosslink density and with the content of poly-sulfidic crosslinks. If the crosslink density is higher and the system is cured by a conventional sulfur system, the crack growth appears at a higher level.

Several models have been developed to describe these crack growth regimes [86]. The determination of the pre-exponential factor B and the exponent β in Equation 11.10 requires considerable experimental effort. The logarithm of deformation cycles, N, in the linear crack propagation regime has to be plotted against the log of tearing energy (i. e. for SEN specimens $T = 2c_0 {}_k W$). The slope of these plots represents the exponent, β. So, crack growth has to be measured at constant temperature for several values of tearing energy (or stress and strains). The procedure is illustrated for EPDM vulcanizates in Figure 11.22 [87].

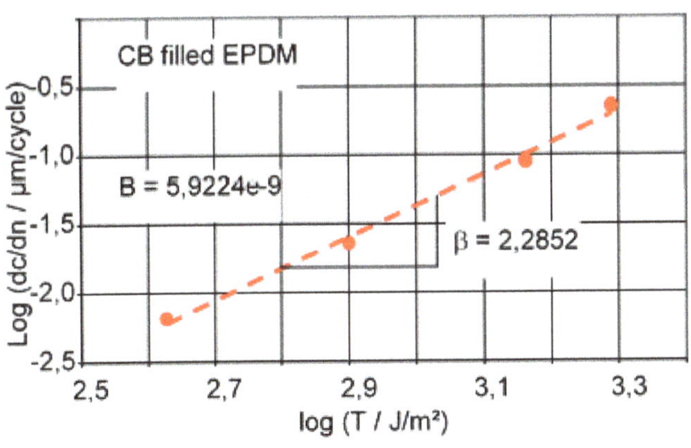


**Figure 11.22** Determination of crack growth parameters B and β for a carbon black-filled EPDM vulcanizate [87]

For adequate accuracy, the measurements have to be performed under mild to severe conditions. There are often considerable differences between the values of β obtained from tear propagation and fatigue life experiments. This reflects the fact that tear propagation in an elastomer during dynamic deformation is described adequately by Equation 11.10 over the whole range of tearing energies. The preexponential factor B and the exponent β for NR, BR and NR/BR blends are given in Table 11.4 [97].

**Table 11.4** Fracture Mechanics Material Parameters B and β for Rubbers

| Rubber | B | β |
|---|---|---|
| NR | $4.46 \times 10^{-12}$ | 1.35 |
| BR | $12.0 \times 10^{-18}$ | 3.44 |
| NR/BR (50/50) | $1.38 \times 10^{-14}$ | 2.12 |

It can be seen that the exponent, β, is polymer specific. NR has the smallest exponent β, due to the fact that it can crystallize under strain and builds up an additional self-reinforcing network that acts in addition to the chemical crosslinking. In the case of cis-BR, the exponent is about three times higher, which also corresponds to the higher crack growth. The blend, composed of equal parts of NR and BR, has an exponent that corresponds to the arithmetic mean of the exponents of the blend constituents if the crosslink density distribution is similar. Based on this fracture mechanics approach, crack growth has been investigated in model vulcanizates of tire sidewall compounds using pure shear test specimens and pulsed cyclic deformations.

The effect of the polymer matrix and of reinforcing blacks (20 and 50 phr) on dynamic crack propagation has been investigated for NR, IR, SBR, BR, CR, IIR and NBR. It has been found that the crack propagation at constant tearing energy (1 kg/cm = 981 J/m$^2$) is by far the lowest in NR. The synthetic rubber shows a crack growth that is roughly two orders of magnitude higher. A comparison of the unfilled and the filled vulcanizates is shown in Figure 11.23 [76].

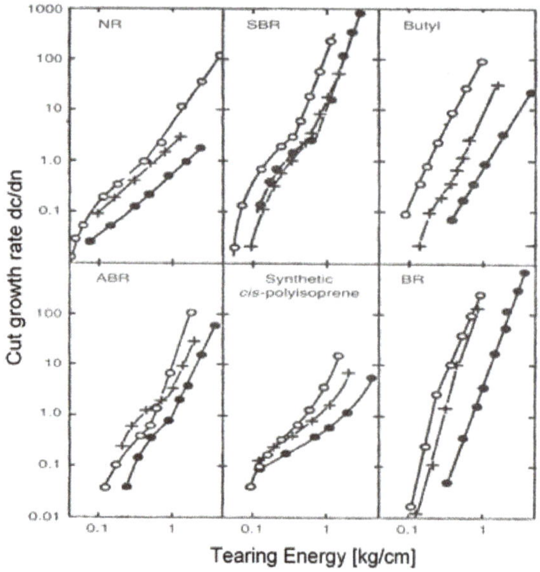


**Figure 11.23** Comparative studies of crack growth in unfilled and filled rubber vulcanizate [76]

The crack growth, dc/dn, indicated by the white dots, is representative of the vulcanized rubber. The curve with the black dots corresponds to the carbon black (50 phr) filled system. The interaction with the carbon black is also very different for the different polymers. For the NR vulcanizate, which shows strain crystallization, the cut growth rate is 10 and for the filled NR it is 0.8. The amorphous SBR shows a cut growth rate at 100 and the filled SBR shows a cut growth rate that is above 10. With butyl rubber, there is a significant increase in tearing energy due to the presence of the carbon black that does not occur to the same extent with unsaturated rubbers. The butyl vulcanizate has a cut growth rate of 100 but the filled vulcanizate demonstrate a very decreased cut growth rate. The synthetic cis-polyisoprene, which has more 3,4-isoprene units and cannot crystallize, has a lower cut growth rate of 6–7, but the filled vulcanizate is similar to the filled NR vulcanizate. The cis-BR shows a similarly high cut growth rate as the SBR but a decreased value for the filled state. The difference from SBR is due to a decreased rubber-filler interaction.

Pulsed cyclic deformation provided by the Tear Analyzer for NR, cis-BR and CIIR shows an interesting trend in terms of tearing energy, as illustrated in Figure 11.24 [88].

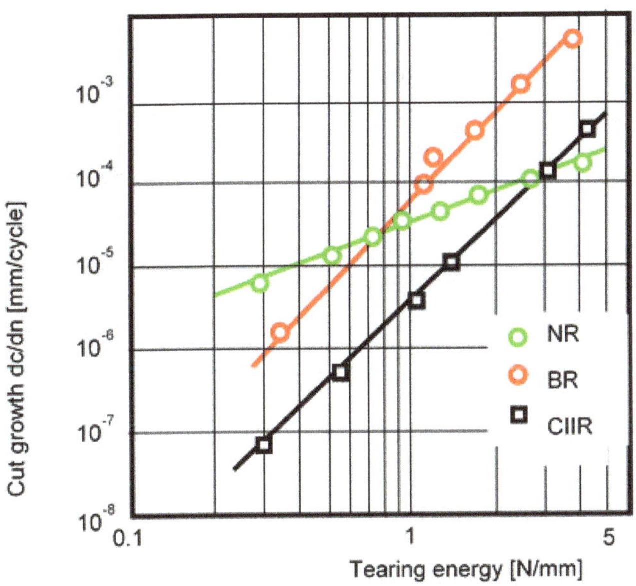


**Figure 11.24** Cut growth versus tearing energy in pulsed cyclic deformation for NR, BR and CIIR vulcanizates (adapted from [88])

The cut growth rate in NR compounds is relatively high at low tearing energies, corresponding to small strains or stresses that cannot provide strain crystallization; but the cut growth values are lowest at high tearing energy, which is the material's response at large stresses or strains under which strain crystallization can take place. The higher rates at low tear energies may be explained by oxidative processes that occur during long-term durability tests, whereas the lower rates at high tearing energy are due to the stabilizing effect of strain-induced crystallization that overcomes the detrimental oxidative effects. The low level of cut growth at high tearing energy is necessary for the abrasion resistance of the NR compounds that are frequently used in the tread of truck tires. The synthetic rubbers show a crack growth rate that is roughly two orders of magnitude higher than that of NR. At high pulsed cyclic deformation, cis-BR and CIIR, known for its damping, have a higher cut growth rate than NR. These two synthetic rubbers are not very suitable for reducing the formation of cracks at high stresses. Elastomers with higher glass transition temperatures dissipate more energy than those with lower $T_g$ due to the higher molecular friction. The large difference in crack growth rate at constant tearing energy also results from the elasticity of the elastomers. The more energy is dissipated, the less energy there is available for crack propagation.

## 11.4.2 Tearing Energy of Unfilled Elastomers

The influence of the viscoelastic material behavior is shown by the increase in tearing energy with the stretching rate at constant temperatures and by the decrease in temperature at constant stretching rate [75]. Using constrained trouser specimens, the effect of temperature and tearing rate has been investigated [89]. The stick-slip phenomenon occurs at high temperatures and at low tear rates. However, the increase in tearing energy with the stretching rate is not dependent on the specimen geometry. Trouser specimens, pure shear specimens and split specimens show the same tearing energy at a certain stretching rate, within error limits [90].

Studies on tear strength have shown that the stretching rate of a vulcanizate can increase non-linearly from smaller stretching rates ($5 \times 10^{-5}$ cm/s) to a high stretching rate of $10^{-2}$ cm/s, which can be attributed to viscoelastic causes. Plotting the tearing strength of a vulcanizate as a function of tearing rate and correcting for the influence of the test temperature according to a WLF-like approach yields a single master-curve that shows a non-linear increase regardless of the geometry of the specimen [79]. This highlights, on one hand, the influence of the viscoelasticity of the specimens and, on the other, the validity of the tearing energy, as a criterion independent of the specimen geometry.

As the example of an SBR vulcanizate shows, the tearing energy of non-crystallizing elastomers decreases with the temperature, while the chain mobility increases with the extension rates in a manner that indicates the dependence of energy dissipation in a viscoelastic material [75].

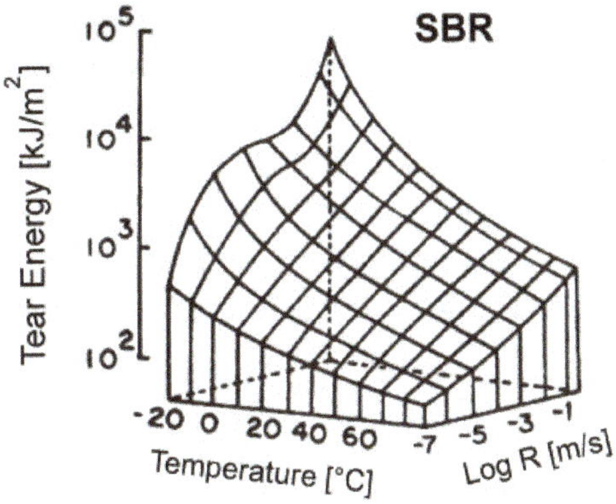


**Figure 11.25**
Development of tearing energy for SBR as a function of temperature and extension rate [75]

This behavior has been qualitatively confirmed for various non-crystallizing elastomers [91]. At temperatures slightly above $T_g$ and high strain rates, the material behaves increasingly rigidly and the chains do not move completely freely. Indeed, at still higher rates the vulcanizate would fracture like a polymer glass [92].

A uniform description based on the temperature-time superposition principle can only be carried out in limited areas of the tearing surface that is formed in the vulcanized rubber. Assuming viscoelastic behavior of amorphous elastomers, it can be concluded that the tearing energy is similar under conditions of equal chain mobility. As the stress is applied, an oriented structure is created at the crack tip that is associated with large local deformations. This oriented structure at the crack tip is parallel with the applied stress and perpendicular to the direction of crack growth. Since more energy is required to propagate the crack perpendicular to the orientation direction, the onset of the crack is delayed [91, 93]. Thus, a large amount of energy is consumed for purposes other than creating new surfaces [94]. The consumption of the energy is attributable to viscoelastic energy loss in the vicinity of the crack. The energy dissipation at the crack tip determines the magnitude of the tearing energy. The more energy dissipated at the crack tip, the higher the energy required to form a new crack surface unit. This dependency is demonstrated in by the proportionality of T and the loss modulus, G″, for six different vulcanizates [91].

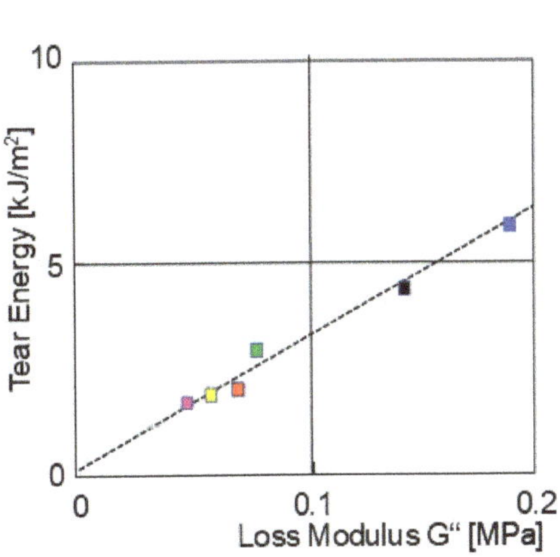


**Figure 11.26**
Increase in tearing energy due to energy dissipation of the rubber [91]

The work required to break the elastomer has been found to increase in proportion to the energy dissipated on stretching [77]. However, since the results concern unfilled systems, the informative value for practical applications is somewhat limited. But the result shows that no new surface can be formed without energy expenditure.

Compared the number of stretches until failure the NR are superior to the amorphous SBR. The depth of the initial cut decreases this number strongly [95]. Also, the number of repeated stretches is much higher for NR than for E-SBR [96] as shown in Figure 11.27.

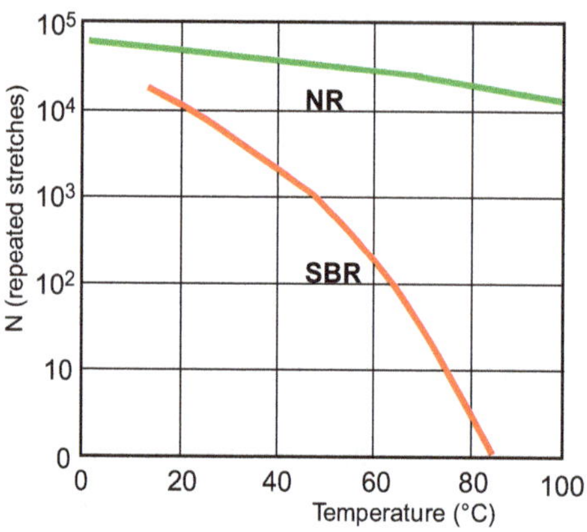


**Figure 11.27** Decrease in the number of stretches to failure as a function of the temperature for NR and E-SBR [96]

Up to 100 °C, the NR shows a small reduction in elongations to failure with less than one order of magnitude, while the E-SBR declines over three orders of magnitude in cyclic deformation [96]. For strain-crystallizing NR, the tearing energy is high and does not decrease strongly with the temperature up to 80 °C. The expected high tearing energy is due to the presence of strain crystallization above the elongation at 300%. The explanation for this high energy can be found in TEM micrographs, which show the crystallite arranged perpendicularly to the direction of the elongation, as shown in Figure 11.28.

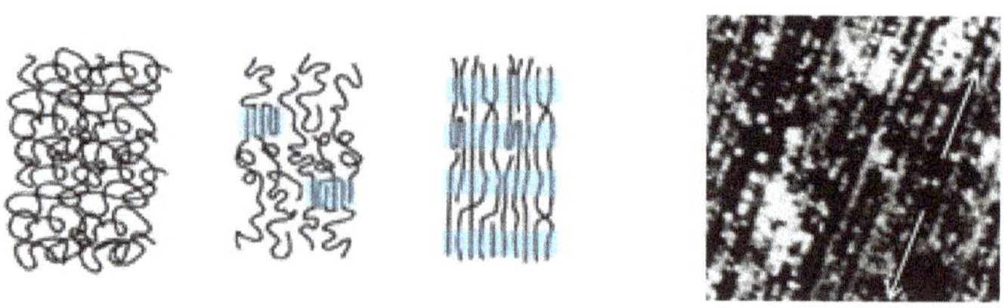

**Figure 11.28** Schematic picture of strain crystallization and the TEM image proof

Strain crystallization of the uniform structure of the polyisoprene chains, which when stretched an elongation of 300%, form crystals that are arranged perpendicularly to the deformation direction and cause the material to strengthen itself through self-reinforcement. This self-reinforcement causes the increase in the stress at high deformations and the appearance of internal cracks appear only at higher stress.

Studies have demonstrated the influence of the viscoelasticity on tearing energy by showing that the equivalence between time and temperature follows a master-curve, with a reference temperature at 25 °C. In the high-rate region, the tearing energy curve exhibits a glassy stage and can be broken. However, over the entire time scale, the tearing energy depends on the crosslink density and the type of micro-structure of the sulfur crosslinks [89]. The longer the sulfur bonds of the crosslinks are, the more the initiation of catastrophic cuts can be prevented.

### 11.4.3 Crack Growth in Filled Elastomers

As far as the effect of reinforcing filler is concerned, it has been found that the presence of filler reduces crack propagation by at least one order of magnitude. This effect of reducing cut growth rate has been covered in the comments given for Figure 11.24. Another example of the cut growth rate can be found in Table 11.5, which illustrates the reduction in cut growth in NR, BR, SBR and NBR due to carbon black of different surface activity [97].

**Table 11.5** Effects of Carbon Black on Cut Growth Rates [91]

| Rubber | Carbon Black | Cut Growth Rate ($10^{-5}$ mm/cycle) |
|---|---|---|
| NR | – | 5.60 |
| | N990 | 2.03 |
| | N330 | 0.182 |
| BR | – | 247 |
| | N990 | 163 |
| | N330 | 1.13 |
| SBR | – | 146 |
| | N990 | 14.3 |
| | N330 | 1.21 |
| NBR | – | 247 |
| | N990 | 3.11 |
| | N330 | 0.87 |

As the surface area of the carbon blacks used increases, the crack growth rate in each rubber vulcanizate decreases progressively. For the development of particularly resistant elastomers, it is advisable to use strain-crystallizing rubber on the one hand [98] and to use a carbon black with a high specific surface on the other, even if this complicates the mixing procedure and makes dispersion more difficult. Note that the improvement in properties is greater for the more reinforcing filler and the amorphous elastomers than for the strain-crystallizing rubbers. In application, low crack growth is desirable in order to minimize premature failure associated with dynamically induced cracking. The phenomenon of deviation of the crack tip from the intended tear path is common for carbon black-filled vulcanizates. When deviation occurs at the crack tip, interpretation of the results is made difficult, since the initial single crack tip can generate branches in different directions. However, this mechanism contributes to prolonging the time to catastrophic failure.

### 11.4.4 Service Life Prediction: Integrated Power-Law Model

In practice, it is important to know the fatigue life of elastomers in order to be able to make predictions about the durability of a component. Tear resistance and dynamic crack growth allows a classification of different materials, but not a determination of the fatigue life. The empirical Equation 11.10 can be used for this ambitious purpose of determining the fatigue life time for a filled vulcanizate.

The fatigue life is determined by integrating the equation for the crack growth rate from its initial size, $c_0$, to its final size, c [88, 99]. If the energy criterion valid for the test specimen is used in the expression of the power law, the number of load cycles that leads to a certain crack propagation can be calculated by integrating Equation 11.10. For a single edge notched specimen with initial flaw size $c_0$, the crack growth equation is given by:

$$dc/dn = B\,(2c_0kW)^{\beta} \tag{11.11}$$

where B is the empirical determined preexponential parameter, $c_0$ is the initial flaw, k is the energy parameter, W is the kinetic energy and β is the exponent

Integrating this differential equation within the initial cut length $c_0$ and the final cut length, c, the number of cycles N to break is obtained:

$$N = 1/(\beta - 1)\,(B2kW)^{\beta}\,(1/{c_0}^{\beta-1} - 1/c^{\beta-1}) \tag{11.12}$$

where c is the final cut length

Integration of Equation 11.12 makes it possible to calculate the service life of a vulcanizate with a specific notch. The results make it clear that the initial flaw size, $c_0$, which determines the number of cycles of the material to final failure, is important.

The smaller the flaw size inherent in a given elastomer, the longer this material resists cyclic dynamic load. Equation 11.12 can be used to evaluate the effect of the stored teasing energy, 2kw, and the real flow size, $c_0$, on the number of cycles of the elastomer and the lifetime of the elastomer. Conversely, the service life, N, of a test specimen can be used to determine the original size of the flaw. Therefore, the pellet fragments with a size of 0.1 mm are the important flaws in the vulcanizate. The loss of pellet fragments is randomly distributed in the vulcanizate. The higher the cut growth rate, the better is the abrasion resistance of the vulcanizate. The wet-mixed vulcanizate offers a reduced number of large agglomerates and that increases the cut growth rate and improves the fatigue resistance of the vulcanizate. The advantage of a wet-mixed vulcanizate is the better filler dispersion obtained by the avoidance of pelletized filler.

Among the different processes for ascertaining the fatigue resistance of elastomers, the Tear Analyzer is a versatile tool that allows a quantitative analysis of the dynamic crack growth of rate single edge notched and pure shear specimens under sine-shaped or pulsed waveforms, followed by arbitrary relaxation times at given pre-stress and a large spectrum of strain amplitudes. A schematic diagram of this complex apparatus is given in Figure 11.29 [83].

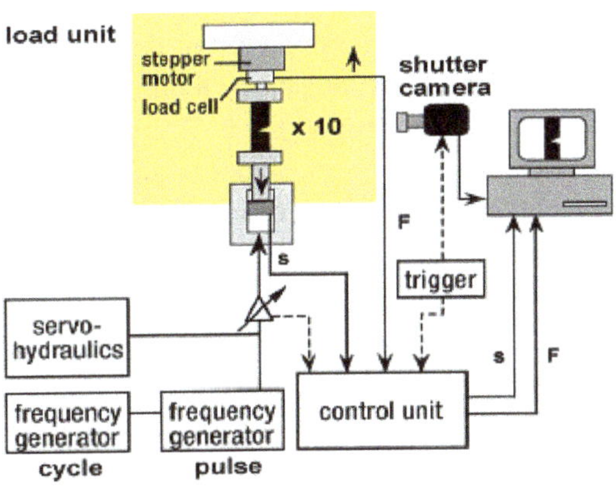


**Figure 11.29**
Schematic model of the Tear Analyzer [83]

The Tear Analyzer contains ten measuring stations with separate load cells to ensure statistical data treatment. Each load cell is connected to a computer-controlled motor that maintains a pre-stress to ensure constant strain during the experiment. The contour length of the crack on each test specimen is recorded by a moving CCD-video camera. The images of the cuts are digitized and the crack propagation is calculated by a computer for each test specimen. The dynamic loading on the notched test specimens to the tearing point is conducted with an elongation of 15 to 30% at a frequency of 10 Hz and a pulse width of several milliseconds. The tear propagation and the tear contour are determined at equidistant intervals.

Data acquisition and analysis of the data from the Tear Analyzer are shown in Figure 11.30. The tear contour length is measured as a function of deformation cycles until the unstable crack regime is reached. In parallel, the dynamic hysteresis and the elastic energy density stored in the material are measured as a function of the cycles of the test specimen. These data are used to plot the cut growth rate as a function of the number of cycles, which is implicitly the fatigue life.

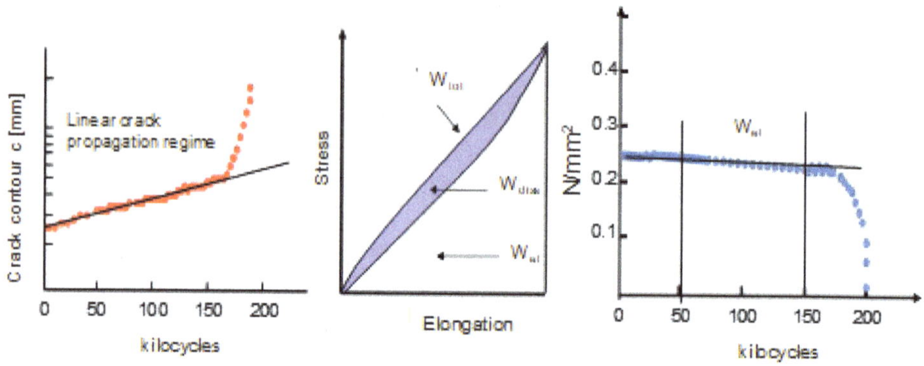


**Figure 11.30** Parameters measured by using a tear analyzer (adapted from [83])

Modern versions of the Tear Analyzer are available with servo-hydraulic drive or two electric drives [100]. Both can produce several pulses (triangle, sawtooth, rectangle etc.) at strain amplitudes up to 50 mm and frequencies from 0.1 to 50 Hz at temperatures from −20 to 150 °C and different environments that can contain air, an ozone-rich atmosphere or simply nitrogen that will not oxidize the specimen during the protracted test.

With the help of the tear analyzer, the experimental conditions can be adapted to the practical requirements when using the filled elastomers. In order to closely approximate the conditions in practice, the materials are not loaded sinusoidally, but in most cases are pulsed with alternating stresses that closely resemble the real deformation pattern. Such conditions can be simulated with a tear analyzer so that of a "predictive" test of the vulcanizate can be carried out. The amplitude of strain has a major influence on the service life of the test specimen. With increase in amplitude, tear propagation enters the unstable and the catastrophic tear regime much earlier.

#### 11.4.4.1 Service Life of Unfilled Vulcanizates

In order to investigate the influence of the crosslink structure in NR vulcanizates with identical crosslink density, crack propagation measurements were carried out using this technique [101]. The results confirm the longer service life for vulcanizates with a higher proportion of poly-sulfidic crosslinks. If the service life cycles are plotted as a function of the strain rate, a significant increase in the service life for CV curing systems (CBS/S) compared with EV curing systems (TMTD/S) is found. Vulcanizates with a higher proportion of poly-sulfidic crosslinks can be exposed to a higher level of strain without any loss of service life. The reversion-stable curing system leads to a lower crack growth rate and or longer service life.

Since fatigue life experiments are subject to environmental conditions over a large number of cycles and increased temperatures occur at the crack tip due to the increased distribution of stress, it is important to see how the service life is influenced by the presence of anti-oxidants. Comparison of vulcanizates of otherwise the same composition shows that the crack contour length in the linear range is increased by a small amount by phenolic antioxidants and TMQ, compared with the reference without antioxidant. A significant improvement occurs when p-phenylenediamines are used. In NR vulcanizates, 6PPD has the greatest effect on the service life of the vulcanizates and prolongs the service life of the vulcanizate [81].

#### 11.4.4.2 Service Life of Filled Vulcanizates

As discussed extensively in previous sections, the type and dosage of reinforcing filler have an impact on dynamic crack propagation and consequently on the service life of the filled elastomer. Tear analyzer tests confirm that the specific surface area and the surface activity of the filler significantly extend the service life of vulcanizates. In any case, the degree of filler dispersion achieved in the mixing process makes a significant contribution to this improvement in performance. Tear propagation per cycle as a function of energy density shows that higher energy densities lead to a reduction of the crack propagation. Therefore, the mixing time after the filler incorporation is important for increasing the fatigue resistance. Particularly in the case of fillers with a high specific surface and surface activity, dispersive mixing is severely restricted by the inter-aggregated forces of the filler interaction in the agglomerate. With reference to strategies that lead to improved filler dispersion (adequate processing parameters, choice of suitable raw materials and surface modification of rubber and filler), it can be stated that the higher degree of filler dispersion achieved during mixing makes it possible to achieve a higher number of cycles, as the filled vulcanizate can withstand the linear cut propagation regime and, therefore, there is an increase in the service life of the vulcanizate.

Tear analyzer investigations on SBR/N330 vulcanizates have shown that an increase in macro-dispersion implicitly also increases micro-dispersion, expressed by a lower Payne effect ($\Delta G'$) or lower dynamic hysteresis, leads to a marked increase in the linear crack propagation regime and prolongs the service life of the vulcanizate. The dynamic cut growth increases with the micro-dispersion of the fillers in any compound. The degree of filler dispersion achieved during mixing allows a higher number of cycles in that the filled vulcanizate can withstand the linear cut propagation regime. Figure 11.31 shows that the mixing time and the rotor speed increase the micro-dispersion of the filler [103]. The resistance to cut growth increases with each mixing action. With the mixing parameters, such as the correct fill factor, rotor speed, ram pressure and mixing time, the erosion of the larger agglomerates can be accelerated and the formation of aggregates in the elastomer can be improved. Thus, more nanoparticles are distributed in the rubber compound.

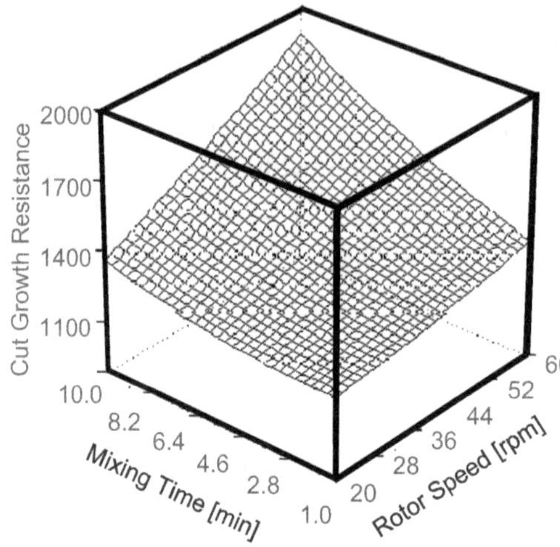


**Figure 11.31**
The influence of mixing time and rotor speed on the cut growth resistance of E-SBR/CB N330 vulcanizates

Comparison of vulcanizates filled with conventional carbon black and carbon black produced in wet-mixing show that the service life of rubber filler composite (RFC) is increased significantly. A fourfold (30% strain) to tenfold (20% strain) lower crack propagation rate has been observed for E-SBR vulcanizates filled with RFC blacks. Vulcanizates that are also further mixed on the two-roll after the mixing process in the internal mixer show an increased degree of dispersion of carbon black (97–98%) and the number of load cycles in the linear range more than doubles accordingly. It is therefore clear that the better filler dispersion results in a longer service life for the vulcanizate. It can be seen that the filler dispersion on the two-roll mill increases the number of cycles in the linear regime and therefore the functionality of the rubber compound is significantly increased. The better dispersion of the filler in the vulcanizates is a prerequisite for a prolonged service life of the vulcanizate.

Another test with the Tear Analyzer has shown that the dynamic crack propagation in a sidewall compound of NR/cis-BR/carbon black depends on the filler distribution in the rubber. The rubber phase of NR/cis-BR was set to 0.45 to 0.55 while the carbon blacks N550 and N762 were varied from 0.40 to 0.60 in the rubber blend. The macrodispersion of the filler was above 95%. The order of mixing was the interesting parameter in this experiment. The dynamic crack propagation differs by three orders of magnitude in its propagation of the length of the created crack and in the number of cycles in the linear regime, where a function of the vulcanizate is shown. The results are shown in Figure 11.32.

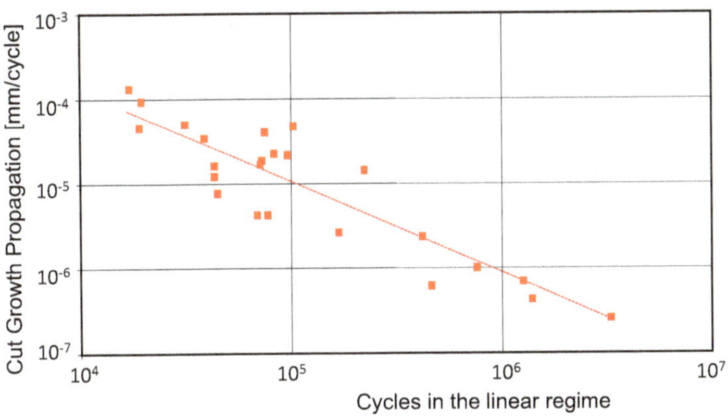


**Figure 11.32** Propagation of dynamic crack growth in a NR/cis-BR/ carbon black vulcanizate

The presence of the more active carbon black N550 and the distribution of the larger amount of carbon blacks N550 in the cis-BR phase are responsible for this large difference in the crack propagation regime. The difference in the dispersion and distribution of a filler in a rubber blend is also shown by an experiment with the Tear Analyzer. The reduction in cut growth propagation and the significant increase in the linear cycles are determined via the setting of the mixing parameter.

For an NR/S-SBR/cis-BR compound fed with a silanized HD silica, the forced distribution of fillers shows an advantage when the silica has been silanized in the S-SBR/cis-BR phase and is subsequently mixed with the NR phase: a longer linear cycle time, which indicates a better service life. If the filler is mixed into an incompatible blend that shows no strain crystallization, it contains a larger amount of the filler and the dynamic crack propagation decreases significantly. This forced distribution leads to a vulcanizate with a longer service life. This influence of the silica distribution in NR/S-SBR/cis-BR can be seen in Figure 11.33 [105].

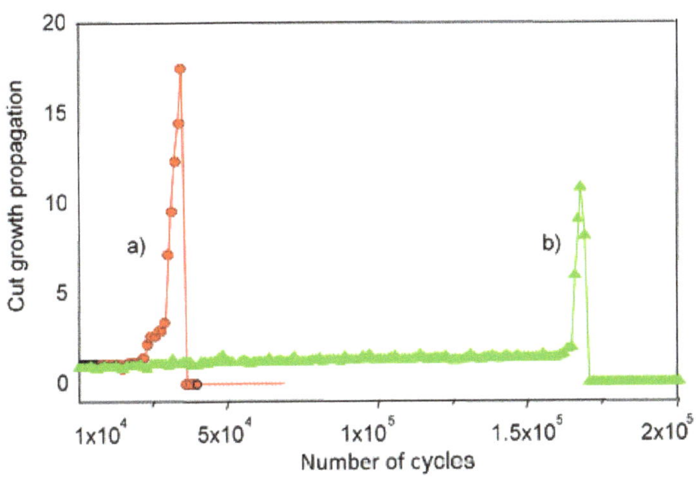


**Figure 11.33** The influence of the silanization procedure of HD silica in NR/S-SBR/cis-BR blend: a) silanization in the NR phase and b) silanization in the S-SBR/cis-BR phase

At this point, it is helpful to emphasize that the proportion of undispersed filler fragments increases the frequency of flaws in the vulcanizate and triggers the initiation of cuts and cracks under dynamic load. Thus, it is essential to reduce the occurrence of pellet fragments by using the right mixing strategy.

Adding a low volume of multi-walled carbon nanotubes to a rubber compound that is normally filled with active carbon or a silanized HD silica increases, the tearing energy of the vulcanizate and significantly reduce crack propagation. The effect of this filler combination on the crack growth propagation is shown in Figure 11.34 [106].

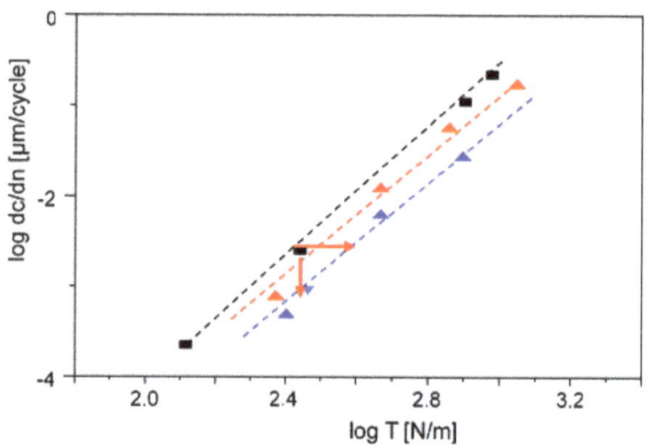


**Figure 11.34** The influence of carbon nanotubes on tearing energy and crack propagation in NR/CB 220

The addition of the multi-walled carbon nanotubes to carbon black or silanized HD silica can deliver lower cut growth and so can prolong the service life of a filled vulcanizate significantly. The micro-dispersion of the filler and the mechanisms for slowing down cut propagation are ultimately responsible for the service life. The contrasting situation between undispersed carbon black and well dispersed filler is shown in Figure 11.35 [106].

The fewer pellet fragments there are in the system, the more likely it is that significant flaws will occur in the vulcanizate after the mixing process is finished and the service life of the vulcanizate will be shortened. However, the deagglomeration of the pellet fragments and the large agglomerates generate aggregates in the nano range. These small particles are crucial for the deflection of the cracks and this leads to a greater lifespan of the vulcanizate. The proportion of sub-micron nanoparticles in the system should be as high as possible in order to prevent the crack from propagating and spreading effectively in the material. Thus, from the fracture mechanics point of view, the micro-dispersion of the filler is the decisive criterion for a longer service life and good fatigue resistance. A perspective for better performance is undoubtedly offered by all measures that enable good filler dispersion, especially the use of innovative filler and advanced mixing technologies that can avoid undispersed filler particles acting as material flaws.

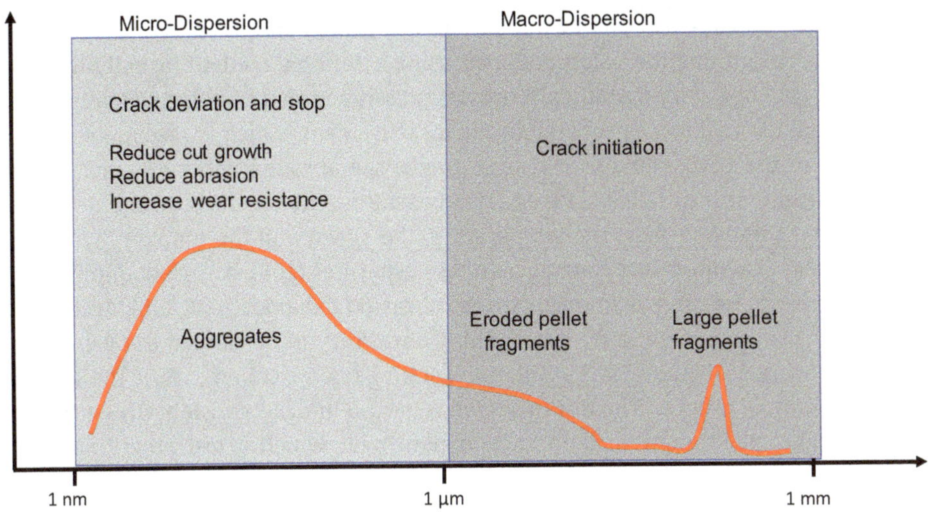


**Figure 11.35** Effects of macro- and micro-dispersion of filler particle on fracture phenomena

A good way of reducing the proportion of pellet fragments is to use wet-compounding, which avoids the use of filler pellets. This technology reduces the number of flaws in the vulcanizate and the dynamic cut growth is significantly reduced, thus significantly improving the abrasion resistance of the vulcanizate.

## 11.5 Abrasion and Wear

In addition to the dynamic properties of elastomers, abrasion and wear are the most important properties of elastomers if a longer service life is the aim. As mentioned in the introduction, the importance of carbon black was discovered in 1904 by Mote who was seeking to increase the abrasion resistance of NR vulcanizate. Since then, high abrasion resistance has always been a sign of strong reinforcement by the carbon black dispersion. For tire treads, abrasion resistance and the skid resistance are important parameters that describe the reinforcement of the vulcanizates.

Abrasion refers to the loss of material on the surface of elastomers due to abrasive mechanical stress that occurs when surfaces interact with one another. The abrasion behavior of elastomers is characterized by the molecular structure of a rubber and the rubber-filler interaction, the mixture composition, the dispersion of the filler and the stress. The temperature at which this process takes place and the nature of the contact surface have a significant influence on the abrasion behavior.

An abrasion experiment is carried out by sliding a rubber specimen over a given distance and determining the volume loss by using a suitable method, usually the difference in weight before and after. It is always assumed that the abraded volume is proportional to the distance covered in the measurement, which is reasonable if the sharpness of the track remains constant. Hence the abraded volume is always referred to as the unit distance covered. This basic quantity depends on the pressure between the sliding surface, the temperature, the speed and the topography of the track, such as sharpness and coarseness of the asperities [106]. A variety of different test methods can be used to characterize the abrasion behavior [107, 108]. According to DIN ISO 4649, the method for determining abrasion can be carried out using two different variants. The first works with non-rotating test specimens, while the second prefers a rotating process. The test specimen is moved under a constant pressure and at a constant speed along a fixed friction path over which a test emery sheet is stretched. The relative abrasion resistance is determined using this method:

$$\Delta V_{rel} = \Delta m_t \Delta m_{const} / \rho \Delta m_r \tag{11.13}$$

where $\Delta V_{rel}$ is the relative volume loss, $\Delta m_t$ is the mass loss of the tested elastomers, $\Delta m_{const}$ is the defined mass loss of the reference elastomers (mg), ρ is the density of the tested elastomer ($g/cm^3$) and $\Delta m_r$ is the mass loss of the reference elastomer (mg)

Abrasion resistance is measured with different abraders, such as the DIN abrader, Akron angle abrader, Pico abrader, Lambourn abrader and Cabot abrader. The abraders have their advantages and also some disadvantages in handling. The attraction of each testing may involve one or more abrasion mechanisms. The ratings for the abrasion resistance of different compounds may vary across the different test devices. However, the real test of abradability are the road conditions, including the texture of the road, the temperature, the weather conditions, tire slip, load on the tires, tire construction and tread compounds.

Abrasion is defined as the weight or volume loss of an elastomer that is guided under a specific contact pressure along a specific path on an abrasive surface of a defined sharpness severity. The abrasion resistance is the ratio of the volume loss of a standard body to the volume loss of the body to be tested in percent. Abrasion damage can occur when there is dynamic motion against an abrasive counterface. Three borderline cases can be defined with regard to the wear of elastomers:

- abrasive stress
- adhesive stress
- deformative stress
- thermal degradation

In practice, several mechanisms are usually involved in abrasion and the wear process at the same time, but they occur in different proportions. Both processes occur during abrasive stress and the quantitative characterization of the abrasion properties is very complex. In this way, the process is associated with cut growth and heating. However, abrasion is also a process in which rubber chains are rupture and this adds an additional level of difficulty that is not only directly related to cut growth propagation.

Closely related to abrasion resistance is tire wear, which is related to the filler loading, specific surface area and the filler dispersion in a tire tread rubber. Wear is usually associated with the detachment of rubber particles, with chain rupture, which in turn means that temperature plays an important role.

### 11.5.1 Abrasion of Unfilled Vulcanizates

In unfilled elastomers the abrasion resistance is a function of chain mobility, which is related to the composition of the rubber and partly to the crosslink density. The increase in the glass transition temperature, $T_g$, indicates an increase in the abrasion loss of the rubber matrix. Therefore, with a low glass transition temperature and a low deformation hysteresis, a high abrasion resistance can be expected. For unfilled non-crystallizing rubber vulcanizates, the temperature and the speed dependence of sliding abrasion is dominated by the viscoelastic nature of the rubber [109]. Consistently, the abrasion loss increases linearly with the glass transition temperature of the elastomer. The lowest abrasion loss is shown by high cis-1,4-BR. The high chain mobility enables rapid adaptation to the high deformation frequency and makes a significant contribution to avoiding chain scissions. As mentioned before, the concentration of the cis-butadiene units is a determining factor in the increase in abrasion resistance [29]. Therefore, Nd-BR with the lowest glass transition temperature shows the highest abrasion resistance. The introduction of styrene groups into the cis-BR increases the glass transition significantly and the abrasion resistance decreases with the content of styrene units. The more styrene groups included in the rubber chains, the higher is the abrasion of the rubber.

In contrast to friction, the resulting abrasion is generally a non-linear function of pressure applied to the test specimen [106]:

$$A = A_{ref}\,(p/p_0)^n \qquad (11.14)$$

where A is the abrasion of the elastomer, p is the pressure on the test specimen and n is a distinct positive power index

For a constant apparent area of contact, a similar relationship results for load dependence. The reference abrasion loss, $A_{ref}$, at the reference pressure, $p_0$, and the power index, n, depend both on the type of track and, of course, on the rubber compound. If the abrasion is related to the frictional energy dissipation, usually referred to as the abradability of the rubber compound, the power index would be modified if there is a load dependence of the friction coefficient but would remain the same if there is not. If the track is smooth the abrasive loss may be so small that it eludes normal measurements, although the frictional force may be very high, whilst on rough sharp surfaces the abrasive loss may be pronounced at a moderate force. In general, the larger the power index, n, the blunter the surface on which the abrasion tacks place.

For non-crystallizing rubber compounds, the temperature and the speed dependence of sliding abrasion on the sharp tracks is dominated by the viscoelastic nature of the rubber, provided that the sliding speed is so low that temperature rise due to friction is small. If the two variables are combined using the WLF transformation, the result is an abrasion master-curve [109]. The Akron abrader has been used to compare E-SBR 1502 with S-SBR [110]. There was less abrasion loss for the E-SBR 1502 at lower revolutions, while at higher revolutions it increased in both cases. The abrasion resistance after one day's aging was lower than that of the unaged specimens.

Strain crystallization is another criterion for reducing abrasion of the unfilled vulcanizate. A significant influence on the level of abrasion is the ability of the rubber to undergo strain crystallization. The tensile strength is consistent with the abrasion resistance of NR, where the crystallization of NR during the elongation is a determining factor [111]. This property is one of the most important reasons for using NR for the tread of truck tires, since the strain crystallization provides a self-reinforcement and increases the tearing energy at deformations higher than 300% and results in an increase in the abrasion resistance.

The type of crosslinking is another important parameter for the abrasion resistance of the rubber vulcanizate. The best choice is to select a conventional curing system for sulfur crosslinking that can react with the rubber chain and can repair the chain fractures that occur during reversion at elevated temperatures [101].

The presence of anti-oxidants is necessary in an elastomer in order to have a choice against oxygen attack in the matrix. Anti-oxidants can be very useful at increasing abrasion resistance. Depending on the abrasive plate, a reduction in abrasion loss of 50 to 60% can be achieved by adjusting the concentration of the anti-oxidants. This knowledge forms the basis for increasing the abrasion resistance of the rubber compound by adding radical scavengers [112]. In practice, 6-PPD is successfully used in rubber compounds to increase resistance to attack by oxygen and ozone in the atmosphere. The importance of the presence of oxygen radical scavengers has been proven and they are present in all vulcanizates that have to endure abrasion. Without a doubt, the p-phenylenediamines are the most commonly used anti-aging agents and are able to increase the abrasion resistance of a rubber compound.

### 11.5.2 Abrasion of Filled Elastomers

A large proportion of the elastomers containing a filler are subject to friction with the substrate, which ultimately leads to abrasion of the material. Therefore, all tire treads filled with carbon black or silanized silica must withstand road abrasion at varying degrees of roughness at different road speeds. The better the abrasion resistance, the longer the tires last and the magic triangle is increased in the direction of better abrasion resistance.

Abrasion resistance is one of the important rubber properties, which is strongly affected by the loading of the fillers added to the formulation. But not only the type and the loading of the active filler are important, the degree of the filler dispersion and distribution in the rubber blends obtained after the mixing process and the vulcanization process that increase the filler-filler interaction and increase the Payne effect, as described in Chapter 10, also play an essential role in abrasion resistance.

The abrasion resistance is determined both by the composition of the elastomers and the conditions of sliding on a rough surface. Furthermore, the temperature plays an important role, since the viscoelastic properties determine the abrasion behavior. In addition, large portions of the elastomer are detached, which requires a chemical process. For these reasons, abrasion resistance is a difficult task for rubber scientists to address. In addition, abrasion is associated with the ecology of road traffic, because it is suspected that the worn parts of a tire can also be carcinogenic.

#### 11.5.2.1 Abrasion of Carbon Black in Filled Vulcanizates

Tires are the only car parts that are in direct contact with the road surface. Friction, especially when a car brakes and accelerates, causes tire wear particles to be released from the tire tread. The increase in temperature can lead to evaporation of volatile tire tread compounds [113]. If the road is rough, the abrasion loss is more pronounced, even at moderate frictional forces, because stress concentrations in the contact area enhance the abrasion process.

Tests of abrasion on tire treads have shown that abrasion loss increases with the load, service temperature and speed, while it decreases with the tire pressure [114]. Abrasion increases with the pressure, the speed of the wheels and the temperature of the wheels on the road. The pressure inside the tire reduces abrasion to a minimum and then increases it again if it increases [115].

Although various test methods are employed in the laboratory, the abrasion of a tire on a rough or smooth surface, in the dry or in the rain, is of great importance for the speed of the tire on the road. Therefore, abrasion wear determined in the laboratory is often over-estimated. Unfortunately, the abrasion tests do not correlate particularly well with application experience. For this reason, tire manufacturers prefer road wear.

#### 11.5.2.1.1 Effect of Filler Loading

The effect of filler loading on the abrasion resistance of vulcanizates used to be an issue in tire technology. For E-SBR vulcanizates filled with different grades of carbon black, it has been shown that a master-curve can be obtained for the abrasion resistance as a function of the effective volume fraction of carbon black, where the volume loss is measured with a DIN abrader. At filler loadings of 0–70 phr for semi-reinforcing carbon blacks and 0–50 phr for reinforcing carbon blacks, abrasion decreases monotonically with the effective volume of the filler [38]. For semi-reinforcing carbon blacks, the abrasion resistance decreases, because the graphitized micro-crystallites on the surface increase and the amorphous carbons decrease, a fact which promotes a decrease in the rubber-filler interaction and an increase in abrasion. The surface activity of the reinforcing carbon blacks increases as the primary particle size decreases in the test. A comparison of E-SBR and S-SBR with carbon black N330 shows that the abrasion resistance improved with the filler loading up to 50 phr. In principle one can state that the higher the filler loading, the more is the decrease of abrasion in the elastomer. However, if E-SBR and S-SBR are compared, the greater abrasion resistance is shown in E-SBR [110].

A more basic approach is to consider the dependence of the abrasion loss not on the pressure but on the energy dissipation, W, caused by the frictional force [116]. For a sliding experiment, it is described by the following equation:

$$W = Fs \tag{11.15}$$

where F is the frictional force and s is the distance covered when sliding

On very sharp roads , the abrasion loss is nearly proportional to the frictional force, but is a power function of the energy dissipation, whereby the power index depends on the sharpness of the sharp roads as well as on the cut growth and fatigue behavior of the filled elastomers [90].

The behavior is similar to the cut growth of the rubber compound. The rate of cut growth is a function of tearing energy, which itself is proportional to the stored elastic energy density in the test piece [117, 118]. Basically, abrasion is a cut growth process that is governed by the tearing energy. The relationship between the abrasion and the energy consumption as a power law is firmly established. The main effect is due to the sharpness of the track rather than its coarseness. However, abrasion is not purely a physical phenomenon; oxidation of the rubber has to be taken in account. For abrasion, this is a much more dominant process than in the case of cut growth, which does not involve chemical reactions on the rubber chains. The main reason for this is that the energy consumption in abrasion processes raises the temperature at the interface between the rubber and the track and thereby modifies the entire process.

If the experiments are carried out s a function of energy dissipation at different speeds, different abrasion results are obtained (mg/km) [119]. This cannot be explained by a purely mechanical cut growth process. In particular, as the energy dissipation in the contact area increases, the temperature is going to rise, as a considerable portion of the energy is turned into heat.

The wet-mixing Cabot elastomer composite (CEC) provides advantages over the dry-mixing rubber compounds of similar composition in terms of abrasion resistance at higher filler loading. The effect of loading for carbon black N134 is compared with dry-mixing in NR SMR 20 and the CEC compound on the abrasion resistance measured by means of the Cabot abrader at a slip ratio of 7%. The excellent carbon black dispersion in CEC compound, even at high loading, accounts for the increase in the abrasion resistance above a loading of 50 phr. The maximum of the abrasion resistance is located at 70 phr with an abrasion index of 180%. The absence of the pelletized carbon black becomes clear at the higher loadings of carbon black, where a significantly higher abrasion resistance is observed [120]. A comparison of dry compounding and wet compounding shows that the latter mixing technology leads to a significant improvement in abrasion properties and that it is definitely worth choosing this method of producing a tire compound.

#### 11.5.2.1.2 Effect of Specific Surface Area

It is generally considered that the most important filler parameter for increasing the abrasion resistance is the specific surface area [121–125]. The abrasion loss of filled elastomers decreases from inactive fillers, such as thermal blacks, to active fillers, such as active furnace blacks [111, 116]. The surface area or the primary particle size is considered to be the most important characteristic for reducing filler abrasion. The higher the rubber-filler interaction, the more efficient the effect on chain rupture, which leads to abrasion. Taking the abrasion index for N330 as a reference, the effect of abrasion can be characterized as shown in Table 11.6.

**Table 11.6** Abrasion Index in SBR Vulcanizate for a Group of Carbon Blacks of Decreasing Specific Surface Area

| Carbon Black Type | Abrasion Index (%) |
|---|---|
| N110 | 1.29 |
| N220 | 1.25 |
| N330 | 1.00 |
| N550 | 0.90 |
| N660 | 0.66 |
| N770 | 0.60 |

The table shows how differently the specific surface area affects the abrasion index. The higher the specific surface area, the greater is the abrasion resistance found in laboratory tests. Note that the carbon black filler reduces the rate of cut growth and increases the abrasion resistance, but has only a small effect on the power index. Generally, it can be seen that abrasion tends to occur more in the region of higher tearing energy, when the abrasive track is very sharp.

The effect of the specific surface area and the structure of carbon black on abrasion behavior of filled NR, SBR and cis-BR compounds was investigated using a modified bladder abrader. Characteristic parameters were introduced via the particle size, with a linear relationship between Young's modulus and the abrasion resistance. The rate of abrasion resistance decreased with the specific surface area and the structure of carbon black. It was shown that cis-BR compounds cause a much slower abrasion than NR and SBR [126].

Rubber filled with high-specific surface-area carbon black generally shows a much higher abrasion resistance than the unfilled elastomers. It has been shown that the product of the filler loading and the specific surface area of carbon black leads to a decrease in the abrasion volume [123, 127]. Since the type and loading of carbon black and its specific surface area make the decisive contribution to abrasion resistance, it has been presented as a product of the two variables. This combined influence is shown in Figure 11.36 [128].

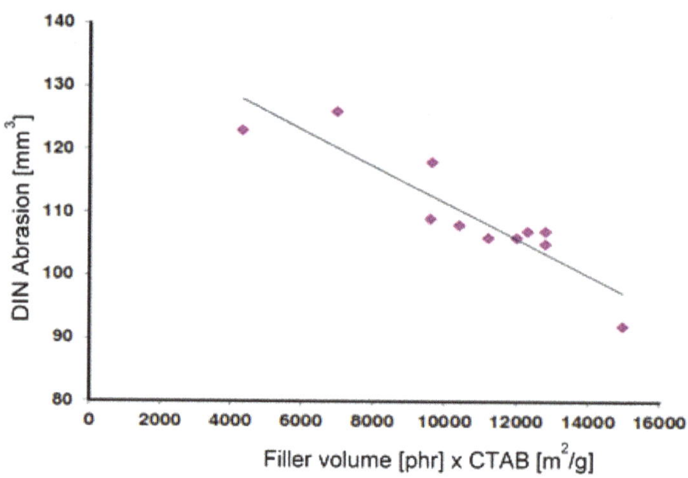


**Figure 11.36** The combination of carbon black loading and specific surface area of carbon black reduces DIN abrasion

A carbon black of high specific surface area and maximum concentration should be used in order to achieve the lowest volume abrasion after successful dispersion. At a constant loading of 65 phr carbon black with 35 phr oil, the optimum surface area for an SBR/BR blend system has bene found between 130 and 150 $m^2/g$ [129].

At lower loading, the increase in abrasion resistance with increase in surface area may be attributed to several abrasion mechanisms related to increasing effective interfacial area between filler and polymer. Given the complexity, it is difficult to identify the main parameter governing the abrasion resistance of filled elastomers. However, there is a limit to the abrasion reduction; the process has a minimum value and an overfilled system increases abrasion, because the filler particles are not attached to the polymer chains.

Wet-mixing of tire tread compounds filled with active carbon black has been found to improve abrasion resistance. Compared with the dry-mixing, the improvements show better abrasion resistance, which is especially important for the carbon blacks of high specific surface area of over 100 $m^2/g$. At a CTAB surface area of 140 $m^2/g$, the dry-mixing carbon blacks show an abrasion index of 115, while the wet-mixing carbon black returns a maximum abrasion index of 130 for the same CTAB surface area. For racing-grade carbon blacks, wet-mixing yields an even higher abrasion index [130].

The negative effect of high surface area carbon blacks on abrasion resistance that is observed in some reports may be related to the difficulty of obtaining a sufficiently high degree of dispersion. This may be due to the stronger filler-filler interaction in the agglomerates, which increases cohesive strength. This is because the surface energy of fillers increases with the specific surface area or the filler activity, as shown in Chapter 3. Wet-mixing of carbon black in NR compounds has shown an increase in abrasion resistance, because the degree of filler dispersion is significantly higher and the dynamic hysteresis is reduced.

#### 11.5.2.1.3 Effect of Filler Structure

The structure of the filler is another important characteristic that has a positive effect on the increase in abrasion resistance. Due to a higher degree of aggregate branching, a larger volume of rubber chains can be included in the hydrodynamic volume of the aggregate, which increases Young's modulus and the Payne effect.

The abrasion resistance for carbon black-filled E-SBR increases with the compressed DBP number at a constant specific surface area. This result is consistent with many reports showing that high structure carbon blacks lead to an increase in abrasion resistance. It has been found that the curves for high and low DBP numbers increase with the severity of the test. The influence of the abrasion loss decreases with the carbon black structure and the reinforcing factor, as shown in Figure 11.37 [131].

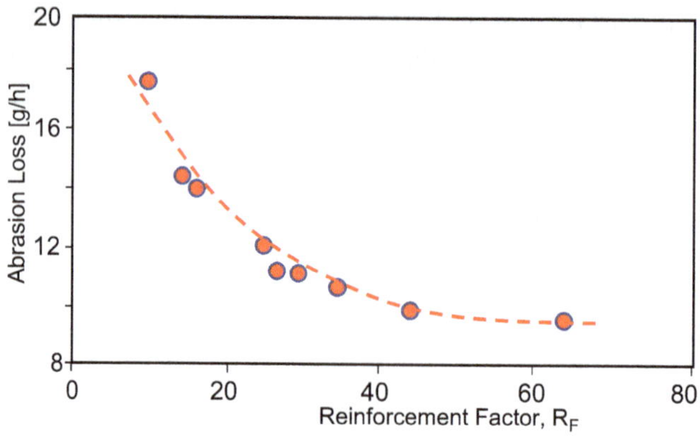


**Figure 11.37** The influence of the carbon black structure on the abrasion loss of an NR vulcanizate with a loading of 50 phr [121]

The abrasion loss values of the carbon blacks from the N300 series show a decrease of more than 30% when the structure of the aggregates increases. To a lesser extent, increasing the structure from N220 to N234 also produces a reduction in abrasion loss values. In most cases, the abrasion tests were carried out at constant loading and similar specific surface area. The effect of higher structure is reflected in the higher effective loading in the branched aggregates and an increased Young's modulus at low deformations of the elastomer [109]. Similar to the specific surface area, the structure of the aggregates also has a positive effect on abrasion resistance. Any change in the structure of the carbon black affects the failure properties of the elastomer.

#### 11.5.2.1.4 Effect of Rubber-Filler Interaction

Very few reports have dealt with the effect of rubber-filler interaction on abrasion resistance, even though this effect has been recognized for a long time. This is mainly due to the fact that the dispersibility of carbon black was considered to be satisfactory.

Elastomers filled with carbon black of high specific surface area exhibit a higher abrasion resistance, but above a certain specific surface area they reach a limit, where the abrasion resistance no longer increases [122, 125]. The lower abrasion resistance of vulcanizates filled with high specific area carbon blacks may be attributed to poor carbon black dispersion in the mixing process, due to increased filler-filler interaction in the elastomer [132].

When carbon black was heated in an inert gas atmosphere, the filler-filler interaction increased substantially because of the increased $sp^2$ hybridization which triggers the filler-filler interaction [133, 134]. It was observed in an academic study that a small amount of CTAB adsorbed on the surface of carbon black determines the surface activity of carbon black, reducing the abrasion resistance, because the adsorptive sites are partially blocked by the CTAB [135]. A consequence of the oxidation of carbon black with nitric acid is a reduction in the abrasion resistance of S-SBR vulcanizate by 38% at a slip ratio of 14% slip [136].

The extensive influence of the carbon black units which increase abrasion resistance can be clearly seen [129]. From the point of view of rubber-filler interaction, the mixing process of carbon black compound should be optimized in such a way that the mineral oil and other ingredients are added after carbon black is incorporated in the rubber compound. For a car tire tread that contains S-SBR/cis-BR (75/25), dispersion of carbon black 234 (75 phr) should occur before the mineral oil (25 phr) is added to the mixer. This process has been found to enhance the physical properties of the rubber and that is expressed also in the abrasion resistance of 19% [130].

The mixing technology can significantly affect the dispersion of carbon black for the purpose of achieving a higher rubber-filler interaction. This is the case when a wet-mixing process is used instead of the usual dry-mixing process in order to eliminate the use of pelletized carbon black. The result is rubber compounds which have significantly higher rubber-filler interaction and which show a strong improvement in dynamic properties and abrasion resistance. Dry-mixed compounds show a maximum abrasion index of 150 at 60 phr, while Cabot elastomer composites (CEC) vulcanizates with 70 phr carbon black achieve a smaller abrasion index and a higher abrasion resistance. Carbon blacks with the same specific surface area and structure of the aggregates show a clear improvement in abrasion behavior when the surface is functionalized during the manufacturing process [130]. In this way, wet-mixing technology in the production of tire compounds is able to achieve not only the advantages in dynamic properties but also a significant increase in abrasion resistance.

In order to improve the abrasion resistance and wear in the tire, the changes in the surface energy distribution of carbon black have to be achieved by functionalizing of the surface obtained during the manufacturing process of the carbon black. The changes obtained by functionalization of the carbon black are not only observed in the tensile-strain behavior, in the strain amplitude dependence of $G'(\gamma)$ and $G''(\gamma)$ or a lower heat build-up, but particularly in an increase in the abrasion resistance which could be achieved with the functionalized carbon blacks. Compared with the ASTM carbon black N115, which has an abrasion index of 100%, a value of 115% is achieved with Ecorax® carbon black. With this advantage, high abrasion resistance can be achieved in the production of tire treads, which cannot otherwise be achieved without functionalization during manufacture [39].

When carbon-silica dual-phase fillers (CSDPFs) are incorporated into hydrocarbon unsaturated rubbers, not only is the filler-filler interaction reduced but higher filler-polymer interaction is achieved, compared with both carbon black and silica [137]. Regardless of the difference in the specific surface area, the CSDP filler leads to significantly higher adsorption energies on heptane at 180 °C, as measured by IGC, for the same silica content. The higher surface activity has been attributed to the change in surface microstructure of the primary particles. The use of CSDPF results in a low dynamic hysteresis and a high abrasion resistance that is comparable to very active carbon blacks, such as N115 [138]. The silica carbon black filler with the higher surface coverage of silica nanoparticles not only is recommended for reducing the rolling resistance and increasing wet-skid resistance, but it also delivers an abrasion resistance on a par with carbon black N110 [139]. CSDPF not only achieves the low Payne effect, but also shows a significant advantage in abrasion resistance in the blend of S-SBR/cis-BR compounds. As environmental protection will become increasingly important in the foreseeable future, it is to be expected that more functionalized carbon blacks will be used and more wet mixing will determine the rubber compound. This contribution can also be understood in the sense of reducing abrasion on the road.

### 11.5.3 Influence of Precipitated Silicas

The use of precipitated silica constitutes a new input to this problem, because its dispersibility is influenced by the strong inter-aggregate interactions that increase the filler-filler interaction and influence not only the dynamic properties but also the abrasion resistance. Compared with carbon black, the interaction between hydrocarbon rubbers and silica is weak and not very important. The abrasion resistance is improved by the larger specific surface area of the precipitated silica, as shown in Figure 11.38 [61]. The lack of strong polymer-filler interaction is the most important factor contributing to the lower abrasion resistance.

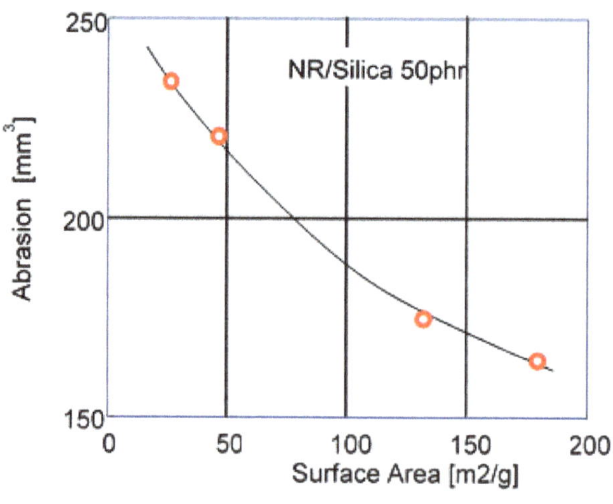


Figure 11.38 Influence of the specific surface area on the abrasion resistance of precipitated silica [61]

Through organo-silane modification, the silica agglomerate can be depressed via a reduction in the silanol groups, leading to improved dynamic properties and an increase in abrasion resistance. The agglomerate of silica particles cannot provide strong interaction to the rubber matrix and detachment of the filler surface from the rubber matrix be observed. Compared with carbon black, the interaction between polar chemicals and silica is very intense, while interaction with alkanes is very low, which also indicate a weak interaction with hydrocarbon rubbers.

Compared with vulcanizates containing precipitated silica, the effect of silanization with various silanes is to increase abrasion resistance. The abrasion resistance is promoted by bifunctional silanes that link the filler to the rubber network and make the filler-rubber network stronger. The silanization efficiency is highest in rubbers polymerized in solution (i. e. S-SBR and cis-BR). In E-SBR, impurities, such as residues from the emulsifier, interfere with the bifunctional silanes and reduce their silanization efficiency.

The lack of strong polymer-filler interaction is the most important factor contributing to the lower abrasion resistance, as shown in Figure 11.39 [140].

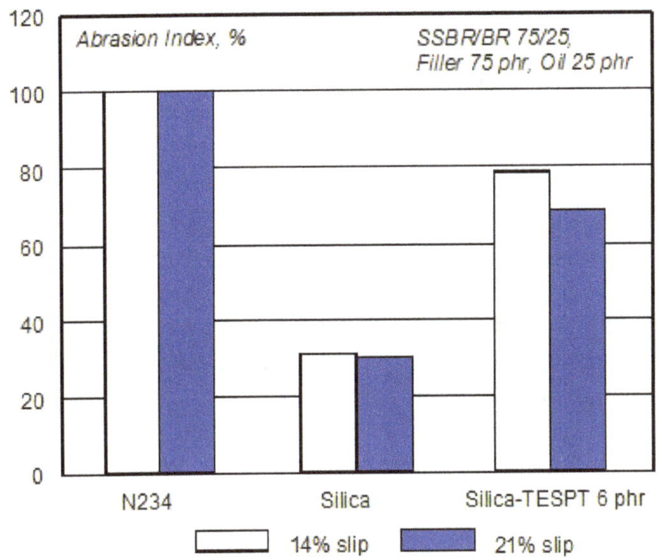


**Figure 11.39** Abrasion index for S-SBR/cis-BR compounds containing carbon black, HD silica and silanized HD silica at different slip percentages

To evaluate the influence of the unsilanized HD silica and the silanized HD silica with TESPT, the abrasion index of S-SBR/BR (75/25) with 50 phr carbon black N234 is taken as reference. Compared with this value, the compound containing HD silica shows only 30% abrasion resistance and the efficiently silanized compound returns an 80% abrasion index [141]. This shows that the silanized HD silica is relatively lower in terms of abrasion resistance of the carbon black N234. One reason for the advantage of furnace black is the higher proportion of $sp^3$-hybridized carbon atoms on the surface of the N234, which ensures significantly better rubber-filler interaction with the rubber matrix, which does not occur with silanized HD silica at the same slip level. However, silanized HD silica is used as the principal filler in car tire treads which exhibit a strong reduction in rolling resistance and a quite high abrasion resistance.

The abrasion resistance measured with an Akron Abrasion tester shows an abrasion index of 105% for wet-mixed silanized HD silica that is higher than that of the dry-mixed silanized HD silica of only 83% and is better than that of carbon black N234, which has an abrasion index of s 100%. It should be pointed out that, because of the different compounding conditions the results show much better adhesion in the interface and a better filler dispersion in the vulcanizate. The higher rubber-filler interaction is obtained in wet-mixing because pelletized silica is not used [142]. Due to the absence of pelletized silica, the rubber-filler interaction can be significantly improved and the distribution in the vulcanizate is increased, which improves the abrasion index.

Great efforts have been made to incorporate E-SBR into silica compounds due to better processability and low cost, but poor abrasion resistance is the main shortcoming of E-SBR/silanized HD silica compounds. However, this inferior property is related to a decrease in the rubber-filler interaction that cannot be compensated for by the use of a bifunctional organo-silane. The 5–8% of non-rubber ingredients, i. e. emulsifiers, modifiers, in E-SBR interfere with the coupling reaction of the organo-silane.

Multiple factors may account for the inferior properties of truck tire treads containing HD silica that are related to the characteristics of the precipitated silica. Again, the non-rubber ingredients, such as proteins, fatty acids, phospholipids that are a part of the NR latex, can interfere with the silanization process [143]. The reduced silanization reduces the rubber-filler interaction, resulting in a decrease in the abrasion resistance. A silanized silica in NR has only 60% of the abrasion resistance of an NR/carbon black N110 with 100% abrasion resistance. Thus, truck tire treads focus primarily on the strain crystallization of the NR, which significantly reduces cut growth propagation, and less on silanization of the silica, which is disadvantaged by the ingredients of the NR latex.

#### 11.5.3.1 Influence of Layered Silicates and Carbon Nanotubes

Elastomer matrices have not been as widely studied as thermoplastic or thermoset matrices, but there are a number of interesting reports on the behavior of layered silicates on abrasion resistance. Recent experiments on frictional and sliding wear performance have shown that incorporation of layered silicates into rubber compounds may improve the wear behavior of the related vulcanizates [144, 145]. The orientation of the layered silicates in the rubber compound increases the resistance to a high degree of wear. The type of HNBR/organo-clay can increase, relative to a non-dispersed MMT, by a factor of 7–8.

Multi-walled carbon nanotubes have been used in abrasion tests to investigate cut growth propagation in different compounds. At small volumes of 1 or 2 vol.%, the tearing energy increases and the cut growth propagation is reduced. Therefore, abrasion is also reduced by the addition of multi-walled carbon nanotubes [146].

### 11.5.4 Mechanisms of Abrasion

Depending on the stress on the elastomer, different processes such as crack formation, crack growth and fatigue are effective for abrasion. Therefore, the amount and type of filler play major roles in tread wear resistance. A comprehensive examination of the effects of specific surface area, aggregate size distribution and aggregate structure has revealed that specific surface area and the degree of dispersion are the most important parameters that affect tread wear. Abrasion is shown schematically in Figure 11.40.

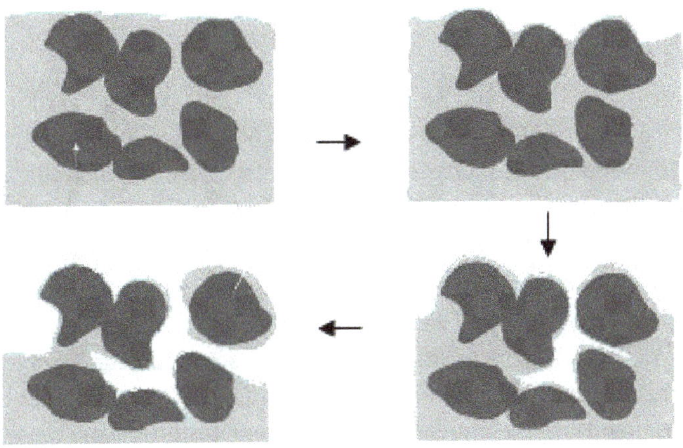

**Figure 11.40** Schematic representation of abrasion loss

Wear involves high temperatures that may preclude strain crystallization. In such a situation, filler reinforcement becomes mandatory for high performance. It is generally recognized that the mechanism of rubber abrasion is highly complex, involving not only mechanical failure of the material, such as fatigue, tearing, etc. but also mechano-chemical and thermo-chemical processes. When referred to in connection with energy dissipation, abrasion loss is the abradability of the elastomer on the track on which it was measured. The rate of the cut growth is a function tearing energy, which itself is proportional to the stored elastic density in the specimen. Several mechanisms have been proposed to explain the very complex abrasion phenomenon. The problem is that abrasion leads to the separation of larger particles of the vulcanizate surface, which is strongly related to chain rupture, filler-rubber interaction and certainly to sharp temperature increases.

During sliding, an asperity on the tracks pull the tip over and extends it until the adhesion force can no longer hold it and it snaps back. However, as the tip extends, the stress at the bottom of the ridge rises and a crack can grow to a limited extent. After a number of passes on an asperity, the lip has become long enough for a particle to become detached [105]. The concepts behind the abrasion mechanism have evolved from different observations of this complex phenomenon that appear during friction between an elastomer and a rough surface. The main mechanisms for rubber abrasion are:

- cutting abrasion
- intrinsic abrasion
- pattern abrasion
- roll formation
- spalling abrasion
- mechano-chemical abrasion

Even if an explanation were offered for the abrasion phenomenon, it is considered to be one of the most complicated areas in rubber science. A lot of work will have to be done here to clarify it. Different mechanisms are presented below. All of them involve detachment of larger parts of the elastomer during a given amount of friction with the abrading surface.

Cutting abrasion is the name given to an abrasion mechanism that regards scratches in an elastomer as a typical defect [108]. The direction of the abrasion process is equated with tensile failure. Detachment in the elastomer gives rise to catastrophic tearing of the rubber surface behind the asperities of the track [106]. The detachment of some of the elastomer depends on the cut growth, which depends on the decrease in stored energy in the elastomer. Thus, the critical tearing energy, $G_c$, depends on the strength of the vulcanizate and on the diameter of the tip of the tear:

$$G_c = W_b d \tag{11.16}$$

where $G_c$ is the critical tearing energy, $W_b$ is the energy at break and d is the diameter of the tip of the tear

Thus, the detachment rate depends on the energy density when the specimen is torn and the diameter of the crack tip, which is able to result in an abrasion during which large particles are detached. The pressure on the test specimen to be tested has a major influence on the result of the abrasion behavior. Up to a certain pressure, the abrasion volume is proportional to the applied pressure. This is characterized by abrasion that is directly proportional to the volumetric coefficient of friction and inversely proportional to the elastic energy during tearing:

$$A = k\,\mu / W_b \quad (11.17)$$

where μ is the coefficient of friction

It is therefore possible to describe the abrasion of an elastomer based on the ratio of the coefficients of friction and the remaining mechanical energy. A fundamental problem lies in the speed below which $W_b$ was used because the tensile strength must be measured on the elastomer surface during the abrasion process. It is recommended to use the WLF equation to solve this problem.

The next abrasion mechanism discussed here is that of intrinsic or small-scale abrasion. This mechanism basically consists of an interaction between the removal of a macroscopic particle from an elastomer by mechanical rupture and the chemical deterioration of the elastomer by mechano-oxidation reaction [147]. This type of abrasion can occur frequently, since all abrasion processes take place in the presence of oxygen. It is obvious that the damage occurs at the weakest point on the elastomer surface as the elastomer is moving over the track asperity, and discrete particles of the surface become detached from the bulk. The probability that a particle will become detached is proportional to the tensile stress. For this reason, the volume abrasion per unit distance of sliding is described by the following equation:

$$A = kF \quad (11.18)$$

where k is a constant and F is the frictional force

In Equation 11.17, the abrasion is proportional to frictional energy dissipation and the influence of the chemical structure of the vulcanizate is summarized in the coefficient k. In the case of intrinsic abrasion, the speed of abrasion is governed by the coefficient of friction, which can be determined. The influence of the elastomer, which is composed according to the type of rubber used, the filler and the crosslink density achieved, is examined in a comparison involving NR, SBR and EPDM vulcanizates filled with carbon black [148]. It is clear on a small scale that the abrasion presents itself as a power law of the frictional force, where the exponent is determined by the type of rubber:

$$A = k'F^n \quad (11.19)$$

For unfilled vulcanizates made from SBR and cis-BR, the exponent will be 2.9 for SBR and 3.5 for cis-BR. This means that the contribution of the cis-BR with the lowest glass transition temperature is higher than that of the strain-crystallizing NR and the amorphous SBR.

It has been found that, with carbon black-filled NR, SBR and EPDM, the abrasion consists of an oily debris, which is believed to reduce the degradation of the rubber chains. Pattern abrasion shows that an array of nearly parallel ridges is formed that are arranged perpendicular to the direction of abrasion [149]. This area of ridges has been called the abrasion pattern mechanism. The physical characteristics of ridges on tires are practically identical with those produced on the roads and in the laboratory [150].

With the help of fracture mechanics, the mechanical energy loss, G, based on the unit area formed is understood to be a function of the thickness of the test specimen and the mechanical energy consumed, U, when the cut grows through an incremental increase, dc. This is represented by the following expression:

$$G = (dU/dc)/W \tag{11.20}$$

As the blade passes over the pattern, the tongue of the rubber is pulled back and then released as the blade passes on. The tearing energy approach has been used to test the crack behavior of a number of rubber compounds, and the results can be expressed in a similar power law to that of crack growth propagation [98]. The researchers who studied crack growth are also working on the mechanism of vulcanizate abrasion and have found a connection between cut growth propagation and abrasion in some vulcanizates. The loss of abrasion depth depends on the crack growth rate and the angle at which the crack grows into the rubber vulcanizate. It has been found that the mechanical energy, G, consumed depends on the cut growth propagation, which increases with the tearing energy. Therefore, if the tearing energy and the cut growth propagation are known, the abrasion can be predicted [76]. The crack growth is therefore determined by the tearing energy and the angle at which crack growth takes place. However, it appears to be closely related to the geometry of the pattern, and much of the loss of rubbers occurs from the steep slope of the pattern [149].

The abrasion mechanism of roll formation postulates that mechano-chemical degraded rubber rolls up into small cylinders that debond from the elastomer [151]. When a smooth abradant presses on a particular rubber surface with a force and moves parallel with the surface of the rubber, a complex deformation takes place. It is known that failure begins with the appearance of a cut perpendicular to the direction of the tensile force. The nature of this abrasion behavior is caused by a stress condition in the rubber which occurs when the elastomer surface is in a maximum strain state. For this reason, larger parts of the vulcanizate will detach. The force that causes this deformation depends on the resistance of the elastomer to the crack propagation at the point where the shred detaches. Failure of the shred at a point where the deformation is critical causes the detachment of the roll from the elastomer and completes the act of frictional abrasion.

The spalling mechanism of abrasion behavior requires that a vulcanizate have small voids and spherical cavities that are under a certain pressure [152]. The predictable diameter size of these voids is on the order of 0.1–1 µm [153]. These voids can cause large cracks when inflation pressure increases, according to reports [154]. Stiffer rubber vulcanizates are less prone to spalling abrasion, which is the case for elastomers with a high loading of carbon black.

One of the obvious abrasion effects in the laboratory that indicates that abrasion is not only a mechanical rupture phenomenon but has also strong chemical aspects is the smearing or mechano-chemical degradation process that occurs in practically all rubber abrasion processes. Abrasion debris on clean surfaces tends to stick together, forming rolls of abraded rubber. They are often quite large, but consist of a number of very small particles that are highly depolymerized and oxidized. Normally, there is sufficient dust on the road, which adsorbs the sticky parts and the rolls disintegrate.

The mechano-chemical degradation process requires a combination of stretching of the elastomer and then a thermal-oxidative process that destroys the materials. For NR/carbon black N330 vulcanizate that does not contain anti-aging agent IPPD, an abrasion rate in $mm^3$/500 cycles of 70 has been determined. In a parallel experiment in which the anti-aging agent IPPD was added, the abrasion rate was significantly lower at 35 [155]. As has been found, the usual effect of mechano-chemical abrasion is that an oily sticky layer appears on the surface of the specimen when abraded under the fatigue conditions. Such behavior is observed in a carbon black-filled vulcanizate of NR and SBR. In cis-BR vulcanizates, however, it can be seen that the radicals formed during mechanical-chemical abrasion initiate crosslinking, such that no oily layer is formed during abrasion behavior [109].

In summary, a large number of attrition mechanisms have been proposed, but no definitive solution to the problem has been found to date. A wide variety of forms of abrasion behavior play a role in the way particles of different sizes are released from the elastomer, and all of them are still exposed to oxidative attack.

### 11.5.5 Tread Wear

Active fillers reduce the abrasion and wear of elastomers, as practice shows. The wear resistance of filled elastomers is determined by the recipe, the type and amount of filler, the type of the rubber but also the dispersion of the filler and the type of stress to which the elastomer is exposed. The question of the influence of the road surface on tire wear cannot be answered unequivocally, because of the different compositions of the elastomer, the conditions of use and weather influences on the abrasive power of an elastomer. There are various test methods for determining wear behavior, which are more or less adapted to the real loads. Hence, reliance has to be placed on the correlation between the laboratory results on an abrasive surface and experience gained from road tests. The basic underlying process concerns the interaction between the filler system and the polymer.

The mechanism of rubber tread wear has been studied for more than a century. In 1929, the first patent was published for a wear tester that correlated the results of tire wear [156]. The slip ratio of tires for normal drying was about 1% and on hilly roads 4%. The slip ratio contributes to wear, with high slip ratios during stops and starts, winding roads, hilly roads contributed the most to increased tire-tread wear. Other parameters, such as the tire temperature, pavement texture, moisture, water, dust feed, compound properties and specimen dimensions, have become important factors in tread wear [157]. It has been observed that the abrasion pattern is perpendicular to the direction of friction [158–160]. This is due to the stick-slip mechanism, when the elastomer deforms, clinging onto the pavement asperities. When the frictional force exceeds the static friction, the rubber suddenly slips at very high speeds and returns to its original shape. This movement is repeated at constant intervals, creating an abrasion pattern; stick is a slow process, while slips is a fast process. Therefore, an asymmetrically patterned cross-section resembling both is obtained.

One of the simplest instruments for measuring wear under controlled load and slip utilizes a two-wheeled trailer with its wheels set at a given slip angle. Both wheels must be fitted with tires of with the same construction, tread patterns and identical composition. Under these conditions, the tires run at the same slip angle [161].

A trailer pulled on a rough road has been used to study, the effect of filler loading on tread wear of tires under various conditions [162, 163]. The index of tread wear resistance increases and reaches a maximum at 75 phr in a compound E-SBR/cis-BR (75/25). Above the maximum tread wear resistance, the decline has been attributed to an interference between neighboring carbon black particles and poorer dispersion in the rubber matrix.

Using a carbon black with a surface area of 130 $m^2/g$ and a DBP number of 124 mL/100 g, it has been found that the optimum filler loadings increase linearly with wear severity. It has been reported that the abrasion of tread of tires increases significantly with the surface temperature. That is especially true of NR tread compounds, as polymer degradation at high temperatures is more severe [164]. At a tire surface temperature of 80 °C, tread wear is 1.4 mm/$10^3$ km for NR compounds and only 0.75 mm/$10^3$ km for E-SBR compounds.

To achieve a minimum wear rate at high wear severity, low specific surface area carbon blacks require a higher filler loading than higher surface area blacks, while no significant difference in wear resistant is found at lower test severities. In radial tires, the tread wear index has been found to be a linear function of the specific surface area [165].

As the DBP number of carbon black decreases, more carbon black must be added to maintain a minimum wear at any severity. For optimum performance, it may be desirable to use a lower loading of a high structure black [108]. Wear measurements at a constant slip angle of the trailer, but changing ambient temperature have revealed a larger temperature dependence of the wear rate for NR and NR [161].

On the road, there are two vibrations at low and high frequencies. The high frequency vibration, initiates cracks on the rubber surface, which has caused the rubber compound to fall as crack propagation during cyclic deformation. The frequency for tire wear is about 40,000 Hz [166]. A much faster wear apparatus has been developed by Coesfield GmbH has been used to generate a chipping and chunking wear pattern and to study the problem of wear [167–171]. For truck tires operating under off-road conditions, which had the same ranking for visible damage or abraded specimen surface after the instrumented chip and cut analysis, the numerically evaluated damage parameters were found to decrease from a pure NR tire to NR/cis-BR and NR/S-SBR tire that are filled with carbon black N234.

In addition to these differences, which are known from chipping and chunking, the type of production of a vulcanizate is also important. With all carbon blacks, a significant improvement in wear resistance can be obtained through wet-mixing, which is particularly advantageous for carbon blacks with a high specific surface area. Off-road tires must meet specific performance requirements as they are subjected to harsh conditions which may cause special fatigue and abrasion wear in the first stage while cutting and chipping will occur in the later stage [172]. Fillers and mixing processes play a visible role in imparting properties to a rubber compound. The effect of carbon black-filled NR, S-SBR and cis-BR has been studied to characterize fracture due to cutting and chipping and the propensity for this phenomenon [168].

Our understanding of the very complex phenomenon of tire abrasion and tire wear and their relation to traction and wear has progressed steadily. Some details may still be missing, but the main points are now clear. It appears that good agreement is reached when the slip conditions are measured. It is therefore not surprising that the laboratory test procedure, which admittedly is more complex than a single point reading, gives a close correlation to this complex phenomenon.

## References


[1] Studebaker, M. L., in "Reinforcement of Elastomers", Kraus, G., Ed., Interscience Publishers, New York (1965) Ch. 12

[2] Griffith, A. A., Philos. Trans. Royal Soc., Ser. A, 1920, 221: 1963

[3] Beurrot S., et al., Int. J. Fatigue, 2013, 52: 82

[4] Hurreau, B., et al., Rubber Chem. Technol., 2016, 89: 126

[5] LeSaux, V., et al., Polym. Eng. Sci., 2011, 51: 1253

[6] Gent, A. N., Park, B., J. Mater. Sci., 1984, 19: 1947

[7] Choi, I. S., Roland, C. M., Rubber Chem. Technol., 1996, 69: 591

[8] Lake, G. J., Lindley, P. B., J. Appl. Polym. Sci., 1965, 9: 348

[9] Fielding-Russel, G. S., Rongone, R. L., Rubber Chem. Technol., 1983, 56: 838

[10] Eldred, R. E., J. Polym. Sci. B, 1973, 10: 391

[11] Abraham, F., PhD Thesis, Coventry University (2002)

[12] Mars, W., Fatemi, A., Rubber Chem. Technol., 2003, 76: 124

[13] Irwin, G. R., J. Appl. Mech., 1957, 24: 361

[14] Clamroth, R., Eisele, U., Kautsch. Gummi Kunstst., 1975, 28: 433

[15] Kainradl, P., Händler F., Rubber Chem. Technol., 1960, 33: 1438

[16] Oberth, A. E., Bruener, R. S., Trans. Soc. Rheol., 1965, 9: 165

[17] Gent, A. N., in "Science and Technology of Rubber", Mark, J. E., Erman, B., Roland, C. M., Eds., Academic Press, 4th ed., (2013)

[18] Gent, A. N., J. Mater. Sci., 1980, 15: 2884

[19] Nicholson, D. W., J. Adher., 1979, 10: 255

[20] Kraus, G., Rubber Chem. Technol., 1965, 38: 1070

[21] Hess, W. M., Lyon, F., Burgess, K. A., Kautsch. Gummi Kunstst., 1967, 20: 135

[22] Gent, A. N., Lindley, P. B., Thomas, A. G., J. Appl. Polym. Sci., 1964, 8: 455

[23] Cadwell, M., et al., Ind. Eng. Chem., 1940, 12: 19

[24] ASTM D 623 Method A

[25] ISO 4666

[26] ASTM D 813–59

[27] Roland, C. M., Smith, C. R., Rubber Chem. Technol., 1985, 58: 806

[28] Eisele, U., in "Introduction to Polymer Physics" Springer Verlag, Berlin (1990) Ch. 10

[29] Obrecht, W., Lecture at the advanced training course Rubber Technology University of Hanover (2013)

[30] Musch, R., ACS Rubber Division Meeting, Detroit (1991)

[31] Coran, A. Y., in "Science and Technology of Rubber", Mark, J. E., Erman, B., Eirich, F. R., Eds., Academic Press, New York (1978) Ch. 7

[32] Dluzneski, P. R., Rubber Chem. Technol., 2001, 74: 451

[33] Hong, S. W., in "Rubber Compounding: Chemistry and Applications", Rodgers, B., Ed., Marcel Dekker, New York (2004) Ch. 10

[34] Barnard, D., Kautsch. Gummi Kunstst., 1982, 35: 747

[35] Kainradl, P., Wolff, F., Kautsch. Gummi Kunstst., 1968, 21: 103

[36] Li, T., Wang, M.-J., Schuster, R. H., Internat. Seminar on Elastomers, Qingdao (2017)

[37] Wolff, S., Panenka, R., IRC'85, Kyoto, Japan (1985)

[38] Wolff, S., Wang, M.-J., Kautsch. Gummi Kunstst., 1994, 47: 17

[39] Niedermeier, W., Fröhlich, J., Educ. Symp. DKG'09, Nuremberg (2009)

[40] Cataldo, F., Ursini, O., Angelini, G., Fullerenes, Nanotubes and Carbon Material, 2009, 17: 18

[41] Meon, W., Blume, A., Luginsland, H.-D., Uhrland, S., in "Rubber Compounding, Chemistry and Applications", Rodgers, B., Ed., Marcel Dekker, Inc., New York, Basel (2004) Ch. 7

[42] Woehler, A., Engineering, 1867, 2: 160

[43] Stahlberg, W., PhD Thesis, Chalmers University (2018)

[44] Marco, Y., et al., Polymer Testing, 2017, 59: 67

[45] Fukahori, Y., et al., Rubber Chem. Technol., 1996, 69: 752

[46] André, N., Cailletaud, G., Piques, R., Kautsch. Gummi Kunstst., 1999, 52: 120

[47] Abraham, F., Alshuth, Th., Jerrams, S., Mater. Design, 2005, 26: 239

[48] Gent, A. N., Mars, W. in "Science and Technology of Rubber", Mark, J. E., Erman, B., Roland, C. M., Eds., Acad. Press, 4th ed., (2013) Ch. 10

[49] Lindley, P. B., J. Strain Anal., 1972, 7: 132

[50] ASTM D 624

[51] DIN 53507

[52] Hofmann, W., Kunstst., 1987, 77: 1057

[53] Kok, A., Yeoh, O. H., Europ. Polym. J., 1986, 22: 341

[54] Elmurasksi, E., Kroon, M., Int. J. Fracture, 2012, 177: 163

[55] Charrier, C., et al., in "Constitutive Models for Rubber". III, Busfield, J., Muhr, A., Eds. (2003)

[56] Timbrell, C., Wiehan, M., Cook, G., Muhr, A., in "Constitutive Models for Rubber", Busfield, J., Muhr, A., Eds. (2003)

[57] Rotten, J., J. Physics: Condensed Matter, 2009, 21: 463101

[58] Higuchi, Y., Kubo, M., Macromolecules, 2017, 50: 3690

[59] Li, H., et al., Polymer, 2020, 199: 122571

[60] David, A., Tartaglino, U., Casagnelo, M., Roos, G., ACS Polymers, 2021, 1: 175

[61] Görl, U., in "Handbuch der Kautschuktechnologie", Hofmann, W., Gupta, H., Eds., Gupta Verlag (2001) Ch. 8.B

[62] Wang, M.-J., et al., Kautsch. Gummi Kunstst., 2020, 73(9): 45

[63] Wang, Y. Z., Zhang, L.-Q., Wang, Y. R., J. Appl. Polym. Sci., 2000, 78: 1879

[64] Wu, Y. P., Zhang, L.-Q., Wang, Y. Q., J. Appl. Polym. Sci., 2001, 82: 2142

[65] Varghese, S., Karger-Kocsis, J., J. Appl. Polym. Sci., 2004, 91: 813

[66] Gatos, K. G., Karger-Kocsis, J., Polymer, 2005, 46: 3069

[67] Gatos, K. G., et al., Macomolecular Mat. Eng., 2004, 289: 1079

[68] Yan, N., et al., Plast, Rubber Compos. 2009, 38: 290

[69] Seppä, PhD Thesis University of Technology Helsinki (2003)

[70] Griffith, A. A., Philos. Trans. Royal Soc., Ser. A, 1920, 221: 1963

[71] Rivlin, R. S., Thomas, A. G., J. Polym. Sci., 1953, 10: 291

[72] Thomas, A. G., J. Polym. Sci., 1958, 31: 123

[73] Thomas, A. G., J. Polym. Sci., 1955, 28: 177

[74] Andrews, E. H., Proc. Phys. Soc., 1961, 77: 983

[75] Greensmith, H. W., Thomas, A. G., J. Polym. Sci., 1955, 18: 189

[76] Lake, G. J., Thomas, A. G., in "Engineering with Rubber", Gent, A. N., Ed., Hanser Publ., Munich (2001) Ch. 1

[77] Lake, G. J., Rubber Chem. Technol., 1995, 68: 435

[78] Oppermann, B., PhD Thesis, University of Hanover (1993)

[79] Hess, W. M., Ford, F. P., Rubber Chem. Technol., 1963, 49: 1175

[80] Greensmith, H. W., Mullins, L., Thomas, A. G., Trans. Soc. Rheol., 1960, 4: 179

[81] Stacer, R. G., Yanyo, L. C., Kelly, F. N., Rubber Chem. Technol., 1985, 58: 422

[82] De, D., Gent, A. N., Rubber Chem. Technol., 1996, 69: 834

[83] Eisele, U., Kelbch, S., Engels, H.-W., Kautsch. Gummi Kunstst., 1992, 45: 1064

[84] Lindley, P. B., Int. J. Fracture, 1973, 9: 449

[85] Lake, G. J., Lindley, P. B., J. Polym. Sci., 1964: 8: 455

[86] Mars, W., Int. J. Fatigue, 2002, 24: 949

[87] Alshuth, Th., Kautsch. Gummi Kunstst.

[88] Young, D. G., Rubber Chem. Technol., 1985, 58: 785

[89] Wang, M.-J., Kelley, F. N., Kautsch. Gummi Kunstst., 2016, 69(1–2): 38

[90] Thomas, A. G., J. Appl. Polym. Sci., 1960, 3: 168

[91] Mullins, L., et al., Trans. Inst. Rubber Ind., 1959, 35: 213

[92] Kambour, R. P., J. Polym. Sci. Macromol. Rev., 1973, 7: 1

[93] Gent, A. N., Kim, H. T., Rubber Chem. Technol., 1978, 51: 35

[94] Nakajima, N., in "Science and Practise of Rubber Mixing", RAPRA Technol. Ltd. (1998)

[95] Lake, G. J., Lindley, J., J. Appl. Polym. Sci., 1965, 9: 1233

[96] Greensmith, H.W., Mullins, L., Thomas, A.G., in "Chemstry and Physics of rubber-like Materials", Bateman, L. Ed., MacLaren and Sons, London (1963) Ch. 9

[97] Gent, A.N., Lindley, B.P., Thomas, A.G., J. Polym. Sci., 1964, 8: 455

[98] Lake, G.J., Lindley, P.B., Rubber J. Inst. Plast., 1964, 146: 30

[99] Lake, G.J., Rubber Chem. Technol., 1995, 68: 435

[100] *www.coesfeld.com*

[101] Schuster, R.H., Kiesekamp, J. Schmidt, H., Weidenhaupt, H.-W., Kleiner, W., ACS Rubber Division Meeting, Chicago (1998)

[102] Nordsiek, K.H., Wolperts, J., Kautsch. Gummi Kunstst., 1992, 45: 792

[103] Klie, B., Teich, S., Haberstroh, E., Giese, U., Kautsch. Gummi Kunstst., 2015, 67: 31

[104] Bogun, M., Abraham, F., Muresan, L. Schuster, R.H., Radusch, H.-J., Kautsch. Gummi Kunstst., 2004, 57: 363

[105] DIK report to Lanxess Co., 2014

[106] Grosch, K.A., in "Current Topics in Elastomer Research", Bhowmick, A. K, Ed., CRC Press 2008, Ch. 26

[107] ASTM D 394

[108] ISO 4649

[109] Medalia, A.I., Kautsch. Gummi Kunstst., 1994, 47: 364

[110] Sun, J.T., Hu, B., Zhang, P., Zhao, S.G., Polymer and Polym. Composit., 2011, 19(8): 647

[111] Grosch, K.A., Schallamach, A., Rubber Chem. Technol., 1966, 39: 287

[112] Hamed, G, Park, B., Rubber Chem. Technol., 1999, 72: 946

[113] Wolff, S., Wang, M.-J., in "Carbon Black, Science and Technology", Donnet, J.-B., Bansal, R.C., Wang, M.-J., Eds., Marcel Dekker, Inc., New York (1993)

[114] Cadle, S.H., Williams, R.L., Rubber Chem. Technol., 1979, 52: 146

[115] Lechtenböhmer, A., Postgraduate Studies in Rubber Technology at the University of Hanover (2014)

[116] Schallamach, A., Grosch, K.A., in "The Mechanics of Pneumatic Tires", Clark, S.K., Ed., US Department of Transportation, p. 407

[117] Rivlin, R.S., Thomas, A.G., J. Polym. Sci., 1953, 10: 291

[118] Thomas, A.G., J. Appl. Sci., 1960, 3: 168

[119] Grosch, K.A., Rubber Chem. Technol., 1996, 69: 496

[120] Wang, M.-J., Wang, T., Wang, Y.L., Kautsch. Gummi Kunstst., 2002, 55: 388

[121] Studebaker, M.L., in "Reinforcement of Elastomers", Kraus, G., Ed., Wiley Interscience (1965)

[122] Dannenberg, E.M., Rubber Age, 1966, 98: 82

[123] Bulgin, D., Composite, 1971, 2: 165, 82

[124] Cotton, G.R., Dannenberg, E.M., Tire Sci. Technol., 1974, 2: 211

[125] Dizon, E.S., Rubber Chem. Technol., 1976, 49: 12

[126] Kang, Sh., et al., J. Mat. Sci., 2007, 42: 8391

[127] Shieh, C.H., et al., ACS Rubber Division Meeting, Toronto (1991)

[128] Warskulat, M., Wehmeier, A., Educational Symp. DKT'15, Nuremberg (2015)

[129] Veith, A.G. Chirico, A.E.A., Rubber Chem. Technol., 1979, 52: 748

[130] Wang, M.-J., Kautsch. Gummi Kunstst., 2007, 59: 438

[131] Byres, J.T., Patel, A.C., Rubber World, 1983, 188: 21

[132] Wang, M.-J., Wolff, S., Donnet, J.-B., Rubber Chem. Technol., 1991, 64: 714

[133] Rivin, D., Rubber Chem. Technol., 1971, 44: 39

[134] Brown, W. A., Patel, A. C., Int. Rubber Conf., (1981)

[135] Dannenberg, E. M., Papirer, E., Donnet, J.-B., Colloques Intern. du C. N.R.S (1975)]

[136] Hess, W. M., Lyon, F., Burgess, K. K., Kautsch. Gummi Kunstst. 1967, 20: 135

[137] Wang, M.-J., Kutsovski, Y., Zhang, P., Rubber Chem. Technol., 2002, 65: 78

[138] Wang, M.-J., Kautsch. Gummi Kunstst., 2005, 58: 626

[139] Bomal, Y., Cochet, P., Dejean, B., Gelling, F., Newell, R., Kautsch. Gummi Kunstst., 1998, 51: 259

[140] Wolff, S., Wang, M.-J., Tan, E.-H., Kautsch. Gummi Kunstst., 1994, 47: 837

[141] Wang, M.-J., Kautsch. Gummi Kunstst., 2008, 61: 159

[142] Wang, M.-J. et al., Kautsch. Gummi Kunstst., 2020, 73(9): 45

[143] Wolff, S., ACS Rubber Division Meeting, New York, 1986

[144] Gatos, K. G., Kamero, K., Karger-Koczis, J., eXPRESS Polym. Lett., 2007, 1: 27

[145] Karger-Koczis, J., Kautsch. Gummi Kunstst., 2006, 59: 537

[146] Schuster, R. H., ACS Rubber Division Meeting, Cincinatti (2012)

[147] Grosch, K. A., Schallamach, A., Inst. Rubber Ind. Trans., IRI, 1965, 41: 80

[148] Schallamach, A., in "The Chemistry and Physics of Rubber-Like Substances", Bateman, L., Ed., MacLaren, London (1963) Ch. 13

[149] Southern, E., Thomas, A. G., Plast. Rubber: Mater. Appl. 1978, 3: 133

[150] Schallamach, A., Trans. IRI, 1952, 28: 256

[151] Reznikovskii, M. M., Brodski, G. I., in "Abrasion of Rubber", James, D. I., Ed., Palmerton Pub., Co., (1964) Ch. 14

[152] Gent, A. N., Lindley, P. B., Proc. R. Soc., London, 1958, Series A 248: 195

[153] Gent, A. N., Tompkins, D. A., J. Polym. Sci., 1969, Part A, 27: 1483

[154] Denecour, R. L., Gent, A. N., J. Polym. Sci., 1968, 40: 2520

[155] Schallamach, A., J. Appl. Polym. Sci., 1968, 12: 281

[156] Lambourn, L. J., Rubber Chem. Technol., 1928, 2: 166

[157] Gent, A. N., Rubber Chem. Technol., 1989, 62: 750

[158] Mouri, H., Rubber World, 2022, 265: 26

[159] Schallamach, A., Rubber Chem. Technol., 1968, 41: 209

[160] Veith, A., Rubber Chem. Technol., 1992, 65: 624

[161] Grosch, K. A., Schallamach, Wear, 1961, 4: 356

[162] Bulgin, D., Walters, M. H., Proc. 5$^{th}$ Intern. Rubber Conf., Brighton (UK) 1967

[163] Shieh, C. H., Mace, M. L., Ouyang, G. B., ACS Rubber Division Meeting, Toronto, (1991)]

[164] Grosch, K. A., Schallamach, A., Wear, 1961, 4: 356]

[165] Hess, W. M., Ayala, J. A., Vegvary, P. C. Kautsch. Gummi Kunstst., 1988, 41: 1215

[166] Grosch, K. A., Schallamach, A., Wear, 1961, 4: 356

[167] Mouri, H., Kautsch. Gummi Kunstst., 2003, 72: 39

[168] Stoceck, R., Mars, W., Kippscholl, R., Robertson, C. G., Plastics, Rubber Comp., 2019, 48: 14

[169] Kippscholl, R., Stoceck, R., Rubber, Filler, Plastics, 2019, 2: 88

[170] Euchler, E., Michael, H., Gehcke, M., Kratina, O., Stoceck R., Kautsch. Gummi Kunstst., 2016, 69: 22

[171] Stoceck, R., et al., Kautsch. Gummi Kunstst., 2020, 73(6)–51

[172] Scherbakov, M., Gurich, M. R. A., J. Elastomer Plast., 2003, 53: 73

# Index

## Symbols



## A

## C

## D

## G



## H

# I

## K



## L

## M

## Q



## R

## T

## U

## V



## W

## X



## Y



## Z

www.ingramcontent.com/pod-product-compliance
Lightning Source LLC
LaVergne TN
LVHW081927200926
848484LV00009B/2389